Atomic Transition Probabilities
Iron through Nickel

Journal of
Physical and Chemical Reference Data

David R. Lide, Jr., Editor

The Journal of Physical and Chemical Reference Data (ISSN 0047-2689) is published quarterly by the American Chemical Society (1155 16th St., N. W., Washington, DC 20036-9976) and the American Institute of Physics (335 E. 45th St., New York, NY 10017-3483) for the National Bureau of Standards. Second-class postage paid at Washington, DC and additional mailing offices. POSTMASTER: Send address changes to *Journal of Physical and Chemical Reference Data*, Membership and Subscription Services, P. O. Box 3337, Columbus, Ohio 43210.

The objective of the Journal is to provide critically evaluated physical and chemical property data, fully documented as to the original sources and the criteria used for evaluation. Critical reviews of measurement techniques, whose aim is to assess the accuracy of available data in a given technical area, are also included. The Journal is not intended as a publication outlet for original experimental measurements such as those that are normally reported in the primary research literature, nor for review articles of a descriptive or primarily theoretical nature.

Supplements to the Journal are published at irregular intervals and are not included in subscriptions to the Journal. They contain compilations which are too lengthy for a journal format.

The Editor welcomes appropriate manuscripts for consideration by the Editorial Board. Potential contributors who are interested in preparing a compilation are invited to submit an outline of the nature and scope of the proposed compilation, with criteria for evaluation of the data and other pertinent factors, to:

David R. Lide, Jr., Editor
J. Phys. Chem. Ref. Data
National Bureau of Standards
Gaithersburg, MD 20899

One source of contributions to the Journal is The National Standard Reference Data System (NSRDS), which was established in 1963 as a means of coordinating on a national scale the production and dissemination of critically evaluated reference data in the physical sciences. Under the Standard Reference Data Act (Public Law 90-396) the National Bureau of Standards of the U.S. Department of Commerce has the primary responsibility in the Federal Government for providing reliable scientific and technical reference data. The Office of Standard Reference Data of NBS coordinates a complex of data evaluation centers, located in university, industrial, and other Government laboratories as well as within the National Bureau of Standards, which are engaged in the compilation and critical evaluation of numerical data on physical and chemical properties retrieved from the world scientific literature. The participants in this NBS-sponsored program, together with similar groups under private or other Government support which are pursuing the same ends, comprise the National Standard Reference Data System.

The primary focus of the NSRDS is on well-defined physical and chemical properties of well-characterized materials or systems. An effort is made to assess the accuracy of data reported in the primary research literature and to prepare compilations of critically evaluated data which will serve as reliable and convenient reference sources for the scientific and technical community.

Information for Contributors

Manuscripts submitted for publication must be prepared in accordance with *Instructions for Preparation of Manuscripts for the Journal of Physical and Chemical Reference Data*, available on request from the Editor.

New and renewal subscriptions should be sent with payment to the Office of the Controller at the American Chemical Society, 1155 Sixteenth Street, N.W., Washington, DC 20036-9976. **Address changes**, with at least six weeks advance notice, should be sent to *Journal of Physical and Chemical Reference Data*, Membership and Subscription Services, American Chemical Society, P.O. Box 3337, Columbus, OH 43210. Changes of address must include both old and new addresses and ZIP codes and, if possible, the address label from the mailing wrapper of a recent issue. Claims for missing numbers will not be allowed: if loss was due to failure of the change-of-address notice to be received in the time specified; if claim is dated (a) North America: more than 90 days beyond issue date, (b) all other foreign: more than one year beyond issue date.

Members of AIP member and affiliate societies requesting member subscription rates should direct subscriptions, renewals, and address changes to American Institute of Physics, Dept. S/F, 335 E. 45th St., NY 10017-3483.

Subscription Prices (1988)
(not including supplements)

	U.S.A.	Foreign (surface mail)	Optional air freight Europe Mideast N. Africa	Asia and Oceania
Members (of ACS, AIP, or affiliated society)	$ 60.00	$ 70.00	$ 80.00	$ 80.00
Regular rate	$265.00	$275.00	$285.00	$285.00

Rates above do not apply to nonmember subscribers in Japan, who must enter subscription orders with Maruzen Company Ltd., 3-10 Nihonbashi 2-chome, Chuo-ku, Tokyo 103, Japan. Tel: (03) 272-7211.

Back numbers are available at a cost of $75 per single copy and $295 per volume.

Orders for reprints, supplements, and back numbers should be addressed to the American Chemical Society, 1155 Sixteenth Street, N. W., Washington, DC 20036-9976. Prices for reprints and supplements are listed at the end of this issue.

Microfilm subscriptions of the *Journal of Physical and Chemical Reference Data* are available on 16 mm and 35 mm. This journal also appears in Sec. I of *Current Physics Microform* (CPM) along with 26 other journals published by the American Institute of Physics and its member societies. A *Microfilm Catalog* is available on request.

Journal of
**Physical and
Chemical
Reference Data**

Volume 17, 1988
Supplement No. 4

Atomic Transition Probabilities
Iron through Nickel

J. R. Fuhr, G. A. Martin, and W. L. Wiese

*National Measurement Laboratory, National Bureau of Standards,
Gaithersburg, Maryland 20899*

Published by the **American Chemical Society**
and the **American Institute of Physics** for
the **National Bureau of Standards**

Library of Congress Catalog Card Number 88-72276

International Standard Book Number
0-88318-586-5

American Institute of Physics, Inc.

335 East 45th Street

New York, New York 10017-3483

Printed in the United States of America

Foreword

The *Journal of Physical and Chemical Reference Data* is published jointly by the American Institute of Physics and the American Chemical Society for the National Bureau of Standards. Its objective is to provide critically evaluated physical and chemical property data, fully documented as to the original sources and the criteria used for evaluation. One of the principal sources of material for the journal is the National Standard Reference Data System (NSRDS), a program coordinated by NBS for the purpose of promoting the compilation and critical evaluation of property data.

The regular issues of the *Journal of Physical and Chemical Reference Data* are published quarterly and contain compilations and critical data reviews of moderate length. Longer monographs, volumes of collected tables, and other material unsuited to a periodical format are published separately as *Supplements to the Journal*. This critical compilation, "Atomic Transition Probabilities—Iron through Nickel," by J. R. Fuhr, G. A. Martin and W. L. Wiese, is presented as Supplement No. 4 to Volume 17 of the *Journal of Physical and Chemical Reference Data*.

David R. Lide, Jr., Editor
Journal of Physical and Chemical Reference Data

Atomic Transition Probabilities
Iron through Nickel

J. R. Fuhr, G. A. Martin, and W. L. Wiese

National Measurement Laboratory, National Bureau of Standards, Gaithersburg, Maryland 20899

Atomic transition probabilities for about 9,500 spectral lines of three iron-group elements, Fe ($Z = 26$) to Ni ($Z = 28$), are critically compiled, based on all available literature sources. The data are presented in separate tables for each element and stage of ionization and are further subdivided into allowed (i.e., electric dipole—E1) and forbidden (magnetic dipole—M1, electric quadrupole—E2, and magnetic quadrupole—M2) transitions. Within each data table the spectral lines are grouped into multiplets, which are in turn arranged according to parent configurations, transition arrays, and ascending quantum numbers. For each line the transition probability for spontaneous emission and the line strength are given, along with the spectroscopic designation, the wavelength, the statistical weights, and the energy levels of the upper and lower states. For allowed lines the absorption oscillator strength is listed, while for forbidden transitions the type of transition is identified (M1, E2, etc.). In addition, the estimated accuracy and the source are indicated. In short introductions, which precede the tables for each ion, the main justifications for the choice of the adopted data and for the accuracy rating are discussed. A general introduction contains a discussion of our method of evaluation and the principal criteria for our judgements.

Key words: Allowed and forbidden transitions; oscillator strengths; transition probabilities; line strengths; iron; cobalt; nickel.

Contents

1. Introductory Remarks

This is the third major critical compilation by the NBS Data Center on Atomic Transition Probabilities. A first tabulation[1] containing transition probabilities for about 4,000 spectral lines of the elements hydrogen through neon, atomic numbers $Z = 1$ through 10, including the neutral atoms as well as their various ions, was published in 1966. A second data volume[2] was issued in 1969, containing data for about 5,000 lines of the elements sodium ($Z = 11$) through calcium ($Z = 20$), again for all stages of ionization for which data were available. The data compilation work then continued with a series of smaller tables for the atoms and ions of the elements of the iron group, i.e., Sc and Ti[3]; V, Cr, and Mn[4]; Fe, Co, and Ni[5]; and the forbidden lines of all these elements.[6] From the beginning, it has been our intention to integrate these smaller tabulations into a single volume for the iron-group elements, in updated and expanded form. Unexpectedly, a great deal of new data were generated for these elements during the past few years, often with much improved accuracy, so that our revisions and additions became very extensive. Thus it took a much longer time than anticipated to complete these largely new data tables, and the greatly expanded tabulations had to be split into two separate volumes. This volume contains the material on the elements Fe ($Z = 26$) through Ni ($Z = 28$), and a companion volume[7] contains the material on Sc ($Z = 21$) through Mn ($Z = 25$).

In the present compilation, we maintain the scope and format of our earlier tabulations, i.e., we present critically evaluated atomic transition probabilities of allowed and forbidden discrete transitions of all stages of ionization for which we have reliable data. We have aimed at listing data for at least the more prominent lines of each spectrum, even if some of these data are of low accuracy. Furthermore, we have also presented transition-probability data for weaker transitions if the accuracy of these data has been estimated to be better than ± 50%.

The original literature is continually monitored by this NBS Data Center, and a master reference list is maintained from which all literature sources for this compilation have been taken.

2. Method of Evaluation

For the compilation of data on a critical basis, the central task is the evaluation of the data accuracy and the subsequent choice of the most accurate material. In order to accomplish this task in a consistent manner, we had established general guideposts for each experimental and theoretical approach in our earlier compilation work, and we have maintained these criteria in this work. Specifically, we judge each original literature source by the following principal criteria:

(1) Our general evaluation of the capabilities and reliability of the applied experimental or theoretical method.

(2) The author's consideration of the major critical factors in his approach that enter into the results.

(3) The degree of agreement and general consistency between the author's results and other reliable data.

(4) The degree of fit of the data into established systematic trends and, if deviations exist, the reasons for such disagreements.

(5) The author's estimate of his uncertainties.

We have discussed our general evaluations of each experimental and theoretical method in considerable detail in the introductions to our previous tabulations.[1-6] Thus, we refer to these publications for further details. However, we should point out that, in this tabulation, we illustrate particularly interesting situations by providing comparison tables or graphs in the introductions to individual spectra. For example, we present graphical comparisons for Fe I.

With respect to error estimates, we should note that the theoretical literature sources, which provide a large part of the data, generally contain no error estimates, since no reliable assessment of the uncertainties introduced by the various approximations is possible. But even for the experimental papers, where error estimates may often readily be made, the statements by some authors are too imprecise and also incomplete, so that they are not particularly useful as presented. Sometimes only statistical measurement errors have been given, without allowance for systematic errors. It therefore became essential to judge each paper by the principal factors 1-4 listed above, in addition to utilizing the author's error estimate (point (5)) whenever appropriate.

3. General Arrangement of the Tables

We have continued to use the same general arrangement of the tables as in our earlier volumes,[1,2] i.e., we have included data which serve to identify the spectral lines, as well as the actual transition probabilities (and related quantities), accuracy estimates, and references to the sources of the compiled material. However, for most of the spectra of neutral and singly ionized atoms of the iron-group elements, the transition array column was dropped. Instead, in order to identify the lower and upper levels of a transition, we adopted the level designation scheme of C. E. Moore,[8] who affixed lower-case letters ($a,b,c,...,x,y,z$) to the term designations. This convention is also retained in the very recent tables of "Atomic Energy Levels" by J. Sugar and C. Corliss.[9] In other special cases, we have adapted our notation to the special coupling situations encountered in those spectra, as, for example, the Jj coupling encountered in Ne-like ions and Jj and $J_1\ell$ coupling for Ar-like ions.

Material pertaining to spectral-line identifications has been taken from the comprehensive wavelength tabulations of Reader and Corliss,[10] Kelly,[11,12] and Kelly and Palumbo,[13] the multiplet tables of C. E. Moore,[14,15] and the recent energy-level compilation of Sugar and

Corliss[9] (this last reference supersedes earlier compilations by Sugar and others[16–18]). We have supplemented the wavelength and energy-level data from these sources with original literature data when needed in the course of preparing our transition-probability tables. A listing of all data sources other than Refs. 9–18 is given in Table 1.

Wavelengths and energy levels which are the results of theoretical calculations, or which were either calculated from experimentally determined data or interpolated or extrapolated from data on similar (e.g.,

isoelectronic) species, are placed in square brackets in order to distinguish them from the usually more accurate experimental material.

For each transition-probability table which contains a minimum of twenty distinct wavelength values, we provide a "list of tabulated lines," i.e., a listing, in ascending order of wavelength, of the spectral lines contained therein, along with an index to the multiplet number (or numbers) in which each is to be found. Wavelengths that are printed in italics in the transition-probability tables are not included in these line lists.

TABLE 1. Special source material for wavelength and energy-level data. Complete citations are given below.

Spectrum	References	Spectrum	References	Spectrum	References
Fe I	1	Co I	75,76,77	Ni I	92,93
Fe II	1,2	Co II	78	Ni II	94
Fe VII	3	Co III	79	Ni IX	80
Fe VIII	4	Co VIII	80	Ni XI	6,95
Fe IX	5,6	Co IX	4,81	Ni XII	11,84,85,86,96
Fe X	5,6,7,8,9,10,11,12	Co X	82,83	Ni XIII	11,14,85,86,97,98
Fe XI	5,7,9,11,13,14,15,16,17	Co XI	82,84,85,86	Ni XIV	7,14,18,20,85,86
Fe XII	5,13,14,18,19,20	Co XII	7,85,86	Ni XV	7,14,21,22,23,85,86
Fe XIII	5,13,14,19,20,21,22,23	Co XIII	7,18,85,86	Ni XVI	22,25,26,85,86
Fe XIV	5,13,20,22,24,25,26,27	Co XIV	7,21,23,85,86	Ni XVII	23,28,29,34,85,86,99
Fe XV	5,13,20,22,23,24,27,28,29,	Co XV	25,85,86	Ni XVIII	28,34,35,36,86
	30,31,32,33,34	Co XVI	23,28,85,87	Ni XIX	38,39,40,42,45,88,100
Fe XVI	5,31,34,35,36,37	Co XVII	28,34,35,36,85	Ni XX	36,40,47,48
Fe XVII	38,39,40,41,42,43,44,45	Co XVIII	38,40,88	Ni XXI	36,40,49,51,53,101
Fe XVIII	36,40,46,47,48	Co XIX	36,46	Ni XXII	36,53
Fe XIX	36,40,47,49,50,51,52	Co XX	36,40	Ni XXIII	36,53
Fe XX	36,51,53,54,55,56	Co XXI	36	Ni XXIV	36
Fe XXI	36,53,57,58	Co XXII	36	Ni XXV	36,62,63,67,89
Fe XXII	36,59,60,61	Co XXIII	36	Ni XXVI	54,67,102
Fe XXIII	36,59,62,63,64,65,66,67,68	Co XXIV	36,62,63	Ni XXVII	73,74
Fe XXIV	65,67,69,70,71,72	Co XXV	60,89,90,91		
Fe XXV	73,74	Co XXVI	73,74		

[1] H. M. Crosswhite, J. Res. Nat. Bur. Stand., Sect. A **79**, 17 (1975).

[2] S. Johansson, Phys. Scr. **18**, 217 (1978).

[3] J. O. Ekberg, Phys. Scr. **23**, 7 (1981).

[4] A. A. Ramonas and A. N. Ryabtsev, Opt. Spectrosc. (USSR) **48**, 348 (1980).

[5] B. C. Fawcett, R. D. Cowan, E. Y. Kononov, and R. W. Hayes, J. Phys. B **5**, 1255 (1972).

[6] B. Edlén and R. Smitt, Sol. Phys. **57**, 329 (1978).

[7] R. Smitt, L. A. Svensson, and M. Outred, Phys. Scr. **13**, 293 (1976).

[8] R. Smitt, Sol. Phys. **51**, 113 (1977).

[9] G. E. Bromage, R. D. Cowan, and B. C. Fawcett, Phys. Scr. **15**, 177 (1977).

[10] B. Edlén, Z. Phys. **104**, 407 (1937).

[11] B. Edlén, Z. Astrophys. **22**, 30 (1942).

[12] G. D. Sandlin and R. Tousey, Astrophys. J. **227**, L107 (1979).

[13] B. C. Fawcett, J. Phys. B **4**, 1577 (1971).

[14] U. Feldman and G. A. Doschek, J. Opt. Soc. Am. **67**, 726 (1977).

[15] B. Edlén, Z. Phys. **104**, 188 (1937).

[16] W. E. Behring, L. Cohen, and U. Feldman, Astrophys. J. **175**, 493 (1972).

[17] B. C. Fawcett, N. J. Peacock, and R. D. Cowan, J. Phys. B **1**, 295 (1968).

[18] K.-N. Huang, At. Data Nucl. Data Tables **30**, 313 (1984).

[19] G. E. Bromage, R. D. Cowan, and B. C. Fawcett, Mon. Not. R. Astron. Soc. **183**, 19 (1978).

[20] W. E. Behring, L. Cohen, U. Feldman, and G. A. Doschek, Astrophys. J. **203**, 521 (1976).

[21] K.-N. Huang, At. Data Nucl. Data Tables **32**, 503 (1985).

[22] J. T. Jefferies, F. Q. Orrall, and J. B. Zirker, Mem. Soc. R. Sci. Liege, Collect. 8°, **17**, 235 (1969).

[23] S. O. Kastner, M. Swartz, A. K. Bhatia, and J. Lapides, J. Opt. Soc. Am. **68**, 1558 (1978).

[24] B. C. Fawcett, J. Phys. B **3**, 1732 (1970).

[25] K.-N. Huang, At. Data Nucl. Data Tables **34**, 1 (1986).

[26] B. C. Fawcett, At. Data Nucl. Data Tables **28**, 557 (1983).

[27] B. Edlén, Z. Phys. **103**, 536 (1936).

[28] B. C. Fawcett, R. D. Cowan, and R. W. Hayes, J. Phys. B **5**, 2143 (1972) and Supplement.

[29] B. C. Fawcett, At. Data Nucl. Data Tables **28**, 579 (1983).

[30] C. Froese Fischer and M. Godefroid, Nucl. Instrum. Methods **202**, 307 (1982).

[31] B. C. Fawcett, A. H. Gabriel, F. E. Irons, N. J. Peacock, and P. A. H. Saunders, Proc. Phys. Soc. (London) **88**, 1051 (1966).

[32] N. J. Peacock, R. D. Cowan, and G. A. Sawyer, *Proceedings of the 7th International Conference on Ionization Phenomena in Gases*, Vol. II, 599–602 (Eds. B. Perovic and D. Tosic, Gradevinska Knjiga Publishing House, Belgrade, 1966).

[33] R. D. Cowan and K. G. Widing, Astrophys. J. **180**, 285 (1973).

[34] U. Feldman, L. Katz, W. Behring, and L. Cohen, J. Opt. Soc. Am. **61**, 91 (1971).

[35] B. Edlén, Z. Phys. **100**, 621 (1936).

[36] K. D. Lawson and N. J. Peacock, J. Phys. B **13**, 3313 (1980).

[37] B. Edlén, Phys. Scr. **17**, 565 (1978).

[38] L. A. Vainshtein and U. I. Safronova, *Spektroskopicheskie Konstanty Atomov*, 5–122 (Ed. V. B. Belyanin, Akad. Nauk SSSR, Ot. Ob. Fiz. Astron., Nauch. Sov. Spektrosk., Moscow, 1977).

[39] M. Finkenthal, P. Mandelbaum, A. Bar-Shalom, M. Klapisch, J. L. Schwob, C. Breton, C. de Michelis, and M. Mattioli, J. Phys. B **18**, L331 (1985).

[40] H. Gordon, M. G. Hobby, and N. J. Peacock, J. Phys. B **13**, 1985 (1980) and Supplement.

[41] B. C. Fawcett, G. E. Bromage, and R. W. Hayes, Mon. Not. R. Astron. Soc. **186**, 113 (1979).

[42] M. Loulergue and H. Nussbaumer, Astron. Astrophys. **45**, 125 (1975).

[43] J.-P. Buchet, M.-C. Buchet-Poulizac, A. Denis, J. Desesquelles, M. Druetta, S. Martin, J. P. Grandin, X. Husson, and I. Lesteven, Phys. Scr. **31**, 364 (1985).

[44] C. Jupén, Mon. Not. R. Astron. Soc. **208**, 1P (1984).

[45] M. Klapisch, A. Bar-Shalom, J. L. Schwob, B. S. Fraenkel, C. Breton, C. de Michelis, M. Finkenthal, and M. Mattioli, Phys. Lett. A **69**, 34 (1978).

[46] U. Feldman, G. A. Doschek, R. D. Cowan, and L. Cohen, J. Opt. Soc. Am. **63**, 1445 (1973).

[47] N. J. Peacock, M. F. Stamp, and J. D. Silver, Phys. Scr. **T8**, 10 (1984).

[48] B. C. Fawcett, At. Data Nucl. Data Tables **31**, 495 (1984).

[49] B. C. Fawcett, At. Data Nucl. Data Tables **34**, 215 (1986).

[50] S. O. Kastner, A. K. Bhatia, and L. Cohen, Phys. Scr. **15**, 259 (1977).

[51] K. G. Widing, Astrophys. J. **222**, 735 (1978).

[52] K. J. H. Phillips, J. W. Leibacher, C. J. Wolfson, J. H. Parkinson, B. C. Fawcett, B. J. Kent, H. E. Mason, L. W. Acton, J. L. Culhane, and A. H. Gabriel, Astrophys. J. **256**, 774 (1982).

[53] E. Hinnov, S. Suckewer, S. Cohen, and K. Sato, Phys. Rev. A **25**, 2293 (1982).

[54] G. D. Sandlin, G. E. Brueckner, V. E. Scherrer, and R. Tousey, Astrophys. J. **205**, L47 (1976).

[55] H. E. Mason and A. K. Bhatia, Astron. Astrophys., Suppl. Ser. **52**, 181 (1983).

[56] K. D. Lawson and N. J. Peacock, Astron. Astrophys., Suppl. Ser. **58**, 475 (1984).

[57] H. E. Mason, G. A. Doschek, U. Feldman, and A. K. Bhatia, Astron. Astrophys. **73**, 74 (1979).

[58] G. A. Doschek, U. Feldman, K. P. Dere, G. D. Sandlin, M. E. VanHoosier, G. E. Brueckner, J. D. Purcell, and R. Tousey, Astrophys. J. **196**, L83 (1975).

[59] G. E. Bromage, R. D. Cowan, B. C. Fawcett, and A. Ridgeley, J. Opt. Soc. Am. **68**, 48 (1978).

[60] N. Spector, A. Zigler, H. Zmora, and J. L. Schwob, J. Opt. Soc. Am. **70**, 857 (1980).

[61] H. E. Mason and P. J. Storey, Mon. Not. R. Astron. Soc. **191**, 631 (1980).

[62] U. I. Safronova and T. G. Lisina, At. Data Nucl. Data Tables **24**, 49 (1979).

[63] V. A. Boiko, S. A. Pikuz, U. I. Safronova, and A. Ya. Faenov, J. Phys. B **10**, 1253 (1977).

[64] K. G. Widing, Astrophys. J. **197**, L33 (1975).

[65] E. Ya. Kononov, K. N. Koshelev, and Yu. V. Sidel'nikov, Sov. J. Plasma Phys. **3**, 375 (1977).

[66] M. Bitter, K. W. Hill, N. R. Sauthoff, P. C. Efthimion, E. Meservey, W. Roney, S. von Goeler, R. Horton, M. Goldman, and W. Stodiek, Phys. Rev. Lett. **43**, 129 (1979).

[67] B. C. Fawcett, A. Ridgeley, and T. P. Hughes, Mon. Not. R. Astron. Soc. **188**, 365 (1979).

[68] R. D. Chapman, Astrophys. J. **156**, 87 (1969).

[69] V. A. Boiko, A. Ya. Faenov, and S. A. Pikuz, J. Quant. Spectrosc. Radiat. Transfer **19**, 11 (1978).

[70] K. G. Widing and J. D. Purcell, Astrophys. J. **204**, L151 (1976).

[71] Yu. I. Grineva, V. I. Karev, V. V. Korneev, V. V. Krutov, S. L. Mandelstam, L. A. Vainshtein, B. N. Vasilyev, and I. A. Zhitnik, Sol. Phys. **29**, 441 (1973).

[72] F. Bely-Dubau, A. H. Gabriel, and S. Volonté, Mon. Not. R. Astron. Soc. **186**, 405 (1979).

[73] L. A. Vainshtein and U. I. Safronova, At. Data Nucl. Data Tables **21**, 49 (1978).

[74] A. M. Ermolaev and M. Jones, J. Phys. B **7**, 199 (1974) and Supplement.

[75] K. Burns and F. Sullivan, Science Studies, St. Bonaventure College **10**, No. 10 (1942).

[76] K. Burns and F. Sullivan, Science Studies, St. Bonaventure College **11**, No. 4 (1942).

[77] B. L. Cardon, P. L. Smith, J. M. Scalo, L. Testerman, and W. Whaling, Astrophys. J. **260**, 395 (1982).

[78] L. Iglesias, Opt. Pura Apl. **12**, 63 (1979).

[79] A. J. J. Raassen and S. Orti Ortin, Physica C (Amsterdam) **123**, 353 (1984).

[80] B. C. Fawcett, A. Ridgeley, and J. O. Ekberg, Phys. Scr. **21**, 155 (1980).

[81] S. Hoory, S. Goldsmith, B. S. Fraenkel, and U. Feldman, Astrophys. J. **160**, 781 (1970).

[82] S. Goldsmith, J. Opt. Soc. Am. **59**, 1678 (1969).

[83] E. Alexander, U. Feldman, and B. S. Fraenkel, J. Opt. Soc. Am. **55**, 650 (1965).

[84] K.-N. Huang, Y.-K. Kim, K. T. Cheng, and J. P. Desclaux, At. Data Nucl. Data Tables **28**, 355 (1983).

[85] B. C. Fawcett and R. W. Hayes, J. Phys. B **5**, 366 (1972).

[86] B. C. Fawcett and A. T. Hatter, Astron. Astrophys. **84**, 78 (1980).

[87] S. O. Kastner and A. K. Bhatia, J. Opt. Soc. Am. **69**, 1391 (1979).

[88] S. O. Kastner, W. E. Behring, and L. Cohen, Astrophys. J. **199**, 777 (1975).

[89] K. T. Cheng, Y.-K. Kim, and J. P. Desclaux, At. Data Nucl. Data Tables **24**, 111 (1979).

[90] J. Hata and I. P. Grant, Mon. Not. R. Astron. Soc. **211**, 549 (1984).

[91] U. I. Safronova, J. Quant. Spectrosc. Radiat. Transfer **14**, 251 (1974).

[92] K. Burns and F. Sullivan, Science Studies, St. Bonaventure College **13**, No. 3 (1947).

[93] K. Burns and F. Sullivan, Science Studies, St. Bonaventure College **14**, No. 3 (1948).

[94] A. G. Shenstone, J. Res. Nat. Bur. Stand., Sect. A **74**, 801 (1970).

[95] M. Even-Zohar and B. S. Fraenkel, J. Opt. Soc. Am. **58**, 1420 (1968).

[96] S. Goldsmith and B. S. Fraenkel, Astrophys. J. **161**, 317 (1970).

[97] G. E. Bromage, Astron. Astrophys., Suppl. Ser. **41**, 79 (1980).

[98] L. A. Svensson, Sol. Phys. **18**, 232 (1971).

[99] M. Finkenthal, E. Hinnov, S. Cohen, and S. Suckewer, Phys. Lett. A **91**, 284 (1982).

[100] R. D. Cowan, Los Alamos Scientific Laboratory Informal Report LA–6679–MS (Jan. 1977).

[101] M. Finkenthal, R. E. Bell, H. W. Moos, and TFR Group, J. Appl. Phys. **56**, 2012 (1984).

[102] U. I. Safronova and Yu. V. Sidel'nikov, Prikl. Spektrosk. 5-7 (1977).

We have denoted the uncertainties in the atomic transition probability data as in our earlier compilations, i.e.,

A . . . . . for uncertainties within 3 percent,
B . . . . . for uncertainties within 10 percent,
C . . . . . for uncertainties within 25 percent,
D . . . . . for uncertainties within 50 percent,
E . . . . . for uncertainties greater than 50 percent.

The word *uncertainty* is used here with the connotation "estimated extent of the deviation from the true value." The estimation procedure is based on our evaluation of random errors *as well as our estimates of the maximum effect of possible systematic errors.* We have often made further distinctions in the uncertainty labels by assigning plus or minus signs to some transitions to indicate that these lines are estimated to be somewhat better or worse than similar lines. These should, therefore, be the first or last choice among similar transitions.

A summary of the abbreviations and special symbols used in the tables is given in Section 4. We have also included there for convenience the relations between line and multiplet values in the case of *LS* coupling. In Table 2, we provide a table of conversion factors, which we have used throughout this compilation to convert from transition probabilities to oscillator strengths and line strengths, and vice versa.

TABLE 2.　Conversion factors

The factor in each box converts by multiplication the quantity above it into the one at its left.

	A_{ki}	f_{ik}	S	
A_{ki}	1	$\dfrac{6.670_3 \times 10^{15} g_i}{g_k \lambda^2}$	E1 $\dfrac{2.026_1 \times 10^{18}}{g_k \lambda^3}$	
			E2 $\dfrac{1.679_9 \times 10^{18}}{g_k \lambda^5}$	
			M1 $\dfrac{2.697_4 \times 10^{13}}{g_k \lambda^3}$	
			M2 $\dfrac{6.626_5 \times 10^{12}}{g_k \lambda^5}$	
f_{ik}	$\dfrac{1.499_2 \times 10^{-16} \lambda^2 g_k}{g_i}$	1	E1 $\dfrac{303.7_6}{g_i \lambda}$	
S	E1 $4.935_5 \times 10^{-19}\, g_k \lambda^3$	E1 $3.292_1 \times 10^{-3}\, g_i \lambda$	1	
	E2 $5.952_6 \times 10^{-19}\, g_k \lambda^5$			
	M1 $3.707_3 \times 10^{-14}\, g_k \lambda^3$			
	M2 $1.509_1 \times 10^{-13}\, g_k \lambda^5$			

The line strength (S) is given in atomic units; formulas and values for these quantities in SI units are as follows:

For E1 transitions, $a_0^2 e^2 = 7.188_3 \times 10^{-59}$ m^2 C^2.
For E2 transitions, $a_0^4 e^2 = 2.012_9 \times 10^{-79}$ m^4 C^2.
For M1 transitions, $\mu_B^2 = (eh/4\pi m_e)^2 = 8.600_7 \times 10^{-47}$ J^2 T^{-2}.
For M2 transitions, $\mu_B^2 a_0^2 = 2.408_5 \times 10^{-67}$ J^2 m^2 T^{-2},

where a_0, e, m_e, and h are the Bohr radius, electron charge, electron mass, and Planck constant, respectively, and μ_B is the Bohr magneton.

The transition probability (A_{ki}) is in units of s^{-1}, and the f-value is dimensionless. The wavelength (λ) is given in Ångström units, and g_i and g_k are the statistical weights of the lower and upper level, respectively.

[Note: the definition of the line strength for E2 transitions which is used by some authors yields an S-value that is 50% higher than that employed here and in earlier NBS transition-probability compilations. We have multiplied such line strengths by $\frac{2}{3}$ before tabulating them here, and have indicated this fact in the short introductions to the pertinent data tables.]

For the atomic constants entering into the relations given in this table, we have used the recommendations of the CODATA Task Group on Fundamental Constants (E. R. Cohen and B. N. Taylor, Rev. Mod. Phys. **59**, 1121 (1987)). The 1987 values were not available at the time we compiled most data for this publication; however, differences between these and the earlier (CODATA Task Group, 1973) values of the fundamental constants, which we utilized, amount to only 0.002% or less for the E1 transitions and 0.05% or less for the M1, E2, and M2 (forbidden) transitions and have therefore not affected our tabulated data.

4. Key to Abbreviations and Symbols Used in the Tables

1. Symbols for indication of accuracy:

A ...... uncertainties within 3 percent,
B ...... uncertainties within 10 percent,
C ...... uncertainties within 25 percent,
D ...... uncertainties within 50 percent,
E ...... uncertainties greater than 50 percent.

2. Abbreviations appearing in the source column of allowed transitions:

ls = LS coupling rules applied
n = normalized to a scale different from that of the author (as explained in the introductory remarks to the pertinent spectrum).
interp. = derived by an interpolation technique, rather than taken directly from the literature.

3. Special symbols used in the wavelength and energy level columns:

The number in parentheses under the multiplet designation refers to the sequence number of Ref. 14 (Revised Multiplet Table). If letters "uv" are added, we refer to the sequence number of Ref. 15 (Ultraviolet Multiplet Table).

Numbers in italics indicate multiplet values, i.e., weighted averages of *line* values.

Numbers in square brackets indicate approximate calculated or extrapolated values.

Useful Relations

(A) Statistical weights:

The statistical weights are related to the inner quantum number J_L (for one-electron spectra: j_l) of a level (i.e., initial or final state of a *line*) by

$$g_L = 2J_L + 1,$$

and to the quantum numbers of a term (initial or final state of a *multiplet*) by

$$g_M = (2L + 1)(2S + 1).$$

(The "multiplet" values g_M may also be obtained by summing over all possible "line" values g_L. S is the resultant spin.)

(B) Relations between the strengths of lines and the total multiplet strength:

1. Line strength S:

$$S(i,k) = \sum_{J_i,J_k} S(J_i,J_k)$$

or

$$S\ (\text{Multiplet}) = \Sigma\ S\ (\text{line})$$

(k denotes the upper and i the lower term).

2. Absorption oscillator strength f_{ik}:

$$f_{ik}^{multiplet} = \frac{1}{\bar{\lambda}_{ik} \sum_{J_i} (2J_i+1)} \sum_{J_i,J_k} (2J_i+1)$$
$$\times \lambda(J_i,J_k) \times f(J_i,J_k).$$

The mean wavelength for the multiplet, $\bar{\lambda}_{ik}$, may be obtained from the *weighted* energy levels. Often the wavelength differences for the lines within a multiplet are small, so that the wavelength factors may be neglected.

3. Transition probability A_{ki}:

$$A_{ki}^{multiplet} = \frac{1}{(\bar{\lambda}_{ik})^3 \sum_{J_k} (2J_k+1)} \sum_{J_i,J_k} (2J_k+1)$$
$$\times \lambda(J_i,J_k)^3 \times A(J_i,J_k).$$

Relative strengths $S(J_i,J_k)$ of the components of a multiplet are listed for the case of LS coupling in C. W. Allen, *Astrophysical Quantities,* 3rd ed. (The Athlone Press, London, 1973); H. E. White and A. Y. Eliason, Phys. Rev. **44**, 753 (1933); B. W. Shore and D. H. Menzel, *Principles of Atomic Structure*, p. 447 (John Wiley & Sons, Inc., New York, 1968); L. Goldberg, Astrophys. J. **82**, 1 (1935) and **84**, 11 (1936).

5. Acknowledgments

We would like to thank A. W. Weiss for his assistance in preparing the data tables for He-like ions, and we would like to express our deep appreciation to Paul Lanthier, Mary Trapane, and Mary Lou Thompson for their competent and untiring efforts to prepare this manuscript for computer typesetting. We would also like to thank Arlene Robey of the NBS Data Center on Atomic Energy Levels for making her bibliographical files available for our extensive use. This compilation has greatly benefited from the preprints and private communications that were provided by many of our colleagues and from their generous cooperation in providing further details of their work on request.

One of us (W.L.W.) performed part of the critical compilation work during his stay at the Ruhr University, Bochum, West Germany, as a recipient of the Humboldt Award. He would like to thank Prof. H.-J. Kunze for his hospitality and the A. von Humboldt Foundation for providing him this opportunity.

This data compilation project was partially supported by the Office of Standard Reference Data of the National Bureau of Standards and the Astronomy Branch of the National Aeronautics and Space Administration.

6. References

[1] W. L. Wiese, M. W. Smith, and B. M. Glennon, "Atomic Transition Probabilities—Hydrogen through Neon," Vol. I, NSRDS–NBS 4 (1966).

[2] W. L. Wiese, M. W. Smith, and B. M. Miles, "Atomic Transition Probabilities—Sodium through Calcium," Vol. II, NSRDS–NBS 22 (1969).

[3] W. L. Wiese and J. R. Fuhr, J. Phys. Chem. Ref. Data **4**, 263 (1975).

[4] S. M. Younger, J. R. Fuhr, G. A. Martin, and W. L. Wiese, J. Phys. Chem. Ref. Data **7**, 495 (1978).

[5] J. R. Fuhr, G. A. Martin, W. L. Wiese, and S. M. Younger, J. Phys. Chem. Ref. Data **10**, 305 (1981).

[6] M. W. Smith and W. L. Wiese, J. Phys. Chem. Ref. Data **2**, 85 (1973).

[7] G. A. Martin, J. R. Fuhr, and W. L. Wiese, "Atomic Transition Probabilities—Scandium through Manganese," J. Phys. Chem. Ref. Data **17**, Suppl. 3 (1988).

[8] C. E. Moore, "Atomic Energy Levels," Vols. I and II, NSRDS–NBS 35 (1971). (Reprints of NBS Circ. 467, originally issued in 1949 and 1952, respectively.)

[9] J. Sugar and C. Corliss, "Atomic Energy Levels of the Iron-Period Elements: Potassium through Nickel," J. Phys. Chem. Ref. Data **14**, Suppl. 2 (1985).

[10] J. Reader and C. Corliss, CRC Handbook of Chemistry and Physics, 67th Ed. (1986–87), pp. E-187 to E-313 (CRC Press, Inc., Boca Raton, FL).

[11] R. L. Kelly, "Atomic Emission Lines in the Near Ultraviolet; Hydrogen through Krypton," Sects. I and II, NASA Techn. Memorand. 80268 (1979).

[12] R. L. Kelly, "Atomic and Ionic Spectrum Lines Below 2000 Ångströms: Hydrogen through Krypton," J. Phys. Chem. Ref. Data **16**, Suppl. 1 (1987).

[13] R. L. Kelly and L. J. Palumbo, "Atomic and Ionic Emission Lines Below 2000 Ångströms—Hydrogen through Krypton," Naval Res. Lab. Rept. 7599 (1973).

[14] C. E. Moore, "A Multiplet Table of Astrophysical Interest," Revised Edition, NSRDS–NBS 40 (1972).

[15] C. E. Moore, "An Ultraviolet Multiplet Table," NBS Circ. 488, U. S. Government Print. Off., Washington, D.C. (1968).

[16] J. Sugar and C. Corliss, J. Phys. Chem. Ref. Data **10**, 1097 (1981).

[17] C. Corliss and J. Sugar, J. Phys. Chem. Ref. Data **10**, 197 (1981); **11**, 135 (1982).

[18] J. Reader and J. Sugar, J. Phys. Chem. Ref. Data **4**, 353 (1975).

Iron

Fe I

Ground State: $1s^2 2s^2 2p^6 3s^2 3p^6 3d^6 4s^2\ ^5D_4$

Ionization Energy: $7.9024\ \text{eV} = 63737\ \text{cm}^{-1}$

Allowed Transitions

List of tabulated lines

Wavelength (Å)	No.	Wavelength (Å)	No.	Wavelength (Å)	No.	Wavelength (Å)	No.
1934.54	31	2501.13	14	2970.12	10	3156.27	354
1937.27	30	2510.83	14	2973.13	9	3160.66	126
1940.66	31	2512.36	15	2973.24	9	3161.95	131
2084.12	29	2518.10	14	2980.53	210	3166.44	183
2102.35	29	2522.85	14	2981.45	10	3168.85	131
2112.97	29	2524.29	14	2983.57	8	3175.45	126
2132.02	28	2527.43	14	2986.46	10	3176.36	182
2138.59	26	2529.13	14	2986.65	150	3193.23	7
2145.19	27	2535.61	14	2987.29	47	3196.93	126
2153.01	26	2540.97	14	2990.39	210	3199.53	127
2161.58	26	2545.98	14	2994.43	8	3205.40	126
2166.77	23	2549.61	14	2994.50	10	3207.07	130
2171.30	26	2584.54	51	2996.39	123	3215.94	127
2173.21	27	2606.83	51	2999.51	47	3217.38	128
2176.84	25	2618.02	51	3000.95	8	3219.58	127
2191.20	24	2623.53	51	3005.31	149	3222.07	127
2191.84	23	2632.59	13	3007.28	10	3225.79	126
2196.04	23	2656.15	151	3008.14	8	3227.80	128
2200.72	23	2669.49	151	3009.09	148	3228.25	128
2228.17	22	2679.06	50	3009.57	47	3229.99	332
2250.79	20	2719.03	12	3011.48	211	3230.21	129
2259.28	19	2720.90	12	3015.92	148	3230.96	128
2259.51	20	2723.58	12	3016.18	47	3233.05	377
2265.05	20	2733.58	49	3017.63	8	3233.97	129
2267.08	21	2735.48	49	3018.14	149	3246.96	97
2272.07	20	2737.31	12	3018.98	47	3248.20	128
2276.03	18	2742.41	12	3021.07	8	3253.60	411
2277.11	52	2744.07	12	3024.03	10	3254.36	377
2287.25	18	2750.14	12	3025.84	8	3257.59	95
2292.52	19	2756.33	12	3026.46	47	3265.62	96
2294.41	18	2788.10	48	3031.63	47	3268.23	97
2300.14	19	2835.46	11	3037.39	8	3271.00	96
2301.68	18	2869.31	11	3039.32	149	3280.26	377
2303.42	20	2874.17	11	3040.43	47	3282.89	410
2303.58	20	2894.50	123	3042.02	47	3284.59	96
2309.00	18	2899.42	122	3042.66	47	3290.99	97
2313.10	18	2912.16	9	3047.60	8	3292.02	410
2320.36	18	2920.69	69	3053.07	121	3292.59	96
2371.43	17	2923.29	378	3057.45	46	3298.13	95
2373.62	17	2925.36	212	3059.09	8	3305.97	96
2374.52	17	2929.01	9	3067.24	46	3306.36	96
2381.83	17	2936.90	9	3068.17	68	3307.23	376
2389.97	17	2941.34	9	3075.72	46	3314.74	410
2462.18	16	2947.88	9	3083.74	46	3317.12	120
2462.65	16	2953.94	9	3091.58	46	3319.25	287
2479.78	16	2954.65	122	3098.19	209	3322.47	262
2483.27	16	2957.36	9	3100.67	46	3323.74	250
2488.14	16	2965.25	9	3119.49	147	3325.46	146
2490.64	16	2966.90	9	3120.43	147	3328.87	376
2491.15	16	2969.36	10	3134.11	46	3337.66	208

List of tabulated lines — Continued

Wavelength (Å)	No.	Wavelength (Å)	No.	Wavelength (Å)	No.	Wavelength (Å)	No.
3347.93	119	3522.27	221	3596.20	142	3669.15	283
3354.06	249	3522.90	224	3597.02	349	3669.52	202
3355.23	376	3523.31	221	3598.72	408	3670.09	282
3369.55	208	3524.08	180	3599.62	480	3670.81	116
3370.78	208	3524.24	115	3602.08	217	3672.69	141
3372.07	93	3527.79	221	3603.20	205	3674.77	248
3380.11	208	3529.82	221	3603.67	177	3676.31	173
3382.40	94	3530.39	221	3603.82	311	3676.88	257
3383.98	93	3531.44	143	3605.45	204	3677.31	458
3392.65	92	3534.53	482	3606.68	204	3677.63	202
3394.58	91	3536.56	221	3608.86	43	3678.86	114
3396.98	45	3537.73	180	3610.16	216	3679.91	5
3399.33	91	3537.90	222	3610.70	218	3681.64	258
3402.26	375	3538.78	482	3612.07	220	3682.24	459
3406.44	409	3540.12	223	3613.15	219	3683.05	5
3407.46	93	3540.71	43	3613.45	406	3684.11	203
3410.17	442	3541.08	221	3614.77	261	3686.00	253
3411.35	207	3542.08	221	3615.19	349	3686.26	114
3413.13	92	3543.39	144	3615.66	66	3687.10	83
3417.27	45	3543.67	441	3616.15	349	3687.46	41
3417.84	91	3544.63	180	3616.32	117	3688.48	405
3418.51	91	3548.02	311	3617.79	311	3688.88	140
3424.28	91	3549.86	67	3618.77	43	3689.02	139
3425.01	331	3551.11	216	3620.24	219	3689.90	330
3427.12	91	3552.11	313	3621.46	204	3690.73	478
3428.19	92	3552.83	216	3622.00	205	3694.01	260
3428.75	496	3553.74	481	3623.19	141	3697.43	257
3440.99	6	3556.88	222	3624.06	350	3698.60	308
3442.36	118	3559.50	312	3624.31	115	3699.15	307
3443.88	6	3560.07	216	3627.05	479	3701.09	253
3445.15	91	3560.70	406	3628.09	86	3702.03	248
3447.28	90	3564.11	67	3628.82	284	3703.69	257
3450.33	90	3565.38	44	3630.35	218	3703.82	248
3462.35	89	3566.31	113	3631.46	43	3704.01	310
3463.30	67	3567.03	220	3632.04	311	3704.46	201
3469.83	181	3567.37	144	3632.55	283	3705.57	5
3476.70	6	3568.42	216	3633.84	285	3707.82	5
3477.85	90	3568.82	407	3635.19	307	3709.25	41
3483.01	44	3568.98	204	3636.99	175	3711.22	173
3485.34	88	3570.10	44	3637.25	141	3711.41	309
3493.28	67	3571.22	66	3637.86	253	3715.91	111
3493.69	206	3572.00	216	3638.30	204	3718.41	203
3495.29	179	3572.59	220	3640.39	205	3719.93	5
3496.19	145	3573.39	407	3641.45	218	3722.56	5
3497.10	88	3576.76	374	3644.58	176	3724.38	110
3497.84	6	3578.38	216	3644.80	350	3725.49	329
3500.57	179	3578.67	113	3645.82	311	3726.93	253
3504.86	114	3581.19	43	3647.84	43	3727.09	255
3505.07	312	3582.20	373	3649.51	202	3727.62	41
3506.50	116	3583.33	353	3650.03	260	3728.67	171
3508.49	286	3585.32	43	3651.47	205	3730.39	330
3509.12	221	3585.71	43	3653.76	141	3730.95	173
3509.87	88	3586.98	43	3654.66	86	3731.37	169
3510.44	120	3589.11	43	3655.46	248	3732.40	85
3511.74	179	3590.08	285	3657.14	116	3733.32	5
3512.22	221	3591.00	352	3657.89	261	3734.86	41
3513.05	67	3591.35	216	3658.55	174	3735.32	256
3513.82	44	3591.48	348	3659.52	141	3737.13	5
3514.63	144	3592.47	178	3661.36	140	3738.31	372
3516.41	286	3592.67	349	3664.54	259	3739.12	83
3516.56	221	3592.89	87	3664.69	258	3739.32	84
3518.68	222	3593.32	351	3666.94	66	3740.24	403
3518.82	88	3594.63	217	3667.25	350	3742.62	255
3520.85	179	3595.30	217	3668.21	348	3743.36	41
3521.84	88	3595.86	142	3668.89	172	3744.10	253

List of tabulated lines — Continued

Wavelength (Å)	No.	Wavelength (Å)	No.	Wavelength (Å)	No.	Wavelength (Å)	No.
3745.56	5	3806.22	440	3890.39	347	3974.40	344
3745.90	5	3806.70	370	3890.84	195	3974.77	81
3746.49	82	3807.54	82	3891.93	439	3975.21	125
3746.93	254	3808.29	306	3893.39	280	3975.85	563
3748.26	5	3808.73	167	3895.66	4	3976.61	438
3749.48	41	3809.04	247	3897.45	279	3977.74	81
3751.06	404	3810.76	401	3899.03	137	3979.65	342
3751.82	199	3811.89	199	3899.71	4	3980.65	125
3753.15	138	3813.63	197	3900.52	345	3981.11	109
3753.61	82	3813.88	504	3902.95	65	3981.77	194
3754.51	254	3814.52	42	3903.90	279	3983.96	193
3756.07	83	3815.84	65	3906.48	4	3985.39	399
3756.94	477	3816.34	82	3906.75	401	3989.86	455
3757.45	402	3817.64	426	3907.47	198	3990.37	326
3758.23	41	3819.50	428	3907.93	195	3994.11	325
3760.05	138	3820.43	40	3909.66	345	3995.20	368
3760.53	85	3821.18	371	3909.83	246	3995.98	192
3761.41	171	3821.83	167	3910.84	198	3996.97	546
3762.21	430	3824.44	4	3911.00	343	3997.39	194
3763.79	41	3825.88	40	3913.63	108	3998.05	191
3765.54	371	3826.84	197	3914.27	347	4000.27	337
3766.09	170	3827.82	65	3916.73	369	4000.46	278
3766.67	254	3829.13	547	3917.18	40	4001.66	81
3767.19	41	3829.77	165	3919.07	280	4003.76	437
3768.03	82	3833.31	165	3920.26	4	4005.24	63
3770.30	199	3834.22	40	3920.84	347	4006.31	367
3771.50	370	3836.33	401	3922.91	4	4007.27	193
3773.36	329	3837.13	167	3925.20	347	4009.71	81
3773.70	254	3839.26	328	3927.92	4	4010.18	530
3774.82	82	3839.61	573	3930.30	4	4011.42	162
3775.86	199	3840.44	40	3931.12	345	4011.71	125
3776.45	83	3841.05	65	3935.31	244	4014.53	475
3777.06	281	3843.26	327	3937.33	194	4016.43	341
3777.45	168	3845.17	112	3940.88	40	4017.15	326
3778.32	247	3845.69	457	3941.28	343	4018.28	341
3778.51	401	3846.00	428	3942.44	246	4019.05	164
3778.70	82	3846.41	476	3943.34	81	4020.49	531
3781.19	83	3846.80	401	3944.75	245	4021.87	194
3781.94	532	3848.29	166	3944.89	280	4022.45	136
3782.45	256	3849.96	40	3945.12	195	4024.11	193
3782.61	308	3850.82	42	3946.99	342	4024.72	341
3785.71	371	3852.57	82	3948.77	368	4030.18	81
3785.95	138	3853.46	279	3949.14	438	4031.24	304
3786.19	247	3856.37	4	3949.95	81	4031.96	398
3786.68	42	3859.21	137	3951.16	399	4032.47	215
3787.16	533	3859.91	4	3952.60	194	4032.63	64
3787.88	41	3863.74	195	3953.15	280	4035.25	495
3789.18	200	3865.52	40	3953.86	244	4036.37	192
3789.82	427	3867.22	305	3955.34	343	4040.64	398
3790.09	42	3867.93	165	3955.96	305	4041.91	366
3791.50	168	3871.75	279	3956.45	368	4042.75	337
3791.73	428	3872.50	40	3957.02	343	4044.61	243
3792.15	199	3872.92	198	3960.28	531	4045.81	63
3792.83	84	3873.76	137	3961.15	245	4049.34	162
3793.87	247	3876.04	42	3962.35	346	4051.92	425
3794.34	138	3878.02	40	3963.10	343	4054.18	338
3795.00	41	3878.57	4	3964.52	245	4054.87	424
3797.95	167	3883.28	400	3966.06	65	4055.03	162
3798.51	41	3884.36	196	3967.42	368	4057.34	193
3799.55	41	3885.15	280	3967.96	342	4058.22	339
3801.68	247	3885.51	111	3969.26	63	4058.75	108
3802.00	429	3886.28	4	3969.63	397	4059.73	454
3802.28	402	3887.05	40	3970.39	305	4062.44	243
3804.01	427	3888.51	65	3971.32	193	4063.59	63
3805.35	371	3888.82	305	3973.65	456	4065.40	424

List of tabulated lines — Continued

Wavelength (Å)	No.	Wavelength (Å)	No.	Wavelength (Å)	No.	Wavelength (Å)	No.
4066.59	277	4134.68	240	4219.36	473	4300.83	562
4067.27	160	4136.51	421	4220.05	572	4302.18	321
4067.49	275	4137.00	436	4220.34	303	4304.54	270
4067.60	398	4137.42	645	4222.21	124	4305.20	450
4067.98	340	4139.93	38	4223.73	272	4305.45	300
4069.08	338	4141.86	275	4224.17	416	4307.90	62
4070.03	215	4142.63	645	4224.51	416	4309.03	502
4070.77	339	4143.87	63	4225.45	420	4309.37	270
4071.74	63	4145.21	190	4225.96	322	4310.37	572
4073.76	339	4146.06	275	4226.42	237	4315.08	80
4074.79	324	4147.67	62	4228.72	417	4317.04	452
4076.23	304	4149.37	421	4229.75	61	4319.46	158
4076.63	339	4150.25	422	4230.58	302	4325.76	62
4078.35	163	4152.17	38	4232.73	3	4326.75	269
4079.18	425	4153.90	422	4233.60	124	4327.09	451
4079.84	243	4154.80	421	4235.94	124	4327.92	363
4080.21	339	4156.80	240	4237.07	39	4337.05	61
4080.89	338	4158.79	422	4237.67	273	4338.26	79
4082.13	424	4160.56	274	4238.81	420	4343.28	393
4082.44	530	4160.78	657	4239.36	528	4343.70	319
4084.49	424	4161.08	416	4240.37	453	4346.55	364
4085.00	242	4161.48	275	4241.11	236	4347.24	2
4085.30	340	4167.86	365	4242.73	395	4347.85	492
4085.98	623	4168.63	416	4243.79	572	4348.94	270
4087.09	421	4168.94	421	4245.26	237	4351.54	269
4088.57	530	4169.78	420	4246.08	527	4352.73	80
4089.22	275	4170.90	303	4247.43	420	4358.50	267
4090.09	425	4171.69	545	4248.22	303	4360.81	525
4090.98	422	4171.90	396	4249.32	107	4365.90	268
4091.55	240	4172.12	395	4250.12	124	4367.58	270
4092.46	38	4172.74	39	4250.79	62	4367.90	61
4095.27	624	4173.32	238	4253.55	731	4369.77	320
4095.97	161	4173.92	39	4256.79	644	4372.99	299
4097.10	339	4174.91	39	4258.31	3	4374.50	394
4098.18	339	4175.64	240	4258.62	236	4375.93	2
4100.74	38	4177.59	38	4258.95	274	4377.80	393
4101.27	424	4180.40	190	4260.47	124	4382.77	472
4101.68	108	4181.75	239	4264.20	419	4383.54	61
4104.97	421	4182.38	301	4264.74	571	4384.68	298
4106.27	160	4182.79	421	4266.96	189	4387.89	300
4106.44	423	4183.03	423	4267.83	303	4388.41	494
4107.49	239	4184.89	238	4268.75	395	4389.24	2
4108.13	340	4187.04	124	4271.15	124	4390.46	269
4109.07	339	4187.79	124	4271.76	62	4390.95	270
4109.80	240	4189.56	544	4275.72	159	4391.87	570
4112.35	422	4191.68	238	4276.68	562	4392.58	559
4112.96	645	4194.50	190	4277.41	158	4395.29	492
4114.45	239	4196.21	420	4278.23	418	4401.29	492
4114.96	422	4196.53	273	4279.48	571	4401.44	235
4116.97	339	4197.38	562	4279.86	236	4404.75	61
4118.54	474	4198.30	124	4280.53	364	4407.71	77
4118.90	340	4198.64	420	4282.40	80	4408.41	77
4120.21	276	4199.09	323	4284.42	272	4409.12	393
4121.80	241	4200.09	571	4285.44	363	4413.40	604
4122.51	241	4200.92	416	4285.83	526	4415.12	61
4124.49	529	4202.03	62	4286.44	270	4419.30	522
4125.88	239	4203.67	731	4288.15	189	4422.57	235
4126.18	422	4203.94	503	4288.96	158	4423.14	267
4126.88	239	4205.54	416	4290.38	271	4423.84	494
4127.61	240	4206.70	3	4290.87	236	4430.19	297
4129.22	424	4207.13	237	4292.13	79	4430.61	77
4129.46	422	4210.34	124	4292.29	79	4432.57	471
4132.06	63	4213.65	238	4294.12	61	4433.22	494
4132.90	239	4216.18	3	4298.04	321	4433.78	490
4133.86	424	4217.55	420	4300.21	561	4435.15	2

List of tabulated lines — Continued

Wavelength (Å)	No.	Wavelength (Å)	No.	Wavelength (Å)	No.	Wavelength (Å)	No.
4436.92	318	4546.48	605	4672.84	60	4804.52	468
4438.34	492	4546.68	569	4673.16	486	4807.71	415
4439.63	317	4547.02	59	4673.28	488	4808.15	386
4439.88	106	4547.85	449	4674.65	60	4809.94	467
4440.48	493	4551.65	558	4677.60	622	4813.11	385
4440.82	570	4554.47	214	4678.85	487	4813.72	730
4442.34	77	4556.93	391	4679.22	415	4817.77	76
4442.83	78	4560.09	489	4680.29	59	4818.66	432
4443.19	235	4561.43	188	4682.56	252	4834.51	105
4445.47	2	4565.31	392	4683.57	229	4835.87	619
4446.83	492	4565.66	335	4685.03	231	4837.65	730
4447.13	78	4566.51	392	4687.39	230	4838.09	386
4447.72	77	4566.99	434	4687.68	231	4838.51	414
4450.32	300	4568.61	569	4690.14	486	4839.55	357
4450.77	558	4571.44	214	4690.38	37	4840.32	619
4452.62	556	4572.86	485	4691.41	265	4841.78	621
4454.38	234	4574.21	335	4700.19	543	4842.79	620
4455.03	560	4574.72	105	4701.05	486	4843.14	414
4456.33	318	4579.82	296	4704.95	487	4844.01	444
4456.63	559	4580.58	491	4705.46	446	4848.90	104
4459.12	77	4581.51	336	4706.31	521	4849.67	467
4461.65	2	4587.13	469	4707.27	335	4854.89	603
4464.77	297	4587.72	557	4707.49	229	4859.13	619
4466.55	233	4592.65	59	4712.10	294	4859.74	213
4466.94	570	4593.53	557	4714.07	715	4860.98	415
4469.37	494	4595.36	362	4714.19	360	4867.53	58
4471.68	2	4596.06	486	4716.85	388	4869.45	445
4478.04	78	4596.41	489	4726.14	252	4870.05	567
4479.97	560	4598.12	335	4727.00	389	4871.32	213
4480.14	317	4598.73	485	4729.02	598	4872.14	213
4481.61	491	4600.93	360	4729.68	415	4872.69	656
4482.17	2	4602.00	59	4734.10	667	4872.91	642
4482.74	492	4603.34	230	4735.84	602	4873.74	385
4483.78	524	4603.95	266	4736.77	335	4874.36	293
4484.22	492	4604.25	231	4737.63	358	4875.87	414
4485.67	494	4607.08	435	4740.34	265	4876.19	384
4485.97	490	4612.64	232	4741.53	229	4877.61	252
4487.74	362	4613.20	335	4745.13	76	4878.21	213
4488.13	485	4614.21	391	4749.95	715	4882.14	414
4489.74	2	4618.76	265	4760.07	252	4887.37	599
4490.08	296	4619.29	487	4765.48	60	4890.75	213
4492.68	556	4620.14	295	4766.87	415	4891.49	213
4493.37	470	4625.04	335	4771.70	76	4892.87	621
4494.05	559	4626.76	266	4776.07	389	4896.44	566
4494.56	77	4630.12	105	4779.44	433	4903.31	213
4495.57	491	4631.48	679	4780.81	385	4905.13	568
4495.95	490	4632.91	59	4782.79	357	4907.73	414
4502.59	470	4633.76	266	4785.96	601	4908.61	105
4504.83	336	4635.62	214	4786.81	292	4911.52	643
4510.82	489	4635.85	232	4787.83	252	4911.78	566
4514.18	316	4636.66	315	4788.76	357	4912.52	600
4515.16	214	4637.50	335	4789.65	447	4916.67	568
4517.53	297	4638.01	488	4790.56	619	4917.23	618
4518.43	361	4643.22	58	4790.75	387	4918.01	621
4518.58	78	4643.46	486	4791.25	386	4918.99	213
4523.40	493	4647.43	265	4793.96	314	4920.50	213
4525.88	214	4649.82	359	4794.36	105	4924.77	104
4527.78	392	4654.50	59	4798.26	602	4925.29	617
4528.61	77	4657.59	229	4798.73	58	4927.42	466
4531.15	59	4658.29	360	4799.06	643	4930.31	567
4533.13	392	4661.33	231	4799.41	520	4935.42	518
4537.67	362	4661.53	716	4800.13	252	4939.69	36
4541.32	390	4661.97	265	4800.65	602	4945.64	654
4541.94	361	4663.18	448	4801.63	656	4946.38	414
4542.41	523	4669.17	487	4802.53	715	4950.10	414

List of tabulated lines — Continued

Wavelength (Å)	No.	Wavelength (Å)	No.	Wavelength (Å)	No.	Wavelength (Å)	No.
4961.91	501	5109.65	635	5253.46	334	5405.77	35
4962.56	642	5110.41	1	5254.96	1	5406.77	676
4966.09	414	5115.78	465	5262.61	677	5409.13	675
4968.69	519	5119.90	551	5262.89	382	5410.91	686
4969.92	618	5121.64	641	5263.30	334	5412.80	683
4970.50	516	5123.72	36	5263.87	464	5415.20	686
4973.10	566	5124.60	356	5266.55	251	5417.03	676
4978.60	554	5125.11	636	5267.28	674	5421.85	698
4979.59	516	5126.19	635	5269.54	35	5422.15	673
4985.98	640	5127.36	36	5270.36	57	5424.07	674
4986.22	621	5127.68	1	5273.37	104	5429.70	35
4986.90	638	5129.63	553	5279.65	355	5432.95	671
4987.62	640	5131.47	75	5280.36	515	5434.52	35
4988.95	618	5133.69	638	5281.79	251	5435.17	682
4991.27	617	5136.09	597	5283.62	334	5436.30	682
4991.86	640	5137.38	636	5284.42	499	5436.59	103
4992.80	651	5141.74	104	5284.62	596	5438.04	726
4993.68	652	5143.73	74	5285.12	687	5441.32	672
4994.13	36	5145.09	75	5288.53	540	5443.41	613
4995.41	654	5145.73	542	5293.03	686	5445.04	684
4999.11	600	5146.30	678	5293.97	595	5446.92	35
5001.86	553	5150.84	36	5294.56	513	5452.12	509
5002.79	414	5151.91	36	5295.32	674	5460.91	290
5004.04	653	5159.06	637	5298.79	513	5461.54	673
5012.07	36	5159.95	641	5300.41	727	5463.27	684
5012.68	639	5162.27	635	5301.33	683	5464.29	594
5014.94	553	5164.55	687	5302.30	334	5466.39	672
5016.48	635	5166.28	1	5307.36	56	5470.17	672
5019.18	729	5167.49	57	5308.71	637	5472.72	649
5021.89	383	5168.90	1	5315.07	675	5473.18	616
5022.24	553	5171.60	56	5319.22	592	5473.90	615
5023.23	641	5177.23	541	5320.05	514	5478.48	615
5023.50	678	5178.80	687	5321.11	686	5480.87	615
5025.08	651	5180.07	687	5322.04	102	5481.25	612
5025.30	655	5184.26	635	5324.18	334	5481.45	614
5027.76	651	5187.91	596	5326.79	675	5482.26	512
5029.62	431	5194.94	56	5328.04	35	5483.11	614
5030.77	356	5197.93	637	5329.99	593	5487.16	671
5031.90	678	5198.71	75	5332.67	595	5487.74	590
5044.21	213	5202.34	75	5332.90	56	5489.85	676
5048.43	566	5204.58	1	5339.93	334	5491.84	595
5049.82	104	5206.80	641	5341.02	57	5493.51	614
5051.63	36	5207.95	515	5349.74	684	5494.46	589
5054.64	517	5208.59	334	5353.39	615	5496.57	747
5056.00	677	5209.90	355	5361.64	671	5497.52	35
5056.86	652	5213.35	686	5364.87	674	5499.60	680
5058.00	555	5213.80	552	5367.47	674	5501.46	35
5058.50	517	5216.27	56	5369.96	674	5506.78	35
5060.08	1	5218.51	728	5371.49	35	5512.28	671
5067.15	638	5223.19	515	5373.71	687	5517.08	650
5068.77	251	5224.30	74	5376.85	666	5521.28	683
5074.75	640	5225.53	1	5379.57	539	5522.46	649
5079.22	75	5228.41	637	5383.37	674	5524.25	613
5079.74	36	5232.94	251	5385.58	538	5525.55	615
5080.95	356	5236.19	596	5386.34	616	5528.89	682
5083.34	36	5236.38	674	5387.51	595	5529.15	511
5085.68	639	5238.25	552	5389.48	673	5531.95	747
5088.16	618	5241.90	678	5393.17	334	5532.75	463
5090.78	636	5242.49	500	5394.68	595	5536.59	228
5099.09	553	5243.78	635	5395.25	671	5539.28	510
5104.04	291	5247.05	1	5397.13	35	5539.83	665
5104.21	638	5249.10	687	5397.62	498	5543.15	537
5104.44	636	5250.21	1	5398.29	673	5543.94	615
5107.45	36	5250.64	75	5400.50	673	5546.51	673
5107.64	56	5253.03	103	5401.27	674	5547.00	614

List of tabulated lines — Continued

Wavelength (Å)	No.	Wavelength (Å)	No.	Wavelength (Å)	No.	Wavelength (Å)	No.
5549.66	756	5679.02	698	5853.18	55	6163.56	73
5549.94	537	5680.26	591	5855.13	694	6165.37	585
5552.70	747	5686.53	697	5856.08	662	6170.49	740
5553.59	682	5691.51	633	5858.77	631	6173.34	71
5554.89	698	5696.10	694	5861.11	631	6180.22	187
5557.95	684	5698.05	505	5864.24	632	6188.04	550
5559.64	748	5698.37	664	5873.21	633	6191.56	134
5560.23	685	5701.54	157	5876.27	631	6199.48	156
5563.60	615	5702.43	507	5877.77	630	6200.32	155
5567.40	157	5705.48	633	5879.49	713	6213.43	71
5568.81	507	5705.99	698	5880.00	713	6215.15	585
5569.62	413	5707.07	508	5881.28	693	6219.28	71
5572.84	413	5707.25	507	5883.84	565	6220.77	549
5576.09	413	5708.11	682	5892.80	72	6226.77	564
5577.03	756	5709.38	413	5898.21	739	6229.23	227
5579.34	614	5709.93	634	5902.52	725	6230.73	155
5583.97	613	5711.87	633	5905.67	696	6240.66	73
5584.77	462	5712.15	413	5909.99	333	6246.32	484
5586.76	413	5715.47	608	5916.25	135	6252.55	134
5587.58	591	5717.85	648	5927.80	690	6253.82	737
5594.66	697	5720.89	693	5929.70	691	6254.26	101
5595.06	756	5724.45	650	5930.17	695	6256.37	134
5598.30	698	5731.77	633	5933.80	711	6265.13	71
5607.66	612	5732.29	755	5934.66	565	6270.24	227
5608.98	649	5732.86	609	5940.97	630	6271.29	412
5609.97	506	5738.22	631	5952.75	550	6280.63	33
5611.35	506	5741.86	632	5955.68	647	6290.55	156
5615.64	413	5742.95	631	5956.70	34	6297.80	71
5617.22	381	5747.95	697	5961.91	627	6303.46	669
5618.65	648	5749.65	681	5969.55	632	6311.51	227
5619.23	535	5753.12	648	6003.03	550	6315.81	582
5619.60	682	5754.41	506	6012.21	73	6322.69	155
5620.53	614	5759.27	699	6015.25	72	6330.86	735
5624.06	681	5760.35	505	6016.66	443	6335.34	71
5624.54	413	5762.43	506	6019.36	461	6336.84	484
5633.97	756	5762.99	648	6020.17	693	6338.90	738
5635.85	634	5778.47	157	6024.07	693	6344.15	134
5636.71	508	5780.62	333	6027.06	585	6355.04	227
5638.27	633	5784.69	413	6032.67	629	6358.69	33
5640.46	714	5787.27	380	6034.04	670	6362.89	586
5641.46	633	5791.04	333	6035.34	660	6364.38	734
5642.75	699	5793.93	632	6054.10	670	6380.75	583
5643.94	587	5798.19	565	6055.99	739	6385.74	734
5644.35	611	5804.06	550	6060.81	628	6392.55	100
5646.70	650	5804.48	633	6062.89	72	6393.60	133
5649.66	497	5805.76	755	6065.48	155	6400.00	484
5650.01	756	5806.73	695	6079.02	691	6411.65	484
5650.71	756	5807.79	333	6082.72	73	6419.98	738
5651.47	682	5807.97	693	6085.27	187	6421.35	101
5652.01	613	5809.25	565	6093.66	692	6430.85	71
5652.32	649	5811.93	588	6094.42	692	6436.43	584
5653.89	680	5814.80	632	6096.69	550	6462.73	133
5655.18	756	5815.16	609	6098.28	712	6469.21	738
5658.82	413	5816.07	661	6105.15	690	6475.63	154
5660.79	507	5816.36	694	6120.25	34	6481.88	100
5661.03	725	5826.64	631	6136.61	134	6494.98	133
5661.36	649	5827.89	333	6137.00	71	6495.78	734
5661.97	650	5833.93	157	6137.69	155	6496.46	738
5662.94	536	5835.10	631	6139.65	156	6498.95	33
5667.67	157	5837.71	663	6141.73	484	6509.56	581
5672.28	725	5838.42	550	6145.42	412	6518.38	227
5677.68	611	5844.88	610	6147.85	584	6533.97	710
5678.04	749	5845.27	755	6151.62	71	6546.24	186
5678.39	565	5849.67	534	6157.73	583	6551.68	33
5678.60	103	5852.19	693	6159.41	690	6569.23	734

List of tabulated lines — Continued

Wavelength (Å)	No.	Wavelength (Å)	No.	Wavelength (Å)	No.	Wavelength (Å)	No.
6574.24	33	6793.26	578	7022.39	626	7401.69	575
6575.02	154	6794.60	745	7022.98	606	7418.32	576
6581.22	54	6796.11	580	7024.08	577	7418.67	575
6591.32	724	6804.02	689	7024.65	701	7420.20	752
6592.91	186	6804.27	719	7038.25	606	7421.60	702
6593.88	133	6806.85	186	7038.82	626	7430.58	152
6597.61	734	6810.28	710	7044.60	743	7443.03	576
6608.03	100	6820.43	710	7057.96	483	7443.26	753
6609.12	154	6824.80	746	7068.02	743	7454.02	574
6625.04	33	6828.61	708	7068.42	575	7461.53	152
6627.56	689	6833.24	707	7069.54	153	7463.38	752
6633.44	738	6837.00	719	7071.88	707	7473.56	702
6633.76	710	6839.83	153	7072.82	577	7476.40	733
6634.10	738	6841.35	708	7079.32	744	7481.74	184
6639.72	708	6842.67	710	7090.40	606	7481.93	732
6639.90	580	6843.67	688	7093.09	703	7484.28	751
6646.98	154	6851.64	54	7094.30	460	7498.56	574
6653.88	607	6854.82	720	7095.43	646	7501.25	576
6663.45	101	6855.74	707	7107.46	578	7540.44	184
6667.42	133	6857.25	579	7109.67	704	7541.61	548
6667.73	723	6858.16	688	7112.18	264	7551.10	750
6677.99	186	6859.49	226	7114.55	185	7582.15	742
6696.32	736	6860.29	153	7118.10	744	7583.80	263
6699.14	723	6861.93	100	7125.00	483	7588.30	751
6703.57	186	6862.48	705	7130.94	606	7723.20	99
6704.48	607	6864.31	700	7132.99	576	7737.67	668
6710.31	54	6880.65	606	7151.50	100	7748.27	263
6712.44	745	6885.77	688	7158.50	483	7820.80	658
6713.76	736	6898.31	626	7180.02	53	7844.55	732
6715.41	689	6911.52	100	7189.17	289	7869.65	668
6716.24	719	6916.70	607	7190.13	289	7879.75	751
6725.39	607	6933.04	606	7213.84	646	7912.87	32
6732.06	719	6936.48	709	7219.69	574	7941.09	379
6733.16	708	6945.20	101	7228.69	185	8075.13	32
6736.56	659	6960.33	717	7256.14	744	8327.05	70
6737.98	706	6971.95	264	7261.02	185	8387.77	70
6739.54	54	6976.93	718	7284.84	574	8468.40	70
6745.11	722	6978.85	101	7285.29	702	8514.07	70
6745.96	578	6988.53	132	7306.61	625	8611.81	225
6746.96	153	6997.13	741	7312.05	754	8674.75	225
6750.15	101	6999.90	606	7330.15	701	8688.62	70
6752.72	708	7000.63	578	7347.16	184	8757.19	225
6753.45	709	7008.01	626	7353.53	733	8804.62	98
6761.07	722	7010.36	718	7359.95	754	8838.43	225
6769.66	721	7014.99	132	7366.37	702	8999.55	225
6783.71	153	7016.08	100	7396.50	744	9088.33	225
6786.88	607	7016.44	606	7400.87	152	10218.36	288

From the large number of articles containing f-value data on Fe I, we have selected most of the recent experiments (Refs. 1–20, 25–32) for this tabulation. Much of the material is taken from two very comprehensive sources, the stabilized-arc emission experiments by May et al.[5] and by Bridges and Kornblith.[4] The most accurate set of oscillator strengths is the one measured by Blackwell and co-workers,[1-3, 27-29] with an absolute scale based on lifetime data. Another reference providing a wealth of data is that of Gurtovenko and Kostik,[30] who derived oscillator strengths from solar spectra.

We established the absolute scale by utilizing accurate data for the principal resonance line at 3719.93 Å. The

atomic beam work by Bell and Tubbs[20] yields the f-value of this transition directly, and lifetime measurements of its upper level, $z\,{}^5F_5^{\circ}$, may also be converted into f-values, since the other downward transitions contribute—at most—a few additional percent to the total lifetime and can be approximately corrected for. Very accurate lifetime measurements of this upper level have been performed by Wagner and Otten,[16] who used the method of optical double resonance; Klose,[17] who used the delayed coincidence technique; Hilborn and de Zafra,[18] who employed the Hanle effect; Brzozowski et al.,[19] who used the high frequency deflection technique; and Marek et al.[31] and Hannaford and Lowe,[32] who employed the method of selective laser excitation. The average f-value resulting from these six lifetime measurements and the atomic beam experiment is $f = 0.0412$, with a standard deviation of the mean of $\pm 0.75\%$ (these lifetime data are given in Table 1). This f-value (obtained by including the effects of the other weak transitions involved) is estimated to have an overall uncertainty not to exceed three percent and forms the basis of the absolute scale for this spectrum, to which all other measurements discussed below were normalized. (For most references, changes (usually small) in the absolute scale had to be made, and we have indicated this by an "n" in the source column.)

TABLE 1. Selected lifetime-oscillator strength data for the Fe I resonance level.

Reference	τ(ns) of the $z\,^5F_5^\circ$ level	Oscillator strength of the 3719.93 Å line
Wagner and Otten[16]	59.5 ± 1.6	0.0425
Klose[17]	61.5 ± 0.4	0.0413
Hilborn and de Zafra[18]	63.2 ± 3.6	0.0400
Brzozowski et al.[19]	60.5 ± 1.5	0.0418
Marek, Richter, and Stahnke[31]	62.4 ± 4.2	0.0405
Hannaford and Lowe[32]	61.0 ± 1.0	0.0414
Bell and Tubbs[20]		0.041 ± 0.003

Avg. f-value $= 0.0412 \pm 0.000311$.

The spectrum of Fe I is very rich in lines of moderate strength in the visible and near uv regions. Two large-scale measurements of relative f-values have been carried out by May et al.[5] and by Bridges and Kornblith[4] for this spectral range. Both experiments were performed in emission with stabilized, steady state arc sources. The most comprehensive set of data on this spectrum is by May et al.,[5] who determined relative oscillator strengths for over 1000 lines with a convection stabilized arc and employed photographic detection. Bridges and Kornblith determined data for 534 lines with a more sophisticated photoelectric data acquisition technnque; this included a self-regulating system for the arc discharge, in which fluctuations in the spectral line signals were monitored and controlled in order to maintain stability in the arc chamber. Since the data of May et al. and of Bridges and Kornblith overlap for 168 lines, we were able to make several graphical comparisons (Figs. 1–3), plotting log gf (May et al.)–log gf (Bridges and Kornblith) (in the graphs denoted by Δlog) vs wavelength, vs log gf (of Ref. 4), and vs upper energy level. These studies show that the mutual scatter is only about ± 0.1 dex and essentially random, i.e., there are no intensity or energy level dependent trends. However, there is some marked disagreement between the f-values of Refs. 4 and 5 for the lines of shortest wavelength, especially $\lambda = 3495.29$, 3699.15, 3540.12, and 3521.84 Å. This may be due to scattered light problems for the radiometric standards at short wavelengths. Since Bridges and Kornblith took this problem into account by application of appropriate filters, we used their data (or more accurate data) in these cases.

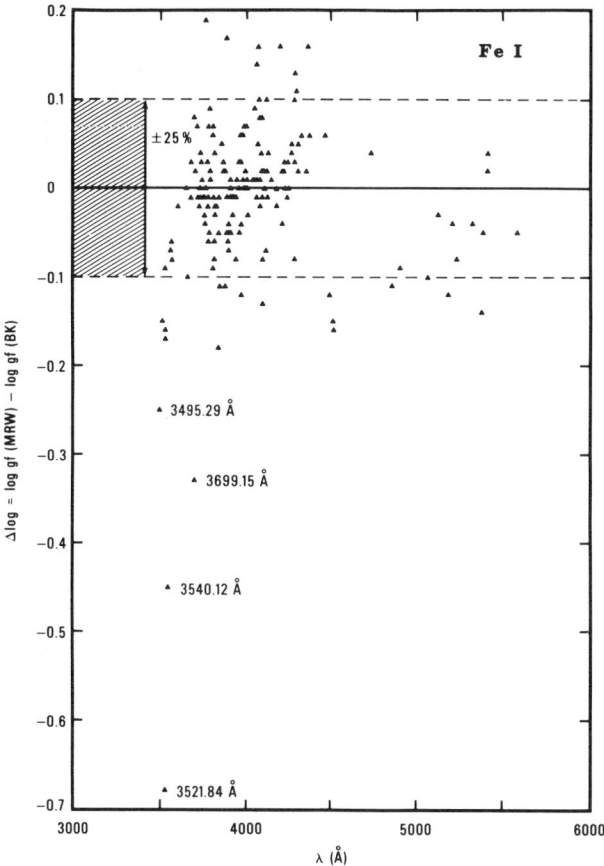

FIG. 1. Plot of Δlog $= \log gf$(May et al.[5]) $- \log gf$(Bridges and Kornblith[4]) vs wavelength (Å).

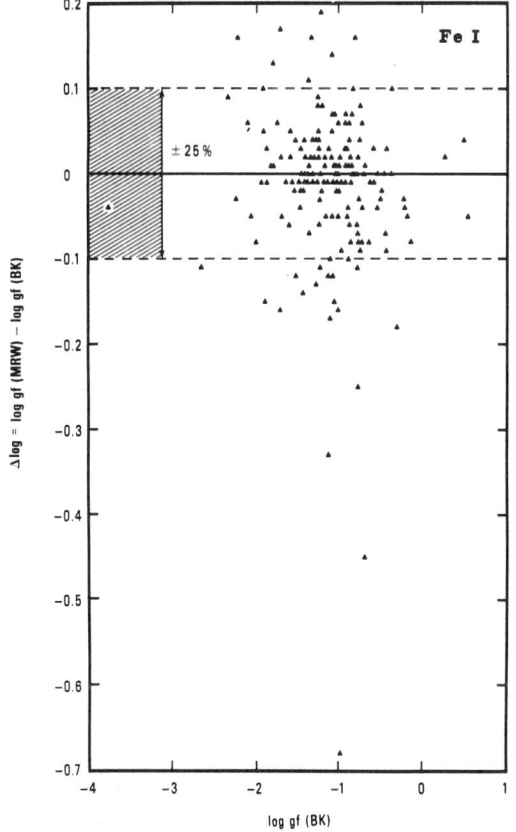

FIG. 2. Plot of Δlog $= \log gf$(May et al.[5]) $- \log gf$(Bridges and Kornblith[4]) vs log gf(Bridges and Kornblith).

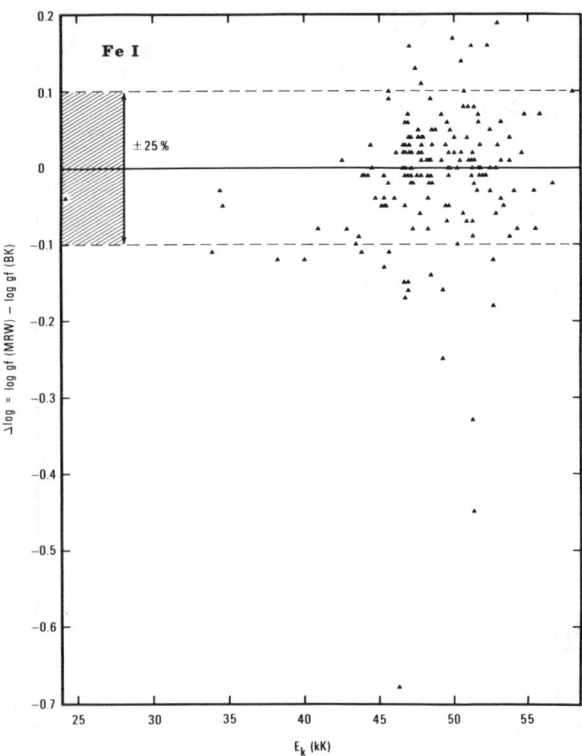

FIG 3. Plot of $\Delta\log = \log gf$(May et al.[5]) $- \log gf$(Bridges and Kornblith[4]) vs upper energy level.

The data of Bridges and Kornblith could be subjected to another important check: They overlap for 106 lines with the data of Blackwell et al.[1,3,27-29] (to be discussed later), which are of outstanding accuracy. The comparison, illustrated in Fig. 4, shows fairly good agreement: for example, 68% of the data are within $\pm 25\%$ of each other. Nevertheless, there are some differences outside the mutually estimated uncertainties. The graphical comparison also indicates: (a) a systematic trend in the data with line intensity (or $\log gf$), (b) a small difference in absolute scales, and (c) a serious disagreement for the 4427.31 Å line. (a) The trend is probably due to two unrelated facts. First, the weak lines measured by Bridges and Kornblith, which have lower accuracy ratings, appear to be systematically too strong, a tendency which has also been observed for some other emission measurements of iron group elements. Secondly, the log gf-values for the strongest lines measured by Bridges and Kornblith may be slightly too small because of undetected minor amounts of self-absorption present (Bridges and Kornblith note that their self-absorption check is good to only a few percent). (b) The small difference in absolute scales is not unexpected, on account of the different normalization procedures employed. Bridges and Kornblith used an average based on various lifetime data involving numerous lines, while Blackwell et al. utilized only the very accurate data for the resonance line at 3719.9 Å. Since the high precision measurements of

Blackwell et al. combined with these resonance line data determine the absolute scale very accurately, we have used that scale to renormalize the data of Bridges and Kornblith. The graphical comparison between Ref. 4 and Blackwell et al. indicates that the data of Bridges and Kornblith suffer a larger degree of scatter and more pronounced shift in absolute scale for weak lines (those lines having log gf-values (Ref. 4) <0.75). As a result of this renormalization, we have lowered the log gf-values of Bridges and Kornblith for weaker and stronger lines by 0.10 and 0.04 respectively. Since May et al. normalized their scale directly to that of Bridges and Kornblith, we have used the same criterion in lowering their log gf-values. (c) A serious disagreement between Blackwell et al. and Bridges and Kornblith is seen in the case of the 4427.31 Å line. This line is completely blended with another line at 4427.30 Å, so we have accordingly omitted it from this compilation.

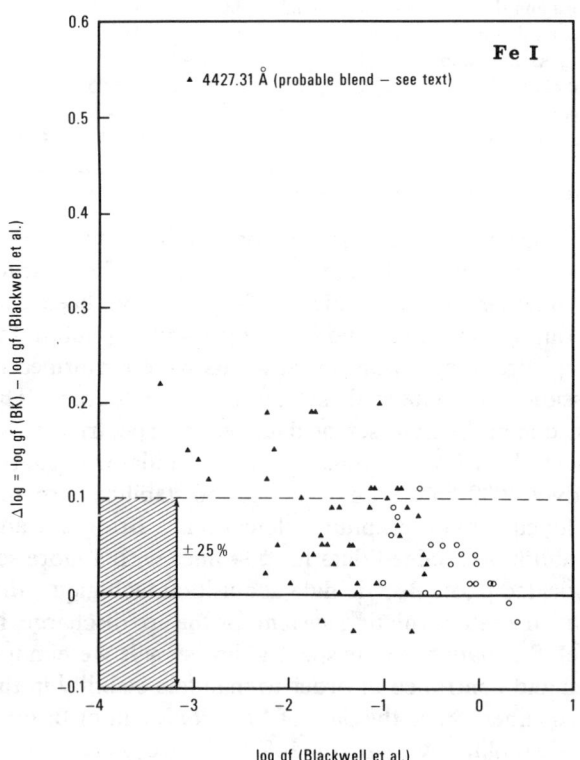

FIG 4. Plot of $\Delta\log = \log gf$(Bridges and Kornblith[4]) $- \log gf$ (Blackwell et al.[1,3]) vs log gf(Blackwell et al.). Open circles are used to represent lines for which the f-values of Bridges and Kornblith are denoted by them to be accurate to within 10% ("a" accuracy, in their notation), while solid triangles are used for lines with uncertainties greater than 10%.

After the few apparently unreliable f-values from Ref. 4 were eliminated, data for over five hundred lines remained. We have utilized these renormalized data as the principal reference source of accurate f-values for Fe I and have normalized and/or compared the other, much less comprehensive data sources (to be discussed later) to it. Our error estimates for the very weak and very strong lines were adjusted to reflect the possible deficiencies detected by the comparison with the data of Blackwell et al., as discussed above. Blackwell et al[3] also suggest a temperature error in the data of Bridges and Kornblith. However, we have found no indication of this from our detailed graphical comparisons. We should also note that temperature errors in the experiment of Bridges and Kornblith are minimized, since their absolute scale is based on numerous lifetime data for levels spanning a large range of excitation energies.

The most accurate relative oscillator strengths for Fe I are provided by the absorption experiments of Blackwell et al.[1-3,27-29] Their work centers on lines originating from the ground state or states of low excitation potential. An extremely stable and well diagnosed King-type furnace was used as the absorption tube, and intensity ratios were determined photoelectrically for various line pairs, which by appropriate overlaps were built up to a network that could be cross-checked and optimized for internal consistency. The relative data thus obtained—which span a large range of gf-values—were estimated to be accurate to within 0.5 percent in most cases. Because of possible blending, Blackwell et al.[29] have lowered their accuracy ratings to ±7 percent for a few of the weakest lines originating from the highest lower level (b ^{3}F).

Another comprehensive source which we utilized for this compilation is the astrophysical work by Gurtovenko and Kostik.[30] These authors derived oscillator strengths from central intensities of Fe I lines, taken from the Liege solar atlas.[33] Gurtovenko and Kostik established their absolute scale by normalizing 20 of their log gf-values to the accurate scale of Blackwell et al. After performing a series of graphical comparisons between Ref. 30 and other data sources, we detected two separate problems in the data of Gurtovenko and Kostik. First, we noticed that the log gf-values of Ref. 30 became increasingly greater than those of several other data sources as the lines become stronger. Since this may be mainly due to the approximate treament of non-LTE effects on the central line intensities, we have omitted all lines from Ref. 30 having log gf-values > -1.5. Secondly, the oscillator strengths of this reference were found to be systematically lower than those of Refs. 4 as well as 5 for lines of short wavelengths. For lines having wavelengths $\leqslant 4500$ Å, we have therefore lowered the accuracy rating of Ref. 30 to D−. We included a total of over 300 lines from the work of Gurtovenko and Kostik in this compilation to supplement the laboratory emission and absorption data. The solar f-values are estimated to be generally accurate to within ±50%.

An additional important data source is the experimental work by Huber and co-workers[6,7,11-13] which makes use of the anomalous dispersion and absorption techniques. Less comprehensive sources of data, which were utilized to supplement this material, are the branching-ratio emission experiments of Martinez-Garcia et al.[8] and of Adams and Whaling,[26] who employed hollow cathode discharges; the shock-tube emission work of Wolnik et

al.[9,14]; and the emission experiments with stabilized arcs by Garz and Kock[10] and Richter and Wulff.[15]

All these data were extensively intercompared in a series of graphic plots to establish their mutual consistency and, if necessary, to find appropriate renormalization factors. Normally, Δlog was plotted vs upper energy level for emission work and vs lower energy level for the anomalous dispersion and absorption experiments. Furthermore, Δlog was also plotted vs wavelength and vs log gf. The material by Bridges and Kornblith or by May et al. served as reference material, since their work covered so many lines. The graphs, of which Figs. 1–3 are samples, are instructive indicators of systematic trends which are dependent on upper or lower energy level, the magnitude of log gf, or the wavelength. Several disagreements in absolute scales were readily detected, and in three cases, an energy level dependent trend was noticed, and a least squares fit was then performed for a renormalization. In other cases, no renormalization was required at all. The resulting renormalization factors are shown in Table 2.

TABLE 2. Renormalization factors.

Reference	Normalization: number to be added to the original log gf-value, as it appeared in the literature.
4	−0.04 (if log gf (Ref. 4) $\geqslant -0.75$) −0.10 (if log gf (Ref. 4) < -0.75)
5	−0.04 (if log gf (Ref. 4) $\geqslant -0.75$) −0.10 (if log gf (Ref. 4) < -0.75)
6	+0.02
9	−0.06
10	$-0.44 + (0.0000101) E_k$[a]
11	+0.24
12	−0.20
13	$-0.11 - (0.00000511) E_i$
14	−0.16
15	$-0.69 + (0.0000129) E_k$

[a]The units of E_k (upper energy level) or E_i (lower energy level) are cm^{-1}.

The graphs are also a very good indicator of the scatter in the various sets of data. By intercomparing all overlapping data, one can readily isolate the principal sources of scatter. Our error estimates take this into account, in addition to an evaluation of the critical factors involved in each method and the error statements provided by the authors. When overlaps in the data occur, we have selected the very precise data of Refs. 1–3, 27–29 as our first choice. Next, we have given equal weight to the data fo Refs. 4–8, 26, averaging them when they overlap. Data from Refs. 9–15, 30 were tabulated with equal weight too, but only in those cases where no material from the earlier cited authors was available. In toto, we have compiled f-value data for about 1950 lines. In this compilation, we have generally omitted blended lines. Wavelengths have been taken from the work of

Crosswhite.[21] Energy level values and term designations as listed in our multiplet column have been taken from the compilation of Corliss and Sugar.[22] Particular attention was paid to the fact that the designations of some energy levels and multiplets have changed from the original classifications by Moore.[23,24] Also, some multiplet designations appear to be identical as we have listed them, for example, Nos. 19 and 26 in our tables, since the present setup does not completely identify the multiplets by their respective transition arrays. For further details on multiplet and term designations, the reader is referred to Ref. 22.

NOTES: After our final data tables were prepared, another paper by Gurtovenko and Kostik[34] became available to us. In this work, the authors determined oscillator strengths for 360 Fe I lines from observed equivalent widths of solar lines. These lines in this new reference overlap with those of Ref. 30 (determined from observed central intensities of solar lines), and the *f*-values agree quite well with one another, particularly for weak lines. We feel that both methods are capable of yielding fairly reliable oscillator strengths.

Dr. Blackwell informed us recently that the *gf*-data for the 3479.83 Å line may be erroneous, due to possible blending. We therefore urge the readers to disregard our tabulated data for this line.

References

[1]D. E. Blackwell, P. A. Ibbetson, A. D. Petford, and M. J. Shallis, Mon. Not. R. Astron. Soc. **186**, 633 (1979).

[2]D. E. Blackwell, P. A. Ibbetson, A. D. Petford, and R. B. Willis, Mon. Not. R. Astron. Soc. **177**, 219 (1976).

[3]D. E. Blackwell, A. D. Petford, and M. J. Shallis, Mon. Not. R. Astron. Soc. **186**, 657 (1979).

[4]J. M. Bridges and R. L. Kornblith, Astrophys. J. **192**, 793 (1974).

[5]M. May, J. Richter, and J. Wichelmann, Astron. Astrophys., Suppl. Ser. **18**, 405 (1974).

[6]F. P. Banfield and M. C. E. Huber, Astrophys. J. **186**, 335 (1973).

[7]M. C. E. Huber, Astrophys. J. **190**, 237 (1974).

[8]M. Martinez-Garcia, W. Whaling, D. L. Mickey, and G. M. Lawrence, Astrophys. J. **165**, 213 (1971).

[9]S. J. Wolnik, R. O. Berthel, and G. W. Wares, Astrophys. J. **162**, 1037 (1970).

[10]T. Garz and M. Kock, Astron. Astrophys. **2**, 274 (1969).

[11]G. L. Grasdalen, M. Huber, and W. H. Parkinson, Astrophys. J. **156**, 1153 (1969).

[12]M. Huber and F. L. Tobey, Jr., Astrophys. J. **152**, 609 (1968).

[13]M. C. E. Huber and W. H. Parkinson, Astrophys. J. **172**, 229 (1972).

[14]S. J. Wolnik, R. O. Berthel, and G. W. Wares, Astrophys. J. **166**, L31 (1971).

[15]J. Richter and P. Wulff, Astron. Astrophys. **9**, 37 (1970).

[16]R. Wagner and E. W. Otten, Z. Phys. **220**, 349 (1969).

[17]J. Z. Klose, Astrophys. J. **165**, 637 (1971).

[18]R. C. Hilborn and R. de Zafra, Astrophys. J. **183**, 347 (1973).

[19]J. Brzozowski, P. Erman, M. Lyyra, and W. H. Smith, Phys. Scr. **14**, 48 (1976).

[20]G. D. Bell and E. F. Tubbs, Astrpohys. J. **159**, 1093 (1970).

[21]H. M. Crosswhite, J. Res. Nat. Bur. Stand., Sect. A **79**, 17 (1975).

[22]C. Corliss and J. Sugar, J. Phys. Chem. Ref. Data **11**, 135 (1982).

[23]C. E. Moore, "Atomic Energy Levels (As derived from the Analyses of Optical Spectra)," Vol. II, Nat. Bur. Stand. (U.S.), Circ. 467, (Aug. 1952).

[24]C. E. Moore, "A Multiplet Table of Astrophysical Interest, Revised Edition," 253 pp., Nat. Stand. Ref. Data Ser., Nat. Bur. Stand. (U.S.), 40 (Feb. 1972).

[25]D. E. Blackwell, private communications.

[26]D. L. Adams and W. Whaling, J. Quant. Spectrosc. Radiat. Transfer **25**, 233 (1981).

[27]D. E. Blackwell, A. D. Petford, M. J. Shallis, and G. J. Simmons, Mon. Not. R. Astron. Soc. **191**, 445 (1980).

[28]D. E. Blackwell, A. D. Petford, M. J. Shallis, and G. J. Simmons, Mon. Not. R. Astron. Soc. **199**, 43 (1982).

[29]D. E. Blackwell, A. D. Petford, and G. J. Simmons, Mon. Not. R. Astron Soc. **210**, 595 (1982).

[30]E. A. Gurtovenko and R. I. Kostik, Astron. Astrophys., Suppl. Ser. **46**, 239 (1981).

[31]J. Marek, J. Richter, and H.-J. Stahnke, Phys. Scr. **19**, 325 (1979).

[32]P. Hannaford and R. M. Lowe, J. Phys. B **14**, L5 (1981).

[33]L. Delbouille, L. Neven, and C. Roland, "Photometric Atlas of the Solar Spectrum from 3000 to 10,000 Å," (Institut d'Astrophysique de l'Université de Liège, Observatoire Royal de Belgique, Liège, Belgique, 1973).

[34]E. A. Gurtovenko and R. I. Kostik, Astron. Astrophys., Suppl. Ser. **47**, 193 (1982).

Fe I: Allowed transitions

No.	Multiplet	λ (Å)	E_i (cm^{-1})	E_k (cm^{-1})	g_i	g_k	A_{ki} (10^8 s^{-1})	f_{ik}	S (at. u.)	log gf	Accuracy	Source
1.	$a\ ^5D - z\ ^7D°$ (1)											
		5166.28	0.0	19351	9	11	1.45(−5)[a]	7.09(−6)	0.00109	−4.195	B+	1
		5247.05	704.0	19757	5	7	3.92(−6)	2.26(−6)	1.96(−4)	−4.946	B+	1
		5254.96	888.1	19912	3	5	8.32(−6)	5.74(−6)	2.98(−4)	−4.764	B+	1
		5250.21	978.1	20020	1	3	9.30(−6)	1.15(−5)	1.99(−4)	−4.938	B+	1
		5110.41	0.0	19562	9	9	4.93(−5)	1.93(−5)	0.00292	−3.760	B+	1
		5168.90	415.9	19757	7	7	3.83(−5)	1.53(−5)	0.00183	−3.969	B+	1
		5204.58	704.0	19912	5	5	2.29(−5)	9.31(−6)	7.98(−4)	−4.332	B+	1
		5225.53	888.1	20020	3	3	1.32(−5)	5.42(−6)	2.80(−4)	−4.789	B+	1
		5060.08	0.0	19757	9	7	1.3(−6)	3.9(−7)	5.8(−5)	−5.46	B+	2
		5127.68	415.9	19912	7	5	3.80(−7)	1.07(−7)	1.27(−5)	−6.125	B+	1

Fe I: Allowed transitions — Continued

No.	Multiplet	λ (Å)	E_i (cm^{-1})	E_k (cm^{-1})	g_i	g_k	A_{ki} (10^8 s^{-1})	f_{ik}	S (at. u.)	log gf	Accuracy	Source
2.	$a\ ^5D - z\ ^7F°$ (2)											
		4375.93	0.0	22846	9	11	2.95(−4)	1.03(−4)	0.0134	−3.031	B+	1
		4461.65	704.0	23111	5	7	2.95(−4)	1.23(−4)	0.00906	−3.210	B+	1
		4482.17	888.1	23192	3	5	2.10(−4)	1.05(−4)	0.00466	−3.501	B+	1
		4489.74	978.1	23245	1	3	1.19(−4)	1.08(−4)	0.00160	−3.966	B+	1
		4347.24	0.0	22997	9	9	1.23(−6)	3.49(−7)	4.49(−5)	−5.503	B+	1
		4445.47	704.0	23192	5	5	2.45(−6)	7.24(−7)	5.30(−5)	−5.441	B+	1
		4471.68	888.1	23245	3	3	1.12(−6)	3.37(−7)	1.49(−5)	−5.995	B+	1
		4389.24	415.9	23192	7	5	1.81(−5)	3.73(−6)	3.77(−4)	−4.583	B+	1
		4435.15	704.0	23245	5	3	4.72(−5)	8.36(−6)	6.10(−4)	−4.379	B+	1
3.	$a\ ^5D - z\ ^7P°$ (3)											
		4216.18	0.0	23711	9	9	1.84(−4)	4.90(−5)	0.00611	−3.356	B+	1
		4206.70	415.9	24181	7	7	7.2(−5)	1.9(−5)	0.0018	−3.88	D	4n,5n
		4258.31	704.0	24181	5	7	2.54(−5)	9.66(−6)	6.77(−4)	−4.316	B+	1
		4232.73	888.1	24507	3	5	8.79(−6)	3.93(−6)	1.64(−4)	−4.928	B+	1
4.	$a\ ^5D - z\ ^5D°$ (4)	*3882.7*	*402.9*	*26151*	*25*	*25*	*0.100*	*0.0226*	*7.23*	*−0.247*	*B*	*1,4n,13n*
		3859.91	0.0	25900	9	9	0.0970	0.0217	2.48	−0.710	B+	1
		3886.28	415.9	26140	7	7	0.0530	0.0120	1.07	−1.076	B+	1
		3899.71	704.0	26340	5	5	0.0258	0.00589	0.378	−1.531	B+	1
		3906.48	888.1	26479	3	3	0.00833	0.00190	0.0735	−2.243	B+	1
		3824.44	0.0	26140	9	7	0.0283	0.00483	0.547	−1.362	B+	1
		3856.37	415.9	26340	7	5	0.0464	0.00739	0.657	−1.286	B+	1
		3878.57	704.0	26479	5	3	0.066	0.0089	0.57	−1.35	D−	13n
		3895.66	888.1	26550	3	1	0.0940	0.00713	0.274	−1.670	B+	1
		3922.91	415.9	25900	7	9	0.0108	0.00319	0.288	−1.651	B+	1
		3930.30	704.0	26140	5	7	0.016	0.0051	0.33	−1.59	C	4n
		3927.92	888.1	26340	3	5	0.022	0.0086	0.33	−1.59	C	4n
		3920.26	978.1	26479	1	3	0.0260	0.0179	0.232	−1.746	B+	1
5.	$a\ ^5D - z\ ^5F°$ (5)											
		3719.93	0.0	26875	9	11	0.162	0.0412	4.54	−0.431	B+	16,17,18, 19,20,31, 32
		3737.13	415.9	27167	7	9	0.142	0.0381	3.28	−0.574	B+	1
		3745.56	704.0	27395	5	7	0.115	0.0339	2.09	−0.771	B+	1
		3748.26	888.1	27560	3	5	0.0915	0.0321	1.19	−1.016	B+	1
		3745.90	978.1	27666	1	3	0.0733	0.0462	0.570	−1.335	B+	1
		3679.91	0.0	27167	9	9	0.0138	0.00280	0.305	−1.599	B+	1
		3705.57	415.9	27395	7	7	0.0322	0.00662	0.565	−1.334	B+	1
		3722.56	704.0	27560	5	5	0.0497	0.0103	0.633	−1.287	B+	1
		3733.32	888.1	27666	3	3	0.062	0.013	0.48	−1.41	C	4n,6n
		3683.05	415.9	27560	7	5	0.0028	4.1(−4)	0.035	−2.54	C−	4n,6n
		3707.82	704.0	27666	5	3	0.0072	8.9(−4)	0.055	−2.35	C	6n
6.	$a\ ^5D - z\ ^5P°$ (6)											
		3440.99	415.9	29469	7	5	0.084	0.011	0.84	−1.13	C	4n
		3443.88	704.0	29733	5	3	0.062	0.0066	0.38	−1.48	C	4n
		3497.84	888.1	29469	3	5	0.026	0.0080	0.28	−1.62	C	4n
		3476.70	978.1	29733	1	3	0.054	0.030	0.34	−1.53	C	4n

Fe I: Allowed transitions — Continued

No.	Multiplet	λ (Å)	E_i (cm^{-1})	E_k (cm^{-1})	g_i	g_k	A_{ki} (10^8 s^{-1})	f_{ik}	S (at. u.)	log gf	Accuracy	Source
7.	$a\,^5D - z\,^3F°$ (7)											
		3193.23	0.0	31307	9	9	0.0053	8.0($-$4)	0.076	-2.14	C	6n
8.	$a\,^5D - y\,^5D°$ (9)											
		3021.07	415.9	33507	7	7	0.456	0.0624	4.34	-0.360	B+	1
		3017.63	888.1	34017	3	3	0.0682	0.00931	0.277	-1.554	B+	1
		2983.57	0.0	33507	9	7	0.280	0.0290	2.57	-0.583	B+	1
		2994.43	415.9	33802	7	5	0.44	0.042	2.9	-0.53	C	6n
		3000.95	704.0	34017	5	3	0.642	0.0520	2.57	-0.585	B+	1
		3008.14	888.1	34122	3	1	1.07	0.0485	1.44	-0.837	B+	1
		3059.09	415.9	33096	7	9	0.17	0.031	2.2	-0.66	C+	4n,6n
		3047.60	704.0	33507	5	7	0.284	0.0553	2.78	-0.558	B+	1
		3037.39	888.1	33802	3	5	0.32	0.074	2.2	-0.65	C+	4n,6n
		3025.84	978.1	34017	1	3	0.348	0.143	1.43	-0.844	B+	1
9.	$a\,^5D - y\,^5F°$ (uv 1)											
		2966.90	0.0	33695	9	11	0.272	0.0438	3.85	-0.404	B+	1
		2973.24	415.9	34040	7	9	0.183	0.0313	2.14	-0.660	B+	1
		2973.13	704.0	34329	5	7	0.135	0.0251	1.23	-0.901	B+	1
		2965.25	978.1	34692	1	3	0.116	0.0460	0.449	-1.337	B+	1
		2936.90	0.0	34040	9	9	0.13	0.017	1.5	-0.82	C+	4n,6n
		2947.88	415.9	34329	7	7	0.20	0.027	1.8	-0.73	C	6n
		2953.94	704.0	34547	5	5	0.189	0.0247	1.20	-0.908	B+	1
		2957.36	888.1	34692	3	3	0.177	0.0232	0.678	-1.157	B+	1
		2912.16	0.0	34329	9	7	0.033	0.0033	0.28	-1.53	C	4n,6n
		2929.01	415.9	34547	7	5	0.073	0.0067	0.45	-1.33	C	6n
		2941.34	704.0	34692	5	3	0.056	0.0044	0.21	-1.66	C	4n
10.	$a\,^5D - z\,^3P°$ (11)											
		2981.45	415.9	33947	7	5	0.0654	0.00622	0.427	-1.361	B+	1
		2970.12	704.0	34363	5	3	0.0273	0.00217	0.106	-1.965	C+	25,26n
		2969.36	888.1	34556	3	1	0.0366	0.00161	0.0473	-2.315	B+	1
		3007.28	704.0	33947	5	5	0.0273	0.00371	0.184	-1.732	B+	1
		2986.46	888.1	34363	3	3	0.00219	2.92(-4)	0.00862	-3.057	B+	1
		3024.03	888.1	33947	3	5	0.0488	0.0111	0.333	-1.476	B+	1
		2994.50	978.1	34363	1	3	0.0149	0.00601	0.0593	-2.221	B+	1
11.	$a\,^5D - z\,^5G°$ (uv 2)											
		2874.17	0.0	34782	9	11	0.013	0.0020	0.17	-1.74	C	6n
		2869.31	415.9	35257	7	9	0.015	0.0023	0.15	-1.79	C	6n
		2835.46	0.0	35257	9	9	0.0090	0.0011	0.091	-2.01	C	6n
12.	$a\,^5D - y\,^5P°$ (uv 5)											
		2719.03	0.0	36767	9	7	1.4	0.12	9.6	0.03	C	6n
		2720.90	415.9	37158	7	5	1.1	0.084	5.3	-0.23	C	6n
		2723.58	704.0	37410	5	3	0.64	0.043	1.9	-0.67	C	6n
		2750.14	415.9	36767	7	7	0.39	0.044	2.8	-0.51	C	6n
		2742.41	704.0	37158	5	5	0.63	0.071	3.2	-0.45	C	6n
		2737.31	888.1	37410	3	3	0.85	0.095	2.6	-0.55	C	6n,7
		2756.33	888.1	37158	3	5	0.20	0.038	1.0	-0.94	C	6n
		2744.07	978.1	37410	1	3	0.35	0.12	1.1	-0.92	C	6n

Fe I: Allowed transitions — Continued

No.	Multiplet	λ (Å)	E_i (cm^{-1})	E_k (cm^{-1})	g_i	g_k	A_{ki} (10^8 s^{-1})	f_{ik}	S (at. u.)	log gf	Accuracy	Source
13.	$a\ ^5D - y\ ^3D°$ (uv 6)											
		2632.59	704.0	38678	5	5	0.015	0.0016	0.067	−2.11	C	6n
14.	$a\ ^5D - x\ ^5D°$ (uv 7)											
		2522.85	0.0	39626	9	9	2.9	0.28	21	0.40	C	6n
		2527.43	415.9	39970	7	7	1.9	0.18	10	0.10	C	6n
		2529.13	704.0	40231	5	5	0.98	0.094	3.9	−0.33	C	6n
		2501.13	0.0	39970	9	7	0.68	0.050	3.7	−0.35	C	6n
		2510.83	415.9	40231	7	5	1.3	0.088	5.1	−0.21	C	6n
		2518.10	704.0	40405	5	3	1.9	0.11	4.5	−0.27	C	6n
		2524.29	888.1	40491	3	1	3.4	0.11	2.7	−0.49	C	6n
		2549.61	415.9	39626	7	9	0.36	0.045	2.7	−0.50	C	6n
		2545.98	704.0	39970	5	7	0.67	0.091	3.8	−0.34	C	6n
		2540.97	888.1	40231	3	5	0.92	0.15	3.7	−0.35	C	6n
		2535.61	978.1	40405	1	3	0.97	0.28	2.4	−0.55	C	6n
15.	$a\ ^5D - y\ ^7P°$ (uv 8)											
		2512.36	415.9	40207	7	7	0.027	0.0025	0.15	−1.75	C	6n
16.	$a\ ^5D - x\ ^5F°$ (uv 9)											
		2483.27	0.0	40257	9	11	4.9	0.56	41	0.70	C	6n
		2488.14	415.9	40594	7	9	4.7	0.56	32	0.59	C	6n
		2490.64	704.0	40842	5	7	3.8	0.49	20	0.39	C	6n
		2491.15	888.1	41018	3	5	3.0	0.47	12	0.15	C	6n
		2462.65	0.0	40594	9	9	0.58	0.053	3.9	−0.32	C	6n
		2479.78	704.0	41018	5	5	1.8	0.17	6.9	−0.07	C	6n
		2462.18	415.9	41018	7	5	0.15	0.0099	0.56	−1.16	C	6n
17.	$a\ ^5D - x\ ^5P°$ (uv 11)											
		2373.62	415.9	42533	7	7	0.067	0.0057	0.31	−1.40	C	6n
		2371.43	704.0	42860	5	5	0.052	0.0044	0.17	−1.66	C	6n
		2389.97	704.0	42533	5	7	0.050	0.0060	0.24	−1.52	C	6n
		2381.83	888.1	42860	3	5	0.054	0.0076	0.18	−1.64	C	6n
		2374.52	978.1	43079	1	3	0.29	0.074	0.58	−1.13	C	6n
18.	$a\ ^5D - w\ ^5D°$ (uv 14)											
		2276.03	0.0	43923	9	7	0.17	0.010	0.68	−1.04	C	6n
		2287.25	704.0	44411	5	3	0.34	0.016	0.60	−1.10	C	6n
		2294.41	888.1	44459	3	1	0.61	0.016	0.36	−1.32	C	6n
		2320.36	415.9	43499	7	9	0.12	0.013	0.68	−1.05	C	6n
		2313.10	704.0	43923	5	7	0.14	0.016	0.59	−1.11	C	6n
		2309.00	888.1	44184	3	5	0.15	0.020	0.46	−1.22	C	6n
		2301.68	978.1	44411	1	3	0.13	0.030	0.23	−1.52	C	6n
19.	$a\ ^5D - \ ^5D°$											
		2259.28	415.9	44664	7	5	0.013	7.0(−4)	0.036	−2.31	C	6n
		2292.52	415.9	44023	7	9	0.043	0.0043	0.23	−1.52	C	6n
		2300.14	704.0	44166	5	7	0.080	0.0089	0.34	−1.35	C	6n

Fe I: Allowed transitions — Continued

No.	Multiplet	λ (Å)	E_i (cm^{-1})	E_k (cm^{-1})	g_i	g_k	A_{ki} (10^8 s^{-1})	f_{ik}	S (at. u.)	log gf	Accuracy	Source
20.	$a\ ^5D - ^5F°$											
		2259.51	0.0	44244	9	11	0.070	0.0065	0.44	−1.23	C	6n
		2272.07	415.9	44415	7	9	0.038	0.0038	0.20	−1.58	C	6n
		2303.58	888.1	44285	3	5	0.076	0.010	0.23	−1.52	C	6n
		2303.42	978.1	44378?	1	3	0.094	0.022	0.17	−1.65	C	6n
		2250.79	0.0	44415	9	9	0.019	0.0014	0.095	−1.89	C	6n
		2265.05	415.9	44551	7	7	0.020	0.0015	0.080	−1.97	C	6n
21.	$a\ ^5D - y\ ^5S°$ (uv 17)											
		2267.08	415.9	44512	7	5	0.071	0.0039	0.21	−1.56	C	6n
22.	$a\ ^5D - (\ °)^b$											
		2228.17	415.9	45282	7	5	0.021	0.0011	0.057	−2.11	C	6n
23.	$a\ ^5D - w\ ^5P°$ (uv 21)											
		2166.77	0.0	46137	9	7	2.7	0.15	9.6	0.13	C	6n
		2191.84	704.0	46314	5	5	1.2	0.083	3.0	−0.38	C	6n
		2196.04	888.1	46410	3	3	1.2	0.086	1.9	−0.59	C	6n
		2200.72	888.1	46314	3	5	0.28	0.034	0.74	−0.99	C	6n
24.	$a\ ^5D - z\ ^3S°$ (uv 22)											
		2191.20	978.1	46601	1	3	0.073	0.016	0.11	−1.80	C	6n
25.	$a\ ^5D - y\ ^3P°$ (uv 23)											
		2176.84	978.1	46902	1	3	0.10	0.022	0.16	−1.66	C	6n
26.	$a\ ^5D - ^5D°$											
		2153.01	704.0	47136	5	5	0.069	0.0048	0.17	−1.62	C	6n
		2138.59	0.0	46745	9	7	0.028	0.0015	0.095	−1.87	C	6n
		2171.30	704.0	46745	5	7	0.051	0.0050	0.18	−1.60	C	6n
		2161.58	888.1	47136	3	5	0.050	0.0058	0.12	−1.76	C	6n
27.	$a\ ^5D - ^3D°$											
		2145.19	415.9	47017	7	7	0.057	0.0039	0.19	−1.56	C	6n
		2173.21	888.1	46889	3	5	0.083	0.0098	0.21	−1.53	C	6n
28.	$a\ ^5D - x\ ^3F°$ (uv 25)											
		2132.02	0.0	46889	9	9	0.076	0.0052	0.33	−1.33	C	6n
29.	$a\ ^5D - v\ ^5P°$ (uv 33)											
		2084.12	0.0	47967	9	7	0.37	0.019	1.2	−0.77	C	6n
		2102.35	415.9	47967	7	7	0.088	0.0058	0.28	−1.39	C	6n
		2112.97	978.1	48290	1	3	0.19	0.038	0.26	−1.42	C	6n

Fe I: Allowed transitions — Continued

No.	Multiplet	λ (Å)	E_i (cm^{-1})	E_k (cm^{-1})	g_i	g_k	A_{ki} (10^8 s^{-1})	f_{ik}	S (at. u.)	log gf	Accuracy	Source
30.	a ^{5}D – u ^{5}F° (uv 35)											
		1937.27	0.0	51619	9	7	0.22	0.0095	0.54	−1.07	C	6n
31.	a ^{5}D – u ^{5}P° (uv 37)											
		1934.54	0.0	51692	9	7	0.25	0.011	0.64	−1.00	C	6n
		1940.66	415.9	51945	7	5	0.26	0.010	0.46	−1.14	C	6n
32.	a ^{5}F – z ^{7}D° (12)											
		7912.87	6928	19562	11	9	1.68(−6)	1.29(−6)	3.70(−4)	−4.848	B+	3
		8075.13	7377	19757	9	7	1.27(−6)	9.63(−7)	2.30(−4)	−5.062	B+	3
33.	a ^{5}F – z ^{7}F° (13)											
		6358.69	6928	22650	11	13	4.32(−6)	3.09(−6)	7.13(−4)	−4.468	B+	3
		6280.63	6928	22846	11	11	6.31(−6)	3.73(−6)	8.48(−4)	−4.387	B+	3
		6498.95	7728	23111	7	7	4.51(−6)	2.86(−6)	4.28(−4)	−4.699	B+	3
		6574.24	7986	23192	5	5	2.8(−6)	1.8(−6)	2.0(−4)	−5.04	D	5n
		6625.04	8155	23245	3	3	2.3(−6)	1.5(−6)	9.7(−5)	−5.35	D	30
		6551.68	7986	23245	5	3	8.4(−7)	3.2(−7)	3.5(−5)	−5.79	D	30
34.	a ^{5}F – z ^{7}P° (14)											
		5956.70	6928	23711	11	9	5.19(−6)	2.26(−6)	4.87(−4)	−4.605	B+	3
		6120.25	7377	23711	9	9	2.2(−7)	1.2(−7)	2.3(−5)	−5.95	D	30
35.	a ^{5}F – z ^{5}D° (15)											
		5269.54	6928	25900	11	9	0.0127	0.00434	0.828	−1.321	B+	3
		5328.04	7377	26140	9	7	0.0115	0.00380	0.600	−1.466	B+	3
		5371.49	7728	26340	7	5	0.0105	0.00324	0.400	−1.645	B+	3
		5405.77	7986	26479	5	3	0.0109	0.00286	0.255	−1.844	B+	3
		5434.52	8155	26550	3	1	0.0171	0.00252	0.135	−2.122	B+	3
		5397.13	7377	25900	9	9	0.00259	0.00113	0.181	−1.993	B+	3
		5429.70	7728	26140	7	7	0.00427	0.00189	0.236	−1.879	B+	3
		5446.92	7986	26340	5	5	0.0053	0.0023	0.21	−1.93	C	4n
		5501.46	7728	25900	7	9	2.7(−4)	1.6(−4)	0.020	−2.95	D	4n
		5506.78	7986	26140	5	7	5.01(−4)	3.19(−4)	0.0289	−2.797	B+	3
		5497.52	8155	26340	3	5	6.25(−4)	4.72(−4)	0.0256	−2.849	B+	3
36.	a ^{5}F – z ^{5}F° (16)											
		5012.07	6928	26875	11	11	5.50(−4)	2.07(−4)	0.0376	−2.642	B+	3
		5051.63	7377	27167	9	9	4.66(−4)	1.78(−4)	0.0267	−2.795	B+	3
		5083.34	7728	27395	7	7	4.06(−4)	1.57(−4)	0.0184	−2.958	B+	3
		5107.45	7986	27560	5	5	4.19(−4)	1.64(−4)	0.0138	−3.087	B+	3
		5123.72	8155	27666	3	3	7.24(−4)	2.85(−4)	0.0144	−3.068	B+	3
		4939.69	6928	27167	11	9	1.39(−4)	4.16(−5)	0.00743	−3.340	B+	3
		4994.13	7377	27395	9	7	3.18(−4)	9.24(−5)	0.0137	−3.080	B+	3
		5079.74	7986	27666	5	3	5.19(−4)	1.21(−4)	0.0101	−3.220	B+	3
		5127.36	7377	26875	9	11	1.14(−4)	5.48(−5)	0.00832	−3.307	B+	3
		5150.84	7986	27395	5	7	3.1(−4)	1.7(−4)	0.014	−3.07	D	4n
		5151.91	8155	27560	3	5	2.39(−4)	1.59(−4)	0.00808	−3.322	B+	3

Fe I:　Allowed transitions — Continued

No.	Multiplet	λ (Å)	E_i (cm^{-1})	E_k (cm^{-1})	g_i	g_k	A_{ki} (10^8 s^{-1})	f_{ik}	S (at. u.)	log gf	Accuracy	Source
37.	$a\,^5F - z\,^5P°$ (17)											
		4690.38	8155	29469	3	5	3.6(−6)	2.0(−6)	9.1(−5)	−5.23	D	30
38.	$a\,^5F - z\,^3F°$ (18)											
		4100.74	6928	31307	11	9	2.92(−4)	6.02(−5)	0.00894	−3.179	B+	3
		4092.46	7377	31805	9	7	2.7(−5)	5.2(−6)	6.3(−4)	−4.33	D	5n
		4177.59	7377	31307	9	9	3.72(−4)	9.72(−5)	0.0120	−3.058	B+	3
		4152.17	7728	31805	7	7	3.24(−4)	8.37(−5)	0.00801	−3.232	B+	3
		4139.93	7986	32134	5	5	1.83(−4)	4.70(−5)	0.00320	−3.629	B+	3
39.	$a\,^5F - z\,^3D°$ (19)											
		4174.91	7377	31323	9	7	5.87(−4)	1.19(−4)	0.0148	−2.969	B+	3
		4172.74	7728	31686	7	5	6.46(−4)	1.20(−4)	0.0116	−3.074	B+	3
		4173.92	7986	31937	5	3	5.3(−4)	8.3(−5)	0.0057	−3.38	D	5n
		4237.07	7728	31323	7	7	2.22(−5)	5.97(−6)	5.83(−4)	−4.379	B+	3
40.	$a\,^5F - y\,^5D°$ (20)											
		3820.43	6928	33096	11	9	0.668	0.120	16.5	0.119	B+	3
		3825.88	7377	33507	9	7	0.598	0.102	11.6	−0.037	B+	3
		3834.22	7728	33802	7	5	0.453	0.0713	6.30	−0.302	B+	3
		3840.44	7986	34017	5	3	0.470	0.0624	3.94	−0.506	B+	3
		3849.96	8155	34122	3	1	0.606	0.0449	1.71	−0.871	B+	3
		3887.05	7377	33096	9	9	0.0352	0.00798	0.919	−1.144	B+	3
		3878.02	7728	33507	7	7	0.0772	0.0174	1.56	−0.914	B+	3
		3872.50	7986	33802	5	5	0.105	0.0236	1.50	−0.928	B+	3
		3865.52	8155	34017	3	3	0.155	0.0347	1.33	−0.982	B+	3
		3940.88	7728	33096	7	9	0.00120	3.59(−4)	0.0326	−2.600	B+	3
		3917.18	7986	33507	5	7	0.00435	0.00140	0.0902	−2.155	B+	3
41.	$a\,^5F - y\,^5F°$ (21)	*3750.2*	*7460*	*34118*	35	35	0.914	0.193	83.3	0.829	B+	3
		3734.86	6928	33695	11	11	0.902	0.189	25.5	0.317	B+	3
		3749.48	7377	34040	9	9	0.764	0.161	17.9	0.161	B+	3
		3758.23	7728	34329	7	7	0.634	0.134	11.6	−0.027	B+	3
		3763.79	7986	34547	5	5	0.544	0.116	7.16	−0.238	B+	3
		3767.19	8155	34692	3	3	0.640	0.136	5.06	−0.389	B+	3
		3687.46	6928	34040	11	9	0.0801	0.0134	1.78	−0.833	B+	3
		3709.25	7377	34329	9	7	0.156	0.0251	2.76	−0.646	B+	3
		3727.62	7728	34547	7	5	0.225	0.0334	2.87	−0.631	B+	3
		3743.36	7986	34692	5	3	0.260	0.0328	2.02	−0.785	B+	3
		3798.51	7377	33695	9	11	0.0323	0.00855	0.962	−1.114	B+	3
		3799.55	7728	34040	7	9	0.0732	0.0204	1.78	−0.846	B+	3
		3795.00	7986	34329	5	7	0.115	0.0347	2.17	−0.761	B+	3
		3787.88	8155	34547	3	5	0.129	0.0461	1.73	−0.859	B+	3
42.	$a\,^5F - z\,^3P°$ (22)											
		3790.09	7986	34363	5	3	0.0268	0.00347	0.216	−1.761	B+	3
		3786.68	8155	34556	3	1	0.0277	0.00199	0.0743	−2.225	B+	3
		3850.82	7986	33947	5	5	0.0166	0.00369	0.234	−1.734	B+	3
		3814.52	8155	34363	3	3	0.00624	0.00136	0.0513	−2.389	B+	25
		3876.04	8155	33947	3	5	0.0013	5.0(−4)	0.019	−2.82	C	4n,5n, 26n

Fe I: Allowed transitions — Continued

No.	Multiplet	λ (Å)	E_i (cm^{-1})	E_k (cm^{-1})	g_i	g_k	A_{ki} (10^8 s^{-1})	f_{ik}	S (at. u.)	log gf	Accu-racy	Source
43.	$a\ ^5F - z\ ^5G°$ (23)											
		3581.19	6928	34844	11	13	1.02	0.232	30.0	0.406	B+	3
		3647.84	7377	34782	9	11	0.292	0.0711	7.68	−0.194	B+	3
		3631.46	7728	35257	7	9	0.517	0.131	11.0	−0.036	B+	3
		3618.77	7986	35612	5	7	0.73	0.20	12	0.00	C+	4n,6n
		3608.86	8155	35856	3	5	0.814	0.265	9.44	−0.100	B+	3
		3589.11	6928	34782	11	11	0.00361	6.98(−4)	0.0907	−2.115	B+	3
		3585.71	7377	35257	9	9	0.0375	0.00722	0.767	−1.187	B+	3
		3585.32	7728	35612	7	7	0.13	0.025	2.1	−0.76	C	6n
		3586.98	7986	35856	5	5	0.16	0.030	1.8	−0.82	C+	4n,6n
		3540.71	7377	35612	9	7	0.0014	2.1(−4)	0.022	−2.73	D	5n
44.	$a\ ^5F - z\ ^3G°$ (24)											
		3513.82	6928	35379	11	11	0.0341	0.00630	0.802	−1.159	B+	3
		3483.01	7377	36079	9	7	0.0010	1.4(−4)	0.015	−2.89	D	12n
		3570.10	7377	35379	9	11	0.677	0.158	16.7	0.153	B+	3
		3565.38	7728	35768	7	9	0.38	0.092	7.6	−0.19	C+	4n
45.	$a\ ^5F - y\ ^5P°$ (26)											
		3396.98	7728	37158	7	5	0.0025	3.1(−4)	0.024	−2.66	D	12n
		3417.27	8155	37410	3	3	7.8(−4)	1.4(−4)	0.0046	−3.39	D	12n
46.	$a\ ^5F - x\ ^5D°$ (28)											
		3057.45	6928	39626	11	9	0.44	0.050	5.5	−0.26	C+	4n
		3067.24	7377	39970	9	7	0.34	0.038	3.4	−0.47	C+	4n
		3075.72	7728	40231	7	5	0.29	0.030	2.1	−0.68	C+	4n
		3083.74	7986	40405	5	3	0.30	0.026	1.3	−0.89	C+	4n
		3091.58	8155	40491	3	1	0.54	0.026	0.79	−1.11	C	4n
		3100.67	7728	39970	7	7	0.14	0.020	1.4	−0.86	C+	4n
		3134.11	7728	39626	7	9	0.012	0.0023	0.17	−1.79	C	4n
47.	$a\ ^5F - x\ ^5F°$ (30)											
		2999.51	6928	40257	11	11	0.23	0.031	3.3	−0.47	C+	4n
		3009.57	7377	40594	9	9	0.17	0.024	2.1	−0.67	C+	4n
		3018.98	7728	40842	7	7	0.13	0.017	1.2	−0.92	C+	4n
		3026.46	7986	41018	5	5	0.11	0.016	0.77	−1.11	C	4n
		3031.63	8155	41131	3	3	0.15	0.021	0.63	−1.20	C	4n
		2987.29	7377	40842	9	7	0.066	0.0069	0.61	−1.21	C	4n
		3016.18	7986	41131	5	3	0.085	0.0069	0.34	−1.46	C	4n
		3040.43	7377	40257	9	11	0.030	0.0051	0.46	−1.34	C	4n
		3042.66	7986	40842	5	7	0.057	0.011	0.55	−1.26	C	4n
		3042.02	8155	41018	3	5	0.049	0.011	0.34	−1.47	C	4n
48.	$a\ ^5F - y\ ^5G°$ (uv 44)											
		2788.10	6928	42784	11	13	0.63	0.087	8.8	−0.02	C	6n

Fe I: Allowed transitions — Continued

No.	Multiplet	λ (Å)	E_i (cm^{-1})	E_k (cm^{-1})	g_i	g_k	A_{ki} (10^8 s^{-1})	f_{ik}	S (at. u.)	log gf	Accuracy	Source
49.	a ^{5}F $- w$ ^{5}D° (uv 46)											
		2733.58	6928	43499	11	9	0.86	0.079	7.8	−0.06	C	7
		2735.48	7377	43923	9	7	0.62	0.054	4.4	−0.31	C	7
50.	a ^{5}F $-$ ^{5}F°											
		2679.06	6928	44244	11	11	0.19	0.021	2.0	−0.64	C	7
51.	a ^{5}F $- x$ ^{5}G° (uv 52)											
		2584.54	6928	45608?	11	13	0.46	0.054	5.1	−0.23	C	6n,7
		2606.83	7377	45726	9	11	0.42	0.052	4.0	−0.33	C	7
		2623.53	7728	45833	7	9	0.33	0.044	2.7	−0.51	C	7
		2618.02	7728	45913	7	7	0.40	0.041	2.5	−0.54	C	7
52.	a ^{5}F $- t$ ^{5}D° (uv 71)											
		2277.11	7728	51630	7	5	37	2.1	110	1.16	C	6n
53.	a ^{3}F $- z$ ^{5}D° (33)											
		7180.02	11976	25900	9	9	2.4(−6)	1.8(−6)	3.9(−4)	−4.78	D	30
54.	a ^{3}F $- z$ ^{5}F° (34)											
		6710.31	11976	26875	9	11	1.8(−6)	1.5(−6)	2.9(−4)	−4.88	D	30
		6581.22	11976	27167	9	9	2.4(−6)	1.5(−6)	3.0(−4)	−4.86	D	5n
		6739.54	12561	27395	7	7	2.4(−6)	1.6(−6)	2.5(−4)	−4.95	D	30
		6851.64	12969	27560	5	5	1.4(−6)	9.6(−7)	1.1(−4)	−5.32	D	30
55.	a ^{3}F $- z$ ^{5}P° (35)											
		5853.18	11976	29056	9	7	1.5(−6)	5.8(−7)	1.0(−4)	−5.28	D	5n
56.	a ^{3}F $- z$ ^{3}F° (36)											
		5171.60	11976	31307	9	9	0.00446	0.00179	0.274	−1.793	B+	27
		5194.94	12561	31805	7	7	0.00287	0.00116	0.139	−2.090	B+	27
		5216.27	12969	32134	5	5	0.00347	0.00142	0.122	−2.150	B+	27
		5107.64	12561	32134	7	5	0.00195	5.46(−4)	0.0642	−2.418	B+	27
		5332.90	12561	31307	7	9	3.0(−4)	1.6(−4)	0.020	−2.94	D	30
		5307.36	12969	31805	5	7	3.49(−4)	2.06(−4)	0.0180	−2.987	B+	27
57.	a ^{3}F $- z$ ^{3}D° (37)											
		5167.49	11976	31323	9	7	0.020	0.0061	0.93	−1.26	C	4n
		5270.36	12969	31937	5	3	0.025	0.0062	0.54	−1.51	C	4n
		5341.02	12969	31686	5	5	0.0041	0.0017	0.15	−2.06	D	9n

Fe I: Allowed transitions — Continued

No.	Multiplet	λ (Å)	E_i (cm^{-1})	E_k (cm^{-1})	g_i	g_k	A_{ki} (10^8 s^{-1})	f_{ik}	S (at. u.)	log gf	Accuracy	Source
58.	$a\ ^3F - y\ ^5D°$ (38)											
		4798.73	12969	33802	5	5	3.3(−5)	1.1(−5)	8.9(−4)	−4.25	D	5n
		4643.22	11976	33507	9	7	3.8(−6)	9.7(−7)	1.3(−4)	−5.06	D	30
		4867.53	12969	33507	5	7	8.0(−6)	4.0(−6)	3.2(−4)	−4.70	D	30
59.	$a\ ^3F - y\ ^5F°$ (39)											
		4654.50	12561	34040	7	9	5.64(−4)	2.35(−4)	0.0253	−2.783	B+	27
		4680.29	12969	34329	5	7	7.32(−5)	3.37(−5)	0.00259	−3.774	B+	27
		4531.15	11976	34040	9	9	0.00253	7.78(−4)	0.104	−2.155	B+	27
		4592.65	12561	34329	7	7	0.00161	5.08(−4)	0.0538	−2.449	B+	27
		4632.91	12969	34547	5	5	7.59(−4)	2.44(−4)	0.0186	−2.913	B+	27
		4547.02	12561	34547	7	5	1.2(−4)	2.7(−5)	0.0028	−3.73	D	5n
		4602.00	12969	34692	5	3	7.36(−4)	1.40(−4)	0.0106	−3.154	B+	27
60.	$a\ ^3F - z\ ^3P°$ (40)											
		4674.65	12561	33947	7	5	9.9(−6)	2.3(−6)	2.5(−4)	−4.79	D	5n
		4672.84	12969	34363	5	3	5.9(−5)	1.2(−5)	8.9(−4)	−4.24	D	30
		4765.48	12969	33947	5	5	5.7(−5)	2.0(−5)	0.0015	−4.01	D	5n
61.	$a\ ^3F - z\ ^5G°$ (41)											
		4383.54	11976	34782	9	11	0.500	0.176	22.9	0.200	B+	27
		4404.75	12561	35257	7	9	0.275	0.103	10.5	−0.142	B+	27
		4415.12	12969	35612	5	7	0.119	0.0485	3.53	−0.615	B+	27
		4294.12	11976	35257	9	9	0.031	0.0086	1.1	−1.11	C	4n
		4337.05	12561	35612	7	7	0.0102	0.00288	0.288	−1.695	B+	27
		4367.90	12969	35856	5	5	0.0016	4.5(−4)	0.032	−2.65	D	5n
		4229.75	11976	35612	9	7	1.99(−4)	4.16(−5)	0.00521	−3.427	B+	27
62.	$a\ ^3F - z\ ^3G°$ (42)	*4293.8*	*12407*	*35690*	21	27	0.39	0.14	41	0.46	C+	4n,13n, 27
		4271.76	11976	35379	9	11	0.228	0.0762	9.64	−0.164	B+	27
		4307.90	12561	35768	7	9	0.34	0.12	12	−0.07	C+	4n
		4325.76	12969	36079	5	7	0.50	0.20	14	−0.01	C+	4n
		4202.03	11976	35768	9	9	0.0822	0.0218	2.71	−0.708	B+	27
		4250.79	12561	36079	7	7	0.10	0.028	2.7	−0.71	D−	13n
		4147.67	11976	36079	9	7	0.00436	8.74(−4)	0.107	−2.104	B+	27
63.	$a\ ^3F - y\ ^3F°$ (43)	*4057.8*	*12407*	*37044*	21	21	1.0	0.25	70	0.72	B	4n,13n, 27
		4045.81	11976	36686	9	9	0.863	0.212	25.4	0.280	B+	27
		4063.59	12561	37163	7	7	0.68	0.17	16	0.07	C+	4n
		4071.74	12969	37521	5	5	0.765	0.190	12.7	−0.022	B+	27
		3969.26	11976	37163	9	7	0.23	0.042	5.0	−0.42	C+	4n
		4005.24	12561	37521	7	5	0.204	0.0351	3.24	−0.610	B+	27
		4143.87	12561	36686	7	9	0.15	0.051	4.8	−0.45	C+	4n
		4132.06	12969	37163	5	7	0.12	0.045	3.0	−0.65	D−	13n
64.	$a\ ^3F - y\ ^5P°$ (44)											
		4032.63	11976	36767	9	7	0.0021	4.0(−4)	0.048	−2.44	D	5n

Fe I: Allowed transitions — Continued

No.	Multiplet	λ (Å)	E_i (cm⁻¹)	E_k (cm⁻¹)	g_i	g_k	A_{ki} (10⁸ s⁻¹)	f_{ik}	S (at. u.)	log gf	Accuracy	Source
65.	$a\ ^3F - y\ ^3D°$ (45)	*3830.3*	*12407*	*38507*	21	15	1.4	0.22	59	0.67	C+	4n,5n,6n, 27
		3815.84	11976	38175	9	7	1.3	0.22	25	0.30	C+	4n,6n
		3827.82	12561	38678	7	5	1.05	0.165	14.5	0.062	B+	27
		3841.05	12969	38996	5	3	1.3	0.18	11	−0.05	C+	4n
		3902.95	12561	38175	7	7	0.214	0.0489	4.39	−0.466	B+	27
		3888.51	12969	38678	5	5	0.26	0.059	3.8	−0.53	C+	4n
		3966.06	12969	38175	5	7	0.014	0.0046	0.30	−1.64	C	4n,5n
66.	$a\ ^3F - x\ ^5D°$ (46)											
		3615.66	11976	39626	9	9	6.4(−4)	1.2(−4)	0.013	−2.95	D	5n
		3666.94	12969	40231	5	5	5.7(−4)	1.2(−4)	0.0069	−3.24	D	5n
		3571.22	11976	39970	9	7	0.0019	2.8(−4)	0.030	−2.60	D	5n
67.	$a\ ^3F - x\ ^5F°$ (48)											
		3493.28	11976	40594	9	9	7.8(−4)	1.4(−4)	0.015	−2.89	D	5n
		3564.11	12969	41018	5	5	0.0014	2.6(−4)	0.015	−2.89	D	5n
		3463.30	11976	40842	9	7	0.0011	1.6(−4)	0.016	−2.84	D	12n
		3513.05	12561	41018	7	5	0.0032	4.2(−4)	0.034	−2.53	D	5n
		3549.86	12969	41131	5	3	0.0051	5.8(−4)	0.034	−2.54	D	5n
68.	$a\ ^3F - (\ °)^b$											
		3068.17	12969	45552	5	3	0.098	0.0083	0.42	−1.38	C	4n
69.	$a\ ^3F - x\ ^3F°$ (uv 87)											
		2920.69	12969	47197	5	5	0.052	0.0066	0.32	−1.48	C	4n
70.	$a\ ^5P - z\ ^5P°$ (60)											
		8688.62	17550	29056	7	7	0.00775	0.00877	1.76	−1.212	B+	28
		8514.07	17727	29469	5	5	0.00109	0.00118	0.165	−2.229	B+	28
		8468.40	17927	29733	3	3	0.00263	0.00282	0.236	−2.072	B+	28
		8387.77	17550	29469	7	5	0.00609	0.00459	0.887	−1.493	B+	28
		8327.05	17727	29733	5	3	0.00957	0.00597	0.818	−1.525	B+	28
71.	$a\ ^5P - y\ ^5D°$ (62)	*6319.4*	*17684*	*33504*	15	25	0.0020	0.0020	0.61	−1.53	B	28,30
		6430.85	17550	33096	7	9	0.00177	0.00141	0.209	−2.006	B+	28
		6335.34	17727	33507	5	7	0.0014	0.0012	0.12	−2.23	D	30
		6297.80	17927	33802	3	5	6.12(−4)	6.07(−4)	0.0377	−2.740	B+	28
		6265.13	17550	33507	7	7	6.84(−4)	4.03(−4)	0.0581	−2.550	B+	28
		6219.28	17727	33802	5	5	0.00127	7.38(−4)	0.0755	−2.433	B+	28
		6213.43	17927	34017	3	3	0.0013	7.3(−4)	0.045	−2.66	D	30
		6151.62	17550	33802	7	5	1.77(−4)	7.18(−5)	0.0102	−3.299	B+	28
		6137.00	17727	34017	5	3	6.62(−4)	2.24(−4)	0.0227	−2.950	B+	28
		6173.34	17927	34122	3	1	0.00231	4.39(−4)	0.0268	−2.880	B+	28
72.	$a\ ^5P - y\ ^5F°$ (63)											
		6062.89	17550	34040	7	9	1.5(−5)	1.0(−5)	0.0014	−4.14	D	5n
		6015.25	17927	34547	3	5	7.7(−6)	7.0(−6)	4.1(−4)	−4.68	D	30
		5892.80	17727	34692	5	3	6.0(−5)	1.9(−5)	0.0018	−4.03	D	5n

Fe I: Allowed transitions — Continued

No.	Multiplet	λ (Å)	E_i (cm^{-1})	E_k (cm^{-1})	g_i	g_k	A_{ki} (10^8 s^{-1})	f_{ik}	S (at. u.)	log gf	Accuracy	Source
73.	$a\ ^5P - z\ ^3P°$ (64)											
		6012.21	17927	34556	3	1	1.2(−4)	2.1(−5)	0.0012	−4.20	D	5n,26n
		6163.56	17727	33947	5	5	5.1(−5)	2.9(−5)	0.0029	−3.84	D	5n
		6082.72	17927	34363	3	3	1.61(−4)	8.91(−5)	0.00535	−3.573	B+	28
		6240.66	17927	33947	3	5	1.4(−4)	1.4(−4)	0.0086	−3.38	D	5n,26n
74.	$a\ ^5P - y\ ^3F°$ (65)											
		5224.30	17550	36686	7	9	2.3(−5)	1.2(−5)	0.0014	−4.08	D	5n
		5143.73	17727	37163	5	7	5.8(−5)	3.2(−5)	0.0027	−3.79	D	5n
75.	$a\ ^5P - y\ ^5P°$ (66)											
		5202.34	17550	36767	7	7	0.00511	0.00207	0.249	−1.838	B+	28
		5145.09	17727	37158	5	5	3.0(−4)	1.2(−4)	0.010	−3.23	D	5n
		5131.47	17927	37410	3	3	0.0023	9.2(−4)	0.047	−2.56	D	30
		5079.22	17727	37410	5	3	0.00739	0.00171	0.143	−2.067	B+	28
		5250.64	17727	36767	5	7	0.0031	0.0018	0.15	−2.05	D	30
		5198.71	17927	37158	3	5	0.00362	0.00244	0.125	−2.135	B+	28
76.	$a\ ^5P - y\ ^3D°$ (67)											
		4771.70	17727	38678	5	5	9.1(−5)	3.1(−5)	0.0024	−3.81	D	5n
		4745.13	17927	38996	3	3	6.7(−5)	2.3(−5)	0.0011	−4.17	D	5n
		4817.77	17927	38678	3	5	1.7(−4)	9.8(−5)	0.0047	−3.53	D	5n
77.	$a\ ^5P - x\ ^5D°$ (68)											
		4528.61	17550	39626	7	9	0.0544	0.0215	2.25	−0.822	B+	28
		4494.56	17727	39970	5	7	0.0345	0.0146	1.08	−1.136	B+	28
		4459.12	17550	39970	7	7	0.0252	0.00751	0.772	−1.279	B+	28
		4442.34	17727	40231	5	5	0.0376	0.0111	0.813	−1.255	B+	28
		4447.72	17927	40405	3	3	0.0511	0.0152	0.666	−1.342	B+	28
		4407.71	17550	40231	7	5	0.0083	0.0017	0.17	−1.92	C	4n
		4408.41	17727	40405	5	3	0.022	0.0039	0.28	−1.71	C	4n
		4430.61	17927	40491	3	1	0.0745	0.00731	0.320	−1.659	B+	28
78.	$a\ ^5P - y\ ^7P°$ (69)											
		4447.13	17727	40207	5	7	0.0012	5.1(−4)	0.038	−2.59	D	5n
		4518.58	17927	40052	3	5	7.5(−5)	3.8(−5)	0.0017	−3.94	D	5n
		4478.04	17727	40052	5	5	1.3(−4)	4.0(−5)	0.0029	−3.70	D	5n
		4442.83	17550	40052	7	5	0.00109	2.31(−4)	0.0236	−2.792	B+	28
79.	$a\ ^5P - x\ ^5F°$ (70)											
		4338.26	17550	40594	7	9	6.5(−4)	2.4(−4)	0.024	−2.78	D	5n
		4292.13	17550	40842	7	7	4.7(−4)	1.3(−4)	0.013	−3.04	D	5n
		4292.29	17727	41018	5	5	0.0012	3.3(−4)	0.023	−2.78	D	5n

Fe I: Allowed transitions — Continued

No.	Multiplet	λ (Å)	E_i (cm^{-1})	E_k (cm^{-1})	g_i	g_k	A_{ki} (10^8 s^{-1})	f_{ik}	S (at. u.)	log gf	Accuracy	Source
80.	$a\ ^5P - z\ ^5S°$ (71)	4307.1	17684	40895	15	5	0.23	0.021	4.5	−0.50	C+	4n,5n
		4282.40	17550	40895	7	5	0.11	0.022	2.2	−0.81	C+	4n,5n
		4315.08	17727	40895	5	5	0.077	0.021	1.5	−0.97	C	4n
		4352.73	17927	40895	3	5	0.039	0.018	0.79	−1.26	C	4n
81.	$a\ ^5P - x\ ^5P°$ (72)	3988.2	17684	42751	15	15	0.068	0.016	3.2	−0.61	C	4n,5n
		4001.66	17550	42533	7	7	0.0079	0.0019	0.18	−1.88	C	4n,5n
		3977.74	17727	42860	5	5	0.070	0.017	1.1	−1.08	C	4n
		3974.77	17927	43079	3	3	0.0035	8.4(−4)	0.033	−2.60	D	5n
		3949.95	17550	42860	7	5	0.059	0.0099	0.90	−1.16	C	5n
		3943.34	17727	43079	5	3	0.0079	0.0011	0.071	−2.26	D	5n
		4030.18	17727	42533	5	7	0.0029	9.8(−4)	0.065	−2.31	D	5n
		4009.71	17927	42860	3	5	0.052	0.021	0.83	−1.20	C	4n
82.	$a\ ^5P - w\ ^5D°$ (73)											
		3852.57	17550	43499	7	9	0.029	0.0082	0.73	−1.24	C	4n
		3816.34	17727	43923	5	7	0.023	0.0069	0.44	−1.46	C	5n
		3807.54	17927	44184	3	5	0.080	0.029	1.1	−1.06	C	4n,5n
		3778.70	17727	44184	5	5	0.0087	0.0019	0.12	−2.03	D	5n
		3774.82	17927	44411	3	3	0.047	0.010	0.38	−1.52	C	4n,5n
		3753.61	17550	44184	7	5	0.093	0.014	1.2	−1.01	C	4n,5n
		3746.49	17727	44411	5	3	0.011	0.0014	0.087	−2.15	D	5n
		3768.03	17927	44459	3	1	0.084	0.0059	0.22	−1.75	C	5n
83.	$a\ ^5P - ^5D°$											
		3776.45	17550	44023	7	9	0.015	0.0041	0.36	−1.54	C	4n,5n
		3781.19	17727	44166	5	7	0.0080	0.0024	0.15	−1.92	C	5n
		3739.12	17927	44664	3	5	0.0060	0.0021	0.078	−2.20	D	5n
		3756.07	17550	44166	7	7	0.0055	0.0012	0.10	−2.09	D	5n
		3687.10	17550	44664	7	5	0.024	0.0034	0.29	−1.62	C	5n
84.	$a\ ^5P - ^5F°$											
		3792.83	17927	44285	3	5	0.0034	0.0012	0.046	−2.43	D	5n
		3739.32	17550	44285	7	5	0.0046	6.8(−4)	0.059	−2.32	D	5n
85.	$a\ ^5P - y\ ^5S°$ (76)											
		3732.40	17727	44512	5	5	0.28	0.058	3.5	−0.54	C+	4n
		3760.53	17927	44512	3	5	0.048	0.017	0.63	−1.29	C	4n,5n
86.	$a\ ^5P - (\ °)^b$											
		3628.09	17727	45282	5	5	0.0052	0.0010	0.061	−2.29	D	5n

Fe I: Allowed transitions — Continued

No.	Multiplet	λ (Å)	E_i (cm^{-1})	E_k (cm^{-1})	g_i	g_k	A_{ki} (10^8 s^{-1})	f_{ik}	S (at. u.)	log gf	Accuracy	Source
88.	$a\ ^5P - w\ ^5P°$ (78)											
		3497.10	17550	46137	7	7	0.14	0.026	2.1	−0.74	C+	4n
		3509.87	17927	46410	3	3	0.015	0.0028	0.098	−2.07	D	5n
		3485.34	17727	46410	5	3	0.14	0.015	0.85	−1.13	C	4n
		3518.82	17727	46137	5	7	0.0063	0.0016	0.094	−2.09	D	5n
		3521.84	17927	46314	3	5	0.096	0.030	1.0	−1.05	C	4n
89.	$a\ ^5P - z\ ^3S°$ (79)											
		3462.35	17727	46601	5	3	0.014	0.0016	0.088	−2.11	D	12n
90.	$a\ ^5P - y\ ^3P°$ (82)											
		3477.85	17927	46673	3	1	0.042	0.0025	0.087	−2.12	D	12n
		3447.28	17727	46727	5	5	0.091	0.016	0.92	−1.09	C	4n
		3450.33	17927	46902	3	3	0.20	0.037	1.2	−0.96	C+	4n
91.	$a\ ^5P - \ ^5D°$											
		3427.12	17550	46721	7	9	0.55	0.12	9.8	−0.06	C+	4n
		3445.15	17727	46745	5	7	0.28	0.069	3.9	−0.46	C+	4n
		3424.28	17550	46745	7	7	0.20	0.036	2.8	−0.60	C+	4n
		3399.33	17727	47136	5	5	0.38	0.066	3.7	−0.48	C+	4n
		3417.84	17927	47177	3	3	0.51	0.090	3.0	−0.57	C+	4n
		3394.58	17727	47177	5	3	0.099	0.010	0.57	−1.29	C	4n
		3418.51	17927	47171?	3	1	1.3	0.076	2.6	−0.64	C+	4n
92.	$a\ ^5P - \ ^3D°$											
		3392.65	17550	47017	7	7	0.26	0.044	3.5	−0.51	C+	4n
		3428.19	17727	46889	5	5	0.21	0.037	2.1	−0.73	C+	4n
		3413.13	17727	47017	5	7	0.36	0.087	4.9	−0.36	C+	4n
93.	$a\ ^5P - x\ ^3F°$ (83)											
		3407.46	17550	46889	7	9	0.58	0.13	10	−0.04	C+	4n
		3383.98	17550	47093	7	7	0.093	0.016	1.2	−0.95	C+	4n
		3372.07	17550	47197	7	5	0.010	0.0012	0.094	−2.07	D	12n
94.	$a\ ^5P - z\ ^3H°$ (84)											
		3382.40	17550	47106	7	9	0.0092	0.0020	0.16	−1.85	D	12n
95.	$a\ ^5P - v\ ^5F°$ (90)											
		3298.13	17927	48239	3	5	0.081	0.022	0.72	−1.18	C	4n
		3257.59	17550	48239	7	5	0.14	0.016	1.2	−0.95	D−	11n

Fe I: Allowed transitions — Continued

No.	Multiplet	λ (Å)	E_i (cm^{-1})	E_k (cm^{-1})	g_i	g_k	A_{ki} (10^8 s^{-1})	f_{ik}	S (at. u.)	log gf	Accuracy	Source
96.	$a\ ^5P - v\ ^5P°$ (91)											
		3284.59	17727	48163	5	5	0.054	0.0087	0.47	−1.36	C	4n
		3292.59	17927	48290	3	3	0.26	0.043	1.4	−0.89	C+	4n
		3265.62	17550	48163	7	5	0.38	0.043	3.2	−0.52	C+	4n
		3271.00	17727	48290	5	3	0.66	0.063	3.4	−0.50	C+	4n
		3305.97	17727	47967	5	7	0.47	0.11	5.8	−0.27	C+	4n
		3306.36	17927	48163	3	5	0.61	0.17	5.5	−0.30	D−	11n
97.	$a\ ^5P - x\ ^3P°$ (95)											
		3246.96	17727	48516	5	3	0.099	0.0094	0.50	−1.33	E	11n
		3268.23	17927	48516	3	3	0.059	0.0094	0.30	−1.55	D	12n
		3290.99	17927	48305	3	5	0.060	0.016	0.53	−1.31	C	4n
98.	$a\ ^3P2 - z\ ^5P°$ (106)											
		8804.62	18378	29733	5	3	1.67(−4)	1.17(−4)	0.0169	−3.234	B+	28
99.	$a\ ^3P2 - z\ ^3D°$ (108)											
		7723.20	18378	31323	5	7	3.86(−5)	4.83(−5)	0.00614	−3.617	B+	28
100.	$a\ ^3P2 - y\ ^5D°$ (109)											
		6608.03	18378	33507	5	7	2.0(−5)	1.9(−5)	0.0020	−4.03	D	5n
		7016.08	19552	33802	3	5	1.7(−4)	2.1(−4)	0.014	−3.21	D	5n
		7151.50	20038	34017	1	3	8.1(−5)	1.9(−4)	0.0044	−3.73	D	30
		6481.88	18378	33802	5	5	3.29(−4)	2.08(−4)	0.0221	−2.984	B+	28
		6911.52	19552	34017	3	3	4.2(−5)	3.0(−5)	0.0021	−4.04	D	30
		6392.55	18378	34017	5	3	5.1(−5)	1.9(−5)	0.0020	−4.03	D	30
		6861.93	19552	34122	3	1	1.8(−4)	4.3(−5)	0.0029	−3.89	D	30
101.	$a\ ^3P2 - z\ ^3P°$ (111)	*6577.6*	*18954*	*34153*	9	9	0.00428	0.00278	0.541	−1.602	B	5n,26n, 28
		6421.35	18378	33947	5	5	0.00304	0.00188	0.199	−2.027	B+	28
		6750.15	19552	34363	3	3	0.00117	7.98(−4)	0.0532	−2.621	B+	28
		6254.26	18378	34363	5	3	0.0019	6.6(−4)	0.068	−2.48	C	5n,26n
		6663.45	19552	34556	3	1	0.00499	0.00111	0.0728	−2.479	B+	28
		6945.20	19552	33947	3	5	9.12(−4)	0.00110	0.0754	−2.482	B+	28
		6978.85	20038	34363	1	3	0.00144	0.00316	0.0727	−2.500	B+	28
102.	$a\ ^3P2 - y\ ^3F°$ (112)											
		5322.04	18378	37163	5	7	3.1(−4)	1.9(−4)	0.016	−3.03	D	5n
103.	$a\ ^3P2 - y\ ^5P°$ (113)											
		5436.59	18378	36767	5	7	1.3(−4)	8.1(−5)	0.0073	−3.39	D	5n
		5678.60	19552	37158	3	5	8.8(−6)	7.1(−6)	4.0(−4)	−4.67	D	30
		5253.03	18378	37410	5	3	9.3(−5)	2.3(−5)	0.0020	−3.94	D	5n

Fe I:· Allowed transitions — Continued

No.	Multiplet	λ (Å)	E_i (cm^{-1})	E_k (cm^{-1})	g_i	g_k	A_{ki} (10^8 s^{-1})	f_{ik}	S (at. u.)	log gf	Accuracy	Source
104.	$a\ ^3P2 - y\ ^3D°$ (114)											
		5049.82	18378	38175	5	7	0.014	0.0076	0.63	−1.42	C	4n
		5273.37	20038	38996	1	3	0.0053	0.0066	0.11	−2.18	D	30
		4924.77	18378	38678	5	5	0.0033	0.0012	0.097	−2.22	D	10n,30
		5141.74	19552	38996	3	3	0.0060	0.0024	0.12	−2.15	D	30
		4848.90	18378	38996	5	3	3.8(−4)	8.0(−5)	0.0064	−3.40	D	5n
105.	$a\ ^3P2 - x\ ^5D°$ (115)											
		4630.12	18378	39970	5	7	0.0011	5.0(−4)	0.038	−2.60	D	5n
		4834.51	19552	40231	3	5	2.2(−4)	1.3(−4)	0.0062	−3.41	D	5n
		4908.61	20038	40405	1	3	6.4(−5)	6.9(−5)	0.0011	−4.16	D	30
		4574.72	18378	40231	5	5	6.8(−4)	2.1(−4)	0.016	−2.97	D	5n
		4794.36	19552	40405	3	3	8.6(−5)	3.0(−5)	0.0014	−4.05	D	5n
106.	$a\ ^3P2 - z\ ^5S°$ (116)											
		4439.88	18378	40895	5	5	6.74(−4)	1.99(−4)	0.0145	−3.002	B+	28
107.	$a\ ^3P2 - x\ ^5P°$ (117)											
		4249.32	19552	43079	3	3	8.9(−4)	2.4(−4)	0.010	−3.14	D	5n
108.	$a\ ^3P2 - w\ ^5D°$ (120)											
		3913.63	18378	43923	5	7	0.0135	0.00435	0.280	−1.663	B+	28
		4058.75	19552	44184	3	5	0.0071	0.0029	0.12	−2.06	D	4n
		4101.68	20038	44411	1	3	0.0034	0.0026	0.035	−2.59	D	5n
109.	$a\ ^3P2 - \ ^5D°$											
		3981.11	19552	44664	3	5	9.4(−4)	3.7(−4)	0.015	−2.95	D	5n
110.	$a\ ^3P2 - (\ °)^b$											
		3724.38	18378	45221	5	7	0.13	0.036	2.2	−0.74	C+	4n
111.	$a\ ^3P2 - (\ °)^b$											
		3885.51	19552	45282	3	5	0.058	0.022	0.84	−1.18	C	4n,5n
		3715.91	18378	45282	5	5	0.0261	0.00540	0.330	−1.569	B+	28
112.	$a\ ^3P2 - (\ °)^b$											
		3845.17	19552	45552	3	3	0.068	0.015	0.57	−1.35	C	4n,5n
113.	$a\ ^3P2 - w\ ^5P°$ (127)											
		3578.67	18378	46314	5	5	0.016	0.0031	0.18	−1.81	C	5n
		3566.31	18378	46410	5	3	0.029	0.0033	0.19	−1.78	C	5n

Fe I: Allowed transitions — Continued

No.	Multiplet	λ (Å)	E_i (cm^{-1})	E_k (cm^{-1})	g_i	g_k	A_{ki} (10^8 s^{-1})	f_{ik}	S (at. u.)	log gf	Accu-racy	Source
114.	$a\ ^3$P2 – $y\ ^3$P° (131)											
		3504.86	18378	46902	5	3	0.017	0.0019	0.11	−2.02	D	5n
		3686.26	19552	46673	3	1	0.12	0.0080	0.29	−1.62	C	5n
		3678.86	19552	46727	3	5	0.041	0.014	0.51	−1.38	C	4n,5n
115.	$a\ ^3$P2 – ^{5}D°											
		3524.24	18378	46745	5	7	0.042	0.011	0.64	−1.26	C	4n,5n
		3624.31	19552	47136	3	5	0.010	0.0034	0.12	−1.99	C	5n
116.	$a\ ^3$P2 – ^{3}D°											
		3657.14	19552	46889	3	5	0.011	0.0038	0.14	−1.94	C	5n
		3670.81	20038	47272	1	3	0.022	0.013	0.16	−1.87	C	5n
		3506.50	18378	46889	5	5	0.071	0.013	0.75	−1.19	C	4n,5n
117.	$a\ ^3$P2 – $x\ ^3$F° (132)											
		3616.32	19552	47197	3	5	0.0072	0.0024	0.084	−2.15	D	5n
118.	$a\ ^3$P2 – ^{1}D°											
		3442.36	18378	47420	5	5	0.0455	0.00809	0.458	−1.393	B+	28
119.	$a\ ^3$P2 – $v\ ^5$F° (138)											
		3347.93	18378	48239	5	5	0.040	0.0068	0.37	−1.47	C	4n
120.	$a\ ^3$P2 – $x\ ^3$P° (139)											
		3317.12	18378	48516	5	3	0.0312	0.00308	0.168	−1.812	B+	28
		3510.44	20038	48516	1	3	0.044	0.025	0.28	−1.61	C	5n
121.	$a\ ^3$P2 – $u\ ^3$D° (146)											
		3053.07	19552	52297	3	5	0.15	0.036	1.1	−0.97	C+	4n
122.	$a\ ^3$P2 – ^{3}D°											
		2954.65	18378	52213	5	7	0.10	0.018	0.89	−1.04	C	4n
		2899.42	18378	52858	5	3	0.59	0.045	2.1	−0.65	C+	4n
123.	$a\ ^3$P2 – ^{3}P°											
		2894.50	18378	52916	5	5	0.62	0.078	3.7	−0.41	C+	4n
		2996.39	19552	52916	3	5	0.16	0.036	1.1	−0.97	C+	4n

Fe I: Allowed transitions — Continued

No.	Multiplet	λ (Å)	E_i (cm^{-1})	E_k (cm^{-1})	g_i	g_k	A_{ki} (10^8 s^{-1})	f_{ik}	S (at. u.)	log gf	Accu-racy	Source
124.	$z\ ^7D° - e\ ^7D$ (152)											
		4260.47	19351	42816	11	11	0.32	0.087	13	−0.02	D	9n
		4235.94	19562	43163	9	9	0.188	0.0507	6.36	−0.341	B+	28
		4222.21	19757	43435	7	7	0.0577	0.0154	1.50	−0.967	B+	28
		4210.34	20020	43764	3	3	0.17	0.045	1.9	−0.87	C+	4n
		4198.30	19351	43163	11	9	0.0803	0.0174	2.64	−0.719	B+	28
		4187.79	19562	43435	9	7	0.152	0.0310	3.85	−0.554	B+	28
		4187.04	19757	43634	7	5	0.215	0.0404	3.90	−0.548	B+	28
		4271.15	19757	43163	7	9	0.182	0.0640	6.30	−0.349	B+	28
		4250.12	19912	43435	5	7	0.208	0.0787	5.51	−0.405	B+	28
		4233.60	20020	43634	3	5	0.185	0.0830	3.47	−0.604	B+	28
125.	$z\ ^7D° - e\ ^5D$ (153)											
		3980.65	19562	44677	9	9	2.2(−4)	5.3(−5)	0.0063	−3.32	D	5n
		4011.71	19757	44677	7	9	9.4(−4)	2.9(−4)	0.027	−2.69	D	5n
		3975.21	19912	45061	5	7	0.0010	3.5(−4)	0.023	−2.76	D	5n
126.	$z\ ^7D° - e\ ^7F$ (155)											
		3225.79	19351	50342	11	13	0.88	0.16	19	0.25	C+	4n
		3196.93	19562	50833	9	11	0.90	0.17	16	0.18	D−	11n
		3175.45	19351	50833	11	11	0.13	0.019	2.2	−0.67	C+	4n
		3160.66	19562	51192	9	9	0.19	0.028	2.6	−0.60	C+	4n
		3205.40	20020	51208	3	3	1.2	0.18	5.7	−0.27	C+	4n
127.	$z\ ^7D° - f\ ^7D$ (156)											
		3222.07	19351	50378	11	11	0.33	0.051	6.0	−0.25	D−	11n
		3199.53	19562	50808	9	9	0.26	0.040	3.8	−0.44	C+	4n
		3215.94	19912	50999	5	5	0.80	0.12	6.5	−0.21	C+	4n
		3219.58	19757	50808	7	9	0.62	0.12	9.2	−0.06	D−	11n
128.	$z\ ^7D° - f\ ^5D$ (157)											
		3217.38	19351	50423	11	9	0.22	0.028	3.3	−0.51	C+	4n
		3227.80	19562	50534	9	7	1.4	0.17	16	0.19	D−	11n
		3230.96	19757	50699	7	5	0.39	0.043	3.2	−0.52	C+	4n
		3228.25	19912	50880	5	3	0.45	0.042	2.2	−0.68	D−	11n
		3248.20	19757	50534	7	7	0.22	0.034	2.6	−0.62	C+	4n
129.	$z\ ^7D° - e\ ^7P$ (158)											
		3233.97	19562	50475	9	9	0.20	0.031	3.0	−0.55	C+	4n
		3230.21	19912	50861	5	5	0.19	0.030	1.6	−0.82	D−	11n
130.	$z\ ^7D° - e\ ^5G$ (159)											
		3207.07	19351	50523	11	13	0.013	0.0023	0.27	−1.60	D	12n

Fe I: Allowed transitions — Continued

No.	Multiplet	λ (Å)	E_i (cm⁻¹)	E_k (cm⁻¹)	g_i	g_k	A_{ki} (10⁸ s⁻¹)	f_{ik}	S (at. u.)	log gf	Accuracy	Source
131.	$z\,^7D° - e\,^7G$ (160)											
		3161.95	19351	50968	11	13	0.12	0.021	2.4	−0.64	C+	4n
		3168.85	19912	51461	5	7	0.057	0.012	0.63	−1.22	D	12n
132.	$a\,^3H - y\,^5F°$ (167)											
		6988.53	19390	33695	13	11	2.7(−5)	1.7(−5)	0.0050	−3.66	D	5n
		7014.99	19788	34040	9	9	8.5(−6)	6.2(−6)	0.0013	−4.25	D	30
133.	$a\,^3H - z\,^5G°$ (168)											
		6593.88	19621	34782	11	11	5.28(−4)	3.44(−4)	0.0822	−2.422	B+	28
		6462.73	19788	35257	9	9	4.5(−4)	2.8(−4)	0.053	−2.60	D	5n
		6494.98	19390	34782	13	11	0.00767	0.00410	1.14	−1.273	B+	28
		6393.60	19621	35257	11	9	0.0044	0.0022	0.51	−1.62	D−	14n,30
		6667.42	19788	34782	9	11	5.4(−6)	4.4(−6)	8.7(−4)	−4.40	D	30
134.	$a\,^3H - z\,^3G°$ (169)											
		6252.55	19390	35379	13	11	0.00319	0.00158	0.423	−1.687	B+	28
		6191.56	19621	35768	11	9	0.0049	0.0023	0.51	−1.60	D−	14n
		6136.61	19788	36079	9	7	0.0101	0.00442	0.804	−1.400	B+	28
		6344.15	19621	35379	11	11	1.80(−4)	1.09(−4)	0.0249	−2.923	B+	28
		6256.37	19788	35768	9	9	4.5(−4)	2.7(−4)	0.049	−2.62	D	5n
135.	$a\,^3H - y\,^3F°$ (170)											
		5916.25	19788	36686	9	9	2.15(−4)	1.13(−4)	0.0197	−2.994	B+	28
136.	$a\,^3H - \,^5F°$											
		4022.45	19390	44244	13	11	9.9(−5)	2.0(−5)	0.0035	−3.58	D−	30
137.	$a\,^3H - y\,^3G°$ (175)											
		3859.21	19390	45295	13	11	0.085	0.016	2.7	−0.68	C+	4n
		3873.76	19621	45428	11	9	0.080	0.015	2.1	−0.79	C+	4n
		3899.03	19788	45428	9	9	0.0083	0.0019	0.22	−1.77	C	4n,5n
138.	$a\,^3H - z\,^3I°$ (177)											
		3760.05	19390	45978?	13	15	0.0447	0.0109	1.76	−0.847	B+	28
		3785.95	19621	46027	11	13	0.042	0.011	1.5	−0.93	C	5n
		3794.34	19788	46136	9	11	0.038	0.0099	1.1	−1.05	C	4n,5n
		3753.15	19390	46027	13	13	0.0010	2.2(−4)	0.035	−2.55	D	5n
139.	$a\,^3H - \,^5D°$											
		3689.02	19621	46721	11	9	0.0040	6.7(−4)	0.090	−2.13	D	5n
140.	$a\,^3H - x\,^3F°$ (179)											
		3661.36	19788	47093	9	7	0.0025	3.9(−4)	0.043	−2.45	D	5n
		3688.88	19788	46889	9	9	0.0039	8.0(−4)	0.088	−2.14	D	5n

Fe I: Allowed transitions — Continued

No.	Multiplet	λ (Å)	E_i (cm^{-1})	E_k (cm^{-1})	g_i	g_k	A_{ki} (10^8 s^{-1})	f_{ik}	S (at. u.)	log gf	Accuracy	Source
141.	$a\ ^3\text{H} - z\ ^3\text{H}°$ (180)											
		3623.19	19390	46982	13	13	0.074	0.015	2.3	−0.72	C+	4n
		3659.52	19788	47106	9	9	0.058	0.012	1.3	−0.98	C+	4n
		3637.25	19621	47106	11	9	0.0071	0.0011	0.15	−1.90	C	5n
		3653.76	19621	46982	11	13	0.0045	0.0011	0.14	−1.93	C	5n
		3672.69	19788	47008	9	11	0.0031	7.7(−4)	0.084	−2.16	D	5n
142.	$a\ ^3\text{H} - w\ ^5\text{G}°$ (181)											
		3596.20	19621	47420	11	11	0.00433	8.39(−4)	0.109	−2.035	B+	28
		3595.86	19788	47590	9	9	0.0020	3.9(−4)	0.041	−2.46	D	5n
143.	$a\ ^3\text{H} - v\ ^5\text{F}°$ (182)											
		3531.44	19621	47930	11	9	0.0024	3.7(−4)	0.047	−2.39	D	5n
144.	$a\ ^3\text{H} - x\ ^3\text{G}°$ (183)											
		3514.63	19390	47835	13	11	0.0037	5.8(−4)	0.088	−2.12	D	5n
		3543.39	19621	47835	11	11	0.0030	5.7(−4)	0.074	−2.20	D	5n
		3567.37	19788	47812	9	9	0.0035	6.7(−4)	0.071	−2.22	D	5n
145.	$a\ ^3\text{H} - z\ ^1\text{H}°$ (186)											
		3496.19	19788	48383	9	11	2.9(−4)	6.5(−5)	0.0068	−3.23	C	8
146.	$a\ ^3\text{H} - v\ ^3\text{G}°$ (191)											
		3325.46	19788	49851	9	7	0.021	0.0027	0.27	−1.61	D	12n
147.	$a\ ^3\text{H} - u\ ^3\text{G}°$ (194)											
		3119.49	19621	51668	11	9	0.082	0.0097	1.1	−0.97	C+	4n
		3120.43	19788	51826	9	7	0.089	0.010	0.94	−1.04	C	4n
148.	$a\ ^3\text{H} - w\ ^3\text{H}°$ (198)											
		3009.09	19390	52613	13	11	0.067	0.0077	0.99	−1.00	C	4n
		3015.92	19621	52769	11	9	0.059	0.0066	0.72	−1.14	C	4n
149.	$a\ ^3\text{H} - y\ ^3\text{I}°$ (199)											
		3005.31	19390	52655?	13	15	0.024	0.0038	0.48	−1.31	C	8
		3039.32	19621	52514	11	13	0.016	0.0026	0.29	−1.54	C	8
		3018.14	19390	52514	13	13	0.012	0.0016	0.21	−1.67	C	8
150.	$a\ ^3\text{H} - z\ ^1\text{I}°$ (200)											
		2986.65	19621	53094	11	13	0.0085	0.0013	0.15	−1.83	C	8

Fe I: Allowed transitions — Continued

No.	Multiplet	λ (Å)	E_i (cm^{-1})	E_k (cm^{-1})	g_i	g_k	A_{ki} (10^8 s^{-1})	f_{ik}	S (at. u.)	log gf	Accuracy	Source
151.	a ^{3}H $-$ x ^{3}I° (uv 156)											
		2656.15	19390	57028?	13	15	0.28	0.034	3.9	-0.35	C	8
		2669.49	19621	57070	11	13	0.17	0.021	2.1	-0.63	C	8
152.	b ^{3}F2 $-$ y ^{5}F° (204)											
		7461.53	20641	34040	9	9	3.5($-$5)	2.9($-$5)	0.0065	-3.58	D	30
		7430.58	20874	34329	7	7	2.4($-$5)	2.0($-$5)	0.0034	-3.86	D	30
		7400.87	21039	34547	5	5	9.5($-$6)	7.8($-$6)	9.5($-$4)	-4.41	D	30
153.	b ^{3}F2 $-$ z ^{5}G° (205)											
		7069.54	20641	34782	9	11	5.5($-$6)	5.1($-$6)	0.0011	-4.34	D	30
		6860.29	21039	35612	5	7	1.4($-$5)	1.4($-$5)	0.0016	-4.15	D	30
		6839.83	20641	35257	9	9	5.6($-$5)	3.9($-$5)	0.0080	-3.45	D	5n
		6783.71	20874	35612	7	7	2.2($-$5)	1.5($-$5)	0.0023	-3.98	D	30
		6746.96	21039	35856	5	5	1.3($-$5)	8.9($-$6)	9.9($-$4)	-4.35	D	30
154.	b ^{3}F2 $-$ z ^{3}G° (206)											
		6646.98	21039	36079	5	7	2.2($-$5)	2.0($-$5)	0.0022	-3.99	D	30
		6609.12	20641	35768	9	9	3.45($-$4)	2.26($-$4)	0.0442	-2.692	B+	29
		6575.02	20874	36079	7	7	3.3($-$4)	2.2($-$4)	0.033	-2.82	D	5n
		6475.63	20641	36079	9	7	2.6($-$4)	1.3($-$4)	0.024	-2.94	D	5n
155.	b ^{3}F2 $-$ y ^{3}F° (207)											
		6230.73	20641	36686	9	9	0.0100	0.00582	1.07	-1.281	B+	29
		6137.69	20874	37163	7	7	0.0100	0.00565	0.799	-1.403	B+	29
		6065.48	21039	37521	5	5	0.0107	0.00590	0.589	-1.530	B+	29
		6322.69	20874	36686	7	9	6.95($-$4)	5.36($-$4)	0.0781	-2.426	B+	29
		6200.32	21039	37163	5	7	9.06($-$4)	7.31($-$4)	0.0746	-2.437	B+	29
156.	b ^{3}F2 $-$ y ^{5}P° (208)											
		6199.48	20641	36767	9	7	9.2($-$6)	4.1($-$6)	7.6($-$4)	-4.43	D	30
		6139.65	20874	37158	7	5	1.1($-$5)	4.5($-$6)	6.4($-$4)	-4.50	D	30
		6290.55	20874	36767	7	7	1.1($-$5)	6.7($-$6)	9.7($-$4)	-4.33	D	30
157.	b ^{3}F2 $-$ y ^{3}D° (209)											
		5701.54	20641	38175	9	7	0.00178	6.76($-$4)	0.114	-2.216	B+	29
		5567.40	21039	38996	5	3	0.0011	3.2($-$4)	0.029	-2.80	D	5n
		5778.47	20874	38175	7	7	7.3($-$5)	3.7($-$5)	0.0049	-3.59	D	5n
		5667.67	21039	38678	5	5	3.9($-$4)	1.9($-$4)	0.017	-3.03	D	5n
		5833.93	21039	38175	5	7	6.1($-$5)	4.4($-$5)	0.0042	-3.66	D	5n
158.	b ^{3}F2 $-$ w ^{5}D° (214)											
		4319.46	21039	44184	5	5	1.8($-$4)	4.9($-$5)	0.0035	-3.61	D$-$	30
		4288.96	20874	44184	7	5	0.0021	4.2($-$4)	0.042	-2.53	D	5n
		4277.41	21039	44411	5	3	0.0012	1.9($-$4)	0.013	-3.02	D	5n

Fe I: Allowed transitions — Continued

No.	Multiplet	λ (Å)	E_i (cm^{-1})	E_k (cm^{-1})	g_i	g_k	A_{ki} (10^8 s^{-1})	f_{ik}	S (at. u.)	log gf	Accuracy	Source
159.	$b\ ^3F2 - ^5D°$											
		4275.72	20641	44023	9	9	5.9(−4)	1.6(−4)	0.020	−2.84	D	5n
160.	$b\ ^3F2 - (\ °)^b$											
		4067.27	20641	45221	9	7	0.0220	0.00423	0.510	−1.419	B+	29
		4106.27	20874	45221	7	7	0.0028	7.2(−4)	0.068	−2.30	D	5n
161.	$b\ ^3F2 - (\ °)^b$											
		4095.97	20874	45282	7	5	0.032	0.0057	0.54	−1.40	C	4n,5n
162.	$b\ ^3F2 - y\ ^3G°$ (218)											
		4055.03	20641	45295	9	11	0.0058	0.0017	0.21	−1.80	C	5n
		4049.34	20874	45563	7	7	0.0025	6.1(−4)	0.057	−2.37	D	4n,5n
		4011.42	20641	45563	9	7	0.0024	4.4(−4)	0.053	−2.40	D	5n
163.	$b\ ^3F2 - (\ °)^b$											
		4078.35	21039	45552	5	3	0.042	0.0063	0.42	−1.50	C	4n,5n
164.	$b\ ^3F2 - x\ ^5G°$ (219)											
		4019.05	21039	45913	5	7	9.8(−4)	3.3(−4)	0.022	−2.78	D	5n
165.	$b\ ^3F2 - ^5D°$											
		3829.77	20641	46745	9	7	0.0066	0.0011	0.13	−1.99	C	5n
		3833.31	20641	46721	9	9	0.0469	0.0103	1.17	−1.032	B+	29
		3867.93	20874	46721	7	9	0.0059	0.0017	0.15	−1.92	C	4n,5n
166.	$b\ ^3F2 - ^3D°$											
		3848.29	21039	47017	5	7	0.0043	0.0013	0.084	−2.18	D	4n
167.	$b\ ^3F2 - x\ ^3F°$ (222)											
		3808.73	20641	46889	9	9	0.0354	0.00770	0.869	−1.159	B+	29
		3821.83	21039	47197	5	5	0.078	0.017	1.1	−1.07	C	4n,5n
		3797.95	20874	47197	7	5	0.018	0.0027	0.24	−1.72	C	5n
		3837.13	21039	47093	5	7	0.013	0.0039	0.25	−1.71	C	4n,5n
168.	$b\ ^3F2 - z\ ^3H°$ (223)											
		3791.50	20641	47008	9	11	0.0034	8.8(−4)	0.099	−2.10	D	5n
		3777.45	20641	47106	9	9	0.00858	0.00184	0.205	−1.782	B+	29
169.	$b\ ^3F2 - w\ ^5G°$ (225)											
		3731.37	21039	47831	5	5	0.0338	0.00705	0.433	−1.453	B+	29
170.	$b\ ^3F2 - ^1D°$											
		3766.09	20874	47420	7	5	0.0068	0.0010	0.090	−2.14	D	5n

Fe I:　Allowed transitions — Continued

No.	Multiplet	λ (Å)	E_i (cm^{-1})	E_k (cm^{-1})	g_i	g_k	A_{ki} (10^8 s^{-1})	f_{ik}	S (at. u.)	log gf	Accuracy	Source
171.	$b\ ^3F2 - z\ ^1G°$ (227)											
		3728.67	20641	47453	9	9	0.011	0.0024	0.26	−1.67	C	5n
		3761.41	20874	47453	7	9	0.0066	0.0018	0.16	−1.90	C	5n
172.	$b\ ^3F2 - v\ ^5F°$ (229)											
		3668.89	20874	48123	7	7	0.0024	4.8(−4)	0.041	−2.47	D	5n
173.	$b\ ^3F2 - x\ ^3G°$ (228)											
		3676.31	20641	47835	9	11	0.0463	0.0115	1.25	−0.986	B+	29
		3711.22	20874	47812	7	9	0.033	0.0088	0.75	−1.21	C	5n
		3730.95	21039	47834	5	7	0.038	0.011	0.68	−1.26	C	4n,5n
174.	$b\ ^3F2 - v\ ^5P°$ (231)											
		3658.55	20641	47967	9	7	0.0023	3.5(−4)	0.038	−2.50	D	5n
175.	$b\ ^3F2 - ^5H°$											
		3636.99	20874	48362	7	9	0.015	0.0038	0.31	−1.58	C	5n
176.	$b\ ^3F2 - x\ ^3P°$ (235)											
		3644.58	20874	48305	7	5	0.0036	5.1(−4)	0.043	−2.45	D	5n
177.	$b\ ^3F2 - z\ ^1H°$											
		3603.67	20641	48383	9	11	0.0023	5.4(−4)	0.058	−2.31	C	8
178.	$b\ ^3F2 - y\ ^1G°$ (237)											
		3592.47	20874	48703	7	9	0.0019	4.8(−4)	0.040	−2.47	D	5n
179.	$b\ ^3F2 - w\ ^3F°$ (238)											
		3511.74	20641	49109	9	9	0.0022	4.1(−4)	0.043	−2.43	D	5n
		3520.85	21039	49433	5	5	0.013	0.0024	0.14	−1.92	C	5n
		3495.29	20641	49243	9	7	0.0946	0.0135	1.40	−0.916	B+	29
		3500.57	20874	49433	7	5	0.029	0.0038	0.30	−1.58	C	5n
180.	$b\ ^3F2 - v\ ^3D°$ (239)											
		3524.08	20874	49243	7	5	0.075	0.010	0.81	−1.15	C	4n,5n
		3537.73	21039	49298	5	3	0.11	0.012	0.70	−1.22	C	5n
		3544.63	21039	49243	5	5	0.015	0.0028	0.16	−1.86	C	5n
181.	$b\ ^3F2 - v\ ^3G°$ (242)											
		3469.83	21039	49851	5	7	0.0184	0.00466	0.266	−1.633	B+	25

Fe I: Allowed transitions — Continued

No.	Multiplet	λ (Å)	E_i (cm^{-1})	E_k (cm^{-1})	g_i	g_k	A_{ki} (10^8 s^{-1})	f_{ik}	S (at. u.)	log gf	Accuracy	Source
182.	$b\ ^3F2 - u\ ^3D°$ (258)											
		3176.36	21039	52512	5	3	0.092	0.0083	0.44	−1.38	D	12n
183.	$b\ ^3F2 - ^3D°$											
		3166.44	20641	52213	9	7	0.114	0.0133	1.25	−0.921	B+	29
184.	$a\ ^3G - z\ ^5G°$ (266)											
		7540.44	21999	35257	9	9	1.8(−5)	1.6(−5)	0.0035	−3.85	D	30
		7481.74	22249	35612	7	7	1.4(−5)	1.1(−5)	0.0020	−4.10	D	30
		7347.16	22249	35856	7	5	1.5(−5)	8.6(−6)	0.0015	−4.22	D	30
185.	$a\ ^3G - z\ ^3G°$ (267)											
		7261.02	21999	35768	9	9	3.5(−5)	2.7(−5)	0.0059	−3.61	D	30
		7228.69	22249	36079	7	7	7.6(−5)	6.0(−5)	0.0099	−3.38	D	30
		7114.55	21716	35768	11	9	1.4(−5)	8.9(−6)	0.0023	−4.01	D	30
186.	$a\ ^3G - y\ ^3F°$ (268)											
		6677.99	21716	36686	11	9	0.0056	0.0031	0.74	−1.47	D−	14n
		6592.91	21999	37163	9	7	0.0055	0.0028	0.55	−1.60	D−	14n
		6546.24	22249	37521	7	5	0.0070	0.0032	0.48	−1.65	D−	14n
		6806.85	21999	36686	9	9	9.9(−5)	6.9(−5)	0.014	−3.21	D	30
		6703.57	22249	37163	7	7	1.5(−4)	9.9(−5)	0.015	−3.16	D	5n
187.	$a\ ^3G - y\ ^3D°$ (269)											
		6180.22	21999	38175	9	7	4.1(−4)	1.8(−4)	0.034	−2.78	D	5n
		6085.27	22249	38678	7	5	2.2(−4)	8.8(−5)	0.012	−3.21	D	5n
188.	$a\ ^3G - ^5D°$											
		4561.43	22249	44166	7	7	3.8(−4)	1.2(−4)	0.012	−3.08	D	30
189.	$a\ ^3G - y\ ^3G°$ (273)											
		4266.96	21999	45428	9	9	0.0085	0.0023	0.29	−1.68	C	5n
		4288.15	22249	45563	7	7	0.0062	0.0017	0.17	−1.92	C	4n,5n
190.	$a\ ^3G - x\ ^5G°$ (274)											
		4194.50	21999	45833	9	9	1.8(−4)	4.9(−5)	0.0060	−3.36	D−	30
		4145.21	21716	45833	11	9	6.8(−4)	1.4(−4)	0.022	−2.80	D	5n
		4180.40	21999	45913	9	7	6.7(−4)	1.4(−4)	0.017	−2.91	D−	30
191.	$a\ ^3G - ^5D°$											
		3998.05	21716	46721	11	9	0.066	0.013	1.9	−0.84	C+	4n,5n
192.	$a\ ^3G - ^3D°$											
		3995.98	21999	47017	9	7	0.021	0.0039	0.46	−1.45	C	4n,5n
		4036.37	22249	47017	7	7	8.5(−4)	2.1(−4)	0.019	−2.84	D	5n

Fe I: Allowed transitions — Continued

No.	Multiplet	λ (Å)	E_i (cm^{-1})	E_k (cm^{-1})	g_i	g_k	A_{ki} (10^8 s^{-1})	f_{ik}	S (at. u.)	log gf	Accuracy	Source
193.	$a\ ^3G - x\ ^3F°$ (277)											
		3971.32	21716	46889	11	9	0.057	0.011	1.6	-0.92	C+	4n,5n
		3983.96	21999	47093	9	7	0.076	0.014	1.7	-0.90	C+	4n,5n
		4007.27	22249	47197	7	5	0.042	0.0072	0.66	-1.30	C	4n
		4024.11	22249	47093	7	7	0.0029	7.2(−4)	0.066	-2.30	D	5n
		4057.34	22249	46889	7	9	0.0044	0.0014	0.13	-2.01	D	4n,5n
194.	$a\ ^3G - z\ ^3H°$ (278)											
		3997.39	21999	47008	9	11	0.15	0.045	5.4	-0.39	C+	4n
		4021.87	22249	47106	7	9	0.10	0.031	2.9	-0.66	C+	4n
		3952.60	21716	47008	11	11	0.041	0.0096	1.4	-0.98	C+	4n,5n
		3981.77	21999	47106	9	9	0.039	0.0092	1.1	-1.08	C	4n,5n
		3937.33	21716	47106	11	9	0.017	0.0032	0.46	-1.45	C	4n,5n
195.	$a\ ^3G - w\ ^5G°$ (280)											
		3945.12	22249	47590	7	9	0.015	0.0046	0.42	-1.49	C	5n
		3863.74	21716	47590	11	9	0.022	0.0040	0.56	-1.36	C	4n,5n
		3890.84	21999	47693	9	7	0.029	0.0051	0.59	-1.34	C	4n,5n
		3907.93	22249	47831	7	5	0.067	0.011	0.99	-1.11	C	4n,5n
196.	$a\ ^3G - z\ ^1G°$ (282)											
		3884.36	21716	47453	11	9	0.035	0.0065	0.91	-1.15	C	4n,5n
197.	$a\ ^3G - v\ ^5F°$ (283)											
		3813.63	21716	47930	11	9	0.014	0.0025	0.35	-1.56	C	4n,5n
		3826.84	21999	48123	9	7	0.015	0.0025	0.29	-1.64	C	5n
198.	$a\ ^3G - x\ ^3G°$ (284)											
		3872.92	21999	47812	9	9	0.0088	0.0020	0.23	-1.75	C	5n
		3907.47	22249	47834	7	7	0.0080	0.0018	0.17	-1.89	C	5n
		3910.84	22249	47812	7	9	0.012	0.0037	0.33	-1.59	C	5n
199.	$a\ ^3G - ^5H°$											
		3770.30	21716	48231	11	11	0.017	0.0037	0.51	-1.39	C	5n
		3792.15	21999	48362	9	9	0.019	0.0042	0.47	-1.42	C	4n,5n
		3811.89	22249	48476	7	7	0.034	0.0073	0.64	-1.29	C	4n,5n
		3751.82	21716	48362	11	9	0.0043	7.4(−4)	0.10	-2.09	D	5n
		3775.86	21999	48476	9	7	0.0022	3.7(−4)	0.041	-2.48	D	5n
200.	$a\ ^3G - z\ ^1H°$ (289)											
		3789.18	21999	48383	9	11	0.023	0.0060	0.67	-1.27	C	4n,5n,8
201.	$a\ ^3G - y\ ^1G°$ (290)											
		3704.46	21716	48703	11	9	0.13	0.022	3.0	-0.61	C+	4n

Fe I: Allowed transitions — Continued

No.	Multiplet	λ (Å)	E_i (cm^{-1})	E_k (cm^{-1})	g_i	g_k	A_{ki} (10^8 s^{-1})	f_{ik}	S (at. u.)	log gf	Accuracy	Source
202.	$a\ ^3G - w\ ^3F°$ (291)											
		3649.51	21716	49109	11	9	0.42	0.069	9.1	−0.12	C+	4n
		3669.52	21999	49243	9	7	0.30	0.046	5.0	−0.38	C+	4n
		3677.63	22249	49433	7	5	0.80	0.12	9.8	−0.09	C+	4n
203.	$a\ ^3G - v\ ^3D°$ (292)											
		3684.11	21999	49135	9	7	0.34	0.053	5.8	−0.32	C+	4n
		3718.41	22249	49135	7	7	0.053	0.011	0.94	−1.11	C	4n,5n
204.	$a\ ^3G - y\ ^3H°$ (294)											
		3606.68	21716	49434	11	13	0.82	0.19	25	0.32	C+	4n
		3621.46	21999	49604	9	11	0.51	0.12	13	0.04	C+	4n
		3638.30	22249	49727	7	9	0.26	0.067	5.6	−0.33	C+	4n
		3605.45	21999	49727	9	9	0.64	0.12	13	0.05	C+	4n
		3568.98	21716	49727	11	9	0.030	0.0047	0.60	−1.29	C	5n
205.	$a\ ^3G - v\ ^3G°$ (295)											
		3603.20	21716	49461	11	11	0.26	0.051	6.7	−0.25	C+	4n
		3622.00	22249	49851	7	7	0.51	0.10	8.4	−0.15	C+	4n
		3640.39	21999	49461	9	11	0.38	0.092	10	−0.08	C+	4n
		3651.47	22249	49628	7	9	0.62	0.16	13	0.05	C+	4n
206.	$a\ ^3G - x\ ^1G°$ (297)											
		3493.69	21999	50614	9	9	0.0046	8.4(−4)	0.087	−2.12	D	5n
207.	$a\ ^3G - v\ ^3F°$ (301)											
		3411.35	21999	51305	9	9	0.055	0.0097	0.98	−1.06	C	4n
208.	$a\ ^3G - u\ ^3G°$ (304)											
		3370.78	21716	51374	11	11	0.33	0.056	6.8	−0.21	C+	4n
		3369.55	21999	51668	9	9	0.24	0.041	4.1	−0.43	C+	4n
		3380.11	22249	51826	7	7	0.24	0.040	3.1	−0.55	C+	4n
		3337.66	21716	51668	11	9	0.057	0.0077	0.94	−1.07	C	4n
209.	$a\ ^3G - t\ ^3G°$ (313)											
		3098.19	21716	53983	11	11	0.11	0.015	1.7	−0.77	C+	4n
210.	$a\ ^3G - ^3G°$											
		2980.53	22249	55791	7	7	0.22	0.029	2.0	−0.69	C+	4n
		2990.39	21999	55430	9	11	0.39	0.064	5.7	−0.24	C+	4n
211.	$a\ ^3G - ^3H°$											
		3011.48	22249	55446	7	9	0.47	0.082	5.7	−0.24	C+	4n

Fe I: Allowed transitions — Continued

No.	Multiplet	λ (Å)	E_i (cm^{-1})	E_k (cm^{-1})	g_i	g_k	A_{ki} (10^8 s^{-1})	f_{ik}	S (at. u.)	log gf	Accuracy	Source
212.	$a\ ^3G - u\ ^3H°$ (uv 167)											
		2925.36	22249	56423	7	9	0.18	0.030	2.0	−0.68	C+	4n
213.	$z\ ^7F° - e\ ^7D$ (318)											
		4920.50	22846	43163	11	9	0.35	0.10	19	0.06	C+	4n
		4891.49	22997	43435	9	7	0.29	0.080	12	−0.14	C+	4n
		4871.32	23111	43634	7	5	0.22	0.056	6.2	−0.41	C+	4n
		4859.74	23192	43764	5	3	0.13	0.028	2.3	−0.85	C+	4n,5n
		4918.99	23111	43435	7	7	0.17	0.061	6.9	−0.37	C+	4n
		4890.75	23192	43634	5	5	0.21	0.074	6.0	−0.43	C+	4n
		4872.14	23245	43764	3	3	0.24	0.084	4.0	−0.60	C+	4n
		5044.21	22997	42816	9	11	0.0017	7.9(−4)	0.12	−2.15	D	30
		4903.31	23245	43634	3	5	0.047	0.028	1.4	−1.08	C	4n,5n
		4878.21	23270	43764	1	3	0.091	0.098	1.6	−1.01	C	4n
214.	$z\ ^7F° - e\ ^5D$ (319)											
		4554.47	23111	45061	7	7	4.1(−4)	1.3(−4)	0.013	−3.05	D	5n
		4515.16	23192	45334	5	5	3.9(−4)	1.2(−4)	0.0088	−3.23	D	5n
		4635.62	23111	44677	7	9	8.9(−5)	3.7(−5)	0.0039	−3.59	D	5n
		4571.44	23192	45061	5	7	2.4(−4)	1.1(−4)	0.0081	−3.27	D	5n
		4525.88	23245	45334	3	5	4.1(−4)	2.1(−4)	0.0094	−3.20	D	5n
215.	$z\ ^7F° - e\ ^5F$ (320)											
		4070.03	23192	47756	5	7	1.9(−4)	6.6(−5)	0.0044	−3.48	D−	30
		4032.47	23245	48037	3	5	0.0059	0.0024	0.096	−2.14	D−	30
216.	$z\ ^7F° - e\ ^7F$ (321)											
		3610.16	22650	50342	13	13	0.48	0.095	15	0.09	C+	4n
		3572.00	22846	50833	11	11	0.24	0.047	6.0	−0.29	C+	4n
		3552.83	23192	51331	5	5	0.15	0.029	1.7	−0.84	C+	4n,5n
		3551.11	22997	51149	9	7	0.0030	4.4(−4)	0.047	−2.40	D	5n
		3568.42	23192	51208	5	3	0.053	0.0060	0.35	−1.52	C	5n
		3591.35	22997	50833	9	11	0.0071	0.0017	0.18	−1.82	C	5n
		3560.07	23111	51192	7	9	0.0034	8.2(−4)	0.067	−2.24	D	5n
		3578.38	23270	51208	1	3	0.063	0.036	0.43	−1.44	C	5n
217.	$z\ ^7F° - f\ ^7D$ (322)											
		3594.63	22997	50808	9	9	0.27	0.053	5.7	−0.32	C+	4n
		3595.30	23192	50999	5	5	0.054	0.010	0.62	−1.28	C	5n
		3602.08	23245	50999	3	5	0.030	0.0096	0.34	−1.54	C	5n
218.	$z\ ^7F° - f\ ^5D$ (323)											
		3630.35	22997	50534	9	7	0.076	0.012	1.3	−0.98	C	5n
		3610.70	23192	50880	5	3	0.071	0.0083	0.50	−1.38	C	5n
		3641.45	23245	50699	3	5	0.0052	0.0017	0.061	−2.29	D	5n

Fe I: Allowed transitions — Continued

No.	Multiplet	λ (Å)	E_i (cm^{-1})	E_k (cm^{-1})	g_i	g_k	A_{ki} (10^8 s^{-1})	f_{ik}	S (at. u.)	log gf	Accuracy	Source
219.	$z\,^7\mathrm{F}° - e\,^7\mathrm{P}$ (324)											
		3620.24	22997	50611	9	7	0.012	0.0018	0.20	−1.78	C	5n
		3613.15	23192	50861	5	5	0.019	0.0038	0.23	−1.72	C	5n
220.	$z\,^7\mathrm{F}° - e\,^5\mathrm{G}$ (325)											
		3572.59	22997	50980	9	9	0.018	0.0035	0.37	−1.50	C	5n
		3612.07	22846	50523	11	13	0.075	0.017	2.3	−0.72	C+	4n
		3567.03	23192	51219	5	7	0.065	0.017	1.0	−1.06	C	5n
221.	$z\,^7\mathrm{F}° - e\,^7\mathrm{G}$ (326)											
		3541.08	22997	51229	9	11	0.62	0.14	15	0.11	C+	4n
		3542.08	23111	51335	7	9	0.74	0.18	15	0.10	C+	4n
		3536.56	23192	51461	5	7	0.78	0.20	12	0.01	C+	4n
		3530.39	22650	50968	13	13	0.032	0.0060	0.90	−1.11	C	5n
		3522.27	22846	51229	11	11	0.032	0.0060	0.77	−1.18	C	5n
		3527.79	22997	51335	9	9	0.20	0.037	3.9	−0.48	C+	4n,5n
		3529.82	23245	51567	3	3	0.76	0.14	5.0	−0.37	C+	4n
		3509.12	22846	51335	11	9	0.0046	6.9(−4)	0.088	−2.12	D	5n
		3512.22	22997	51461	9	7	0.021	0.0030	0.31	−1.57	C	5n
		3516.56	23111	51540	7	5	0.037	0.0050	0.40	−1.46	C	5n
		3523.31	23192	51567	5	3	0.076	0.0085	0.49	−1.37	C	5n
222.	$z\,^7\mathrm{F}° - f\,^5\mathrm{F}$ (327)											
		3537.90	22846	51103	11	11	0.084	0.016	2.0	−0.76	C	5n
		3556.88	22997	51103	9	11	0.44	0.10	11	−0.04	C+	4n
		3518.68	23192	51604	5	7	0.016	0.0041	0.24	−1.69	C	5n
223.	$z\,^7\mathrm{F}° - g\,^5\mathrm{D}$ (329)											
		3540.12	23111	51350	7	9	0.12	0.029	2.3	−0.70	C+	4n
224.	$z\,^7\mathrm{F}° - e\,^7\mathrm{S}$ (330)											
		3522.90	23192	51570	5	7	0.021	0.0055	0.32	−1.56	C	5n
225.	$b\,^3\mathrm{P} - z\,^3\mathrm{P}°$ (339)	*8882.5*	*22898*	*34153*	9	9	0.010	0.012	3.2	−0.96	C	26n
		8999.55	22838	33947	5	5	0.0082	0.010	1.5	−1.30	C	26n
		8757.19	22947	34363	3	3	0.00273	0.00314	0.271	−2.026	C	26n
		8674.75	22838	34363	5	3	0.00421	0.00285	0.407	−1.85	C	26n
		8611.81	22947	34556	3	1	0.0114	0.00423	0.359	−1.90	C	26n
		9088.33	22947	33947	3	5	0.0021	0.0043	0.39	−1.89	C	26n
		8838.43	23052	34363	1	3	0.00297	0.0104	0.304	−1.98	C	26n
226.	$b\,^3\mathrm{P} - y\,^3\mathrm{F}°$ (340)											
		6859.49	22947	37521	3	5	8.6(−6)	1.0(−5)	6.8(−4)	−4.52	D	30

Fe I: Allowed transitions — Continued

No.	Multiplet	λ (Å)	E_i (cm^{-1})	E_k (cm^{-1})	g_i	g_k	A_{ki} (10^8 s^{-1})	f_{ik}	S (at. u.)	log gf	Accuracy	Source
227.	$b\ ^3P - y\ ^3D°$ (342)											
		6518.38	22838	38175	5	7	4.0(−4)	3.6(−4)	0.038	−2.75	D	5n
		6355.04	22947	38678	3	5	0.0013	0.0013	0.080	−2.42	D	5n
		6270.24	23052	38996	1	3	0.0011	0.0019	0.040	−2.71	D	5n
		6311.51	22838	38678	5	5	2.0(−4)	1.2(−4)	0.012	−3.23	D	5n
		6229.23	22947	38996	3	3	6.1(−4)	3.6(−4)	0.022	−2.97	D	5n
228.	$b\ ^3P - z\ ^5S°$ (345)											
		5536.59	22838	40895	5	5	6.7(−5)	3.1(−5)	0.0028	−3.81	D	5n
229.	$b\ ^3P - w\ ^5D°$ (346)											
		4741.53	22838	43923	5	7	0.0042	0.0020	0.16	−2.00	D	10n,30
		4707.49	22947	44184	3	5	0.0028	0.0015	0.071	−2.34	D	30
		4683.57	22838	44184	5	5	0.0018	5.9(−4)	0.046	−2.53	D	5n
		4657.59	22947	44411	3	3	0.0013	4.2(−4)	0.019	−2.90	D	5n
230.	$b\ ^3P - ^5D°$											
		4687.39	22838	44166	5	7	6.7(−4)	3.1(−4)	0.024	−2.81	D	5n
		4603.34	22947	44664	3	5	4.6(−4)	2.4(−4)	0.011	−3.14	D	5n
231.	$b\ ^3P - ^5F°$											
		4604.25	22838	44551	5	7	5.2(−5)	2.3(−5)	0.0017	−3.94	D	30
		4685.03	22947	44285	3	5	2.8(−4)	1.5(−4)	0.0070	−3.34	D	5n
		4687.68	23052	44378?	1	3	1.7(−4)	1.7(−4)	0.0026	−3.78	D	30
		4661.33	22838	44285	5	5	5.2(−5)	1.7(−5)	0.0013	−4.07	D	30
232.	$b\ ^3P - y\ ^5S°$ (349)											
		4612.64	22838	44512	5	5	1.0(−4)	3.3(−5)	0.0025	−3.78	D	30
		4635.85	22947	44512	3	5	0.0024	0.0013	0.058	−2.42	D	5n
233.	$b\ ^3P - (\ °)^b$											
		4466.55	22838	45221	5	7	0.12	0.051	3.8	−0.59	C+	4n
234.	$b\ ^3P - (\ °)^b$											
		4454.38	22838	45282	5	5	0.038	0.011	0.82	−1.25	C	4n
235.	$b\ ^3P - (\ °)^b$											
		4443.19	23052	45552	1	3	0.11	0.095	1.4	−1.02	C	4n
		4422.57	22947	45552	3	3	0.088	0.026	1.1	−1.11	C	4n
		4401.44	22838	45552	5	3	0.026	0.0045	0.32	−1.65	C	5n
236.	$b\ ^3P - w\ ^5P°$ (351)											
		4290.87	22838	46137	5	7	0.0044	0.0017	0.12	−2.07	D	5n
		4279.86	23052	46410	1	3	0.0057	0.0047	0.066	−2.33	D	5n
		4258.62	22838	46314	5	5	0.0070	0.0019	0.13	−2.02	D	5n
		4241.11	22838	46410	5	3	0.0038	6.2(−4)	0.043	−2.51	D	5n

Fe I: Allowed transitions — Continued

No.	Multiplet	λ (Å)	E_i (cm^{-1})	E_k (cm^{-1})	g_i	g_k	A_{ki} (10^8 s^{-1})	f_{ik}	S (at. u.)	log gf	Accuracy	Source
237.	$b\ ^3$P – $z\ ^3$S° (352)	4217.7	22898	46601	9	3	0.16	0.014	1.8	−0.89	C	4n,5n
		4207.13	22838	46601	5	3	0.043	0.0069	0.48	−1.46	C	4n,5n
		4226.42	22947	46601	3	3	0.037	0.010	0.42	−1.52	C	4n,5n
		4245.26	23052	46601	1	3	0.083	0.067	0.94	−1.17	C	4n,5n
238.	$b\ ^3$P – $y\ ^3$P° (355)											
		4184.89	22838	46727	5	5	0.11	0.028	1.9	−0.86	C+	4n
		4173.32	22947	46902	3	3	0.021	0.0054	0.22	−1.79	C	5n
		4213.65	22947	46673	3	1	0.19	0.017	0.71	−1.29	C	4n,5n
		4191.68	23052	46902	1	3	0.048	0.038	0.52	−1.42	C	5n
239.	$b\ ^3$P – ^{5}D°											
		4181.75	22838	46745	5	7	0.36	0.13	9.1	−0.18	D−	13n
		4132.90	22947	47136	3	5	0.094	0.040	1.6	−0.92	C+	4n
		4114.45	22838	47136	5	5	0.047	0.012	0.81	−1.22	C	4n,5n
		4125.88	22947	47177	3	3	0.015	0.0038	0.16	−1.94	C	5n
		4107.49	22838	47177	5	3	0.25	0.037	2.5	−0.73	C+	4n
		4126.88	22947	47171?	3	1	0.011	9.6(−4)	0.039	−2.54	D	5n
240.	$b\ ^3$P – ^{3}D°	4143.6	22898	47025	9	15	0.27	0.11	14	0.01	C+	4n,5n
		4134.68	22838	47017	5	7	0.18	0.065	4.4	−0.49	C+	4n
		4175.64	22947	46889	3	5	0.16	0.071	2.9	−0.67	C+	4n
		4127.61	23052	47272	1	3	0.13	0.10	1.4	−0.99	C+	4n
		4156.80	22838	46889	5	5	0.19	0.048	3.3	−0.62	C+	4n
		4109.80	22947	47272	3	3	0.16	0.041	1.7	−0.91	C+	4n
		4091.55	22838	47272	5	3	0.010	0.0015	0.10	−2.12	D	4n,5n
241.	$b\ ^3$P – $x\ ^3$F° (356)											
		4121.80	22838	47093	5	7	0.028	0.010	0.68	−1.30	C	4n,5n
		4122.51	22947	47197	3	5	0.029	0.012	0.50	−1.43	C	5n
242.	$b\ ^3$P – ^{1}D°											
		4085.00	22947	47420	3	5	0.042	0.017	0.71	−1.28	C	5n
243.	$b\ ^3$P – $y\ ^3$S° (359)	4054.3	22898	47556	9	3	0.40	0.032	3.9	−0.53	C	4n,5n
		4044.61	22838	47556	5	3	0.11	0.017	1.1	−1.08	C	5n
		4062.44	22947	47556	3	3	0.22	0.055	2.2	−0.78	C+	4n
		4079.84	23052	47556	1	3	0.063	0.047	0.63	−1.33	C	4n,5n
244.	$b\ ^3$P – $v\ ^5$F° (362)											
		3953.86	22838	48123	5	7	0.0057	0.0019	0.12	−2.03	D	5n
		3935.31	22947	48351	3	3	0.019	0.0045	0.17	−1.87	C	5n
245.	$b\ ^3$P – $v\ ^5$P° (361)											
		3964.52	22947	48163	3	5	0.024	0.0094	0.37	−1.55	C	4n,5n
		3961.15	23052	48290	1	3	0.023	0.016	0.21	−1.79	C	5n
		3944.75	22947	48290	3	3	0.012	0.0027	0.11	−2.09	D	5n

Fe I: Allowed transitions — Continued

No.	Multiplet	λ (Å)	E_i (cm^{-1})	E_k (cm^{-1})	g_i	g_k	A_{ki} (10^8 s^{-1})	f_{ik}	S (at. u.)	log gf	Accuracy	Source
246.	$b\ ^3P - x\ ^3P°$ (364)											
		3909.83	22947	48516	3	3	0.065	0.015	0.57	−1.35	C	5n
		3942.44	22947	48305	3	5	0.090	0.035	1.4	−0.98	C+	4n,5n
247.	$b\ ^3P - v\ ^3D°$ (367)											
		3801.68	22838	49135	5	7	0.066	0.020	1.3	−1.00	C	5n
		3809.04	23052	49298	1	3	0.016	0.010	0.13	−1.99	C	5n
		3786.19	22838	49243	5	5	0.12	0.025	1.6	−0.90	C	5n
		3793.87	22947	49298	3	3	0.074	0.016	0.60	−1.32	C	5n
		3778.32	22838	49298	5	3	0.024	0.0030	0.19	−1.82	C	5n
248.	$b\ ^3P - w\ ^3P°$ (369)											
		3655.46	22838	50187	5	5	0.10	0.020	1.2	−1.00	C	4n,5n
		3674.77	22838	50043	5	3	0.067	0.0081	0.49	−1.39	C	5n
		3702.03	22947	49951	3	1	0.35	0.024	0.88	−1.14	C	4n,5n
		3703.82	23052	50043	1	3	0.12	0.072	0.88	−1.14	C	5n
249.	$b\ ^3P - \ ^3D°$											
		3354.06	23052	52858	1	3	0.077	0.039	0.43	−1.41	C	12n
250.	$b\ ^3P - \ ^3P°$											
		3323.74	22838	52916	5	5	0.30	0.050	2.7	−0.60	C+	4n
251.	$z\ ^7P° - e\ ^7D$ (383)											
		5232.94	23711	42816	9	11	0.14	0.072	11	−0.19	C+	4n,5n
		5266.55	24181	43163	7	9	0.086	0.046	5.6	−0.49	C+	4n
		5281.79	24507	43435	5	7	0.033	0.019	1.7	−1.02	C	4n
		5068.77	23711	43435	9	7	0.022	0.0066	0.99	−1.23	C	4n,5n
252.	$z\ ^7P° - e\ ^5D$ (384)											
		4787.83	24181	45061	7	7	7.1(−4)	2.4(−4)	0.027	−2.77	D	5n
		4800.13	24507	45334	5	5	0.0011	3.6(−4)	0.029	−2.74	D	5n
		4682.56	23711	45061	9	7	3.2(−4)	8.2(−5)	0.011	−3.13	D	5n
		4726.14	24181	45334	7	5	3.4(−4)	8.0(−5)	0.0087	−3.25	D	5n
		4760.07	24507	45509	5	3	2.6(−4)	5.4(−5)	0.0042	−3.57	D	30
		4877.61	24181	44677	7	9	2.2(−4)	1.0(−4)	0.011	−3.15	D	5n
253.	$z\ ^7P° - e\ ^7F$ (385)											
		3686.00	23711	50833	9	11	0.26	0.064	7.0	−0.24	C+	4n
		3701.09	24181	51192	7	9	0.48	0.13	11	−0.05	C+	4n
		3637.86	23711	51192	9	9	0.055	0.011	1.2	−1.01	C	5n
		3726.93	24507	51331	5	5	0.46	0.096	5.9	−0.32	C	5n
		3744.10	24507	51208	5	3	0.36	0.046	2.8	−0.64	C+	4n,5n

Fe I: Allowed transitions — Continued

No.	Multiplet	λ (Å)	E_i (cm^{-1})	E_k (cm^{-1})	g_i	g_k	A_{ki} (10^8 s^{-1})	f_{ik}	S (at. u.)	log gf	Accuracy	Source
254.	$z\,^7P° - f\,^7D$ (386)											
		3754.51	24181	50808	7	9	0.024	0.0065	0.56	−1.34	C	5n
		3746.93	24181	50862	7	7	0.22	0.047	4.1	−0.48	C	5n
		3773.70	24507	50999	5	5	0.033	0.0071	0.44	−1.45	C	5n
		3766.67	24507	51048	5	3	0.097	0.012	0.76	−1.21	C	5n
255.	$z\,^7P° - f\,^5D$ (387)											
		3742.62	23711	50423	9	9	0.10	0.022	2.4	−0.70	C+	4n,5n
		3727.09	23711	50534	9	7	0.20	0.032	3.5	−0.54	C	5n
256.	$z\,^7P° - e\,^7P$ (388)											
		3735.32	23711	50475	9	9	0.24	0.050	5.5	−0.35	C	5n
		3782.45	24181	50611	7	7	0.012	0.0027	0.23	−1.73	C	5n
257.	$z\,^7P° - e\,^5G$ (389)											
		3703.69	23711	50704	9	11	0.053	0.013	1.5	−0.92	C	5n
		3697.43	24181	51219	7	7	0.21	0.042	3.6	−0.53	C+	4n
		3676.88	24181	51370	7	5	0.023	0.0033	0.28	−1.64	C	5n
258.	$z\,^7P° - e\,^7G$ (390)											
		3681.64	24181	51335	7	9	0.013	0.0034	0.29	−1.62	C	5n
		3664.69	24181	51461	7	7	0.021	0.0043	0.36	−1.52	C	5n
259.	$z\,^7P° - f\,^5F$ (391)											
		3664.54	24181	51462	7	9	0.034	0.0088	0.74	−1.21	C	5n
260.	$z\,^7P° - e\,^7S$ (394)											
		3650.03	24181	51570	7	7	0.099	0.020	1.7	−0.86	C	5n
		3694.01	24507	51570	5	7	0.68	0.20	12	−0.01	C+	4n
261.	$z\,^7P° - e\,^5P$ (395)											
		3614.77	24181	51837	7	7	0.034	0.0067	0.56	−1.33	C	5n
		3657.89	24507	51837	5	7	0.033	0.0094	0.56	−1.33	C	5n
262.	$z\,^7P° - g\,^7D$ (396)											
		3322.47	23711	53801	9	11	0.062	0.012	1.2	−0.95	D	12n
263.	$b\,^3G - y\,^3F°$ (402)											
		7748.27	23784	36686	11	9	0.0021	0.0016	0.44	−1.76	D	30
		7583.80	24339	37521	7	5	0.0024	0.0015	0.26	−1.99	D	30

Fe I:　Allowed transitions — Continued

No.	Multiplet	λ (Å)	E_i (cm^{-1})	E_k (cm^{-1})	g_i	g_k	A_{ki} (10^8 s^{-1})	f_{ik}	S (at. u.)	log gf	Accuracy	Source
264.	$b\ ^3G - y\ ^3D°$ (404)											
		7112.18	24119	38175	9	7	1.5(−4)	9.0(−5)	0.019	−3.09	D	5n
		6971.95	24339	38678	7	5	8.9(−5)	4.6(−5)	0.0074	−3.49	D	30
265.	$b\ ^3G - y\ ^3G°$ (409)											
		4647.43	23784	45295	11	11	0.014	0.0045	0.75	−1.31	D−	14n
		4691.41	24119	45428	9	9	0.012	0.0039	0.55	−1.45	D−	14n
		4618.76	23784	45428	11	9	0.0016	4.2(−4)	0.070	−2.34	D	5n
		4661.97	24119	45563	9	7	0.0015	3.9(−4)	0.053	−2.46	D	5n
		4740.34	24339	45428	7	9	6.9(−4)	3.0(−4)	0.033	−2.68	D	5n
266.	$b\ ^3G - x\ ^5G°$ (410)											
		4626.76	24119	45726	9	11	4.9(−5)	1.9(−5)	0.0026	−3.76	D	30
		4603.95	24119	45833	9	9	5.1(−4)	1.6(−4)	0.022	−2.84	D	5n
		4633.76	24339	45913	7	7	4.1(−4)	1.3(−4)	0.014	−3.03	D	5n
267.	$b\ ^3G - ^5D°$											
		4358.50	23784	46721	11	9	0.0090	0.0021	0.33	−1.64	C	4n,5n
		4423.14	24119	46721	9	9	0.0012	3.4(−4)	0.045	−2.51	D	5n
268.	$b\ ^3G - ^3D°$											
		4365.90	24119	47017	9	7	0.0031	6.8(−4)	0.088	−2.21	D	4n,5n
269.	$b\ ^3G - x\ ^3F°$ (413)											
		4326.75	23784	46889	11	9	0.0049	0.0011	0.18	−1.91	C	5n
		4351.54	24119	47093	9	7	0.014	0.0031	0.40	−1.55	C	5n
		4390.46	24119	46889	9	9	0.0012	3.5(−4)	0.046	−2.50	D	5n
270.	$b\ ^3G - z\ ^3H°$ (414)	*4349.6*	*24040*	*47024*	*27*	*33*	0.019	0.0065	2.5	−0.76	C	4n,5n
		4309.37	23784	46982	11	13	0.018	0.0060	0.94	−1.18	C	4n
		4367.58	24119	47008	9	11	0.017	0.0060	0.77	−1.27	C	5n
		4390.95	24339	47106	7	9	0.013	0.0050	0.50	−1.46	C	5n
		4304.54	23784	47008	11	11	0.0032	8.9(−4)	0.14	−2.01	D	5n
		4348.94	24119	47106	9	9	0.0029	8.2(−4)	0.11	−2.13	D	5n
		4286.44	23784	47106	11	9	0.0015	3.3(−4)	0.051	−2.44	D	5n
271.	$b\ ^3G - w\ ^5G°$ (416)											
		4290.38	24119	47420	9	11	0.0053	0.0018	0.23	−1.79	C	4n,5n
272.	$b\ ^3G - z\ ^1G°$ (417)											
		4223.73	23784	47453	11	9	5.1(−4)	1.1(−4)	0.017	−2.91	D−	30
		4284.42	24119	47453	9	9	0.0010	2.8(−4)	0.035	−2.60	D	5n

Fe I: Allowed transitions — Continued

No.	Multiplet	λ (Å)	E_i (cm^{-1})	E_k (cm^{-1})	g_i	g_k	A_{ki} (10^8 s^{-1})	f_{ik}	S (at. u.)	log gf	Accuracy	Source
273.	$b\ ^3G - v\ ^5F°$ (418)											
		4196.53	23784	47606	11	11	0.0027	7.1(−4)	0.11	−2.11	D	5n
		4237.67	24339	47930	7	9	0.0018	6.1(−4)	0.060	−2.37	D	5n
274.	$b\ ^3G - x\ ^3G°$ (419)											
		4160.56	23784	47812	11	9	5.6(−4)	1.2(−4)	0.018	−2.88	D−	30
		4258.95	24339	47812	7	9	0.0040	0.0014	0.14	−2.01	D	5n
275.	$b\ ^3G - ^5H°$											
		4146.06	24119	48231	9	11	0.0051	0.0016	0.20	−1.84	C	4n,5n
		4161.48	24339	48362	7	9	0.0027	9.0(−4)	0.086	−2.20	D	5n
		4089.22	23784	48231	11	11	0.0040	0.0010	0.15	−1.96	C	4n,5n
		4141.86	24339	48476	7	7	0.0070	0.0018	0.17	−1.90	C	5n
		4067.49	23784	48362	11	9	3.3(−4)	6.7(−5)	0.0099	−3.13	D−	30
276.	$b\ ^3G - z\ ^1H°$ (423)											
		4120.21	24119	48383	9	11	0.024	0.0075	0.92	−1.17	C	4n,5n,8
277.	$b\ ^3G - y\ ^1G°$ (424)											
		4066.59	24119	48703	9	9	0.011	0.0027	0.33	−1.61	C	4n,5n
278.	$b\ ^3G - w\ ^3F°$ (426)											
		4000.46	24119	49109	9	9	0.011	0.0026	0.31	−1.63	C	5n
279.	$b\ ^3G - y\ ^3H°$ (429)											
		3897.45	23784	49434	11	13	0.019	0.0051	0.72	−1.25	C	4n,5n
		3871.75	23784	49604	11	11	0.067	0.015	2.1	−0.78	C+	4n,5n
		3903.90	24119	49727	9	9	0.096	0.022	2.5	−0.70	C+	4n,5n
		3853.46	23784	49727	11	9	0.0055	0.0010	0.14	−1.96	C	4n,5n
280.	$b\ ^3G - v\ ^3G°$ (430)											
		3893.39	23784	49461	11	11	0.13	0.030	4.2	−0.48	C+	4n,5n
		3919.07	24119	49628	9	9	0.039	0.0089	1.0	−1.10	C	4n,5n
		3885.15	24119	49851	9	7	0.013	0.0023	0.26	−1.68	C	4n,5n
		3944.89	24119	49461	9	11	0.014	0.0039	0.45	−1.46	C	5n
		3953.15	24339	49628	7	9	0.037	0.011	1.0	−1.11	C	4n,5n
281.	$b\ ^3G - z\ ^1F°$ (432)											
		3777.06	24119	50587	9	7	0.014	0.0023	0.26	−1.68	C	4n,5n
282.	$b\ ^3G - x\ ^3H°$ (435)											
		3670.09	23784	51023	11	13	0.076	0.018	2.4	−0.70	C+	4n

Fe I: Allowed transitions — Continued

No.	Multiplet	λ (Å)	E_i (cm^{-1})	E_k (cm^{-1})	g_i	g_k	A_{ki} (10^8 s^{-1})	f_{ik}	S (at. u.)	log gf	Accuracy	Source
283.	$b\ ^3G - v\ ^3F°$ (437)											
		3632.55	23784	51305	11	9	0.052	0.0085	1.1	−1.03	C	5n
		3669.15	24119	51365	9	7	0.074	0.012	1.3	−0.98	C	5n
284.	$b\ ^3G - u\ ^3G°$ (438)											
		3628.82	24119	51668	9	9	0.0030	5.8(−4)	0.063	−2.28	D	5n
285.	$b\ ^3G - {}^1H°$											
		3590.08	23784	51630	11	11	0.010	0.0019	0.25	−1.67	C	5n
		3633.84	24119	51630	9	11	0.017	0.0042	0.45	−1.42	C	5n
286.	$b\ ^3G - w\ ^3H°$ (442)											
		3508.49	24119	52613	9	11	0.057	0.013	1.3	−0.94	C	5n
		3516.41	24339	52769	7	9	0.034	0.0080	0.65	−1.25	C	5n
287.	$b\ ^3G - t\ ^3G°$ (449)											
		3319.25	24119	54237	9	9	0.034	0.0056	0.55	−1.30	D	12n
288.	$c\ ^3P - z\ ^3P°$ (461)											
		10218.36	24772	34556	3	1	0.0011	5.8(−4)	0.058	−2.76	C	26n
289.	$c\ ^3P - y\ ^3D°$ (463)											
		7189.17	24772	38678	3	5	3.8(−4)	4.9(−4)	0.035	−2.83	D	30
		7190.13	25092	38996	1	3	1.6(−4)	3.8(−4)	0.0090	−3.42	D	30
290.	$c\ ^3P - x\ ^5P°$ (464)											
		5460.91	24772	43079	3	3	2.0(−4)	8.8(−5)	0.0047	−3.58	D	30
291.	$c\ ^3P - w\ ^5D°$ (465)											
		5104.04	24336	43923	5	7	4.9(−4)	2.7(−4)	0.023	−2.87	D	5n
292.	$c\ ^3P - (\ °)^b$											
		4786.81	24336	45221	5	7	0.011	0.0052	0.41	−1.59	D−	14n,30
293.	$c\ ^3P - (\ °)^b$											
		4874.36	24772	45282	3	5	5.2(−4)	3.1(−4)	0.015	−3.03	D	30
294.	$c\ ^3P - (\ °)^b$											
		4712.10	24336	45552	5	3	7.4(−4)	1.5(−4)	0.011	−3.13	D	5n

Fe I: Allowed transitions — Continued

No.	Multiplet	λ (Å)	E_i (cm^{-1})	E_k (cm^{-1})	g_i	g_k	A_{ki} (10^8 s^{-1})	f_{ik}	S (at. u.)	log gf	Accuracy	Source
295.	$c\ ^3P - w\ ^5P°$ (468)											
		4620.14	24772	46410	3	3	1.9(−4)	5.9(−5)	0.0027	−3.75	D	30
296.	$c\ ^3P - z\ ^3S°$ (469)											
		4490.08	24336	46601	5	3	0.029	0.0053	0.39	−1.58	C	5n
		4579.82	24772	46601	3	3	0.0016	4.9(−4)	0.022	−2.83	D	5n
297.	$c\ ^3P - y\ ^3P°$ (472)											
		4464.77	24336	46727	5	5	0.011	0.0033	0.24	−1.78	C	4n,5n
		4517.53	24772	46902	3	3	0.016	0.0048	0.21	−1.84	C	4n,5n
		4430.19	24336	46902	5	3	0.012	0.0021	0.16	−1.97	C	5n
298.	$c\ ^3P - \ ^5D°$											
		4384.68	24336	47136	5	5	0.0048	0.0014	0.10	−2.16	D	5n
299.	$c\ ^3P - x\ ^3F°$ (473)											
		4372.99	24336	47197	5	5	0.0016	4.7(−4)	0.034	−2.63	D	5n
300.	$c\ ^3P - y\ ^3S°$ (476)	*4348.3*	*24565*	47556	9	3	0.11	0.010	1.3	−1.04	C	4n,5n
		4305.45	24336	47556	5	3	0.060	0.010	0.71	−1.30	C	4n,5n
		4387.89	24772	47556	3	3	0.039	0.011	0.49	−1.47	C	5n
		4450.32	25092	47556	1	3	0.010	0.0089	0.13	−2.05	D	5n
301.	$c\ ^3P - v\ ^5F°$ (476a)											
		4182.38	24336	48239	5	5	0.049	0.013	0.89	−1.19	C	5n
302.	$c\ ^3P - v\ ^5P°$ (478)											
		4230.58	24336	47967	5	7	0.0013	5.0(−4)	0.035	−2.60	D	5n
303.	$c\ ^3P - x\ ^3P°$ (482)											
		4170.90	24336	48305	5	5	0.061	0.016	1.1	−1.10	C	4n,5n
		4220.34	24772	48460	3	1	0.19	0.017	0.71	−1.29	C	4n,5n
		4248.22	24772	48305	3	5	0.035	0.016	0.67	−1.32	C	5n
		4267.83	25092	48516	1	3	0.094	0.077	1.1	−1.11	C	4n,5n
304.	$c\ ^3P - v\ ^3D°$ (486)											
		4031.24	24336	49135	5	7	0.0016	5.4(−4)	0.036	−2.57	D−	30
		4076.23	24772	49298	3	3	0.012	0.0031	0.13	−2.03	D	5n

Fe I: Allowed transitions — Continued

No.	Multiplet	λ (Å)	E_i (cm^{-1})	E_k (cm^{-1})	g_i	g_k	A_{ki} (10^8 s^{-1})	f_{ik}	S (at. u.)	log gf	Accuracy	Source
305.	$c\ ^3P - w\ ^3P°$ (488)											
		3867.22	24336	50187	5	5	0.34	0.076	4.8	−0.42	C+	4n
		3955.96	24772	50043	3	3	0.057	0.013	0.52	−1.40	C	5n
		3888.82	24336	50043	5	3	0.27	0.036	2.3	−0.74	C	5n
		3970.39	24772	49951	3	1	0.35	0.028	1.1	−1.08	C	5n
306.	$c\ ^3P - z\ ^1F°$ (489)											
		3808.29	24336	50587	5	7	0.012	0.0036	0.22	−1.75	C	5n
307.	$c\ ^3P - t\ ^5D°$ (490)											
		3699.15	24336	51361	5	7	0.045	0.013	0.79	−1.19	C	4n
		3635.19	24336	51828	5	3	0.14	0.016	0.97	−1.09	C	5n
308.	$c\ ^3P - v\ ^3F°$ (491)											
		3698.60	24336	51365	5	7	0.038	0.011	0.67	−1.26	C	4n,5n
		3782.61	24772	51201	3	5	0.013	0.0046	0.17	−1.86	C	5n
309.	$c\ ^3P - y\ ^1D°$ (494)											
		3711.41	24772	51708	3	5	0.073	0.025	0.93	−1.12	C	5n
310.	$c\ ^3P - x\ ^1D°$ (495)											
		3704.01	24772	51762	3	5	0.015	0.0053	0.19	−1.80	C	5n
311.	$c\ ^3P - u\ ^3D°$ (496)											
		3617.79	24336	51969	5	7	0.65	0.18	11	−0.05	C+	4n
		3632.04	24772	52297	3	5	0.48	0.16	5.7	−0.32	C+	4n
		3645.82	25092	52512	1	3	0.57	0.34	4.1	−0.47	C+	4n,5n
		3603.82	24772	52512	3	3	0.17	0.033	1.2	−1.01	C	5n
		3548.02	24336	52512	5	3	0.097	0.011	0.64	−1.26	D	12n
312.	$c\ ^3P - ^3D°$											
		3559.50	24772	52858	3	3	0.19	0.036	1.3	−0.97	C+	4n,5n
		3505.07	24336	52858	5	3	0.099	0.011	0.63	−1.26	C	5n
313.	$c\ ^3P - ^3P°$											
		3552.11	24772	52916	3	5	0.045	0.014	0.50	−1.37	C	5n
314.	$a\ ^1G - y\ ^3G°$ (512)											
		4793.96	24575	45428	9	9	9.5(−5)	3.3(−5)	0.0047	−3.53	D	5n
315.	$a\ ^1G - z\ ^3I°$ (513)											
		4636.66	24575	46136	9	11	4.8(−5)	1.9(−5)	0.0026	−3.77	D	15n,30

Fe I: Allowed transitions — Continued

No.	Multiplet	λ (Å)	E_i (cm^{-1})	E_k (cm^{-1})	g_i	g_k	A_{ki} (10^8 s^{-1})	f_{ik}	S (at. u.)	log gf	Accuracy	Source
316.	$a\ ^1G - \ ^5D°$											
		4514.18	24575	46721	9	9	0.0033	0.0010	0.13	-2.05	D	$4n,5n$
317.	$a\ ^1G - x\ ^3F°$ (515)											
		4480.14	24575	46889	9	9	0.0034	0.0010	0.13	-2.04	D$-$	30
		4439.63	24575	47093	9	7	7.0(-4)	1.6(-4)	0.021	-2.84	D	$5n$
318.	$a\ ^1G - z\ ^3H°$ (516)											
		4456.33	24575	47008	9	11	0.0021	7.5(-4)	0.099	-2.17	D	$5n$
		4436.92	24575	47106	9	9	0.0029	8.6(-4)	0.11	-2.11	D	$5n$
319.	$a\ ^1G - w\ ^5G°$ (517)											
		4343.70	24575	47590	9	9	0.0052	0.0015	0.19	-1.88	C	$5n$
320.	$a\ ^1G - z\ ^1G°$ (518)	4369.77	24575	47453	9	9	0.072	0.021	2.7	-0.73	C$+$	$4n$
321.	$a\ ^1G - x\ ^3G°$ (520)											
		4298.04	24575	47835	9	11	0.014	0.0047	0.60	-1.37	C	$4n,5n$
		4302.18	24575	47812	9	9	0.0072	0.0020	0.25	-1.74	C	$4n,5n$
322.	$a\ ^1G - \ ^5H°$											
		4225.96	24575	48231	9	11	0.014	0.0045	0.57	-1.39	C	$4n$
323.	$a\ ^1G - z\ ^1H°$ (522)	4199.09	24575	48383	9	11	0.61	0.20	25	0.25	C	8
324.	$a\ ^1G - w\ ^3F°$ (524)											
		4074.79	24575	49109	9	9	0.048	0.012	1.4	-0.97	C$+$	$4n,5n$
325.	$a\ ^1G - y\ ^3H°$ (526)											
		3994.11	24575	49604	9	11	0.013	0.0038	0.45	-1.47	C	$4n,5n$
326.	$a\ ^1G - v\ ^3G°$ (527)											
		4017.15	24575	49461	9	11	0.045	0.013	1.6	-0.92	C$+$	$4n$
		3990.37	24575	49628	9	9	0.016	0.0039	0.46	-1.45	C	$4n,5n$
327.	$a\ ^1G - z\ ^1F°$ (528)	3843.26	24575	50587	9	7	0.47	0.080	9.2	-0.14	C$+$	$4n$
328.	$a\ ^1G - x\ ^1G°$ (529)	3839.26	24575	50614	9	9	0.28	0.062	7.1	-0.25	C$+$	$4n$

Fe I: Allowed transitions — Continued

No.	Multiplet	λ (Å)	E_i (cm^{-1})	E_k (cm^{-1})	g_i	g_k	A_{ki} (10^8 s^{-1})	f_{ik}	S (at. u.)	log gf	Accuracy	Source
329.	$a\ ^1G - x\ ^3H°$ (531)											
		3773.36	24575	51069	9	11	0.0028	7.5(−4)	0.084	−2.17	D	5n
		3725.49	24575	51409	9	9	0.015	0.0032	0.35	−1.54	C	4n,5n
330.	$a\ ^1G - u\ ^3G°$ (533)											
		3730.39	24575	51374	9	11	0.13	0.032	3.5	−0.54	C+	4n,5n
		3689.90	24575	51668	9	9	0.017	0.0035	0.38	−1.50	C	5n
331.	$a\ ^1G - x\ ^1F°$ (541)	3425.01	24575	53763	9	7	0.28	0.039	3.9	−0.46	C+	4n
332.	$a\ ^1G - \ ^3H°$											
		3229.99	24575	55526	9	11	0.45	0.086	8.3	−0.11	D−	11n
333.	$z\ ^5D° - e\ ^7D$ (552)											
		5909.99	25900	42816	9	11	2.9(−4)	1.8(−4)	0.032	−2.78	D	5n
		5827.89	26479	43634	3	5	1.5(−4)	1.3(−4)	0.0075	−3.41	D	5n
		5807.79	26550	43764	1	3	2.6(−4)	3.9(−4)	0.0074	−3.41	D	30
		5791.04	25900	43163	9	9	7.7(−4)	3.9(−4)	0.066	−2.46	D	5n
		5780.62	26140	43435	7	7	6.5(−4)	3.3(−4)	0.044	−2.64	D	5n
334.	$z\ ^5D° - e\ ^5D$ (553)											
		5324.18	25900	44677	9	9	0.15	0.064	10	−0.24	C+	4n,5n
		5283.62	26140	45061	7	7	0.080	0.033	4.1	−0.63	D	9n
		5263.30	26340	45334	5	5	0.052	0.021	1.9	−0.97	C+	4n
		5253.46	26479	45509	3	3	0.017	0.0071	0.37	−1.67	C	4n
		5208.59	26140	45334	7	5	0.052	0.015	1.8	−0.98	C+	4n,5n
		5393.17	26140	44677	7	9	0.031	0.018	2.2	−0.91	C+	4n
		5339.93	26340	45061	5	7	0.070	0.042	3.7	−0.68	C+	4n
		5302.30	26479	45334	3	5	0.063	0.044	2.3	−0.88	C+	4n
335.	$z\ ^5D° - e\ ^5F$ (554)											
		4736.77	25900	47006	9	11	0.049	0.020	2.8	−0.74	C+	4n,5n
		4707.27	26140	47378	7	9	0.028	0.012	1.3	−1.08	D−	14n
		4637.50	26479	48037	3	5	0.025	0.014	0.62	−1.39	C	5n
		4613.20	26550	48221	1	3	0.022	0.021	0.32	−1.67	C	5n
		4625.04	26140	47756	7	7	0.020	0.0065	0.70	−1.34	D−	14n
		4598.12	26479	48221	3	3	0.028	0.0090	0.41	−1.57	D	30
		4574.21	25900	47756	9	7	0.0014	3.5(−4)	0.048	−2.50	D	5n
		4565.66	26140	48037	7	5	0.0036	8.0(−4)	0.085	−2.25	D	5n
336.	$z\ ^5D° - e\ ^3F$ (555)											
		4581.51	26140	47961	7	9	0.0052	0.0021	0.22	−1.83	C	5n
		4504.83	26340	48532	5	7	0.0025	0.0011	0.080	−2.27	D	5n
337.	$z\ ^5D° - e\ ^7F$ (556)											
		4000.27	26340	51331	5	5	0.020	0.0048	0.32	−1.62	C	5n
		4042.75	26479	51208	3	3	7.3(−4)	1.8(−4)	0.0071	−3.27	D−	30

Fe I: Allowed transitions — Continued

No.	Multiplet	λ (Å)	E_i (cm^{-1})	E_k (cm^{-1})	g_i	g_k	A_{ki} (10^8 s^{-1})	f_{ik}	S (at. u.)	log gf	Accu-racy	Source
338.	$z\,^5D° - f\,^7D$ (557)											
		4080.89	26550	51048	1	3	0.021	0.016	0.21	−1.80	C	5n
		4054.18	26340	50999	5	5	0.0071	0.0017	0.12	−2.06	D	5n
		4069.08	26479	51048	3	3	0.017	0.0043	0.17	−1.89	C	5n
339.	$z\,^5D° - f\,^5D$ (558)											
		4076.63	25900	50423	9	9	0.19	0.049	5.9	−0.36	C+	4n
		4098.18	26140	50534	7	7	0.068	0.017	1.6	−0.92	C+	4n,5n
		4097.10	26479	50880	3	3	0.027	0.0068	0.28	−1.69	C	5n
		4058.22	25900	50534	9	7	0.049	0.0094	1.1	−1.07	C	4n,5n
		4070.77	26140	50699	7	5	0.13	0.023	2.2	−0.79	C+	4n,5n
		4073.76	26340	50880	5	3	0.16	0.024	1.6	−0.92	C+	4n,5n
		4080.21	26479	50981	3	1	0.24	0.020	0.81	−1.22	C	4n,5n
		4116.97	26140	50423	7	9	0.0011	3.5(−4)	0.033	−2.61	D−	30
		4109.07	26550	50880	1	3	0.045	0.034	0.46	−1.47	C	4n,5n
340.	$z\,^5D° - e\,^7P$ (559)											
		4067.98	25900	50475	9	9	0.17	0.041	5.0	−0.43	C+	4n
		4085.30	26140	50611	7	7	0.11	0.028	2.6	−0.71	C+	4n,5n
		4108.13	26140	50475	7	9	0.0032	0.0010	0.098	−2.14	D	5n
		4118.90	26340	50611	5	7	0.017	0.0062	0.42	−1.51	C	5n
341.	$z\,^5D° - e\,^5G$ (560)											
		4024.72	26140	50980	7	9	0.089	0.028	2.6	−0.71	C	5n
		4018.28	26340	51219	5	7	0.026	0.0087	0.58	−1.36	C	5n
		4016.43	26479	51370	3	5	0.021	0.0084	0.33	−1.60	D−	30
342.	$z\,^5D° - e\,^7G$ (561)											
		3946.99	25900	51229	9	11	0.044	0.012	1.5	−0.95	C	5n
		3967.96	26140	51335	7	9	0.063	0.019	1.8	−0.87	C	5n
		3979.65	26340	51461	5	7	0.0064	0.0021	0.14	−1.97	C	5n
343.	$z\,^5D° - f\,^5F$ (562)											
		3957.02	26340	51604	5	7	0.16	0.051	3.3	−0.59	C	5n
		3963.10	26479	51705	3	5	0.17	0.067	2.6	−0.70	C+	4n,5n
		3911.00	25900	51462	9	9	0.010	0.0023	0.27	−1.68	C	5n
		3941.28	26340	51705	5	5	0.084	0.020	1.3	−1.01	C	5n
		3955.34	26479	51754	3	3	0.14	0.033	1.3	−1.01	C	5n
344.	$z\,^5D° - e\,^3D$ (564)											
		3974.40	26140	51294	7	7	0.0076	0.0018	0.16	−1.90	C	5n
345.	$z\,^5D° - g\,^5D$ (565)											
		3900.52	26140	51771	7	7	0.075	0.017	1.5	−0.92	C+	4n,5n
		3931.12	26340	51771	5	7	0.045	0.014	0.94	−1.14	C	5n
		3909.66	26479	52050	3	5	0.053	0.020	0.78	−1.22	C	5n

Fe I: Allowed transitions — Continued

No.	Multiplet	λ (Å)	E_i (cm^{-1})	E_k (cm^{-1})	g_i	g_k	A_{ki} (10^8 s^{-1})	f_{ik}	S (at. u.)	log gf	Accuracy	Source
346.	$z\ ^5D° - e\ ^7S$ (566)											
		3962.35	26340	51570	5	7	0.011	0.0037	0.24	−1.73	C	5n
347.	$z\ ^5D° - e\ ^5P$ (567)											
		3890.39	26140	51837	7	7	0.015	0.0033	0.30	−1.63	C	5n
		3914.27	26479	52020	3	3	0.054	0.012	0.48	−1.43	C	5n
		3920.84	26340	51837	5	7	0.018	0.0059	0.38	−1.53	C	5n
		3925.20	26550	52020	1	3	0.057	0.040	0.51	−1.40	C	5n
348.	$z\ ^5D° - g\ ^5F$ (568)											
		3668.21	26140	53394	7	9	0.030	0.0079	0.66	−1.26	C	5n
		3591.48	26550	54386	1	3	0.060	0.035	0.41	−1.46	C	5n
349.	$z\ ^5D° - h\ ^5D$ (569)											
		3615.19	26479	54133	3	3	0.058	0.011	0.40	−1.47	C	5n
		3616.15	25900	53546	9	7	0.030	0.0046	0.50	−1.38	C	5n
		3592.67	26140	53967	7	5	0.040	0.0056	0.46	−1.41	C	5n
		3597.02	26340	54133	5	3	0.17	0.020	1.2	−1.00	C	5n
350.	$z\ ^5D° - f\ ^5P$ (570)											
		3667.25	25900	53160	9	7	0.14	0.022	2.4	−0.71	C	5n
		3644.80	26140	53569	7	5	0.078	0.011	0.93	−1.11	C	5n
		3624.06	26340	53925	5	3	0.054	0.0063	0.38	−1.50	C	5n
351.	$z\ ^5D° - f\ ^5G$ (571)											
		3593.32	26340	54161	5	7	0.029	0.0078	0.46	−1.41	C	5n
352.	$z\ ^5D° - e\ ^3G$ (573)											
		3591.00	25900	53739	9	11	0.0075	0.0018	0.19	−1.80	C	5n
353.	$z\ ^5D° - f\ ^3D$ (574)											
		3583.33	26550	54449	1	3	0.23	0.13	1.6	−0.88	C	5n
354.	$z\ ^5D° - i\ ^5D$ (578)											
		3156.27	26140	57814	7	7	0.54	0.080	5.8	−0.25	D	12n
355.	$b\ ^3H - y\ ^3G°$ (584)											
		5209.90	26106	45295	13	11	1.2(−4)	4.2(−5)	0.0094	−3.26	D	30
		5279.65	26628	45563	9	7	1.2(−4)	4.0(−5)	0.0063	−3.44	D	5n

Fe I: Allowed transitions — Continued

No.	Multiplet	λ (Å)	E_i (cm^{-1})	E_k (cm^{-1})	g_i	g_k	A_{ki} (10^8 s^{-1})	f_{ik}	S (at. u.)	log gf	Accuracy	Source
356.	$b\ ^3H - z\ ^3I°$ (585)											
		5030.77	26106	45978?	13	15	2.0(−4)	8.6(−5)	0.019	−2.95	D	5n
		5080.95	26351	46027	11	13	1.6(−4)	7.4(−5)	0.014	−3.09	D	5n
		5124.60	26628	46136	9	11	2.5(−4)	1.2(−4)	0.018	−2.96	D	30
357.	$b\ ^3H - z\ ^3H°$ (588)											
		4788.76	26106	46982	13	13	0.0035	0.0012	0.24	−1.81	C	5n
		4839.55	26351	47008	11	11	0.0039	0.0014	0.24	−1.82	C	5n
		4782.79	26106	47008	13	11	5.5(−5)	1.6(−5)	0.0033	−3.68	D	30
358.	$b\ ^3H - z\ ^1G°$ (590)											
		4737.63	26351	47453	11	9	9.5(−4)	2.6(−4)	0.045	−2.54	D	5n
359.	$b\ ^3H - v\ ^5F°$ (592)											
		4649.82	26106	47606	13	11	5.7(−4)	1.6(−4)	0.031	−2.69	D	5n
360.	$b\ ^3H - x\ ^3G°$ (591)											
		4600.93	26106	47835	13	11	7.7(−4)	2.1(−4)	0.041	−2.57	D	5n
		4658.29	26351	47812	11	9	3.1(−4)	8.3(−5)	0.014	−3.04	D	5n
		4714.19	26628	47834	9	7	9.2(−4)	2.4(−4)	0.033	−2.67	D	5n
361.	$b\ ^3H - ^5H°$											
		4518.43	26106	48231	13	11	1.9(−4)	4.9(−5)	0.0094	−3.20	D	5n
		4541.94	26351	48362	11	9	2.7(−4)	6.7(−5)	0.011	−3.13	D	5n
362.	$b\ ^3H - z\ ^1H°$ (594)											
		4487.74	26106	48383	13	11	4.3(−4)	1.1(−4)	0.021	−2.84	C−	5n,8
		4537.67	26351	48383	11	11	3.9(−4)	1.2(−4)	0.020	−2.88	C−	5n,8
		4595.36	26628	48383	9	11	0.0054	0.0021	0.29	−1.72	C	5n,8
363.	$b\ ^3H - y\ ^3H°$ (597)											
		4285.44	26106	49434	13	13	0.018	0.0050	0.92	−1.19	C	4n,5n
		4327.92	26628	49727	9	9	0.0079	0.0022	0.28	−1.70	C	5n
364.	$b\ ^3H - v\ ^3G°$ (598)											
		4280.53	26106	49461	13	11	0.0028	6.5(−4)	0.12	−2.07	D	5n
		4346.55	26628	49628	9	9	0.011	0.0032	0.41	−1.54	C	5n
365.	$b\ ^3H - x\ ^1G°$ (599)											
		4167.86	26628	50614	9	9	0.0052	0.0014	0.17	−1.91	C	5n

Fe I: Allowed transitions — Continued

No.	Multiplet	λ (Å)	E_i (cm^{-1})	E_k (cm^{-1})	g_i	g_k	A_{ki} (10^8 s^{-1})	f_{ik}	S (at. u.)	log gf	Accuracy	Source
366.	$b\ ^3H - t\ ^5D°$ (602)											
		4041.91	26628	51361	9	7	5.2(−4)	9.9(−5)	0.012	−3.05	D−	30
367.	$b\ ^3H - v\ ^3F°$ (603)											
		4006.31	26351	51305	11	9	0.047	0.0093	1.3	−0.99	C	5n
368.	$b\ ^3H - u\ ^3G°$ (604)											
		3956.45	26106	51374	13	11	0.21	0.042	7.2	−0.26	C	5n
		3948.77	26351	51668	11	9	0.22	0.042	5.9	−0.34	C	5n
		3967.42	26628	51826	9	7	0.23	0.043	5.1	−0.41	C+	4n,5n
		3995.20	26351	51374	11	11	0.0083	0.0020	0.29	−1.66	C	5n
369.	$b\ ^3H - ^1H°$											
		3916.73	26106	51630	13	11	0.12	0.023	3.9	−0.52	C+	4n,5n
370.	$b\ ^3H - w\ ^3H°$ (607)											
		3806.70	26351	52613	11	11	0.54	0.12	16	0.11	C+	4n
		3771.50	26106	52613	13	11	0.0060	0.0011	0.18	−1.85	C	5n
371.	$b\ ^3H - y\ ^3I°$ (608)											
		3765.54	26106	52655?	13	15	0.98	0.24	39	0.49	C+	4n,8
		3821.18	26351	52514	11	13	0.70	0.18	25	0.30	C+	4n,8
		3805.35	26628	52899	9	11	0.98	0.26	29	0.37	C+	4n
		3785.71	26106	52514	13	13	0.014	0.0030	0.48	−1.41	C	8
372.	$b\ ^3H - z\ ^1I°$ (609)											
		3738.31	26351	53094	11	13	0.38	0.093	13	0.01	C+	4n,8
373.	$b\ ^3H - ^5F°$											
		3582.20	26106	54014	13	11	0.25	0.040	6.2	−0.28	C+	4n
374.	$b\ ^3H - ^5D°$											
		3576.76	26351	54301	11	9	0.096	0.015	2.0	−0.78	C	5n
375.	$b\ ^3H - ^3H°$											
		3402.26	26106	55490	13	13	0.28	0.049	7.1	−0.20	C+	4n
376.	$b\ ^3H - u\ ^3H°$ (617)											
		3307.23	26106	56334	13	13	0.20	0.033	4.6	−0.37	C+	4n
		3328.87	26351	56383	11	11	0.27	0.045	5.4	−0.31	C+	4n
		3355.23	26628	56423	9	9	0.32	0.054	5.4	−0.31	C+	4n

Fe I: Allowed transitions — Continued

No.	Multiplet	λ (Å)	E_i (cm^{-1})	E_k (cm^{-1})	g_i	g_k	A_{ki} (10^8 s^{-1})	f_{ik}	S (at. u.)	log gf	Accuracy	Source
377.	$b\ ^3$H $-\ x\ ^3$I° (620)											
		3233.05	26106	57028?	13	15	0.54	0.098	14	0.11	C+	4n,8
		3254.36	26351	57070	11	13	0.51	0.095	11	0.02	C+	4n,8
		3280.26	26628	57104	9	11	0.54	0.11	10	−0.02	C+	4n
378.	$b\ ^3$H $-\ t\ ^3$H° (uv 182)											
		2923.29	26351	60549	11	11	1.6	0.21	22	0.36	C+	4n
379.	$a\ ^3$D $-\ y\ ^3$D° (623)											
		7941.09	26406	38996	3	3	9.3(−4)	8.8(−4)	0.069	−2.58	D	30
380.	$a\ ^3$D $-\ w\ ^5$D° (625)											
		5787.27	26225	43499	7	9	1.6(−5)	1.1(−5)	0.0014	−4.13	D	30
381.	$a\ ^3$D $-\ ^5$D°											
		5617.22	26225	44023	7	9	3.1(−4)	1.9(−4)	0.024	−2.88	D	5n
382.	$a\ ^3$D $-\ (\ °)^b$											
		5262.89	26225	45221	7	7	7.5(−4)	3.1(−4)	0.038	−2.66	D	5n
383.	$a\ ^3$D $-\ w\ ^5$P° (629)											
		5021.89	26406	46314	3	5	0.0039	0.0025	0.12	−2.13	D	5n
384.	$a\ ^3$D $-\ y\ ^3$P° (631)											
		4876.19	26225	46727	7	5	2.3(−4)	6.0(−5)	0.0067	−3.38	D	5n
385.	$a\ ^3$D $-\ ^5$D°											
		4873.74	26624	47136	5	5	4.9(−4)	1.7(−4)	0.014	−3.06	D	5n
		4813.11	26406	47177	3	3	0.0012	4.3(−4)	0.020	−2.89	D	5n
		4780.81	26225	47136	7	5	2.6(−4)	6.3(−5)	0.0069	−3.36	D	15n,30
386.	$a\ ^3$D $-\ ^3$D°											
		4808.15	26225	47017	7	7	6.7(−4)	2.3(−4)	0.026	−2.79	D	5n
		4791.25	26406	47272	3	3	0.0030	0.0010	0.049	−2.51	D	5n
		4838.09	26225	46889	7	5	3.4(−4)	8.6(−5)	0.0096	−3.22	D	30
387.	$a\ ^3$D $-\ x\ ^3$F° (632)											
		4790.75	26225	47093	7	7	2.4(−4)	8.2(−5)	0.0091	−3.24	D	5n
388.	$a\ ^3$D $-\ ^1$D°											
		4716.85	26225	47420	7	5	2.3(−4)	5.6(−5)	0.0060	−3.41	D	30

J. Phys. Chem. Ref. Data, Vol. 17, Suppl. 4, 1988

Fe I: Allowed transitions — Continued

No.	Multiplet	λ (Å)	E_i (cm^{-1})	E_k (cm^{-1})	g_i	g_k	A_{ki} (10^8 s^{-1})	f_{ik}	S (at. u.)	log gf	Accuracy	Source
389.	$a\ ^3D - y\ ^3S°$ (635)											
		4776.07	26624	47556	5	3	0.0019	4.0(−4)	0.031	−2.70	D	5n
		4727.00	26406	47556	3	3	4.7(−4)	1.6(−4)	0.0073	−3.33	D	30
390.	$a\ ^3D - v\ ^5F°$ (640)											
		4541.32	26225	48239	7	5	7.3(−4)	1.6(−4)	0.017	−2.95	D	30
391.	$a\ ^3D - v\ ^5P°$ (638)											
		4556.93	26225	48163	7	5	0.0013	2.8(−4)	0.029	−2.71	D	5n
		4614.21	26624	48290	5	3	0.0025	4.8(−4)	0.036	−2.62	D	5n
392.	$a\ ^3D - x\ ^3P°$ (641)											
		4527.78	26225	48305	7	5	0.0012	2.6(−4)	0.027	−2.74	D	5n
		4566.51	26624	48516	5	3	0.0060	0.0011	0.085	−2.25	D	5n
		4533.13	26406	48460	3	1	0.037	0.0038	0.17	−1.94	C	5n
		4565.31	26406	48305	3	5	0.0020	0.0010	0.046	−2.51	D	5n
393.	$a\ ^3D - v\ ^3D°$ (645)											
		4343.28	26225	49243	7	5	0.014	0.0029	0.29	−1.70	C	5n
		4409.12	26624	49298	5	3	0.0067	0.0012	0.085	−2.23	D	5n
		4377.80	26406	49243	3	5	0.0034	0.0016	0.071	−2.31	D	5n
394.	$a\ ^3D - z\ ^1D°$ (648)											
		4374.50	26624	49477	5	5	0.0049	0.0014	0.10	−2.15	D	4n,5n
395.	$a\ ^3D - w\ ^3P°$ (649)											
		4172.12	26225	50187	7	5	0.097	0.018	1.7	−0.90	C+	4n,5n
		4268.75	26624	50043	5	3	0.042	0.0069	0.48	−1.46	C	4n,5n
		4242.73	26624	50187	5	5	0.018	0.0048	0.34	−1.62	C	5n
396.	$a\ ^3D - z\ ^1F°$ (650)											
		4171.90	26624	50587	5	7	0.013	0.0046	0.31	−1.64	C	5n
397.	$a\ ^3D - x\ ^3H°$											
		3969.63	26225	51409	7	9	0.026	0.0080	0.73	−1.25	C	5n
398.	$a\ ^3D - v\ ^3F°$ (655)											
		4040.64	26624	51365	5	7	0.044	0.015	1.0	−1.12	C	4n,5n
		4031.96	26406	51201	3	5	0.071	0.029	1.2	−1.06	C	4n,5n
		4067.60	26624	51201	5	5	0.0013	3.2(−4)	0.022	−2.79	D−	30

Fe I: Allowed transitions — Continued

No.	Multiplet	λ (Å)	E_i (cm^{-1})	E_k (cm^{-1})	g_i	g_k	A_{ki} (10^8 s^{-1})	f_{ik}	S (at. u.)	log gf	Accuracy	Source
399.	$a\ ^3D - y\ ^1D°$ (661)											
		3985.39	26624	51708	5	5	0.067	0.016	1.0	−1.10	C	4n,5n
		3951.16	26406	51708	3	5	0.36	0.14	5.5	−0.38	C+	4n,5n
400.	$a\ ^3D - u\ ^3D°$ (663)											
		3883.28	26225	51969	7	7	0.16	0.036	3.2	−0.60	C+	4n,5n
401.	$a\ ^3D - ^3D°$											
		3846.80	26225	52213	7	7	0.66	0.15	13	0.01	C+	4n
		3836.33	26624	52683	5	5	0.37	0.081	5.1	−0.39	C+	4n,5n
		3778.51	26225	52683	7	5	0.12	0.019	1.6	−0.88	C	5n
		3810.76	26624	52858	5	3	0.20	0.026	1.6	−0.89	C+	4n,5n
		3906.75	26624	52213	5	7	0.067	0.021	1.4	−0.97	C	5n
402.	$a\ ^3D - ^3P°$											
		3757.45	26624	53230	5	3	0.12	0.015	0.92	−1.13	C	5n
		3802.28	26624	52916	5	5	0.050	0.011	0.67	−1.27	C	5n
403.	$a\ ^3D - ^5F°$											
		3740.24	26225	52954	7	9	0.14	0.038	3.3	−0.58	C+	4n,5n
404.	$a\ ^3D - ^5F°$											
		3751.06	26624	53275	5	5	0.012	0.0025	0.16	−1.90	C	5n
405.	$a\ ^3D - ^5D°$											
		3688.48	26225	53329	7	9	0.069	0.018	1.5	−0.90	C	5n
406.	$a\ ^3D - ^5D°$											
		3613.45	26225	53892	7	7	0.067	0.013	1.1	−1.04	C	5n
		3560.70	26225	54301	7	9	0.065	0.016	1.3	−0.95	C+	4n,5n
407.	$a\ ^3D - t\ ^3G°$ (673)											
		3568.82	26225	54237	7	9	0.056	0.014	1.1	−1.02	C	5n
		3573.39	26624	54600	5	7	0.075	0.020	1.2	−1.00	C	5n
408.	$a\ ^3D - ^5P°$											
		3598.72	26225	54005	7	7	0.029	0.0057	0.47	−1.40	C	5n
409.	$a\ ^3D - w\ ^1D°$ (676)											
		3406.44	26406	55754	3	5	0.30	0.086	2.9	−0.59	C+	4n
410.	$a\ ^3D - u\ ^3F°$ (680)											
		3292.02	26225	56593	7	9	0.61	0.13	9.7	−0.05	C+	4n
		3314.74	26624	56783	5	7	0.69	0.16	8.7	−0.10	C+	4n
		3282.89	26406	56859	3	5	0.30	0.082	2.7	−0.61	C+	4n

Fe I: Allowed transitions — Continued

No.	Multiplet	λ (Å)	E_i (cm^{-1})	E_k (cm^{-1})	g_i	g_k	A_{ki} (10^8 s^{-1})	f_{ik}	S (at. u.)	log gf	Accuracy	Source
411.	$a\ ^3D - v\ ^1G°$ (681)											
		3253.60	26225	56951	7	9	0.18	0.037	2.8	−0.59	C+	4n
412.	$z\ ^5F° - e\ ^7D$ (685)											
		6271.29	26875	42816	11	11	1.7(−4)	1.0(−4)	0.023	−2.95	D	5n
		6145.42	27167	43435	9	7	5.0(−5)	2.2(−5)	0.0040	−3.70	D	30
413.	$z\ ^5F° - e\ ^5D$ (686)											
		5615.64	26875	44677	11	9	0.17	0.066	13	−0.14	C+	4n
		5586.76	27167	45061	9	7	0.19	0.068	11	−0.21	C+	4n,5n
		5572.84	27395	45334	7	5	0.21	0.070	9.0	−0.31	C+	4n
		5569.62	27560	45509	5	3	0.21	0.058	5.3	−0.54	C+	4n
		5576.09	27666	45595	3	1	0.21	0.033	1.8	−1.00	C	5n
		5709.38	27167	44677	9	9	0.013	0.0064	1.1	−1.24	C	4n
		5658.82	27395	45061	7	7	0.036	0.017	2.2	−0.92	C+	4n
		5624.54	27560	45334	5	5	0.053	0.025	2.3	−0.90	C+	4n
		5784.69	27395	44677	7	9	4.7(−4)	3.1(−4)	0.041	−2.67	D	5n
		5712.15	27560	45061	5	7	0.0025	0.0017	0.16	−2.06	D	5n
414.	$z\ ^5F° - e\ ^5F$ (687)											
		4966.09	26875	47006	11	11	0.032	0.012	2.1	−0.89	C+	4n
		4946.38	27167	47378	9	9	0.020	0.0075	1.1	−1.17	D−	14n
		4882.14	27560	48037	5	5	0.013	0.0046	0.37	−1.64	D	30
		4875.87	26875	47378	11	9	0.0030	8.7(−4)	0.15	−2.02	D	5n
		4843.14	27395	48037	7	5	0.0082	0.0021	0.23	−1.84	C	5n
		4838.51	27560	48221	5	3	0.011	0.0022	0.18	−1.95	C	5n
		5002.79	27395	47378	7	9	0.0078	0.0038	0.43	−1.58	C	5n
		4950.10	27560	47756	5	7	0.0084	0.0043	0.35	−1.67	D	10n,30
		4907.73	27666	48037	3	5	0.0080	0.0048	0.23	−1.84	D	10n,30
415.	$z\ ^5F° - e\ ^3F$ (688)											
		4679.22	27167	48532	9	7	0.0032	8.0(−4)	0.11	−2.14	D	5n
		4807.71	27167	47961	9	9	0.0020	7.0(−4)	0.10	−2.20	D	5n
		4729.68	27395	48532	7	7	0.0014	4.8(−4)	0.053	−2.47	D	5n
		4860.98	27395	47961	7	9	0.0012	5.3(−4)	0.059	−2.43	D	5n
		4766.87	27560	48532	5	7	0.0021	9.8(−4)	0.077	−2.31	D	5n
416.	$z\ ^5F° - e\ ^7F$ (689)											
		4224.17	27167	50833	9	11	0.13	0.043	5.4	−0.41	C+	4n
		4200.92	27395	51192	7	9	0.042	0.014	1.4	−1.00	C	5n
		4224.51	27666	51331	3	5	0.071	0.032	1.3	−1.02	C	5n
		4161.08	27167	51192	9	9	0.0087	0.0023	0.28	−1.69	C	5n
		4205.54	27560	51331	5	5	0.036	0.0096	0.66	−1.32	C	5n
		4168.63	27167	51149	9	7	0.0063	0.0013	0.16	−1.94	C	5n
417.	$z\ ^5F° - f\ ^7D$ (690)											
		4228.72	27167	50808	9	9	0.0012	3.2(−4)	0.040	−2.54	D−	30

Fe I: Allowed transitions — Continued

No.	Multiplet	λ (Å)	E_i (cm^{-1})	E_k (cm^{-1})	g_i	g_k	A_{ki} (10^8 s^{-1})	f_{ik}	S (at. u.)	log gf	Accuracy	Source
418.	$z\,^5F° - f\,^5D$ (691)											
		4278.23	27167	50534	9	7	0.0095	0.0020	0.26	−1.74	C	5n
419.	$z\,^5F° - e\,^7P$ (692)											
		4264.20	27167	50611	9	7	0.017	0.0037	0.46	−1.48	C	5n
420.	$z\,^5F° - e\,^5G$ (693)											
		4247.43	27167	50704	9	11	0.20	0.065	8.2	−0.23	C+	4n
		4238.81	27395	50980	7	9	0.22	0.075	7.3	−0.28	C+	4n
		4225.45	27560	51219	5	7	0.17	0.063	4.4	−0.50	C+	4n
		4217.55	27666	51370	3	5	0.23	0.10	4.3	−0.51	C+	4n
		4196.21	27395	51219	7	7	0.098	0.026	2.5	−0.74	C+	4n,5n
		4198.64	27560	51370	5	5	0.13	0.035	2.4	−0.76	C	5n
		4169.78	27395	51370	7	5	0.0094	0.0018	0.17	−1.91	C	5n
421.	$z\,^5F° - e\,^7G$ (694)											
		4149.37	26875	50968	11	13	0.036	0.011	1.6	−0.92	C+	4n,5n
		4154.80	27167	51229	9	11	0.15	0.047	5.8	−0.37	C+	4n
		4182.79	27560	51461	5	7	0.012	0.0044	0.30	−1.66	C	5n
		4104.97	26875	51229	11	11	0.0024	6.0(−4)	0.089	−2.18	D	5n
		4136.51	27167	51335	9	9	0.013	0.0033	0.40	−1.53	C	5n
		4168.94	27560	51540	5	5	0.017	0.0045	0.31	−1.65	C	5n
		4087.09	26875	51335	11	9	0.018	0.0036	0.53	−1.40	C	4n,5n
422.	$z\,^5F° - f\,^5F$ (695)											
		4126.18	26875	51103	11	11	0.039	0.010	1.5	−0.96	C	5n
		4114.96	27167	51462	9	9	0.010	0.0025	0.31	−1.64	C	5n
		4129.46	27395	51604	7	7	0.0060	0.0015	0.15	−1.97	C	5n
		4150.25	27666	51754	3	3	0.071	0.018	0.75	−1.26	C	5n
		4090.98	27167	51604	9	7	0.0099	0.0019	0.23	−1.76	C	5n
		4112.35	27395	51705	7	5	0.014	0.0025	0.24	−1.75	C	5n
		4153.90	27395	51462	7	9	0.23	0.077	7.3	−0.27	C+	4n
		4158.79	27666	51705	3	5	0.16	0.071	2.9	−0.67	C	5n
423.	$z\,^5F° - e\,^3D$ (697)											
		4106.44	27395	51740	7	5	0.024	0.0044	0.42	−1.51	C	5n
		4183.03	27395	51294	7	7	0.0037	9.7(−4)	0.093	−2.17	D	5n
424.	$z\,^5F° - g\,^5D$ (698)											
		4084.49	26875	51350	11	9	0.11	0.023	3.5	−0.59	C+	4n
		4054.87	27560	52214	5	3	0.16	0.023	1.5	−0.94	C	5n
		4065.40	27666	52257	3	1	0.19	0.016	0.64	−1.32	C	4n,5n
		4133.86	27167	51350	9	9	0.022	0.0056	0.68	−1.30	C	4n
		4101.27	27395	51771	7	7	0.024	0.0060	0.57	−1.38	C	4n,5n
		4082.13	27560	52050	5	5	0.023	0.0058	0.39	−1.54	C	5n
		4129.22	27560	51771	5	7	0.0052	0.0019	0.13	−2.03	D	5n

Fe I: Allowed transitions — Continued

No.	Multiplet	λ (Å)	E_i (cm^{-1})	E_k (cm^{-1})	g_i	g_k	A_{ki} (10^8 s^{-1})	f_{ik}	S (at. u.)	log gf	Accuracy	Source
425.	$z\ ^5F° - e\ ^5P$ (700)											
		4051.92	27395	52067	7	5	0.030	0.0053	0.50	−1.43	C	5n
		4090.09	27395	51837	7	7	0.0095	0.0024	0.22	−1.78	C	5n
		4079.18	27560	52067	5	5	0.051	0.013	0.85	−1.20	C	5n
426.	$z\ ^5F° - g\ ^5F$ (701)											
		3817.64	26875	53061	11	11	0.083	0.018	2.5	−0.70	C	5n
427.	$z\ ^5F° - h\ ^5D$ (702)											
		3804.01	26875	53155	11	9	0.047	0.0083	1.1	−1.04	C	4n,5n
		3789.82	27167	53546	9	7	0.039	0.0065	0.73	−1.23	C	5n
428.	$z\ ^5F° - f\ ^5P$ (703)											
		3846.00	27167	53160	9	7	0.043	0.0073	0.84	−1.18	C	5n
		3819.50	27395	53569	7	5	0.046	0.0072	0.63	−1.30	C	5n
		3791.73	27560	53925	5	3	0.063	0.0081	0.51	−1.39	C	5n
429.	$z\ ^5F° - f\ ^5G$ (704)											
		3802.00	26875	53169	11	13	0.035	0.0089	1.2	−1.01	C	5n
430.	$z\ ^5F° - e\ ^3G$ (705)											
		3762.21	27167	53739	9	11	0.029	0.0074	0.82	−1.18	C	4n,5n
431.	$a\ ^1P - ^1D°$	5029.62	27543	47420	3	5	0.0047	0.0030	0.15	−2.05	D	5n
432.	$a\ ^1P - v\ ^5P°$ (719)											
		4818.66	27543	48290	3	3	1.7(−4)	5.8(−5)	0.0028	−3.76	D	30
433.	$a\ ^1P - x\ ^3P°$ (720)											
		4779.44	27543	48460	3	1	0.014	0.0016	0.077	−2.31	D	5n
434.	$a\ ^1P - w\ ^3F°$ (723)											
		4566.99	27543	49433	3	5	0.0053	0.0028	0.13	−2.08	D	5n
435.	$a\ ^1P - v\ ^3D°$ (724)											
		4607.08	27543	49243	3	5	1.9(−4)	9.8(−5)	0.0045	−3.53	D	30
436.	$a\ ^1P - y\ ^1D°$ (726)	4137.00	27543	51708	3	5	0.22	0.096	3.9	−0.54	C+	4n

Fe I: Allowed transitions — Continued

No.	Multiplet	λ (Å)	E_i (cm^{-1})	E_k (cm^{-1})	g_i	g_k	A_{ki} (10^8 s^{-1})	f_{ik}	S (at. u.)	log gf	Accuracy	Source
437.	$a\ ^1P - u\ ^3D°$ (728)											
		4003.76	27543	52512	3	3	0.071	0.017	0.67	−1.29	C	4n,5n
438.	$a\ ^1P - {}^3D°$											
		3976.61	27543	52683	3	5	0.18	0.071	2.8	−0.67	C	5n
		3949.14	27543	52858	3	3	0.039	0.0092	0.36	−1.56	C	5n
439.	$a\ ^1P - {}^3P°$											
		3891.93	27543	53230	3	3	0.40	0.090	3.5	−0.57	C+	4n,5n
440.	$a\ ^1P - {}^5D°$											
		3806.22	27543	53808	3	3	0.23	0.051	1.9	−0.82	C+	4n,5n
441.	$a\ ^1P - w\ ^1D°$ (734)	3543.67	27543	55754	3	5	0.18	0.057	2.0	−0.77	C	5n
442.	$a\ ^1P - u\ ^3F°$ (735)											
		3410.17	27543	56859	3	5	0.47	0.14	4.6	−0.39	C+	4n
443.	$a\ ^1D - (\ °)^b$											
		6016.66	28605	45221	5	7	0.0040	0.0030	0.30	−1.82	C	5n
444.	$a\ ^1D - w\ ^3F°$ (750)											
		4844.01	28605	49243	5	7	0.0038	0.0019	0.15	−2.03	D	5n
445.	$a\ ^1D - v\ ^3D°$ (751)											
		4869.45	28605	49135	5	7	0.0012	6.0(−4)	0.048	−2.52	D	5n
446.	$a\ ^1D - v\ ^3G°$ (752)											
		4705.46	28605	49851	5	7	0.0021	9.8(−4)	0.076	−2.31	D	5n
447.	$a\ ^1D - z\ ^1D°$ (753)	4789.65	28605	49477	5	5	0.072	0.025	1.9	−0.91	C+	4n
448.	$a\ ^1D - w\ ^3P°$ (754)											
		4663.18	28605	50043	5	3	0.0039	7.6(−4)	0.058	−2.42	D	5n
449.	$a\ ^1D - z\ ^1F°$ (755)	4547.85	28605	50587	5	7	0.076	0.033	2.5	−0.78	C+	4n
450.	$a\ ^1D - u\ ^3G°$ (760)											
		4305.20	28605	51826	5	7	0.0044	0.0017	0.12	−2.07	D	5n

Fe I: Allowed transitions — Continued

No.	Multiplet	λ (Å)	E_i (cm^{-1})	E_k (cm^{-1})	g_i	g_k	A_{ki} (10^8 s^{-1})	f_{ik}	S (at. u.)	log gf	Accuracy	Source
451.	$a\ ^1D - y\ ^1D°$ (761)	4327.09	28605	51708	5	5	0.078	0.022	1.6	−0.96	C+	4n,5n
452.	$a\ ^1D - x\ ^1D°$ (762)	4317.04	28605	51762	5	5	0.0048	0.0014	0.096	−2.17	D	5n
453.	$a\ ^1D - ^1P°$	4240.37	28605	52181	5	3	0.057	0.0092	0.64	−1.34	C	4n,5n
454.	$a\ ^1D - ^3P°$	4059.73	28605	53230	5	3	0.081	0.012	0.80	−1.22	C	4n,5n
455.	$a\ ^1D - ^5F°$	3989.86	28605	53661	5	7	0.050	0.017	1.1	−1.08	C	5n
456.	$a\ ^1D - x\ ^1F°$ (769)	3973.65	28605	53763	5	7	0.066	0.022	1.4	−0.96	C+	4n,5n
457.	$a\ ^1D - t\ ^3G°$ (771)	3845.69	28605	54600	5	7	0.049	0.015	0.96	−1.12	C	5n
458.	$a\ ^1D - ^3G°$	3677.31	28605	55791	5	7	0.31	0.087	5.3	−0.36	C	5n
459.	$a\ ^1D - w\ ^1D°$ (772)	3682.24	28605	55754	5	5	1.7	0.35	21	0.24	C+	4n
460.	$a\ ^1H - y\ ^5G°$ (778)	7094.30	28820	42912	11	11	4.1(−5)	3.1(−5)	0.0079	−3.47	D	30
461.	$a\ ^1H - y\ ^3G°$ (780)	6019.36	28820	45428	11	9	8.9(−5)	4.0(−5)	0.0087	−3.36	D	30
462.	$a\ ^1H - ^5D°$	5584.77	28820	46721	11	9	0.0011	4.4(−4)	0.088	−2.32	D	5n
463.	$a\ ^1H - x\ ^3F°$ (783)	5532.75	28820	46889	11	9	0.0017	6.4(−4)	0.13	−2.15	D	5n
464.	$a\ ^1H - x\ ^3G°$ (788)	5263.87	28820	47812	11	9	0.0020	6.7(−4)	0.13	−2.13	D	5n
465.	$a\ ^1H - ^5H°$	5115.78	28820	48362	11	9	5.2(−4)	1.7(−4)	0.031	−2.74	D	5n

Fe I: Allowed transitions — Continued

No.	Multiplet	λ (Å)	E_i (cm^{-1})	E_k (cm^{-1})	g_i	g_k	A_{ki} (10^8 s^{-1})	f_{ik}	S (at. u.)	log gf	Accuracy	Source
466.	$a\,^1H - w\,^3F°$ (792)											
		4927.42	28820	49109	11	9	0.0031	9.3(−4)	0.17	−1.99	C	5n
467.	$a\,^1H - y\,^3H°$ (793)											
		4849.67	28820	49434	11	13	4.6(−4)	1.9(−4)	0.033	−2.68	D	5n
		4809.94	28820	49604	11	11	5.0(−4)	1.7(−4)	0.030	−2.72	D	5n
468.	$a\,^1H - v\,^3G°$ (794)											
		4804.52	28820	49628	11	9	8.3(−4)	2.3(−4)	0.041	−2.59	D	30
469.	$a\,^1H - x\,^1G°$ (795)	4587.13	28820	50614	11	9	0.0058	0.0015	0.25	−1.78	C	5n
470.	$a\,^1H - x\,^3H°$ (796)											
		4502.59	28820	51023	11	13	0.0011	4.1(−4)	0.066	−2.35	D	5n
		4493.37	28820	51069	11	11	2.7(−4)	8.3(−5)	0.013	−3.04	D−	30
471.	$a\,^1H - u\,^3G°$ (797)											
		4432.57	28820	51374	11	11	0.0078	0.0023	0.37	−1.60	C	5n
472.	$a\,^1H - \,^1H°$	4382.77	28820	51630	11	11	0.011	0.0033	0.52	−1.44	C	5n
473.	$a\,^1H - y\,^3I°$ (800)											
		4219.36	28820	52514	11	13	0.38	0.12	18	0.12	C+	4n,8
474.	$a\,^1H - z\,^1I°$ (801)	4118.54	28820	53094	11	13	0.58	0.17	26	0.28	C	8
475.	$a\,^1H - y\,^1H°$ (802)	4014.53	28820	53722	11	11	0.24	0.057	8.3	−0.20	C+	4n
476.	$a\,^1H - w\,^1G°$ (804)	3846.41	28820	54811	11	9	0.19	0.034	4.7	−0.43	C	5n
477.	$a\,^1H - \,^3G°$											
		3756.94	28820	55430	11	11	0.24	0.051	6.9	−0.25	C+	4n,5n
478.	$a\,^1H - (\,°)^b$											
		3690.73	28820	55907	11	11	0.27	0.056	7.5	−0.21	C+	4n
479.	$a\,^1H - u\,^3H°$ (808)											
		3627.05	28820	56383	11	11	0.023	0.0045	0.58	−1.31	C	5n

Fe I: Allowed transitions — Continued

No.	Multiplet	λ (Å)	E_i (cm^{-1})	E_k (cm^{-1})	g_i	g_k	A_{ki} (10^8 s^{-1})	f_{ik}	S (at. u.)	log gf	Accuracy	Source
480.	a ^{1}H – u ^{3}F° (809)											
		3599.62	28820	56593	11	9	0.18	0.029	3.8	−0.50	C+	4n,5n
481.	a ^{1}H – v ^{1}G° (810)	3553.74	28820	56951	11	9	0.81	0.13	16	0.14	C+	4n
482.	a ^{1}H – x ^{3}I° (811)											
		3538.78	28820	57070	11	13	0.0076	0.0017	0.22	−1.73	C	8
		3534.53	28820	57104	11	11	0.019	0.0035	0.45	−1.41	C	5n
483.	z ^{5}P° – e ^{7}D (815)											
		7158.50	29469	43435	5	7	2.4(−4)	2.6(−4)	0.030	−2.89	D	30
		7057.96	29469	43634	5	5	1.1(−4)	8.3(−5)	0.0097	−3.38	D	30
		7125.00	29733	43764	3	3	1.4(−4)	1.1(−4)	0.0076	−3.49	D	30
484.	z ^{5}P° – e ^{5}D (816)											
		6400.00	29056	44677	7	9	0.055	0.043	6.4	−0.52	D−	14n
		6411.65	29469	45061	5	7	0.035	0.030	3.2	−0.82	D−	14n
		6246.32	29056	45061	7	7	0.027	0.016	2.3	−0.96	D−	14n
		6336.84	29733	45509	3	3	0.049	0.030	1.9	−1.05	D−	14n
		6141.73	29056	45334	7	5	0.0087	0.0035	0.50	−1.61	C	5n
485.	z ^{5}P° – e ^{7}F (819)											
		4572.86	29469	51331	5	5	9.9(−4)	3.1(−4)	0.023	−2.81	D	5n
		4488.13	29056	51331	7	5	0.013	0.0027	0.28	−1.72	C	5n
		4598.73	29469	51208	5	3	0.0023	4.4(−4)	0.033	−2.66	D	30
486.	z ^{5}P° – f ^{7}D (820)											
		4596.06	29056	50808	7	9	0.0080	0.0033	0.35	−1.64	C	5n
		4673.16	29469	50862	5	7	0.046	0.021	1.6	−0.98	D−	14n
		4701.05	29733	50999	3	5	0.0066	0.0037	0.17	−1.96	C	5n
		4643.46	29469	50999	5	5	0.032	0.010	0.78	−1.29	C	5n
		4690.14	29733	51048	3	3	0.021	0.0070	0.32	−1.68	C	5n
487.	z ^{5}P° – f ^{5}D (821)											
		4678.85	29056	50423	7	9	0.074	0.031	3.4	−0.66	D	9n
		4619.29	29056	50699	7	5	0.047	0.011	1.2	−1.12	C	5n
		4669.17	29469	50880	5	3	0.040	0.0078	0.60	−1.41	C	5n
		4704.95	29733	50981	3	1	0.081	0.0090	0.42	−1.57	C	5n
488.	z ^{5}P° – e ^{7}P (822)											
		4638.01	29056	50611	7	7	0.034	0.011	1.2	−1.12	C	5n
		4673.28	29469	50861	5	5	0.034	0.011	0.87	−1.25	C	5n

Fe I: Allowed transitions — Continued

No.	Multiplet	λ (Å)	E_i (cm^{-1})	E_k (cm^{-1})	g_i	g_k	A_{ki} (10^8 s^{-1})	f_{ik}	S (at. u.)	log gf	Accuracy	Source
489.	$z\ ^5P° - e\ ^5G$ (823)											
		4560.09	29056	50980	7	9	0.0043	0.0017	0.18	−1.92	C	5n
		4596.41	29469	51219	5	7	0.0022	9.6(−4)	0.072	−2.32	D	5n
		4510.82	29056	51219	7	7	6.0(−4)	1.8(−4)	0.019	−2.89	D	30
490.	$z\ ^5P° - f\ ^5F$ (825)											
		4433.78	29056	51604	7	7	0.026	0.0077	0.78	−1.27	C	5n
		4495.95	29469	51705	5	5	0.013	0.0038	0.28	−1.72	C	5n
		4485.97	29469	51754	5	3	0.0049	8.9(−4)	0.066	−2.35	D	5n
491.	$z\ ^5P° - e\ ^3D$ (827)											
		4495.57	29056	51294	7	7	0.0036	0.0011	0.11	−2.12	D	5n
		4481.61	29733	52040	3	3	0.042	0.013	0.56	−1.42	C	5n
		4580.58	29469	51294	5	7	0.0037	0.0016	0.12	−2.09	D	5n
492.	$z\ ^5P° - g\ ^5D$ (828)											
		4484.22	29056	51350	7	9	0.070	0.027	2.8	−0.72	D	9n
		4482.74	29469	51771	5	7	0.021	0.0089	0.66	−1.35	C	5n
		4401.29	29056	51771	7	7	0.059	0.017	1.7	−0.92	C+	4n
		4446.83	29733	52214	3	3	0.053	0.016	0.68	−1.33	C	5n
		4347.85	29056	52050	7	5	0.015	0.0031	0.31	−1.66	C	5n
		4395.29	29469	52214	5	3	0.017	0.0030	0.21	−1.83	C	5n
		4438.34	29733	52257	3	1	0.079	0.0078	0.34	−1.63	C	5n
493.	$z\ ^5P° - e\ ^7S$ (829)											
		4440.48	29056	51570	7	7	0.0041	0.0012	0.12	−2.07	D	5n
		4523.40	29469	51570	5	7	0.0048	0.0020	0.15	−1.99	C	5n
494.	$z\ ^5P° - e\ ^5P$ (830)											
		4388.41	29056	51837	7	7	0.13	0.037	3.7	−0.59	C+	4n
		4423.84	29469	52067	5	5	0.017	0.0049	0.36	−1.61	C	5n
		4485.67	29733	52020	3	3	0.11	0.032	1.4	−1.02	C	5n
		4433.22	29469	52020	5	3	0.23	0.040	2.9	−0.70	C	5n
		4469.37	29469	51837	5	7	0.26	0.11	8.1	−0.26	C+	4n
495.	$z\ ^5P° - g\ ^5F$ (831)											
		4035.25	29056	53831	7	7	0.0021	5.1(−4)	0.047	−2.45	D−	30
496.	$z\ ^5P° - 4$ (836)											
		3428.75	29056	58213	7	5	0.27	0.033	2.6	−0.63	D	12n
497.	$a\ ^1I - z\ ^3H°$ (838)											
		5649.66	29313	47008	13	11	3.2(−4)	1.3(−4)	0.032	−2.77	D	5n

Fe I: Allowed transitions — Continued

No.	Multiplet	λ (Å)	E_i (cm⁻¹)	E_k (cm⁻¹)	g_i	g_k	A_{ki} (10⁸ s⁻¹)	f_{ik}	S (at. u.)	log gf	Accuracy	Source
498.	$a\ ^1I - x\ ^3G°$ (841)											
		5397.62	29313	47835	13	11	6.9(−4)	2.5(−4)	0.059	−2.48	D	30
499.	$a\ ^1I - \ ^5H°$											
		5284.42	29313	48231	13	11	5.7(−4)	2.0(−4)	0.046	−2.58	D	5n
500.	$a\ ^1I - z\ ^1H°$ (843)	5242.49	29313	48383	13	11	0.032	0.011	2.5	−0.84	C+	4n,8
501.	$a\ ^1I - v\ ^3G°$ (845)											
		4961.91	29313	49461	13	11	0.0013	3.9(−4)	0.084	−2.29	D	5n
502.	$a\ ^1I - y\ ^3I°$ (849)											
		4309.03	29313	52514	13	13	0.022	0.0061	1.1	−1.10	C	4n,5n,8
503.	$a\ ^1I - z\ ^1I°$ (850)	4203.94	29313	53094	13	13	0.13	0.034	6.2	−0.35	C	8
504.	$a\ ^1I - \ ^3H°$											
		3813.88	29313	55526	13	11	0.087	0.016	2.6	−0.68	C+	4n,5n
505.	$b\ ^3D - y\ ^3P°$ (867)											
		5760.35	29372	46727	7	5	0.0013	4.6(−4)	0.061	−2.49	D	5n
		5698.05	29357	46902	5	3	0.0014	4.2(−4)	0.039	−2.68	D	5n
506.	$b\ ^3D - \ ^5D°$											
		5762.43	29372	46721	7	9	0.0012	7.5(−4)	0.10	−2.28	D	5n
		5611.35	29320	47136	3	5	4.3(−4)	3.4(−4)	0.019	−2.99	D	30
		5754.41	29372	46745	7	7	5.7(−4)	2.9(−4)	0.038	−2.70	D	5n
		5609.97	29357	47177	5	3	4.1(−4)	1.2(−4)	0.011	−3.24	D	30
507.	$b\ ^3D - \ ^3D°$											
		5702.43	29357	46889	5	5	5.5(−4)	2.7(−4)	0.025	−2.87	D	5n
		5568.81	29320	47272	3	3	8.0(−4)	3.7(−4)	0.021	−2.95	D	5n
		5707.25	29372	46889	7	5	1.6(−4)	5.7(−5)	0.0075	−3.40	D	30
		5660.79	29357	47017	5	7	3.9(−4)	2.6(−4)	0.025	−2.88	D	5n
508.	$b\ ^3D - x\ ^3F°$ (868)											
		5707.07	29372	46889	7	9	9.1(−4)	5.7(−4)	0.075	−2.40	D	5n
		5636.71	29357	47093	5	7	7.4(−4)	4.9(−4)	0.046	−2.61	D	5n
509.	$b\ ^3D - w\ ^5G°$ (870)											
		5452.12	29357	47693	5	7	4.4(−4)	2.8(−4)	0.025	−2.86	D	30
510.	$b\ ^3D - \ ^1D°$											
		5539.28	29372	47420	7	5	9.5(−4)	3.1(−4)	0.040	−2.66	D	5n

Fe I: Allowed transitions — Continued

No.	Multiplet	λ (Å)	E_i (cm^{-1})	E_k (cm^{-1})	g_i	g_k	A_{ki} (10^8 s^{-1})	f_{ik}	S (at. u.)	log gf	Accuracy	Source
511.	$b\ ^3$D $- z\ ^1$G° (872)											
		5529.15	29372	47453	7	9	4.5(−4)	2.7(−4)	0.034	−2.73	D	5n
512.	$b\ ^3$D $- y\ ^3$S° (873)											
		5482.26	29320	47556	3	3	2.1(−4)	9.6(−5)	0.0052	−3.54	D	30
513.	$b\ ^3$D $- v\ ^5$F° (875)											
		5294.56	29357	48239	5	5	6.6(−4)	2.8(−4)	0.024	−2.86	D	5n
		5298.79	29372	48239	7	5	0.0033	9.9(−4)	0.12	−2.16	D	5n
514.	$b\ ^3$D $- v\ ^5$P° (877)											
		5320.05	29372	48163	7	5	0.0014	4.1(−4)	0.051	−2.54	D	5n
515.	$b\ ^3$D $- x\ ^3$P° (880)											
		5280.36	29372	48305	7	5	0.0045	0.0013	0.16	−2.03	D	5n
		5223.19	29320	48460	3	1	0.010	0.0014	0.070	−2.39	D	5n
		5207.95	29320	48516	3	3	0.0029	0.0012	0.061	−2.45	D	5n
516.	$b\ ^3$D $- w\ ^3$F° (883)											
		4970.50	29320	49433	3	5	0.011	0.0067	0.33	−1.70	C	5n
		4979.59	29357	49433	5	5	0.0014	5.3(−4)	0.043	−2.58	D	30
517.	$b\ ^3$D $- v\ ^3$D° (884)											
		5058.50	29372	49135	7	7	5.5(−4)	2.1(−4)	0.024	−2.83	D	15n,30
		5054.64	29357	49135	5	7	0.0027	0.0014	0.12	−2.14	D	5n
518.	$b\ ^3$D $- v\ ^3$G° (886)											
		4935.42	29372	49628	7	9	1.1(−4)	5.3(−5)	0.0060	−3.43	D	30
519.	$b\ ^3$D $- z\ ^1$D° (887)											
		4968.69	29357	49477	5	5	0.0090	0.0033	0.27	−1.78	C	5n
520.	$b\ ^3$D $- w\ ^3$P° (888)											
		4799.41	29357	50187	5	5	0.0034	0.0012	0.093	−2.23	D	5n
521.	$b\ ^3$D $- x\ ^1$G° (890)											
		4706.31	29372	50614	7	9	3.8(−4)	1.6(−4)	0.017	−2.95	D	30

Fe I: Allowed transitions — Continued

No.	Multiplet	λ (Å)	E_i (cm^{-1})	E_k (cm^{-1})	g_i	g_k	A_{ki} (10^8 s^{-1})	f_{ik}	S (at. u.)	log gf	Accuracy	Source
522.	$b\ ^3D - t\ ^5D°$ (893)											
		4419.30	29320	51942	3	1	0.0057	5.5(−4)	0.024	−2.78	D−	30
523.	$b\ ^3D - v\ ^3F°$ (894)											
		4542.41	29357	51365	5	7	0.0041	0.0018	0.13	−2.05	D	5n
524.	$b\ ^3D - u\ ^3G°$ (898)											
		4483.78	29372	51668	7	9	0.0012	4.8(−4)	0.050	−2.47	D	5n
525.	$b\ ^3D - u\ ^3D°$ (903)											
		4360.81	29372	52297	7	5	0.0095	0.0019	0.19	−1.87	C	5n
526.	$b\ ^3D - ^3D°$											
		4285.83	29357	52683	5	5	0.012	0.0033	0.23	−1.78	C	5n
527.	$b\ ^3D - ^3P°$											
		4246.08	29372	52916	7	5	0.057	0.011	1.1	−1.11	C	4n,5n
528.	$b\ ^3D - ^5F°$											
		4239.36	29372	52954	7	9	0.012	0.0043	0.42	−1.52	C	5n
529.	$b\ ^3D - ^5G°$											
		4124.49	29372	53610	7	9	0.0025	8.2(−4)	0.078	−2.24	D−	30
530.	$b\ ^3D - ^5D°$											
		4010.18	29372	54301	7	9	0.0075	0.0023	0.21	−1.79	C	5n
		4082.44	29320	53808	3	3	0.038	0.0094	0.38	−1.55	C	5n
		4088.57	29357	53808	5	3	0.039	0.0059	0.40	−1.53	C	5n
531.	$b\ ^3D - t\ ^3G°$ (913)											
		4020.49	29372	54237	7	9	0.0078	0.0024	0.22	−1.77	C	5n
		3960.28	29357	54600	5	7	0.042	0.014	0.90	−1.16	C	5n
532.	$b\ ^3D - ^3G°$											
		3781.94	29357	55791	5	7	0.037	0.011	0.68	−1.26	C	5n
533.	$b\ ^3D - w\ ^1D°$ (916)											
		3787.16	29357	55754	5	5	0.10	0.022	1.4	−0.96	C+	4n,5n
534.	$b\ ^1G2 - x\ ^3F°$ (922)											
		5849.67	29799	46889	9	9	2.2(−4)	1.1(−4)	0.020	−2.99	D	5n

Fe I: Allowed transitions — Continued

No.	Multiplet	λ (Å)	E_i (cm^{-1})	E_k (cm^{-1})	g_i	g_k	A_{ki} (10^8 s^{-1})	f_{ik}	S (at. u.)	log gf	Accuracy	Source
535.	$b\ ^1G2 - w\ ^5G°$ (923)											
		5619.23	29799	47590	9	9	1.3(−4)	6.0(−5)	0.0099	−3.27	D	30
536.	$b\ ^1G2 - z\ ^1G°$ (924)	5662.94	29799	47453	9	9	0.0010	5.0(−4)	0.083	−2.35	D	5n
537.	$b\ ^1G2 - x\ ^3G°$ (926)											
		5549.94	29799	47812	9	9	3.0(−4)	1.4(−4)	0.022	−2.91	D	5n
		5543.15	29799	47834	9	7	0.0083	0.0030	0.49	−1.57	C	5n
538.	$b\ ^1G2 - ^5H°$											
		5385.58	29799	48362	9	9	2.7(−4)	1.2(−4)	0.019	−2.97	D	5n
539.	$b\ ^1G2 - z\ ^1H°$ (928)	5379.57	29799	48383	9	11	0.0070	0.0037	0.59	−1.48	C	4n,5n,8
540.	$b\ ^1G2 - y\ ^1G°$ (929)	5288.53	29799	48703	9	9	0.0057	0.0024	0.38	−1.67	D	15n,30
541.	$b\ ^1G2 - w\ ^3F°$ (930)											
		5177.23	29799	49109	9	9	0.0011	4.2(−4)	0.065	−2.42	D	5n
542.	$b\ ^1G2 - ^1F°$											
		5145.73	29799	49227	9	7	1.9(−4)	6.0(−5)	0.0091	−3.27	D	30
543.	$b\ ^1G2 - x\ ^3H°$ (935)											
		4700.19	29799	51069	9	11	0.0057	0.0023	0.32	−1.68	C	5n
544.	$b\ ^1G2 - ^5F°$											
		4189.56	29799	53661	9	7	0.025	0.0052	0.65	−1.33	C	5n
545.	$b\ ^1G2 - x\ ^1F°$ (941)	4171.69	29799	53763	9	7	0.029	0.0058	0.72	−1.28	C	5n
546.	$b\ ^1G2 - w\ ^1G°$ (945)	3996.97	29799	54811	9	9	0.067	0.016	1.9	−0.84	C+	4n,5n
547.	$b\ ^1G2 - (\ °)^b$											
		3829.13	29799	55907	9	11	0.027	0.0072	0.81	−1.19	C	5n
548.	$z\ ^3F° - e\ ^5D$ (957)											
		7541.61	31805	45061	7	7	1.3(−4)	1.1(−4)	0.019	−3.12	D	30
549.	$z\ ^3F° - e\ ^5F$ (958)											
		6220.77	31307	47378	9	9	6.6(−4)	3.9(−4)	0.071	−2.46	D	30

Fe I:　Allowed transitions — Continued

No.	Multiplet	λ (Å)	E_i (cm^{-1})	E_k (cm^{-1})	g_i	g_k	A_{ki} (10^8 s^{-1})	f_{ik}	S (at. u.)	$\log gf$	Accuracy	Source
550.	$z\ ^3F^\circ - e\ ^3F$ (959)											
		6003.03	31307	47961	9	9	0.016	0.0084	1.5	−1.12	D−	14n
		5952.75	32134	48928	5	5	0.014	0.0073	0.71	−1.44	C	5n
		5804.06	31307	48532	9	7	0.0015	5.7(−4)	0.098	−2.29	D	5n
		5838.42	31805	48928	7	5	0.0018	6.5(−4)	0.088	−2.34	D	5n
		6188.04	31805	47961	7	9	0.0037	0.0027	0.39	−1.72	C	5n
		6096.69	32134	48532	5	7	0.0030	0.0023	0.24	−1.93	C	5n
551.	$z\ ^3F^\circ - e\ ^7F$ (960)											
		5119.90	31307	50833	9	11	2.1(−4)	9.9(−5)	0.015	−3.05	D	30
552.	$z\ ^3F^\circ - e\ ^5G$ (962)											
		5213.80	31805	50980	7	9	4.7(−4)	2.5(−4)	0.030	−2.76	D	30
		5238.25	32134	51219	5	7	2.5(−4)	1.4(−4)	0.012	−3.15	D	30
553.	$z\ ^3F^\circ - e\ ^3D$ (965)											
		5001.86	31307	51294	9	7	0.39	0.11	17	0.01	C+	4n
		5014.94	31805	51740	7	5	0.30	0.080	9.3	−0.25	C+	4n
		5022.24	32134	52040	5	3	0.26	0.059	4.9	−0.53	C+	4n
		5129.63	31805	51294	7	7	0.0051	0.0020	0.24	−1.85	C	5n
		5099.09	32134	51740	5	5	0.015	0.0059	0.50	−1.53	C	5n
554.	$z\ ^3F^\circ - g\ ^5D$ (966)											
		4978.60	32134	52214	5	3	0.11	0.023	1.9	−0.93	D−	14n
555.	$z\ ^3F^\circ - e\ ^7S$ (967)											
		5058.00	31805	51570	7	7	0.0011	4.2(−4)	0.049	−2.53	D	5n
556.	$z\ ^3F^\circ - g\ ^5F$ (969)											
		4452.62	31805	54258	7	5	0.0079	0.0017	0.17	−1.93	C	5n
		4492.68	32134	54386	5	3	0.025	0.0045	0.33	−1.65	C	5n
557.	$z\ ^3F^\circ - f\ ^5P$ (971)											
		4593.53	31805	53569	7	5	0.0055	0.0012	0.13	−2.06	D	5n
		4587.72	32134	53925	5	3	0.0075	0.0014	0.11	−2.15	D	5n
558.	$z\ ^3F^\circ - f\ ^5G$ (972)											
		4551.65	31805	53769	7	9	0.0031	0.0012	0.13	−2.06	D	5n
		4450.77	31307	53769	9	9	0.0022	6.4(−4)	0.084	−2.24	D	5n

Fe I: Allowed transitions — Continued

No.	Multiplet	λ (Å)	E_i (cm^{-1})	E_k (cm^{-1})	g_i	g_k	A_{ki} (10^8 s^{-1})	f_{ik}	S (at. u.)	log gf	Accuracy	Source
559.	$z\ ^3F^\circ - e\ ^3G$ (973)											
		4456.63	31307	53739	9	11	0.0064	0.0023	0.31	−1.68	C	5n
		4494.05	32134	54379	5	7	0.0073	0.0031	0.23	−1.81	C	5n
		4392.58	31307	54067	9	9	0.0038	0.0011	0.14	−2.00	D	5n
560.	$z\ ^3F^\circ - f\ ^3D$ (974)											
		4455.03	31307	53748	9	7	0.039	0.0090	1.2	−1.09	C	5n
		4479.97	32134	54449	5	3	0.033	0.0060	0.45	−1.52	D−	30
561.	$z\ ^3F^\circ - e\ ^3H$ (975)											
		4300.21	31307	54555?	9	9	0.0062	0.0017	0.22	−1.81	C	5n
562.	$z\ ^3F^\circ - f\ ^3F$ (976)											
		4276.68	31307	54683	9	9	0.025	0.0069	0.87	−1.21	C	5n
		4300.83	32134	55379	5	5	0.047	0.013	0.91	−1.19	C	5n
		4197.38	31307	55125	9	7	0.0020	4.0(−4)	0.050	−2.44	D−	30
563.	$z\ ^3F^\circ - 2$ (977)											
		3975.85	31307	56452	9	9	0.025	0.0060	0.70	−1.27	C	5n
564.	$z\ ^3D^\circ - e\ ^5F$ (981)											
		6226.77	31323	47378	7	9	0.0012	8.6(−4)	0.12	−2.22	D	5n
565.	$z\ ^3D^\circ - e\ ^3F$ (982)											
		5934.66	31686	48532	5	7	0.018	0.014	1.3	−1.17	C	5n
		5883.84	31937	48928	3	5	0.017	0.015	0.85	−1.36	C	5n
		5809.25	31323	48532	7	7	0.0041	0.0021	0.28	−1.84	C	5n
		5798.19	31686	48928	5	5	0.0051	0.0026	0.25	−1.89	C	5n
		5678.39	31323	48928	7	5	4.0(−4)	1.4(−4)	0.018	−3.02	D	30
566.	$z\ ^3D^\circ - e\ ^3D$ (984)											
		4973.10	31937	52040	3	3	0.10	0.037	1.8	−0.95	C+	4n
		4896.44	31323	51740	7	5	0.0050	0.0013	0.14	−2.05	D	5n
		4911.78	31686	52040	5	3	0.015	0.0032	0.26	−1.79	C	5n
		5048.43	31937	51740	3	5	0.029	0.018	0.91	−1.26	C	4n
567.	$z\ ^3D^\circ - g\ ^5D$ (985)											
		4930.31	31937	52214	3	3	0.041	0.015	0.73	−1.35	C	5n
		4870.05	31686	52214	5	3	0.0043	9.1(−4)	0.073	−2.34	D	5n
568.	$z\ ^3D^\circ - e\ ^5P$ (986)											
		4905.13	31686	52067	5	5	0.0049	0.0018	0.14	−2.05	D	5n
		4916.67	31686	52020	5	3	0.0010	2.2(−4)	0.018	−2.96	D	30

Fe I: Allowed transitions — Continued

No.	Multiplet	λ (Å)	E_i (cm^{-1})	E_k (cm^{-1})	g_i	g_k	A_{ki} (10^8 s^{-1})	f_{ik}	S (at. u.)	log gf	Accuracy	Source
569.	$z\ ^3$D° $- f\ ^5$P (989)											
		4568.61	31686	53569	5	5	0.0024	7.6($-$4)	0.057	-2.42	D	30
		4546.68	31937	53925	3	3	0.0026	8.2($-$4)	0.037	-2.61	D	30
570.	$z\ ^3$D° $- f\ ^3$D (992)											
		4466.94	31686	54067	5	5	0.030	0.0089	0.66	-1.35	C	5n
		4440.82	31937	54449	3	3	0.028	0.0084	0.37	-1.60	C	5n
		4391.87	31686	54449	5	3	0.011	0.0018	0.13	-2.04	D	5n
571.	$z\ ^3$D° $- f\ ^3$F (993)											
		4279.48	31323	54683	7	9	0.014	0.0050	0.49	-1.46	C	5n
		4264.74	31937	55379	3	5	0.026	0.012	0.50	-1.45	C	5n
		4200.09	31323	55125	7	7	0.040	0.011	1.0	-1.13	C	5n
572.	$z\ ^3$D° $- e\ ^3$P (994)											
		4243.79	31323	54880	7	5	0.023	0.0045	0.44	-1.50	C	5n
		4220.05	31686	55376	5	3	0.027	0.0044	0.30	-1.66	C	5n
		4310.37	31686	54880	5	5	0.023	0.0063	0.45	-1.50	C	5n
573.	$z\ ^3$D° $- i\ ^5$D (995)											
		3839.61	31937	57974	3	5	0.39	0.15	5.5	-0.36	C	5n
574.	$c\ ^3$F $- ^5$D°											
		7219.69	32874	46721	9	9	0.0029	0.0023	0.49	-1.69	D	30
		7498.56	33413	46745	7	7	9.5($-$4)	8.0($-$4)	0.14	-2.25	D	30
		7284.84	33413	47136	7	5	0.0045	0.0025	0.43	-1.75	D	30
		7454.02	33765	47177	5	3	0.0016	7.8($-$4)	0.095	-2.41	D	30
575.	$c\ ^3$F $- ^3$D°											
		7068.42	32874	47017	9	7	0.0080	0.0046	0.97	-1.38	C	5n
		7418.67	33413	46889	7	5	0.0062	0.0037	0.63	-1.59	D	30
		7401.69	33765	47272	5	3	0.0083	0.0041	0.50	-1.69	D	30
576.	$c\ ^3$F $- x\ ^3$F° (1002)											
		7132.99	32874	46889	9	9	0.0026	0.0020	0.42	-1.75	C	5n
		7443.03	33765	47197	5	5	0.0036	0.0030	0.37	-1.82	D	30
		7418.32	33413	46889	7	9	1.6($-$4)	1.7($-$4)	0.029	-2.93	D	30
		7501.25	33765	47093	5	7	2.2($-$4)	2.6($-$4)	0.032	-2.89	D	30
577.	$c\ ^3$F $- z\ ^3$H° (1003)											
		7072.82	32874	47008	9	11	1.8($-$4)	1.6($-$4)	0.034	-2.84	D	30
		7024.08	32874	47106	9	9	0.0012	9.0($-$4)	0.19	-2.09	D	5n

Fe I: Allowed transitions — Continued

No.	Multiplet	λ (Å)	E_i (cm^{-1})	E_k (cm^{-1})	g_i	g_k	A_{ki} (10^8 s^{-1})	f_{ik}	S (at. u.)	log gf	Accuracy	Source
578.	$c\ ^3F - w\ ^5G°$ (1005)											
		6793.26	32874	47590	9	9	5.4(−4)	3.8(−4)	0.076	−2.47	D	30
		7000.63	33413	47693	7	7	0.0012	8.8(−4)	0.14	−2.21	D	5n
		7107.46	33765	47831	5	5	0.0024	0.0018	0.21	−2.04	D	5n
		6745.96	32874	47693	9	7	3.6(−4)	1.9(−4)	0.038	−2.77	D	30
579.	$c\ ^3F - z\ ^1G°$ (1006)											
		6857.25	32874	47453	9	9	0.0011	7.9(−4)	0.16	−2.15	D	5n
580.	$c\ ^3F - v\ ^5F°$ (1007)											
		6639.90	32874	47930	9	9	5.8(−4)	3.9(−4)	0.076	−2.46	D	30
		6796.11	33413	48123	7	7	6.1(−4)	4.2(−4)	0.066	−2.53	D	30
581.	$c\ ^3F - ^5H°$											
		6509.56	32874	48231	9	11	1.5(−4)	1.2(−4)	0.023	−2.97	D	30
582.	$c\ ^3F - y\ ^1G°$ (1014)											
		6315.81	32874	48703	9	9	0.0036	0.0022	0.41	−1.71	C	5n
583.	$c\ ^3F - w\ ^3F°$ (1015)											
		6157.73	32874	49109	9	9	0.011	0.0061	1.1	−1.26	C	5n
		6380.75	33765	49433	5	5	0.013	0.0080	0.84	−1.40	C	5n
584.	$c\ ^3F - v\ ^3D°$ (1016)											
		6147.85	32874	49135	9	7	0.0050	0.0022	0.40	−1.70	C	5n
		6436.43	33765	49298	5	3	0.0019	6.9(−4)	0.073	−2.46	D	30
585.	$c\ ^3F - v\ ^3G°$ (1018)											
		6027.06	32874	49461	9	11	0.010	0.0069	1.2	−1.21	C	5n
		6165.37	33413	49628	7	9	0.0055	0.0040	0.57	−1.55	C	5n
		6215.15	33765	49851	5	7	0.0090	0.0073	0.74	−1.44	C	5n
586.	$c\ ^3F - z\ ^1D°$ (1019)											
		6362.89	33765	49477	5	5	0.0035	0.0021	0.22	−1.97	C	5n
587.	$c\ ^3F - z\ ^1F°$ (1021)											
		5643.94	32874	50587	9	7	0.0027	9.9(−4)	0.17	−2.05	D	5n
588.	$c\ ^3F - x\ ^1G°$ (1022)											
		5811.93	33413	50614	7	9	8.2(−4)	5.3(−4)	0.071	−2.43	D	5n

Fe I: Allowed transitions — Continued

No.	Multiplet	λ (Å)	E_i (cm^{-1})	E_k (cm^{-1})	g_i	g_k	A_{ki} (10^8 s^{-1})	f_{ik}	S (at. u.)	log gf	Accuracy	Source
589.	$c\ ^3F - x\ ^3H°$ (1024)											
		5494.46	32874	51069	9	11	0.0016	9.0(-4)	0.15	-2.09	D	$5n$
590.	$c\ ^3F - t\ ^5D°$ (1025)											
		5487.74	33413	51630	7	5	0.086	0.028	3.5	-0.71	D$-$	$14n$
591.	$c\ ^3F - v\ ^3F°$ (1026)											
		5587.58	33413	51305	7	9	0.0034	0.0020	0.26	-1.85	C	$5n$
		5680.26	33765	51365	5	7	7.8(-4)	5.3(-4)	0.049	-2.58	D	$5n$
592.	$c\ ^3F - u\ ^3G°$ (1029)											
		5319.22	32874	51668	9	9	0.0011	4.7(-4)	0.075	-2.37	D	$5n$
593.	$c\ ^3F - ^1H°$											
		5329.99	32874	51630	9	11	0.011	0.0056	0.88	-1.30	C	$5n$
594.	$c\ ^3F - y\ ^1D°$ (1030)											
		5464.29	33413	51708	7	5	0.0085	0.0027	0.34	-1.72	C	$5n$
595.	$c\ ^3F - u\ ^3D°$ (1031)											
		5293.97	33413	52297	7	5	0.0064	0.0019	0.24	-1.87	C	$5n$
		5332.67	33765	52512	5	3	0.0064	0.0016	0.14	-2.09	D	$5n$
		5387.51	33413	51969	7	7	0.0024	0.0010	0.13	-2.14	D	$5n$
		5394.68	33765	52297	5	5	0.011	0.0048	0.43	-1.62	C	$5n$
		5491.84	33765	51969	5	7	0.0013	8.0(-4)	0.072	-2.40	D	$5n$
596.	$c\ ^3F - ^3D°$											
		5187.91	33413	52683	7	5	0.027	0.0079	0.94	-1.26	C	$4n,5n$
		5236.19	33765	52858	5	3	0.015	0.0038	0.33	-1.72	C	$5n$
		5284.62	33765	52683	5	5	0.0037	0.0016	0.14	-2.11	D	$5n$
597.	$c\ ^3F - ^3P°$											
		5136.09	33765	53230	5	3	0.0064	0.0015	0.13	-2.12	D	$5n$
598.	$c\ ^3F - ^5F°$											
		4729.02	32874	54014	9	11	0.0059	0.0024	0.34	-1.66	C	$5n$
599.	$c\ ^3F - ^5D°$											
		4887.37	32874	53329	9	9	0.0016	5.8(-4)	0.084	-2.28	D	30
600.	$c\ ^3F - x\ ^1F°$ (1040)											
		4912.52	33413	53763	7	7	0.0015	5.4(-4)	0.061	-2.42	D	30
		4999.11	33765	53763	5	7	0.0069	0.0036	0.30	-1.74	C	$5n$

Fe I: Allowed transitions — Continued

No.	Multiplet	λ (Å)	E_i (cm^{-1})	E_k (cm^{-1})	g_i	g_k	A_{ki} (10^8 s^{-1})	f_{ik}	S (at. u.)	log gf	Accuracy	Source
601.	$c\ ^3F - ^5D^\circ$											
		4785.96	33413	54301	7	9	0.0038	0.0017	0.19	−1.93	C	5n
602.	$c\ ^3F - t\ ^3G^\circ$ (1042)											
		4735.84	32874	53983	9	11	0.016	0.0067	0.94	−1.22	C	5n
		4800.65	33413	54237	7	9	0.018	0.0079	0.87	−1.26	C	5n
		4798.26	33765	54600	5	7	0.012	0.0056	0.45	−1.55	C	5n
603.	$c\ ^3F - ^5P^\circ$											
		4854.89	33413	54005	7	7	0.0037	0.0013	0.15	−2.04	D	5n
604.	$c\ ^3F - ^3H^\circ$											
		4413.40	32874	55526	9	11	0.0092	0.0033	0.43	−1.53	C	5n
605.	$c\ ^3F - w\ ^1D^\circ$ (1047)											
		4546.48	33765	55754	5	5	0.0020	6.2(−4)	0.046	−2.51	D	30
606.	$y\ ^5D^\circ - e\ ^5F$ (1051)											
		7130.94	34017	48037	3	5	0.043	0.054	3.8	−0.79	C	5n
		7090.40	34122	48221	1	3	0.027	0.062	1.4	−1.21	C	5n
		6999.90	33096	47378	9	9	0.0042	0.0031	0.63	−1.56	C	5n
		7016.44	33507	47756	7	7	0.011	0.0079	1.3	−1.26	C	5n
		7022.98	33802	48037	5	5	0.015	0.011	1.3	−1.25	C	5n
		7038.25	34017	48221	3	3	0.022	0.017	1.2	−1.30	C	5n
		6880.65	33507	48037	7	5	0.0012	6.1(−4)	0.097	−2.37	D	30
		6933.04	33802	48221	5	3	0.0021	9.1(−4)	0.10	−2.34	D	30
607.	$y\ ^5D^\circ - e\ ^3F$ (1052)											
		6725.39	33096	47961	9	9	8.2(−4)	5.6(−4)	0.11	−2.30	D	30
		6653.88	33507	48532	7	7	6.5(−4)	4.3(−4)	0.066	−2.52	D	30
		6916.70	33507	47961	7	9	0.0055	0.0051	0.81	−1.45	C	5n
		6786.88	33802	48532	5	7	0.0018	0.0017	0.19	−2.07	D	5n
		6704.48	34017	48928	3	5	6.5(−4)	7.3(−4)	0.048	−2.66	D	30
608.	$y\ ^5D^\circ - f\ ^7D$ (1054)											
		5715.47	33507	50999	7	5	4.3(−4)	1.5(−4)	0.020	−2.98	D	30
609.	$y\ ^5D^\circ - f\ ^5D$ (1055)											
		5732.86	33096	50534	9	7	2.6(−4)	1.0(−4)	0.017	−3.04	D	30
		5815.16	33507	50699	7	5	9.5(−4)	3.4(−4)	0.046	−2.62	D	5n
610.	$y\ ^5D^\circ - e\ ^7P$ (1056)											
		5844.88	33507	50611	7	7	3.2(−4)	1.6(−4)	0.022	−2.94	D	30

Fe I: Allowed transitions — Continued

No.	Multiplet	λ (Å)	E_i (cm^{-1})	E_k (cm^{-1})	g_i	g_k	A_{ki} (10^8 s^{-1})	f_{ik}	S (at. u.)	log gf	Accuracy	Source
611.	$y\ ^5D° - e\ ^5G$ (1057)											
		5677.68	33096	50704	9	11	3.8(−4)	2.2(−4)	0.037	−2.70	D	30
		5644.35	33507	51219	7	7	2.7(−4)	1.3(−4)	0.017	−3.04	D	30
612.	$y\ ^5D° - e\ ^7G$ (1058)											
		5607.66	33507	51335	7	9	0.0013	7.7(−4)	0.099	−2.27	D	30
		5481.25	33096	51335	9	9	0.0098	0.0044	0.72	−1.40	C	5n
613.	$y\ ^5D° - f\ ^5F$ (1059)											
		5652.01	34017	51705	3	5	3.2(−4)	2.5(−4)	0.014	−3.12	D	30
		5443.41	33096	51462	9	9	2.8(−4)	1.2(−4)	0.020	−2.95	D	30
		5524.25	33507	51604	7	7	4.6(−4)	2.1(−4)	0.027	−2.83	D	30
		5583.97	33802	51705	5	5	7.3(−4)	3.4(−4)	0.031	−2.77	D	30
614.	$y\ ^5D° - e\ ^3D$ (1061)											
		5493.51	33096	51294	9	7	0.0046	0.0016	0.26	−1.84	C	5n
		5483.11	33507	51740	7	5	0.012	0.0038	0.47	−1.58	C	5n
		5481.45	33802	52040	5	3	0.026	0.0071	0.64	−1.45	C	5n
		5620.53	33507	51294	7	7	0.0049	0.0023	0.30	−1.79	C	5n
		5547.00	34017	52040	3	3	0.0089	0.0041	0.22	−1.91	C	5n
		5579.34	34122	52040	1	3	0.0028	0.0040	0.073	−2.40	D	30
615.	$y\ ^5D° - g\ ^5D$ (1062)											
		5473.90	33507	51771	7	7	0.055	0.025	3.1	−0.76	C+	4n
		5478.48	33802	52050	5	5	0.0063	0.0028	0.25	−1.85	C	5n
		5353.39	33096	51771	9	7	0.048	0.016	2.5	−0.84	D−	14n
		5480.87	34017	52257	3	1	0.12	0.018	0.99	−1.26	C	5n
		5563.60	33802	51771	5	7	0.032	0.020	1.9	−0.99	C	5n
		5543.94	34017	52050	3	5	0.031	0.024	1.3	−1.14	C	5n
		5525.55	34122	52214	1	3	0.034	0.047	0.85	−1.33	C	5n
616.	$y\ ^5D° - e\ ^5P$ (1064)											
		5386.34	33507	52067	7	5	0.0078	0.0024	0.30	−1.77	C	5n
		5473.18	33802	52067	5	5	0.0032	0.0014	0.13	−2.14	D	5n
617.	$y\ ^5D° - g\ ^5F$ (1065)											
		4991.27	33802	53831	5	7	0.082	0.043	3.5	−0.67	D−	14n
		4925.29	33096	53394	9	9	0.0023	8.4(−4)	0.12	−2.12	D	30
618.	$y\ ^5D° - h\ ^5D$ (1066)											
		4988.95	33507	53546	7	7	0.049	0.018	2.1	−0.89	C+	4n
		4969.92	34017	54133	3	3	0.18	0.065	3.2	−0.71	D+	10n
		4917.23	33802	54133	5	3	0.061	0.013	1.1	−1.18	C	5n
		5088.16	33507	53155	7	9	0.0048	0.0024	0.28	−1.78	C	5n

Fe I: Allowed transitions — Continued

No.	Multiplet	λ (Å)	E_i (cm^{-1})	E_k (cm^{-1})	g_i	g_k	A_{ki} (10^8 s^{-1})	f_{ik}	S (at. u.)	log gf	Accuracy	Source
619.	$y\,^5D° - f\,^5G$ (1068)											
		4835.87	33096	53769	9	9	0.010	0.0035	0.50	−1.50	C	5n
		4840.32	33507	54161	7	7	0.017	0.0058	0.65	−1.39	C	5n
		4859.13	33802	54376	5	5	0.011	0.0037	0.30	−1.73	C	5n
		4790.56	33507	54376	7	5	0.0013	3.3(−4)	0.036	−2.64	D	5n
620.	$y\,^5D° - e\,^3G$ (1069)											
		4842.79	33096	53739	9	11	0.0071	0.0031	0.44	−1.56	C	5n
621.	$y\,^5D° - f\,^3D$ (1070)											
		4841.78	33802	54449	5	3	0.013	0.0026	0.21	−1.88	C	5n
		4892.87	34017	54449	3	3	0.048	0.017	0.83	−1.29	D	15n
		4986.22	34017	54067	3	5	0.022	0.014	0.67	−1.39	C	5n
		4918.01	34122	54449	1	3	0.040	0.044	0.71	−1.36	C	5n
622.	$y\,^5D° - e\,^3P$ (1072)											
		4677.60	33507	54880	7	5	0.0032	7.5(−4)	0.081	−2.28	D	30
623.	$y\,^5D° - i\,^5D$ (1073)											
		4085.98	33507	57974	7	5	0.050	0.0090	0.85	−1.20	C	5n
624.	$y\,^5D° - 4$ (1075)											
		4095.27	33802	58213	5	5	0.032	0.0080	0.54	−1.40	C	5n
625.	$y\,^5F° - e\,^5F$ (1077)											
		7306.61	33695	47378	11	9	0.0025	0.0017	0.44	−1.74	D	30
626.	$y\,^5F° - e\,^3F$ (1078)											
		7008.01	33695	47961	11	9	0.0013	7.9(−4)	0.20	−2.06	D	5n
		6898.31	34040	48532	9	7	0.0012	6.5(−4)	0.13	−2.23	D	30
		7038.82	34329	48532	7	7	0.0020	0.0015	0.24	−1.99	C	5n
		7022.39	34692	48928	3	5	0.0014	0.0017	0.12	−2.29	D	30
627.	$y\,^5F° - f\,^7D$ (1080)											
		5961.91	34040	50808	9	9	1.4(−4)	7.7(−5)	0.014	−3.16	D	30
628.	$y\,^5F° - f\,^5D$ (1081)											
		6060.81	34040	50534	9	7	1.9(−4)	8.0(−5)	0.014	−3.14	D	30
629.	$y\,^5F° - e\,^7P$ (1082)											
		6032.67	34040	50611	9	7	1.2(−4)	5.2(−5)	0.0093	−3.33	D	30

Fe I: Allowed transitions — Continued

No.	Multiplet	λ (Å)	E_i (cm^{-1})	E_k (cm^{-1})	g_i	g_k	A_{ki} (10^8 s^{-1})	f_{ik}	S (at. u.)	$\log gf$	Accuracy	Source
630.	$y\ ^5F^\circ - e\ ^5G$ (1083)											
		5940.97	33695	50523	11	13	0.0010	6.4(−4)	0.14	−2.15	D	5n
		5877.77	33695	50704	11	11	0.0010	5.4(−4)	0.11	−2.23	D	5n
631.	$y\ ^5F^\circ - f\ ^5F$ (1084)											
		5742.95	33695	51103	11	11	5.7(−4)	2.8(−4)	0.058	−2.51	D	5n
		5738.22	34040	51462	9	9	0.0010	5.1(−4)	0.086	−2.34	D	30
		5826.64	34547	51705	5	5	4.5(−4)	2.3(−4)	0.022	−2.94	D	30
		5858.77	34040	51103	9	11	9.7(−4)	6.1(−4)	0.11	−2.26	D	5n
		5835.10	34329	51462	7	9	9.3(−4)	6.1(−4)	0.082	−2.37	D	5n
		5861.11	34547	51604	5	7	9.8(−4)	7.1(−4)	0.068	−2.45	D	30
		5876.27	34692	51705	3	5	8.6(−4)	7.5(−4)	0.043	−2.65	D	30
632.	$y\ ^5F^\circ - e\ ^3D$ (1086)											
		5793.93	34040	51294	9	7	0.0057	0.0022	0.38	−1.70	C	5n
		5741.86	34329	51740	7	5	0.0075	0.0027	0.35	−1.73	C	5n
		5814.80	34547	51740	5	5	0.0042	0.0021	0.21	−1.97	C	5n
		5969.55	34547	51294	5	7	5.0(−4)	3.7(−4)	0.037	−2.73	D	30
		5864.24	34692	51740	3	5	0.0012	0.0010	0.058	−2.52	D	30
633.	$y\ ^5F^\circ - g\ ^5D$ (1087)											
		5638.27	34040	51771	9	7	0.040	0.015	2.5	−0.87	C	5n
		5641.46	34329	52050	7	5	0.028	0.0094	1.2	−1.18	C	5n
		5691.51	34692	52257	3	1	0.062	0.010	0.57	−1.52	C	5n
		5731.77	34329	51771	7	7	0.015	0.0072	0.95	−1.30	C	5n
		5711.87	34547	52050	5	5	0.014	0.0069	0.65	−1.46	C	5n
		5705.48	34692	52214	3	3	0.017	0.0084	0.47	−1.60	C	5n
		5873.21	34329	51350	7	9	0.0016	0.0010	0.14	−2.14	D	5n
		5804.48	34547	51771	5	7	0.0026	0.0018	0.17	−2.04	D	5n
634.	$y\ ^5F^\circ - e\ ^5P$ (1088)											
		5635.85	34329	52067	7	5	0.0054	0.0018	0.24	−1.89	C	5n
		5709.93	34329	51837	7	7	0.0013	6.5(−4)	0.086	−2.34	D	30
635.	$y\ ^5F^\circ - g\ ^5F$ (1089)											
		5162.27	33695	53061	11	11	0.24	0.095	18	0.02	D	9n
		5126.19	34329	53831	7	7	0.030	0.012	1.4	−1.08	C	5n
		5016.48	34329	54258	7	5	0.011	0.0029	0.34	−1.69	D	30
		5243.78	34329	53394	7	9	0.019	0.010	1.2	−1.15	C	5n
		5184.26	34547	53831	5	7	0.035	0.020	1.7	−1.00	C	5n
		5109.65	34692	54258	3	5	0.054	0.035	1.8	−0.98	C−	15n
636.	$y\ ^5F^\circ - h\ ^5D$ (1090)											
		5137.38	33695	53155	11	9	0.11	0.036	6.7	−0.40	C+	4n
		5125.11	34040	53546	9	7	0.26	0.080	12	−0.14	D	9n
		5090.78	34329	53967	7	5	0.20	0.057	6.7	−0.40	C+	4n
		5104.44	34547	54133	5	3	0.017	0.0041	0.34	−1.69	C	5n

Fe I: Allowed transitions — Continued

No.	Multiplet	λ (Å)	E_i (cm^{-1})	E_k (cm^{-1})	g_i	g_k	A_{ki} (10^8 s^{-1})	f_{ik}	S (at. u.)	log gf	Accuracy	Source
637.	$y\,^5F° - f\,^5P$ (1091)											
		5228.41	34040	53160	9	7	0.018	0.0057	0.88	−1.29	C	5n
		5159.06	34547	53925	5	3	0.13	0.030	2.5	−0.82	D−	14n,15n
		5308.71	34329	53160	7	7	0.0011	4.5(−4)	0.055	−2.50	D	30
		5197.93	34692	53925	3	3	0.019	0.0076	0.39	−1.64	C	5n
638.	$y\,^5F° - f\,^5G$ (1092)											
		5133.69	33695	53169	11	13	0.27	0.13	23	0.14	D	9
		5104.21	33695	53282	11	11	0.0025	9.7(−4)	0.18	−1.97	C	5n
		5067.15	34040	53769	9	9	0.031	0.012	1.8	−0.97	C−	15n,30
		4986.90	34329	54376	7	5	0.0044	0.0012	0.13	−2.09	D	30
639.	$y\,^5F° - e\,^5H$ (1093)											
		5085.68	33695	53353?	11	13	4.9(−4)	2.2(−4)	0.041	−2.61	D	30
		5012.68	34547	54491	5	7	0.0062	0.0032	0.27	−1.79	C	5n
640.	$y\,^5F° - e\,^3G$ (1094)											
		4987.62	33695	53739	11	11	4.6(−4)	1.7(−4)	0.031	−2.72	D	30
		4991.86	34040	54067	9	9	0.0037	0.0014	0.20	−1.91	C	5n
		4985.98	34329	54379	7	7	0.0048	0.0018	0.21	−1.90	D	30
		5074.75	34040	53739	9	11	0.15	0.070	11	−0.20	C+	4n
641.	$y\,^5F° - f\,^3D$ (1095)											
		5023.23	34547	54449	5	3	0.022	0.0050	0.42	−1.60	C	5n
		5121.64	34547	54067	5	5	0.079	0.031	2.6	−0.81	C+	4n,5n
		5206.80	34547	53748	5	7	0.0010	5.9(−4)	0.051	−2.53	D	30
		5159.95	34692	54067	3	5	0.0012	8.2(−4)	0.042	−2.61	D	30
642.	$y\,^5F° - e\,^3H$ (1097)											
		4962.56	33695	53841?	11	13	0.011	0.0047	0.84	−1.29	C	5n
		4872.91	34040	54555?	9	9	0.0016	5.6(−4)	0.080	−2.30	D	30
643.	$y\,^5F° - f\,^3F$ (1098)											
		4799.06	34547	55379	5	5	9.8(−4)	3.4(−4)	0.027	−2.77	D	30
		4911.52	34329	54683	7	9	0.0018	8.2(−4)	0.093	−2.24	D	5n
644.	$y\,^5F° - i\,^5D$ (1102)											
		4256.79	34329	57814	7	7	0.014	0.0039	0.39	−1.56	D−	30
645.	$y\,^5F° - g\,^5G$ (1103)											
		4112.96	33695	58002	11	13	0.14	0.043	6.4	−0.33	C+	4n,5n
		4137.42	34547	58710?	5	7	0.061	0.022	1.5	−0.96	C	5n
		4142.63	34692	58825	3	5	0.074	0.032	1.3	−1.02	C	5n

Fe I: Allowed transitions — Continued

No.	Multiplet	λ (Å)	E_i (cm^{-1})	E_k (cm^{-1})	g_i	g_k	A_{ki} (10^8 s^{-1})	f_{ik}	S (at. u.)	log gf	Accuracy	Source
646.	$z\ ^3P^\circ - e\ ^5F$ (1105)											
		7095.43	33947	48037	5	5	0.0025	0.0019	0.22	−2.02	D	5n
		7213.84	34363	48221	3	3	9.8(−4)	7.6(−4)	0.054	−2.64	D	30
647.	$z\ ^3P^\circ - e\ ^5S$ (1106)											
		5955.68	34363	51149	3	5	1.8(−4)	1.6(−4)	0.0094	−3.32	D	30
648.	$z\ ^3P^\circ - e\ ^3D$ (1107)											
		5762.99	33947	51294	5	7	0.10	0.071	6.7	−0.45	C+	4n
		5753.12	34363	51740	3	5	0.070	0.058	3.3	−0.76	C+	4n
		5717.85	34556	52040	1	3	0.050	0.074	1.4	−1.13	C	5n
		5618.65	33947	51740	5	5	0.018	0.0083	0.77	−1.38	C	5n
649.	$z\ ^3P^\circ - g\ ^5D$ (1108)											
		5608.98	33947	51771	5	7	0.0012	8.0(−4)	0.074	−2.40	D	30
		5652.32	34363	52050	3	5	0.0047	0.0037	0.21	−1.95	C	5n
		5661.36	34556	52214	1	3	0.0066	0.0095	0.18	−2.02	D	5n
		5522.46	33947	52050	5	5	0.012	0.0056	0.51	−1.55	C	5n
		5472.72	33947	52214	5	3	0.014	0.0038	0.34	−1.72	C	5n
650.	$z\ ^3P^\circ - e\ ^5P$ (1109)											
		5646.70	34363	52067	3	5	0.0013	0.0011	0.059	−2.50	D	30
		5724.45	34556	52020	1	3	0.0016	0.0023	0.043	−2.64	D	30
		5517.08	33947	52067	5	5	0.0019	8.5(−4)	0.077	−2.37	D	5n
		5661.97	34363	52020	3	3	0.0013	6.2(−4)	0.035	−2.73	D	30
651.	$z\ ^3P^\circ - g\ ^5F$ (1110)											
		5027.76	33947	53831	5	7	0.021	0.011	0.93	−1.25	C	5n
		5025.08	34363	54258	3	5	0.0054	0.0034	0.17	−1.99	D	30
		4992.80	34363	54386	3	3	0.0040	0.0015	0.074	−2.35	D	15n,30
652.	$z\ ^3P^\circ - h\ ^5D$ (1111)											
		4993.68	33947	53967	5	5	0.018	0.0068	0.56	−1.47	C	5n
		5056.86	34363	54133	3	3	0.0095	0.0037	0.18	−1.96	C	5n
653.	$z\ ^3P^\circ - f\ ^5P$ (1112)											
		5004.04	33947	53925	5	3	0.035	0.0080	0.66	−1.40	C	5n
654.	$z\ ^3P^\circ - f\ ^5G$ (1113)											
		4945.64	33947	54161	5	7	0.012	0.0062	0.50	−1.51	C	5n
		4995.41	34363	54376	3	5	0.0069	0.0043	0.21	−1.89	C	5n
655.	$z\ ^3P^\circ - f\ ^3D$											
		5025.30	34556	54449	1	3	0.0080	0.0091	0.15	−2.04	D	30

Fe I: Allowed transitions — Continued

No.	Multiplet	λ (Å)	E_i (cm^{-1})	E_k (cm^{-1})	g_i	g_k	A_{ki} (10^8 s^{-1})	f_{ik}	S (at. u.)	log gf	Accuracy	Source
656.	$z\,^3P° - e\,^3P$ (1115)											
		4872.69	34363	54880	3	5	0.0017	0.0010	0.050	−2.51	D	30
		4801.63	34556	55376	1	3	0.0022	0.0022	0.035	−2.65	D	30
657.	$z\,^3P° - i\,^5D$ (1116)											
		4160.78	33947	57974	5	5	0.0084	0.0022	0.15	−1.96	D−	30
658.	$b\,^1D2 - {}^1D°$	7820.80	34637	47420	5	5	5.0(−4)	4.6(−4)	0.059	−2.64	D	30
659.	$b\,^1D2 - z\,^1D°$ (1122)	6736.56	34637	49477	5	5	3.1(−4)	2.1(−4)	0.023	−2.98	D	30
660.	$b\,^1D2 - v\,^3F°$ (1125)											
		6035.34	34637	51201	5	5	9.4(−4)	5.1(−4)	0.051	−2.59	D	30
661.	$b\,^1D2 - u\,^3G°$ (1127)											
		5816.07	34637	51826	5	7	0.0014	0.0010	0.096	−2.30	D	30
662.	$b\,^1D2 - y\,^1D°$ (1128)	5856.08	34637	51708	5	5	0.0089	0.0046	0.44	−1.64	C	5n
663.	$b\,^1D2 - x\,^1D°$ (1129)	5837.71	34637	51762	5	5	0.0018	9.1(−4)	0.088	−2.34	D	5n
664.	$b\,^1D2 - {}^1P°$	5698.37	34637	52181	5	3	0.0048	0.0014	0.13	−2.15	D	5n
665.	$b\,^1D2 - {}^3D°$											
		5539.83	34637	52683	5	5	0.0013	5.9(−4)	0.054	−2.53	D	5n
666.	$b\,^1D2 - {}^3P°$											
		5376.85	34637	53230	5	3	0.0038	9.8(−4)	0.087	−2.31	D	5n
667.	$b\,^1D2 - w\,^1D°$ (1133)	4734.10	34637	55754	5	5	0.016	0.0053	0.41	−1.58	C	5n
668.	$z\,^5G° - e\,^3F$ (1137)											
		7869.65	35257	47961	9	9	0.0016	0.0015	0.34	−1.88	D	30
		7737.67	35612	48532	7	7	2.8(−4)	2.5(−4)	0.045	−2.75	D	30
669.	$z\,^5G° - e\,^5G$ (1140)											
		6303.46	34844	50704	13	11	3.3(−4)	1.7(−4)	0.045	−2.66	D	30
670.	$z\,^5G° - g\,^5D$ (1142)											
		6034.04	34782	51350	11	9	7.7(−4)	3.5(−4)	0.076	−2.42	D	30
		6054.10	35257	51771	9	7	0.0013	5.4(−4)	0.098	−2.31	D	30

Fe I: Allowed transitions — Continued

No.	Multiplet	λ (Å)	E_i (cm^{-1})	E_k (cm^{-1})	g_i	g_k	A_{ki} (10^8 s^{-1})	f_{ik}	S (at. u.)	log gf	Accuracy	Source
671.	$z\,^5G° - g\,^5F$ (1143)											
		5361.64	35612	54258	7	5	0.017	0.0053	0.66	−1.43	C	5n
		5395.25	35856	54386	5	3	0.0052	0.0014	0.12	−2.17	D	5n
		5512.28	35257	53394	9	9	0.0093	0.0042	0.69	−1.42	C	5n
		5487.16	35612	53831	7	7	0.0093	0.0042	0.53	−1.53	C	5n
		5432.95	35856	54258	5	5	0.041	0.018	1.6	−1.04	C	5n
672.	$z\,^5G° - h\,^5D$ (1144)											
		5441.32	34782	53155	11	9	0.0047	0.0017	0.33	−1.73	C	5n
		5466.39	35257	53546	9	7	0.075	0.026	4.2	−0.63	D−	14n
		5470.17	35856	54133	5	3	0.012	0.0031	0.28	−1.81	C	5n
673.	$z\,^5G° - f\,^5G$ (1145)											
		5400.50	35257	53769	9	9	0.18	0.077	12	−0.16	D	9n
		5389.48	35612	54161	7	7	0.13	0.056	6.9	−0.41	D−	14n
		5398.29	35856	54376	5	5	0.098	0.043	3.8	−0.67	C	5n
		5422.15	34844	53282	13	11	0.0011	4.2(−4)	0.098	−2.26	D	30
		5546.51	35257	53282	9	11	0.0097	0.0054	0.89	−1.31	C	5n
		5461.54	35856	54161	5	7	0.0040	0.0025	0.23	−1.90	C	5n
674.	$z\,^5G° - e\,^5H$ (1146)											
		5424.07	34844	53275?	13	15	0.50	0.25	59	0.52	D	9n
		5383.37	34782	53353?	11	13	0.56	0.29	57	0.50	C+	4n,5n
		5369.96	35257	53874?	9	11	0.47	0.25	40	0.35	C+	4n
		5367.47	35612	54237	7	9	0.58	0.32	40	0.35	C+	4n
		5364.87	35856	54491	5	7	0.55	0.33	29	0.22	D	9n
		5401.27	34844	53353?	13	13	0.0021	9.2(−4)	0.21	−1.92	C	5n
		5236.38	34782	53874?	11	11	8.2(−4)	3.4(−4)	0.064	−2.43	D	30
		5267.28	35257	54237	9	9	0.0045	0.0019	0.29	−1.77	D	30
		5295.32	35612	54491	7	7	0.0069	0.0029	0.36	−1.69	C	5n
675.	$z\,^5G° - e\,^3G$ (1147)											
		5315.07	35257	54067	9	9	0.0074	0.0031	0.49	−1.55	C	5n
		5326.79	35612	54379	7	7	0.0027	0.0011	0.14	−2.10	D	30
		5409.13	35257	53739	9	11	0.010	0.0056	0.89	−1.30	C	5n
676.	$z\,^5G° - f\,^3D$ (1148)											
		5406.77	35257	53748	9	7	0.0062	0.0021	0.34	−1.72	C	5n
		5417.03	35612	54067	7	5	0.0095	0.0030	0.37	−1.68	C	5n
		5489.85	35856	54067	5	5	0.0025	0.0012	0.10	−2.24	D	5n
677.	$z\,^5G° - e\,^3H$ (1149)											
		5262.61	34844	53841?	13	13	9.7(−4)	4.0(−4)	0.091	−2.28	D	30
		5056.00	34782	54555?	11	9	0.0028	8.9(−4)	0.16	−2.01	D	5n

Fe I: Allowed transitions — Continued

No.	Multiplet	λ (Å)	E_i (cm^{-1})	E_k (cm^{-1})	g_i	g_k	A_{ki} (10^8 s^{-1})	f_{ik}	S (at. u.)	log gf	Accuracy	Source
678.	$z\,^5G° - f\,^3F$ (1150)											
		5023.50	34782	54683	11	9	0.0057	0.0018	0.32	-1.71	C	5n
		5031.90	35257	55125	9	7	0.0080	0.0024	0.35	-1.67	C	5n
		5146.30	35257	54683	9	9	0.0026	0.0010	0.16	-2.03	D	5n
		5241.90	35612	54683	7	9	0.0058	0.0031	0.37	-1.67	C	5n
679.	$z\,^5G° - 3$ (1152)											
		4631.48	35257	56843	9	9	0.0034	0.0011	0.15	-2.01	D	5n
680.	$z\,^3G° - g\,^5F$ (1159)											
		5653.89	35379	53061	11	11	0.0043	0.0021	0.43	-1.64	C	5n
		5499.60	36079	54258	7	5	9.4(-4)	3.1(-4)	0.039	-2.67	D	30
681.	$z\,^3G° - h\,^5D$ (1160)											
		5624.06	35379	53155	11	9	0.0078	0.0030	0.61	-1.48	C	5n
		5749.65	35768	53155	9	9	2.6(-4)	1.3(-4)	0.022	-2.94	D	30
682.	$z\,^3G° - f\,^5G$ (1161)											
		5619.60	35379	53169	11	13	0.0032	0.0018	0.37	-1.70	C	5n
		5708.11	35768	53282	9	11	0.0050	0.0030	0.51	-1.57	C	5n
		5651.47	36079	53769	7	9	0.0023	0.0014	0.19	-2.00	D	5n
		5553.59	35768	53769	9	9	0.0093	0.0043	0.71	-1.41	C	5n
		5528.89	36079	54161	7	7	0.0030	0.0014	0.17	-2.02	D	5n
		5436.30	35379	53769	11	9	0.0072	0.0026	0.52	-1.54	C	5n
		5435.17	35768	54161	9	7	0.0021	7.2(-4)	0.12	-2.19	D	30
683.	$z\,^3G° - e\,^5H$ (1162)											
		5521.28	35768	53874?	9	11	8.1(-4)	4.5(-4)	0.074	-2.39	D	30
		5412.80	35768	54237	9	9	0.0033	0.0014	0.23	-1.89	D	30
		5301.33	35379	54237	11	9	4.7(-4)	1.6(-4)	0.031	-2.75	D	30
684.	$z\,^3G° - e\,^3G$ (1163)											
		5445.04	35379	53739	11	11	0.20	0.087	17	-0.02	D	9n
		5463.27	35768	54067	9	9	0.32	0.14	23	0.11	C+	4n
		5349.74	35379	54067	11	9	0.013	0.0046	0.88	-1.30	C	5n
		5557.95	36079	54067	7	9	0.013	0.0075	0.96	-1.28	C	5n
685.	$z\,^3G° - f\,^3D$ (1164)											
		5560.23	35768	53748	9	7	0.020	0.0072	1.2	-1.19	C	5n

Fe I: Allowed transitions — Continued

No.	Multiplet	λ (Å)	E_i (cm^{-1})	E_k (cm^{-1})	g_i	g_k	A_{ki} (10^8 s^{-1})	f_{ik}	S (at. u.)	log gf	Accuracy	Source
686.	$z\ ^3G° - e\ ^3H$ (1165)											
		5415.20	35379	53841?	11	13	0.56	0.29	57	0.50	C+	4n,5n
		5410.91	36079	54555?	7	9	0.48	0.27	34	0.28	C+	4n,5n
		5293.03	35379	54267?	11	11	9.7(−4)	4.1(−4)	0.078	−2.35	D	5n
		5321.11	35768	54555?	9	9	0.0095	0.0040	0.64	−1.44	C	5n
		5213.35	35379	54555?	11	9	0.0016	5.2(−4)	0.099	−2.24	D	30
687.	$z\ ^3G° - f\ ^3F$ (1166)	*5178.3*	*35690*	*54996*	27	21	0.037	0.012	5.3	−0.51	C	5n
		5178.80	35379	54683	11	9	0.0040	0.0013	0.25	−1.84	C	5n
		5164.55	35768	55125	9	7	0.016	0.0049	0.74	−1.36	C	5n
		5180.07	36079	55379	7	5	0.027	0.0079	0.94	−1.26	C	5n
		5285.12	35768	54683	9	9	0.0061	0.0025	0.40	−1.64	C	5n
		5249.10	36079	55125	7	7	0.011	0.0047	0.57	−1.48	C	5n
		5373.71	36079	54683	7	9	0.035	0.020	2.4	−0.86	C	5n
688.	$y\ ^3F° - e\ ^3D$ (1173)											
		6843.67	36686	51294	9	7	0.024	0.013	2.6	−0.93	C	5n
		6858.16	37163	51740	7	5	0.025	0.012	2.0	−1.06	C	5n
		6885.77	37521	52040	5	3	0.020	0.0083	0.94	−1.38	C	5n
689.	$y\ ^3F° - g\ ^5D$ (1174)											
		6627.56	36686	51771	9	7	0.0045	0.0023	0.46	−1.68	C	5n
		6715.41	37163	52050	7	5	0.0068	0.0033	0.51	−1.64	C	5n
		6804.02	37521	52214	5	3	0.010	0.0043	0.48	−1.67	D	30
690.	$y\ ^3F° - g\ ^5F$ (1175)											
		6105.15	36686	53061	9	11	0.0014	9.9(−4)	0.18	−2.05	D	30
		6159.41	37163	53394	7	9	0.0021	0.0015	0.22	−1.97	D	30
		5927.80	37521	54386	5	3	0.051	0.016	1.6	−1.09	C	5n
691.	$y\ ^3F° - h\ ^5D$ (1176)											
		6079.02	37521	53967	5	5	0.027	0.015	1.5	−1.12	C	5n
		5929.70	36686	53546	9	7	0.011	0.0043	0.76	−1.41	C	5n
692.	$y\ ^3F° - f\ ^5P$ (1177)											
		6093.66	37163	53569	7	5	0.011	0.0045	0.63	−1.50	C	5n
		6094.42	37521	53925	5	3	0.0069	0.0023	0.23	−1.94	C	5n
693.	$y\ ^3F° - f\ ^5G$ (1178)											
		6024.07	36686	53282	9	11	0.13	0.084	15	−0.12	D−	14n
		6020.17	37163	53769	7	9	0.11	0.077	11	−0.27	D−	14n
		5852.19	36686	53769	9	9	0.010	0.0052	0.90	−1.33	C	5n
		5881.28	37163	54161	7	7	0.0040	0.0021	0.28	−1.84	C	5n
		5720.89	36686	54161	9	7	0.0033	0.0012	0.21	−1.95	D	30
		5807.97	37163	54376	7	5	0.0013	4.8(−4)	0.065	−2.47	D	30

Fe I: Allowed transitions — Continued

No.	Multiplet	λ (Å)	E_i (cm^{-1})	E_k (cm^{-1})	g_i	g_k	A_{ki} (10^8 s^{-1})	f_{ik}	S (at. u.)	log gf	Accuracy	Source
694.	$y\ ^3F° - e\ ^5H$ (1179)											
		5816.36	36686	53874?	9	11	0.037	0.023	4.0	−0.68	D−	14n
		5855.13	37163	54237	7	9	0.0038	0.0025	0.33	−1.76	C	5n
		5696.10	36686	54237	9	9	0.0023	0.0011	0.19	−1.99	C	5n
695.	$y\ ^3F° - e\ ^3G$ (1180)											
		5930.17	37521	54379	5	7	0.16	0.12	11	−0.23	D−	14n
		5806.73	37163	54379	7	7	0.025	0.013	1.7	−1.05	C	5n
696.	$y\ ^3F° - f\ ^3D$ (1181)											
		5905.67	37521	54449	5	3	0.12	0.037	3.6	−0.73	C	5n
697.	$y\ ^3F° - e\ ^3H$ (1182)											
		5686.53	36686	54267?	9	11	0.044	0.026	4.4	−0.63	C	5n
		5747.95	37163	54555?	7	9	0.0083	0.0053	0.70	−1.43	C	5n
		5594.66	36686	54555?	9	9	0.030	0.014	2.3	−0.90	C	5n
698.	$y\ ^3F° - f\ ^3F$ (1183)											
		5554.89	36686	54683	9	9	0.087	0.040	6.6	−0.44	D−	14n
		5598.30	37521	55379	5	5	0.18	0.085	7.9	−0.37	D−	14n
		5421.85	36686	55125	9	7	0.0054	0.0018	0.30	−1.78	C	5n
		5705.99	37163	54683	7	9	0.067	0.042	5.5	−0.53	C	5n
		5679.02	37521	55125	5	7	0.036	0.024	2.2	−0.92	C	5n
699.	$y\ ^3F° - e\ ^3P$ (1184)											
		5642.75	37163	54880	7	5	0.0032	0.0011	0.14	−2.12	D	5n
		5759.27	37521	54880	5	5	0.0034	0.0017	0.16	−2.07	D	30
700.	$y\ ^5P° - e\ ^7F$ (1186)											
		6864.31	36767	51331	7	5	0.0014	6.8(−4)	0.11	−2.32	D	30
701.	$y\ ^5P° - f\ ^7D$ (1187)											
		7330.15	37410	51048	3	3	0.0039	0.0031	0.23	−2.03	D	30
		7024.65	36767	50999	7	5	0.012	0.0061	0.99	−1.37	C	5n
702.	$y\ ^5P° - f\ ^5D$ (1188)											
		7473.56	37158	50534	5	7	0.0023	0.0027	0.33	−1.87	D	30
		7421.60	37410	50880	3	3	0.0064	0.0053	0.39	−1.80	D	30
		7285.29	37158	50880	5	3	0.0084	0.0040	0.48	−1.70	D	30
		7366.37	37410	50981	3	1	0.015	0.0042	0.31	−1.90	D	30
703.	$y\ ^5P° - e\ ^7P$ (1189)											
		7093.09	36767	50861	7	5	0.0025	0.0014	0.22	−2.02	D	30

Fe I: Allowed transitions — Continued

No.	Multiplet	λ (Å)	E_i (cm^{-1})	E_k (cm^{-1})	g_i	g_k	A_{ki} (10^8 s^{-1})	f_{ik}	S (at. u.)	log gf	Accu-racy	Source
704.	$y\ ^5P° - e\ ^5G$ (1190)											
		7109.67	37158	51219	5	7	3.2(−4)	3.4(−4)	0.040	−2.77	D	30
705.	$y\ ^5P° - e\ ^7G$ (1191)											
		6862.48	36767	51335	7	9	0.0042	0.0038	0.61	−1.57	C	5n
706.	$y\ ^5P° - f\ ^5F$ (1192)											
		6737.98	36767	51604	7	7	0.0037	0.0025	0.39	−1.75	D	30
707.	$y\ ^5P° - e\ ^3D$ (1194)											
		6855.74	37158	51740	5	5	0.0043	0.0030	0.34	−1.82	D	30
		6833.24	37410	52040	3	3	0.0040	0.0028	0.19	−2.08	D	30
		7071.88	37158	51294	5	7	0.0038	0.0040	0.46	−1.70	D	30
708.	$y\ ^5P° - g\ ^5D$ (1195)											
		6841.35	37158	51771	5	7	0.036	0.036	4.0	−0.75	C	5n
		6828.61	37410	52050	3	5	0.034	0.040	2.7	−0.92	C	5n
		6752.72	37410	52214	3	3	0.021	0.015	0.97	−1.36	C	5n
		6639.72	37158	52214	5	3	0.0096	0.0038	0.42	−1.72	D	30
		6733.16	37410	52257	3	1	0.039	0.0088	0.58	−1.58	C	5n
709.	$y\ ^5P° - e\ ^7S$ (1196)											
		6753.45	36767	51570	7	7	0.0011	7.3(−4)	0.11	−2.29	D	30
		6936.48	37158	51570	5	7	0.0011	0.0011	0.13	−2.25	D	30
710.	$y\ ^5P° - e\ ^5P$ (1197)											
		6633.76	36767	51837	7	7	0.036	0.024	3.6	−0.78	C	5n
		6842.67	37410	52020	3	3	0.023	0.016	1.1	−1.32	C	5n
		6533.97	36767	52067	7	5	0.011	0.0050	0.75	−1.46	C	5n
		6810.28	37158	51837	5	7	0.016	0.015	1.7	−1.12	C	5n
		6820.43	37410	52067	3	5	0.014	0.016	1.1	−1.32	C	5n
711.	$y\ ^5P° - g\ ^5F$ (1198)											
		5933.80	37410	54258	3	5	0.0022	0.0020	0.12	−2.23	D	30
712.	$y\ ^5P° - f\ ^5P$ (1200)											
		6098.28	36767	53160	7	7	0.0034	0.0019	0.26	−1.88	D	30
713.	$y\ ^5P° - f\ ^5G$ (1201)											
		5880.00	36767	53769	7	9	0.0025	0.0016	0.22	−1.94	C	5n
		5879.49	37158	54161	5	7	0.0020	0.0014	0.14	−2.14	D	5n

Fe I: Allowed transitions — Continued

No.	Multiplet	λ (Å)	E_i (cm^{-1})	E_k (cm^{-1})	g_i	g_k	A_{ki} (10^8 s^{-1})	f_{ik}	S (at. u.)	log gf	Accuracy	Source
714.	$y\,^5P° - e\,^5H$ (1202)											
		5640.46	36767	54491	7	7	0.0056	0.0027	0.35	−1.73	C	5n
715.	$y\,^5P° - i\,^5D$ (1206)											
		4749.95	36767	57814	7	7	0.019	0.0065	0.71	−1.34	C	5n
		4802.53	37158	57974	5	5	0.0088	0.0030	0.24	−1.82	D	30
		4714.07	36767	57974	7	5	0.017	0.0041	0.45	−1.54	C	5n
716.	$y\,^5P° - 4$ (1207)											
		4661.53	36767	58213	7	5	0.033	0.0077	0.82	−1.27	C	5n
717.	$d\,^3F - x\,^3H°$ (1222)											
		6960.33	37046	51409	9	9	0.0016	0.0012	0.24	−1.98	D	30
718.	$d\,^3F - v\,^3F°$ (1221)											
		7010.36	36941	51201	5	5	0.0027	0.0020	0.23	−2.01	D	30
		6976.93	36976	51305	7	9	0.0022	0.0020	0.32	−1.85	D	30
719.	$d\,^3F - u\,^3G°$ (1225)											
		6804.27	36976	51668	7	9	0.0019	0.0017	0.27	−1.92	D	30
		6716.24	36941	51826	5	7	0.0025	0.0024	0.27	−1.92	D	30
		6837.00	37046	51668	9	9	0.0025	0.0017	0.35	−1.81	C	5n
		6732.06	36976	51826	7	7	0.0013	8.8(−4)	0.14	−2.21	D	30
720.	$d\,^3F - {}^1H°$											
		6854.82	37046	51630	9	11	0.0014	0.0012	0.24	−1.98	D	30
721.	$d\,^3F - y\,^1D°$ (1226)											
		6769.66	36941	51708	5	5	6.4(−4)	4.4(−4)	0.049	−2.66	D	30
722.	$d\,^3F - x\,^1D°$ (1227)											
		6761.07	36976	51762	7	5	8.6(−4)	4.2(−4)	0.066	−2.53	D	30
		6745.11	36941	51762	5	5	0.0020	0.0014	0.15	−2.16	D	30
723.	$d\,^3F - u\,^3D°$ (1228)											
		6699.14	37046	51969	9	7	0.0014	7.2(−4)	0.14	−2.19	D	30
		6667.73	36976	51969	7	7	0.0015	0.0010	0.16	−2.15	D	30
724.	$d\,^3F - {}^3D°$											
		6591.32	37046	52213	9	7	0.0019	9.5(−4)	0.18	−2.07	D	30

Fe I:　Allowed transitions — Continued

No.	Multiplet	λ (Å)	E_i (cm^{-1})	E_k (cm^{-1})	g_i	g_k	A_{ki} (10^8 s^{-1})	f_{ik}	S (at. u.)	$\log gf$	Accuracy	Source
725.	$d\ ^3F - t\ ^3G°$ (1234)											
		5902.52	37046	53983	9	11	0.0027	0.0017	0.30	-1.81	C	$5n$
		5661.03	36941	54600	5	7	0.0011	7.4(-4)	0.069	-2.43	D	30
		5672.28	36976	54600	7	7	4.7(-4)	2.3(-4)	0.030	-2.80	D	30
726.	$d\ ^3F - {}^3G°$											
		5438.04	37046	55430	9	11	5.2(-4)	2.8(-4)	0.045	-2.60	D	30
727.	$d\ ^3F - (\ °)^b$											
		5300.41	37046	55907	9	11	0.0038	0.0020	0.31	-1.75	C	$5n$
728.	$d\ ^3F - {}^1F°$											
		5218.51	36941	56098	5	7	8.8(-4)	5.0(-4)	0.043	-2.60	D	30
729.	$d\ ^3F - u\ ^3F°$ (1242)											
		5019.18	36941	56859	5	5	0.0044	0.0017	0.14	-2.08	D	30
730.	$d\ ^3F - t\ ^3F°$ (1243)											
		4837.65	36976	57641	7	7	0.0013	4.4(-4)	0.049	-2.51	D	30
		4813.72	36941	57709	5	5	0.0021	7.4(-4)	0.059	-2.43	D	30
731.	$d\ ^3F - t\ ^3H°$ (1245)											
		4253.55	37046	60549	9	11	0.022	0.0072	0.90	-1.19	C	$5n$
		4203.67	36976	60758	7	9	0.086	0.029	2.8	-0.69	C	$5n$
732.	$y\ ^3D° - e\ ^3D$ (1250)											
		7481.93	38678	52040	5	3	0.0063	0.0032	0.39	-1.80	D	30
		7844.55	38996	51740	3	5	0.0034	0.0052	0.40	-1.81	D	30
733.	$y\ ^3D° - g\ ^5D$ (1251)											
		7353.53	38175	51771	7	7	0.0047	0.0038	0.65	-1.57	D	30
		7476.40	38678	52050	5	5	0.0050	0.0042	0.51	-1.68	D	30
734.	$y\ ^3D° - g\ ^5F$ (1253)											
		6569.23	38175	53394	7	9	0.065	0.054	8.2	-0.42	C	$5n$
		6597.61	38678	53831	5	7	0.019	0.017	1.8	-1.07	C	$5n$
		6385.74	38175	53831	7	7	0.0029	0.0018	0.26	-1.91	D	30
		6495.78	38996	54386	3	3	0.060	0.038	2.5	-0.94	C	$5n$
		6364.38	38678	54386	5	3	0.020	0.0074	0.78	-1.43	C	$5n$
735.	$y\ ^3D° - h\ ^5D$ (1254)											
		6330.86	38175	53967	7	5	0.0061	0.0026	0.38	-1.74	C	$5n$

Fe I: Allowed transitions — Continued

No.	Multiplet	λ (Å)	E_i (cm^{-1})	E_k (cm^{-1})	g_i	g_k	A_{ki} (10^8 s^{-1})	f_{ik}	S (at. u.)	log gf	Accuracy	Source
736.	$y\ ^3$D° – $f\ ^5$P (1255)											
		6713.76	38678	53569	5	5	0.0074	0.0050	0.56	−1.60	C	5n
		6696.32	38996	53925	3	3	0.011	0.0071	0.47	−1.67	D	30
737.	$y\ ^3$D° – $f\ ^5$G (1256)											
		6253.82	38175	54161	7	7	0.0053	0.0031	0.45	−1.66	D	30
738.	$y\ ^3$D° – $f\ ^3$D (1258)											
		6419.98	38175	53748	7	7	0.13	0.082	12	−0.24	C	5n
		6496.46	38678	54067	5	5	0.085	0.054	5.8	−0.57	C	5n
		6469.21	38996	54449	3	3	0.090	0.057	3.6	−0.77	C	5n
		6338.90	38678	54449	5	3	0.048	0.017	1.8	−1.06	C	5n
		6634.10	38678	53748	5	7	0.0080	0.0074	0.81	−1.43	C	5n
		6633.44	38996	54067	3	5	0.0098	0.011	0.71	−1.49	C	5n
739.	$y\ ^3$D° – $f\ ^3$F (1259)											
		6055.99	38175	54683	7	9	0.070	0.050	6.9	−0.46	D−	14n
		5898.21	38175	55125	7	7	0.0041	0.0021	0.29	−1.83	C	5n
740.	$y\ ^3$D° – $e\ ^3$P (1260)											
		6170.49	38678	54880	5	5	0.13	0.073	7.4	−0.44	D−	14n
741.	$x\ ^5$D° – $g\ ^5$F (1273)											
		6997.13	39970	54258	7	5	0.0014	7.2(−4)	0.12	−2.30	D	30
742.	$x\ ^5$D° – $h\ ^5$D (1274)											
		7582.15	39970	53155	7	9	0.0023	0.0025	0.44	−1.75	D	30
743.	$x\ ^5$D° – $f\ ^5$G (1276)											
		7044.60	39970	54161	7	7	0.0044	0.0033	0.53	−1.64	D	30
		7068.02	40231	54376	5	5	0.0028	0.0021	0.24	−1.98	D	30
744.	$x\ ^5$D° – $f\ ^3$D (1278)											
		7079.32	39626	53748	9	7	0.0016	9.5(−4)	0.20	−2.07	D	30
		7256.14	39970	53748	7	7	0.0047	0.0037	0.61	−1.59	D	30
		7118.10	40405	54449	3	3	0.012	0.0090	0.63	−1.57	D	30
		7396.50	40231	53748	5	7	0.0040	0.0046	0.56	−1.64	D	30
745.	$x\ ^5$D° – $f\ ^3$F (1279)											
		6794.60	39970	54683	7	9	0.0012	0.0011	0.17	−2.11	D	30
		6712.44	40231	55125	5	7	0.0015	0.0014	0.15	−2.16	D	30

Fe I: Allowed transitions — Continued

No.	Multiplet	λ (Å)	E_i (cm⁻¹)	E_k (cm⁻¹)	g_i	g_k	A_{ki} (10^8 s⁻¹)	f_{ik}	S (at. u.)	log gf	Accuracy	Source
746.	$x\ ^5D° - e\ ^3P$ (1280)											
		6824.80	40231	54880	5	5	0.0021	0.0015	0.17	−2.13	D	30
747.	$x\ ^5D° - i\ ^5D$ (1281)											
		5531.95	39626	57698	9	9	0.0059	0.0027	0.45	−1.61	C	5n
		5496.57	39626	57814	9	7	0.0059	0.0021	0.34	−1.73	D	30
		5552.70	39970	57974	7	5	0.0044	0.0015	0.19	−1.99	C	5n
748.	$x\ ^5D° - 4$ (1282)											
		5559.64	40231	58213	5	5	0.0064	0.0030	0.27	−1.83	C	5n
749.	$y\ ^7P° - i\ ^5D$ (1290)											
		5678.04	40207	57814	7	7	7.1(−4)	3.4(−4)	0.045	−2.62	D	30
750.	$x\ ^5F° - g\ ^5F$ (1303)											
		7551.10	41018	54258	5	5	0.0055	0.0047	0.58	−1.63	D	30
751.	$x\ ^5F° - f\ ^5G$ (1306)											
		7879.75	40594	53282	9	11	0.0022	0.0025	0.58	−1.65	D	30
		7588.30	40594	53769	9	9	0.0011	9.7(−4)	0.22	−2.06	D	30
		7484.28	41018	54376	5	5	0.0048	0.0040	0.49	−1.70	D	30
752.	$x\ ^5F° - e\ ^5H$ (1307)											
		7463.38	40842	54237	7	9	0.0025	0.0027	0.47	−1.72	D	30
		7420.20	41018	54491	5	7	0.0016	0.0018	0.22	−2.04	D	30
753.	$x\ ^5F° - f\ ^3D$ (1309)											
		7443.26	41018	54449	5	3	0.0047	0.0023	0.29	−1.93	D	30
754.	$x\ ^5F° - e\ ^3H$ (1310)											
		7359.95	40257	53841?	11	13	0.0013	0.0013	0.34	−1.85	D	30
		7312.05	40594	54267?	9	11	0.0015	0.0015	0.32	−1.87	D	30
755.	$x\ ^5F° - i\ ^5D$ (1313)											
		5732.29	40257	57698	11	9	0.0062	0.0025	0.52	−1.56	C	5n
		5805.76	40594	57814	9	7	0.0073	0.0029	0.49	−1.59	C	5n
		5845.27	40594	57698	9	9	0.0033	0.0017	0.29	−1.82	D	30

Fe I: Allowed transitions — Continued

No.	Multiplet	λ (Å)	E_i (cm^{-1})	E_k (cm^{-1})	g_i	g_k	A_{ki} (10^8 s^{-1})	f_{ik}	S (at. u.)	log gf	Accuracy	Source
756.	$x\ ^5F° - g\ ^5G$ (1314)											
		5633.97	40257	58002	11	13	0.087	0.049	10	−0.27	C	5n
		5655.18	40842	58520?	7	9	0.053	0.033	4.3	−0.64	C	5n
		5650.71	41018	58710?	5	7	0.033	0.022	2.0	−0.96	C	5n
		5650.01	41131	58825	3	5	0.050	0.040	2.2	−0.92	C	5n
		5549.66	40257	58271?	11	11	0.0040	0.0019	0.37	−1.69	C	5n
		5577.03	40594	58520?	9	9	0.0067	0.0031	0.52	−1.55	D	30
		5595.06	40842	58710?	7	7	0.0052	0.0024	0.31	−1.77	D	30

[a]The number in parentheses following the tabulated value indicates the power of ten by which this value has to be multiplied.
[b]The *LS*-coupling designation of the upper term of this multiplet was not provided in the NBS energy level compilation [22], so we have accordingly omitted it from this work.

Fe I

Forbidden Transitions

List of tabulated lines

Wavelength (Å)	No.	Wavelength (Å)	No.	Wavelength (Å)	No.	Wavelength (Å)	No.
3403.65	9	5156.3	13	5936.99	2	10056.0	10
3452.54	9	5160.6	4	5968.87	2	10075.0	23
3454.34	9	5220.56	3	5999.99	2	10229.8	10
3487.23	9	5290.75	12	6231.27	18	10235.2	22
3489.07	9	5303.99	3	6393.72	18	10262.8	10
3511.64	9	5304.06	12	6760.61	11	10264.7	14
3516.17	9	5356.32	3	6836.94	11	10315.0	23
3527.33	9	5363.91	12	6884.50	11	10318.7	14
3812.07	8	5382.26	3	6972.07	11	10916.6	24
3856.98	8	5412.97	12	7005.23	11	11202.1	24
3917.64	8	5427.17	12	7008.89	11	11237.0	20
3931.50	8	5439.72	3	7016.21	17	11524.5	20
4153.72	7	5477.40	12	7316.44	17	11764.2	20
4458.57	6	5565.68	3	7321.23	17	12124	21
4494.57	6	5639.55	2	7439.58	17	13207	1
4510.63	6	5656.39	3	7935.32	16	13419	1
4789.19	5	5696.36	2	8321.51	16	13552	1
4843.34	5	5708.96	2	9093.67	22	13672	1
4847.58	5	5715.94	3	9106.17	22	13731	1
4886.56	5	5745.49	3	9203.8	15	13759	1
4916.26	5	5775.05	2	9411.9	10	13954	19
4942.95	5	5804.45	2	9659.0	10	14072	25
4956.35	5	5834.64	2	9801.9	10	14430	1
4961.18	5	5867.17	2	9826.83	10	14586	21
5052.1	4	5872.77	2	9974.41	14	14657	25
5111.8	13	5934.41	2	9998.31	10		

For this spectrum, we have selected the calculations of Grevesse et al.[1] These authors employed analytic wavefunctions calculated in a Thomas-Fermi potential. These wavefunctions were then mixed in configuration interaction. Spin-orbit interaction was then used to derive empirical parameters for intermediate coupling by fitting to observed energy levels. The reliability of the M1 transition probabilities calculated in this way depend upon configuration mixing but in general are expected to be accurate to within 40 – 80 percent.

The situation concerning the E2 lines is not as clear. Grevesse et al. employed limited configuration interaction but did not specify the choice of basis functions. They used a Racah representation (employed by Garstang[2,3]) to obtain angular factors but were obliged to use their own wavefunctions to calculate the radial matrix elements. For this complex a spectrum, it is likely that the Thomas-Fermi potential will lead to unreliable E2 transition probabilities, since cancellation effects due to different configuration mixing may not be accounted for. Therefore, the uncertainties for the E2 lines tabulated here are expected to be quite large.

In this compilation, we have tabulated data for lines of at least moderate strength (which comprise about half the lines listed in Ref. 1). For most lines, we have assigned only one type of radiation if this is the predominant contribution to the strength of the line, i.e. if A_{ki} (specific radiation) $> (0.99)A_{ki}$(total line). However, for some lines, it was necessary to tabulate both the magnetic dipole and the electric quadrupole contributions, when they are comparable in strength.

References

[1]N. Grevesse, H. Nussbaumer, and J. P. Swings, Mon. Not. R. Astron. Soc. **151**, 239 (1971).

[2]R. H. Garstang, Proc. Cambridge Philos. Soc. **53**, 214 (1957).

[3]R. H. Garstang, Proc. Cambridge Philos. Soc. **54**, 383 (1958).

Fe I: Forbidden transitions

No.	Multiplet	λ (Å)	E_i (cm^{-1})	E_k (cm^{-1})	g_i	g_k	Type of transition	A_{ki} (s^{-1})	S (at. u.)	Accuracy	Source
1.	$a\,^5D - a\,^5F$										
		[14430]	0.0	6928	9	11	E2	0.0020	8.2	E	1
		[13552]	0.0	7377	9	9	E2	0.0015	3.7	E	1
		[13672]	415.9	7728	7	7	E2	0.0017	3.4	E	1
		[13731]	704.0	7986	5	5	E2	0.0016	2.3	E	1
		[13759]	888.1	8155	3	3	E2	0.0019	1.7	E	1
		[13207]	415.9	7986	7	5	E2	0.0011	1.3	E	1
		[13419]	704.0	8155	5	3	E2	0.0015	1.2	E	1
2.	$a\,^5D - a\,^5P$ (2F)										
		5639.55	0.0	17727	9	5	E2	0.14	2.4	E	1
		5708.96	415.9	17927	7	3	E2	0.15	1.6	E	1
		5696.36	0.0	17550	9	7	E2	0.12	3.0	E	1
		5775.05	415.9	17727	7	5	M1	0.0096	3.4(−4)^a	E	1
		5804.45	704.0	17927	5	3	E2	0.088	1.0	E	1
		5834.64	415.9	17550	7	7	E2	0.090	2.5	E	1
		5872.77	704.0	17727	5	5	E2	0.034	0.71	E	1
		5867.17	888.1	17927	3	3	E2	0.021	0.26	E	1
		5934.41	704.0	17550	5	7	E2	0.039	1.2	E	1
		5936.99	888.1	17727	3	5	E2	0.053	1.2	E	1
		5999.99	888.1	17550	3	7	E2	0.0085	0.28	E	1
		5968.87	978.1	17727	1	5	E2	0.024	0.54	E	1
3.	$a\,^5D - a\,^3P2$ (3F)										
		5439.72	0.0	18378	9	5	E2	0.0053	0.075	E	1
		5565.68	415.9	18378	7	5	M1	0.36	0.012	E	1
		5303.99	704.0	19552	5	3	M1	0.46	0.0076	E	1
		5220.56	888.1	20038	3	1	M1	0.57	0.0030	E	1
		5656.39	704.0	18378	5	5	E2	0.0018	0.031	E	1
		5356.32	888.1	19552	3	3	E2	0.0010	0.0079	E	1
		5715.94	888.1	18378	3	5	M1	0.034	0.0012	E	1
		"	"	"	3	5	E2	0.0038	0.069	E	1
		5382.26	978.1	19552	1	3	M1	0.079	0.0014	E	1
		5745.49	978.1	18378	1	5	E2	0.0020	0.037	E	1

Fe I: Forbidden transitions — Continued

No.	Multiplet	λ (Å)	E_i (cm^{-1})	E_k (cm^{-1})	g_i	g_k	Type of transition	A_{ki} (s^{-1})	S (at. u.)	Accuracy	Source
4.	$a\,^5D - a\,^3H$										
		[5160.6]	415.9	19788	7	9	M1	0.0015	6.9(−5)	E	1
		[5052.1]	0.0	19788	9	9	M1	0.0082	3.5(−4)	E	1
5.	$a\,^5D - b\,^3F2$ (4F)										
		4843.34	0.0	20641	9	9	M1	0.42	0.016	E	1
		4886.56	415.9	20874	7	7	M1	0.23	0.0070	E	1
		4916.26	704.0	21039	5	5	M1	0.092	0.0020	E	1
		4789.19	0.0	20874	9	7	M1	0.039	0.0011	E	1
		4847.58	415.9	21039	7	5	M1	0.025	5.3(−4)	E	1
		4942.95	415.9	20641	7	9	M1	0.077	0.0031	E	1
		4956.35	704.0	20874	5	7	M1	0.079	0.0025	E	1
		4961.18	888.1	21039	3	5	M1	0.045	0.0010	E	1
6.	$a\,^5D - b\,^3P$ (6F)										
		4458.57	415.9	22838	7	5	M1	0.030	4.9(−4)	E	1
		"	"	"	7	5	E2	0.003	0.02	E	1
		4494.57	704.0	22947	5	3	M1	0.044	4.4(−4)	E	1
		"	"	"	5	3	E2	0.004	0.01	E	1
		4510.63	888.1	23052	3	1	M1	0.10	3.4(−4)	E	1
7.	$a\,^5D - c\,^3P$ (8F)										
		4153.72	704.0	24772	5	3	M1	0.016	1.3(−4)	E	1
8.	$a\,^5D - a\,^3D$ (9F)										
		3812.07	0.0	26225	9	7	M1	0.014	2.0(−4)	E	1
		"	"	"	9	7	E2	0.007	0.02	E	1
		3856.98	704.0	26624	5	5	M1	0.011	1.2(−4)	E	1
		3917.64	888.1	26406	3	3	M1	0.012	8.0(−5)	E	1
		"	"	"	3	3	E2	0.004	0.007	E	1
		3931.50	978.1	26406	1	3	M1	0.011	7.4(−5)	E	1
9.	$a\,^5D - b\,^3D$ (10F)										
		3403.65	0.0	29372	9	7	M1	0.18	0.0018	D	1
		3454.34	415.9	29357	7	5	M1	0.021	1.6(−4)	D	1
		3452.54	415.9	29372	7	7	M1	0.052	5.6(−4)	D	1
		3489.07	704.0	29357	5	5	M1	0.083	6.5(−4)	D	1
		3516.17	888.1	29320	3	3	M1	0.10	4.8(−4)	D	1
		3487.23	704.0	29372	5	7	M1	0.037	4.1(−4)	D	1
		"	"	"	5	7	E2	0.002	0.004	E	1
		3511.64	888.1	29357	3	5	M1	0.070	5.6(−4)	D	1
		3527.33	978.1	29320	1	3	M1	0.089	4.3(−4)	D	1

Fe I: Forbidden transitions — Continued

No.	Multiplet	λ (Å)	E_i (cm^{-1})	E_k (cm^{-1})	g_i	g_k	Type of transition	A_{ki} (s^{-1})	S (at. u.)	Accu-racy	Source
10.	$a\,^5F - a\,^5P$ (11F)										
		[9411.9]	6928	17550	11	7	E2	0.010	3.1	E	1
		[9659.0]	7377	17727	9	5	E2	0.0054	1.4	E	1
		[9801.9]	7728	17927	7	3	E2	0.0029	0.47	E	1
		9826.83	7377	17550	9	7	E2	0.0029	1.1	E	1
		9998.31	7728	17727	7	5	E2	0.0039	1.2	E	1
		10056.0	7986	17927	5	3	E2	0.0043	0.79	E	1
		10262.8	7986	17727	5	5	E2	0.016	0.54	E	1
		10229.8	8155	17927	3	3	E2	0.0035	0.70	E	1
11.	$a\,^5F - a\,^3G$ (15F)										
		6760.61	6928	21716	11	11	M1	0.13	0.016	E	1
		6836.94	7377	21999	9	9	M1	0.072	0.0077	E	1
		6884.50	7728	22249	7	7	M1	0.028	0.0024	E	1
		6972.07	7377	21716	9	11	M1	0.026	0.0036	E	1
		7005.23	7728	21999	7	9	M1	0.032	0.0037	E	1
		7008.89	7986	22249	5	7	M1	0.022	0.0020	E	1
12.	$a\,^5F - a\,^3D$ (20F)										
		5304.06	7377	26225	9	7	M1	0.18	0.0070	E	1
		5290.75	7728	26624	7	5	M1	0.22	0.0060	E	1
		5427.17	7986	26406	5	3	M1	0.17	0.0030	E	1
		5363.91	7986	26624	5	5	M1	0.020	5.7(−4)	E	1
		5477.40	8155	26406	3	3	M1	0.082	0.0015	E	1
		5412.97	8155	26624	3	5	M1	0.022	6.5(−4)	E	1
13.	$a\,^5F - a\,^1P$										
		[5111.8]	7986	27543	5	3	M1	0.023	3.4(−4)	E	1
		[5156.3]	8155	27543	3	3	M1	0.011	1.7(−4)	E	1
14.	$a\,^3F - a\,^3G$ (23F)										
		10264.7	11976	21716	9	11	M1	0.011	0.0049	E	1
		9974.41	11976	21999	9	9	M1	0.015	0.0050	E	1
		10318.7	12561	22249	7	7	M1	0.012	0.0034	E	1
15.	$a\,^3F - b\,^3P$										
		[9203.8]	11976	22838	9	5	E2	0.013	2.6	E	1
16.	$a\,^3F - a\,^1G$ (26F)										
		7935.32	11976	24575	9	9	M1	0.064	0.011	E	1
		8321.51	12561	24575	7	9	M1	0.041	0.0079	E	1
17.	$a\,^3F - a\,^3D$ (28F)										
		7016.21	11976	26225	9	7	M1	0.033	0.0030	E	1
		7439.58	12969	26406	5	3	M1	0.016	7.3(−4)	E	1
		7316.44	12561	26225	7	7	M1	0.020	0.0020	E	1
		7321.23	12696	26624	5	5	M1	0.011	8.0(−4)	E	1

Fe I: Forbidden transitions — Continued

No.	Multiplet	λ (Å)	E_i (cm^{-1})	E_k (cm^{-1})	g_i	g_k	Type of transition	A_{ki} (s^{-1})	S (at. u.)	Accu-racy	Source
18.	$a\ ^3F - a\ ^1D$ (29F)										
		6231.27	12561	28605	7	5	M1	0.17	0.0076	E	1
		6393.72	12969	28605	5	5	M1	0.093	0.0045	E	1
19.	$a\ ^5P - c\ ^3P$										
		[13954]	17927	25092	3	1	M1	0.020	0.0020	D	1
20.	$a\ ^5P - a\ ^3D$ (32F)										
		11524.5	17550	26225	7	7	M1	0.068	0.027	E	1
		11237.0	17727	26624	5	5	M1	0.025	0.0066	E	1
		11764.2	17727	26225	5	7	M1	0.015	0.0063	E	1
21.	$a\ ^3P2 - a\ ^3D$										
		[12124]	18378	26624	5	5	M1	0.020	0.0066	E	1
		[14586]	19552	26406	3	3	M1	0.014	0.0048	E	1
22.	$a\ ^3P2 - b\ ^3D$ (36F)										
		9093.67	18378	29372	5	7	M1	0.034	0.0066	E	1
		"	"	"	5	7	E2	0.003	0.8	E	1
		9106.17	18378	29357	5	5	M1	0.028	0.0039	E	1
		"	"	"	5	5	E2	0.002	0.4	E	1
		10235.2	19552	29320	3	3	M1	0.029	0.0035	E	1
		"	"	"	3	3	E2	0.001	0.2	E	1
23.	$a\ ^3H - a\ ^1I$ (38F)										
		10075.0	19390	29313	13	13	M1	0.079	0.039	E	1
		10315.0	19621	29313	11	13	M1	0.052	0.028	E	1
24.	$b\ ^3F2 - b\ ^1G2$ (41F)										
		10916.6	20641	29799	9	9	M1	0.19	0.082	E	1
		11202.1	20874	29799	7	9	M1	0.092	0.043	E	1
25.	$a\ ^3G - a\ ^1H$										
		[14072]	21716	28820	11	11	M1	0.033	0.038	E	1
		[14657]	21999	28820	9	11	M1	0.015	0.019	E	1

[a]The number in parentheses following the tabulated value indicates the power of ten by which this value has to be multiplied.

Fe II

Mn Isoelectronic Sequence

Ground State: $1s^2 2s^2 2p^6 3s^2 3p^6 3d^6 4s\ ^6D_{9/2}$

Ionization Energy: 16.1879 eV = 130563 cm^{-1}

Allowed Transitions

List of tabulated lines

Wavelength (Å)	No.	Wavelength (Å)	No.	Wavelength (Å)	No.	Wavelength (Å)	No.
1144.94	6	2388.63	2	2535.49	54	2566.62	51
1635.40	17	2390.10	226	2536.67	96	2566.91	16
1641.76	17	2390.77	203	2537.14	188	2568.41	33
1647.16	17	2391.48	9	2538.20	158	2568.88	52
2208.41	189	2395.42	2	2538.39	55	2569.78	180
2213.66	43	2395.62	2	2538.50	40	2570.53	218
2218.27	189	2399.24	2	2538.80	38	2570.85	127
2249.18	5	2400.06	226	2538.91	38	2571.54	51
2250.18	4	2401.29	203	2538.99	38	2572.97	61
2250.93	4	2402.60	10	2539.80	53	2573.21	76
2253.12	4	2404.43	2	2540.52	180	2573.75	127
2260.08	4	2404.89	2	2541.10	56	2574.36	32
2260.86	4	2406.66	2	2541.84	38	2576.86	168
2262.69	5	2410.52	2	2542.32	8	2577.43	52
2267.58	4	2411.07	2	2542.73	85	2577.92	16
2268.65	5	2413.31	2	2543.38	39	2580.72	169
2279.92	4	2416.45	198	2543.43	56	2581.11	61
2327.40	3	2418.44	198	2544.97	35	2582.41	151
2331.31	9	2423.21	143	2545.22	39	2582.58	16
2332.80	3	2428.36	142	2545.44	115	2583.05	51
2338.01	3	2432.87	159	2545.51	55	2583.34	114
2343.49	3	2434.06	191	2546.67	54	2585.63	168
2343.96	9	2434.24	194	2547.34	38	2585.88	1
2344.28	3	2434.73	159	2548.33	34	2587.95	168
2348.11	10	2439.30	77	2548.59	38	2588.18	33
2348.30	3	2445.11	191	2548.74	33	2588.79	113
2351.67	192	2445.80	142	2548.92	158	2590.55	33
2352.31	192	2446.47	42	2549.08	127	2591.54	16
2353.68	192	2447.20	142	2549.40	56	2592.78	157
2354.89	9	2453.98	202	2549.46	56	2593.72	16
2360.00	9	2455.71	197	2549.77	114	2594.96	151
2360.29	10	2458.78	77	2550.03	95	2595.29	49
2362.02	9	2458.97	141	2550.15	188	2598.37	1
2363.86	192	2460.44	197	2550.58	38	2599.40	1
2364.83	3	2461.28	77	2550.68	95	2604.05	209
2365.77	195	2461.86	77	2551.21	170	2604.66	113
2366.59	9	2466.52	213	2554.95	76	2605.04	209
2368.60	10	2469.51	141	2555.07	54	2605.34	177
2369.95	192	2472.61	197	2555.45	56	2605.42	75
2370.50	9	2475.12	197	2557.08	38	2605.90	187
2373.74	2	2475.54	197	2557.51	52	2606.51	177
2375.19	10	2481.05	97	2559.27	114	2607.09	1
2379.27	10	2484.44	201	2559.77	76	2608.85	48
2380.76	3	2492.34	97	2559.92	115	2609.13	151
2382.04	2	2493.26	41	2560.28	84	2609.44	113
2382.36	9	2501.31	201	2561.58	76	2609.87	75
2382.90	25	2503.87	128	2562.09	84	2611.07	16
2383.25	10	2508.34	222	2562.54	16	2611.34	50
2384.39	10	2533.63	39	2563.48	16	2611.87	1
2385.01	9	2534.42	39	2566.22	209	2613.58	49
2388.37	25	2535.36	210	2566.40	210	2613.82	1

List of tabulated lines — Continued

Wavelength (Å)	No.	Wavelength (Å)	No.	Wavelength (Å)	No.	Wavelength (Å)	No.
2614.18	112	2712.39	72	2796.63	190	2888.10	79
2614.87	48	2714.41	15	2797.91	91	2892.82	13
2617.62	1	2716.22	109	2799.29	90	2894.78	88
2619.07	48	2716.43	175	2799.71	68	2895.22	139
2620.17	50	2716.56	225	2804.02	107	2897.27	102
2620.41	1	2716.70	14	2805.32	140	2902.32	104
2620.70	48	2717.87	224	2805.79	107	2902.46	122
2621.67	1	2718.64	221	2809.78	193	2906.12	79
2623.11	157	2719.30	174	2811.27	66	2907.85	12
2623.73	48	2721.81	71	2812.49	79	2910.76	122
2625.49	157	2722.06	108	2813.61	68	2916.15	12
2625.67	1	2722.74	219	2817.09	193	2917.09	173
2626.50	50	2724.88	14	2819.33	66	2917.47	13
2626.70	74	2726.51	109	2826.02	103	2922.02	138
2628.29	1	2727.38	70	2827.43	89	2926.59	12
2628.57	74	2727.54	15	2831.56	81	2934.49	122
2629.59	48	2728.91	108	2833.09	193	2939.51	12
2630.07	48	2730.73	14	2835.71	80	2944.40	23
2631.05	1	2732.01	93	2836.19	139	2945.26	12
2631.32	1	2732.44	7	2836.51	139	2947.66	23
2631.61	48	2732.94	221	2837.30	89	2949.18	121
2633.20	187	2736.50	83	2838.22	193	2953.77	12
2636.69	187	2739.55	15	2839.51	196	2954.05	101
2637.50	212	2741.40	108	2839.80	193	2959.60	102
2637.64	84	2743.20	14	2840.34	65	2959.84	208
2639.56	84	2746.48	14	2840.65	81	2961.27	12
2641.12	32	2746.98	15	2840.76	124	2964.13	100
2642.01	150	2749.18	15	2841.35	66	2964.63	23
2646.21	94	2749.32	14	2842.08	66	2965.04	23
2649.47	223	2749.49	15	2843.32	89	2965.40	99
2650.48	212	2750.00	71	2843.49	139	2968.74	101
2651.27	94	2753.29	92	2844.96	200	2969.93	121
2652.57	94	2754.91	190	2847.21	67	2970.52	12
2654.63	212	2755.73	14	2847.77	193	2970.68	120
2657.92	126	2756.51	70	2848.11	200	2975.94	12
2658.25	150	2757.03	69	2848.32	196	2979.36	12
2662.56	212	2759.34	7	2848.90	156	2980.96	101
2664.66	111	2761.81	15	2849.61	66	2982.06	172
2666.64	111	2762.34	190	2853.20	67	2984.82	23
2667.22	212	2762.44	71	2855.69	66	2985.55	23
2669.93	219	2763.66	227	2856.15	65	2997.30	172
2670.38	186	2763.91	71	2856.38	193	2998.86	100
2671.40	212	2764.79	68	2856.91	200	3000.06	120
2682.51	223	2765.13	228	2857.17	139	3002.33	149
2683.00	219	2767.50	92	2857.42	65	3002.65	23
2684.75	126	2768.93	15	2861.19	13	3021.41	99
2684.96	72	2769.15	70	2864.97	139	3036.96	211
2686.11	73	2769.36	68	2868.87	13	3044.84	149
2686.39	110	2771.18	125	2869.16	106	3048.99	211
2691.74	73	2771.55	67	2869.69	106	3056.80	155
2692.60	126	2772.72	15	2871.06	65	3062.23	154
2692.83	14	2774.69	82	2871.13	88	3065.32	148
2693.86	109	2775.34	7	2872.39	88	3070.69	119
2697.33	176	2776.18	69	2873.40	123	3071.12	211
2697.46	176	2776.91	190	2875.35	105	3076.44	211
2697.72	167	2779.30	91	2876.80	104	3077.17	154
2699.20	219	2780.04	179	2879.24	122	3078.68	211
2703.99	109	2783.69	91	2880.76	13	3089.39	185
2704.57	73	2784.28	140	2883.71	88	3096.30	148
2707.13	175	2785.19	190	2884.77	200	3105.17	137
2709.05	82	2787.24	193	2885.93	156	3105.55	137
2709.37	14	2790.56	125	2886.23	87	3106.56	119
2711.84	72	2793.89	68	2887.31	106	3114.30	137

List of tabulated lines — Continued

Wavelength (Å)	No.	Wavelength (Å)	No.	Wavelength (Å)	No.	Wavelength (Å)	No.
3114.68	137	3281.29	11	3945.21	18	4923.93	59
3116.59	137	3285.43	11	3974.16	31	4993.35	45
3131.72	153	3289.35	116	4024.55	171	5000.73	28
3133.05	137	3295.24	133	4075.95	27	5018.45	59
3135.36	137	3295.81	11	4087.27	29	5100.66	44
3144.75	137	3297.89	145	4122.64	29	5132.66	44
3146.75	118	3302.86	11	4124.79	26	5136.80	44
3154.20	117	3303.47	11	4128.74	30	5169.00	59
3162.80	166	3314.00	11	4173.45	30	5197.56	63
3163.09	21	3323.07	147	4178.86	29	5234.62	63
3166.67	22	3360.10	152	4180.97	181	5247.95	229
3167.86	117	3366.98	207	4233.17	30	5264.81	64
3170.34	22	3381.00	207	4258.16	29	5272.40	214
3177.54	137	3395.34	163	4273.32	30	5276.00	63
3179.50	184	3398.36	152	4296.57	29	5284.10	58
3180.16	184	3416.05	24	4303.17	30	5316.62	63
3183.12	21	3425.58	20	4351.76	30	5325.56	63
3185.32	21	3436.11	145	4369.40	29	5414.05	64
3186.74	22	3442.24	144	4384.33	37	5425.25	63
3187.29	166	3456.93	132	4385.38	30	5506.20	230
3192.06	117	3463.97	19	4413.60	37	5534.83	86
3192.92	22	3464.50	162	4416.82	30	5627.49	98
3193.81	22	3468.68	162	4472.92	46	5813.68	204
3196.08	21	3475.74	19	4489.19	46	5823.15	205
3210.45	22	3487.99	19	4491.40	46	5961.71	231
3213.31	22	3493.47	162	4508.28	47	5991.37	62
3227.73	22	3494.67	24	4515.34	46	6084.10	62
3231.70	135	3495.62	161	4520.23	46	6113.33	62
3232.79	165	3499.88	161	4522.63	47	6129.69	62
3237.40	136	3503.47	19	4534.17	46	6149.25	131
3237.82	136	3507.39	24	4541.52	47	6179.39	204
3241.69	135	3508.21	19	4549.21	215	6239.91	131
3243.72	165	3614.87	160	4549.47	47	6247.55	131
3247.18	136	3621.27	178	4555.89	46	6369.46	57
3247.39	165	3624.89	178	4576.33	47	6383.72	199
3249.66	136	3632.29	160	4582.84	46	6416.92	131
3255.88	11	3711.97	217	4583.83	47	6432.68	57
3257.36	146	3748.49	183	4620.51	47	6446.40	220
3258.77	134	3759.46	183	4629.34	46	6456.39	131
3259.05	136	3814.12	182	4635.33	215	6516.08	57
3266.94	116	3824.91	31	4656.97	60	7222.39	130
3267.04	135	3827.08	182	4666.75	46	7224.47	130
3268.51	164	3870.61	65	4670.17	28	7301.56	129
3269.77	164	3906.04	206	4720.15	78	7479.69	129
3273.50	164	3914.48	18	4731.44	60	7515.79	130
3276.61	147	3935.94	206	4833.19	36	7711.71	130
3277.35	11	3938.29	18	4840.00	36		
3279.65	164	3938.97	216	4893.83	45		

Since the publication of our last set of critically evaluated oscillator strengths for Fe II,[8] the data situation has improved considerably. Recent theoretical and experimental work has increased the quality and quantity of fairly reliable data. In addition, a better absolute scale is now available, because of a recent lifetime experiment which employed selective laser excitation.

For this compilation, we have chosen primarily experimental data from a variety of sources: emission data of Whaling,[1] Bridges,[2] Moity,[3] Baschek et al.,[4] and Kroll and Kock[17] (emission and hook measurements), as well as solar data from Blackwell et al.[5] and Kostyk and Orlova,[6] who derived oscillator strengths from equivalent widths of Fe II lines taken from the Liege solar atlas.[9] Furthermore, for some additional prominent lines, we have tabulated log gf-values from Kurucz,[7] who used a semiempirical, scaled Thomas-Fermi-Dirac approach which included configuration mixing.

Whaling[1] measured branching ratios using a hollow cathode discharge as a source, and a Fourier transform spectrometer for the spectral recordings. For lines arising from twelve different upper levels, he normalized his branching ratios to very accurate lifetimes measured by Hannaford and Lowe,[10] using the method of laser-excited fluorescence, which is estimated to be accurate to within ten percent. For several additional lines, Whaling normalized his branching ratios to the lifetime data of Brzozowski et al.,[11] who used the high-frequency-deflection, or pulsed-electron technique. Since these lifetimes were obtained by means of non-selective excitation, they are subject to cascading effects—and therefore not as reliable as those of Ref. 10. These factors were considered in Ref. 1; therefore, we have used Whaling's own error estimates in this compilation.

Another reference providing reliable oscillator strengths is the work of Kroll and Kock.[17] These authors measured oscillator strengths by a combined hook-emission technique. The "hook" data were taken by using a wall-stabilized arc in conjunction with a Michelson interferometer, while the emission data were measured with a hollow cathode discharge. These relative oscillator strengths (branching ratios) were then normalized to an absolute scale via lifetime data. The data of Kroll and Kock overlap with those of Whaling[1] for 58 lines. For strong lines (log $gf > -0.5$), we found excellent agreement between the two sources: for the 19 overlapping lines, Refs. 1 and 17 agreed within 12%. With the exception of four lines, the f-value data for the remaining weaker lines agreed within 50%. We estimate the data of Ref. 17 to be generally reliable to within ±25 percent.

The experiment by Bridges[2] is also a source of reliable f-value data. In this work, relative oscillator strengths for 14 lines were measured photoelectrically with a wall-stabilized arc, and effects of self-absorption were accounted for. Bridges used a calibrated tungsten strip lamp as a radiometric standard and a predisperser to reduce scattered light. He normalized his data to an absolute scale by using the phase-shift lifetime of the z $^6D^\circ_{9/2}$ level, as measured by Assousa and Smith.[12] This lifetime (3.9 ns) is in perfect agreement with that measured by Hannaford and Lowe[10]. We estimate that the data tabulated in Ref. 2 are accurate to within 25% or better, except for the weakest lines (log $gf < -2.00$).

In evaluating data for this spectrum, we found one procedure to be particularly valuable. This involved comparing the various data sources by graphical analysis. As in the case of Fe I, we plotted the difference, $\Delta \log = \log gf$ (source A) $- \log gf$ (source B), either versus log gf (source B), upper energy level, or wavelength. These plots enabled us to detect systematic trends or errors, differences in scale, and random scatter. As a result of our evaluation and analyses, we recommend that the data sources listed in Table 1 be renormalized as follows:

TABLE 1. Renormalization factors.

Reference	Normalization factor (logarithmic): number to be added to the original log gf-value, as it appeared in the literature.
3	−0.06 (if $E_k < 48000$ cm^{-1})
	−0.24 (if $E_k > 48000$ cm^{-1})
4	+0.16
6	−0.06

The experimental source providing by far the largest quantity of data is the paper by Moity.[3] In this work, Moity measured relative oscillator strengths photographically for 494 Fe II lines by using a wall-stabilized arc. He used a carbon arc as a reference light source in calibrating the spectral efficiency of the system. Moity placed all of his relative data on a common absolute scale by use of the Saha equation and Fe II/Fe I intensity ratios, where the Fe I f-values are precisely known.[8] As a result of our evaluation of Moity's work, we arrived at the following conclusions. In general, Moity's f-values for resonance lines arising from low-lying upper levels ($E_k < 40000$ cm^{-1}) are fairly reliable (accuracies vary from 25-50%), are slightly too high (by about 0.06 dex), and do not suffer from serious scatter. On comparing Moity's results to those of Whaling, however, we found that for lines arising from higher upper levels ($E_k > 40000$ cm^{-1}, and especially for $E_k > 60000$ cm^{-1}), there is appreciable scatter in Moity's data, and an energy-level-dependent trend as well as a possible wavelength-dependent trend is apparent. For such lines arising from high upper levels, Moity's log gf-values are consistently too large—by about 75%—suggesting a temperature error. A comparison with the theoretical data of Kurucz suggests similar problems with Moity's data.

Moity was aware that for high-lying upper levels, his f-values are systematically greater than those of Smith and Whaling.[13] However, he reasoned that Smith and Whaling's data were likely to be too small, since they had normalized their f-values to lifetimes that were affected by cascading problems. At this time, the cause of this disagreement is not settled.

The only theoretical data source selected is the work of Kurucz, who calculated f-values from scaled Thomas-Fermi-Dirac wavefunctions and used for his semiempirical approach the excellent, very complete lists of observed energy levels of Johansson[14] and Dobbie.[15] Our criterion for selecting data from Ref. 7 was to consider only strong (log $gf > -1.0$), observed lines having relatively pure upper and lower energy levels (purity of level $\geqslant$ 75%, as taken from the principal eigenvector components of Kurucz). We found that for this small, select group of lines, the data of Whaling[1] and Kurucz[7] agreed quite well. (For the 42 overlapping lines, 64% of the data agreed within 25 percent, and 88% of the data agreed within 50 percent.)

In our previous compilation,[8] we had included the data of Phillips,[16] who derived oscillator strengths from solar spectra. We have omitted Ref. 16 from this compilation because our new comparisons revealed severe scatter in Phillips' data. A possible explanation may rest with Phillips' choice of the model atmosphere and use of not too accurate equivalent width data. Finally, we have generally omitted blended lines from this compilation.

References

[1]W. Whaling, private communication.

[2]J. M. Bridges, "Contributed Papers—International Conference on Phenomena in Ionized Gases, 11th," 418 (Ed. I. Stoll, Czech. Acad. Sci., Inst. Phys., Prague, Czech., 1973).

[3]J. Moity, Astron. Astrophys., Suppl. Ser. **52**, 37 (1983).

[4]B. Baschek, T. Garz, H. Holweger, and J. Richter, Astron. Astrophys. **4**, 229 (1970).

[5]D. E. Blackwell, M. J. Shallis, and G. J. Simmons, Astron. Astrophys. **81**, 340 (1980).

[6]R. I. Kostyk, and T. V. Orlova, Astrometriya Astrofiz. **47**, 32 (1982).

[7]R. L. Kurucz, Smithsonian Astrophysical Observatory Special Report 390 (1981).

[8]J. R. Fuhr, G. A. Martin, W. L. Wiese, and S. M. Younger, J. Phys. Chem. Ref. Data **10**, 305 (1981).

[9]L. Delbouille, L. Neven, and C. Roland, "Photometric Atlas of the Solar Spectrum from 3000 to 10,000 Å," (Institut d'Astrophysique de l'Université de Liege, Observatoire Royal de Belgique, Liege, Belgique, 1973).

[10]P. Hannaford and R. M. Lowe, J. Phys. B **16**, L43 (1983).

[11]J. Brzozowski, P. Erman, M. Lyyra, and W. H. Smith, Phys. Scr. **14**, 48 (1976).

[12]G. E. Assousa and W. H. Smith, Astrophys. J. **176**, 259 (1972).

[13]P. L. Smith and W. Whaling, Astrophys. J. **183**, 313 (1973).

[14]S. Johansson, Phys. Scr. **18**, 217 (1978).

[15]J. C. Dobbie, Ann. Solar Phys. Obs. Cambridge **5**, 1 (1938).

[16]M. M. Phillips, Astrophys. J., Suppl. Ser. **39**, 377 (1979).

[17]S. Kroll and M. Kock, Astron. Astrophys., Suppl. Ser. **67**, 225 (1987).

Fe II: Allowed transitions

No.	Multiplet	λ (Å)	E_i (cm^{-1})	E_k (cm^{-1})	g_i	g_k	A_{ki} (10^8 s^{-1})	f_{ik}	S (at. u.)	log gf	Accuracy	Source
1.	$a\,^6D - z\,^6D°$ (uv 1)	*2610.6*	*416.3*	*38710*	30	30	2.5	0.26	66	0.89	B	1
		2599.40	0.0	38459	10	10	2.2	0.22	19	0.35	B	1
		2611.87	384.8	38660	8	8	1.1	0.11	7.7	−0.05	B	1
		2617.62	667.7	38859	6	6	0.44	0.045	2.3	−0.57	B	1
		2620.41	862.6	39013	4	4	0.036	0.0037	0.13	−1.83	B	1
		2621.67	977.1	39109	2	2	0.49	0.050	0.86	−1.00	B	1
		2585.88	0.0	38660	10	8	0.81	0.065	5.5	−0.19	B	1
		2598.37	384.8	38859	8	6	1.3	0.099	6.8	−0.10	B	1
		2607.09	667.7	39013	6	4	1.7	0.11	5.8	−0.17	B	1
		2613.82	862.6	39109	4	2	2.0	0.10	3.5	−0.39	B	1
		2625.67	384.8	38459	8	10	0.34	0.043	3.0	−0.46	B	1
		2631.32	667.7	38660	6	8	0.60	0.084	4.3	−0.30	B	1
		2631.05	862.6	38859	4	6	0.77	0.12	4.1	−0.32	B	1
		2628.29	977.1	39013	2	4	0.86	0.18	3.1	−0.45	B	1
2.	$a\,^6D - z\,^6F°$ (uv 2)											
		2382.04	0.0	41968	10	12	3.8	0.39	31	0.59	D	7
		2395.62	384.8	42115	8	10	2.5	0.27	17	0.33	B	1
		2404.89	667.7	42237	6	8	1.7	0.20	9.3	0.07	C+	17
		2410.52	862.6	42335	4	6	1.5	0.19	6.2	−0.11	C+	17
		2413.31	977.1	42401	2	4	1.1	0.19	3.0	−0.43	C+	17
		2373.74	0.0	42115	10	10	0.33	0.028	2.2	−0.55	B	1
		2388.63	384.8	42237	8	8	1.0	0.089	5.6	−0.15	C+	17
		2399.24	667.7	42335	6	6	1.4	0.12	5.6	−0.15	C+	17
		2406.66	862.6	42401	4	4	1.6	0.14	4.5	−0.25	C+	17
		2411.07	977.1	42440	2	2	2.4	0.21	3.3	−0.38	C+	17
		2395.42	667.7	42401	6	4	0.33	0.019	0.89	−0.95	C	17
		2404.43	862.6	42440	4	2	0.71	0.031	0.97	−0.91	C	17

Fe II: Allowed transitions — Continued

No.	Multiplet	λ (Å)	E_i (cm^{-1})	E_k (cm^{-1})	g_i	g_k	A_{ki} (10^8 s^{-1})	f_{ik}	S (at. u.)	log gf	Accuracy	Source
3.	$a\,^6D - z\,^6P°$ (uv 3)											
		2343.49	0.0	42658	10	8	1.7	0.11	8.5	0.04	C+	17
		2332.80	384.8	43239	8	6	1.5	0.091	5.6	−0.14	D	7
		2327.40	667.7	43621	6	4	0.59	0.032	1.5	−0.72	B	1
		2364.83	384.8	42658	8	8	0.61	0.051	3.2	−0.39	C+	17
		2348.30	667.7	43239	6	6	1.2	0.10	4.8	−0.21	D	7
		2338.01	862.6	43621	4	4	1.1	0.087	2.7	−0.46	B	1
		2380.76	667.7	42658	6	8	0.31	0.036	1.7	−0.67	C	17
		2344.28	977.1	43621	2	4	0.82	0.13	2.1	−0.57	B	1
4.	$a\,^6D - z\,^4F°$ (uv 4)											
		2260.08	0.0	44233	10	10	0.049	0.0037	0.28	−1.43	D	1
		2253.12	384.8	44754	8	8	0.051	0.0039	0.23	−1.51	D	1
		2250.93	667.7	45080	6	6	0.031	0.0024	0.11	−1.85	C	17
		2250.18	862.6	45290	4	4	0.015	0.0011	0.034	−2.34	C	17
		2279.92	384.8	44233	8	10	0.039	0.0038	0.23	−1.52	C	1
		2267.58	667.7	44754	6	8	0.031	0.0032	0.14	−1.72	D	1
		2260.86	862.6	45080	4	6	0.021	0.0024	0.073	−2.01	C	17
5.	$a\,^6D - z\,^4D°$ (uv 5)											
		2249.18	0.0	44447	10	8	0.041	0.0025	0.19	−1.60	C	17
		2262.69	862.6	45044	4	4	0.018	0.0014	0.042	−2.25	C	17
		2268.56	977.1	45044	2	4	0.0055	8.5(−4)[a]	0.013	−2.77	C	17
6.	$a\,^6D - y\,^6F°$ (uv 10)											
		1144.94	0.0	87341	10	12	4.8	0.11	4.2	0.05	D	7
7.	$a\,^4F - z\,^6D°$ (uv 32)											
		2732.44	1873	38459	10	10	9.1(−4)	1.0(−4)	0.0092	−2.99	C	17
		2759.34	2430	38660	8	8	5.0(−4)	5.7(−5)	0.0042	−3.34	D	3n
		2775.34	2838	38859	6	6	2.8(−4)	3.2(−5)	0.0017	−3.72	D	3n
8.	$a\,^4F - z\,^6F°$ (uv 33)											
		2542.32	3117	42440	4	2	0.0068	3.3(−4)	0.011	−2.88	D	3n
9.	$a\,^4F - z\,^4F°$ (uv 35)	*2363.5*	*2417*	*44714*	28	28	0.37	0.031	6.8	−0.06	C	1
		2360.00	1873	44233	10	10	0.24	0.020	1.6	−0.70	C	1
		2362.02	2430	44754	8	8	0.13	0.011	0.66	−1.07	C	1
		2366.59	2838	45080	6	6	0.099	0.0084	0.39	−1.30	B	1
		2370.50	3117	45290	4	4	0.14	0.012	0.37	−1.33	C+	1
		2331.31	1873	44754	10	8	0.29	0.019	1.5	−0.72	C	1
		2343.96	2430	45080	8	6	0.29	0.018	1.1	−0.85	B	1
		2354.89	2838	45290	6	4	0.24	0.013	0.62	−1.10	B	1
		2391.48	2430	44233	8	10	0.027	0.0029	0.18	−1.64	D	1
		2385.01	2838	44754	6	8	0.034	0.0038	0.18	−1.64	D	1
		2382.36	3117	45080	4	6	0.040	0.0051	0.16	−1.69	C+	1

Fe II: Allowed transitions — Continued

No.	Multiplet	λ (Å)	E_i (cm^{-1})	E_k (cm^{-1})	g_i	g_k	A_{ki} (10^8 s^{-1})	f_{ik}	S (at. u.)	log gf	Accuracy	Source
10.	$a\ ^4F - z\ ^4D°$ (uv 36)											
		2348.11	1873	44447	10	8	0.51	0.034	2.6	−0.47	C+	1
		2360.29	2430	44785	8	6	0.59	0.037	2.3	−0.53	C	17
		2368.60	2838	45044	6	4	0.59	0.033	1.6	−0.70	B	1
		2375.19	3117	45206	4	2	0.98	0.041	1.3	−0.78	D	7
		2379.27	2430	44447	8	8	0.15	0.013	0.78	−1.00	C+	1
		2383.25	2838	44785	6	6	0.34	0.029	1.4	−0.76	C	17
		2384.39	3117	45044	4	4	0.23	0.020	0.62	−1.10	C+	1
		2402.60	2838	44447	6	8	0.019	0.0022	0.10	−1.88	D	1
11.	$a\ ^4D - z\ ^6D°$ (1)											
		3277.35	7955	38459	8	10	0.0040	4.2(−4)	0.027	−2.47	C	1
		3302.86	8392	38660	6	8	2.4(−4)	5.2(−5)	0.0034	−3.51	C	17
		3314.00	8847	39013	2	4	1.8(−4)	5.9(−5)	0.0013	−3.93	D	3n
		3255.88	7955	38660	8	8	0.0025	4.0(−4)	0.034	−2.50	D	1
		3281.29	8392	38859	6	6	0.0021	3.4(−4)	0.022	−2.69	C	17
		3295.81	8680	39013	4	4	0.0019	3.2(−4)	0.014	−2.90	C	17
		3303.47	8847	39109	2	2	0.0061	0.0010	0.022	−2.70	C	17
12.	$a\ ^4D - z\ ^6F°$ (uv 60)											
		2926.59	7955	42115	8	10	0.046	0.0074	0.57	−1.23	C	1
		2953.77	8392	42237	6	8	0.047	0.0082	0.48	−1.31	C	17
		2970.52	8680	42335	4	6	0.028	0.0055	0.21	−1.66	C	17
		2979.36	8847	42401	2	4	0.018	0.0047	0.092	−2.03	C	17
		2916.15	7955	42237	8	8	4.3(−4)	5.5(−5)	0.0042	−3.36	D	3n
		2945.26	8392	42335	6	6	5.7(−4)	7.4(−5)	0.0043	−3.35	D	3n
		2975.94	8847	42440	2	2	0.0086	0.0011	0.022	−2.64	D	3n
		2907.85	7955	42335	8	6	0.0012	1.1(−4)	0.0087	−3.04	D	3n
		2939.51	8392	42401	6	4	0.0034	3.0(−4)	0.017	−2.75	D	3n
		2961.27	8680	42440	4	2	0.0078	5.1(−4)	0.020	−2.69	D	3n
13.	$a\ ^4D - z\ ^6P°$ (uv 61)											
		2880.76	7955	42658	8	8	0.022	0.0027	0.21	−1.66	C	17
		2868.87	8392	43239	6	6	0.0069	8.5(−4)	0.048	−2.29	D	3n
		2861.19	8680	43621	4	4	0.0019	2.3(−4)	0.0088	−3.03	D	3n
		2917.47	8392	42658	6	8	0.0015	2.6(−4)	0.015	−2.81	D	3n
		2892.82	8680	43239	4	6	0.0018	3.3(−4)	0.013	−2.88	D	3n
14.	$a\ ^4D - z\ ^4F°$ (uv 62)	*2746.9*	*8320*	*44714*	*20*	*28*	*2.1*	*0.33*	*60*	*0.82*	*B*	*1,3n*
		2755.73	7955	44233	8	10	2.1	0.30	22	0.38	B	1
		2749.32	8392	44754	6	8	2.1	0.32	17	0.28	B	1
		2746.48	8680	45080	4	6	1.9	0.33	12	0.12	B	1
		2743.20	8847	45290	2	4	1.8	0.41	7.3	−0.09	B	1
		2716.70	7955	44754	8	8	9.6(−4)	1.1(−4)	0.0076	−3.07	C	17
		2724.88	8392	45080	6	6	0.97	0.011	0.58	−1.19	B	1
		2730.73	8680	45290	4	4	0.25	0.028	1.0	−0.95	B	1
		2692.83	7955	45080	8	6	0.012	0.0010	0.072	−2.09	D	1
		2709.37	8392	45290	6	4	0.0026	1.9(−4)	0.010	−2.95	C	17

Fe II: Allowed transitions — Continued

No.	Multiplet	λ (Å)	E_i (cm^{-1})	E_k (cm^{-1})	g_i	g_k	A_{ki} (10^8 s^{-1})	f_{ik}	S (at. u.)	log gf	Accuracy	Source
15.	$a\ ^4D - z\ ^4D°$ (uv 63)											
		2739.55	7955	44447	8	8	1.9	0.22	16	0.24	B	1
		2746.98	8392	44785	6	6	1.6	0.18	9.7	0.03	C+	17
		2749.18	8680	45044	4	4	1.1	0.12	4.3	−0.32	B	1
		2749.49	8847	45206	2	2	1.1	0.12	2.2	−0.61	D	3n
		2714.41	7955	44785	8	6	0.55	0.045	3.2	−0.44	C+	17
		2727.54	8392	45044	6	4	0.85	0.063	3.4	−0.42	B	1
		2772.72	8392	44447	6	8	5.2(−4)	8.0(−5)	0.0044	−3.32	D	3n
		2768.93	8680	44785	4	6	0.045	0.0077	0.28	−1.51	C	17
		2761.81	8847	45044	2	4	0.11	0.026	0.48	−1.28	B	1
16.	$a\ ^4D - z\ ^4P°$ (uv 64)	*2570.1*	*8320*	*47218*	*20*	*12*	*2.2*	*0.13*	*22*	*0.42*	*C*	*1,2,3n,17*
		2562.54	7955	46967	8	6	1.5	0.11	7.5	−0.05	C+	17
		2563.48	8392	47390	6	4	1.3	0.085	4.3	−0.29	B	1
		2566.91	8680	47626	4	2	1.1	0.056	1.9	−0.65	D	3n
		2591.54	8392	46967	6	6	0.51	0.052	2.6	−0.51	C	2
		2582.58	8680	47390	4	4	0.77	0.077	2.6	−0.51	B	1
		2577.92	8847	47626	2	2	1.3	0.13	2.1	−0.60	D	3n
		2611.07	8680	46967	4	6	0.061	0.0093	0.32	−1.43	C	17
		2593.72	8847	47390	2	4	0.13	0.026	0.44	−1.29	B	1
17.	$a\ ^4D - x\ ^4P°$ (uv 68)											
		1635.40	7955	69102	8	6	2.4	0.072	3.1	−0.24	D	7
		1641.76	8392	69302	6	4	1.8	0.049	1.6	−0.53	D	7
		1647.16	8392	69102	6	6	0.52	0.021	0.68	−0.90	D	7
18.	$a\ ^4P - z\ ^6D°$ (3)											
		3938.29	13474	38859	6	6	9.2(−5)	2.1(−5)	0.0017	−3.89	D	3n
		3945.21	13673	39013	4	4	6.0(−5)	1.4(−5)	7.3(−4)	−4.25	D	3n
		3914.48	13474	39013	6	4	9.7(−5)	1.5(−5)	0.0011	−4.05	D	3n
19.	$a\ ^4P - z\ ^6F°$ (4)											
		3475.74	13474	42237	6	8	1.2(−4)	3.0(−5)	0.0020	−3.75	D	3n
		3487.99	13673	42335	4	6	1.2(−4)	3.2(−5)	0.0015	−3.89	D	3n
		3508.21	13905	42401	2	4	8.4(−5)	3.1(−5)	7.1(−4)	−4.21	D	3n
		3463.97	13474	42335	6	6	4.9(−5)	8.7(−6)	6.0(−4)	−4.28	D	3n
		3503.47	13905	42440	2	2	4.3(−4)	7.9(−5)	0.0018	−3.80	D	3n
20.	$a\ ^4P - z\ ^6P°$ (5)											
		3425.58	13474	42658	6	8	1.3(−4)	3.1(−5)	0.0021	−3.73	D	3n
21.	$a\ ^4P - z\ ^4F°$ (7)											
		3196.08	13474	44754	6	8	0.018	0.0036	0.23	−1.66	C+	1
		3183.12	13673	45080	4	6	0.010	0.0023	0.096	−2.04	C	1
		3185.32	13905	45290	2	4	0.0027	8.3(−4)	0.017	−2.78	C	17
		3163.09	13474	45080	6	6	0.0017	2.5(−4)	0.016	−2.82	C	17

Fe II: Allowed transitions — Continued

No.	Multiplet	λ (Å)	E_i (cm^{-1})	E_k (cm^{-1})	g_i	g_k	A_{ki} (10^8 s^{-1})	f_{ik}	S (at. u.)	log gf	Accuracy	Source
22.	$a\ ^4P - z\ ^4D°$ (6)	*3215.5*	*13612*	*44702*	12	20	0.065	0.0168	2.14	−0.69	C	1,2,3n,17
		3227.73	13474	44447	6	8	0.059	0.012	0.79	−1.13	B	1
		3213.31	13673	44785	4	6	0.063	0.015	0.62	−1.23	C	2
		3210.45	13905	45044	2	4	0.026	0.0081	0.17	−1.79	C	1
		3192.92	13474	44785	6	6	0.012	0.0019	0.12	−1.95	C	17
		3186.74	13673	45044	4	4	0.032	0.0049	0.20	−1.71	C	1
		3193.81	13905	45206	2	2	0.062	0.0095	0.20	−1.72	D	3n
		3166.67	13474	45044	6	4	0.0014	1.4(−4)	0.0087	−3.08	D	3n
		3170.34	13673	45206	4	2	0.0091	6.9(−4)	0.029	−2.56	D	3n
23.	$a\ ^4P - z\ ^4P°$ (uv 78)	*2974.8*	*13612*	*47218*	12	12	0.49	0.066	7.7	−0.10	C	1,3n,17
		2984.82	13474	46967	6	6	0.36	0.048	2.8	−0.54	C	17
		2965.04	13673	47390	4	4	0.060	0.0079	0.31	−1.50	C+	1
		2964.63	13905	47626	2	2	0.093	0.012	0.24	−1.61	D	3n
		2947.66	13474	47390	6	4	0.20	0.017	1.0	−0.98	B	1
		2944.40	13673	47626	4	2	0.46	0.030	1.2	−0.92	D	3n
		3002.65	13673	46967	4	6	0.14	0.029	1.2	−0.93	C	17
		2985.55	13905	47390	2	4	0.18	0.048	0.94	−1.02	B	1
24.	$a\ ^2P - z\ ^4P°$ (16)											
		3494.67	18361	46967	4	6	7.6(−4)	2.1(−4)	0.0096	−3.08	D	3n
		3507.39	18887	47390	2	4	6.3(−4)	2.3(−4)	0.0054	−3.33	D	3n
		3416.05	18361	47626	4	2	0.0032	2.8(−4)	0.013	−2.95	D	3n
25.	$a\ ^2H - z\ ^2I°$ (uv 117)											
		2382.90	20340	62293	12	14	0.22	0.021	2.0	−0.59	D	7
		2388.37	20806	62662	10	12	0.22	0.022	1.8	−0.65	D	7
26.	$a\ ^2D2 - z\ ^4F°$ (22)											
		4124.79	20517	44754	6	8	3.1(−5)	1.1(−5)	8.6(−4)	−4.20	D	5
27.	$a\ ^2D2 - z\ ^4D°$ (21)											
		4075.95	20517	45044	6	4	4.2(−4)	6.9(−5)	0.0056	−3.38	D	3n
28.	$b\ ^4P - z\ ^6F°$ (25)											
		4670.17	20831	42237	6	8	3.0(−5)	1.3(−5)	0.0012	−4.10	E	4n
		5000.73	22410	42401	2	4	1.2(−5)	9.1(−6)	3.0(−4)	−4.74	D	5
29.	$b\ ^4P - z\ ^4F°$ (28)											
		4178.86	20831	44754	6	8	0.0016	5.5(−4)	0.046	−2.48	C	17
		4296.57	21812	45080	4	6	5.9(−4)	2.4(−4)	0.014	−3.01	D	3n
		4369.40	22410	45290	2	4	1.9(−4)	1.1(−4)	0.0031	−3.67	D	3n
		4122.64	20831	45080	6	6	2.7(−4)	6.9(−5)	0.0057	−3.38	D	3n
		4258.16	21812	45290	4	4	3.7(−4)	1.0(−4)	0.0056	−3.40	D	3n
		4087.27	20831	45290	6	4	1.9(−5)	3.2(−6)	2.6(−4)	−4.71	D	6n

Fe II: Allowed transitions — Continued

No.	Multiplet	λ (Å)	E_i (cm^{-1})	E_k (cm^{-1})	g_i	g_k	A_{ki} (10^8 s^{-1})	f_{ik}	S (at. u.)	log gf	Accuracy	Source
30.	b ^{4}P – z ^{4}D° (27)	*4294.1*	*21421*	*44702*	12	20	0.0060	0.0028	0.47	−1.48	C−	1,3n,17
		4233.17	20831	44447	6	8	0.0047	0.0017	0.14	−2.00	C	1
		4351.76	21812	44785	4	6	0.0047	0.0020	0.011	−2.10	C	17
		4416.82	22410	45044	2	4	0.0021	0.0013	0.037	−2.60	D	3n
		4173.45	20831	44785	6	6	0.0042	0.0011	0.091	−2.18	C	17
		4303.17	21812	45044	4	4	0.0029	8.1(−4)	0.046	−2.49	C	17
		4385.38	22410	45206	2	2	0.0047	0.0013	0.039	−2.57	D	3n
		4128.74	20831	45044	6	4	1.7(−4)	2.8(−5)	0.0023	−3.77	D	3n
		4273.32	21812	45026	4	2	8.3(−4)	1.1(−4)	0.0064	−3.34	D	3n
31.	b ^{4}P – z ^{4}P° (29)											
		3824.91	20831	46967	6	6	3.0(−4)	6.5(−5)	0.0049	−3.41	D	3n
		3974.16	21812	46967	4	6	2.2(−4)	7.7(−5)	0.0040	−3.51	D	3n
32.	b ^{4}P – z ^{4}S° (uv 144)											
		2574.36	20831	59663	6	4	1.6	0.10	5.2	−0.21	D−	3n
		2641.12	21812	59663	4	4	0.054	0.0056	0.19	−1.65	D−	3n
33.	b ^{4}P – y ^{4}P° (uv 145)											
		2588.18	22410	61035	2	2	0.16	0.016	0.27	−1.50	D−	3n
		2548.74	21812	61035	4	2	1.7	0.081	2.7	−0.49	D−	3n
		2590.55	21812	60402	4	6	0.091	0.014	0.47	−1.26	D−	3n
		2568.41	22410	61333	2	4	0.44	0.087	1.5	−0.76	D−	3n
34.	b ^{4}P – z ^{4}G° (uv 146)											
		2548.33	21812	61042	4	6	0.20	0.029	0.96	−0.94	D−	3n
35.	b ^{4}P – z ^{2}D° (uv 147)											
		2544.97	21812	61093	4	6	0.40	0.059	2.0	−0.63	D−	3n
36.	a ^{4}H – z ^{6}F° (30)											
		4833.19	21430	42115	12	10	4.7(−6)	1.4(−6)	2.6(−4)	−4.78	D	6n
		4840.00	21582	42237	10	8	4.5(−6)	1.3(−6)	2.0(−4)	−4.90	D	6n
37.	a ^{4}H – z ^{4}F° (32)											
		4384.33	21430	44233	12	10	1.1(−4)	2.6(−5)	0.0046	−3.50	D	3n
		4413.60	21582	44233	10	10	4.6(−5)	1.3(−5)	0.0020	−3.87	D	5

Fe II: Allowed transitions — Continued

No.	Multiplet	λ (Å)	E_i (cm^{-1})	E_k (cm^{-1})	g_i	g_k	A_{ki} (10^8 s^{-1})	f_{ik}	S (at. u.)	log gf	Accuracy	Source
38.	$a\ ^4$H – $z\ ^4$G° (uv 158)											
		2538.99	21252	60625	14	12	1.2	0.10	12	0.16	D−	3n
		2538.80	21430	60807	12	10	0.82	0.066	6.6	−0.10	D−	3n
		2538.91	21582	60957	10	8	0.78	0.060	5.0	−0.22	D−	3n
		2541.84	21712	61042	8	6	0.77	0.056	3.7	−0.35	D−	3n
		2550.58	21430	60625	12	12	0.020	0.0020	0.20	−1.63	D−	3n
		2548.59	21582	60807	10	10	0.19	0.018	1.5	−0.74	D−	3n
		2547.34	21712	60957	8	8	0.20	0.019	1.3	−0.81	D−	3n
		2557.08	21712	60807	8	10	0.021	0.0026	0.17	−1.69	D−	3n
39.	$a\ ^4$H – $z\ ^4$H° (uv 159)											
		2533.63	21430	60888	12	12	1.3	0.13	13	0.18	D−	3n
		2534.42	21712	61157	8	8	1.2	0.11	7.4	−0.05	D−	3n
		2543.38	21582	60888	10	12	0.44	0.051	4.3	−0.29	D−	3n
		2545.22	21712	60989	8	10	0.33	0.040	2.7	−0.49	D−	3n
40.	$a\ ^4$H – $z\ ^2$D° (uv 160)											
		2538.50	21712	61093	8	6	0.33	0.024	1.6	−0.72	D−	3n
41.	$a\ ^4$H – $z\ ^4$I° (uv 161)											
		2493.26	21252	61348	14	16	3.4	0.37	42	0.71	D	7
42.	$a\ ^4$H – $z\ ^2$I° (uv 164)											
		2446.47	21430	62293	12	14	0.29	0.030	2.9	−0.44	D	7
43.	$a\ ^4$H – $y\ ^4$H° (uv 168)											
		2213.66	21252	66412	14	14	0.44	0.033	3.3	−0.34	D	7
44.	$b\ ^4$F – $z\ ^6$F° (35)											
		5132.66	22637	42115	10	10	1.7(−5)	6.6(−6)	0.0011	−4.18	D	5
		5100.66	22637	42237	10	8	1.4(−5)	4.3(−6)	7.2(−4)	−4.37	D	5
		5136.80	22939	42401	6	4	2.0(−5)	5.4(−6)	5.5(−4)	−4.49	D	5
45.	$b\ ^4$F – $z\ ^6$P° (36)											
		4993.35	22637	42658	10	8	7.5(−5)	2.2(−5)	0.0037	−3.65	E	4n
		4893.83	22810	43239	8	6	1.6(−5)	4.4(−6)	5.7(−4)	−4.45	D	6n

Fe II: Allowed transitions — Continued

No.	Multiplet	λ (Å)	E_i (cm^{-1})	E_k (cm^{-1})	g_i	g_k	A_{ki} (10^8 s^{-1})	f_{ik}	S (at. u.)	log gf	Accuracy	Source
46.	$b\ ^4F - z\ ^4F°$ (37)	*4563.5*	*22807*	*44714*	28	28	0.0023	7.1(−4)	0.30	−1.70	D	2,3n,17
		4629.34	22637	44233	10	10	0.0013	4.3(−4)	0.065	−2.37	D	2
		4555.89	22810	44754	8	8	0.0021	6.4(−4)	0.077	−2.29	D	3n
		4515.34	22939	45080	6	6	0.0018	5.5(−4)	0.049	−2.48	D	2
		4491.40	23031	45290	4	4	0.0017	5.0(−4)	0.030	−2.70	C	17
		4520.23	22637	44754	10	8	0.0010	2.5(−4)	0.037	−2.60	D	2
		4489.19	22810	45080	8	6	5.9(−4)	1.3(−4)	0.016	−2.97	D	3n
		4472.92	22939	45290	6	4	3.1(−4)	6.2(−5)	0.0055	−3.43	D	3n
		4666.75	22810	44233	8	10	1.4(−4)	5.8(−5)	0.0072	−3.33	D	3n
		4582.84	22939	44754	6	8	3.2(−4)	1.3(−4)	0.012	−3.10	C	17
		4534.17	23031	45080	4	6	1.8(−4)	8.5(−5)	0.0051	−3.47	D	3n
47.	$b\ ^4F - z\ ^4D°$ (38)											
		4583.83	22637	44447	10	8	0.0038	9.5(−4)	0.14	−2.02	D	1
		4549.47	22810	44785	8	6	0.0096	0.0022	0.27	−1.75	C	17
		4522.63	22939	45044	6	4	0.0076	0.0016	0.14	−2.03	C	17
		4508.28	23031	45206	4	2	0.010	0.0015	0.092	−2.21	D	2
		4620.51	22810	44447	8	8	2.0(−4)	6.6(−5)	0.0080	−3.28	D	3n
		4576.33	22939	44785	6	6	4.8(−4)	1.5(−4)	0.014	−3.04	D	3n
		4541.52	23031	45044	4	4	7.2(−4)	2.2(−4)	0.013	−3.05	D	3n
48.	$b\ ^4F - z\ ^4G°$ (uv 171)											
		2631.61	22637	60625	10	12	0.53	0.066	5.7	−0.18	D−	3n
		2629.59	22939	60957	6	8	0.62	0.085	4.4	−0.29	D−	3n
		2630.07	23031	61042	4	6	0.57	0.089	3.1	−0.45	D−	3n
		2619.07	22637	60807	10	10	0.27	0.028	2.4	−0.55	D−	3n
		2620.70	22810	60957	8	8	0.33	0.034	2.4	−0.56	D−	3n
		2623.73	22939	61042	6	6	0.22	0.023	1.2	−0.86	D−	3n
		2608.85	22637	60957	10	8	0.044	0.0036	0.31	−1.44	D−	3n
		2614.87	22810	61042	8	6	0.037	0.0029	0.20	−1.64	D−	3n
49.	$b\ ^4F - z\ ^4H°$ (uv 172)											
		2613.58	22637	60888	10	12	0.031	0.0038	0.33	−1.42	D−	3n
		2595.29	22637	61157	10	8	0.022	0.0017	0.15	−1.76	D−	3n
50.	$b\ ^4F - z\ ^2D°$ (uv 173)											
		2611.34	22810	61093	8	6	0.036	0.0027	0.19	−1.66	D−	3n
		2620.17	22939	61093	6	6	0.13	0.014	0.72	−1.08	D−	3n
		2626.50	23031	61093	4	6	0.34	0.052	1.8	−0.68	D−	3n
51.	$b\ ^4F - z\ ^4I°$ (uv 174)											
		2566.62	22637	61587	10	12	0.063	0.0074	0.63	−1.13	D−	3n
		2583.05	22810	61513	8	10	0.022	0.0027	0.19	−1.66	D−	3n
		2571.54	22637	61513	10	10	0.027	0.0026	0.22	−1.58	D−	3n

Fe II: Allowed transitions — Continued

No.	Multiplet	λ (Å)	E_i (cm^{-1})	E_k (cm^{-1})	g_i	g_k	A_{ki} (10^8 s^{-1})	f_{ik}	S (at. u.)	log gf	Accuracy	Source
52.	b ^{4}F – y ^{4}D° (uv 175)											
		2557.51	22637	61726	10	8	0.13	0.010	0.88	−0.98	D−	3n
		2568.88	22810	61726	8	8	0.040	0.0040	0.27	−1.50	D−	3n
		2577.43	22939	61726	6	8	0.0091	0.0012	0.061	−2.14	D−	3n
53.	b ^{4}F – y ^{6}P° (uv 176)											
		2539.80	22810	62172	8	8	0.025	0.0024	0.16	−1.72	D−	3n
54.	b ^{4}F – (°)^b											
		2535.49	22637	62066	10	8	0.54	0.042	3.5	−0.38	D−	3n
		2546.67	22810	62066	8	8	0.62	0.060	4.0	−0.32	D−	3n
		2555.07	22939	62066	6	8	0.18	0.024	1.2	−0.84	D−	3n
55.	b ^{4}F – z ^{2}G° (uv 178)											
		2545.51	22810	62083	8	10	0.026	0.0031	0.21	−1.60	D−	3n
		2538.39	22939	62322	6	8	0.028	0.0036	0.18	−1.66	D−	3n
56.	b ^{4}F – y ^{4}F° (uv 177)											
		2549.46	22939	62152	6	6	0.80	0.078	3.9	−0.33	D−	3n
		2549.40	23031	62245	4	4	1.3	0.13	4.3	−0.29	D−	3n
		2541.10	22810	62152	8	6	0.73	0.053	3.6	−0.37	D−	3n
		2543.43	22939	62245	6	4	0.71	0.046	2.3	−0.56	D−	3n
		2555.45	23031	62152	4	6	0.25	0.036	1.2	−0.84	D−	3n
57.	a ^{6}S – z ^{6}D° (40)											
		6516.08	23318	38660	6	8	7.0(−5)	5.9(−5)	0.0076	−3.45	D	5
		6432.68	23318	38859	6	6	4.9(−5)	3.0(−5)	0.0039	−3.74	D	5
		6369.46	23318	39013	6	4	1.8(−5)	7.3(−6)	9.2(−4)	−4.36	D	5
58.	a ^{6}S – z ^{6}F° (41)											
		5284.10	23318	42237	6	8	1.9(−4)	1.1(−4)	0.011	−3.19	D	5
59.	a ^{6}S – z ^{6}P° (42)	5062.4	23318	43066	6	18	0.036	0.041	4.1	−0.61	C	1,2,17
		5169.00	23318	42658	6	8	0.042	0.023	2.3	−0.87	C	17
		5018.45	23318	43239	6	6	0.027	0.010	1.0	−1.22	C	2
		4923.93	23318	43621	6	4	0.033	0.0080	0.78	−1.32	C	1
60.	a ^{6}S – z ^{4}D° (43)											
		4731.44	23318	44447	6	8	1.6(−4)	7.3(−5)	0.0068	−3.36	D	3n
		4656.97	23318	44785	6	6	1.2(−4)	3.9(−5)	0.0036	−3.63	E	4n

Fe II: Allowed transitions — Continued

No.	Multiplet	λ (Å)	E_i (cm^{-1})	E_k (cm^{-1})	g_i	g_k	A_{ki} (10^8 s^{-1})	f_{ik}	S (at. u.)	log gf	Accuracy	Source
61.	$a\ ^6S - y\ ^6P°$ (uv 190)											
		2572.97	23318	62172	6	8	0.037	0.0049	0.25	−1.53	D−	3n
		2581.11	23318	62049	6	6	0.033	0.0033	0.17	−1.70	D−	3n
62.	$a\ ^4G - z\ ^6F°$ (46)											
		6129.69	25805	42115	10	10	3.4(−6)	1.9(−6)	3.8(−4)	−4.72	D	6n
		5991.37	25429	42115	12	10	3.4(−5)	1.5(−5)	0.0036	−3.74	D	5
		6084.10	25805	42237	10	8	2.4(−5)	1.0(−5)	0.0021	−3.98	D	5
		6113.33	25982	42335	8	6	1.5(−5)	6.1(−6)	9.9(−4)	−4.31	D	5
63.	$a\ ^4G - z\ ^4F°$ (49)											
		5316.62	25429	44233	12	10	0.0033	0.0012	0.25	−1.85	C	17
		5276.00	25805	44754	10	8	0.0034	0.0012	0.20	−1.94	C	17
		5234.62	25982	45080	8	6	0.0036	0.0011	0.15	−2.05	C	17
		5197.56	26055	45290	6	4	0.0049	0.0013	0.14	−2.10	C	17
		5425.25	25805	44233	10	10	9.9(−5)	4.4(−5)	0.0078	−3.36	D	5
		5325.56	25982	44754	8	8	7.4(−4)	3.1(−4)	0.044	−2.60	C	17
64.	$a\ ^4G - z\ ^4D°$ (48)											
		5264.81	26055	45044	6	4	3.9(−4)	1.1(−4)	0.011	−3.19	D	5
		5414.05	25982	44447	8	8	4.6(−5)	2.0(−5)	0.0029	−3.79	D	5
65.	$a\ ^4G - z\ ^4G°$ (uv 195)											
		2840.34	25429	60625	12	12	0.056	0.0068	0.76	−1.09	D−	3n
		2856.15	25805	60807	10	10	0.038	0.0047	0.44	−1.33	D−	3n
		2857.42	26055	61042	6	6	0.028	0.0035	0.20	−1.68	D−	3n
		2871.06	25805	60625	10	12	0.021	0.0032	0.30	−1.50	D−	3n
		2870.61	25982	60807	8	10	0.0071	0.0011	0.082	−2.06	D−	3n
66.	$a\ ^4G - z\ ^4H°$ (uv 196)											
		2849.61	25805	60888	10	12	0.041	0.0060	0.57	−1.22	D−	3n
		2855.69	25982	60989	8	10	0.10	0.016	1.2	−0.90	D−	3n
		2819.33	25429	60888	12	12	0.0094	0.0011	0.13	−1.87	D−	3n
		2841.35	25805	60989	10	10	0.0039	4.7(−4)	0.044	−2.33	D−	3n
		2842.08	25982	61157	8	8	0.012	0.0014	0.11	−1.94	D−	3n
		2811.27	25429	60989	12	10	0.012	0.0012	0.13	−1.86	D−	3n
67.	$a\ ^4G - z\ ^2D°$ (uv 197)											
		2847.21	25982	61093	8	6	0.019	0.0017	0.13	−1.86	D−	3n
		2771.55	26055	62126	6	4	0.018	0.0014	0.074	−2.09	D−	3n
		2853.20	26055	61093	6	6	0.0072	8.7(−4)	0.049	−2.28	D−	3n

Fe II: Allowed transitions — Continued

No.	Multiplet	λ (Å)	E_i (cm^{-1})	E_k (cm^{-1})	g_i	g_k	A_{ki} (10^8 s^{-1})	f_{ik}	S (at. u.)	log gf	Accuracy	Source
68.	a ^{4}G – z ^{4}I° (uv 198)											
		2769.36	25429	61528	12	14	0.16	0.021	2.3	−0.59	D−	3n
		2793.89	25805	61587	10	12	0.096	0.013	1.2	−0.87	D−	3n
		2813.61	25982	61513	8	10	0.034	0.0051	0.38	−1.39	D−	3n
		2764.79	25429	61587	12	12	0.0082	9.4(−4)	0.10	−1.95	D−	3n
		2799.71	25805	61513	10	10	0.0035	4.2(−4)	0.038	−2.38	D−	3n
69.	a ^{4}G – (°)^b											
		2757.03	25805	62066	10	8	0.054	0.0049	0.44	−1.31	D−	3n
		2776.18	26055	62066	6	8	0.020	0.0031	0.17	−1.73	D−	3n
70.	a ^{4}G – z ^{2}G° (uv 200)											
		2727.38	25429	62083	12	10	0.32	0.030	3.2	−0.45	D−	3n
		2769.15	25982	62083	8	10	0.051	0.0074	0.54	−1.23	D−	3n
		2756.51	26055	62322	6	8	0.062	0.0094	0.51	−1.25	D−	3n
71.	a ^{4}G – y ^{4}F° (uv 199)											
		2721.81	25429	62158	12	10	0.033	0.0031	0.33	−1.43	D−	3n
		2763.91	25982	62152	8	6	0.032	0.0027	0.20	−1.66	D−	3n
		2762.44	26055	62245	6	4	0.031	0.0024	0.13	−1.85	D−	3n
		2750.00	25805	62158	10	10	0.012	0.0013	0.12	−1.87	D−	3n
72.	a ^{4}G – z ^{2}I° (uv 201)											
		2711.84	25429	62293	12	14	0.38	0.049	5.3	−0.23	D−	3n
		2712.39	25805	62662	10	12	0.13	0.017	1.6	−0.76	D−	3n
		2684.96	25429	62662	12	12	0.0097	0.0010	0.11	−1.90	D−	3n
73.	a ^{4}G – x ^{4}D° (uv 202)											
		2691.74	25805	62945	10	8	0.047	0.0041	0.36	−1.39	D−	3n
		2704.57	25982	62945	8	8	0.013	0.0014	0.10	−1.95	D−	3n
		2686.11	26055	63273	6	6	0.010	0.0011	0.058	−2.18	D−	3n
74.	a ^{4}G – y ^{4}G° (uv 203)											
		2626.70	25982	64041	8	8	0.019	0.0020	0.14	−1.80	D−	3n
		2628.57	26055	64087	6	6	0.063	0.0065	0.34	−1.41	D−	3n
75.	a ^{4}G – z ^{2}F° (uv 204)											
		2609.87	25982	64286	8	8	0.18	0.018	1.2	−0.84	D−	3n
		2605.42	26055	64425	6	6	0.26	0.026	1.4	−0.80	D−	3n
76.	a ^{4}G – y ^{2}G° (uv 205)											
		2561.58	25805	64832	10	10	0.010	0.0010	0.084	−2.00	D−	3n
		2554.95	25982	65110	8	8	0.022	0.0022	0.15	−1.76	D−	3n
		2573.21	25982	64832	8	10	0.14	0.018	1.2	−0.85	D−	3n
		2559.77	26055	65110	6	8	0.24	0.031	1.6	−0.73	D−	3n

Fe II: Allowed transitions — Continued

No.	Multiplet	λ (Å)	E_i (cm^{-1})	E_k (cm^{-1})	g_i	g_k	A_{ki} (10^8 s^{-1})	f_{ik}	S (at. u.)	log gf	Accuracy	Source
77.	$a\ ^4G - y\ ^4H°$ (uv 209)											
		2439.30	25429	66412	12	14	2.8	0.29	28	0.54	D	7
		2458.78	25805	66464	10	12	2.7	0.29	23	0.46	D	7
		2461.86	25982	66589	8	10	2.6	0.30	19	0.38	D	7
		2461.28	26055	66672	6	8	2.6	0.31	15	0.27	D	7
78.	$b\ ^2P - z\ ^4P°$ (54)											
		4720.15	25788	46967	4	6	8.9(−6)	4.4(−6)	2.8(−4)	−4.75	D	6n
79.	$b\ ^2P - y\ ^4P°$ (uv 215)											
		2888.10	25788	60402	4	6	0.036	0.0067	0.26	−1.57	D−	3n
		2906.12	26933	61333	2	4	0.029	0.0074	0.14	−1.83	D−	3n
		2812.49	25788	61333	4	4	0.024	0.0028	0.10	−1.95	D−	3n
80.	$b\ ^2P - z\ ^4G°$ (uv 216)											
		2835.71	25788	61042	4	6	0.31	0.056	2.1	−0.65	D−	3n
81.	$b\ ^2P - z\ ^2D°$ (uv 217)											
		2831.56	25788	61093	4	6	0.58	0.10	3.9	−0.38	D−	3n
		2840.65	26933	62126	2	4	0.53	0.13	2.4	−0.59	D−	3n
82.	$b\ ^2P - y\ ^4D°$ (uv 218)											
		2709.05	25788	62690	4	6	0.37	0.061	2.2	−0.61	D−	3n
		2774.69	26933	62962	2	4	0.24	0.055	1.0	−0.96	D−	3n
83.	$b\ ^2P - x\ ^4D°$ (uv 220)											
		2736.50	26933	63465	2	4	0.012	0.0026	0.047	−2.28	D−	3n
84.	$b\ ^2P - z\ ^2P°$ (uv 221)	*2586.2*	*26170*	*64825*	6	6	2.3	0.23	12	0.15	D−	3n
		2560.28	25788	64834	4	4	1.5	0.15	5.0	−0.23	D−	3n
		2639.56	26933	64806	2	2	1.1	0.11	1.9	−0.65	D−	3n
		2562.09	25788	64806	4	2	1.5	0.072	2.4	−0.54	D−	3n
		2637.64	26933	64834	2	4	0.83	0.17	3.0	−0.46	D−	3n
85.	$b\ ^2P - z\ ^2S°$ (uv 223)											
		2542.73	26933	66249	2	2	1.9	0.18	3.0	−0.44	D−	3n
86.	$b\ ^2H - z\ ^4F°$ (55)											
		5534.83	26170	44233	12	10	2.6(−4)	9.8(−5)	0.021	−2.93	D	5

J. Phys. Chem. Ref. Data, Vol. 17, Suppl. 4, 1988

Fe II: Allowed transitions — Continued

No.	Multiplet	λ (Å)	E_i (cm^{-1})	E_k (cm^{-1})	g_i	g_k	A_{ki} (10^8 s^{-1})	f_{ik}	S (at. u.)	log gf	Accuracy	Source
87.	$b\ ^2H - z\ ^4G°$ (uv 229)											
		2886.23	26170	60807	12	10	0.0055	5.8(−4)	0.066	−2.16	D−	3n
88.	$b\ ^2H - z\ ^4H°$ (uv 230)											
		2883.71	26170	60838	12	14	0.10	0.014	1.6	−0.76	D−	3n
		2894.78	26353	60888	10	12	0.040	0.0060	0.57	−1.22	D−	3n
		2871.13	26170	60989	12	10	0.027	0.0028	0.31	−1.48	D−	3n
		2872.39	26353	61157	10	8	0.15	0.014	1.4	−0.84	D−	3n
89.	$b\ ^2H - z\ ^4I°$ (uv 231)											
		2827.43	26170	61528	12	14	0.024	0.0034	0.38	−1.39	D−	3n
		2837.30	26353	61587	10	12	0.019	0.0027	0.25	−1.57	D−	3n
		2843.32	26353	61513	10	10	0.015	0.0018	0.17	−1.75	D−	3n
90.	$b\ ^2H - (\ °)^b$											
		2799.29	26353	62066	10	8	0.11	0.010	0.92	−1.00	D−	3n
91.	$b\ ^2H - z\ ^2G°$ (uv 234)	*2781.9*	*26253*	*62189*	22	18	0.73	0.069	14	0.18	D−	3n
		2783.69	26170	62083	12	10	0.70	0.068	7.4	−0.09	D−	3n
		2779.30	26353	62322	10	8	0.76	0.071	6.5	−0.15	D−	3n
		2797.91	26353	62083	10	10	0.028	0.0033	0.31	−1.48	D−	3n
92.	$b\ ^2H - z\ ^2I°$ (uv 235)											
		2767.50	26170	62293	12	14	1.9	0.26	28	0.49	D	7
		2753.29	26353	62662	10	12	1.2	0.16	15	0.21	D−	3n
93.	$b\ ^2H - x\ ^4D°$ (uv 236)											
		2732.01	26353	62945	10	8	0.056	0.0050	0.45	−1.30	D−	3n
94.	$b\ ^2H - y\ ^4G°$ (uv 237)											
		2651.27	26170	63876	12	12	0.0042	4.5(−4)	0.047	−2.27	D−	3n
		2646.21	26170	63949	12	10	0.014	0.0012	0.13	−1.84	D−	3n
		2652.57	26353	64041	10	8	0.046	0.0039	0.34	−1.41	D−	3n
95.	$b\ ^2H - z\ ^2H°$ (uv 240)											
		2550.68	26170	65364	12	12	0.89	0.087	8.8	0.02	D−	3n
		2550.03	26353	65556	10	10	1.2	0.12	10	0.08	D−	3n
96.	$b\ ^2H - x\ ^4G°$ (uv 241)											
		2536.67	26170	65580	12	12	0.40	0.039	3.9	−0.33	D−	3n

Fe II: Allowed transitions — Continued

No.	Multiplet	λ (Å)	E_i (cm^{-1})	E_k (cm^{-1})	g_i	g_k	A_{ki} (10^8 s^{-1})	f_{ik}	S (at. u.)	log gf	Accuracy	Source
97.	$b\ ^2$H – $y\ ^4$H° (uv 243)											
		2492.34	26353	66464	10	12	0.16	0.018	1.5	−0.74	D	7
		2481.05	26170	66464	12	12	0.19	0.017	1.7	−0.68	D	7
98.	$a\ ^2$F – $z\ ^4$F° (57)											
		5627.49	27315	45080	8	6	1.5(−5)	5.5(−6)	8.1(−4)	−4.36	D	5
99.	$a\ ^2$F – $y\ ^4$P° (uv 251)											
		3021.41	27315	60402	8	6	0.0022	2.2(−4)	0.018	−2.75	D−	3n
		2965.40	27620	61333	6	4	0.0087	7.6(−4)	0.045	−2.34	D−	3n
100.	$a\ ^2$F – $z\ ^4$G° (uv 252)											
		2998.86	27620	60957	6	8	0.0034	6.2(−4)	0.037	−2.43	D−	3n
		2964.13	27315	61042	8	6	0.049	0.0049	0.38	−1.41	D−	3n
101.	$a\ ^2$F – $z\ ^4$H° (uv 253)											
		2968.74	27315	60989	8	10	0.0027	3.6(−4)	0.028	−2.54	D−	3n
		2980.96	27620	61157	6	8	0.0092	0.0016	0.096	−2.01	D−	3n
		2954.05	27315	61157	8	8	0.0096	0.0013	0.097	−2.00	D−	3n
102.	$a\ ^2$F – $z\ ^2$D° (uv 254)											
		2959.60	27315	61093	8	6	0.064	0.0063	0.49	−1.30	D−	3n
		2897.27	27620	62126	6	4	0.14	0.012	0.69	−1.14	D−	3n
103.	$a\ ^2$F – $y\ ^4$D° (uv 255)											
		2826.02	27315	62690	8	6	0.039	0.0035	0.26	−1.55	D−	3n
104.	$a\ ^2$F – (°)^b											
		2902.32	27620	62066	6	8	0.0050	8.4(−4)	0.048	−2.30	D−	3n
		2876.80	27315	62066	8	8	0.059	0.0074	0.56	−1.23	D−	3n
105.	$a\ ^2$F – $z\ ^2$G° (uv 258)											
		2875.35	27315	62083	8	10	0.095	0.015	1.1	−0.93	D−	3n
106.	$a\ ^2$F – $y\ ^4$F° (uv 257)											
		2869.16	27315	62158	8	10	0.018	0.0028	0.21	−1.65	D−	3n
		2869.69	27315	62152	8	6	0.0072	6.7(−4)	0.051	−2.27	D−	3n
		2887.31	27620	62245	6	4	0.019	0.0016	0.089	−2.03	D−	3n

Fe II: Allowed transitions — Continued

No.	Multiplet	λ (Å)	E_i (cm^{-1})	E_k (cm^{-1})	g_i	g_k	A_{ki} (10^8 s^{-1})	f_{ik}	S (at. u.)	log gf	Accuracy	Source
107.	a ^{2}F – x ^{4}D° (uv 259)											
		2805.79	27315	62945	8	8	0.022	0.0026	0.19	−1.68	D−	3n
		2804.02	27620	63273	6	6	0.013	0.0015	0.084	−2.04	D−	3n
108.	a ^{2}F – y ^{4}G° (uv 260)											
		2728.91	27315	63949	8	10	0.088	0.012	0.88	−1.01	D−	3n
		2722.06	27315	64041	8	8	0.11	0.012	0.86	−1.02	D−	3n
		2741.40	27620	64087	6	6	0.17	0.020	1.1	−0.93	D−	3n
109.	a ^{2}F – z ^{2}F° (uv 261)	*2709.2*	*27446*	*64346*	14	14	1.2	0.14	17	0.28	D−	3n
		2703.99	27315	64286	8	8	1.2	0.13	9.5	0.03	D−	3n
		2716.22	27620	64425	6	6	1.1	0.12	6.5	−0.14	D−	3n
		2693.86	27315	64425	8	6	0.052	0.0042	0.30	−1.47	D−	3n
		2726.51	27620	64286	6	8	0.046	0.0068	0.37	−1.39	D−	3n
110.	a ^{2}F – z ^{2}P° (uv 262)											
		2686.39	27620	64834	6	4	0.023	0.0017	0.088	−2.00	D−	3n
111.	a ^{2}F – y ^{2}G° (uv 263)											
		2664.66	27315	64832	8	10	1.5	0.20	14	0.21	D−	3n
		2666.64	27620	65110	6	8	1.7	0.25	13	0.17	D−	3n
112.	a ^{2}F – z ^{2}H° (uv 264)											
		2614.18	27315	65556	8	10	0.062	0.0079	0.54	−1.20	D−	3n
113.	a ^{2}F – x ^{4}G° (uv 265)											
		2604.66	27315	65696	8	10	0.0092	0.0012	0.080	−2.03	D−	3n
		2609.44	27620	65931	6	8	0.077	0.011	0.54	−1.20	D−	3n
		2588.79	27315	65931	8	8	0.077	0.0077	0.53	−1.21	D−	3n
114.	a ^{2}F – x ^{4}F° (uv 266)											
		2583.34	27315	66013	8	10	0.0074	9.3(−4)	0.063	−2.13	D−	3n
		2559.27	27315	66377	8	8	0.067	0.0066	0.44	−1.28	D−	3n
		2549.77	27315	66522	8	6	0.25	0.018	1.2	−0.84	D−	3n
115.	a ^{2}F – y ^{4}H° (uv 267)											
		2545.44	27315	66589	8	10	0.14	0.017	1.1	−0.87	D−	3n
		2559.92	27620	66672	6	8	0.24	0.031	1.6	−0.73	D−	3n
116.	b ^{2}G – z ^{4}H° (65)											
		3266.94	30389	60989	10	10	0.0043	6.9(−4)	0.074	−2.16	D−	3n
		3289.35	30764	61157	8	8	0.018	0.0030	0.26	−1.62	D−	3n

Fe II: Allowed transitions — Continued

No.	Multiplet	λ (Å)	E_i (cm^{-1})	E_k (cm^{-1})	g_i	g_k	A_{ki} (10^8 s^{-1})	f_{ik}	S (at. u.)	log gf	Accuracy	Source
117.	$b\,^2G - z\,^2G°$ (66)											
		3154.20	30389	62083	10	10	0.15	0.022	2.3	−0.66	D−	3n
		3167.86	30764	62322	8	8	0.13	0.019	1.6	−0.82	D−	3n
		3192.06	30764	62083	8	10	0.0042	8.1(−4)	0.068	−2.19	D−	3n
118.	$b\,^2G - y\,^4F°$ (67)											
		3146.75	30389	62158	10	10	0.0010	1.5(−4)	0.016	−2.82	D−	3n
119.	$b\,^2G - x\,^4D°$ (68)											
		3070.69	30389	62945	10	8	0.0075	8.5(−4)	0.086	−2.07	D−	3n
		3106.56	30764	62945	8	8	0.012	0.0018	0.15	−1.84	D−	3n
120.	$b\,^2G - y\,^4G°$ (uv 276)											
		2970.68	30389	64041	10	8	0.022	0.0023	0.23	−1.63	D−	3n
		3000.06	30764	64087	8	6	0.029	0.0029	0.23	−1.63	D−	3n
121.	$b\,^2G - z\,^2F°$ (uv 277)											
		2949.18	30389	64286	10	8	0.20	0.021	2.0	−0.68	D−	3n
		2969.93	30764	64425	8	6	0.18	0.018	1.4	−0.85	D−	3n
122.	$b\,^2G - y\,^2G°$ (uv 278)	*2906.1*	*30556*	*64956*	18	18	0.030	0.0038	0.65	−1.17	D−	3n
		2902.46	30389	64832	10	10	0.022	0.0028	0.26	−1.56	D−	3n
		2910.76	30764	65110	8	8	0.0082	0.0010	0.080	−2.08	D−	3n
		2879.24	30389	65110	10	8	0.025	0.0025	0.24	−1.60	D−	3n
		2934.49	30764	64832	8	10	0.0055	8.8(−4)	0.068	−2.15	D−	3n
123.	$b\,^2G - z\,^2H°$ (uv 279)											
		2873.40	30764	65556	8	10	0.34	0.053	4.0	−0.37	D−	3n
124.	$b\,^2G - x\,^4G°$ (uv 280)											
		2840.76	30389	65580	10	12	0.11	0.017	1.6	−0.78	D−	3n
125.	$b\,^2G - y\,^4H°$ (uv 282)											
		2771.18	30389	66464	10	12	0.036	0.0050	0.46	−1.30	D−	3n
		2790.56	30764	66589	8	10	0.018	0.0026	0.19	−1.68	D−	3n
126.	$b\,^2G - y\,^2H°$ (uv 283)	*2688.8*	*30556*	*67736*	18	22	1.3	0.18	28	0.50	D−	3n
		2692.60	30389	67516	10	12	1.2	0.16	14	0.21	D−	3n
		2684.75	30764	68001	8	10	1.4	0.19	14	0.19	D−	3n
		2657.92	30389	68001	10	10	0.040	0.0043	0.37	−1.37	D−	3n

Fe II: Allowed transitions — Continued

No.	Multiplet	λ (Å)	E_i (cm^{-1})	E_k (cm^{-1})	g_i	g_k	A_{ki} (10^8 s^{-1})	f_{ik}	S (at. u.)	log gf	Accuracy	Source
127.	b ^{2}G – y ^{2}F° (uv 284)	2558.8	30556	69625	18	14	1.6	0.13	19	0.35	D–	3n
		2549.08	30389	69607	10	8	1.5	0.12	10	0.08	D–	3n
		2570.85	30764	69650	8	6	1.7	0.13	8.7	0.01	D–	3n
		2573.75	30764	69607	8	8	0.020	0.0020	0.13	−1.80	D–	3n
128.	b ^{2}G – x ^{2}G° (uv 285)											
		2503.87	30389	70315	10	10	2.4	0.23	19	0.36	D	7
129.	b ^{4}D – z ^{4}F° (72)											
		7479.69	31388	44754	6	8	2.0(−5)	2.2(−5)	0.0032	−3.88	D	5
		7301.56	31388	45080	6	6	2.7(−5)	2.2(−5)	0.0032	−3.88	D	5
130.	b ^{4}D – z ^{4}D° (73)											
		7711.71	31483	44447	8	8	2.6(−4)	2.3(−4)	0.046	−2.74	D	5
		7224.47	31368	45206	2	2	2.5(−4)	2.0(−4)	0.0095	−3.40	D	5
		7515.79	31483	44785	8	6	4.1(−5)	2.6(−5)	0.0052	−3.68	D	5
		7222.39	31364	45206	4	2	2.1(−4)	8.1(−5)	0.0077	−3.49	D	6n
131.	b ^{4}D – z ^{4}P° (74)											
		6456.39	31483	46967	8	6	0.0013	6.3(−4)	0.11	−2.30	D	5
		6247.55	31388	47390	6	4	0.0013	5.2(−4)	0.064	−2.51	D	5
		6416.92	31388	46967	6	6	3.8(−4)	2.4(−4)	0.030	−2.85	D	5
		6149.25	31368	47626	2	2	0.0011	6.0(−4)	0.024	−2.92	D	5
		6239.91	31368	47390	2	4	8.9(−5)	1.0(−4)	0.0043	−3.68	D	6n
132.	b ^{4}D – y ^{4}P° (76)											
		3456.93	31483	60402	8	6	0.0049	6.6(−4)	0.060	−2.28	D–	3n
133.	b ^{4}D – y ^{4}D° (79)											
		3295.24	31388	61726	6	8	0.0042	9.2(−4)	0.060	−2.26	D–	3n
134.	b ^{4}D – (°)^b											
		3258.77	31388	62066	6	8	0.052	0.011	0.71	−1.18	D–	3n
135.	b ^{4}D – z ^{2}G° (80)											
		3267.04	31483	62083	8	10	0.0020	4.0(−4)	0.034	−2.50	D–	3n
		3231.70	31388	62322	6	8	0.012	0.0025	0.16	−1.83	D–	3n
		3241.69	31483	62322	8	8	0.0021	3.4(−4)	0.029	−2.57	D–	3n

Fe II: Allowed transitions — Continued

No.	Multiplet	λ (Å)	E_i (cm⁻¹)	E_k (cm⁻¹)	g_i	g_k	A_{ki} (10⁸ s⁻¹)	f_{ik}	S (at. u.)	log gf	Accuracy	Source
136.	$b\ ^4D - y\ ^4F°$ (81)											
		3259.05	31483	62158	8	10	0.057	0.011	0.98	−1.04	D−	3n
		3247.18	31364	62152	4	6	0.064	0.015	0.64	−1.22	D−	3n
		3237.82	31368	62245	2	4	0.066	0.021	0.44	−1.38	D−	3n
		3249.66	31388	62152	6	6	0.0090	0.0014	0.091	−2.07	D−	3n
		3237.40	31364	62245	4	4	0.016	0.0026	0.11	−1.99	D−	3n
137.	$b\ ^4D - x\ ^4D°$ (82)											
		3177.54	31483	62945	8	8	0.081	0.012	1.0	−1.01	D−	3n
		3135.36	31388	63273	6	6	0.084	0.012	0.77	−1.13	D−	3n
		3114.30	31364	63465	4	4	0.064	0.0093	0.38	−1.43	D−	3n
		3105.55	31368	63559	2	2	0.076	0.011	0.22	−1.66	D−	3n
		3144.75	31483	63273	8	6	0.020	0.0023	0.19	−1.74	D−	3n
		3116.59	31388	63465	6	4	0.070	0.0068	0.42	−1.39	D−	3n
		3105.17	31364	63559	4	2	0.079	0.0057	0.23	−1.64	D−	3n
		3133.05	31364	63273	4	6	0.016	0.0035	0.15	−1.85	D−	3n
		3114.68	31368	63465	2	4	0.028	0.0081	0.17	−1.79	D−	3n
138.	$b\ ^4D - x\ ^4G°$ (uv 293)											
		2922.02	31483	65696	8	10	0.028	0.0044	0.34	−1.45	D−	3n
139.	$b\ ^4D - x\ ^4F°$ (uv 294)											
		2895.22	31483	66013	8	10	0.080	0.013	0.95	−1.00	D−	3n
		2857.17	31388	66377	6	8	0.095	0.016	0.88	−1.03	D−	3n
		2843.49	31364	66522	4	6	0.072	0.013	0.49	−1.28	D−	3n
		2836.51	31368	66613	2	4	0.077	0.019	0.35	−1.43	D−	3n
		2864.97	31483	66377	8	8	0.035	0.0043	0.33	−1.46	D−	3n
		2836.19	31364	66613	4	4	0.058	0.0070	0.26	−1.55	D−	3n
140.	$b\ ^4D - y\ ^2D°$ (uv 295)											
		2805.32	31364	67001	4	6	0.024	0.0042	0.16	−1.77	D−	3n
		2784.28	31368	67274	2	4	0.031	0.0072	0.13	−1.84	D−	3n
141.	$b\ ^4D - ^4P°$											
		2469.51	31483	71965	8	6	2.8	0.19	13	0.19	D	7
		2458.97	31388	72043	6	4	2.0	0.12	5.9	−0.14	D	7
142.	$b\ ^4D - w\ ^4F°$ (uv 300)											
		2428.36	31483	72651	8	10	2.7	0.30	19	0.38	D	7
		2445.80	31364	72239	4	6	1.5	0.20	6.5	−0.09	D	7
		2447.20	31388	72239	6	6	1.2	0.11	5.2	−0.19	D	7
143.	$b\ ^4D - w\ ^4D°$ (uv 301)											
		2423.21	31364	72619	4	6	1.4	0.19	6.1	−0.12	D	7

Fe II: Allowed transitions — Continued

No.	Multiplet	λ (Å)	E_i (cm^{-1})	E_k (cm^{-1})	g_i	g_k	A_{ki} (10^8 s^{-1})	f_{ik}	S (at. u.)	log gf	Accuracy	Source
144.	$b\ ^2$F $-\ z\ ^4$G° (89)											
		3442.24	31999	61042	8	6	0.0033	4.4(−4)	0.040	−2.45	D−	3n
145.	$b\ ^2$F $-\ z\ ^2$D° (91)											
		3436.11	31999	61093	8	6	0.0051	6.7(−4)	0.061	−2.27	D−	3n
		3297.89	31812	62126	6	4	0.013	0.0014	0.092	−2.07	D−	3n
146.	$b\ ^2$F $-\ y\ ^4$D° (94)											
		3257.36	31999	62690	8	6	0.0025	2.9(−4)	0.025	−2.63	D−	3n
147.	$b\ ^2$F $-\ z\ ^2$G° (92)											
		3323.07	31999	62083	8	10	0.014	0.0029	0.26	−1.63	D−	3n
		3276.61	31812	62332	6	8	0.011	0.0025	0.16	−1.83	D−	3n
148.	$b\ ^2$F $-\ z\ ^2$F° (97)											
		3096.30	31999	64286	8	8	0.025	0.0035	0.29	−1.55	D−	3n
		3065.32	31812	64425	6	6	0.037	0.0053	0.32	−1.50	D−	3n
149.	$b\ ^2$F $-\ y\ ^2$G° (98)											
		3044.84	31999	64832	8	10	0.014	0.0024	0.19	−1.72	D−	3n
		3002.33	31812	65110	6	8	0.023	0.0041	0.24	−1.61	D−	3n
150.	$b\ ^2$F $-\ y\ ^2$F° (uv 309)											
		2658.25	31999	69607	8	8	0.32	0.034	2.4	−0.56	D−	3n
		2642.01	31812	69650	6	6	0.36	0.037	1.9	−0.65	D−	3n
151.	$b\ ^2$F $-\ x\ ^2$G° (uv 310)	*2597.4*	*31919*	*70408*	14	18	0.32	0.042	5.0	−0.23	D−	3n
		2609.13	31999	70315	8	10	0.30	0.038	2.6	−0.52	D−	3n
		2582.41	31812	70524	6	8	0.24	0.032	1.7	−0.71	D−	3n
		2594.96	31999	70524	8	8	0.10	0.010	0.69	−1.09	D−	3n
152.	$a\ ^2$I $-\ z\ ^2$I° (105)											
		3398.36	32876	62293	14	14	0.0014	2.4(−4)	0.037	−2.48	D−	3n
		3360.10	32910	62662	12	12	0.0011	1.8(−4)	0.024	−2.67	D−	3n
153.	$a\ ^2$I $-\ y\ ^2$G° (107)											
		3131.72	32910	64832	12	10	0.0094	0.0012	0.14	−1.86	D−	3n

Fe II: Allowed transitions — Continued

No.	Multiplet	λ (Å)	E_i (cm^{-1})	E_k (cm^{-1})	g_i	g_k	A_{ki} (10^8 s^{-1})	f_{ik}	S (at. u.)	log gf	Accuracy	Source
154.	$a\ ^2$I – $z\ ^2$H° (108)											
		3077.17	32876	65364	14	12	0.11	0.014	1.9	−0.72	D−	3n
		3062.23	32910	65556	12	10	0.12	0.014	1.7	−0.78	D−	3n
155.	$a\ ^2$I – $x\ ^4$G° (109)											
		3056.80	32876	65580	14	12	0.018	0.0022	0.30	−1.52	D−	3n
156.	$a\ ^2$I – $y\ ^2$H° (uv 317)											
		2885.93	32876	67516	14	12	0.035	0.0037	0.50	−1.28	D−	3n
		2848.90	32910	68001	12	10	0.054	0.0055	0.62	−1.18	D−	3n
157.	$a\ ^2$I – $z\ ^2$K° (uv 318)	*2607.9*	*32892*	*71225*	26	30	2.2	0.26	58	0.83	D−	3n
		2592.78	32876	71433	14	16	2.1	0.24	29	0.53	D−	3n
		2625.49	32910	70987	12	14	2.2	0.27	28	0.51	D−	3n
		2623.11	32876	70987	14	14	0.11	0.012	1.4	−0.78	D−	3n
158.	$a\ ^2$I – $x\ ^2$H° (uv 319)											
		2538.20	32876	72262	14	12	1.2	0.099	12	0.14	D−	3n
		2548.92	32910	72130	12	10	0.48	0.039	3.9	−0.33	D−	3n
159.	$a\ ^2$I – $y\ ^2$I° (uv 321)											
		2432.87	32876	73967	14	14	3.2	0.28	32	0.60	D	7
		2434.73	32910	73970	12	12	3.2	0.28	27	0.53	D	7
160.	$c\ ^2$G – $z\ ^4$H° (112)											
		3632.29	33466	60989	10	10	0.0010	2.0(−4)	0.024	−2.69	D−	3n
		3614.87	33501	61157	8	8	0.0029	5.7(−4)	0.054	−2.34	D−	3n
161.	$c\ ^2$G – (°)^b											
		3499.88	33501	62066	8	8	0.0026	4.9(−4)	0.045	−2.41	D−	3n
		3495.62	33466	62066	10	8	0.0017	2.5(−4)	0.028	−2.61	D−	3n
162.	$c\ ^2$G – $z\ ^2$G° (114)											
		3493.47	33466	62083	10	10	0.024	0.0044	0.50	−1.36	D−	3n
		3468.68	33501	62322	8	8	0.014	0.0026	0.23	−1.69	D−	3n
		3464.50	33466	62322	10	8	0.0018	2.6(−4)	0.029	−2.59	D−	3n
163.	$c\ ^2$G – $x\ ^4$D° (117)											
		3395.34	33501	62945	8	8	0.0015	2.7(−4)	0.024	−2.67	D−	3n

Fe II: Allowed transitions — Continued

No.	Multiplet	λ (Å)	E_i (cm^{-1})	E_k (cm^{-1})	g_i	g_k	A_{ki} (10^8 s^{-1})	f_{ik}	S (at. u.)	log gf	Accuracy	Source
164.	$c\,^2G - y\,^4G°$ (118)											
		3279.65	33466	63949	10	10	0.0028	4.6(−4)	0.049	−2.34	D−	3n
		3273.50	33501	64041	8	8	0.0037	6.0(−4)	0.052	−2.32	D−	3n
		3269.77	33466	64041	10	8	0.0030	3.9(−4)	0.042	−2.41	D−	3n
		3268.51	33501	64087	8	6	0.0063	7.5(−4)	0.065	−2.22	D−	3n
165.	$c\,^2G - z\,^2F°$ (119)	*3239.1*	*33482*	*64346*	18	14	0.043	0.0052	1.0	−1.03	D−	3n
		3243.72	33466	64286	10	8	0.040	0.0050	0.54	−1.30	D−	3n
		3232.79	33501	64425	8	6	0.039	0.0045	0.39	−1.44	D−	3n
		3247.39	33501	64286	8	8	0.0056	8.8(−4)	0.076	−2.15	D−	3n
166.	$c\,^2G - y\,^2G°$ (120)											
		3187.29	33466	64832	10	10	0.035	0.0054	0.56	−1.27	D−	3n
		3162.80	33501	65110	8	8	0.039	0.0058	0.49	−1.33	D−	3n
167.	$c\,^2G - x\,^2G°$ (uv 325)											
		2697.72	33466	70524	10	8	0.024	0.0021	0.19	−1.67	D−	3n
168.	$c\,^2G - x\,^2H°$ (uv 326)	*2581.9*	*33482*	*72202*	18	22	1.4	0.17	26	0.49	D−	3n
		2576.86	33466	72262	10	12	1.1	0.13	11	0.12	D−	3n
		2587.95	33501	72130	8	10	1.4	0.18	12	0.16	D−	3n
		2585.63	33466	72130	10	10	0.36	0.036	3.1	−0.44	D−	3n
169.	$c\,^2G - w\,^4F°$ (uv 327)											
		2580.72	33501	72239	8	6	0.021	0.0015	0.10	−1.91	D−	3n
170.	$c\,^2G - w\,^4D°$ (uv 328)											
		2551.21	33466	72652	10	8	0.32	0.025	2.1	−0.60	D−	3n
171.	$b\,^2D - z\,^2D°$ (127)											
		4024.55	36253	61093	6	6	0.0023	5.5(−4)	0.044	−2.48	D−	3n
172.	$b\,^2D - y\,^2F°$ (uv 335)											
		2997.30	36253	69607	6	8	0.083	0.015	0.88	−1.05	D−	3n
		2982.06	36126	69650	4	6	0.21	0.041	1.6	−0.78	D−	3n
173.	$b\,^2D - x\,^2G°$ (uv 336)											
		2917.09	36253	70524	6	8	0.019	0.0032	0.19	−1.71	D−	3n
174.	$b\,^2D - (\quad°)^b$											
		2719.30	36253	73016	6	8	0.37	0.055	3.0	−0.48	D−	3n

Fe II: Allowed transitions — Continued

No.	Multiplet	λ (Å)	E_i (cm⁻¹)	E_k (cm⁻¹)	g_i	g_k	A_{ki} (10⁸ s⁻¹)	f_{ik}	S (at. u.)	log gf	Accuracy	Source
175.	$b\ ^2D$ – (°)[b]											
		2707.13	36126	73055	4	6	0.85	0.14	5.0	−0.25	D−	3n
		2716.43	36253	73055	6	6	0.048	0.0053	0.28	−1.50	D−	3n
176.	$b\ ^2D$ – $y\ ^2P°$ (uv 341)											
		2697.46	36126	73187	4	2	1.8	0.10	3.5	−0.40	D−	3n
		2697.33	36126	73189	4	4	0.27	0.029	1.0	−0.93	D−	3n
177.	$b\ ^2D$ – $x\ ^2D°$ (uv 342)											
		2606.51	36253	74607	6	6	1.8	0.18	9.4	0.04	D−	3n
		2605.34	36126	74498	4	4	1.6	0.17	5.7	−0.18	D−	3n
178.	$a\ ^2S$ – $z\ ^2P°$ (144)	*3622.4*	37227	*64825*	2	6	0.014	0.0084	0.20	−1.78	D−	3n
		3621.27	37227	64834	2	4	0.013	0.0051	0.12	−1.99	D−	3n
		3624.89	37227	64806	2	2	0.018	0.0035	0.083	−2.16	D−	3n
179.	$a\ ^2S$ – $y\ ^2P°$ (uv 348)	*2780.0*	37227	*73188*	2	6	0.25	0.087	1.6	−0.76	D−	3n
		2779.91	37227	73189	2	4	0.23	0.054	0.98	−0.97	D−	3n
		2780.04	37227	73187	2	2	0.29	0.034	0.62	−1.17	D−	3n
180.	$a\ ^2S$ – $x\ ^2P°$ (uv 349)	*2560.0*	37227	*76278*	2	6	1.3	0.39	6.5	−0.11	D−	3n
		2569.78	37227	76129	2	4	1.2	0.23	4.0	−0.33	D−	3n
		2540.52	37227	76577	2	2	1.5	0.15	2.5	−0.53	D−	3n
181.	$c\ ^2D$ – $z\ ^2D°$ (148)											
		4180.97	38215	62126	4	4	0.014	0.0036	0.20	−1.84	D−	3n
182.	$c\ ^2D$ – $z\ ^2F°$ (153)											
		3827.08	38164	64286	6	8	0.0013	3.8(−4)	0.029	−2.64	D−	3n
		3814.12	38215	64425	4	6	0.0030	9.7(−4)	0.049	−2.41	D−	3n
183.	$c\ ^2D$ – $z\ ^2P°$ (154)											
		3748.49	38164	64834	6	4	0.041	0.0058	0.43	−1.46	D−	3n
		3759.46	38215	64806	4	2	0.016	0.0017	0.086	−2.16	D−	3n
184.	$c\ ^2D$ – $y\ ^2F°$ (157)											
		3179.50	38164	69607	6	8	0.099	0.020	1.3	−0.92	D−	3n
		3180.16	38215	69650	4	6	0.071	0.016	0.68	−1.19	D−	3n
185.	$c\ ^2D$ – $x\ ^2G°$ (158)											
		3089.39	38164	70524	6	8	0.022	0.0042	0.26	−1.60	D−	3n

Fe II: Allowed transitions — Continued

No.	Multiplet	λ (Å)	E_i (cm^{-1})	E_k (cm^{-1})	g_i	g_k	A_{ki} (10^8 s^{-1})	f_{ik}	S (at. u.)	log gf	Accuracy	Source
186.	$c\ ^2$D $-\ w\ ^2$F° (uv 355)											
		2670.38	38164	75601	6	8	0.060	0.0085	0.45	-1.29	D$-$	$3n$
187.	$c\ ^2$D $-\ x\ ^2$P° (uv 356)	*2624.3*	*38184*	*76278*	10	6	1.6	0.098	8.5	-0.01	D$-$	$3n$
		2633.20	38164	76129	6	4	1.7	0.12	6.0	-0.16	D$-$	$3n$
		2605.90	38215	76577	4	2	1.2	0.061	2.1	-0.61	D$-$	$3n$
		2636.69	38215	76129	4	4	0.12	0.012	0.43	-1.31	D$-$	$3n$
188.	$z\ ^6$D° $-\ e\ ^6$D (uv 363)											
		2537.14	38459	77862	10	10	1.4	0.13	11	0.13	D	7
		2550.15	38660	77862	8	10	0.40	0.049	3.3	-0.41	D	7
189.	$z\ ^6$D° $-\ ^6$D											
		2208.41	38459	83726	10	10	1.8	0.13	9.8	0.13	D	7
		2218.27	38660	83726	8	10	1.9	0.17	10	0.14	D	7
190.	$z\ ^6$F° $-\ e\ ^6$D (uv 373)											
		2785.19	41968	77862	12	10	1.0	0.098	11	0.07	D$-$	$3n$
		2754.91	42237	78525	8	6	0.84	0.072	5.2	-0.24	D$-$	$3n$
		2796.63	42115	77862	10	10	0.10	0.012	1.1	-0.92	D$-$	$3n$
		2776.91	42237	78238	8	8	0.30	0.034	2.5	-0.56	D$-$	$3n$
		2762.34	42335	78525	6	6	0.37	0.043	2.3	-0.59	D$-$	$3n$
191.	$z\ ^6$F° $-\ e\ ^6$F (uv 375)											
		2445.11	41968	82854	12	12	1.9	0.17	16	0.31	D	7
		2434.06	42237	83308	8	6	0.70	0.046	3.0	-0.43	D	7
192.	$z\ ^6$F° $-\ e\ ^6$G (uv 379)											
		2369.95	42115	84297	10	12	5.7	0.58	45	0.76	D	7
		2363.86	42237	84528	8	10	5.1	0.53	33	0.63	D	7
		2352.31	42440	84938	2	4	4.2	0.69	11	0.14	D	7
		2353.68	42237	84711	8	8	1.3	0.11	6.6	-0.07	D	7
		2351.67	42335	84845	6	6	1.7	0.14	6.6	-0.07	D	7
193.	$z\ ^6$P° $-\ e\ ^6$D (uv 380)											
		2839.80	42658	77862	8	10	0.41	0.063	4.7	-0.30	D$-$	$3n$
		2856.38	43239	78238	6	8	0.27	0.044	2.5	-0.58	D$-$	$3n$
		2809.78	42658	78238	8	8	0.16	0.019	1.4	-0.81	D$-$	$3n$
		2833.09	43239	78525	6	6	0.27	0.032	1.8	-0.71	D$-$	$3n$
		2847.77	43621	78726	4	4	0.33	0.040	1.5	-0.80	D$-$	$3n$
		2787.24	42658	78525	8	6	0.13	0.011	0.82	-1.05	D$-$	$3n$
		2817.09	43239	78726	6	4	0.21	0.017	0.93	-1.00	D$-$	$3n$
		2838.22	43621	78844	4	2	0.42	0.026	0.96	-0.99	D$-$	$3n$

Fe II: Allowed transitions — Continued

No.	Multiplet	λ (Å)	E_i (cm^{-1})	E_k (cm^{-1})	g_i	g_k	A_{ki} (10^8 s^{-1})	f_{ik}	S (at. u.)	log gf	Accuracy	Source
194.	$z\ ^6P^\circ - \ ^6D$											
		2434.24	42658	83726	8	10	2.0	0.22	14	0.24	D	7
195.	$z\ ^6P^\circ - \ ^6S$											
		2365.77	43239	85495	6	6	2.1	0.17	8.2	0.02	D	7
196.	$z\ ^4F^\circ - e\ ^4D$ (uv 391)											
		2839.51	44233	79439	10	8	0.99	0.095	8.9	−0.02	D−	3n
		2848.32	45080	80178	6	4	1.1	0.087	4.9	−0.28	D−	3n
197.	$z\ ^4F^\circ - e\ ^4G$ (uv 395)											
		2460.44	44233	84863	10	12	5.3	0.58	47	0.76	D	7
		2472.61	44754	85185	8	10	3.7	0.42	28	0.53	D	7
		2475.54	45080	85463	6	8	3.5	0.43	21	0.41	D	7
		2475.12	45290	85680	4	6	3.9	0.53	17	0.33	D	7
		2455.71	44754	85463	8	8	1.0	0.095	6.1	−0.12	D	7
198.	$z\ ^4F^\circ - e\ ^4F$ (uv 396)											
		2416.45	44754	86124	8	10	1.6	0.17	11	0.14	D	7
		2418.44	45080	86416	6	8	1.6	0.19	8.9	0.05	D	7
199.	$z\ ^4D^\circ - c\ ^4D$											
		6383.72	44785	60445	6	6	0.0015	9.0(−4)	0.11	−2.27	D	5
200.	$z\ ^4D^\circ - e\ ^4D$ (uv 399)											
		2856.91	44447	79439	8	8	0.87	0.11	8.0	−0.07	D−	3n
		2848.11	44785	79885	6	6	0.70	0.085	4.8	−0.29	D	7
		2844.96	45206	80346	2	2	0.45	0.055	1.0	−0.96	D−	3n
		2884.77	44785	79439	6	8	0.14	0.024	1.4	−0.84	D−	3n
201.	$z\ ^4D^\circ - f\ ^4D$ (uv 400)											
		2484.44	44447	84685	8	8	2.3	0.21	14	0.23	D	7
		2501.31	45206	85173	2	2	1.4	0.13	2.1	−0.59	D	7
202.	$z\ ^4D^\circ - e\ ^4G$ (uv 401)											
		2453.98	44447	85185	8	10	0.73	0.083	5.3	−0.18	D	7
203.	$z\ ^4D^\circ - e\ ^4F$ (uv 402)											
		2401.29	44785	86416	6	8	2.5	0.29	14	0.24	D	7
		2390.77	44785	86600	6	6	0.93	0.080	3.8	−0.32	D	7

Fe II: Allowed transitions — Continued

No.	Multiplet	λ (Å)	E_i (cm^{-1})	E_k (cm^{-1})	g_i	g_k	A_{ki} (10^8 s^{-1})	f_{ik}	S (at. u.)	log gf	Accuracy	Source
204.	$c\ ^2F - z\ ^2D°$ (163)											
		6179.39	44915	61093	8	6	4.5(−4)	1.9(−4)	0.032	−2.81	D	6n
		5813.68	44930	62126	6	4	8.8(−4)	3.0(−4)	0.034	−2.75	D	6n
205.	$c\ ^2F - z\ ^2G°$ (164)											
		5823.15	44915	62083	8	10	1.7(−4)	1.1(−4)	0.016	−3.07	D	6n
206.	$c\ ^2F - x\ ^2G°$ (173)											
		3935.94	44915	70315	8	10	0.0059	0.0017	0.18	−1.86	D−	3n
		3906.04	44930	70524	6	8	0.0081	0.0025	0.19	−1.83	D−	3n
207.	$c\ ^2F - x\ ^2D°$ (177)											
		3366.98	44915	74607	8	6	0.019	0.0024	0.22	−1.71	D−	3n
		3381.00	44930	74498	6	4	0.020	0.0022	0.15	−1.87	D−	3n
208.	$c\ ^2F - w\ ^2D°$ (uv 403)											
		2959.84	44915	78691	8	6	0.16	0.016	1.2	−0.90	D−	3n
209.	$c\ ^2F - v\ ^2G°$ (uv 404)	*2583.3*	*44921*	*83619*	14	18	2.4	0.31	37	0.64	D−	3n
		2566.22	44915	83871	8	10	2.5	0.31	21	0.39	D−	3n
		2605.04	44930	83305	6	8	2.1	0.28	15	0.23	D−	3n
		2604.05	44915	83305	8	8	0.11	0.012	0.80	−1.03	D−	3n
210.	$c\ ^2F - v\ ^2D°$ (uv 405)											
		2566.40	44915	83868	8	6	2.1	0.16	11	0.10	D−	3n
		2535.36	44930	84360	6	4	3.3	0.21	11	0.10	D−	3n
211.	$z\ ^4P° - e\ ^4D$ (181)											
		3078.68	46967	79439	6	8	0.42	0.080	4.9	−0.32	D−	3n
		3076.44	47390	79885	4	6	0.28	0.060	2.4	−0.62	D−	3n
		3071.12	47626	80178	2	4	0.19	0.052	1.1	−0.98	D−	3n
		3036.96	46967	79885	6	6	0.16	0.022	1.3	−0.87	D−	3n
		3048.99	47390	80178	4	4	0.28	0.040	1.6	−0.80	D−	3n
212.	$z\ ^4P° - f\ ^4D$ (uv 410)											
		2650.48	46967	84685	6	8	1.6	0.22	12	0.12	D−	3n
		2667.22	47390	84871	4	6	0.92	0.15	5.2	−0.23	D−	3n
		2671.40	47626	85049	2	4	0.56	0.12	2.1	−0.62	D−	3n
		2637.50	46967	84871	6	6	0.52	0.054	2.8	−0.49	D−	3n
		2654.63	47390	85049	4	4	0.77	0.081	2.8	−0.49	D−	3n
		2662.56	47626	85173	2	2	0.96	0.10	1.8	−0.69	D−	3n
213.	$z\ ^4P° - ^4P$											
		2466.52	47626	88157	2	4	2.1	0.39	6.3	−0.11	D	7

Fe II: Allowed transitions — Continued

No.	Multiplet	λ (Å)	E_i (cm^{-1})	E_k (cm^{-1})	g_i	g_k	A_{ki} (10^8 s^{-1})	f_{ik}	S (at. u.)	log gf	Accuracy	Source
214.	d ^{2}D1 $-$ y ^{2}D° (185)											
		5272.40	48039	67001	6	6	0.0037	0.0016	0.16	-2.03	D	6n
215.	d ^{2}D1 $-$ y ^{2}F° (186)											
		4635.33	48039	69607	6	8	0.0087	0.0037	0.34	-1.65	D$-$	3n
		4549.21	47675	69650	4	6	0.0072	0.0034	0.20	-1.87	D$-$	3n
216.	d ^{2}D1 $-$ (°)^b											
		3938.97	47675	73055	4	6	0.010	0.0035	0.18	-1.85	D$-$	3n
217.	d ^{2}D1 $-$ x ^{2}D° (192)											
		3711.97	47675	74607	4	6	0.0097	0.0030	0.15	-1.92	D$-$	3n
218.	d ^{2}D1 $-$ v ^{4}D° (uv 412)											
		2570.53	48039	86930	6	8	1.2	0.16	8.3	-0.01	D$-$	3n
219.	c ^{4}P $-$ v ^{4}D° (uv 416)											
		2722.74	50213	86930	6	8	0.78	0.12	6.2	-0.16	D$-$	3n
		2683.00	49507	86768	4	6	0.64	0.10	3.7	-0.38	D$-$	3n
		2669.93	49101	86544	2	4	0.47	0.10	1.8	-0.70	D$-$	3n
		2699.20	49507	86544	4	4	0.66	0.072	2.6	-0.54	D$-$	3n
220.	c ^{4}F $-$ x ^{4}G° (199)											
		6446.40	50188	65696	8	10	0.0011	8.6($-$4)	0.15	-2.16	D	5
221.	c ^{4}F $-$ v ^{4}D° (uv 417)											
		2718.64	50157	86930	10	8	1.3	0.11	10	0.06	D$-$	3n
		2732.94	50188	86768	8	6	0.78	0.066	4.7	-0.28	D$-$	3n
222.	c ^{4}F $-$ ^{4}G°											
		2508.34	50188	90043	8	10	2.7	0.32	21	0.41	D	7
223.	d ^{2}F $-$ u ^{2}G°											
		2682.51	54904	92172	8	10	0.70	0.095	6.7	-0.12	D$-$	3n
		2649.47	54871	92603	6	8	1.8	0.26	14	0.19	D$-$	3n
224.	z ^{4}I° $-$ e ^{4}H (uv 431)											
		2717.87	61348	98130	16	14	1.4	0.13	19	0.33	D$-$	3n
225.	z ^{2}I° $-$ e ^{2}H (uv 434)											
		2716.56	62293	99093	14	12	1.6	0.16	20	0.34	D$-$	3n

Fe II: Allowed transitions — Continued

No.	Multiplet	λ (Å)	E_i (cm^{-1})	E_k (cm^{-1})	g_i	g_k	A_{ki} (10^8 s^{-1})	f_{ik}	S (at. u.)	log gf	Accuracy	Source
226.	z ^{2}I° – ^{2}K											
		2390.10	62293	104120	14	16	5.5	0.54	60	0.88	D	7
		2400.06	62662	104315	12	14	5.2	0.52	50	0.80	D	7
227.	y ^{4}H° – f ^{4}G (uv 440)											
		2763.66	66412	102585	14	12	1.3	0.12	16	0.24	D	7
228.	w ^{4}F° – ^{4}D											
		2765.13	72651	108805	10	8	1.2	0.11	9.8	0.03	D	7
229.	e ^{6}G – ^{2}[3]°											
		5247.95	84938	103988	4	6	1.7	1.1	74	0.63	D	7
230.	e ^{4}G – ^{2}[7]°											
		5506.20	84863	103020	12	14	1.4	0.74	160	0.95	D	7
231.	e ^{4}F – ^{2}[6]°											
		5961.71	86124	102893	10	12	0.77	0.49	96	0.69	D	7

[a]The number in parentheses following the tabulated value indicates the power of ten by which this value has to be multiplied.
[b]The LS-coupling designation of the upper term of this multiplet was not provided in the NBS energy level compilation (C. Corliss and J. Sugar, J. Phys. Chem. Ref. Data **14**, Suppl. 2 (1985)), so we have accordingly omitted it from this work.

Fe II

Forbidden Transitions

List of tabulated lines

Wavelength (Å)	No.	Wavelength (Å)	No.	Wavelength (Å)	No.	Wavelength (Å)	No.
3175.38	8	3440.99	20	3579.81	19	4231.56	16
3214.67	8	3450.39	21	3588.2	31	4234.81	30
3224.54	8	3452.30	20	3604.6	31	4243.98	16
3244.18	8	3455.11	20	3625.8	31	4244.81	16
3254.24	8	3484.01	21	3628.65	19	4251.4	18
3275.02	8	3489.98	20	3642.5	31	4266.34	29
3277.12	8	3501.62	20	3664.7	31	4276.83	16
3277.55	8	3504.02	20	4082.0	36	4280.1	17
3289.46	8	3504.51	20	4083.78	18	4287.40	7
3289.89	8	3505.81	19	4114.48	18	4305.90	16
3318.38	21	3528.28	19	4149.1	17	4319.62	16
3376.20	20	3532.9	31	4157.89	30	4329.43	29
3380.95	21	3536.25	20	4177.21	16	4346.85	16
3387.10	20	3538.69	20	4178.95	18	4351.80	29
3402.50	21	3539.19	20	4197.81	17	4352.78	16
3428.24	21	3575.72	19	4211.1	18	4356.14	17

List of tabulated lines — Continued

Wavelength (Å)	No.	Wavelength (Å)	No.	Wavelength (Å)	No.	Wavelength (Å)	No.
4358.37	16	5433.15	13	7432.23	46	10708	34
4359.34	7	5477.25	26	7452.50	12	10789	34
4372.43	16	5527.61	26	7613.15	24	10796.48	45
4382.75	6	5551.31	35	7637.52	4	10874	71
4383.0	28	5556.31	13	7665.29	4	10887	63
4402.60	29	5580.82	35	7686.90	4	10911	34
4409.86	17	5586.9	35	7720.2	54	10941	34
4413.78	7	5587.5	47	7733.12	4	10954	34
4416.27	6	5588.15	35	7734.8	46	11044	63
4432.45	6	5613.3	35	7764.69	24	11066	34
4452.11	7	5627.2	47	7874.23	4	11159	59
4457.95	6	5649.67	35	7975.2	52	11160	63
4468.5	28	5650.94	35	8044.8	73	11165	45
4470.29	6	5673.2	42	8077.5	73	11352	63
4474.91	7	5718.2	35	8199.0	54	11447	45
4479.1	44	5725.92	35	8259.4	52	11655	67
4488.75	6	5746.96	26	8306.1	65	12485	3
4492.64	6	5756.7	47	8387.2	65	12521	3
4509.61	6	5767.5	62	8411.8	65	12567	3
4514.90	6	5799.0	47	8479.9	65	12569	66
4528.39	6	5835.4	42	8489.7	38	12703	3
4533.00	6	5843.90	26	8574.9	51	12776	66
4576.4	44	5847.3	42	8575.2	64	12788	3
4639.68	5	5870.0	41	8600.5	64	12897	66
4664.45	5	5901.26	26	8616.96	11	12943	3
4728.07	5	5913.3	57	8708.8	64	12978	3
4745.49	15	5982.7	56	8715.84	38	13192	66
4772.07	5	6044.1	41	8734.8	64	13206	3
4774.74	15	6095.0	62	8739.1	72	13278	3
4798.28	5	6188.55	40	8825.1	64	13702	76
4814.55	15	6261.1	40	8861.4	73	13718	3
4852.73	15	6280.0	56	8885.66	38	13985	76
4874.49	15	6353.1	55	8891.88	11	14054	76
4889.63	5	6396.30	40	8931.47	51	14603	79
4898.6	43	6404.6	55	9033.45	11	14833	75
4905.35	15	6473.86	40	9051.92	11	14909	75
4947.38	15	6482.3	60	9083.4	51	14964	79
4950.74	15	6485.3	61	9133.63	38	15246	75
4973.39	15	6511.2	69	9202.1	68	15335	10
5005.52	15	6544.8	69	9226.60	11	15995	10
5006.65	5	6566.4	69	9231.7	68	16160	77
5020.24	15	6584.4	69	9267.54	11	16252	77
5027.9	43	6671.90	25	9351.2	68	16435	10
5035.4	48	6689.4	55	9381.7	68	16638	10
5043.53	15	6700.68	39	9384.8	59	16769	10
5048.2	48	6729.85	25	9436.6	78	17000	77
5060.1	43	6746.5	55	9465.4	68	17111	10
5072.40	14	6746.9	61	9469.5	59	17148	49
5107.95	13	6809.21	25	9470.93	11	17449	10
5111.63	14	6829.01	25	9490.6	59	17484	23
5158.00	13	6872.17	25	9513.87	37	17851	70
5158.81	14	6873.87	39	9552.7	71	17971	10
5163.94	27	6896.18	12	9590.4	50	18000	10
5172.5	48	6922.9	60	9669.7	71	18023	49
5181.97	13	6933.67	25	9682.13	37	18094	10
5184.80	14	6944.91	39	9711.2	72	18114	23
5186.0	48	6966.32	25	9949.32	50	18134	23
5220.06	14	7011.24	25	10013.88	37	19136	23
5261.61	14	7047.99	25	10038.79	50	19523	58
5268.88	13	7131.13	39	10127	53	19670	23
5273.38	13	7155.14	12	10432.60	50	20024	23
5296.84	14	7171.98	12	10461	59	20067	33
5333.65	14	7330.2	46	10466	59	20168	58
5347.67	13	7370.94	24	10572	53	20460	33
5376.47	14	7388.16	12	10683	34	20714	23

List of tabulated lines — Continued

Wavelength (Å)	No.	Wavelength (Å)	No.	Wavelength (Å)	No.	Wavelength (Å)	No.
20854	23	25147	58	48879	2	346510	22
21328	33	29597	58	50610	2	353390	1
22104	58	40753	2	53388	2	357670	9
22436	33	40809	2	56725	2	431590	32
22559	70	41139	2	179310	9	502950	32
22661	74	44337	2	228960	22	512860	1
23699	74	46065	2	245120	9	601110	22
24772	74	46707	2	259810	1	873580	1

For this spectrum, we have chosen the calculations of Nussbaumer and Storey,[1] Garstang,[2] and Nussbaumer *et al.*[3] In his calculations, Garstang used a three-configuration basis ($3d^7$, $3d^64s$, and $3d^54s^2$), plus spin-orbit interaction. The authors of Refs. 1 and 3 also employed configuration interaction, utilizing a four-configuration basis and a 17-configuration basis, respectively. In these two more recent works, radial wavefunctions were obtained via a scaled Thomas-Fermi potential. In general, the data of Refs. 1 and 3 should be more accurate than those of Ref. 2, since in the twenty years elapsed since Garstang's work, much more sophisticated theoretical and computational techniques had become available.

In evaluating data for this spectrum, we were able to compare some results of Refs. 1 or 3 to those of Ref. 2. For magnetic dipole transitions within the same configuration and term, the data of Refs. 1 and 2 agree within five percent. However, for M1 transitions between different configurations, the agreement between Refs. 1 and 2 is decidedly worse. The agreement is especially bad (up a factor of six) for weak M1 lines and/or M1 lines involving spin (intercombination) plus configuration changes.

As in the case of [Fe I], the uncertainties for the E2 lines tabulated here are expected to be quite large. For this complex a spectrum, it is likely that the Thomas-Fermi potential (used in Refs. 1 and 3) will lead to unreliable E2 transition probabilities, since cancellation effects due to different configuration mixing may not be accounted for. Nevertheless, for E2 lines, the agreement between Refs. 1 and 2 is fairly satisfactory—within a factor of two.

In selecting data from Garstang's work, we chose lines which are observed and are fairly strong. These lines generally produce about 95 percent of the intensity of a multiplet and also about 95 percent of the radiative de-excitation from a given upper level. The predominant type of radiation for each line listed is either electric quadrupole or magnetic dipole. However, if both of these radiation mechanisms contribute comparably to the total intensity of a line, then we have listed each component separately.

References

[1] H. Nussbaumer and P. J. Storey, Astron. Astrophys. **89**, 308 (1980).
[2] R. H. Garstang, Mon. Not. R. Astron. Soc. **124**, 321 (1962).
[3] H. Nussbaumer, M. Pettini, and P. J. Storey, Astron. Astrophys. **102**, 351 (1981).

Fe II: Forbidden transitions

No.	Multiplet	λ (Å)	E_i (cm^{-1})	E_k (cm^{-1})	g_i	g_k	Type of transition	A_{ki} (s^{-1})	S (at. u.)	Accuracy	Source
1.	$a\,^6D - a\,^6D$										
		[259810]	0.0	384.8	10	8	M1	0.00213	11.1	C+	1
		[353390]	384.8	667.7	8	6	M1	0.00157	15.4	C+	1
		[512860]	667.7	862.6	6	4	M1	7.18(−4)ᵃ	14.4	C+	1
		[873580]	862.6	977.1	4	2	M1	1.88(−4)	9.29	C+	1

Fe II: Forbidden transitions — Continued

No.	Multiplet	λ (Å)	E_i (cm^{-1})	E_k (cm^{-1})	g_i	g_k	Type of transition	A_{ki} (s^{-1})	S (at. u.)	Accuracy	Source
2.	$a\,^6D - a\,^4F$										
		[53388]	0.0	1873	10	10	M1	2.8(−4)	0.016	E	1
		[48879]	384.8	2430	8	8	M1	1.8(−4)	0.0062	E	1
		[46065]	667.7	2838	6	6	M1	1.0(−4)	0.022	E	1
		[44337]	862.6	3117	4	4	M1	4.0(−5)	5.2(−4)	E	1
		[41139]	0.0	2430	10	8	M1	1.0(−4)	0.0021	E	1
		[40753]	384.8	2838	8	6	M1	6.9(−4)	0.0010	E	1
		[40809]	667.7	3117	6	4	M1	2.8(−5)	2.8(−4)	E	1
		[56725]	667.7	2430	6	8	M1	3.4(−5)	0.0018	E	1
		[50610]	862.6	2838	4	6	M1	2.5(−5)	7.2(−4)	E	1
		[46707]	977.1	3117	2	4	M1	9.9(−6)	1.5(−4)	E	1
3.	$a\,^6D - a\,^4D$										
		[12567]	0.0	7955	10	8	M1	0.0056	0.0033	D	1
		[12485]	384.8	8392	8	6	M1	4.8(−4)	2.1(−4)	E	1
		[12521]	862.6	8847	4	2	M1	7.8(−4)	1.1(−4)	E	1
		[13206]	384.8	7955	8	8	M1	0.0016	0.0011	D	1
		[12943]	667.7	8392	6	6	M1	0.0024	0.0012	D	1
		[12788]	862.6	8680	4	4	M1	0.0030	9.3(−4)	D	1
		[12703]	977.1	8847	2	2	M1	0.0040	6.1(−4)	D	1
		[13718]	667.7	7955	6	8	M1	9.7(−4)	7.4(−4)	E	1
		[13278]	862.6	8392	4	6	M1	0.0014	7.3(−4)	D	1
		[12978]	977.1	8680	2	4	M1	0.0013	4.2(−4)	D	1
4.	$a\,^6D - a\,^4P$ (1F)										
		7637.52	384.8	13474	8	6	M1	0.0015	1.5(−4)	E	1
		7686.90	667.7	13673	6	4	M1	0.0018	1.2(−4)	E	1
		7665.29	862.6	13905	4	2	M1	0.0018	6.0(−5)	E	1
		7733.12	977.1	13905	2	2	M1	5.2(−4)	1.8(−5)	E	1
		7874.23	977.1	13673	2	4	M1	2.2(−4)	1.6(−5)	E	1
5.	$a\,^6D - b\,^4P$ (4F)										
		4889.63	384.8	20831	8	6	M1	0.36	0.0094	E	2
		4728.07	667.7	21812	6	4	M1	0.48	0.0075	E	2
		4639.68	862.6	22410	4	2	M1	0.49	0.0036	E	2
		4772.07	862.6	21812	4	4	M1	0.026	4.2(−4)	E	2
		4664.45	977.1	22410	2	2	M1	0.15	0.0011	E	2
		5006.65	862.6	20831	4	6	M1	0.027	7.5(−4)	E	2
		4798.28	977.1	21812	2	4	M1	0.082	0.0013	E	2
6.	$a\,^6D - b\,^4F$ (6F)										
		4416.27	0.0	22637	10	10	M1	0.46	0.015	E	2
		4457.95	384.8	22810	8	8	M1	0.29	0.0076	E	2
		4488.75	667.7	22939	6	6	M1	0.15	0.0030	E	2
		4509.61	862.6	23031	4	4	M1	0.058	7.9(−4)	E	2
		4382.75	0.0	22810	10	8	M1	0.055	0.0014	E	2
		4432.45	384.8	22939	8	6	M1	0.054	0.0010	E	2
		4470.29	667.7	23031	6	4	M1	0.029	3.8(−4)	E	2
		4492.64	384.8	22637	8	10	M1	0.060	0.0020	E	2
		4514.90	667.7	22810	6	8	M1	0.066	0.0018	E	2
		4528.39	862.6	22939	4	6	M1	0.046	9.5(−4)	E	2
		4533.00	977.1	23031	2	4	M1	0.016	2.2(−4)	E	2

Fe II: Forbidden transitions — Continued

No.	Multiplet	λ (Å)	E_i (cm^{-1})	E_k (cm^{-1})	g_i	g_k	Type of transition	A_{ki} (s^{-1})	S (at. u.)	Accuracy	Source
7.	$a\ ^6D - a\ ^6S$ (7F)										
		4287.40	0.0	23318	10	6	E2	1.5	7.8	E	3
		4359.34	384.8	23318	8	6	E2	1.1	6.2	E	3
		4413.78	667.7	23318	6	6	E2	0.81	4.8	E	3
		4452.11	862.6	23318	4	6	E2	0.52	3.2	E	3
		4474.91	977.1	23318	2	6	E2	0.26	1.7	E	3
8.	$a\ ^6D - b\ ^4D$ (11F)										
		3175.38	0.0	31483	10	8	M1	0.22	0.0021	D	2
		3224.54	384.8	31388	8	6	M1	0.046	3.4(−4)	D	2
		3277.12	862.6	31368	4	2	M1	0.041	1.1(−4)	D	2
		3214.67	384.8	31483	8	8	M1	0.069	6.8(−4)	D	2
		3254.24	667.7	31388	6	6	M1	0.12	9.2(−4)	D	2
		3277.55	862.6	31364	4	4	M1	0.16	8.4(−4)	D	2
		3289.46	977.1	31368	2	2	M1	0.22	5.8(−4)	D	2
		3244.18	667.7	31483	6	8	M1	0.029	2.9(−4)	D	2
		3275.02	862.6	31388	4	6	M1	0.056	4.4(−4)	D	2
		3289.89	977.1	31364	2	4	M1	0.061	3.2(−4)	D	2
9.	$a\ ^4F - a\ ^4F$										
		[179310]	1873	2430	10	8	M1	0.00612	10.5	C+	1
		[245120]	2430	2838	8	6	M1	0.00366	12.0	C+	1
		[357670]	2838	3117	6	4	M1	0.00141	9.57	C+	1
10.	$a\ ^4F - a\ ^4D$										
		[15335]	1873	8392	10	6	E2	0.0010	3.0	E	1
		[15995]	2430	8680	8	4	E2	0.0014	3.5	E	1
		[16638]	2838	8847	6	2	E2	0.0016	2.4	E	1
		[16435]	1873	7955	10	8	E2	0.0019	11	E	1
		[16769]	2430	8392	8	6	E2	8.0(−4)	3.8	E	1
		[17111]	2838	8680	6	4	E2	3.8(−4)	1.3	E	1
		[17449]	3117	8847	4	2	E2	8.0(−4)	1.5	E	1
		[18094]	2430	7955	8	8	E2	4.2(−4)	3.9	E	1
		[18000]	2838	8392	6	6	E2	5.8(−4)	3.9	E	1
		[17971]	3117	8680	4	4	E2	6.8(−4)	3.0	E	1
11.	$a\ ^4F - a\ ^4P$ (13F)										
		8616.96	1873	13474	10	6	E2	0.019	3.2	E	1
		8891.88	2430	13673	8	4	E2	0.011	1.5	E	1
		9033.45	2838	13905	6	2	E2	0.0082	0.59	E	1
		9051.92	2430	13474	8	6	E2	0.0049	1.1	E	1
		9226.60	2838	13673	6	4	E2	0.0069	1.1	E	1
		9267.54	3117	13905	4	2	E2	0.011	0.90	E	1
		9470.93	3117	13673	4	4	E2	0.0020	0.36	E	1
12.	$a\ ^4F - a\ ^2G$ (14F)										
		7155.14	1873	15845	10	10	M1	0.15	0.020	E	2
		7171.98	2430	16369	8	8	M1	0.056	0.0061	E	2
		6896.18	1873	16369	10	8	M1	0.0052	5.1(−4)	E	2
		7452.50	2430	15845	8	10	M1	0.048	0.0074	E	2
		7388.16	2838	16369	6	8	M1	0.043	0.0051	E	2

Fe II: Forbidden transitions — Continued

No.	Multiplet	λ (Å)	E_i (cm^{-1})	E_k (cm^{-1})	g_i	g_k	Type of transition	A_{ki} (s^{-1})	S (at. u.)	Accu-racy	Source
13.	$a\ ^4F - b\ ^4P$ (18F)										
		5273.38	1873	20831	10	6	E2	0.37	5.4	E	2
		5158.00	2430	21812	8	4	E2	0.30	2.6	E	2
		5107.95	2838	22410	6	2	E2	0.24	0.99	E	2
		5433.15	2430	20831	8	6	E2	0.11	1.9	E	2
		5268.88	2838	21812	6	4	E2	0.19	1.8	E	2
		5181.97	3117	22410	4	2	E2	0.34	1.5	E	2
		5556.31	2838	20831	6	6	E2	0.022	0.42	E	2
		5347.67	3117	21812	4	4	E2	0.057	0.59	E	2
14.	$a\ ^4F - a\ ^4H$ (19F)										
		5158.81	1873	21252	10	14	E2	0.44	13	E	2
		5261.61	2430	21430	8	12	E2	0.31	8.9	E	2
		5333.65	2838	21582	6	10	E2	0.26	6.7	E	2
		5376.47	3117	21712	4	8	E2	0.26	5.6	E	2
		5111.63	1873	21430	10	12	E2	0.10	2.5	E	2
		5220.06	2430	21582	8	10	E2	0.11	2.5	E	2
		5296.84	2838	21712	6	8	E2	0.091	1.8	E	2
		5072.40	1873	21582	10	10	E2	0.022	0.44	E	2
		5184.80	2430	21712	8	8	E2	0.021	0.37	E	2
15.	$a\ ^4F - b\ ^4F$ (20F)										
		4814.55	1873	22637	10	10	E2	0.40	6.2	E	2
		4905.35	2430	22810	8	8	E2	0.22	3.0	E	2
		4973.39	2838	22939	6	6	E2	0.14	1.5	E	2
		5020.24	3117	23031	4	4	E2	0.18	1.4	E	2
		4774.74	1873	22810	10	8	E2	0.13	1.5	E	2
		4874.49	2430	22939	8	6	E2	0.17	1.7	E	2
		4950.74	2838	23031	6	4	E2	0.17	1.2	E	2
		4947.38	2430	22637	8	10	E2	0.050	0.88	E	2
		5005.52	2838	22810	6	8	E2	0.071	1.1	E	2
		5043.53	3117	22939	4	6	E2	0.065	0.76	E	2
		4745.49	1873	22939	10	6	E2	0.013	0.11	E	2
		4852.73	2430	23031	8	4	E2	0.022	0.14	E	2
16.	$a\ ^4F - a\ ^4G$ (21F)										
		4346.85	2430	25429	8	12	E2	0.21	2.3	E	2
		4352.78	2838	25805	6	10	E2	0.31	2.9	E	2
		4372.43	3117	25982	4	8	E2	0.28	2.1	E	2
		4243.98	1873	25429	10	12	E2	0.90	8.9	E	2
		4276.83	2430	25805	8	10	E2	0.65	5.5	E	2
		4319.62	2838	25982	6	8	E2	0.53	3.8	E	2
		4358.37	3117	26055	4	6	E2	0.73	4.1	E	2
		4177.21	1873	25805	10	10	E2	0.14	1.1	E	2
		4244.81	2430	25982	8	8	E2	0.25	1.6	E	2
		4305.90	2838	26055	6	6	E2	0.31	1.6	E	2
		4231.56	2430	26055	8	6	E2	0.024	0.12	E	2

Fe II: Forbidden transitions — Continued

No.	Multiplet	λ (Å)	E_i (cm^{-1})	E_k (cm^{-1})	g_i	g_k	Type of transition	A_{ki} (s^{-1})	S (at. u.)	Accuracy	Source
17.	$a\ ^4F - b\ ^2P$ (22F)										
		[4280.1]	2430	25788	8	4	E2	0.0017	0.0058	E	2
		[4149.1]	2838	26933	6	2	E2	0.0021	0.0031	E	2
		4356.14	2838	25788	6	4	E2	0.0080	0.030	E	2
		4197.81	3117	26933	4	2	E2	0.0099	0.015	E	2
		4409.86	3117	25788	4	4	E2	0.0042	0.017	E	2
18.	$a\ ^4F - b\ ^2H$ (23F)										
		[4211.1]	2430	26170	8	12	E2	0.024	0.23	E	2
		[4251.4]	2838	26353	6	10	E2	0.0087	0.072	E	2
		4114.48	1873	26170	10	12	E2	0.045	0.38	E	2
		4178.95	2430	26353	8	10	E2	0.0051	0.039	E	2
		4083.78	1873	26353	10	10	E2	0.0030	0.020	E	2
19.	$a\ ^4F - b\ ^2G$ (25F)										
		3505.81	1873	30389	10	10	E2	0.0032	0.010	E	2
		3528.28	2430	30764	8	8	E2	0.0022	0.0057	E	2
		3575.72	2430	30389	8	10	E2	0.0028	0.0097	E	2
		3579.81	2838	30764	6	8	E2	0.0016	0.0045	E	2
		3628.65	2838	30389	6	10	E2	0.0011	0.0041	E	2
20.	$a\ ^4F - b\ ^4D$ (26F)										
		3387.10	1873	31388	10	6	E2	0.20	0.32	E	2
		3455.11	2430	31364	8	4	E2	0.36	0.42	E	2
		3504.02	2838	31368	6	2	E2	0.52	0.33	E	2
		3376.20	1873	31483	10	8	E2	0.73	1.5	E	2
		3452.30	2430	31388	8	6	E2	0.37	0.65	E	2
		3504.51	2838	31364	6	4	E2	0.21	0.26	E	2
		3538.69	3117	31368	4	2	E2	0.40	0.26	E	2
		3440.99	2430	31483	8	8	E2	0.24	0.55	E	2
		3501.62	2838	31388	6	6	E2	0.34	0.64	E	2
		3539.19	3117	31364	4	4	E2	0.38	0.50	E	2
		3489.98	2838	31483	6	8	E2	0.035	0.086	E	2
		3536.25	3117	31388	4	6	E2	0.063	0.12	E	2
21.	$a\ ^4F - b\ ^2F$ (27F)										
		3318.38	1873	31999	10	8	M1	0.042	4.6(−4)	D	2
		3402.50	2430	31812	8	6	M1	0.012	1.1(−4)	D	2
		3380.95	2430	31999	8	8	M1	0.0062	7.1(−5)	D	2
		3450.39	2838	31812	6	6	M1	0.0087	7.9(−5)	D	2
		3428.24	2838	31999	6	8	M1	0.017	2.0(−4)	D	2
		3484.01	3117	31812	4	6	M1	0.046	4.3(−4)	D	2
22.	$a\ ^4D - a\ ^4D$										
		[228960]	7955	8392	8	6	M1	0.00256	6.83	C+	1
		[346510]	8392	8680	6	4	M1	0.00136	8.39	C+	1
		[601110]	8680	8847	4	2	M1	3.71(−4)	5.97	C+	1

Fe II: Forbidden transitions — Continued

No.	Multiplet	λ (Å)	E_i (cm^{-1})	E_k (cm^{-1})	g_i	g_k	Type of transition	A_{ki} (s^{-1})	S (at. u.)	Accuracy	Source
23.	$a\ ^4D - a\ ^4P$										
		[17484]	7955	13673	8	4	E2	6.1(−4)	2.4	E	1
		[18134]	8392	13905	6	2	E2	7.6(−4)	1.8	E	1
		[18114]	7955	13474	8	6	M1	8.3(−5)	1.1(−4)	E	1
		"	"	"	8	6	E2	5.3(−4)	3.7	E	1
		[19136]	8680	13905	4	2	E2	2.5(−4)	0.76	E	1
		[19670]	8392	13474	6	6	E2	2.7(−4)	2.8	E	1
		[20024]	8680	13673	4	4	E2	1.6(−4)	1.2	E	1
		[20854]	8680	13474	4	6	E2	7.7(−5)	1.1	E	1
		[20714]	8847	13673	2	4	E2	1.5(−4)	1.4	E	1
24.	$a\ ^4D - b\ ^4P$ (30F)										
		7764.69	7955	20831	8	6	M1	0.028	0.0029	E	2
		7613.15	8680	21812	4	4	M1	0.012	7.9(−4)	E	2
		7370.94	8847	22410	2	2	M1	0.018	5.3(−4)	E	2
25.	$a\ ^4D - b\ ^4F$ (31F)										
		6809.21	7955	22637	8	10	M1	0.025	0.0029	E	2
		6933.67	8392	22810	6	8	M1	0.0017	1.7(−4)	E	2
		7011.24	8680	22939	4	6	M1	0.0017	1.3(−4)	E	2
		7047.99	8847	23031	2	4	M1	0.016	8.3(−4)	E	2
		6729.85	7955	22810	8	8	M1	0.017	0.0015	E	2
		6872.17	8392	22939	6	6	M1	0.022	0.0016	E	2
		6966.32	8680	23031	4	4	M1	0.026	0.0013	E	2
		6671.90	7955	22939	8	6	M1	0.0044	2.9(−4)	E	2
		6829.01	8392	23031	6	4	M1	0.0060	2.8(−4)	E	2
26.	$a\ ^4D - b\ ^2P$ (34F)										
		5746.96	8392	25788	6	4	M1	0.37	0.010	E	2
		5477.25	8680	26933	4	2	M1	0.44	0.0054	E	2
		5843.90	8680	25788	4	4	M1	0.015	4.4(−4)	E	2
		5527.61	8847	26933	2	2	M1	0.12	0.0015	E	2
		5901.26	8847	25788	2	4	M1	0.040	0.0012	E	2
27.	$a\ ^4D - a\ ^2F$ (35F)										
		5163.94	7955	27315	8	8	M1	0.32	0.013	E	2
28.	$a\ ^4D - b\ ^2G$										
		[4468.5]	8392	30764	6	8	M1	0.0032	8.5(−5)	E	2
		[4383.0]	7955	30764	8	8	M1	0.0050	1.2(−4)	E	2
29.	$a\ ^4D - b\ ^4D$ (36F)										
		4266.34	7955	31388	8	6	M1	0.024	4.2(−4)	C	2
		4351.80	8392	31364	6	4	M1	0.014	1.7(−4)	C	2
		4329.43	8392	31483	6	8	M1	0.017	4.1(−4)	C	2
		4402.60	8680	31388	4	6	M1	0.013	2.5(−4)	C	2

Fe II: Forbidden transitions — Continued

No.	Multiplet	λ (Å)	E_i (cm^{-1})	E_k (cm^{-1})	g_i	g_k	Type of transition	A_{ki} (s^{-1})	S (at. u.)	Accu-racy	Source
30.	$a\,^4D - b\,^2F$ (37F)										
		4157.89	7955	31999	8	8	M1	0.018	3.8($-$4)	E	2
		4234.81	8392	31999	6	8	M1	0.0082	1.8($-$4)	E	2
31.	$a\,^4D - b\,^2D$										
		[3532.9]	7955	36253	8	6	M1	0.20	0.0020	D	2
		[3604.6]	8392	36126	6	4	M1	0.014	9.7($-$5)	D	2
		[3588.2]	8392	36253	6	6	M1	0.038	3.9($-$4)	D	2
		[3642.5]	8680	36126	4	4	M1	0.071	5.1($-$4)	D	2
		[3625.8]	8680	36253	4	6	M1	0.058	6.1($-$4)	D	2
		[3664.7]	8847	36126	2	4	M1	0.14	0.0010	D	2
32.	$a\,^4P - a\,^4P$										
		[502950]	13474	13673	6	4	M1	1.9($-$4)	3.6	D	1
		[431590]	13673	13905	4	2	M1	5.5($-$4)	3.3	D	1
33.	$a\,^4P - a\,^2P$										
		[20460]	13474	18361	6	4	M1	0.063	0.080	D	2
		[21328]	13673	18361	4	4	M1	0.032	0.046	D	2
		[20067]	13905	18887	2	2	M1	0.081	0.049	D	2
		[22436]	13905	18361	2	4	M1	0.014	0.023	D	2
34.	$a\,^4P - b\,^4F$										
		[10911]	13474	22637	6	10	E2	0.0039	3.6	E	2
		[10941]	13673	22810	4	8	E2	0.0025	1.9	E	2
		[11066]	13905	22939	2	6	E2	0.0013	0.77	E	2
		[10708]	13474	22810	6	8	E2	0.0016	1.1	E	2
		[10789]	13673	22939	4	6	E2	0.0025	1.3	E	2
		[10954]	13905	23031	2	4	E2	0.0031	1.2	E	2
		[10683]	13673	23031	4	4	E2	0.0013	0.43	E	2
35.	$a\,^4P - b\,^4D$ (39F)										
		[5613.3]	13673	31483	4	8	E2	0.073	1.9	E	2
		[5718.2]	13905	31388	2	6	E2	0.066	1.4	E	2
		5551.31	13474	31483	6	8	E2	0.13	3.3	E	2
		5725.92	13905	31364	2	4	E2	0.039	0.57	E	2
		5580.82	13474	31388	6	6	E2	0.13	2.5	E	2
		5650.94	13673	31364	4	4	E2	0.074	1.0	E	2
		5588.15	13474	31364	6	4	E2	0.072	0.93	E	2
		5649.67	13673	31368	4	2	E2	0.16	1.1	E	2
		[5586.9]	13474	31368	6	2	E2	0.020	0.13	E	2
36.	$a\,^4P - c\,^2D$										
		[4082.0]	13673	38164	4	6	E2	0.033	0.13	E	2
37.	$a\,^2G - b\,^2H$ (41F)										
		9682.13	15845	26170	10	12	E2	0.013	7.9	E	2
		10013.88	16369	26353	8	10	E2	0.010	6.0	E	2
		9513.87	15845	26353	10	10	M1	0.0012	3.8($-$4)	E	2
		″	″	″	10	10	E2	0.0079	3.7	E	2

Fe II: Forbidden transitions — Continued

No.	Multiplet	λ (Å)	E_i (cm^{-1})	E_k (cm^{-1})	g_i	g_k	Type of transition	A_{ki} (s^{-1})	S (at. u.)	Accu-racy	Source
38.	$a\,^2G - a\,^2F$ (42F)										
		[8489.7]	15845	27620	10	6	E2	0.0049	0.77	E	2
		8715.84	15845	27315	10	8	M1	0.0022	4.3(−4)	E	2
		"	"	"	10	8	E2	0.050	12	E	2
		8885.66	16369	27620	8	6	E2	0.012	2.4	E	2
		9133.63	16369	27315	8	8	M1	0.0030	6.8(−4)	E	2
		"	"	"	8	8	E2	0.0039	1.2	E	2
39.	$a\,^2G - b\,^2G$ (43F)										
		6873.87	15845	30389	10	10	E2	0.098	9.0	E	2
		6944.91	16369	30764	8	8	E2	0.087	6.7	E	2
		6700.68	15845	30764	10	8	E2	0.012	0.77	E	2
		7131.13	16369	30389	8	10	E2	0.011	1.2	E	2
40.	$a\,^2G - b\,^2F$ (44F)										
		[6261.1]	15845	31812	10	6	E2	0.014	0.48	E	2
		6188.55	15845	31999	10	8	M1	0.029	0.0020	E	2
		"	"	"	10	8	E2	0.10	4.3	E	2
		6473.86	16369	31812	8	6	M1	0.028	0.0017	E	2
		"	"	"	8	6	E2	0.035	1.4	E	2
		6396.30	16369	31999	8	8	M1	0.045	0.0035	E	2
41.	$a\,^2G - a\,^2I$										
		[5870.0]	15845	32876	10	14	E2	0.14	8.1	E	2
		[6044.1]	16369	32910	8	12	E2	0.11	6.3	E	2
42.	$a\,^2G - c\,^2G$										
		[5673.2]	15845	33466	10	10	E2	0.30	10	E	2
		[5835.4]	16369	33501	8	8	E2	0.32	10	E	2
		[5847.3]	16369	33466	8	10	E2	0.033	1.3	E	2
43.	$a\,^2G - b\,^2D$										
		[4898.6]	15845	36253	10	6	E2	0.82	8.3	E	2
		[5060.1]	16369	36126	8	4	E2	0.63	5.0	E	2
		[5027.9]	16369	36253	8	6	E2	0.087	1.0	E	2
44.	$a\,^2G - c\,^2D$										
		[4479.1]	15845	38164	10	6	E2	0.24	1.5	E	2
		[4576.4]	16369	38215	8	4	E2	0.48	2.3	E	2
45.	$a\,^2P - a\,^2F$ (45F)										
		[11165]	18361	27315	4	8	E2	0.0070	5.8	E	2
		[11447]	18887	27620	2	6	E2	0.0026	1.8	E	2
		10796.48	18361	27620	4	6	E2	0.0021	1.1	E	2
46.	$a\,^2P - b\,^2F$ (47F)										
		[7330.2]	18361	31999	4	8	E2	0.0014	0.14	E	2
		[7734.8]	18887	31812	2	6	E2	0.0044	0.44	E	2
		7432.23	18361	31812	4	6	E2	0.0059	0.48	E	2

Fe II: Forbidden transitions — Continued

No.	Multiplet	λ (Å)	E_i (cm^{-1})	E_k (cm^{-1})	g_i	g_k	Type of transition	A_{ki} (s^{-1})	S (at. u.)	Accu-racy	Source
47.	$a\,^2P - b\,^2D$										
		[5756.7]	18887	36253	2	6	E2	0.018	0.41	E	2
		[5587.5]	18361	36253	4	6	E2	0.036	0.70	E	2
		[5799.0]	18887	36126	2	4	E2	0.081	1.3	E	2
		[5627.2]	18361	36126	4	4	E2	0.15	2.0	E	2
48.	$a\,^2P - c\,^2D$										
		[5186.0]	18887	38164	2	6	E2	0.11	1.5	E	2
		[5048.2]	18361	38164	4	6	E2	0.42	4.9	E	2
		[5172.5]	18887	38215	2	4	E2	0.21	1.9	E	2
		[5035.4]	18361	38215	4	4	E2	0.11	0.85	E	2
49.	$a\,^2H - b\,^2H$										
		[17148]	20340	26170	12	12	E2	0.0026	28	E	2
		[18023]	20806	26353	10	10	E2	0.0020	23	E	2
50.	$a\,^2H - b\,^2G$ (48F)										
		[9590.4]	20340	30764	12	8	E2	0.0016	0.62	E	2
		9949.32	20340	30389	12	10	E2	0.021	12	E	2
		10038.79	20806	30764	10	8	E2	0.019	9.2	E	2
		10432.60	20806	30389	10	10	M1	0.0016	6.7(−4)	E	2
51.	$a\,^2H - b\,^2F$ (49F)										
		[8574.9]	20340	31999	12	8	E2	0.031	6.8	E	2
		[9083.4]	20806	31812	10	6	E2	0.030	6.6	E	2
		8931.47	20806	31999	10	8	E2	0.0034	0.92	E	2
52.	$a\,^2H - a\,^2I$										
		[7975.2]	20340	32876	12	14	E2	0.069	19	E	2
		[8259.4]	20806	32910	10	12	E2	0.060	16	E	2
53.	$a\,^2D2 - b\,^2G$ (50F)										
		[10127]	20517	30389	6	10	E2	0.0076	4.8	E	2
		[10572]	21308	30764	4	8	E2	0.0067	4.2	E	2
54.	$a\,^2D2 - c\,^2G$										
		[7720.2]	20517	33466	6	10	E2	0.029	4.7	E	2
		[8199.0]	21308	33501	4	8	E2	0.019	3.4	E	2
55.	$a\,^2D2 - b\,^2D$										
		[6353.1]	20517	36253	6	6	E2	0.17	6.3	E	2
		[6746.5]	21308	36126	4	4	E2	0.056	1.9	E	2
		[6404.6]	20517	36126	6	4	E2	0.062	1.6	E	2
		[6689.4]	21308	36253	4	6	E2	0.043	2.1	E	2
56.	$a\,^2D2 - a\,^2S$										
		[5982.7]	20517	37227	6	2	E2	0.25	2.3	E	2
		[6280.0]	21308	37227	4	2	E2	0.17	2.0	E	2

Fe II: Forbidden transitions — Continued

No.	Multiplet	λ (Å)	E_i (cm^{-1})	E_k (cm^{-1})	g_i	g_k	Type of transition	A_{ki} (s^{-1})	S (at. u.)	Accuracy	Source
57.	$a\ ^2$D2 – $c\ ^2$D										
		[5913.3]	21308	38215	4	4	E2	0.090	1.5	E	2
58.	$b\ ^4$P – $b\ ^2$P										
		[20168]	20831	25788	6	4	M1	0.043	0.052	D	2
		[19523]	21812	26933	4	2	M1	0.0049	0.0027	D	2
		[25147]	21812	25788	4	4	M1	0.015	0.035	D	2
		[22104]	22410	26933	2	2	M1	0.024	0.019	D	2
		[29597]	22410	25788	2	4	M1	0.0064	0.025	D	2
59.	$b\ ^4$P – $b\ ^4$D										
		[9384.8]	20831	31483	6	8	M1	0.055	0.013	E	2
		[9469.5]	20831	31388	6	6	M1	0.030	0.0057	E	2
		[10466]	21812	31364	4	4	M1	0.043	0.0073	E	2
		[11159]	22410	31368	2	2	M1	0.057	0.0059	E	2
		[9490.6]	20831	31364	6	4	M1	0.021	0.0027	E	2
		[10461]	21812	31368	4	2	M1	0.037	0.0031	E	2
60.	$b\ ^4$P – $b\ ^2$D										
		[6482.3]	20831	36253	6	6	M1	0.042	0.0025	E	2
		[6922.9]	21812	36253	4	6	M1	0.010	7.4(−4)	E	2
61.	$b\ ^4$P – $a\ ^2$S										
		[6485.3]	21812	37227	4	2	M1	0.73	0.015	E	2
		[6746.9]	22410	37227	2	2	M1	0.20	0.0046	E	2
62.	$b\ ^4$P – $c\ ^2$D										
		[5767.5]	20831	38164	6	6	M1	0.082	0.0035	E	2
		[6095.0]	21812	38215	4	4	M1	0.047	0.0016	E	2
63.	$a\ ^4$H – $b\ ^2$G										
		[11160]	21430	30389	12	10	M1	0.020	0.010	E	2
		[10887]	21582	30764	10	8	M1	0.020	0.0077	E	2
		[11352]	21582	30389	10	10	M1	0.0045	0.0024	E	2
		[11044]	21712	30764	8	8	M1	0.014	0.0056	E	2
64.	$a\ ^4$H – $a\ ^2$I										
		[8600.5]	21252	32876	14	14	M1	0.097	0.032	E	2
		[8708.8]	21430	32910	12	12	M1	0.038	0.011	E	2
		[8575.2]	21252	32910	14	12	M1	0.0014	3.9(−4)	E	2
		[8734.8]	21430	32876	12	14	M1	0.039	0.013	E	2
		[8825.1]	21582	32910	10	12	M1	0.038	0.012	E	2
65.	$a\ ^4$H – $c\ ^2$G										
		[8306.1]	21430	33466	12	10	M1	0.15	0.032	E	2
		[8387.2]	21582	33501	10	8	M1	0.15	0.026	E	2
		[8411.8]	21582	33466	10	10	M1	0.019	0.0042	E	2
		[8479.9]	21712	33501	8	8	M1	0.10	0.018	E	2

Fe II: Forbidden transitions — Continued

No.	Multiplet	λ (Å)	E_i (cm^{-1})	E_k (cm^{-1})	g_i	g_k	Type of transition	A_{ki} (s^{-1})	S (at. u.)	Accuracy	Source
66.	$b\,^4F - b\,^2G$										
		[12897]	22637	30389	10	10	M1	0.020	0.016	E	2
		[12569]	22810	30764	8	8	M1	0.0065	0.0038	E	2
		[13192]	22810	30389	8	10	M1	0.0070	0.0060	E	2
		[12776]	22939	30764	6	8	M1	0.0073	0.0045	E	2
67.	$b\,^4F - b\,^4D$										
		[11655]	22810	31388	8	6	M1	0.048	0.017	E	2
68.	$b\,^4F - c\,^2G$										
		[9231.7]	22637	33466	10	10	M1	0.21	0.061	E	2
		[9351.2]	22810	33501	8	8	M1	0.10	0.024	E	2
		[9202.1]	22637	33501	10	8	M1	0.013	0.0030	E	2
		[9381.7]	22810	33466	8	10	M1	0.070	0.021	E	2
		[9465.4]	22939	33501	6	8	M1	0.065	0.016	E	2
69.	$b\,^4F - c\,^2D$										
		[6511.2]	22810	38164	8	6	M1	0.16	0.0098	E	2
		[6544.8]	22939	38215	6	4	M1	0.19	0.0079	E	2
		[6566.4]	22939	38164	6	6	M1	0.022	0.0014	E	2
		[6584.4]	23031	38215	4	4	M1	0.11	0.0047	E	2
70.	$b\,^2P - b\,^4D$										
		[17851]	25788	31388	4	6	M1	0.011	0.014	E	2
		[22559]	26933	31364	2	4	M1	0.014	0.024	E	2
71.	$b\,^2P - b\,^2D$										
		[9552.7]	25788	36253	4	6	M1	0.022	0.0043	E	2
		[10874]	26933	36126	2	4	M1	0.015	0.0029	E	2
		[9669.7]	25788	36126	4	4	M1	0.075	0.010	E	2
72.	$b\,^2P - a\,^2S$										
		[8739.1]	25788	37227	4	2	M1	0.23	0.011	E	2
		[9711.2]	26933	37227	2	2	M1	0.23	0.016	E	2
73.	$b\,^2P - c\,^2D$										
		[8077.5]	25788	38164	4	6	M1	0.021	0.0025	E	2
		[8861.4]	26933	38215	2	4	M1	0.013	0.0013	E	2
		[8044.8]	25788	38215	4	4	M1	0.030	0.0023	E	2
74.	$b\,^2H - b\,^2G$										
		[23699]	26170	30389	12	10	M1	0.019	0.094	E	2
		[22661]	26353	30764	10	8	M1	0.019	0.066	E	2
		[24772]	26353	30389	10	10	M1	0.027	0.15	E	2
75.	$b\,^2H - a\,^2I$										
		[14909]	26170	32876	12	14	M1	0.016	0.028	E	2
		[15246]	26353	32910	10	12	M1	0.013	0.020	E	2
		[14833]	26170	32910	12	12	M1	0.031	0.045	E	2

Fe II: Forbidden transitions — Continued

No.	Multiplet	λ (Å)	E_i (cm^{-1})	E_k (cm^{-1})	g_i	g_k	Type of transition	A_{ki} (s^{-1})	S (at. u.)	Accuracy	Source
76.	$b\,^2$H – $c\,^2$G										
		[13702]	26170	33466	12	10	M1	0.034	0.032	E	2
		[13985]	26353	33501	10	8	M1	0.036	0.029	E	2
		[14054]	26353	33466	10	10	M1	0.065	0.067	E	2
77.	$a\,^2$F – $c\,^2$G										
		[16252]	27315	33466	8	10	M1	0.037	0.059	E	2
		[17000]	27620	33501	6	8	M1	0.012	0.017	E	2
		[16160]	27315	33501	8	8	M1	0.073	0.091	E	2
78.	$a\,^2$F – $c\,^2$D										
		[9436.6]	27620	38215	6	4	M1	0.045	0.0056	E	2
79.	$b\,^4$D – $c\,^2$D										
		[14964]	31483	38164	8	6	M1	0.029	0.022	E	2
		[14603]	31368	38215	2	4	M1	0.027	0.012	E	2

[a]The number in parentheses following the tabulated value indicates the power of ten by which this value has to be multiplied.

Fe III

Cr Isoelectronic Sequence

Ground State: $1s^2 2s^2 2p^6 3s^2 3p^6 3d^6\,^5D_4$

Ionization Energy: 30.652 eV = 247220 cm^{-1}

Allowed Transitions

List of tabulated lines

Wavelength (Å)	No.	Wavelength (Å)	No.	Wavelength (Å)	No.	Wavelength (Å)	No.
1843.4	15	1930.39	2	1961.23	7	2058.2	9
1844.3	15	1931.51	7	1962.72	7	2058.56	11
1845.0	15	1937.35	2	1964.26	7	2059.68	9
1846.9	15	1943.48	2	1966.20	7	2061.75	9
1849.41	12	1945.34	7	1987.50	1	2084.97	8
1854.38	12	1950.33	13	1991.61	1	2087.13	8
1865.20	17	1951.01	5	1994.07	1	2087.91	8
1893.98	10	1951.3	5	1995.27	1	2088.63	4
1896.80	10	1952.3	5	1995.56	1	2089.09	8
1898.9	6	1952.65	5	1996.42	1	2090.1	8
1903.3	6	1953.32	5	2002.5	18	2090.14	4
1904.3	19	1953.5	5	2039.51	14	2091.31	8
1907.58	10	1954.22	7	2053.5	9	2097.48	4
1915.08	2	1954.98	13	2057.06	9	2103.80	3
1922.79	2	1959.32	7	2057.9	16	2107.32	3

For this spectrum we have chosen the calculations of Biemont[1] and of Kurucz and Peytremann.[2] Biemont obtained radial wavefunctions by means of the scaled Thomas-Fermi method and calculated individual line strengths in intermediate coupling. Similarly, Kurucz and Peytremann used a semiempirical scaled Thomas-Fermi-Dirac approach, with very limited configuration interaction. Generally, the log gf-values of Refs. 1 and 2 are in quite good agreement, particularly for strong lines; e.g., 68% of the data for common lines agree within $\pm 50\%$. In this compilation, we have included only those lines showing 50% or better agreement between Refs. 1 and 2.

We were able to assess the reliability of Kurucz and Peytremann's (or Biemont's) absolute scale by comparing reciprocals of sums of the calculated transition probabilities to beam-foil lifetimes for four excited levels measured by Anderson et al.[3] We considered only Ref. 2 for this study because its branching ratio data are fairly complete, while those of Biemont are not. The comparison shows that the beam-foil lifetimes are, on the average, about 14% longer than the corresponding $(\sum_i A_{ki})^{-1}$ values of Kurucz and Peytremann.

References

[1]E. Biemont, J. Quant. Spectrosc. Radiat. Transfer **16**, 137 (1976).
[2]R. L. Kurucz and E. Peytremann, Smithsonian Astrophysical Observatory Special Report 362 (1975).
[3]T. Andersen, P. Petersen, and E. Biemont, J. Quant. Spectrosc. Radiat. Transfer **17**, 389 (1977).

Fe III: Allowed transitions

No.	Transition Array	Multiplet	λ (Å)	E_i (cm^{-1})	E_k (cm^{-1})	g_i	g_k	A_{ki} (10^8 s^{-1})	f_{ik}	S (at. u.)	log gf	Accuracy	Source
1.	$3d^5(^4G)4s-$ $3d^5(^4G)4p$	$^5G - ^5G°$ (uv 50)											
			1987.50	63425	113740	13	13	4.9	0.29	25	0.58	D	1,2
			1991.61	63466	113677	11	11	4.2	0.25	18	0.44	D	1,2
			1994.07	63487	113635	9	9	3.5	0.21	12	0.28	D	1,2
			1995.56	63494	113605	7	7	3.7	0.22	10	0.19	D	1,2
			1996.42	63495	113584	5	5	4.2	0.25	8.2	0.10	D	1,2
			1995.27	63487	113605	9	7	1.0	0.048	2.8	−0.36	D	1,2
			1996.42	63494	113584	7	5	0.96	0.041	1.9	−0.54	D	1,2
2.		$^5G - ^5H°$ (uv 51)											
			1915.08	63425	115642	13	15	6.0	0.38	31	0.69	D	1,2
			1922.79	63466	115474	11	13	5.5	0.36	25	0.60	D	1,2
			1930.39	63487	115290	9	11	5.1	0.35	20	0.50	D	1,2
			1937.35	63494	115111	7	9	5.1	0.37	17	0.41	D	1,2
			1943.48	63495	114949	5	7	5.0	0.40	13	0.30	D	1,2
3.		$^3G - ^3F°$ (uv 66)											
			2103.80	70729	118247	9	7	2.9	0.15	9.3	0.13	D	1,2
			2107.32	70725	118164	7	5	3.8	0.18	8.7	0.10	D	1,2
4.		$^3G - ^3H°$ (uv 67)											
			2097.48	70694	118355	11	13	4.5	0.35	27	0.59	D	1,2
			2090.14	70729	118557	9	11	4.4	0.35	22	0.50	D	1,2
			2088.63	70694	118557	11	11	0.17	0.011	0.83	−0.92	D	1,2
5.		$^3G - ^3G°$ (uv 68)											
			1951.01	70694	121950	11	11	5.3	0.30	21	0.52	D	1,2
			1952.65	70729	121941	9	9	4.9	0.28	16	0.40	D	1,2
			1953.32	70725	121920	7	7	5.1	0.29	13	0.31	D	1,2
			[1951.3]	70694	121941	11	9	0.34	0.016	1.1	−0.75	D	1,2
			[1953.5]	70729	121920	9	7	0.40	0.018	1.0	−0.79	D	1,2
			[1952.3]	70729	121950	9	11	0.20	0.014	0.81	−0.90	D	1,2

Fe III: Allowed transitions — Continued

No.	Transition Array	Multiplet	λ (Å)	E_i (cm^{-1})	E_k (cm^{-1})	g_i	g_k	A_{ki} (10^8 s^{-1})	f_{ik}	S (at. u.)	log gf	Accuracy	Source
6.	$3d^5(^4\text{P})4s-$ $3d^5(^4\text{P})4p$	^{3}P – ^{3}S°											
			[1898.9]	73728	126391	5	3	3.7	0.12	3.8	−0.22	D	1,2
			[1903.3]	73849	126391	3	3	1.8	0.099	1.9	−0.53	D	1,2
7.	$3d^5(^4\text{D})4s-$ $3d^5(^4\text{D})4p$	^{5}D – ^{5}F° (uv 61)											
			1931.51	69696	121469	9	11	5.3	0.36	21	0.51	D	1,2
			1945.34	69837	121242	7	9	3.7	0.27	12	0.28	D	1,2
			1954.22	69838	121009	5	7	3.5	0.28	9.0	0.15	D	1,2
			1959.32	69788	120826	3	5	2.8	0.27	5.2	−0.09	D	1,2
			1962.72	69747	120697	1	3	2.3	0.39	2.5	−0.41	D	1,2
			1954.22	69837	121009	7	7	1.3	0.074	3.3	−0.29	D	1,2
			1961.23	69838	120826	5	5	1.7	0.10	3.2	−0.30	D	1,2
			1964.26	69788	120697	3	3	2.2	0.13	2.5	−0.42	D	1,2
			1966.20	69838	120697	5	3	0.28	0.0099	0.32	−1.31	D	1,2
8.		^{3}D – ^{3}D° (uv 77)											
			2087.13	76957	124854	7	7	3.1	0.20	9.6	0.15	D	1,2
			2091.31	77102	124904	5	5	2.6	0.17	5.9	−0.07	D	1,2
			2087.91	77075	124955	3	3	2.9	0.19	3.9	−0.24	D	1,2
			2084.97	76957	124904	7	5	0.75	0.035	1.7	−0.61	D	1,2
			2089.09	77102	124955	5	3	1.1	0.043	1.5	−0.67	D	1,2
			[2090.1]	77075	124904	3	5	0.64	0.070	1.4	−0.68	D	1,2
9.		^{3}D – ^{3}F° (uv 78)											
			2061.75	76957	125444	7	9	4.4	0.36	17	0.40	D	1,2
			2059.68	77102	125638	5	7	3.9	0.35	12	0.24	D	1,2
			2057.06	77075	125673	3	5	3.7	0.39	7.9	0.07	D	1,2
			[2053.5]	76957	125638	7	7	0.44	0.028	1.3	−0.71	D	1,2
			[2058.2]	77102	125673	5	5	0.76	0.048	1.6	−0.62	D	1,2
10.	$3d^5(^2\text{I})4s-$ $3d^5(^2\text{I})4p$	^{3}I – ^{3}H° (uv 83)											
			1907.58	79840	132263	15	13	5.3	0.25	24	0.57	D	1,2
			1896.80	79845	132565	13	11	5.0	0.23	19	0.48	D	1,2
			1893.98	79860	132659	11	9	5.5	0.24	16	0.42	D	1,2
11.		^{1}I – ^{1}K° (uv 100)	2058.56	83430	131992	13	15	4.5	0.33	29	0.63	D	1,2
12.	$3d^5(^4\text{F})4s-$ $3d^5(^4\text{F})4p$	^{5}F – ^{5}D° (uv 97)											
			1849.41	83138	137210	11	9	4.3	0.18	12	0.30	D	1,2
			1854.38	83647	137573	3	1	5.7	0.098	1.8	−0.53	D	1,2

Fe III:　Allowed transitions — Continued

No.	Transition Array	Multiplet	λ (Å)	E_i (cm^{-1})	E_k (cm^{-1})	g_i	g_k	A_{ki} (10^8 s^{-1})	f_{ik}	S (at. u.)	log gf	Accuracy	Source
13.	$3d^5(^2$H$)4s-$ $3d^5(^2$H$)4p$	^{3}H – ^{3}I° (uv 116)											
			1950.33	88923	140196	13	15	5.5	0.36	30	0.67	D	1,2
			1954.98	88695	139846	11	13	4.3	0.29	21	0.50	D	1,2
14.		^{1}H – ^{1}I° (uv 134)	2039.51	92524	141540	11	13	4.3	0.32	24	0.55	D	1,2
15.	$3d^5(^2$F1$)4s-$ $3d^5(b\ ^2$F$)4p$	^{3}F – ^{3}D°											
			[1843.4]	93389	147636	9	7	4.8	0.19	10	0.23	D	1,2
			[1844.3]	93392	147615	7	5	4.9	0.18	7.7	0.10	D	1,2
			[1846.9]	93413	147556	5	3	5.5	0.17	5.2	−0.07	D	1,2
			[1845.0]	93413	147615	5	5	0.78	0.040	1.2	−0.70	D	1,2
16.		^{1}F – ^{1}D°	[2057.9]	97041	145618	7	5	3.7	0.17	8.1	0.08	D	1,2
17.		^{1}F – ^{1}F° (uv 154)	1865.20	97041	150655	7	7	6.1	0.32	14	0.35	D	1,2
18.	$3d^5(^2$D2$)4s-$ $3d^5(b\ ^2$D$)4p$	^{1}D – ^{1}F°	[2002.5]	109571	159493	5	7	4.3	0.36	12	0.26	D	1,2
19.		^{1}D – ^{1}D°	[1904.3]	109571	162085	5	5	5.7	0.31	9.7	0.19	D	1,2

Fe III

Forbidden Transitions

List of tabulated lines

Wavelength (Å)	No.	Wavelength (Å)	No.	Wavelength (Å)	No.	Wavelength (Å)	No.
3236.7	6	3356.6	5	4667.0	3	5011.3	1
3239.7	5	3366.2	5	4701.62	3	5084.8	1
3286.2	5	4008.3	4	4733.93	3	5270.3	1
3301.6	5	4046.4	4	4754.83	3	5412.0	1
3319.2	5	4079.7	4	4769.60	3	6096.3	8
3333.8	5	4096.6	4	4777.88	3	6614.0	8
3334.9	5	4607.13	3	4881.11	2	7078.2	7
3355.5	5	4658.10	3	4930.5	1		

We have compiled the data of Garstang,[1] who calculated transition probabilities for 186 [Fe III] lines between levels of the $3d^6$ configuration. The lines tabulated here represent a small subset of Garstang's complete list; i.e., we have selected only the relatively strong transitions, each one being a magnetic dipole line.

As usual, it is difficult to assess the uncertainties of these theoretical data. Furthermore, comparisons are not possible for this spectrum, since Ref. 1 is the only data source available. Nevertheless, for this spectrum, Garstang could assemble in his paper some astrophysical observational intensity data for the lines of a fairly prominent multiplet (our running number 3), and he has found a suprisingly close agreement between the astrophysical intensities and his calculated line intensities. This good agreement, as well as similar good agreement with astrophysical observations for the spectra of [Fe II] and [Ni II] (see the General Introduction) and our previous experience with the evaluation of forbidden line data on simpler atoms, suggests that the strengths of the listed stronger lines should be accurate to within fifty percent. We expect, on the other hand, that larger uncertainties will be encountered in the weaker lines which are not included in our tabulation.

Reference

[1]R. H. Garstang, Mon. Not. R. Astron. Soc. **117**, 393 (1957).

Fe III: Forbidden transitions

No.	Transition Array	Multiplet	λ (Å)	E_i (cm^{-1})	E_k (cm^{-1})	g_i	g_k	Type of transition	A_{ki} (s^{-1})	S (at. u.)	Accuracy	Source
1.	$3d^6$–$3d^6$	^{5}D – ^{3}P2 (1F)										
			5270.3	436.2	19405	7	5	M1	0.40	0.011	D	1
			5011.3	738.9	20688	5	3	M1	0.53	0.0074	D	1
			4930.5	932.4	21209	3	1	M1	0.67	0.0030	D	1
			5412.0	932.4	19405	3	5	M1	0.038	0.0011	D	1
			5084.8	1027.3	20688	1	3	M1	0.091	0.0013	D	1
2.		^{5}D – ^{3}H (2F)										
			4881.11	0.0	20482	9	9	M1	0.0048	1.9(−4)^a	D	1
3.		^{5}D – ^{3}F2 (3F)										
			4658.10	0.0	21462	9	9	M1	0.44	0.015	D	1
			4701.62	436.2	21700	7	7	M1	0.27	0.0073	D	1
			4733.93	738.9	21857	5	5	M1	0.10	0.0020	D	1
			4607.13	0.0	21700	9	7	M1	0.038	9.6(−4)	D	1
			4667.0	436.2	21857	7	5	M1	0.026	4.9(−4)	D	1
			4754.83	436.2	21462	7	9	M1	0.081	0.0029	D	1
			4769.60	738.9	21700	5	7	M1	0.087	0.0024	D	1
			4777.88	932.4	21857	3	5	M1	0.049	9.9(−4)	D	1
4.		^{5}D – ^{3}G (4F)										
			4079.7	436.2	24941	7	9	M1	0.0037	8.4(−5)	D	1
			4096.6	738.9	25142	5	7	M1	0.0027	4.8(−5)	D	1
			4008.3	0.0	24941	9	9	M1	0.019	4.1(−4)	D	1
			4046.4	436.2	25142	7	7	M1	0.0080	1.4(−4)	D	1
5.		^{5}D – ^{3}D (6F)										
			3239.7	0.0	30858	9	7	M1	0.23	0.0020	D	1
			3301.6	436.2	30716	7	5	M1	0.027	1.8(−4)	D	1
			3333.8	738.9	30726	5	3	M1	0.0013	5.4(−6)	D	1
			3286.2	436.2	30858	7	7	M1	0.047	4.3(−4)	D	1
			3334.9	738.9	30716	5	5	M1	0.11	7.6(−4)	D	1
			3355.5	932.4	30726	3	3	M1	0.015	6.3(−5)	D	1
			3319.2	738.9	30858	5	7	M1	0.044	4.2(−4)	D	1
			3356.6	932.4	30716	3	5	M1	0.095	6.7(−4)	D	1
			3366.2	1027.3	30726	1	3	M1	0.13	5.5(−4)	D	1

Fe III: Forbidden transitions — Continued

No.	Transition Array	Multiplet	λ (Å)	E_i (cm^{-1})	E_k (cm^{-1})	g_i	g_k	Type of transition	A_{ki} (s^{-1})	S (at. u.)	Accuracy	Source
6.		^{5}D – ^{1}G2 (7F)										
			3236.7	0.0	30886	9	9	M1	0.0022	2.5(−5)	D	1
7.		^{3}P2 – ^{1}S2 (9F)										
			7078.2	20688	34812	3	1	M1	1.5	0.020	D	1
8.		^{3}P2 – ^{1}D2 (10F)										
			6096.3	19405	35804	5	5	M1	0.096	0.0040	D	1
			6614.0	20688	35804	3	5	M1	0.033	0.0018	D	1

^{a}The number in parentheses following the tabulated value indicates the power of ten by which this value has to be multiplied.

Fe IV

V Isoelectronic Sequence

Ground State: $1s^2 2s^2 2p^6 3s^2 3p^6 3d^5$ ^{6}S$_{5/2}$

Ionization Energy: 54.8 eV = 442000 cm^{-1}

Forbidden Transitions

List of tabulated lines

Wavelength (Å)	No.	Wavelength (Å)	No.	Wavelength (Å)	No.	Wavelength (Å)	No.
2567.4	2	3980.5	6	5236.1	3	10775	14
2567.6	2	4144.2	5	5237.3	3	11001	18
2791.0	16	4152.3	5	5911.0	8	11133	14
2791.2	16	4198.2	5	5938.9	8	11579	20
2792.1	16	4206.6	5	7111.1	11	11680	20
2827.7	13	4208.9	5	7171.4	11	11816	20
2829.4	1	4598.4	24	7184.0	11	11878	20
2835.7	1	4619.2	24	7190.8	11	12017	20
2840.1	13	4619.7	24	7191.2	11	12018	20
2840.2	13	4866.0	4	7192.4	11	16624	17
2840.5	13	4868.0	4	7192.8	11	19074	17
2843.2	13	4868.2	4	7222.8	11	19601	19
2843.4	13	4869.0	4	7555.6	10	19892	19
3185.1	9	4869.2	4	7557.4	10	21214	19
3192.6	9	4888.6	4	7704.1	23	21876	19
3593.1	12	4900.0	4	7725.8	23	27134	7
3598.4	12	4903.1	4	7762.6	23	27157	7
3922.1	22	4906.6	4	7925.0	10	27732	7
3922.5	22	4918.0	4	8024.1	10	28055	7
3933.5	6	5032.5	3	9393.6	15	28356	7
3970.9	22	5033.6	3	10241	18	28645	7
3971.3	22	5233.8	3	10761	14	60120	21

The data for this spectrum were taken from the calculations of Garstang.[1] We have selected all lines of at least moderate strength, the majority of which are due to magnetic dipole radiation. At the time of Garstang's calculations, no observational analysis of the [Fe IV] spectrum existed, and therefore, no precise wavelengths for the forbidden lines were available. In the meantime, Edlén[2] and Ekberg and Edlén[3] have completed an analysis of this spectrum, so that accurate wavelengths and energy levels are now known, which we we have utilized in this compilation. For the lines tabulated here, the differences between the calculated wavelengths of Ref. 1 and the data derived from the observational analyses in Refs. 2 and 3 turned out to be quite small, the largest being approximately four percent. Therefore, we have not made any corrections in Garstang's transition probabilities due to the slight changes in the wavelengths.

References

[1]R. H. Garstang, Mon. Not. R. Astron. Soc. **118**, 572 (1958).
[2]B. Edlén, Mon. Not. R. Astron. Soc. **144**, 391 (1969).
[3]J. O. Ekberg and B. Edlén, Phys. Scr. **18**, 107 (1978).

Fe IV: Forbidden transitions

No.	Transition Array	Multiplet	λ (Å)	E_i (cm^{-1})	E_k (cm^{-1})	g_i	g_k	Type of transition	A_{ki} (s^{-1})	S (at. u.)	Accuracy	Source
1.	$3d^5$–$3d^5$	^6S – ^4P										
			[2835.7]	0	35254	6	6	M1	1.4	0.0071	D	1
			[2829.4]	0	35333	6	4	M1	0.88	0.0030	D	1
2.		^6S – ^4D										
			[2567.6]	0	38935	6	6	M1	0.051	1.9(−4)^a	D	1
			[2567.4]	0	38938	6	4	M1	0.038	9.5(−5)	D	1
3.		^4G – $^2F2$										
			[5233.8]	32293	51394	10	8	M1	0.50	0.021	D	1
			[5033.6]	32306	52167	8	6	M1	0.47	0.013	D	1
			[5237.3]	32306	51394	8	8	M1	0.080	0.0034	D	1
			[5032.5]	32301	52167	6	6	M1	0.21	0.0060	D	1
			[5236.1]	32301	51394	6	8	M1	0.015	6.4(−4)	D	1
4.		^4G – ^4F										
			[4888.6]	32246	52695	12	8	E2	0.040	0.53	E	1
			[4866.0]	32293	52838	10	6	E2	0.061	0.59	E	1
			[4869.2]	32306	52837	8	4	E2	0.071	0.46	E	1
			[4906.6]	32246	52621	12	10	M1	0.11	0.0048	D	1
			"	"	"	12	10	E2	0.21	3.6	E	1
			[4900.0]	32293	52695	10	8	M1	0.039	0.0014	D	1
			"	"	"	10	8	E2	0.15	2.0	E	1
			[4869.0]	32306	52838	8	6	E2	0.11	1.1	E	1
			[4868.2]	32301	52837	6	4	M1	0.14	0.0024	D	1
			"	"	"	6	4	E2	0.18	1.2	E	1
			[4918.0]	32293	52621	10	10	M1	0.073	0.0032	D	1
			"	"	"	10	10	E2	0.045	0.77	E	1
			[4903.1]	32306	52695	8	8	M1	0.13	0.0045	D	1
			"	"	"	8	8	E2	0.067	0.90	E	1
			[4868.0]	32301	52838	6	6	M1	0.19	0.0049	D	1
			"	"	"	6	6	E2	0.064	0.62	E	1
5.		^4G – ^2H										
			[4144.2]	32246	56369	12	12	M1	0.47	0.015	D	1
			[4206.6]	32293	56058	10	10	M1	0.042	0.0012	D	1
			[4198.2]	32246	56058	12	10	M1	0.56	0.015	D	1
			[4152.3]	32293	56369	10	12	M1	0.61	0.019	D	1
			[4208.9]	32306	56058	8	10	M1	0.13	0.0036	D	1

Fe IV: Forbidden transitions — Continued

No.	Transition Array	Multiplet	λ (Å)	E_i (cm^{-1})	E_k (cm^{-1})	g_i	g_k	Type of transition	A_{ki} (s^{-1})	S (at. u.)	Accuracy	Source
6.		$^4G - ^2G2$										
			[3980.5]	32293	57408	10	8	M1	0.077	0.0014	D	1
			[3933.5]	32306	57721	8	10	M1	0.064	0.0014	D	1
7.		$^4P - ^4D$										
			[28356]	35254	38779	6	8	M1	0.038	0.26	D	1
			[27157]	35254	38935	6	6	M1	0.022	0.098	D	1
			[27732]	35333	38938	4	4	M1	0.039	0.12	D	1
			[28645]	35407	38897	2	2	M1	0.058	0.10	D	1
			[27134]	35254	38938	6	4	M1	0.018	0.053	D	1
			[28055]	35333	38897	4	2	M1	0.034	0.056	D	1
8.		$^4P - ^2F2$										
			[5938.9]	35333	52167	4	6	M1	0.013	6.1(−4)	D	1
			[5911.0]	35254	52167	6	6	M1	0.087	0.0040	D	1
9.		$^4P - ^2S$										
			[3185.1]	35333	66720	4	2	M1	0.73	0.0017	D	1
			[3192.6]	35407	66720	2	2	M1	0.11	2.7(−4)	D	1
10.		$^4D - ^2F2$										
			[7925.0]	38779	51394	8	8	M1	0.10	0.015	D	1
			[7555.6]	38935	52167	6	6	M1	0.022	0.0021	D	1
			[8024.1]	38935	51394	6	8	M1	0.034	0.0052	D	1
			[7557.4]	38938	52167	4	6	M1	0.021	0.0020	D	1
11.		$^4D - ^4F$										
			[7222.8]	38779	52621	8	10	M1	0.14	0.020	D	1
			"	"	"	8	10	E2	0.023	2.7	E	1
			[7192.4]	38938	52838	4	6	M1	0.023	0.0019	D	1
			[7171.4]	38897	52837	2	4	M1	0.10	0.055	D	1
			[7184.0]	38779	52695	8	8	M1	0.10	0.011	D	1
			"	"	"	8	8	E2	0.011	1.0	E	1
			[7190.8]	38935	52838	6	6	M1	0.18	0.015	D	1
			"	"	"	6	6	E2	0.012	0.82	E	1
			[7192.8]	38938	52837	4	4	M1	0.16	0.0088	D	1
			"	"	"	4	4	E2	0.016	0.73	E	1
			[7111.1]	38779	52838	8	6	M1	0.028	0.0022	D	1
			[7191.2]	38935	52837	6	4	M1	0.039	0.0022	D	1
12.		$^4D - ^2S$										
			[3598.4]	38938	66720	4	2	M1	0.010	3.5(−5)	D	1
			[3593.1]	38897	66720	2	2	M1	0.078	2.7(−4)	D	1
13.		$^4D - ^2D2$										
			[2827.7]	38779	74133	8	6	M1	0.72	0.0036	D	1
			[2843.2]	38935	74097	6	4	M1	0.090	3.1(−4)	D	1
			[2840.2]	38935	74133	6	6	M1	0.16	8.2(−4)	D	1
			[2843.4]	38938	74097	4	4	M1	0.22	7.5(−4)	D	1
			[2840.5]	38938	74133	4	6	M1	0.027	1.4(−4)	D	1
			[2840.1]	38897	74097	2	4	M1	0.63	0.0021	D	1

Fe IV: Forbidden transitions — Continued

No.	Transition Array	Multiplet	λ (Å)	E_i (cm^{-1})	E_k (cm^{-1})	g_i	g_k	Type of transition	A_{ki} (s^{-1})	S (at. u.)	Accu-racy	Source
14.		^{2}I – ^{2}H										
			[10775]	47091	56369	14	12	M1	0.094	0.052	D	1
			[11133]	47079	56058	12	10	M1	0.060	0.031	D	1
			[10761]	47079	56369	12	12	M1	0.18	0.10	D	1
15.		^{2}I – ^{2}G2										
			[9393.6]	47079	57721	12	10	M1	0.036	0.011	D	1
16.		^{2}I – ^{2}G1										
			[2792.1]	47091	82895	14	10	E2	4.1	4.1	E	1
			[2791.0]	47079	82897	12	8	E2	4.2	3.4	E	1
			[2791.2]	47079	82895	12	10	E2	0.12	0.12	E	1
17.		^{2}F2 – ^{2}G2										
			[19074]	52167	57408	6	8	M1	0.013	0.027	D	1
			[16624]	51394	57408	8	8	M1	0.026	0.035	D	1
18.		^{2}F2 – ^{2}F1										
			[10241]	51394	61157	8	6	M1	0.047	0.011	D	1
			[11001]	52167	61254	6	8	M1	0.051	0.020	D	1
19.		^{4}F – ^{2}G2										
			[19601]	52621	57721	10	10	M1	0.048	0.13	D	1
			[21214]	52695	57408	8	8	M1	0.020	0.057	D	1
			[19892]	52695	57721	8	10	M1	0.017	0.050	D	1
			[21876]	52838	57408	6	8	M1	0.013	0.040	D	1
20.		^{4}F – ^{2}F1										
			[11579]	52621	61254	10	8	M1	0.11	0.051	D	1
			[11816]	52695	61157	8	6	M1	0.027	0.0099	D	1
			[11680]	52695	61254	8	8	M1	0.013	0.0061	D	1
			[12018]	52838	61157	6	6	M1	0.016	0.0062	D	1
			[11878]	52838	61254	6	8	M1	0.026	0.013	D	1
			[12017]	52837	61157	4	6	M1	0.098	0.038	D	1
21.		^{2}H – ^{2}G2										
			[60120]	56058	57721	10	10	M1	0.013	1.0	D	1
22.		^{2}G2 – ^{2}G1										
			[3971.3]	57721	82895	10	10	M1	0.017	3.9(−4)	D	1
			"	"	"	10	10	E2	0.29	1.7	E	1
			[3922.1]	57408	82897	8	8	E2	0.40	1.8	E	1
			[3970.9]	57721	82897	10	8	E2	0.028	0.13	E	1
			[3922.5]	57408	82895	8	10	M1	0.029	6.5(−4)	D	1
			"	"	"	8	10	E2	0.033	0.18	E	1
23.		^{2}F1 – ^{2}D2										
			[7762.6]	61254	74133	8	6	M1	0.10	0.010	D	1
			[7725.8]	61157	74097	6	4	M1	0.10	0.0068	D	1
			[7704.1]	61157	74133	6	6	M1	0.18	0.018	D	1

Fe IV: Forbidden transitions — Continued

No.	Transition Array	Multiplet	λ (Å)	E_i (cm^{-1})	E_k (cm^{-1})	g_i	g_k	Type of transition	A_{ki} (s^{-1})	S (at. u.)	Accuracy	Source
24.		^{2}F1 – ^{2}G1										
			[4619.7]	61254	82895	8	10	M1	0.10	0.0037	D	1
			″	″	″	8	10	E2	0.058	0.73	E	1
			[4598.4]	61157	82897	6	8	M1	0.091	0.0026	D	1
			″	″	″	6	8	E2	0.055	0.54	E	1
			[4619.2]	61254	82897	8	8	M1	0.22	0.0064	D	1

^{a}The number in parentheses following the tabulated value indicates the power of ten by which this value has to be multiplied.

Fe V

Ti Isoelectronic Sequence

Ground State: $1s^2 2s^2 2p^6 3s^2 3p^6 3d^4\ ^5D_0$

Ionization Energy: 75.0 eV $= 605000$ cm^{-1}

Forbidden Transitions

List of tabulated lines

Wavelength (Å)	No.	Wavelength (Å)	No.	Wavelength (Å)	No.	Wavelength (Å)	No.
2707.4	5	3755.7	3	4362.6	11	10488	12
2717.8	5	3783.2	3	4426.3	11	10674	12
2730.2	5	3794.9	3	5118.8	10	18187	6
2738.3	5	3797.4	1	5140.3	10	19174	6
2750.9	5	3820.0	3	6208.5	13	19211	6
2760.6	5	3839.3	3	6258.4	13	20398	6
2780.4	5	3851.3	3	8137.2	7	20469	6
2790.4	5	3891.3	3	8342.7	7	28927	8
2828.2	5	3895.2	1	9835.3	9	31507	8
3400.4	4	3911.3	3	9999.5	9	32714	8
3406.9	4	4003.2	1	10081	9	33604	8
3445.6	4	4071.2	1	10214	9		
3463.5	4	4180.6	1	10348	12		
3503.6	4	4227.2	2	10353	9		

For this spectrum, we have tabulated the data of Garstang.[1] We have selected the most prominent lines, all of which are due to magnetic dipole radiation. According to Garstang's calculations, electric quadrupole contributions are essentially negligible for these lines.

Reference

[1]R. H. Garstang, Mon. Not. R. Astron. Soc. **117**, 393 (1957).

Fe v: Forbidden transitions

No.	Transition Array	Multiplet	λ (Å)	E_i (cm^{-1})	E_k (cm^{-1})	g_i	g_k	Type of transition	A_{ki} (s^{-1})	S (at. u.)	Accuracy	Source
1.	$3d^4$–$3d^4$	^5D – $^3P2$ (1F)										
			[3895.2]	803.1	26468	7	5	M1	0.71	0.0078	D	1
			[4071.2]	417.3	24973	5	3	M1	1.1	0.0083	D	1
			[4180.6]	142.1	24055	3	1	M1	1.3	0.0035	D	1
			[3797.4]	142.1	26468	3	5	M1	0.036	3.7(−4)[a]	D	1
			[4003.2]	0.0	24973	1	3	M1	0.13	9.3(−4)	D	1
2.		^5D – ^3H (2F)										
			[4227.2]	1282.8	24933	9	9	M1	0.0011	2.8(−5)	D	1
3.		^5D – $^3F2$ (3F)										
			[3891.3]	1282.8	26974	9	9	M1	0.74	0.015	D	1
			[3839.3]	803.1	26842	7	7	M1	0.40	0.0059	D	1
			[3794.9]	417.3	26761	5	5	M1	0.20	0.0020	D	1
			[3911.3]	1282.8	26842	9	7	M1	0.066	0.0010	D	1
			[3851.3]	803.1	26761	7	5	M1	0.047	5.0(−4)	D	1
			[3820.0]	803.1	26974	7	9	M1	0.16	0.0030	D	1
			[3783.2]	417.3	26842	5	7	M1	0.16	0.0022	D	1
			[3755.7]	142.1	26761	3	5	M1	0.10	9.8(−4)	D	1
4.		^5D – ^3G (4F)										
			[3406.9]	803.1	30147	7	9	M1	0.0078	1.0(−4)	D	1
			[3400.4]	417.3	29817	5	7	M1	0.0070	7.1(−5)	D	1
			[3463.5]	1282.8	30147	9	9	M1	0.032	4.4(−4)	D	1
			[3445.6]	803.1	29817	7	7	M1	0.017	1.8(−4)	D	1
			[3503.6]	1282.8	29817	9	7	M1	0.0026	2.9(−5)	D	1
5.		^5D – ^3D										
			[2828.2]	1282.8	36630	9	7	M1	0.37	0.0022	D	1
			[2780.4]	803.1	36759	7	5	M1	0.11	4.4(−4)	D	1
			[2738.3]	417.3	36925	5	3	M1	0.0019	4.3(−6)	D	1
			[2790.4]	803.1	36630	7	7	M1	0.089	5.0(−4)	D	1
			[2750.9]	417.3	36759	5	5	M1	0.18	6.9(−4)	D	1
			[2717.8]	142.1	36925	3	3	M1	0.19	4.2(−4)	D	1
			[2760.6]	417.3	36630	5	7	M1	0.097	5.3(−4)	D	1
			[2730.2]	142.1	36759	3	5	M1	0.20	7.5(−4)	D	1
			[2707.4]	0.0	36925	1	3	M1	0.22	4.9(−4)	D	1
6.		^3H – ^3G										
			[20398]	25529	30430	13	11	M1	0.041	0.14	D	1
			[20469]	24933	29817	9	7	M1	0.036	0.080	D	1
			[19211]	25226	30430	11	11	M1	0.041	0.12	D	1
			[19174]	24933	30147	9	9	M1	0.033	0.078	D	1
			[18187]	24933	30430	9	11	M1	0.0012	0.0029	D	1
7.		^3H – ^1I										
			[8342.7]	25529	37512	13	13	M1	0.14	0.039	D	1
			[8137.2]	25226	37512	11	13	M1	0.11	0.029	D	1

J. Phys. Chem. Ref. Data, Vol. 17, Suppl. 4, 1988

Fe V: Forbidden transitions — Continued

No.	Transition Array	Multiplet	λ (Å)	E_i (cm^{-1})	E_k (cm^{-1})	g_i	g_k	Type of transition	A_{ki} (s^{-1})	S (at. u.)	Accuracy	Source
8.		^{3}F2 – ^{3}G										
			[28927]	26974	30430	9	11	M1	0.037	0.37	D	1
			[32714]	26761	29817	5	7	M1	0.030	0.27	D	1
			[31507]	26974	30147	9	9	M1	0.027	0.28	D	1
			[33604]	26842	29817	7	7	M1	0.037	0.36	D	1
9.		^{3}F2 – ^{3}D										
			[10353]	26974	36630	9	7	M1	0.0069	0.0020	D	1
			[10081]	26842	36759	7	5	M1	0.0016	3.0(−4)	D	1
			[9835.3]	26761	36925	5	3	M1	0.014	0.0015	D	1
			[10214]	26842	36630	7	7	M1	0.0064	0.0018	D	1
			[9999.5]	26761	36759	5	5	M1	0.017	0.0032	D	1
10.		^{3}F2 – ^{1}D2										
			[5140.3]	26842	46291	7	5	M1	0.42	0.011	D	1
			[5118.8]	26761	46291	5	5	M1	0.21	0.0052	D	1
11.		^{3}G – ^{1}F										
			[4426.3]	30147	52733	9	7	M1	0.17	0.0038	D	1
			[4362.6]	29817	52733	7	7	M1	0.12	0.0026	D	1
12.		^{3}D – ^{1}D2										
			[10348]	36630	46291	7	5	M1	0.090	0.018	D	1
			[10488]	36759	46291	5	5	M1	0.017	0.0036	D	1
			[10674]	36925	46291	3	5	M1	0.080	0.018	D	1
13.		^{3}D – ^{1}F										
			[6208.5]	36630	52733	7	7	M1	0.15	0.0093	D	1
			[6258.4]	36759	52733	5	7	M1	0.070	0.0045	D	1

[a]The number in parentheses following the tabulated value indicates the power of ten by which this value has to be multiplied.

Fe VI

Sc Isoelectronic Sequence

Ground State: $1s^2 2s^2 2p^6 3s^2 3p^6 3d^3$ $^4F_{3/2}$

Ionization Energy: 99.1 eV = 799000 cm^{-1}

Forbidden Transitions

List of tabulated lines

Wavelength (Å)	No.	Wavelength (Å)	No.	Wavelength (Å)	No.	Wavelength (Å)	No.
1387.9	8	2294.7	23	3928.9	28	10109	11
1394.6	8	2302.3	23	3982.3	28	10321	11
1397.9	8	2312.9	23	3994.6	4	10477	11
1404.6	8	2320.6	23	4014.6	16	11087	11
1411.2	8	2325.8	26	4903.3	20	11266	11
1418.1	8	3492.1	5	4967.1	3	12330	15
1434.6	8	3509.7	5	4971.7	20	12674	15
1875.8	13	3555.6	5	4972.5	3	12888	10
1883.0	13	3573.9	5	4998.0	20	13371	10
1887.9	13	3587.7	12	5097.8	2	13493	15
1895.2	13	3614.1	12	5145.8	3	13746	10
1907.0	13	3630.6	6	5176.0	3	15138	10
1919.5	13	3643.3	5	5234.3	2	41431	19
1944.3	17	3662.5	5	5277.8	2	44052	19
1957.3	17	3665.3	12	5335.2	2	46883	19
1984.4	17	3675.1	6	5370.3	3	123070	1
2145.1	7	3703.6	12	5424.2	2	143100	14
2163.0	7	3740.9	6	5426.6	2	147670	1
2168.9	7	3757.4	12	5484.8	2	149480	9
2181.1	21	3773.2	4	5517.4	22	195530	1
2187.2	7	3774.8	5	5561.5	22	208890	24
2194.5	21	3813.5	4	5591.5	25	258730	27
2197.5	21	3846.9	16	5631.1	2	292920	29
2201.2	7	3847.4	4	5637.6	22	356280	18
2211.1	21	3870.1	28	5677.0	2	490780	9
2220.1	7	3889.4	4	5683.7	22		
2241.3	7	3905.0	16	5715.0	25		
2260.9	7	3922.0	28	5875.7	25		

For this ion, we selected the work of Nussbaumer and Storey,[1] who calculated magnetic dipole and electric quadrupole transition probabilities for radiative transitions between the 19 levels of the $3d^3$ (ground) configuration. These authors employed a single configuration approximation and calculated radial wavefunctions via adjustable Thomas-Fermi potentials. Nussbaumer and Storey then applied additional corrections to their coupling coefficients so that calculated eigenenergies are in close agreement with observed energy levels.

Other data on this spectrum are provided by the work of Garstang et al.[2] The agreement between Refs. 1 and 2 is quite good for M1 data within the same term, i.e., for ^4F - ^4F or ^4P - ^4P transitions, where no radial wavefunc-

tions are required. However, for some M1 transitions between different terms, the A-values for Refs. 1 and 2 disagree by a factor of three or worse. We estimate that the M1 transition probabilities of Ref. 1 are more accurate than those of Ref. 2 because the above cited corrections should lead to a somewhat better representation of intermediate coupling than that found in Ref. 2. In general, the accuracies of the E2 transition probabilities are estimated to be not better than 50 percent.

References

[1]H. Nussbaumer and P. J. Storey, Astron. Astrophys. **70**, 37 (1978).
[2]R. H. Garstang, W. D. Robb, and S. P. Rountree, Astrophys. J. **222**, 384 (1978).

Fe VI: Forbidden transitions

No.	Transition Array	Multiplet	λ (Å)	E_i (cm^{-1})	E_k (cm^{-1})	g_i	g_k	Type of transition	A_{ki} (s^{-1})	S (at. u.)	Accuracy	Source
1.	$3d^3$–$3d^3$	^{4}F – ^{4}F										
			[123070]	1188	2001	8	10	M1	0.0145	10.0	C+	1
			[147670]	511.3	1188	6	8	M1	0.0134	12.8	C+	1
			[195530]	0.0	511.3	4	6	M1	0.00574	9.54	C+	1
2.		^{4}F – ^{4}P (1F)										
			[5677.0]	2001	19611	10	6	E2	0.052	1.1	E	1
			[5631.1]	1188	18942	8	4	E2	0.038	0.51	E	1
			[5484.8]	511.3	18738	6	2	E2	0.034	0.20	E	1
			[5426.6]	1188	19611	8	6	M1	0.0026	9.2(−5)[a]	E	1
			"	"	"	8	6	E2	0.021	0.35	E	1
			[5424.2]	511.3	18942	6	4	M1	0.0018	4.3(−5)	E	1
			"	"	"	6	4	E2	0.033	0.37	E	1
			[5335.2]	0.0	18738	4	2	M1	3.3(−4)	3.7(−6)	E	1
			"	"	"	4	2	E2	0.060	0.31	E	1
			[5234.3]	511.3	19611	6	6	M1	0.0014	4.5(−5)	E	1
			"	"	"	6	6	E2	0.0056	0.079	E	1
			[5277.8]	0.0	18942	4	4	M1	0.0041	8.9(−5)	E	1
			"	"	"	4	4	E2	0.013	0.13	E	1
			[5097.8]	0.0	19611	4	6	M1	2.7(−4)	8.0(−6)	E	1
			"	"	"	4	6	E2	7.0(−4)	0.0086	E	1
3.		^{4}F – ^{2}G (2F)										
			[5176.0]	2001	21315	10	10	M1	0.62	0.032	D	1
			[5145.8]	1188	20616	8	8	M1	0.26	0.011	D	1
			[5370.3]	2001	20616	10	8	M1	0.013	6.0(−4)	D	1
			[4967.1]	1188	21315	8	10	M1	0.25	0.011	D	1
			[4972.5]	511.3	20616	6	8	M1	0.24	0.0088	D	1
4.		^{4}F – ^{2}P (3F)										
			[3994.6]	1188	26215	8	4	E2	0.0041	0.0099	E	1
			[3847.4]	511.3	26496	6	2	E2	0.0028	0.0028	E	1
			[3889.4]	511.3	26215	6	4	M1	0.58	0.0051	E	1
			"	"	"	6	4	E2	0.0028	0.0059	E	1
			[3773.2]	0.0	26496	4	2	M1	0.0020	8.0(−6)	E	1
			"	"	"	4	2	E2	0.0015	0.0014	E	1
			[3813.5]	0.0	26215	4	4	M1	0.36	0.0030	E	1
			"	"	"	4	4	E2	5.4(−4)	0.0010	E	1
5.		^{4}F – ^{2}D2 (4F)										
			[3774.8]	2001	28484	10	6	E2	6.1(−4)	0.0017	E	1
			[3643.3]	1188	28628	8	4	E2	0.0017	0.0026	E	1
			[3662.5]	1188	28484	8	6	M1	1.1	0.012	E	1
			"	"	"	8	6	E2	3.4(−4)	8.0(−4)	E	1
			[3555.6]	511.3	28628	6	4	M1	0.73	0.0049	E	1
			"	"	"	6	4	E2	5.4(−4)	7.3(−4)	E	1
			[3573.9]	511.3	28484	6	6	M1	0.14	0.0014	E	1
			[3492.1]	0.0	28628	4	4	M1	0.39	0.0025	E	1
			"	"	"	4	4	E2	2.7(−4)	3.3(−4)	E	1
			[3509.7]	0.0	28484	4	6	M1	0.043	4.1(−4)	E	1

Fe VI: Forbidden transitions — Continued

No.	Transition Array	Multiplet	λ (Å)	E_i (cm^{-1})	E_k (cm^{-1})	g_i	g_k	Type of transition	A_{ki} (s^{-1})	S (at. u.)	Accuracy	Source
6.		$^4F - ^2H$ (5F)										
			[3675.1]	2001	29203	10	12	M1	0.0010	2.2(−5)	D−	1
			"	"	"	10	12	E2	1.7(−4)	8.1(−4)	E	1
			[3630.6]	1188	28724	8	10	M1	0.0041	7.3(−5)	D−	1
			[3740.9]	2001	28724	10	10	M1	0.0069	1.3(−4)	D−	1
7.		$^4F - ^2F$										
			[2241.3]	2001	46604	10	6	E2	2.0(−4)	4.0(−5)	E	1
			[2260.9]	2001	46217	10	8	M1	0.26	8.9(−4)	D−	1
			"	"	"	10	8	E2	0.0050	0.0014	E	1
			[2201.2]	1188	46604	8	6	M1	0.038	9.0(−5)	D−	1
			"	"	"	8	6	E2	0.0011	2.0(−4)	E	1
			[2220.1]	1188	46217	8	8	M1	0.017	5.5(−5)	D−	1
			"	"	"	8	8	E2	6.0(−4)	1.5(−4)	E	1
			[2168.9]	511.3	46604	6	6	M1	0.031	7.0(−5)	D−	1
			"	"	"	6	6	E2	0.0011	1.9(−4)	E	1
			[2187.2]	511.3	46217	6	8	M1	0.10	3.1(−4)	D−	1
			"	"	"	6	8	E2	0.0017	4.1(−4)	E	1
			[2145.1]	0.0	46604	4	6	M1	0.22	4.8(−4)	D−	1
			"	"	"	4	6	E2	0.0045	7.3(−4)	E	1
			[2163.0]	0.0	46217	4	8	E2	6.6(−4)	1.5(−4)	E	1
8.		$^4F - ^2D1$										
			[1434.6]	2001	71708	10	6	E2	0.064	0.0014	E	1
			[1411.2]	1188	72049	8	4	E2	0.016	2.1(−4)	E	1
			[1418.1]	1188	71078	8	6	M1	0.25	1.6(−4)	E	1
			"	"	"	8	6	E2	0.0049	1.0(−4)	E	1
			[1397.9]	511.3	72049	6	4	M1	0.25	1.0(−4)	E	1
			[1404.6]	511.3	71708	6	6	M1	0.024	1.5(−5)	E	1
			"	"	"	6	6	E2	0.0021	4.1(−5)	E	1
			[1387.9]	0.0	72049	4	4	M1	0.13	5.2(−5)	E	1
			"	"	"	4	4	E2	0.0041	5.0(−5)	E	1
			[1394.6]	0.0	71708	4	6	M1	0.0094	5.7(−6)	E	1
			"	"	"	4	6	E2	6.3(−4)	1.2(−5)	E	1
9.		$^4P - ^4P$										
			[149480]	18942	19611	4	6	M1	0.00473	3.51	C+	1
			[490780]	18738	18942	2	4	M1	1.87(−4)	3.28	C+	1
10.		$^4P - ^2P$										
			[15138]	19611	26215	6	4	M1	0.10	0.051	D	1
			[13746]	18942	26215	4	4	M1	0.21	0.081	D	1
			[12888]	18738	26496	2	2	M1	0.38	0.060	D	1
			[13371]	18738	26215	2	4	M1	0.093	0.033	D	1
11.		$^4P - ^2D2$										
			[11266]	19611	28484	6	6	M1	0.067	0.021	D−	1
			[10321]	18942	28628	4	4	M1	0.0067	0.0011	D−	1
			[11087]	19611	28628	6	4	M1	0.18	0.036	D−	1
			[10477]	18942	28484	4	6	M1	0.0023	5.9(−4)	D−	1
			[10109]	18738	28628	2	4	M1	0.015	0.0023	D−	1

Fe VI:　Forbidden transitions — Continued

No.	Transition Array	Multiplet	λ (Å)	E_i (cm^{-1})	E_k (cm^{-1})	g_i	g_k	Type of transition	A_{ki} (s^{-1})	S (at. u.)	Accu-racy	Source
12.		$^4P - {}^2F$										
			[3665.3]	18942	46217	4	8	E2	0.0022	0.0069	E	1
			[3587.7]	18738	46604	2	6	E2	5.8(−4)	0.0012	E	1
			[3757.4]	19611	46217	6	8	M1	5.0(−4)	7.9(−6)	D−	1
			"	"	"	6	8	E2	2.7(−4)	9.6(−4)	E	1
			[3614.1]	18942	46604	4	6	M1	0.0010	1.1(−5)	D−	1
			[3703.6]	19611	46604	6	6	M1	0.0055	6.2(−5)	D−	1
13.		$^4P - {}^2D1$										
			[1919.5]	19611	71708	6	6	M1	1.4	0.022	E	1
			"	"	"	6	6	E2	0.0012	1.1(−4)	E	1
			[1883.0]	18942	72049	4	4	M1	0.52	5.1(−4)	E	1
			"	"	"	4	4	E2	0.081	0.0046	E	1
			[1907.0]	19611	72049	6	4	M1	0.12	1.2(−4)	E	1
			"	"	"	6	4	E2	8.0(−4)	4.8(−5)	E	1
			[1895.2]	18942	71708	4	6	M1	0.24	3.6(−4)	E	1
			"	"	"	4	6	E2	0.077	0.0067	E	1
			[1875.8]	18738	72049	2	4	M1	0.14	1.4(−4)	E	1
			"	"	"	2	4	E2	0.030	0.0017	E	1
			[1887.9]	18738	71708	2	6	E2	0.0072	6.2(−4)	E	1
14.		$^2G - {}^2G$										
			[143100]	20616	21315	8	10	M1	0.00401	4.36	C	1
15.		$^2G - {}^2H$										
			[12674]	21315	29203	10	12	M1	0.12	0.11	D	1
			"	"	"	10	12	E2	1.1(−4)	0.26	E	1
			[12330]	20616	28724	8	10	M1	0.12	0.083	D	1
			"	"	"	8	10	E2	1.6(−4)	0.27	E	1
			[13493]	21315	28724	10	10	M1	0.21	0.19	D	1
16.		$^2G - {}^2F$										
			[4014.6]	21315	46217	10	8	M1	0.12	0.0023	D−	1
			"	"	"	10	8	E2	0.13	0.65	E	1
			[3846.9]	20616	46604	8	6	M1	0.15	0.0019	D−	1
			"	"	"	8	6	E2	0.15	0.45	E	1
			[3905.0]	20616	46217	8	8	M1	0.25	0.0044	D−	1
			"	"	"	8	8	E2	0.012	0.052	E	1
17.		$^2G - {}^2D1$										
			[1984.4]	21315	71708	10	6	E2	10	1.1	E	1
			[1944.3]	20616	72049	8	4	E2	12	0.79	E	1
			[1957.3]	20616	71708	8	6	M1	0.0022	3.7(−6)	E	1
			"	"	"	8	6	E2	0.91	0.093	E	1
18.		$^2P - {}^2P$										
			[356280]	26215	26496	4	2	M1	2.30(−4)	0.771	C+	1
19.		$^2P - {}^2D2$										
			[44052]	26215	28484	4	6	M1	0.056	1.1	E	1
			[46883]	26496	28628	2	4	M1	0.038	0.58	E	1
			[41431]	26215	28628	4	4	M1	0.10	1.1	E	1

Fe VI: Forbidden transitions — Continued

No.	Transition Array	Multiplet	λ (Å)	E_i (cm^{-1})	E_k (cm^{-1})	g_i	g_k	Type of transition	A_{ki} (s^{-1})	S (at. u.)	Accuracy	Source
20.		^{2}P – ^{2}F										
			[4998.0]	26215	46217	4	8	E2	0.021	0.31	E	1
			[4971.7]	26496	46604	2	6	E2	0.014	0.15	E	1
			[4903.3]	26215	46604	4	6	M1	0.0081	2.1(−4)	D−	1
						4	6	E2	0.011	0.11	E	1
21.		^{2}P – ^{2}D1										
			[2211.1]	26496	71708	2	6	E2	0.59	0.11	E	1
			[2197.5]	26215	71708	4	6	M1	0.35	8.3(−4)	E	1
			"	"	"	4	6	E2	0.66	0.12	E	1
			[2194.5]	26496	72049	2	4	M1	0.0022	3.4(−6)	E	1
			"	"	"	2	4	E2	1.4	0.17	E	1
			[2181.1]	26215	72049	4	4	M1	0.011	1.7(−5)	E	1
			"	"	"	4	4	E2	2.1	0.25	E	1
22.		^{2}D2 – ^{2}F										
			[5683.7]	28628	46217	4	8	E2	4.4(−4)	0.012	E	1
			[5637.6]	28484	46217	6	8	M1	0.014	7.4(−4)	E	1
			"	"	"	6	8	E2	0.031	0.84	E	1
			[5561.5]	28628	46604	4	6	M1	0.0077	2.9(−4)	E	1
			"	"	"	4	6	E2	0.023	0.44	E	1
			[5517.4]	28484	46604	6	6	M1	0.036	0.0013	E	1
			"	"	"	6	6	E2	0.0077	0.14	E	1
23.		^{2}D2 – ^{2}D1										
			[2312.9]	28484	71708	6	6	M1	0.0048	1.3(−5)	E	1
			"	"	"	6	6	E2	0.62	0.15	E	1
			[2302.3]	28628	72049	4	4	M1	0.0038	6.9(−6)	E	1
			"	"	"	4	4	E2	0.026	0.0040	E	1
			[2294.7]	28484	72049	6	4	M1	0.86	0.0015	E	1
			"	"	"	6	4	E2	0.30	0.045	E	1
			[2320.6]	28628	71708	4	6	M1	0.31	8.6(−4)	E	1
			"	"	"	4	6	E2	1.5	0.36	E	1
24.		^{2}H – ^{2}H										
			[208890]	28724	29203	10	12	M1	0.00132	5.35	C+	1
25.		^{2}H – ^{2}F										
			[5875.7]	29203	46217	12	8	E2	0.050	1.7	E	1
			[5591.5]	28724	46604	10	6	E2	0.068	1.3	E	1
			[5715.0]	28724	46217	10	8	M1	8.1(−4)	4.5(−5)	D−	1
			"	"	"	10	8	E2	8.9(−4)	0.026	E	1
26.		^{2}H – ^{2}D1										
			[2325.8]	28724	71708	10	6	E2	0.15	0.036	E	1
27.		^{2}F – ^{2}F										
			[258730]	46217	46604	8	6	M1	8.86(−4)	3.41	C+	1

Fe VI: Forbidden transitions — Continued

No.	Transition Array	Multiplet	λ (Å)	E_i (cm^{-1})	E_k (cm^{-1})	g_i	g_k	Type of transition	A_{ki} (s^{-1})	S (at. u.)	Accu-racy	Source
28.		^{2}F – ^{2}D1										
			[3870.1]	46217	72049	8	4	E2	0.097	0.20	E	1
			[3922.0]	46217	71708	8	6	M1	0.35	0.0047	E	1
			"	"	"	8	6	E2	0.48	1.6	E	1
			[3928.9]	46604	72049	6	4	M1	0.37	0.0033	E	1
			"	"	"	6	4	E2	0.47	1.0	E	1
			[3982.3]	46604	71708	6	6	M1	0.60	0.0084	E	1
			"	"	"	6	6	E2	0.083	0.30	E	1
29.		^{2}D1 – ^{2}D1										
			[292920]	71708	72049	6	4	M1	6.41(−4)	2.39	C+	1

[a]The number in parentheses following the tabulated value indicates the power of ten by which this value has to be multiplied.

Fe VII

Ca Isoelectronic Sequence

Ground State: $1s^2 2s^2 2p^6 3s^2 3p^6 3d^2$ ^{3}F$_2$

Ionization Energy: 124.98 eV = 1008000 cm^{-1}

Allowed Transitions

List of tabulated lines

Wavelength (Å)	No.	Wavelength (Å)	No.	Wavelength (Å)	No.	Wavelength (Å)	No.
150.186	16	154.921	18	232.946	5	246.859	9
150.403	16	154.941	18	233.015	5	247.098	9
150.530	16	154.949	18	233.308	4	265.697	13
150.807	15	155.124	18	233.762	4	1073.95	29
150.852	15	155.150	18	234.337	4	1080.64	26
151.023	15	155.994	22	234.757	8	1080.74	26
151.046	15	157.112	21	235.221	4	1087.86	26
151.145	15	158.481	20	236.778	7	1095.34	26
151.432	14	165.087	23	239.734	11	1117.58	28
151.512	14	165.919	3	239.860	11	1141.44	25
151.675	14	166.365	3	240.053	11	1154.99	25
151.754	14	173.441	1	240.083	11	1163.88	24
151.782	14	176.744	2	240.223	11	1166.18	25
151.971	14	176.928	2	240.572	11	1173.92	25
154.271	19	177.172	2	243.379	12	1180.82	24
154.307	19	231.044	5	243.705	10	1208.38	24
154.335	17	231.693	5	244.030	10	1226.65	24
154.363	19	231.728	5	244.098	9	1239.69	24
154.447	19	232.047	4	244.541	10	1332.38	27
154.565	19	232.256	5	245.153	6		
154.650	19	232.442	5	245.488	9		
154.848	18	232.613	4	246.000	9		

For this spectrum, we have chosen the data of Fawcett and Cowan,[1] who used self-consistent-field calculations with exchange and correlation (the Hartree-X method) to determine oscillator strengths for six lines. These data should be reasonably accurate, since the authors included the dominant contributing configurations in their calculations. For the remaining lines tabulated here, we have selected the data of Warner and Kirkpatrick,[2] who used the single-configuration scaled Thomas-Fermi approximation and calculated individual line strengths in intermediate coupling. One criterion of selecting data was that all lines had to be experimentally observed, i.e., they appear in the comprehensive line list of Ekberg.[3] In this compilation, we have omitted all intercombination (spin-forbidden) transitions and have assigned accuracies of "E" to the weakest lines.

We estimate that for the stronger lines of this relatively simple spectrum, Warner and Kirkpatrick's data should be fairly reliable (except when configuration interaction effects become appreciable). There is indirect support for this estimate from the good consistency between similarly calculated values and lifetime measurements for the isoelectronic ion Ti III.

References

[1]B. C. Fawcett and R. D. Cowan, Sol. Phys. **31**, 339 (1973).

[2]B. Warner and R. Kirkpatrick, Publications of the Department of Astronomy, University of Texas at Austin, Vol. 3, No. 2 (1969).

[3]J. O. Ekberg, Phys. Scr. **23**, 7 (1981).

Fe VII: Allowed transitions

No.	Transition Array	Multiplet	λ (Å)	E_i (cm^{-1})	E_k (cm^{-1})	g_i	g_k	A_{ki} (10^8 s^{-1})	f_{ik}	S (at. u.)	log gf	Accuracy	Source
1.	$3p^63d^2$– $3p^5(^2P°)3d^3(^2H)$	^1G – $^1G°$	173.441	28927	605489	9	9	3600	1.6	8.4	1.17	D–	1
2.	$3p^63d^2$– $3p^5(^2P°)3d^3(^4F)$	^3F – $^3F°$											
			176.744	2332	568118	9	9	2700	1.2	6.5	1.05	D–	1
			176.928	1052	566256	7	7	2400	1.1	4.6	0.90	D–	1
			177.172	0	564425	5	5	1500	0.69	2.0	0.54	D–	1
3.		^3F – $^3D°$											
			166.365	2332	603419	9	7	2900	0.95	4.7	0.93	D–	1
			165.919	1052	603757	7	5	2800	0.82	3.1	0.76	D–	1
4.	$3d^2$–$3d4p$	^3F – $^3D°$	*233.98*	*1350*	*428730*	21	15	160	0.093	1.5	0.29	D–	2
			233.308	2332	430949	9	7	100	0.064	0.44	−0.24	D–	2
			234.337	1052	427785	7	5	110	0.067	0.36	−0.33	D–	2
			235.221	0	425129	5	3	170	0.085	0.33	−0.37	D–	2
			232.613	1052	430949	7	7	45	0.037	0.20	−0.59	D–	2
			233.762	0	427785	5	5	34	0.028	0.11	−0.86	D–	2
			232.047	0	430949	5	7	2.6	0.0029	0.011	−1.84	D–	2
5.		^3F – $^3F°$	*232.07*	*1350*	*432246*	21	21	73	0.059	0.95	0.09	D–	2
			231.728	2332	433871	9	9	60	0.049	0.33	−0.36	D–	2
			232.256	1052	431610	7	7	21	0.017	0.092	−0.92	D–	2
			232.442	0	430213	5	5	21	0.017	0.065	−1.07	D–	2
			232.946	2332	431610	9	7	67	0.042	0.29	−0.42	D–	2
			233.015	1052	430213	7	5	46	0.027	0.14	−0.73	D–	2
			231.044	1052	433871	7	9	4.1	0.0042	0.022	−1.53	D–	2
			231.693	0	431610	5	7	2.8	0.0032	0.012	−1.80	D–	2
6.		^1D – $^1D°$	245.153	17476	425386	5	5	70	0.063	0.26	−0.50	D–	2
7.		^1D – $^1F°$	236.778	17476	439812	5	7	6.8	0.0080	0.031	−1.40	D–	2
8.		^1D – $^1P°$	234.757	17476	443447	5	3	86	0.043	0.17	−0.67	D–	2

Fe VII: Allowed transitions — Continued

No.	Transition Array	Multiplet	λ (Å)	E_i (cm^{-1})	E_k (cm^{-1})	g_i	g_k	A_{ki} (10^8 s^{-1})	f_{ik}	S (at. u.)	log gf	Accuracy	Source
9.		^{3}P – ^{3}D°											
			244.098	21279	430949	5	7	16	0.020	0.082	−0.99	D−	2
			245.488	20430	427785	3	5	23	0.035	0.085	−0.98	D−	2
			246.859	20040	425129	1	3	19	0.051	0.042	−1.29	D−	2
			246.000	21279	427785	5	5	1.9	0.0017	0.0069	−2.07	E	2
			247.098	20430	425129	3	3	6.5	0.0059	0.014	−1.75	D−	2
10.		^{3}P – ^{3}F°											
			243.705	21279	431610	5	7	5.6	0.0069	0.028	−1.46	D−	2
			244.030	20430	430213	3	5	1.9	0.0028	0.0068	−2.07	E	2
			244.541	21279	430213	5	5	0.024	2.2(−5)[a]	8.8(−5)	−3.96	E	2
11.		^{3}P – ^{3}P°	*240.13*	*20858*	*437294*	9	9	120	0.11	0.75	−0.02	D−	2
			240.223	21279	437558	5	5	100	0.087	0.35	−0.36	D−	2
			240.083	20430	436952	3	3	35	0.030	0.072	−1.04	D−	2
			240.572	21279	436952	5	3	40	0.021	0.083	−0.98	D−	2
			240.053	20430	437001	3	1	130	0.037	0.089	−0.95	D−	2
			239.734	20430	437558	3	5	25	0.037	0.087	−0.96	D−	2
			239.860	20040	436952	1	3	34	0.089	0.070	−1.05	D−	2
12.		^{1}G – ^{1}F°	243.379	28927	439812	9	7	210	0.15	1.1	0.12	D−	2
13.		^{1}S – ^{1}P°	265.697	67078	443447	1	3	41	0.13	0.11	−0.89	D−	2
14.	3d^2–3d4f	^{3}F – ^{3}F°											
			151.782	2332	661169	9	9	240	0.082	0.37	−0.13	D−	2
			151.675	1052	660358	7	7	390	0.13	0.47	−0.03	D−	2
			151.512	0	660015	5	5	530	0.18	0.45	−0.04	D−	2
			151.971	2332	660358	9	7	29	0.0077	0.035	−1.16	D−	2
			151.754	1052	660015	7	5	50	0.012	0.044	−1.06	D−	2
			151.432	0	660358	5	7	220	0.10	0.26	−0.28	D−	2
15.		^{3}F – ^{3}G°											
			151.023	2332	664482	9	11	1600	0.67	3.0	0.78	D−	2
			150.852	1052	663950	7	9	1300	0.58	2.0	0.61	D−	2
			150.807	0	663097	5	7	1300	0.62	1.5	0.49	D−	2
			151.145	2332	663950	9	9	210	0.072	0.32	−0.19	D−	2
			151.046	1052	663097	7	7	220	0.077	0.27	−0.27	D−	2
16.		^{3}F – ^{3}D°											
			150.530	2332	666651	9	7	68	0.018	0.080	−0.79	D−	2
			150.403	1052	665923	7	5	73	0.018	0.061	−0.91	D−	2
			150.186	0	665832	5	3	75	0.015	0.038	−1.12	D−	2
17.		^{1}D – ^{1}F°	154.335	17476	665417	5	7	1200	0.58	1.5	0.46	D−	2
18.		^{3}P – ^{3}D°	*154.95*	*20858*	*666245*	9	15	980	0.59	2.7	0.72	D−	2
			154.949	21279	666651	5	7	1000	0.53	1.3	0.42	D−	2
			154.921	20430	665923	3	5	970	0.58	0.89	0.24	D−	2
			154.848	20040	665832	1	3	770	0.83	0.42	−0.08	D−	2
			155.124	21279	665923	5	5	8.2	0.0030	0.0076	−1.83	D−	2
			154.941	20430	665832	3	3	240	0.088	0.13	−0.58	D−	2
			155.150	21279	665832	5	3	0.019	4.2(−6)	1.1(−5)	−4.68	E	2

Fe VII: Allowed transitions — Continued

No.	Transition Array	Multiplet	λ (Å)	E_i (cm^{-1})	E_k (cm^{-1})	g_i	g_k	A_{ki} (10^8 s^{-1})	f_{ik}	S (at. u.)	log gf	Accuracy	Source
19.		^{3}P – ^{3}P°	*154.51*	*20858*	*668083*	9	9	850	0.31	1.4	0.44	D–	2
			154.650	21279	667899	5	5	880	0.32	0.81	0.20	D–	2
			154.363	20430	668253	3	3	420	0.15	0.23	−0.35	D–	2
			154.565	21279	668253	5	3	350	0.076	0.19	−0.42	D–	2
			154.307	20430	668489	3	1	890	0.11	0.16	−0.50	D–	2
			154.447	20430	667899	3	5	1.5	8.8(−4)	0.0013	−2.58	E	2
			154.271	20040	668253	1	3	81	0.087	0.044	−1.06	D–	2
20.		^{1}G – ^{1}G°	158.481	28927	659917	9	9	230	0.086	0.40	−0.11	D–	2
21.		^{1}G – ^{1}F°	157.112	28927	665417	9	7	18	0.0052	0.024	−1.33	D–	2
22.		^{1}G – ^{1}H°	155.994	28927	669978	9	11	1800	0.80	3.7	0.86	D–	2
23.		^{1}S – ^{1}P°	165.087	67078	672820	1	3	690	0.85	0.46	−0.07	D–	2
24.	$3d\,4s$–$3d\,4p$	^{3}D – ^{3}D°											
			1180.82	346262	430949	7	7	10	0.22	5.9	0.18	D–	2
			1208.38	345029	427785	5	5	5.8	0.13	2.5	−0.20	D–	2
			1239.69	344463	425129	3	3	6.2	0.14	1.7	−0.37	D–	2
			1226.65	346262	427785	7	5	2.7	0.044	1.2	−0.51	D–	2
			1163.88	345029	430949	5	7	0.21	0.0060	0.12	−1.52	D–	2
25.		^{3}D – ^{3}F°											
			1141.44	346262	433871	7	9	12	0.31	8.2	0.34	D–	2
			1154.99	345029	431610	5	7	11	0.32	6.0	0.20	D–	2
			1166.18	344463	430213	3	5	9.8	0.33	3.8	−0.00	D–	2
			1173.92	345029	430213	5	5	0.017	3.6(−4)	0.0069	−2.75	E	2
26.		^{3}D – ^{3}P°											
			1095.34	346262	437558	7	5	9.9	0.13	3.2	−0.05	D–	2
			1087.86	345029	436952	5	3	9.0	0.096	1.7	−0.32	D–	2
			1080.64	344463	437001	3	1	15	0.086	0.91	−0.59	D–	2
			1080.74	345029	437558	5	5	3.1	0.054	0.96	−0.57	D–	2
27.		^{1}D – ^{1}D°	1332.38	350333	425386	5	5	4.9	0.13	2.8	−0.19	D–	2
28.		^{1}D – ^{1}F°	1117.58	350333	439812	5	7	12	0.32	6.0	0.21	D–	2
29.		^{1}D – ^{1}P°	1073.95	350333	443447	5	3	15	0.15	2.7	−0.12	D–	2

[a]The number in parentheses following the tabulated value indicates the power of ten by which the value has to be multiplied.

Fe VII

Forbidden Transitions

List of tabulated lines

Wavelength (Å)	No.	Wavelength (Å)	No.	Wavelength (Å)	No.	Wavelength (Å)	No.
285.44	12	303.90	16	353.04	20	5276.4	3
286.30	12	304.15	13	359.78	19	5720.7	2
287.36	12	305.29	13	1490.8	5	6087.0	2
288.80	11	305.82	13	2015.4	8	6601.5	2
289.68	11	306.91	15	2143.0	10	8729.9	7
289.83	11	307.70	15	2182.7	10	26287	6
290.31	11	307.71	15	3456.0	4	33836	6
290.72	11	308.07	15	3586.3	4	78104	1
290.76	11	308.61	15	3758.9	4	95076	1
291.20	11	308.88	15	4698.2	3	117820	9
291.80	11	309.42	15	4893.4	3	256470	9
300.43	14	311.13	18	4942.5	3		
302.76	16	315.12	17	4988.6	3		
303.12	16	316.35	17	5158.9	3		

For this ion, we selected the work of Nussbaumer and Storey[1] and by Warner and Kirkpatrick.[2] Nussbaumer and Storey calculated magnetic dipole and electric quadrupole transition probabilities for radiative transitions between levels of the $3d^2$ (ground) configuration. They used a 17-configuration basis set to represent the eigenfunctions of the $3d^2$ levels. They calculated radial wavefunctions either by adjustable Thomas-Fermi potentials or by a hydrogenic potential, depending upon the value of the principal quantum number, n. Furthermore, these authors applied additional corrections to their coupling coefficients, so that calculated eigenenergies are in close agreement with observed energy levels. In general, the accuracies of the E2 transition probabilities are estimated to be no better than 50 percent.

For lines of the $3d^2$–$3d4s$ transition array, we chose the work of Warner and Kirkpatrick. These authors used a single configuration approximation and calculated radial integrals with scaled Thomas-Fermi wavefunctions. Although this work employed a much less sophisicated theoretical approach than that of Ref. 1, the data for Refs. 1 and 2 agreed reasonably well (within 50 percent) for M1 and E2 transitions within the $3d^2$ configuration.

In converting line strengths to transition probabilities, Warner and Kirkpatrick used their calculated energy-level data. When more accurate experimental data are employed, the changes in the A-values were generally found to be quite small—of the order of 15 percent or less. We have therefore retained the original data of this reference.

References

[1]H. Nussbaumer and P. J. Storey, Astron. Astrophys. **113**, 21 (1982).
[2]B. Warner and R. C. Kirkpatrick, Mon. Not. R. Astron. Soc. **144**, 397 (1969).

Fe VII: Forbidden transitions

No.	Transition Array	Multiplet	λ (Å)	E_i (cm^{-1})	E_k (cm^{-1})	g_i	g_k	Type of transition	A_{ki} (s^{-1})	S (at. u.)	Accuracy	Source
1.	$3d^2$–$3d^2$	^3F – ^3F										
			[78104]	1051.5	2331.5	7	9	M1	0.0424	6.74	C+	1
			[95076]	0.0	1051.5	5	7	M1	0.0298	6.65	C+	1

Fe VII: Forbidden transitions – Continued

No.	Transition Array	Multiplet	λ (Å)	E_i (cm^{-1})	E_k (cm^{-1})	g_i	g_k	Type of transition	A_{ki} (s^{-1})	S (at. u.)	Accuracy	Source
2.		^{3}F – ^{1}D (1F)										
			[6601.5]	2331.5	17476	9	5	E2	0.0016	0.060	E	1
			[6087.0]	1051.5	17476	7	5	M1	0.58	0.024	D	1
			"	"	"	7	5	E2	4.3(−4)^a	0.011	E	1
			[5720.7]	0.0	17476	5	5	M1	0.36	0.012	D	1
			"	"	"	5	5	E2	2.9(−4)	0.0053	E	1
3.		^{3}F – ^{3}P (2F)										
			[5276.4]	2331.5	21279	9	5	E2	0.050	0.61	E	1
			[5158.9]	1051.5	20430	7	3	E2	0.053	0.35	E	1
			[4988.6]	0.0	20040	5	1	E2	0.094	0.17	E	1
			[4942.5]	1051.5	21279	7	5	M1	0.059	0.0013	D	1
			"	"	"	7	5	E2	0.018	0.16	E	1
			[4893.4]	0.0	20430	5	3	M1	0.0017	2.2(−5)	D	1
			"	"	"	5	3	E2	0.034	0.17	E	1
			[4698.2]	0.0	21279	5	5	M1	0.016	3.1(−4)	D	1
			"	"	"	5	5	E2	0.0030	0.020	E	1
4.		^{3}F – ^{1}G (3F)										
			[3758.9]	2331.5	28927	9	9	M1	0.45	0.0080	D	1
			"	"	"	9	9	E2	2.2(−4)	8.8(−4)	E	1
			[3586.3]	1051.5	28927	7	9	M1	0.31	0.0048	D	1
			[3456.0]	0.0	28927	5	9	E2	5.6(−4)	0.0015	E	1
5.		^{3}F – ^{1}S										
			[1490.8]	0.0	67078	5	1	E2	0.17	7.5(−4)	E	1
6.		^{1}D – ^{3}P										
			[26287]	17476	21279	5	5	M1	0.16	0.54	D	1
			[33836]	17476	20430	5	3	M1	0.044	0.19	D	1
7.		^{1}D – ^{1}G (4F)	[8729.9]	17476	28927	5	9	E2	0.0012	0.33	E	1
8.		^{1}D – ^{1}S	[2015.4]	17476	67078	5	1	E2	22	0.44	E	1
9.		^{3}P – ^{3}P										
			[117820]	20430	21279	3	5	M1	0.00762	2.31	C+	1
			[256470]	20040	20430	1	3	M1	0.00106	1.99	C+	1
10.		^{3}P – ^{1}S										
			[2182.2]	21279	67078	5	1	E2	1.4	0.041	E	1
			[2143.0]	20430	67078	3	1	M1	6.9	0.0025	D	1

Fe VII: Forbidden transitions — Continued

No.	Transition Array	Multiplet	λ (Å)	E_i (cm^{-1})	E_k (cm^{-1})	g_i	g_k	Type of transition	A_{ki} (s^{-1})	S (at. u.)	Accuracy	Source
11.	$3d^2$–$3d\,4s$	^3F – ^3D										
			[291.80]	2331.5	345029	9	5	E2	3.8(+4)	0.24	E	2
			[291.20]	1051.5	344463	7	3	E2	6.1(+4)	0.23	E	2
			[290.76]	2331.5	346262	9	7	E2	1.4(+5)	1.2	E	2
			[290.72]	1051.5	345029	7	5	E2	9.0(+4)	0.56	E	2
			[290.31]	0.0	344463	5	3	E2	1.2(+5)	0.44	E	2
			[289.68]	1051.5	346262	7	7	E2	4.0(+4)	0.34	E	2
			[289.83]	0.0	345029	5	5	E2	5.4(+4)	0.33	E	2
			[288.80]	0.0	346262	5	7	E2	3800	0.032	E	2
12.		^3F – ^1D										
			[287.36]	2331.5	350333	9	5	E2	460	0.0027	E	2
			[286.30]	1051.5	350333	7	5	E2	2700	0.015	E	2
			[285.44]	0.0	350333	5	5	E2	320	0.0018	E	2
13.		^1D – ^3D										
			[304.15]	17476	346262	5	7	E2	3000	0.033	E	2
			[305.29]	17476	345029	5	5	E2	7600	0.060	E	2
			[305.82]	17476	344463	5	3	E2	2400	0.011	E	2
14.		^1D – ^1D	[300.43]	17476	350333	5	5	E2	9.8(+4)	0.71	E	2
15.		^3P – ^3D										
			[306.91]	20430	346262	3	7	E2	2.0(+4)	0.23	E	2
			[307.70]	20040	345029	1	5	E2	2.0(+4)	0.16	E	2
			[307.71]	21279	346262	5	7	E2	3.7(+4)	0.43	E	2
			[308.07]	20430	235029	3	5	E2	4800	0.040	E	2
			[308.88]	21279	345029	5	5	E2	2.8(+4)	0.23	E	2
			[308.61]	20430	344463	3	3	E2	4.4(+4)	0.22	E	2
			[309.42]	21279	344463	5	3	E2	1.3(+4)	0.066	E	2
16.		^3P – ^1D										
			[303.90]	21279	350333	5	5	E2	1.3(+4)	0.10	E	2
			[303.12]	20430	350333	3	5	E2	140	0.0011	E	2
			[302.76]	20040	350333	1	5	E2	360	0.0027	E	2
17.		^1G – ^3D										
			[316.35]	28927	345029	9	5	E2	3700	0.035	E	2
			[315.12]	28927	346262	9	7	E2	56	7.3(−4)	E	2
18.		^1G – ^1D	[311.13]	28927	350333	9	5	E2	1.7(+5)	1.5	E	2
19.		^1S – ^3D										
			[359.78]	67078	345029	1	5	E2	160	0.0029	E	2
20.		^1S – ^1D	[353.04]	67078	350333	1	5	E2	1.0(+4)	0.16	E	2

[a]The number in parentheses following the tabulated value indicates the power of ten by which this value has to be multiplied.

Fe VIII

K Isoelectronic Sequence

Ground State: $1s^2 2s^2 2p^6 3s^2 3p^6 3d\ ^2D_{3/2}$

Ionization Energy: $151.061\ eV = 1218380\ cm^{-1}$

Allowed Transitions

List of tabulated lines

Wavelength (Å)	No.	Wavelength (Å)	No.	Wavelength (Å)	No.	Wavelength (Å)	No.
112.472	8	118.907	6	168.545	5	197.362	2
112.486	8	119.380	6	168.929	5	217.691	1
112.932	9	167.486	4	185.213	3	218.564	1
116.196	7	167.656	4	186.601	3	224.305	1
116.442	7	168.002	4	187.237	3		
117.197	7	168.024	5	192.004	2		
118.648	6	168.172	4	196.650	2		

For this spectrum, we have chosen the data of Tiwary,[1,2] who calculated absolute multiplet oscillator strengths for the $3p^6 3d$–$3p^5 3d^2$ and $3p^6 3d$–$3p^5 3d\,4s$ arrays by using configuration interaction wavefunctions. For the $3p^6 3d$–$3p^5 3d\,4s$ array, the LS-coupling line strengths generally agree quite well with the intermediate coupling calculations of Cowan.[3] Where this agreement is not good (worse than $\pm 50\%$), we have omitted the lines from this compilation. Within this transition array, we have normalized Cowan's line strengths to the multiplet strengths of Ref. 2.

For lines within the $3p^6 3d$–$3p^5 3d^2$ transition array, we have obtained line strengths from Tiwary's multiplet strengths by applying LS-coupling rules. We estimate these data to be accurate within fifty percent for stronger lines.

References

[1] S. N. Tiwary, Chem. Phys. Lett. **93**, 47 (1982).
[2] S. N. Tiwary, Astrophys. J. **269**, 803 (1983).
[3] R. D. Cowan, Astrophys. J. **147**, 377 (1967).

Fe VIII: Allowed transitions

No.	Transition Array	Multiplet	λ (Å)	E_i (cm^{-1})	E_k (cm^{-1})	g_i	g_k	A_{ki} (10^8 s^{-1})	f_{ik}	S (at. u.)	log gf	Accuracy	Source
1.	$3p^6 3d$–$3p^5(^2P°)3d^2(^1D)$	^2D – $^2F°$	*221.45*	*1102*	*452676*	10	14	36	0.037	0.27	−0.43	D−	1
			224.305	1836	447658	6	8	34	0.034	0.15	−0.69	D−	*ls*
			217.691	0	459367	4	6	36	0.038	0.11	−0.81	D−	*ls*
			218.564	1836	459367	6	6	2.5	0.0018	0.0077	−1.97	E	*ls*

Fe VIII: Allowed transitions — Continued

No.	Transition Array	Multiplet	λ (Å)	E_i (cm^{-1})	E_k (cm^{-1})	g_i	g_k	A_{ki} (10^8 s^{-1})	f_{ik}	S (at. u.)	log gf	Accuracy	Source
2.	$3p^6 3d-$ $3p^5(^2P°)3d^2(^1S)$	$^2D - ^2P°$	*195.50*	*1102*	*512619*	10	6	17	0.0058	0.037	−1.24	D−	1
			197.362	1836	508518	6	4	14	0.0056	0.022	−1.47	D−	*ls*
			192.004?	0	520822?	4	2	17	0.0047	0.012	−1.72	D−	*ls*
			196.650	0	508518	4	4	1.7	9.7(−4)[a]	0.0025	−2.41	E	*ls*
3.	$3p^6 3d-$ $3p^5(^2P°)3d^2(^3F)$	$^2D - ^2F°$	*185.82*	*1102*	*539250*	10	14	1000	0.76	4.6	0.88	D−	1
			185.213	1836	541755	6	8	1000	0.71	2.6	0.63	D−	*ls*
			186.601	0	535909	4	6	940	0.73	1.8	0.47	D−	*ls*
			187.237	1836	535909	6	6	67	0.035	0.13	−0.68	E	*ls*
4.		$^2D - ^2D°$	*167.90*	*1102*	*596704*	10	10	3300	1.4	7.7	1.15	D−	1
			168.172	1836	596463	6	6	3100	1.3	4.3	0.89	D−	*ls*
			167.486	0	597065	4	4	3000	1.3	2.8	0.71	D−	*ls*
			168.002	1836	597065	6	4	330	0.093	0.31	−0.25	E	*ls*
			167.656	0	596463	4	6	220	0.14	0.31	−0.25	E	*ls*
5.	$3p^6 3d-$ $3p^5(^2P°)3d^2(^3P)$	$^2D - ^2P°$	*168.64*	*1102*	*594089*	10	6	2200	0.56	3.1	0.75	D−	1
			168.545	1836	595152	6	4	2000	0.57	1.9	0.53	D−	*ls*
			168.929	0	591964	4	2	2100	0.45	1.0	0.25	D−	*ls*
			168.024	0	595152	4	4	220	0.095	0.21	−0.42	E	*ls*
6.	$3p^6 3d-$ $3p^5 3d(^3P°)4s$	$^2D - ^2P°$	*119.05*	*1102*	*841106*	10	6	340	0.043	0.17	−0.37	D	2
			118.907	1836	842829	6	4	300	0.042	0.098	−0.60	D	*3n*
			119.380	0	837661	4	2	340	0.036	0.057	−0.84	D	*3n*
			118.648	0	842829	4	4	45	0.0096	0.015	−1.42	E	*3n*
7.	$3p^6 3d-$ $3p^5 3d(^3F°)4s$	$^2D - ^2F°$	*116.77*	*1102*	*857464*	10	14	420	0.12	0.46	0.08	D	2
			117.197	1836	855100	6	8	380	0.10	0.24	−0.21	D	*3n*
			116.196	0	860615	4	6	450	0.14	0.21	−0.26	D	*3n*
			116.442	1836	860615	6	6	26	0.0052	0.012	−1.50	E	*3n*
8.	$3p^6 3d-$ $3p^5 3d(^3D°)4s$	$^2D - ^2D°$	*112.48*	*1102*	*890152*	10	10	490	0.092	0.34	−0.04	D	2
			112.486	1836	890845	6	6	430	0.081	0.18	−0.31	D	*3n*
			112.472	0	889113	4	4	360	0.068	0.10	−0.57	D	*3n*
9.	$3p^6 3d-$ $3p^5 3d(^1F°)4s$	$^2D - ^2F°$	*113.00*	*1102*	*886042*	10	14	180	0.048	0.18	−0.32	D	2
			112.932	1836	887325	6	8	210	0.054	0.12	−0.49	D	*3n*

[a]The number in parentheses following the tabulated value indicates the power of ten by which this value has to be multiplied.

Fe VIII

Forbidden Transitions

List of tabulated lines

Wavelength (Å)	No.	Wavelength (Å)	No.	Wavelength (Å)	No.	Wavelength (Å)	No.
178.70	6	326.55	9	612.00	7	1117.7	10
180.35	6	326.62	9	612.16	7	1117.8	10
180.40	6	326.72	9	612.36	7	1870.2	12
197.52	5	394.59	3	612.52	7	1870.6	12
199.59	5	395.46	8	699.45	11	1875.7	12
199.60	5	395.48	8	699.66	11	1876.1	12
239.93	4	395.62	8	700.21	11	18959	2
242.98	4	395.63	8	700.42	11	54450	1
243.01	4	402.82	3	1117.2	10		
326.45	9	402.98	3	1117.3	10		

For this spectrum, we have tabulated the data of Czyzak and Krueger,[1] who calculated radial wavefunctions by the Hartree-Fock self-consistent field method (with exchange). These authors used *LS* coupling to calculate M1 and E2 transition probabilities. Data for magnetic dipole transitions within the same term are expected to be fairly accurate (better than 25 percent), while data for E2 transitions are much more uncertain.

Reference

[1] S. J. Czyzak and T. K. Krueger, Astrophys. J. **144**, 381 (1966).

Fe VIII: Forbidden transtions

No.	Transition Array	Multiplet	λ (Å)	E_i (cm^{-1})	E_k (cm^{-1})	g_i	g_k	Type of transition	A_{ki} (s^{-1})	S (at. u.)	Accuracy	Source
1.	3d–3d	^{2}D – ^{2}D										
			[54450]	0	1836	4	6	M1	0.0705	2.53	C+	1
2.	4p–4p	^{2}P° – ^{2}P°										
			[18959]	510277	515550	2	4	M1	0.339	0.343	C+	1
3.	4p–4f	^{2}P° – ^{2}F°										
			[402.82]	515550	763799	4	8	E2	4.6(+5)^a	23	E	1
			[394.59]	510277	763703	2	6	E2	3.8(+5)	13	E	1
			[402.98]	515550	763703	4	6	E2	1.0(+5)	3.8	E	1
4.	4p–5f	^{2}P° – ^{2}F°										
			[242.98]	515550	927102	4	8	E2	3.9(+5)	1.6	E	1
			[239.93]	510277	927059	2	6	E2	3.2(+5)	0.91	E	1
			[243.01]	515550	927059	4	6	E2	8.7(+4)	0.26	E	1
5.	4p–6f	^{2}P° – ^{2}F°										
			[199.59]	515550	1016570	4	8	E2	5.4(+5)	0.81	E	1
			[197.52]	510277	1016560	2	6	E2	4.4(+5)	0.47	E	1
			[199.60]	515550	1016560	4	6	E2	1.2(+5)	0.14	E	1
6.	4p–7f	^{2}P° – ^{2}F°										
			[180.35]	515550	1070029	4	8	E2	4.6(+5)	0.42	E	1
			[178.70]	510277	1069873	2	6	E2	3.7(+5)	0.24	E	1
			[180.40]	515550	1069873	4	6	E2	1.0(+5)	0.068	E	1

Fe VIII: Forbidden transitons — Continued

No.	Transition Array	Multiplet	λ (Å)	E_i (cm^{-1})	E_k (cm^{-1})	g_i	g_k	Type of transition	A_{ki} (s^{-1})	S (at. u.)	Accu-racy	Source
7.	4f–5f	^{2}F° – ^{2}F°										
			[612.36]	763799	927102	8	8	E2	2.2(+4)	9.0	E	1
			[612.16]	763703	927059	6	6	E2	3.5(+4)	11	E	1
			[612.52]	763799	927059	8	6	E2	5700	1.8	E	1
			[612.00]	763703	927102	6	8	E2	4300	1.8	E	1
8.	4f–6f	^{2}F° – ^{2}F°										
			[395.62]	763799	1016570	8	8	E2	1.1(+4)	0.51	E	1
			[395.48]	763703	1016560	6	6	E2	1.7(+4)	0.59	E	1
			[395.63]	763799	1016560	8	6	E2	2900	0.10	E	1
			[395.46]	763703	1016570	6	8	E2	2100	0.097	E	1
9.	4f–7f	^{2}F° – ^{2}F°										
			[326.55]	763799	1070029	8	8	E2	6600	0.12	E	1
			[326.62]	763703	1069873	6	6	E2	1.1(+4)	0.15	E	1
			[326.72]	763799	1069873	8	6	E2	1800	0.024	E	1
			[326.45]	763703	1070029	6	8	E2	1300	0.023	E	1
10.	5f–6f	^{2}F° – ^{2}F°										
			[1117.7]	927102	1016570	8	8	E2	8500	71	E	1
			[1117.3]	927059	1016560	6	6	E2	1.4(+4)	87	E	1
			[1117.8]	927102	1016560	8	6	E2	2300	14	E	1
			[1117.2]	927509	1016570	6	8	E2	1700	14	E	1
11.	5f–7f	^{2}F° – ^{2}F°										
			[699.66]	927102	1070029	8	8	E2	5800	4.6	E	1
			[700.21]	927059	1069873	6	6	E2	9300	5.6	E	1
			[700.42]	927102	1069873	8	6	E2	1500	0.90	E	1
			[699.45]	927059	1070029	6	8	E2	1200	0.96	E	1
12.	6f–7f	^{2}F° – ^{2}F°										
			[1870.6]	1016570	1070029	8	8	E2	3200	350	E	1
			[1875.7]	1016560	1069873	6	6	E2	5100	420	E	1
			[1876.1]	1016570	1069873	8	6	E2	850	71	E	1
			[1870.2]	1016560	1070029	6	8	E2	640	70	E	1

[a]The number in parentheses following the tabulated value indicates the power of ten by which this value has to be multiplied.

Fe IX

Ar Isoelectronic Sequence

Ground State: $1s^2 2s^2 2p^6 3s^2 3p^6\ ^1S_0$

Ionization Energy: 233.6 eV = 1884000 cm^{-1}

Allowed Transitions

Line strengths for the first three multiplets of this argon-like ion are from the superposition-of-configurations (SOC) calculations of Weiss,[1] which are expected to be fairly accurate. Lin *et al.*[2] have computed transitions to 4s and 4d states by using the Dirac-Hartree-Fock method, but they have omitted correlation in excited states. Oscillator strengths for 3d–4f transitions have been calculated by Fawcett *et al.*[3] using Cowan's HX (Hartree-Fock with statistical exchange) method. Transitions involving the $^3D_2$ and $^1D_2$ levels of the $3p^5 3d$ configuration are omitted from this compilation, since they are indicated by Wagner and House[4] to be of low purity in *LS* coupling.

References

[1] A. W. Weiss, private communication.
[2] D. L. Lin, W. Fielder, Jr., and L. Armstrong, Jr., Phys. Rev. A **16**, 589 (1977).
[3] B. C. Fawcett, R. D. Cowan, E. Y. Kononov, and R. W. Hayes, J. Phys. B **5**, 1255 (1972).
[4] W. J. Wagner and L. L. House, Astrophys. J. **155**, 677 (1969).

Fe IX: Allowed transitions

No.	Transition Array	Multiplet	λ (Å)	E_i (cm^{-1})	E_k (cm^{-1})	g_i	g_k	A_{ki} (10^8 s^{-1})	f_{ik}	S (at. u.)	log gf	Accuracy	Source
1.	$3p^6$–$3p^5 3d$	^1S – $^3P°$											
			244.911	0	408315	1	3	0.087	2.4(−4)^a	1.9(−4)	−3.63	E	1
2.		^1S – $^3D°$											
			217.100	0	460616	1	3	2.0	0.0043	0.0031	−2.36	E	1
3.		^1S – $^1P°$	171.073	0	584546	1	3	2010	2.65	1.49	0.423	C+	1
4.	$3p^6$– $3p^5(^2P°_{3/2})4s$	^1S – $(^3/_2,^1/_2)°$											
			105.208	0	950498	1	3	320	0.16	0.055	−0.80	D	2
5.	$3p^6$– $3p^5(^2P°_{1/2})4s$	^1S – $(^1/_2,^1/_2)°$											
			103.566	0	965568	1	3	520	0.25	0.085	−0.60	D	2
6.	$3p^6$– $3p^5(^2P°_{3/2})4d$	^1S – $^2[^3/_2]°$											
			83.457	0	1198200	1	3	990	0.31	0.085	−0.51	D	2
7.	$3p^6$– $3p^5(^2P°_{1/2})4d$	^1S – $^2[^3/_2]°$											
			82.430	0	1213200	1	3	560	0.17	0.046	−0.77	D	2
8.	$3p^5 3d$– $3p^5(^2P°_{3/2})4f$	$^3P°$ – $^2[^3/_2]$											
			111.791	408315	1302841	3	5	1200	0.39	0.43	0.07	E	3
			111.713	405772	1300923	1	3	1000	0.56	0.21	−0.25	E	3

Fe IX: Allowed transitions — Continued

No.	Transition Array	Multiplet	λ (Å)	E_i (cm^{-1})	E_k (cm^{-1})	g_i	g_k	A_{ki} (10^8 s^{-1})	f_{ik}	S (at. u.)	log gf	Accuracy	Source
9.		^{3}P° – ^{2}[5/2]											
			112.096	413669	1305761	5	7	1600	0.41	0.76	0.31	E	3
10.		^{3}F° – ^{2}[9/2]											
			113.793	425810	1304599	9	11	2000	0.48	1.6	0.64	E	3
			114.024	429311	1306319	7	9	1600	0.40	1.1	0.45	E	3
11.		^{3}F° – ^{2}[7/2]											
			114.111	433819	1310159	5	7	1400	0.37	0.69	0.27	E	3
12.		^{3}D° – ^{2}[7/2]											
			116.803	455616	1311758	7	9	1600	0.41	1.1	0.46	E	3

[a]The number in parentheses following the tabulated value indicates the power of ten by which this value has to be multiplied.

Fe x

Cl Isoelectronic Sequence

Ground State: $1s^2 2s^2 2p^6 3s^2 3p^5\ ^2P^°_{3/2}$

Ionization Energy: 262.1 eV = 2114000 cm^{-1}

Allowed Transitions

List of tabulated lines

Wavelength (Å)	No.	Wavelength (Å)	No.	Wavelength (Å)	No.	Wavelength (Å)	No.
75.685	17	96.122	10	104.638	26	207.6	6
76.006	16	96.788	10	137.1	18	220.1	4
76.495	17	97.122	9	139.868	20	227.3	3
76.822	16	97.591	10	140.296	19	229.99	3
77.627	14	100.026	22	140.678	20	234.356	2
77.728	13	101.733	23	144.2	21	235.7	3
77.812	12	101.846	23	170.58	5	238.60	3
77.865	12	102.095	23	174.534	5	242.34	2
78.151	15	102.192	25	175.266	5	345.723	1
78.769	12	102.829	27	184.542	7	365.543	1
94.012	11	103.319	27	190.044	7		
95.338	10	103.724	24	195.399	8		
95.374	11	104.248	26	201.556	8		

Line strengths for transitions of the arrays $3s^23p^5$–$3s3p^6$ and $3p^5$–$3p^43d$ are the results of the multiconfiguration Dirac-Fock (MCDF) calculations of Huang et al.[1] These relativistic calculations include a perturbative treatment of the Breit interaction and the Lamb shift. Configuration mixing was limited to some configurations within the $n=3$ complex. Those configurations which were assumed to lie far above $3p^5$ or $3p^43d$ in energy were excluded, as were all configurations outside the complex.

According to the semi-empirical HX (Hartree-Fock with statistical allowance for exchange) calculations of Bromage et al.,[2] some levels of the $3p^43d$ configuration are strongly mixed in the LS basis, and in a few cases the LS designations given in Ref. 2 differed from those of Huang et al. The level designations used in this compilation are in accord with the theoretical results of Refs. 1 and 2. Transitions involving highly mixed levels have been excluded, as have the very weak transitions.

The calculated wavelengths of Huang et al. differ appreciably from the observed ones found in the literature. Thus the available experimentally determined wavelengths were used in making the conversion from line strengths to f- and A-values. (Otherwise, the calculated

wavelengths of Bromage et al. were used.) Bromage et al. indicate that it was necessary to scale down some configuration-interaction parameters by a greater amount than usual in order to fit their calculated energy levels to the experimental data. This could be an indication that, in purely ab initio calculations, configuration interaction should be treated on a larger scale to produce accurate calculated energy levels and f-values.

Oscillator strengths for transitions involving a single electron in the $n=4$ shell are the results of earlier HX calculations published by Fawcett et al.[3] An accuracy rating of "E" has been assigned to weak transitions, as well as to those involving a level for which the purity in the LS basis is less than 60%. Lines which are very weak, or which involve a level that is severely mixed, have been omitted.

References

[1] K.-N. Huang, Y.-K. Kim, K. T. Cheng, and J. P. Desclaux, At. Data Nucl. Data Tables **28**, 355 (1983).

[2] G. E. Bromage, R. D. Cowan, and B. C. Fawcett, Phys. Scr. **15**, 177 (1977).

[3] B. C. Fawcett, R. D. Cowan, E. Y. Kononov, and R. W. Hayes, J. Phys. B **5**, 1255 (1972).

Fe x: Allowed transitions

No.	Transition Array	Multiplet	λ (Å)	E_i (cm^{-1})	E_k (cm^{-1})	g_i	g_k	A_{ki} (10^8 s^{-1})	f_{ik}	S (at. u.)	log gf	Accuracy	Source
1.	$3s^23p^5$–$3s3p^6$	^{2}P° – ^{2}S	352.09	5228	289249	6	2	56	0.0345	0.240	−0.68	C−	1
			345.723	0	289249	4	2	39.0	0.0349	0.159	−0.85	C−	1
			365.543	15683	289249	2	2	17	0.034	0.081	−1.17	C−	1
2.	$3p^5$–$3p^4(^3$P$)3d$	^{2}P° – ^{4}F											
			234.356	0	426701	4	6	0.22	2.7(−4)^a	8.2(−4)	−2.97	E	1
			[242.34]	15683	428330	2	4	0.12	2.2(−4)	3.5(−4)	−3.36	E	1
3.		^{2}P° – ^{4}P											
			[235.7]			2	4	0.10	1.7(−4)	2.6(−4)	−3.47	E	1
			[227.3]			4	4	0.26	2.0(−4)	6.0(−4)	−3.10	E	1
			[238.60]	15683	434800	2	2	0.51	4.4(−4)	6.9(−4)	−3.06	E	1
			229.99	0	434800	4	2	2.1	8.3(−4)	0.0025	−2.48	E	1
4.		^{2}P° – ^{2}F											
			[220.1]			4	6	0.089	9.7(−5)	2.8(−4)	−3.41	E	1
5.		^{2}P° – ^{2}D	174.51	5228	578270	6	10	1800	1.37	4.72	0.91	C−	1
			174.534	0	572954	4	6	1800	1.24	2.84	0.69	C	1
			175.266	15683	586244	2	4	1720	1.59	1.83	0.50	C	1
			170.58	0	586244	4	4	54	0.024	0.053	−1.03	D	1
6.	$3p^5$–$3p^4(^1$D$)3d$	^{2}P° – ^{2}F											
			[207.6]			4	6	1.1	0.0011	0.0029	−2.37	E	1

Fe x: Allowed transitions — Continued

No.	Transition Array	Multiplet	λ (Å)	E_i (cm^{-1})	E_k (cm^{-1})	g_i	g_k	A_{ki} (10^8 s^{-1})	f_{ik}	S (at. u.)	log gf	Accuracy	Source
7.		^{2}P° – ^{2}S	*186.34*	*5228*	541882	6	2	1580	0.274	1.01	0.217	C–	1
			184.542	0	541882	4	2	1200	0.30	0.73	0.08	C–	1
			190.044	15683	541882	2	2	418	0.226	0.283	−0.345	C–	1
8.	$3p^5$–$3p^4(^1$S)$3d$	^{2}P° – ^{2}D											
			201.556	15683	511773	2	4	6.8	0.0083	0.011	−1.78	E	1
			195.399	0	511773	4	4	1.8	0.0010	0.0026	−2.39	E	1
9.	$3p^5$–$3p^4(^3$P)$4s$	^{2}P° – ^{4}P											
			97.122	0	1029600	4	4	350	0.050	0.064	−0.70	D	3
10.		^{2}P° – ^{2}P	*96.339*	*5228*	*1043200*	6	6	1100	0.15	0.28	−0.05	E	3
			96.122	0	1040300	4	4	870	0.12	0.15	−0.32	D	3
			96.788	15683	1048900	2	2	780	0.11	0.070	−0.66	D	3
			95.338	0	1048900	4	2	590	0.040	0.050	−0.80	D	3
			97.591	15683	1040300	2	4	70	0.02	0.01	−1.4	E	3
11.	$3p^5$–$3p^4(^1$D)$4s$	^{2}P° – ^{2}D											
			94.012	0	1063700	4	6	470	0.093	0.12	−0.43	D	3
			95.374	15683	1064200	2	4	550	0.15	0.094	−0.52	D	3
12.	$3p^5$–$3p^4(^3$P)$4d$	^{2}P° – ^{2}D	*78.162*	*5228*	*1284600*	6	10	1400	0.22	0.34	0.12	E	3
			77.865	0	1284300	4	6	1600	0.22	0.23	−0.06	D	3
			78.769	15683	1285100	2	4	400	0.075	0.039	−0.82	E	3
			77.812	0	1285100	4	4	800	0.073	0.075	−0.53	E	3
13.		^{2}P° – ^{4}F											
			77.728	0	1286500	4	6	280	0.038	0.039	−0.82	D	3
14.		^{2}P° – ^{2}F											
			77.627	0	1288200	4	6	480	0.065	0.066	−0.59	D	3
15.		^{2}P° – ^{2}P											
			78.151	15683	1295300	2	4	440	0.080	0.041	−0.80	D	3
16.	$3p^5$–$3p^4(^1$D)$4d$	^{2}P° – ^{2}P											
			76.006	0	1315700	4	4	1300	0.11	0.11	−0.36	D	3
			76.822	15683	1317400	2	2	1800	0.16	0.081	−0.49	D	3
17.		^{2}P° – ^{2}D											
			75.685	0	1321300	4	6	780	0.10	0.10	−0.40	D	3
			76.495	15683	1323000	2	4	1400	0.24	0.12	−0.32	D	3
18.	$3p^4(^3$P)$3d$ – $3p^4(^3$P)$4p$	^{4}D – ^{4}P°											
			[137.1]			8	6	150	0.031	0.11	−0.61	D	3
19.		^{4}F – ^{4}D°											
			140.296	417652	1130431	10	8	220	0.052	0.24	−0.28	D	3

Fe x: Allowed transitions — Continued

No.	Transition Array	Multiplet	λ (Å)	E_i (cm^{-1})	E_k (cm^{-1})	g_i	g_k	A_{ki} (10^8 s^{-1})	f_{ik}	S (at. u.)	log gf	Accuracy	Source
20.	$3p^4(^1D)3d-$ $3p^4(^1D)4p$	$^2G - ^2F°$											
			139.868	450750	1165710	10	8	220	0.052	0.24	−0.28	D	3
			140.678	451083	1161926	8	6	170	0.038	0.14	−0.52	E	3
21.		$^2F - ^2D°$											
			[144.2]			8	6	140	0.033	0.13	−0.58	D	3
22.	$3p^4(^3P)3d-$ $3p^4(^3P)4f$	$^4D - ^4F°$											
			100.026	388708	1388448	8	10	2600	0.49	1.3	0.59	D	3
23.		$^4F - ^4G°$											
			102.095	417652	1397132	10	12	2900	0.55	1.8	0.74	D	3
			101.733	426701	1409666	6	8	1800	0.38	0.76	0.36	D	3
			101.846	428330	1410200	4	6	1700	0.39	0.52	0.19	E	3
24.		$^2F - ^2G°$											
			103.724			6	8	1700	0.36	0.74	0.33	E	3
25.	$3p^4(^1D)3d-$ $3p^4(^1D)4f$	$^2G - ^2H°$											
			102.192	450750	1429300	10	12	2900	0.55	1.9	0.74	D	3
26.		$^2F - ^2G°$											
			104.638	485982	1441658	8	10	2100	0.43	1.2	0.54	D	3
			104.248			6	8	1400	0.31	0.64	0.27	D	3
27.	$3p^4(^1S)3d-$ $3p^4(^1S)4f$	$^2D - ^2F°$											
			103.319			6	8	2600	0.55	1.1	0.52	D	3
			102.829	511773	1484261	4	6	2100	0.49	0.66	0.29	D	3

[a]The number in parentheses following the tabulated value indicates the power of ten by which this value has to be multiplied.

Fe x

Forbidden Transitions

Line strengths for the magnetic dipole and electric quadrupole contributions to the transition between the two levels of the $3p^5$ configuration are the results of the multiconfiguration Dirac-Fock (MCDF) calculations of Huang *et al.*[1] These relativistic calculations included a perturbative treatment of the Breit interaction and the Lamb shift. Allowance for mixing among odd-parity configurations was limited to the set $3s^23p^5$, $3s3p^53d$, $3p^53d^2$, and $3s^23p^33d^2$. The strength of the electric quadrupole transition as defined in Ref. 1 was multiplied by the factor $^2/_3$ which is needed to bring this value into conformance with the definition of quadrupole strengths used in the NBS tables.

A-values for a number of transitions within the $3p^43d$ configuration were calculated by Mason and Nussbaumer[2] using the scaled Thomas-Fermi method with limited allowance for configuration interaction. Al-though each of the A-values reported in Ref. 2 is due to the stronger of the magnetic dipole and electric quadru-pole contributions, and not to the sum of the two,[3] there is no indication as to which type of transition is to be attributed to each of the values given there. Thus no conversion of transition probabilities to line strengths could be carried out. Moreover, the A-values themselves could not be corrected for errors in the calculated wave-lengths used in their determination, since the correction procedure is dependent on the type of transition to which it is applied. Transitions involving the $3p^4(^3P)3d\ ^2F_{7/2}$ level are excluded here, since this level is indicated by Bromage et al.[4] to be of low purity in LS coupling.

Mason and Nussbaumer have also calculated A-values for two magnetic quadrupole lines of the array $3p^5$–$3p^43d$, which were first corrected for differences in theo-retically and experimentally determined wavelengths and then quoted here.

References

[1] K.-N. Huang, Y.-K. Kim, K. T. Cheng, and J. P. Desclaux, At. Data Nucl. Data Tables **28**, 355 (1983).
[2] H. E. Mason and H. Nussbaumer, Astron. Astrophys. **54**, 547 (1977).
[3] H. E. Mason, private communication (1982).
[4] G. E. Bromage, R. D. Cowan, and B. C. Fawcett, Phys. Scr. **15**, 177 (1977).

Fe x: Forbidden transitions

No.	Transition Array	Multiplet	λ (Å)	E_i (cm^{-1})	E_k (cm^{-1})	g_i	g_k	Type of transition	A_{ki} (s^{-1})	S (at. u.)	Accu-racy	Source
1.	$3p^5$–$3p^5$	$^2P°$ – $^2P°$										
			6374.51	0	15683	4	2	M1	69.2	1.33	B	1
			"	"	"	4	2	E2	0.017	0.21	D–	1
2.	$3p^4(^3P)3d$ – $3p^4(^3P)3d$	^4D – ^4F										
			3454.2	388708	417652	8	10		12		E	2
			[2932.9]	388708	422794	8	8		8.5		E	2
3.		^4F – ^4F										
			[19440]	417652	422794	10	8		3.3		E	2
4.	$3p^4(^3P)3d$ – $3p^4(^1D)3d$	^4D – ^2G										
			1611.70	388708	450750	8	10		4.0		E	2
			1603.35	388714	451083	6	8		8.9		E	2
			1603.21	388708	451083	8	8		20		E	2
5.		^4D – ^2F										
			[1028.0]	388708	485982	8	8		65		E	2
			[1028.1]	388714	485982	6	8		18		E	2
6.		^4F – ^2G										
			3020.1	417652	450750	10	10		55		E	2
			3533.6	422794	451083	8	8		14		E	2
			[2990.4]	417652	451083	10	8		5.5		E	2
			3577.1	422794	450750	8	10		6.2		E	2
7.		^4F – ^2F										
			1463.49	417652	485982	10	8		70		E	2
8.	$3p^5$–$3p^4(^3P)3d$	$^2P°$ – ^4D										
			257.262	0	388708	4	8	M2	74	100	E	2

Fe X: Forbidden transitions — Continued

No.	Transition Array	Multiplet	λ (Å)	E_i (cm^{-1})	E_k (cm^{-1})	g_i	g_k	Type of transition	A_{ki} (s^{-1})	S (at. u.)	Accuracy	Source
9.		$^2P^\circ$ – ^4F										
			[236.52]	0	422794	4	8	M2	3.6	3.2	E	2

Fe XI

S Isoelectronic Sequence

Ground State: $1s^2 2s^2 2p^6 3s^2 3p^4 \ ^3P_2$

Ionization Energy: 290.3 eV $= 2341000$ cm^{-1}

Allowed Transitions

List of tabulated lines

Wavelength (Å)	No.	Wavelength (Å)	No.	Wavelength (Å)	No.	Wavelength (Å)	No.
72.166	20	89.863	17	123.49	22	202.4	11
72.310	19	90.205	13	123.572	24	276.35	2
72.635	18	90.345	13	123.822	22	308.61	4
73.2	18	91.394	26	124.725	23	341.113	1
86.513	16	91.472	26	176.620	7	349.046	1
86.772	14	91.63	25	179.762	9	352.661	1
87.025	14	91.733	25	184.41	12	355.837	5
87.995	14	92.81	29	184.800	8	356.519	1
88.029	14	92.87	27,29	188.219	6	358.621	1
88.167	14	93.433	28	189.017	6	369.154	1
89.104	15	121.419	21	192.819	6	406.81	3
89.185	13	121.747	21	201.575	10		

Oscillator strengths for transitions of the $3s^2 3p^4$–$3s 3p^5$ array are the results of the multiconfiguration scaled Thomas-Fermi calculations of Mason.[1] She has published a table of f-values which were obtained by including various sets of configurations in her calculations, as well as the data which result from neglect of configuration interaction. It is not surprising that the results vary considerably, given the complexity of the Fe XI ion. Low accuracy ratings are assigned to the values tabulated here, since even the largest basis set used by Mason does not include all configurations within the $n = 3$ complex, and all configurations outside the complex are excluded.

Results of the HX (Hartree-Fock with statistical allowance for exchange) calculations of Bromage et al.[2] are quoted for the $3p^4$–$3p^3 3d$ array. Correlation with a few configurations, all of which are within the $n = 3$ complex, is included. They had to scale down the configuration-interaction parameters considerably in order to bring the theoretically predicted energy levels into agreement with experiment—an indication that, in purely *ab initio* calculations, configuration interaction should be treated on a much larger scale to produce accurate calculated energy levels and oscillator strengths.

The percentage compositions published by Bromage et al. for the $3p^3 3d$ configuration indicate that many levels are severely mixed in the *LS* basis. For this reason, transitions involving a number of levels have been excluded from this compilation. For transitions involving the remaining levels, the term designations of Bromage et al. have been used.

Results of the earlier HX calculations of Fawcett *et al.*[3,4] are quoted for transitions to configurations in which one electron occupies the $n = 4$ shell. Transitions involving levels which are indicated to be of low purity in LS coupling are omitted here.

Lines which are characterized by very small f-values are assigned lower accuracy ratings; the weakest lines have been excluded.

References

[1] H. E. Mason, Mon. Not. R. Astron. Soc. **170**, 651 (1975).
[2] G. E. Bromage, R. D. Cowan, and B. C. Fawcett, Phys. Scr. **15**, 177 (1977).
[3] B. C. Fawcett, R. D. Cowan, E. Y. Kononov, and R. W. Hayes, J. Phys. B **5**, 1255 (1972).
[4] B. C. Fawcett, N. J. Peacock, and R. D. Cowan, J. Phys. B **1**, 295 (1968).

Fe XI: Allowed transitions

No.	Transition Array	Multiplet	λ (Å)	E_i (cm^{-1})	E_k (cm^{-1})	g_i	g_k	A_{ki} (10^8 s^{-1})	f_{ik}	S (at. u.)	log gf	Accuracy	Source
1.	$3s^2 3p^4$–$3s 3p^5$	^3P – $^3P°$	*353.76*	*5813*	*288492*	9	9	23	0.043	0.45	−0.41	D	1
			352.661	0	283558	5	5	17	0.032	0.19	−0.80	D	1
			356.519	12668	293158	3	3	5.2	0.010	0.035	−1.52	D	1
			341.113	0	293158	5	3	11	0.012	0.067	−1.22	D	1
			349.046	12668	299163	3	1	23	0.014	0.048	−1.38	D	1
			369.154	12668	283558	3	5	5.3	0.018	0.066	−1.27	D	1
			358.621	14312	293158	1	3	7.1	0.041	0.048	−1.39	D	1
2.		^3P – $^1P°$	[276.35]	0	361859	5	3	2.0	0.0014	0.0064	−2.15	E	1
3.		^1D – $^3P°$	[406.81]	37743	283558	5	5	0.40	0.0010	0.0067	−2.30	E	1
4.		^1D – $^1P°$	308.61	37743	361859	5	3	75	0.064	0.33	−0.49	D	1
5.		^1S – $^1P°$	355.837	80831	361859	1	3	1.7	0.0094	0.011	−2.03	D−	1
6.	$3p^4$–$3p^3(^2D°)3d$	^3P – $^3P°$	188.219	0	531296	5	5	1100	0.59	1.8	0.47	D	2
			189.017	12668	541721	3	1	1400	0.25	0.47	−0.12	D	2
			192.819	12668	531296	3	5	220	0.20	0.38	−0.22	D	2
7.		^3P – $^1D°$	176.620	12668	578855	3	5	86	0.067	0.12	−0.70	D	2
8.		^1D – $^1D°$	184.800	37743	578855	5	5	1200	0.63	1.9	0.50	C	2
9.		^1D – $^1F°$	179.762	37743	594034	5	7	1670	1.13	3.34	0.75	C	2
10.	$3p^4$–$3p^3(^2P°)3d$	^3P – $^3P°$	201.575	0	496093	5	5	36	0.022	0.073	−0.96	D	2
11.		^1D – $^3D°$	[202.4]			5	3	170	0.062	0.21	−0.51	E	2
12.		^1S – $^1P°$	184.41	80831	623100	1	3	1430	2.19	1.33	0.340	C	2
13.	$3p^4$–$3p^3(^4S°)4s$	^3P – $^3S°$	*89.646*	*5813*	1121300	9	3	2100	0.083	0.22	−0.13	D−	3
			89.185	0	1121300	5	3	1300	0.092	0.14	−0.34	D	3
			90.205	12668	1121300	3	3	550	0.067	0.060	−0.70	D	3
			90.345	14312	1121300	1	3	200	0.08	0.02	−1.1	E	3

Fe XI: Allowed transitions — Continued

No.	Transition Array	Multiplet	λ (Å)	E_i (cm^{-1})	E_k (cm^{-1})	g_i	g_k	A_{ki} (10^8 s^{-1})	f_{ik}	S (at. u.)	log gf	Accuracy	Source
14.	$3p^4$–$3p^3(^2D°)4s$	^3P – $^3D°$											
			86.772	0	1152400	5	7	540	0.086	0.12	−0.37	D	3
			87.995	12668	1149100	3	5	220	0.043	0.037	−0.89	D	3
			88.167	14312	1148500	1	3	200	0.06	0.02	−1.2	E	3
			87.025	0	1149100	5	5	350	0.040	0.057	−0.70	D	3
			88.029	12668	1148500	3	3	400	0.047	0.041	−0.85	D	3
15.		^1D – $^1D°$	89.104	37743	1160000	5	5	1300	0.16	0.23	−0.10	D	3
16.	$3p^4$–$3p^3(^2P°)4s$	^1D – $^1P°$	86.513	37743	1193600	5	3	830	0.056	0.080	−0.55	D	3
17.		^1S – $^1P°$	[89.863]	80831	1193600	1	3	690	0.25	0.074	−0.60	D	3
18.	$3p^4$–$3p^3(^4S°)4d$	^3P – $^3D°$											
			72.635	0	1376700	5	7	1600	0.18	0.22	−0.05	D	3
			[73.2]			3	5	820	0.11	0.080	−0.48	D	4
19.	$3p^4$–$3p^3(^2D°)4d$	^1D – $^1D°$	72.310	37743	1420600	5	5	1500	0.12	0.14	−0.22	D	3
20.		^1D – $^1F°$	72.166	37743	1423400	5	7	2900	0.32	0.38	0.20	D	3
21.	$3p^3(^4S°)3d$ – $3p^3(^4S°)4p$	$^5D°$ – ^5P											
			121.419			9	7	290	0.050	0.18	−0.35	D	3
			121.747			7	5	210	0.033	0.093	−0.64	D	3
22.	$3p^3(^2D°)3d$ – $3p^3(^2D°)4p$	$^3G°$ – ^3F											
			123.49			11	9	270	0.050	0.22	−0.26	D	3
			123.49			9	7	170	0.030	0.11	−0.57	E	3
			123.822			7	5	220	0.036	0.10	−0.60	E	3
23.		$^1G°$ – ^1F	124.725			9	7	220	0.040	0.15	−0.44	D	3
24.	$3p^3(^2P°)3d$ – $3p^3(^2P°)4p$	$^3F°$ – ^3D											
			123.572			7	5	360	0.059	0.17	−0.38	E	3
25.	$3p^3(^4S°)3d$ – $3p^3(^4S°)4f$	$^5D°$ – ^5F											
			91.733			9	11	4100	0.63	1.7	0.75	D	3
			91.63			7	9	3400	0.55	1.2	0.59	D	3
			91.63			5	7	2800	0.49	0.74	0.39	D	3
			91.63			3	5	2300	0.48	0.43	0.16	D	3
26.	$3p^3(^2D°)3d$ – $3p^3(^2D°)4f$	$^3F°$ – ^3G											
			91.472			7	9	2500	0.41	0.86	0.46	D	3
			91.394			5	7	2600	0.45	0.68	0.35	D	3
27.		$^3G°$ – ^3H											
			92.87			11	13	3900	0.60	2.0	0.82	D	3
			92.87			7	9	3400	0.57	1.2	0.60	D	3

Fe XI: Allowed transitions — Continued

No.	Transition Array	Multiplet	λ (Å)	E_i (cm^{-1})	E_k (cm^{-1})	g_i	g_k	A_{ki} (10^8 s^{-1})	f_{ik}	S (at. u.)	log gf	Accuracy	Source
28.		^{1}G° – ^{1}H	93.433			9	11	3200	0.51	1.4	0.66	D	3
29.	$3p^3(^2P°)3d$– $3p^3(^2P°)4f$	^{3}F° – ^{3}G											
			92.81			9	11	3700	0.59	1.6	0.73	D	3
			92.87			7	9	2800	0.47	1.0	0.52	D	3

Fe XI

Forbidden Transitions

Transition probabilities for magnetic dipole and electric quadrupole lines within the $3p^4$ configuration are the results of the scaled Thomas-Fermi calculations of Mendoza and Zeippen.[1] They included a number of correlation configurations in their basis set and introduced Breit-Pauli relativistic corrections as a perturbation to the nonrelativistic Hamiltonian.

Mason and Nussbaumer[2] calculated A-values for a number of transitions within the $3p^33d$ configuration. They applied the scaled Thomas-Fermi method with very limited allowance for configuration interaction. Although each of the values reported in Ref. 2 for transitions within the $3p^33d$ configuration is due to the stronger of the magnetic dipole and electric quadrupole contributions, and not to the sum of the two,[3] there is no indication as to which type of transition is to be at-

tributed to each of the values given there. The transition probability for the $3p^3(^2D°)3d$ $^3D_3°$ – $3p^3(^2P°)3d$ $^3F_4°$ transition was excluded from this compilation, since the $^3D_3°$ level is indicated[4] to be of low purity in LS coupling. A-values reported in Ref. 2 for electric quadrupole lines of the $3s3p^5$–$3s^23p^33d$ array and magnetic quadrupole lines of the $3p^4$–$3p^33d$ array are tabulated here as well.

References

[1] C. Mendoza and C. J. Zeippen, Mon. Not. R. Astron. Soc. **202**, 981 (1983).
[2] H. E. Mason and H. Nussbaumer, Astron. Astrophys. **54**, 547 (1977).
[3] H. E. Mason, private communication (1982).
[4] G. E. Bromage, R. D. Cowan, and B. C. Fawcett, Phys. Scr. **15**, 177 (1977).

Fe XI: Forbidden transitions

No.	Transition Array	Multiplet	λ (Å)	E_i (cm^{-1})	E_k (cm^{-1})	g_i	g_k	Type of transition	A_{ki} (s^{-1})	S (at. u.)	Accuracy	Source
1.	$3p^4$–$3p^4$	^{3}P – ^{3}P										
			7891.94	0	12668	5	3	M1	43.6	2.38	C+	1
			"	"	"	5	3	E2	0.0035	0.19	D–	1
			[60810]	12668	14312	3	1	M1	0.226	1.88	C+	1
			[6985.2]	0	14312	5	1	E2	0.0099	0.098	E	1
2.		^{3}P – ^{1}D										
			2648.73	0	37743	5	5	M1	92	0.32	D–	1
			"	"	"	5	5	E2	0.15	0.058	E	1
			3986.9	12668	37743	3	5	M1	9.5	0.11	D–	1
			"	"	"	3	5	E2	0.0030	0.0090	E	1
			[4266.6]	14312	37743	1	5	E2	0.0014	0.0059	E	1

Fe XI: Forbidden transitions — Continued

No.	Transition Array	Multiplet	λ (Å)	E_i (cm^{-1})	E_k (cm^{-1})	g_i	g_k	Type of transition	A_{ki} (s^{-1})	S (at. u.)	Accuracy	Source
3.		^{3}P – ^{1}S										
			[1237.1]	0	80831	5	1	E2	1.7	0.0029	E	1
			1467.08	12668	80831	3	1	M1	980	0.11	D–	1
4.		^{1}D – ^{1}S	[2320.1]	37743	80831	5	1	E2	8.3	0.33	D–	1
5.	$3p^3(^4$S$^\circ)3d$ – $3p^3(^2$D$^\circ)3d$	^{5}D$^\circ$ – ^{3}F$^\circ$										
						9	9		28		E	2
6.		^{5}D$^\circ$ – ^{3}G$^\circ$										
						9	11		1.7		E	2
						9	9		6.7		E	2
7.		^{5}D$^\circ$ – ^{1}G$^\circ$										
						9	9		2.0		E	2
						7	9		4.6		E	2
8.	$3p^3(^4$S$^\circ)3d$ – $3p^3(^2$P$^\circ)3d$	^{5}D$^\circ$ – ^{3}F$^\circ$										
						9	9		260		E	2
						7	9		50		E	2
9.	$3p^3(^2$D$^\circ)3d$ – $3p^3(^2$D$^\circ)3d$	^{3}F$^\circ$ – ^{3}G$^\circ$										
						9	11		6.8		E	2
						9	9		4.5		E	2
10.		^{3}F$^\circ$ – ^{1}G$^\circ$										
						9	9		3.3		E	2
						7	9		2.9		E	2
11.	$3p^3(^2$D$^\circ)3d$ – $3p^3(^2$P$^\circ)3d$	^{3}F$^\circ$ – ^{3}F$^\circ$										
						9	9		21		E	2
						7	9		10		E	2
12.		^{3}G$^\circ$ – ^{3}F$^\circ$										
						11	9		51		E	2
						9	9		42		E	2
13.		^{1}G$^\circ$ – ^{3}F$^\circ$										
						9	9		45		E	2
14.	$3s3p^5$ – $3s^23p^3(^2$D$^\circ)3d$	^{3}P$^\circ$ – ^{3}F$^\circ$										
			[658]			5	9	E2	30	0.020	E	2
15.		^{3}P$^\circ$ – ^{1}G$^\circ$										
			[562]			5	9	E2	1.8	5.4(−4)[a]	E	2

Fe XI: Forbidden transitions — Continued

No.	Transition Array	Multiplet	λ (Å)	E_i (cm^{-1})	E_k (cm^{-1})	g_i	g_k	Type of transition	A_{ki} (s^{-1})	S (at. u.)	Accu-racy	Source
16.	$3s3p^5-$ $3s^23p^3(^2P°)3d$	$^3P° - {}^3F°$										
			[485]			5	9	E2	96	0.014	E	2
17.	$3p^4-3p^3(^4S°)3d$	$^3P - {}^5D°$										
			[253]			5	9	M2	25	35	E	2
18.	$3p^4-3p^3(^2D°)3d$	$^3P - {}^1G°$										
			[216]			5	9	M2	1.3	0.83	E	2
19.		$^1D - {}^3F°$										
			[251]			5	9	M2	40	54	E	2
20.		$^1D - {}^3G°$										
			[240]			5	9	M2	1.4	1.5	E	2
21.		$^1D - {}^1G°$	[236]			5	9	M2	6.0	6.0	E	2

[a]The number in parentheses following the tabulated value indicates the power of ten by which this value has to be multiplied.

Fe XII

P Isoelectronic Sequence

Ground State: $1s^22s^22p^63s^23p^3\,{}^4S°_{3/2}$

Ionization Energy: 330.8 eV = 2668000 cm^{-1}

Allowed Transitions

List of tabulated lines

Wavelength (Å)	No.	Wavelength (Å)	No.	Wavelength (Å)	No.	Wavelength (Å)	No.
65.805	26	84.48	30	195.119	5	229.22	10
65.905	21	84.52	30	195.18	13	244	7
66.526	25	84.85	31	196.640	13	246	7
66.960	23	85.14	32	196.923	13	256	6
67.164	27	85.477	32	198.555	16	333.15	2
67.821	22	108.440	28	201.121	16	335.06	2
68.382	24	108.605	28	208.42	15	338.263	2
79.488	17	108.862	28	209.11	8	340.23	2
80.022	17	110.591	29	210.932	15	346.852	1
80.160	19	110.732	29	211.24	15	352.107	1
80.5	17	180.30	11	212.34	8	364.468	1
80.541	19	186.51	14	212.46	8	376.08	3
81.651	18	186.880	9	214.40	8	382.83	3
81.943	18	188.10	14	217	4	385.36	3
82.226	18	192.394	5	218	12		
82.744	20	193.509	5	219	4		
82.837	20	194.920	13	224.39	10		

Line strengths for transitions of the arrays $3s^2 3p^3$–$3s 3p^4$ and $3p^3$–$3p^2 3d$ are the results of the multiconfiguration Dirac-Fock (MCDF) calculations of Huang.[1] These relativistic calculations included a perturbative treatment of the Breit interaction and the Lamb shift. Allowance for configuration mixing was limited to configurations within the $n = 3$ complex having no more than two electrons in the $3d$ subshell.

Huang published diagrams of energy levels (designated in LS coupling) in the $3s^2 3p^3$, $3s 3p^4$, and $3s^2 3p^2 3d$ configurations of Fe XII, but he has not provided percentage compositions. We have used the percentages given by Bromage $et\ al.$[2] as a guide to naming the levels; their values resulted from Hartree-Fock calculations with relativistic effects and statistical allowance for exchange (HXR), and incorporated correlation effects due to a few configurations within the $n = 3$ complex. Whenever a term designation of a level, as given in

Ref. 1, is different from that indicated in Ref. 2, all transitions involving that level are omitted from this compilation.

Results of the earlier HX calculations of Fawcett $et\ al.$[3] are quoted for transitions to configurations in which one electron occupies the $n = 4$ shell.

Transitions involving levels which are indicated to be of low purity in LS coupling are omitted here. Lines which are characterized by very small f-values are assigned lower accuracy ratings; the weakest lines have been excluded.

References

[1] K.-N. Huang, At. Data Nucl. Data Tables **30**, 313 (1984).
[2] G. E. Bromage, R. D. Cowan, and B. C. Fawcett, Mon. Not. R. Astron. Soc. **183**, 19 (1978).
[3] B. C. Fawcett, R. D. Cowan, E. Y. Kononov, and R. W. Hayes, J. Phys. B **5**, 1255 (1972).

Fe XII: Allowed transitions

No.	Transition Array	Multiplet	λ (Å)	E_i (cm^{-1})	E_k (cm^{-1})	g_i	g_k	A_{ki} (10^8 s^{-1})	f_{ik}	S (at. u.)	log gf	Accuracy	Source
1.	$3s^2 3p^3$–$3s 3p^4$	^{4}S° – ^{4}P	357.26	0	279906	4	12	17	0.098	0.46	−0.41	D	1
			364.468	0	274373	4	6	16	0.048	0.23	−0.72	D	1
			352.107	0	284005	4	4	17	0.032	0.15	−0.89	D	1
			346.852	0	288307	4	2	18	0.016	0.075	−1.18	D	1
2.		^{2}D° – ^{2}D	336.97	44275	341040	10	10	32	0.054	0.60	−0.27	E	1
			338.263	46089	341717	6	6	29	0.049	0.33	−0.53	D	1
			335.06	41555	340010	4	4	35	0.059	0.26	−0.63	D	1
			[340.23]	46089	340010	6	4	0.49	5.7(−4)^a	0.0038	−2.47	E	1
			[333.15]	41555	341717	4	6	0.23	5.7(−4)	0.0025	−2.64	E	1
3.		^{2}P° – ^{2}D	380.72	78378	341040	6	10	4.8	0.017	0.13	−0.98	E	1
			382.83	80513	341717	4	6	5.8	0.019	0.097	−1.11	D	1
			[376.08]	74108	340010	2	4	3.0	0.013	0.032	−1.59	D	1
			[385.36]	80513	340010	4	4	0.19	4.1(−4)	0.0021	−2.78	E	1
4.	$3p^3$–$3p^2(^3$P)$3d$	^{4}S° – ^{4}D	[217]			4	6	2.6	0.0028	0.0080	−1.95	E	1
			[219]			4	4	2.6	0.0018	0.0053	−2.13	E	1
5.		^{4}S° – ^{4}P	194.12	0	515139	4	12	880	1.5	3.8	0.77	D	1
			195.119	0	512508	4	6	860	0.74	1.9	0.47	D	1
			193.509	0	516772	4	4	910	0.51	1.3	0.31	D	1
			192.394	0	519767	4	2	900	0.25	0.63	−0.00	D	1
6.		^{2}D° – ^{4}F	[256]			4	4	1.5	0.0015	0.0049	−2.24	E	1
7.		^{2}D° – ^{4}D	[246]			6	4	3.1	0.0019	0.0091	−1.95	E	1
			[244]			4	2	4.0	0.0018	0.0057	−2.15	E	1

Fe XII: Allowed transitions — Continued

No.	Transition Array	Multiplet	λ (Å)	E_i (cm^{-1})	E_k (cm^{-1})	g_i	g_k	A_{ki} (10^8 s^{-1})	f_{ik}	S (at. u.)	log gf	Accuracy	Source
8.		^{2}D° – ^{4}P											
			[214.40]	46089	512508	6	6	11	0.0076	0.032	−1.34	E	1
			[212.46]	46089	516772	6	4	2.5	0.0011	0.0047	−2.17	E	1
			[209.11]	41555	519767	4	2	18	0.0058	0.016	−1.63	E	1
			[212.34]	41555	512508	4	6	4.2	0.0043	0.012	−1.77	E	1
9.		^{2}D° – ^{2}F											
			186.880	46089	581192	6	8	1000	0.73	2.7	0.64	E	1
10.		^{2}P° – ^{4}P											
			[229.22]	80513	516772	4	4	1.7	0.0013	0.0040	−2.28	E	1
			[224.39]	74108	519767	2	2	2.3	0.0018	0.0026	−2.45	E	1
11.	3p^3–3p^2(^{1}D)3d	^{4}S° – ^{2}D											
			[180.30]	0	554633	4	6	1.7	0.0013	0.0030	−2.30	E	1
12.		^{2}D° – ^{2}G											
			[218]			6	8	2.4	0.0023	0.010	−1.86	E	1
13.		^{2}D° – ^{2}D	*196.05*	*44275*	*554341*	10	10	590	0.34	2.2	0.53	D	1
			196.640	46089	554633	6	6	490	0.28	1.1	0.23	D	1
			[195.18]	41555	553902	4	4	610	0.35	0.89	0.14	D	1
			196.923	46089	553902	6	4	110	0.041	0.16	−0.61	D	1
			194.920	41555	554633	4	6	17	0.014	0.037	−1.24	D	1
14.		^{2}D° – ^{2}P											
			[188.10]	46089	577726	6	4	9.9	0.0035	0.013	−1.68	E	1
			[186.51]	41555	577726	4	4	14	0.0073	0.018	−1.53	E	1
15.		^{2}P° – ^{2}D	*210.10*	*78378*	*554341*	6	10	70	0.077	0.32	−0.33	E	1
			210.932	80513	554633	4	6	68	0.068	0.19	−0.56	D	1
			[208.42]	74108	553902	2	4	67	0.087	0.12	−0.76	D	1
			[211.24]	80513	553902	4	4	3.6	0.0024	0.0067	−2.02	E	1
16.		^{2}P° – ^{2}P											
			201.121	80513	577726	4	4	510	0.31	0.82	0.09	E	1
			198.555	74108	577726	2	4	160	0.19	0.25	−0.42	E	1
17.	3p^3–3p^2(^{3}P)4s	^{4}S° – ^{4}P	*80.0*	*0*	*1250000*	4	12	660	0.19	0.20	−0.12	D	3
			79.488	0	1258100	4	6	670	0.095	0.099	−0.42	D	3
			80.022	0	1249700	4	4	680	0.065	0.068	−0.59	D	3
			80.5	0	1240000	4	2	720	0.035	0.037	−0.85	D	3
18.		^{2}D° – ^{2}P	*82.014*	*44275*	*1263600*	10	6	1700	0.10	0.27	0.00	E	3
			81.943	46089	1266500	6	4	1400	0.097	0.16	−0.24	D	3
			82.226	41555	1257800	4	2	1900	0.095	0.10	−0.42	D	3
			81.651	41555	1266500	4	4	100	0.01	0.01	−1.4	E	3
19.	3p^3–3p^2(^{1}D)4s	^{2}D° – ^{2}D											
			[80.541]	46089	1287700	6	6	870	0.085	0.14	−0.29	D	3
			80.160	41555	1289000	4	4	600	0.058	0.061	−0.63	D	3

Fe XII:　Allowed transitions — Continued

No.	Transition Array	Multiplet	λ (Å)	E_i (cm^{-1})	E_k (cm^{-1})	g_i	g_k	A_{ki} (10^8 s^{-1})	f_{ik}	S (at. u.)	log gf	Accuracy	Source
20.		^{2}P° – ^{2}D											
			82.837	80513	1287700	4	6	190	0.030	0.033	−0.92	D	3
			82.744	80513	1289000	4	4	760	0.078	0.085	−0.51	D	3
21.	$3p^3$–$3p^2(^3$P$)4d$	^{4}S° – ^{4}P											
			65.905	0	1517300	4	4	2000	0.13	0.11	−0.28	D	3
22.		^{2}D° – ^{2}F											
			67.821	41555	1516100	4	6	1400	0.14	0.13	−0.25	D	3
23.		^{2}D° – ^{2}D											
			66.960	41555	1535000	4	6	1600	0.16	0.14	−0.19	D	3
24.		^{2}P° – ^{2}D											
			68.382	74108	1536500	2	4	1700	0.24	0.11	−0.32	D	3
25.	$3p^3$–$3p^2(^1$D$)4d$	^{2}D° – ^{2}F											
			66.526	46089	1549300	6	8	1700	0.15	0.20	−0.05	D	3
26.		^{2}D° – ^{2}P											
			65.805	46089	1565700	6	4	510	0.022	0.029	−0.88	D	3
27.		^{2}P° – ^{2}S											
			67.164	80513	1569400	4	2	1100	0.038	0.034	−0.82	D	3
28.	$3p^2(^3$P$)3d$– $3p^2(^3$P$)4p$	^{4}F – ^{4}D°											
			108.440			10	8	330	0.047	0.17	−0.33	D	3
			108.605			8	6	330	0.044	0.13	−0.45	D	3
			108.862			6	4	320	0.038	0.082	−0.64	D	3
29.	$3p^2(^1$D$)3d$– $3p^2(^1$D$)4p$	^{2}G – ^{2}F°											
			110.591			10	8	310	0.046	0.17	−0.34	D	3
			110.732			8	6	130	0.018	0.052	−0.84	D	3
30.	$3p^2(^3$P$)3d$– $3p^2(^3$P$)4f$	^{4}F – ^{4}G°											
			84.52			10	12	5200	0.67	1.9	0.83	D	3
			84.48			8	10	4900	0.66	1.5	0.72	D	3
			84.52			6	8	4000	0.57	0.95	0.53	D	3
			84.48			4	6	4500	0.72	0.80	0.46	D	3
31.		^{4}D – ^{4}F°											
			84.85			6	8	2300	0.33	0.55	0.30	D	3
32.	$3p^2(^1$D$)3d$– $3p^2(^1$D$)4f$	^{2}G – ^{2}H°											
			85.477			10	12	4600	0.60	1.7	0.78	D	3
			85.14			8	10	3400	0.46	1.0	0.57	D	3

[a]The number in parentheses following the tabulated value indicates the power of ten by which this value has to be multiplied.

J. Phys. Chem. Ref. Data, Vol. 17, Suppl. 4, 1988

Fe XII

Forbidden Transitions

Line strengths for magnetic dipole and electric quadrupole transitions within the $3p^3$ configuration are the results of the multiconfiguration Dirac-Fock (MCDF) calculations of Huang.[1] These relativistic calculations included a perturbative treatment of the Breit interaction and the Lamb shift. Allowance for configuration mixing was limited to configurations within the $n = 3$ complex having no more than two electrons in the $3d$ subshell. Strengths of electric quadrupole transitions as defined in Ref. 1 were multiplied by the factor $^2/_3$ which is needed to bring these values into conformance with the definition of quadrupole strengths used in the NBS tables. We have excluded from this compilation the electric quadrupole contributions to the $^4S^\circ_{3/2} - {}^2P^\circ_{3/2}$ and $^4S^\circ_{3/2} - {}^2P^\circ_{1/2}$ transitions, since their strengths are very small and thus subject to considerable uncertainty.

Data for these same transitions calculated by Mendoza and Zeippen[2] with the scaled Thomas-Fermi approach with allowance for correlation are generally in very good agreement with the results of Ref. 1. These latter calculations treated relativistic effects by introducing Breit-Pauli corrections as a perturbation to the nonrelativistic Hamiltonian.

References

[1]K.-N. Huang, At. Data Nucl. Data Tables **30**, 313 (1984).
[2]C. Mendoza and C. J. Zeippen, Mon. Not. R. Astron. Soc. **198**, 127 (1982).

Fe XII: Forbidden transitions

No.	Transition Array	Multiplet	λ (Å)	E_i (cm^{-1})	E_k (cm^{-1})	g_i	g_k	Type of transition	A_{ki} (s^{-1})	S (at. u.)	Accuracy	Source
1.	$3p^3-3p^3$	$^4S^\circ - {}^2D^\circ$										
			2169.03	0	46089	4	6	M1	1.7	0.0038	E	1
			"	"	"	4	6	E2	0.099	0.017	E	1
			2405.71	0	41555	4	4	M1	48	0.099	D	1
			"	"	"	4	4	E2	0.040	0.0077	E	1
2.		$^4S^\circ - {}^2P^\circ$										
			1242.03	0	80513	4	4	M1	320	0.090	D	1
			1349.38	0	74108	4	2	M1	180	0.032	D	1
3.		$^2D^\circ - {}^2D^\circ$										
			[22050]	41555	46089	4	6	M1	0.88	2.11	C+	1
			"	"	"	4	6	E2	4.4(−6)^a	0.081	E	1
4.		$^2D^\circ - {}^2P^\circ$										
			[3568.0]	46089	74108	6	2	E2	0.31	0.21	D−	1
			[2904.1]	46089	80513	6	4	M1	80	0.29	C	1
			"	"	"	6	4	E2	1.3	0.62	D−	1
			[3071.0]	41555	74108	4	2	M1	70	0.15	C	1
			"	"	"	4	2	E2	0.83	0.27	D−	1
			2565.99	41555	80513	4	4	M1	200	0.50	C	1
			"	"	"	4	4	E2	0.64	0.17	D−	1
5.		$^2P^\circ - {}^2P^\circ$										
			[15608]	74108	80513	2	4	M1	2.04	1.15	C+	1
			"	"	"	2	4	E2	1.5(−5)	0.033	E	1

^{a}The number in parentheses following the tabulated value indicates the power of ten by which this value has to be multiplied.

Fe XIII

Si Isoelectronic Sequence

Ground State: $1s^2 2s^2 2p^6 3s^2 3p^2\ {}^3P_0$

Ionization Energy: $361.0\,\mathrm{eV} = 2912000\,\mathrm{cm}^{-1}$

Allowed Transitions

List of tabulated lines

Wavelength (Å)	No.	Wavelength (Å)	No.	Wavelength (Å)	No.	Wavelength (Å)	No.
62.353	16	84.270	25	208.679	13	320.800	2
62.46	17	98.128	21	216.87	11	321.45	2
62.699	16	98.523	21	234	5	348.184	1
63.188	19	98.826	21	236	5	359.63	1
64.139	20	107.384	22	240	5	359.837	1
74.327	14	175.22	9	260	10	368.12	1
74.845	14	185.75	8	303.35	2	372.02	1
75.892	14	196.525	12	311	2	372.24	1
76.117	15	202.424	6	311.552	2	412.98	3
78.452	23	203.826	7	312.164	2	419.92	4

Line strengths for transitions of the arrays $3s^2 3p^2$–$3s 3p^3$ and $3p^2$–$3p\,3d$ are the results of the multiconfiguration Dirac-Fock (MCDF) calculations of Huang.[1] These relativistic calculations included a perturbative treatment of the Breit interaction and the Lamb shift. Allowance for configuration mixing included all configurations within the $n=3$ complex.

Huang published a diagram of energy levels (designated in LS coupling) in the $3s^2 3p^2$, $3s 3p^3$, and $3s^2 3p\,3d$ configurations of Fe XIII, but he has not provided percentage compositions. We have used the percentages given by Bromage et al.[2] as a guide to naming the levels; their values resulted from Hartree-Fock calculations with relativistic effects and statistical allowance for exchange (HXR), and incorporated correlation effects due to a few configurations within the $n=3$ complex. Whenever the term designation of a level, as given in Ref. 1, is different from that indicated in Ref. 2, all transitions involving that level are omitted from this compilation.

Results of the earlier HX calculations of Fawcett et al.[3] and of the more recent multiconfiguration scaled Thomas-Fermi approach of Kastner et al.[4] are quoted for transitions to configurations in which one electron occupies the $n=4$ shell. The results have been averaged for lines common to the two sources.

Transitions involving levels which are indicated to be of low purity in LS coupling are omitted here. Lines which are characterized by very small f-values are assigned lower accuracy ratings; the weakest lines have been excluded.

The lifetime of the $3s 3p^3\ {}^3S_1^\circ$ level was measured by Träbert et al.[5] using the beam-foil technique. This datum does not provide a transition probability, as there are five branches through which this level can be depopulated by electric dipole radiation. A comparison of their measured lifetime to the result derived from the theoretically determined transition probabilities of Huang would be of questionable value, since this is one of the levels whose term designation according to Ref. 1 is in disagreement with the designation given by Bromage et al.

References

[1]K.-N. Huang, At. Data Nucl. Data Tables **32**, 503 (1985).

[2]G. E. Bromage, R. D. Cowan, and B. C. Fawcett, Mon. Not. R. Astron. Soc. **183**, 19 (1978).

[3]B. C. Fawcett, R. D. Cowan, E. Y. Kononov, and R. W. Hayes, J. Phys. B **5**, 1255 (1972).

[4]S. O. Kastner, M. Swartz, A. K. Bhatia, and J. Lapides, J. Opt. Soc. Am. **68**, 1558 (1978).

[5]E. Träbert, K. W. Jones, B. M. Johnson, D. C. Gregory, and T. H. Kruse, Phys. Lett. A **87**, 336 (1982).

Fe XIII:　Allowed transitions

No.	Transition Array	Multiplet	λ (Å)	E_i (cm^{-1})	E_k (cm^{-1})	g_i	g_k	A_{ki} (10^8 s^{-1})	f_{ik}	S (at. u.)	log gf	Accuracy	Source
1.	$3s^2 3p^2 - 3s 3p^3$	^{3}P – ^{3}D°	*363.31*	*13413*	*288660*	9	15	14	0.046	0.50	−0.38	D−	1
			368.12	18561	290210	5	7	13	0.036	0.22	−0.74	D	1
			359.63	9303	287360	3	5	15	0.048	0.17	−0.84	D	1
			348.184	0	287204	1	3	13	0.069	0.079	−1.16	D	1
			[372.02]	18561	287360	5	5	0.50	0.0010	0.0064	−2.28	D−	1
			359.837	9303	287204	3	3	3.3	0.0065	0.023	−1.71	D−	1
			[372.24]	18561	287204	5	3	0.12	1.5(−4)[a]	8.9(−4)	−3.14	E	1
2.		^{3}P – ^{3}P°	*316*			9	9	38	0.057	0.53	−0.29	D	1
			320.800	18561	330282	5	5	32	0.049	0.26	−0.61	D	1
			312.164	9303	329647	3	3	18	0.027	0.083	−1.09	D	1
			[321.45]	18561	329647	5	3	8.9	0.0083	0.044	−1.38	D−	1
			[311]			3	1	41	0.020	0.061	−1.22	C−	1
			311.552	9303	330282	3	5	4.2	0.010	0.031	−1.52	D	1
			[303.35]	0	329647	1	3	12	0.051	0.051	−1.29	D	1
3.		^{1}D – ^{3}D°											
			[412.98]	48069	290210	5	7	0.78	0.0028	0.019	−1.85	E	1
4.		^{1}S – ^{3}P°											
			[419.92]	91508	329647	1	3	0.21	0.0017	0.0023	−2.78	E	1
5.	$3p^2 - 3p 3d$	^{3}P – ^{3}F°											
			[236]			5	7	3.1	0.0036	0.014	−1.74	E	1
			[234]			3	5	1.0	0.0014	0.0032	−2.38	E	1
			[240]			5	5	1.2	0.0011	0.0042	−2.27	E	1
6.		^{3}P – ^{3}P°											
			202.424	9303	503316	3	1	460	0.095	0.19	−0.54	D	1
7.		^{3}P – ^{3}D°											
			203.826	18561	509176	5	7	650	0.57	1.9	0.45	D	1
8.		^{3}P – ^{1}F°											
			[185.75]	18561	556910	5	7	30	0.022	0.066	−0.97	E	1
9.		^{3}P – ^{1}P°											
			[175.22]	0	570713	1	3	4.4	0.0061	0.0035	−2.22	E	1
10.		^{1}D – ^{3}F°											
			[260]			5	5	3.2	0.0033	0.014	−1.79	E	1
11.		^{1}D – ^{3}D°											
			[216.87]	48069	509176	5	7	22	0.022	0.077	−0.97	E	1
12.		^{1}D – ^{1}F°	196.525	48069	556910	5	7	680	0.55	1.79	0.442	C	1
13.		^{1}S – ^{1}P°	208.679	91508	570713	1	3	560	1.1	0.75	0.04	D	1

Fe XIII: Allowed transitions — Continued

No.	Transition Array	Multiplet	λ (Å)	E_i (cm^{-1})	E_k (cm^{-1})	g_i	g_k	A_{ki} (10^8 s^{-1})	f_{ik}	S (at. u.)	log gf	Accuracy	Source
14.	$3p^2$–$3p\,4s$	^{3}P – ^{3}P°											
			74.845	18561	1354700	5	5	1000	0.088	0.11	−0.36	D	3
			75.892	18561	1336300	5	3	770	0.040	0.050	−0.70	D	3
			74.327	9303	1354700	3	5	410	0.057	0.042	−0.77	D	3
15.		^{1}D – ^{1}P°	76.117	48069	1361900	5	3	2100	0.11	0.14	−0.26	D	3
16.	$3p^2$–$3p\,4d$	^{3}P – ^{3}D°											
			62.699	9303	1604200	3	5	2300	0.23	0.14	−0.16	E	3,4
			62.353	0	1603800	1	3	2000	0.35	0.072	−0.46	D	4
17.		^{3}P – ^{3}F°											
			62.46	18561	1620000	5	7	1200	0.098	0.10	−0.31	D	4
18.		^{3}P – ^{3}P°											
						5	5	1400				D	4
						3	1	1600				D	4
19.		^{1}D – ^{1}F°	63.188	48069	1630700	5	7	3900	0.33	0.34	0.22	D	3,4
20.		^{1}S – ^{1}P°	64.139	91508	1650600	1	3	2100	0.39	0.082	−0.41	D	4
21.	$3p\,3d$–$3p\,4p$	^{3}F° – ^{3}D											
			98.128			9	7	410	0.046	0.13	−0.38	D	3
			98.523			7	5	380	0.040	0.091	−0.55	D	3
			98.826			5	3	390	0.034	0.055	−0.77	E	3
22.		^{1}F° – ^{1}D	107.384	556910	1488147	7	5	1800	0.22	0.54	0.19	D	3
23.	$3p\,3d$–$3p\,4f$	^{3}F° – ^{3}G											
			78.452			9	11	6300	0.71	1.7	0.81	D	3,4
24.		^{3}P° – ^{3}D											
						1	3	2300				D	4
25.		^{1}F° – ^{1}G	84.270	556910	1743600	7	9	5500	0.75	1.5	0.72	D	3,4
26.		^{1}P° – ^{1}D				3	5	3400				D	4

[a]The number in parentheses following the tabulated value indicates the power of ten by which this value has to be multiplied.

Fe XIII

Forbidden Transitions

List of tabulated lines

Wavelength (Å)	No.	Wavelength (Å)	No.	Wavelength (Å)	No.	Wavelength (Å)	No.
456.68	11	830	6	2320.7	8	3388.5	2
463.05	10	1200	5	2329	8	5386.1	1
610	9	1216.46	3	2355.4	8	10746.8	1
620	9	1300	5	2364	8	10797.9	1
650	9	1370.9	3	2495	8	33200	7
680	9	2300	8	2535	8	35100	7
820	6	2301.4	4	2578.77	2	62000	7

Line strengths for magnetic dipole and electric quadrupole transitions are the results of the multiconfiguration Dirac-Fock (MCDF) calculations of Huang.[1] These relativistic calculations included a perturbative treatment of the Breit interaction and the Lamb shift. Allowance for configuration interaction encompassed all configurations within the $n = 3$ complex. Huang calculated line strengths for transitions within the $3p^2$ configuration, as well as for transitions between pairs of odd-parity levels whose lower level is one of the four lowest-lying odd-parity levels in the $n = 3$ complex. Transitions involving odd-parity levels which are indicated by Bromage *et al.*[2] (for Fe XIII) or Bromage[3] (for V X and Ni XV) to be of low purity in *LS* coupling in

Fe-group species are omitted here, as are lines whose strengths are very small. Strengths of electric quadrupole transitions as reported in Ref. 1 were multiplied by the factor $2/3$ which is needed to bring these values into conformance with the definition of quadrupole strengths used in the NBS tables.

References

[1] K.-N. Huang, At. Data Nucl. Data Tables **32**, 503 (1985).
[2] G. E. Bromage, R. D. Cowan, and B. C. Fawcett, Mon. Not. R. Astron. Soc. **183**, 19 (1978).
[3] G. E. Bromage, Astron. Astrophys., Suppl. Ser. **41**, 79 (1980).

Fe XIII: Forbidden transitions

No.	Transition Array	Multiplet	λ (Å)	E_i (cm^{-1})	E_k (cm^{-1})	g_i	g_k	Type of transition	A_{ki} (s^{-1})	S (at. u.)	Accuracy	Source
1.	$3p^2$–$3p^2$	^3P – ^3P										
			10797.9	9303	18561	3	5	M1	9.86	2.30	C+	1
			"	"	"	3	5	E2	3.7(−4)^a	0.16	D−	1
			10746.8	0	9303	1	3	M1	14.0	1.93	C+	1
			[5386.1]	0	18561	1	5	E2	0.0063	0.085	E	1
2.		^3P – ^1D										
			3388.5	18561	48069	5	5	M1	75	0.54	E	1
			"	"	"	5	5	E2	0.067	0.089	E	1
			2578.77	9303	48069	3	5	M1	63	0.20	E	1
			"	"	"	3	5	E2	0.038	0.013	E	1
3.		^3P – ^1S										
			[1370.9]	18561	91508	5	1	E2	3.8	0.011	E	1
			1216.46	9303	91508	3	1	M1	1000	0.068	E	1
4.		^1D – ^1S	[2301.4]	48069	91508	5	1	E2	8.1	0.31	D−	1

Fe XIII: Forbidden transitions — Continued

No.	Transition Array	Multiplet	λ (Å)	E_i (cm^{-1})	E_k (cm^{-1})	g_i	g_k	Type of transition	A_{ki} (s^{-1})	S (at. u.)	Accuracy	Source
5.	$3s\,3p^3$–$3s\,3p^3$	$^5S^\circ$ – $^3D^\circ$										
			[1200]			5	7	E2	0.62	0.0064	E	1
			[1300]			5	5	M1	39	0.016	E	1
			"	"	"	5	5	E2	0.38	0.0042	E	1
			[1300]			5	3	M1	12	0.0029	E	1
			"	"	"	5	3	E2	0.17	0.0011	E	1
6.		$^5S^\circ$ – $^3P^\circ$										
			[820]			5	5	M1	620	0.063	E	1
			[830]			5	3	M1	350	0.022	E	1
7.		$^3D^\circ$ – $^3D^\circ$										
			[35100]	287360	290210	5	7	M1	0.39	4.4	D+	1
			"	"	"	5	7	E2	2.0(−7)	0.044	E	1
			[62000]	287204	287360	3	5	M1	1.0(−4)	4.4	E	1
			[33200]	287204	290210	3	7	E2	7.1(−8)	0.012	E	1
8.		$^3D^\circ$ – $^3P^\circ$										
			[2535]	290210	329647	7	3	E2	1.7	0.32	D−	1
			[2300]			5	1	E2	5.5	0.21	D−	1
			[2495]	290210	330282	7	5	M1	83	0.24	E	1
			"	"	"	7	5	E2	2.0	0.57	D−	1
			[2364]	287360	329647	5	3	E2	0.37	0.049	E	1
			[2300]			3	1	M1	140	0.061	E	1
			[2329]	287360	330282	5	5	M1	73	0.17	E	1
			"	"	"	5	5	E2	1.6	0.32	D−	1
			[2355.4]	287204	329647	3	3	M1	120	0.18	E	1
			"	"	"	3	3	E2	2.1	0.27	D−	1
			[2320.7]	287204	330282	3	5	M1	20	0.047	E	1
			"	"	"	3	5	E2	0.55	0.11	D−	1
9.	$3s\,3p^3$–$3s^2\,3p\,3d$	$^3D^\circ$ – $^3F^\circ$										
			[610]			5	9	E2	13	0.0057	E	1
			[650]			3	7	E2	5.4	0.0026	E	1
			[620]			7	9	M1	770	0.061	E	1
			[680]			5	5	M1	19	0.0011	E	1
10.		$^3D^\circ$ – $^3P^\circ$										
			[463.05]	287360	503316	5	1	E2	300	0.0038	E	1
11.		$^3D^\circ$ – $^3D^\circ$										
			[456.68]	290210	509176	7	7	M1	61	0.0015	E	1

aThe number in parentheses following the tabulated value indicates the power of ten by which this value has to be multiplied.

Fe XIV

Al Isoelectronic Sequence

Ground State: $1s^2 2s^2 2p^6 3s^2 3p\ ^2P^\circ_{1/2}$

Ionization Energy: $392.2\ \text{eV} = 3163000\ \text{cm}^{-1}$

Allowed Transitions

List of tabulated lines

Wavelength (Å)	No.	Wavelength (Å)	No.	Wavelength (Å)	No.	Wavelength (Å)	No.
58.963	42	206	36	240	7	288.45	6
59.579	42	207	36	243	20,31	289.160	3
69.176	41	210	21	245	7,20	290	5
69.386	41	211.316	19	248	27	292	25
69.66	40,41	213	21,22	252.197	4	294	25
70.251	41	214	22	257	26	299	15
70.613	40	216	28	257.392	4	301	15
72.80	45	217	22	264.787	4	334.171	2
76.022	44	218	22,39	265	18	342	9
76.152	44	219	21,38	266	18	344	8
91.009	43	219.123	19	267	17	345	14
91.273	43	220	39	268	17,18	347	8
171	32	220.082	19	269	17	348	11
172	24	221	30,38	270	17	352	11
181	35	222	28,39	270.524	4	353.833	2
182	35	223	28,38	274	6	356.60	2
183	34,35	224	21	274.203	3	361	14
184	34	226	30	280	16	364	14
188	33	230	37	280.69	6	384	13
189	33	234	31	281	16	391	13
190	33	235	7	283	10,16	396	13
192	23	238	31	285	10	443	1
193	29	239	31	286	10	513	12

Line strengths for transitions of the arrays $3s^2 3p$–$3s 3p^2$, $3s 3p^2$–$3p^3$, $3s^2 3d$–$3s 3p 3d$, $3s^2 3p$–$3s^2 3d$, and $3s 3p^2$–$3s 3p 3d$ are the results of the multiconfiguration Dirac-Fock (MCDF) calculations of Huang.[1] These relativistic calculations included a perturbative treatment of the Breit interaction. Allowance for configuration mixing included all configurations within the $n = 3$ complex.

Huang published a diagram of energy levels (designated in LS coupling) in the $3s^2 3p$, $3s 3p^2$, $3s^2 3d$, $3p^3$, and $3s 3p 3d$ configurations of Fe XIV, but he has not provided percentage compositions. We have used the percentages given by Fawcett[2] as a guide to naming the levels; the latter's values resulted from Hartree-Fock calculations with relativistic effects and statistical allowance for exchange (HXR), and incorporated correlation effects due to all configurations within the $n = 3$ complex.

Oscillator strengths resulting from earlier HX calculations of Fawcett et al.[3] are quoted for a few transitions to configurations in which one electron occupies the $n = 4$ shell.

Transitions involving levels which are indicated to be of low purity in LS coupling are omitted here. Lines which are characterized by very small f-values are assigned lower accuracy ratings; the weakest lines have been excluded. A few wavelengths computed by Huang for transitions differ significantly from those which resulted from the fitting and scaling procedure applied by Fawcett[2]; lines for which the wavelengths are in serious disagreement have been omitted.

Lifetimes of the $3s 3p^2\ ^2P_{1/2}$ and $3p^3\ ^4S^\circ_{3/2}$ levels measured by Träbert et al.[4] using the beam-foil method are greater, by factors of 2.5 and 3.0, respectively, than values derived from the theoretically determined transition probabilities calculated by Huang.

References

[1] K.-N. Huang, At. Data Nucl. Data Tables **34**, 1 (1986) and private communication.

[2] B. C. Fawcett, At. Data Nucl. Data Tables **28**, 557 (1983).

[3] B. C. Fawcett, R. D. Cowan, E. Y. Kononov, and R. W. Hayes, J. Phys. B **5**, 1255 (1972).

[4] E. Träbert, K. W. Jones, B. M. Johnson, D. C. Gregory, and T. H. Kruse, Phys. Lett. **87A**, 336 (1982).

Fe XIV: Allowed transitions

No.	Transition Array	Multiplet	λ (Å)	E_i (cm^{-1})	E_k (cm^{-1})	g_i	g_k	A_{ki} (10^8 s^{-1})	f_{ik}	S (at. u.)	log gf	Accuracy	Source
1.	$3s^23p$–$3s3p^2$	$^2P°$ – ^4P											
			[443]			4	6	0.26	0.0011	0.0067	−2.34	E	1
2.		$^2P°$ – ^2D	347.21	12567	300581	6	10	21	0.063	0.43	−0.42	E	1
			353.833	18851	301470	4	6	19	0.054	0.25	−0.67	D	1
			334.171	0	299248	2	4	23	0.077	0.17	−0.81	D	1
			356.60	18851	299248	4	4	0.75	0.0014	0.0067	−2.24	E	1
3.		$^2P°$ – ^2S	283.99	12567	364693	6	2	170	0.070	0.39	−0.38	E	1
			289.160	18851	364693	4	2	12	0.0074	0.028	−1.53	E	1
			274.203	0	364693	2	2	180	0.20	0.36	−0.40	D	1
4.		$^2P°$ – ^2P	262.27	12567	393847	6	6	390	0.41	2.1	0.39	D	1
			264.787	18851	396515	4	4	338	0.356	1.24	0.153	C−	1
			257.392	0	388512	2	2	140	0.14	0.23	−0.57	D	1
			270.524	18851	388512	4	2	210	0.12	0.42	−0.33	D	1
			252.197	0	396515	2	4	76	0.145	0.240	−0.54	C−	1
5.	$3s3p^2$–$3p^3$	^4P – $^2D°$											
			[290]			6	6	0.90	0.0011	0.0065	−2.17	E	1
6.		^4P – $^4S°$	283			12	4	360	0.14	1.6	0.23	D	1
			288.45			6	4	160	0.14	0.77	−0.09	D	1
			280.69			4	4	120	0.14	0.52	−0.25	D	1
			[274]			2	4	64	0.14	0.26	−0.54	D	1
7.		^4P – $^2P°$											
			[245]			6	4	2.1	0.0012	0.0060	−2.13	E	1
			[240]			4	4	4.8	0.0041	0.013	−1.78	E	1
			[235]			2	4	2.1	0.0034	0.0053	−2.16	E	1
8.		^2D – $^2D°$											
			[347]			6	6	33	0.060	0.41	−0.45	E	1
			[344]			4	6	3.4	0.0091	0.041	−1.44	E	1
9.		^2D – $^4S°$											
			[342]			4	4	1.2	0.0021	0.0093	−2.08	E	1
10.		^2D – $^2P°$	285			10	6	120	0.090	0.84	−0.05	D	1
			[285]			6	4	100	0.083	0.47	−0.30	D	1
			[286]			4	2	130	0.082	0.31	−0.48	D	1
			[283]			4	4	13	0.016	0.060	−1.19	D	1
11.		^2S – $^2P°$	349			2	6	14	0.078	0.18	−0.81	E	1
			[348]			2	4	20	0.074	0.17	−0.83	D	1
			[352]			2	2	1.4	0.0025	0.0059	−2.29	E	1
12.		^2P – $^4S°$											
			[513]			4	4	0.68	0.0027	0.018	−1.97	E	1

Fe xiv: Allowed transitions — Continued

No.	Transition Array	Multiplet	λ (Å)	E_i (cm^{-1})	E_k (cm^{-1})	g_i	g_k	A_{ki} (10^8 s^{-1})	f_{ik}	S (at. u.)	log gf	Accuracy	Source
13.		^{2}P – ^{2}P°											
			[391]			4	4	30	0.068	0.35	−0.57	D	1
			[384]			2	2	36	0.079	0.20	−0.81	D	1
			[396]			4	2	7.7	0.0090	0.047	−1.44	E	1
14.	$3s^23d$ – $3s3p(^3P°)3d$	^{2}D – ^{2}F°	352			10	14	21	0.054	0.63	−0.26	E	1
			[345]			6	8	23	0.054	0.37	−0.49	E	1
			[361]			4	6	15	0.044	0.21	−0.75	E	1
			[364]			6	6	3.6	0.0072	0.052	−1.36	E	1
15.		^{2}D – ^{2}P°											
			[301]			6	4	4.1	0.0037	0.022	−1.65	E	1
			[299]			4	4	5.7	0.0076	0.030	−1.52	E	1
16.	$3s^23d$ – $3s3p(^1P°)3d$	^{2}D – ^{2}F°	282			10	14	280	0.46	4.3	0.67	E	1
			[283]			6	8	270	0.43	2.4	0.41	E	1
			[280]			4	6	280	0.49	1.8	0.29	E	1
			[281]			6	6	10	0.012	0.067	−1.14	E	1
17.		^{2}D – ^{2}D°	268			10	10	220	0.24	2.1	0.38	E	1
			[268]			6	6	210	0.23	1.2	0.13	E	1
			[269]			4	4	83	0.090	0.32	−0.44	E	1
			[270]			6	4	140	0.10	0.55	−0.21	E	1
			[267]			4	6	3.5	0.0057	0.020	−1.64	E	1
18.		^{2}D – ^{2}P°	267			10	6	320	0.20	1.8	0.31	D	1
			[266]			6	4	170	0.12	0.65	−0.13	D	1
			[268]			4	2	330	0.18	0.62	−0.15	D	1
			[265]			4	4	150	0.15	0.54	−0.21	D	1
19.	$3p$–$3d$	^{2}P° – ^{2}D	216.52	12567	474420	6	10	400	0.47	2.0	0.45	D	1
			219.123	18851	475216	4	6	390	0.42	1.2	0.22	D	1
			211.316	0	473225	2	4	360	0.48	0.67	−0.02	D	1
			220.082	18851	473225	4	4	81	0.059	0.17	−0.63	D	1
20.	$3s3p^2$ – $3s3p(^3P°)3d$	^{4}P – ^{4}F°											
			[245]			6	8	2.8	0.0033	0.016	−1.70	E	1
			[243]			4	6	1.5	0.0020	0.0064	−2.10	E	1
21.		^{4}P – ^{4}P°											
			[224]			6	6	30	0.023	0.10	−0.87	E	1
			[210]			2	2	2.3	0.0015	0.0021	−2.52	E	1
			[213]			4	2	280	0.096	0.27	−0.41	D	1
			[219]			4	6	240	0.26	0.76	0.02	E	1

Fe XIV: Allowed transitions — Continued

No.	Transition Array	Multiplet	λ (Å)	E_i (cm^{-1})	E_k (cm^{-1})	g_i	g_k	A_{ki} (10^8 s^{-1})	f_{ik}	S (at. u.)	log gf	Accuracy	Source
22.		^{4}P – ^{4}D°											
			[217]			6	8	404	0.380	1.63	0.358	C–	1
			[213]			4	6	110	0.11	0.32	−0.34	D	1
			[217]			6	6	260	0.19	0.80	0.05	D	1
			[214]			2	2	400	0.28	0.39	−0.26	D	1
			[218]			4	2	4.4	0.0016	0.0045	−2.20	E	1
23.		^{4}P – ^{2}F°											
			[192]			6	8	4.3	0.0032	0.012	−1.72	E	1
24.		^{4}P – ^{2}P°											
			[172]			2	4	3.1	0.0027	0.0031	−2.26	E	1
25.		^{2}D – ^{4}F°											
			[292]			4	4	0.96	0.0012	0.0047	−2.31	E	1
			[294]			6	4	1.2	0.0010	0.0060	−2.21	E	1
26.		^{2}D – ^{4}P°											
			[257]			6	6	9.0	0.0089	0.045	−1.27	E	1
27.		^{2}D – ^{4}D°											
			[248]			6	8	3.3	0.0041	0.020	−1.61	E	1
28.		^{2}D – ^{2}F°	219			10	14	170	0.17	1.2	0.22	E	1
			[216]			6	8	170	0.15	0.66	−0.03	E	1
			[222]			4	6	130	0.15	0.43	−0.23	E	1
			[223]			6	6	27	0.020	0.088	−0.92	E	1
29.		^{2}D – ^{2}P°											
			[193]			4	2	1.1	3.0(−4)[a]	7.6(−4)	−2.92	E	1
30.		^{2}S – ^{2}P°	224			2	6	300	0.68	1.0	0.13	D	1
			[226]			2	4	390	0.60	0.90	0.08	D	1
			[221]			2	2	130	0.096	0.14	−0.72	D	1
31.		^{2}P – ^{2}P°	240			6	6	180	0.15	0.73	−0.03	E	1
			[243]			4	4	110	0.094	0.30	−0.43	D	1
			[234]			2	2	280	0.23	0.36	−0.33	D	1
			[238]			4	2	47	0.020	0.062	−1.10	D	1
			[239]			2	4	4.1	0.0070	0.011	−1.85	E	1
32.	$3s3p^2$– $3s3p(^1$P°$)3d$	^{4}P – ^{2}F°											
			[171]			6	8	3.4	0.0020	0.0068	−1.92	E	1
33.		^{2}D – ^{2}F°	189			10	14	280	0.21	1.3	0.32	E	1
			[190]			6	8	280	0.21	0.77	0.09	E	1
			[188]			4	6	270	0.22	0.54	−0.06	E	1
			[189]			6	6	15	0.0080	0.030	−1.32	E	1

J. Phys. Chem. Ref. Data, Vol. 17, Suppl. 4, 1988

Fe XIV: Allowed transitions — Continued

No.	Transition Array	Multiplet	λ (Å)	E_i (cm^{-1})	E_k (cm^{-1})	g_i	g_k	A_{ki} (10^8 s^{-1})	f_{ik}	S (at. u.)	log gf	Accuracy	Source
34.		^{2}D – ^{2}D°											
			[183]			4	4	0.91	4.6(−4)	0.0011	−2.74	E	1
			[184]			6	4	2.0	6.6(−4)	0.0024	−2.40	E	1
35.		^{2}D – ^{2}P°	182			10	6	4.5	0.0013	0.0080	−1.87	E	1
			[182]			6	4	2.9	9.5(−4)	0.0034	−2.25	E	1
			[183]			4	2	1.5	3.7(−4)	9.0(−4)	−2.83	E	1
			[181]			4	4	3.2	0.0016	0.0037	−2.21	E	1
36.		^{2}S – ^{2}P°	206			2	6	120	0.22	0.30	−0.35	D	1
			[206]			2	4	70	0.088	0.12	−0.75	D	1
			[207]			2	2	210	0.13	0.18	−0.58	D	1
37.		^{2}P – ^{2}F°											
			[230]			4	6	3.3	0.0040	0.012	−1.80	E	1
38.		^{2}P – ^{2}D°	220			6	10	570	0.69	3.0	0.62	E	1
			[221]			4	6	590	0.65	1.9	0.42	E	1
			[219]			2	4	480	0.69	1.0	0.14	E	1
			[223]			4	4	23	0.017	0.051	−1.16	E	1
39.		^{2}P – ^{2}P°											
			[220]			4	4	320	0.23	0.67	−0.03	D	1
			[218]			2	2	62	0.044	0.063	−1.06	D	1
			[222]			4	2	94	0.0345	0.101	−0.86	C−	1
40.	$3p$–$4s$	^{2}P° – ^{2}S	70.299	12567	1435100	6	2	2600	0.064	0.089	−0.42	D	3
			70.613	18851	1435100	4	2	1700	0.063	0.059	−0.60	D	3
			69.66	0	1435100	2	2	890	0.065	0.030	−0.89	D	3
41.	$3s3p^2$– $3s3p(^3$P°$)4s$	^{4}P – ^{4}P°											
			69.66			6	6	1300	0.092	0.13	−0.26	D	3
			70.251			6	4	810	0.040	0.056	−0.62	D	3
			69.176			4	6	560	0.060	0.055	−0.62	D	3
			69.386			2	4	760	0.11	0.050	−0.66	D	3
42.	$3p$–$4d$	^{2}P° – ^{2}D											
			59.579	18851	1697300	4	6	3100	0.25	0.20	0.00	C	3
			58.963	0	1696000	2	4	2700	0.28	0.11	−0.25	C	3
43.	$3d$–$4p$	^{2}D – ^{2}P°											
			91.009	475216	1574000	6	4	510	0.042	0.076	−0.60	D	3
			91.273	473225	1568800	4	2	560	0.035	0.042	−0.85	D	3
44.	$3d$–$4f$	^{2}D – ^{2}F°											
			76.152	475216	1788400	6	8	7000	0.81	1.2	0.69	C	3
			76.022	473225	1788600	4	6	6600	0.86	0.86	0.54	C	3

Fe XIV: Allowed transitions — Continued

No.	Transition Array	Multiplet	λ (Å)	E_i (cm^{-1})	E_k (cm^{-1})	g_i	g_k	A_{ki} (10^8 s^{-1})	f_{ik}	S (at. u.)	log gf	Accuracy	Source
45.	$3s3p(^3P°)3d-$ $3s3p(^3P°)4f$	$^4F° - {^4}G$											
			72.80			9	11	8600	0.84	1.8	0.88	D	3

[a]The number in parentheses following the tabulated value indicates the power of ten by which this value has to be multiplied.

Fe XIV

Forbidden Transitions

Line strengths for magnetic dipole and electric quadrupole transitions within the $3s^23p$ $^2P°$ and $3s3p^2$ ^4P terms are the results of the multiconfiguration Dirac-Fock (MCDF) calculations of Huang.[1] These relativistic calculations included a perturbative treatment of the Breit interaction and the Lamb shift. Allowance for configuration mixing included all configurations within the $n=3$ complex. Strengths of electric quadrupole transitions as reported in Ref. 1 were multiplied by the factor $^2/_3$ which is needed to bring these values into conformance with the definition of quadrupole strengths used in the NBS tables.

A-values for transitions between pairs of levels of the $3s3p^2$ and $3s^23d$ configurations were calculated by Garstang.[2] He incorporated a Hartree-Fock radial integral published by Froese[3] for electric quadrupole transitions within the $3s3p^2$ configuration, and he calculated a radial integral for the $3s^23d$ configuration by applying the Coulomb approximation. Garstang's theoretical model allowed for mixing between the $3s3p^2$ and $3s^23d$ configurations, as well as for spin-orbit interaction. We quote his results for prominent transitions within the $3s3p^2$ and $3s^23d$ configurations, but we have modified his A-values by incorporating wavelengths computed from differences of experimentally determined energies rather than from differences of his theoretically derived energies.

References

[1]K.-N. Huang, At. Data Nucl. Data Tables **34**, 1 (1986).
[2]R. H. Garstang, Ann. Astrophys. **25**, 109 (1962).
[3]C. Froese, Mon. Not. R. Astron. Soc. **117**, 615 (1957).

Fe XIV: Forbidden transitions

No.	Transition Array	Multiplet	λ (Å)	E_i (cm^{-1})	E_k (cm^{-1})	g_i	g_k	Type of transition	A_{ki} (s^{-1})	S (at. u.)	Accuracy	Source
1.	$3p-3p$	$^2P° - {^2}P°$										
			5303.4	0	18851	2	4	M1	60.1	1.33	C+	1
			"	"	"	2	4	E2	0.015	0.15	D−	1
2.	$3s3p^2-3s3p^2$	$^4P - {^4}P$										
			[10400]			4	6	M1	14.2	3.56	C	1
			"			4	6	E2	5.1(−4)[a]	0.22	D−	1
			[13100]			2	4	M1	9.9	3.29	C	1
			"			2	4	E2	2.0(−5)	0.018	E	1
			[5800]			2	6	E2	0.0068	0.16	D−	1

Fe XIV: Forbidden transitions — Continued

No.	Transition Array	Multiplet	λ (Å)	E_i (cm^{-1})	E_k (cm^{-1})	g_i	g_k	Type of transition	A_{ki} (s^{-1})	S (at. u.)	Accuracy	Source
3.		^{2}D – ^{2}D										
			[44990]	299248	301470	4	6	M1	0.11	2.3	C	2
			"	"	"	4	6	E2	1.8(−7)	0.12	E	2
4.		^{2}D – ^{2}S										
			[1581.7]	301470	364693	6	2	E2	22	0.26	E	2
			[1528.0]	299248	364693	4	2	E2	19	0.19	E	2
5.		^{2}P – ^{2}P										
			[12492]	388512	396515	2	4	M1	3.5	1.0	D	2
			"	"	"	2	4	E2	1.9(−4)	0.14	E	2
6.	3d–3d	^{2}D – ^{2}D										
			[50210]	473225	475216	4	6	M1	0.082	2.3	C−	2

[a]The number in parentheses following the tabulated value indicates the power of ten by which this value has to be multiplied.

Fe XV

Mg Isoelectronic Sequence

Ground State: $1s^2 2s^2 2p^6 3s^2$ ^{1}S$_0$

Ionization Energy: 457.0 eV = 3686000 cm^{-1}

Allowed Transitions

List of tabulated lines

Wavelength (Å)	No.	Wavelength (Å)	No.	Wavelength (Å)	No.	Wavelength (Å)	No.
38.95	16	191.41	18	258	23	333	10,13
52.911	15	196	24	284.160	2	334	10
59.404	30	196.74	18	292.36	3	372.78	9
63.959	33	224.76	17	299	26	387.00	9
65.370	28	227.208	17	302.45	3	389.48	9
65.612	28	227.70	17	305.00	3,19	400.65	9
66.238	28	233	22	305.86	19	402.16	9
68.860	34	233.857	17	307.78	3	404.84	9
69.7	29	234.76	17	312.55	4	417.258	1
69.942	31	235	21	317.62	3	435.18	6
69.989	31	235.27	17	318	11	470.23	6
70.052	31	237	5	319	11	481.52	7
70.224	36	241	21	320	10,11	493.61	6
70.53	35	243	21,22,25	321.76	3	540	12
70.59	35	243.790	20	322	10		
73.199	37	248	27	324	8,14		
73.473	32	251	21	327.03	4		

Oscillator strengths selected for the three transitions $3s^2\,{}^1S_0 - 3snp\,{}^1P_1^\circ$ ($n = 3$–5) are the results of the relativistic random phase approximation (RRPA) calculations of Shorer et al.,[1] who allowed for correlation within the context of a frozen core. Line strengths tabulated for the intercombination line $3s^2\,{}^1S_0 - 3s\,3p\,{}^3P_1^\circ$ and most lines of the $3s\,3p - 3p^2$ array were calculated by Cheng and Johnson[2] using the relativistic multiconfiguration Hartree-Fock (MCHF) method. The two approaches, RRPA and MCHF, provided nearly identical results for the $3s^2\,{}^1S - 3s\,3p\,{}^1P^\circ$ transition. Transition probabilities were computed by Anderson and Anderson[3] using a simplified relativistic self-consistent-field (SCF) approach in which a common set of radial orbitals was generated for all configurations of the $n = 3$ complex, as compared to the more sophisticated approach of Cheng and Johnson, which allowed for variation of a given radial orbital from state to state. Line strengths derived from the data of Ref. 3 are in excellent agreement with those tabulated in Ref. 2, and are quoted here for transitions not treated in either Ref. 1 or Ref. 2, namely, all lines of the $3s\,3p - 3s\,3d$ array and two lines of the $3s\,3p - 3p^2$ array. For nearly all remaining transitions within the complex, we have tabulated the oscillator strength data of Fawcett,[4] which were derived by means of Hartree-Fock calculations that included relativistic effects and statistical allowance for exchange (HXR). These calculations as well as those of Refs. 2 and 3 allowed for all electron correlations within the $n = 3$ complex. Froese Fischer and Godefroid[5] determined f-values for singlet-singlet transitions within this same complex by applying a nonrelativistic MCHF technique with large-scale allowance for configuration interaction; their results are quoted for two transitions of the $3p\,3d - 3d^2$ array for which we estimate the contribution of singlet-triplet mixing to the f-value to be insignificant.

Data are tabulated for a number of additional transitions in which a single electron jumps from the $n = 3$ shell to an $n = 4$ orbital. Oscillator strengths quoted for lines of arrays in which the upper-state configuration is of type $3s\,4l$ ($l = 0,2,3$) were reported by Cowan and Widing,[6] who applied an improved version of Cowan's earlier HX method which takes relativistic and correlation effects into account. A-values calculated by Kastner

et al.[7] using a multiconfiguration STF approach are tabulated for a few transitions of the $3p\,3d - 3p\,4f$ array. Oscillator strengths for two additional lines of this array are the results of earlier HX calculations reported by Fawcett et al.[8]

Transitions involving levels which are indicated in Ref. 4, 7, or 8 to be of low purity in LS coupling are omitted here. Lines which are characterized by very small f-values are assigned lower accuracy ratings.

Träbert et al.[9] measured the lifetimes of four levels by applying the beam-foil method. Their analysis of the experimental results for the $3p^2\,{}^3P_1$ and ${}^3P_2$ levels encompassed both multiexponential fitting techniques and consideration of cascading effects. Interpretation of data for the $3s\,3p\,{}^3P_1^\circ$ and $3p^2\,{}^1D_2$ levels was limited to cascade analysis only. In the case of the two $J = 1$ levels, the sum of our tabulated, theoretically determined A-values for all possible (downward) transitions from the given level, together with our estimate of their uncertainties, lies within the range of the measured lifetime of that level. For the two $J = 2$ levels, however, it is difficult to make a definitive comparison between theory and experiment, since the A-value for at least one transition out of each of these two levels is highly uncertain; moreover, in the case of the lifetime of the $3p^2\,{}^1D_2$ level the A-values of the intercombination lines are comparable in magnitude to that of the singlet–singlet transition from that level.

References

[1] P. Shorer, C. D. Lin, and W. R. Johnson, Phys. Rev. A **16**, 1109 (1977).
[2] K. T. Cheng and W. R. Johnson, Phys. Rev. A **16**, 263 (1977).
[3] E. K. Anderson and E. M. Anderson, Opt. Spectrosc. (USSR) **55**, 500 (1983).
[4] B. C. Fawcett, At. Data Nucl. Data Tables **28**, 579 (1983).
[5] C. Froese Fischer and M. Godefroid, Nucl. Instrum. Methods **202**, 307 (1982).
[6] R. D. Cowan and K. G. Widing, Astrophys. J. **180**, 285 (1973).
[7] S. O. Kastner, M. Swartz, A. K. Bhatia, and J. Lapides, J. Opt. Soc. Am. **68**, 1558 (1978).
[8] B. C. Fawcett, R. D. Cowan, E. Y. Kononov, and R. W. Hayes, J. Phys. B **5**, 1255 (1972).
[9] E. Träbert, K. W. Jones, B. M. Johnson, D. C. Gregory, and T. H. Kruse, Phys. Lett. A **87**, 336 (1982).

Fe xv: Allowed transitions

No.	Transition Array	Multiplet	λ (Å)	E_i (cm^{-1})	E_k (cm^{-1})	g_i	g_k	A_{ki} (10^8 s^{-1})	f_{ik}	S (at. u.)	log gf	Accuracy	Source
1.	$3s^2 - 3s\,3p$	${}^1S - {}^3P^\circ$											
			417.258	0	239660	1	3	0.41	0.0032	0.0044	−2.49	E	2
2.		${}^1S - {}^1P^\circ$	284.160	0	351914	1	3	228	0.827	0.774	−0.082	B	1

Fe xv: Allowed transitions — Continued

No.	Transition Array	Multiplet	λ (Å)	E_i (cm^{-1})	E_k (cm^{-1})	g_i	g_k	A_{ki} (10^8 s^{-1})	f_{ik}	S (at. u.)	log gf	Accuracy	Source
3.	$3s3p$–$3p^2$	$^3P^\circ$ – ^3P	306.67	246890	572970	9	9	180	0.25	2.3	0.36	D	2
			305.00	253823	581700	5	5	130	0.18	0.89	−0.05	D	2
			307.78	239660	564570	3	3	49.1	0.070	0.212	−0.68	C	2
			321.76	253823	564570	5	3	71	0.066	0.351	−0.480	C	2
			317.62	239660	554500	3	1	177	0.089	0.280	−0.57	C	2
			292.36	239660	581700	3	5	45	0.097	0.28	−0.54	D	2
			302.45	233940	564570	1	3	69	0.285	0.284	−0.54	C	2
4.		$^3P^\circ$ – ^1D											
			327.03	253823	559590	5	5	20	0.032	0.17	−0.80	E	2
			312.55	239660	559590	3	5	11	0.027	0.083	−1.09	E	2
5.		$^3P^\circ$ – ^1S											
			[237]			3	1	3.2	9.0(−4)[a]	0.0021	−2.57	E	3
6.		$^1P^\circ$ – ^3P											
			[435.18]	351914	581700	3	5	4.7	0.022	0.096	−1.17	E	2
			[470.23]	351914	564570	3	3	0.084	2.8(−4)	0.0013	−3.08	E	2
			[493.61]	351914	554500	3	1	0.64	7.8(−4)	0.0038	−2.63	E	2
7.		$^1P^\circ$ – ^1D	481.52	351914	559590	3	5	16	0.090	0.43	−0.57	E	2
8.		$^1P^\circ$ – ^1S	[324]			3	1	197	0.103	0.330	−0.51	C	3
9.	$3s3d$–$3p3d$	^3D – $^3F^\circ$	383.98	680370	940800	15	21	53	0.16	3.1	0.39	D	4
			372.78	681435	949690	7	9	60	0.160	1.37	0.049	C	4
			387.00	679785	938190	5	7	41	0.13	0.83	−0.19	C	4
			400.65	678860	928450	3	5	32	0.13	0.51	−0.41	D	4
			[389.48]	681435	938190	7	7	10	0.023	0.21	−0.79	C	4
			[402.16]	679785	928450	5	5	9.1	0.022	0.15	−0.96	D	4
			[404.84]	681435	928450	7	5	0.11	2.0(−4)	0.0019	−2.85	E	4
10.		^3D – $^3D^\circ$											
			[322]			7	7	77	0.12	0.89	−0.08	C	4
			[333]			3	3	20	0.033	0.11	−1.00	E	4
			[334]			5	3	66	0.066	0.36	−0.48	E	4
			[320]			5	7	19	0.040	0.21	−0.70	C	4
11.		^3D – $^3P^\circ$											
			[320]			5	3	28	0.026	0.14	−0.89	E	4
			[319]			3	1	100	0.053	0.17	−0.80	C	4
			[318]			3	3	79	0.12	0.38	−0.44	E	4
12.		^1D – $^1D^\circ$	[540]			5	5	7.3	0.032	0.28	−0.80	D	4
13.		^1D – $^1F^\circ$	[333]			5	7	180	0.42	2.3	0.32	D	4
14.		^1D – $^1P^\circ$	[324]			5	3	130	0.12	0.64	−0.22	D	4
15.	$3s^2$–$3s4p$	^1S – $^1P^\circ$	52.911	0	1890000	1	3	2940	0.370	0.064	−0.432	C	1
16.	$3s^2$–$3s5p$	^1S – $^1P^\circ$	38.95	0	2567000	1	3	1690	0.115	0.0147	−0.94	C	1

Fe xv: Allowed transitions — Continued

No.	Transition Array	Multiplet	λ (Å)	E_i (cm^{-1})	E_k (cm^{-1})	g_i	g_k	A_{ki} (10^8 s^{-1})	f_{ik}	S (at. u.)	log gf	Accuracy	Source
17.	$3s3p$–$3s3d$	^{3}P° – ^{3}D	230.69	246890	680370	9	15	230	0.306	2.09	0.440	C	3
			233.857	253823	681435	5	7	220	0.25	0.98	0.10	C	3
			227.208	239660	679785	3	5	180	0.23	0.52	−0.16	C	3
			224.76	233940	678860	1	3	138	0.314	0.232	−0.50	C	3
			234.76	253823	679785	5	5	55	0.0453	0.175	−0.65	C	3
			227.70	239660	678860	3	3	98	0.076	0.172	−0.64	C	3
			[235.27]	253823	678860	5	3	6.2	0.0031	0.012	−1.81	D	3
18.		^{3}P° – ^{1}D											
			[196.74]	253823	762103	5	5	0.16	9.3(−5)	3.0(−4)	−3.33	E	3
			[191.41]	239660	762103	3	5	3.5	0.0032	0.0061	−2.01	E	3
19.		^{1}P° – ^{3}D											
			[305.00]	351914	679785	3	5	0.30	7.0(−4)	0.0021	−2.68	E	3
			[305.86]	351914	678860	3	3	0.26	3.6(−4)	0.0011	−2.96	E	3
20.		^{1}P° – ^{1}D	243.790	351914	762103	3	5	420	0.62	1.5	0.27	D	3
21.	$3p^2$–$3p3d$	^{3}P – ^{3}D°											
			[243]			5	7	230	0.28	1.1	0.15	D	4
			[235]			1	3	250	0.63	0.49	−0.20	E	4
			[241]			3	3	42	0.037	0.088	−0.95	E	4
			[251]			5	3	2.8	0.0016	0.0066	−2.10	E	4
22.		^{3}P – ^{3}P°											
			[233]			3	3	150	0.12	0.28	−0.44	E	4
			[243]			5	3	64	0.034	0.14	−0.77	E	4
			[233]			3	1	210	0.057	0.13	−0.77	C	4
23.		^{1}D – ^{1}D°	[258]			5	5	140	0.14	0.59	−0.15	E	4
24.		^{1}D – ^{1}P°	[196]			5	3	3.8	0.0013	0.0042	−2.19	E	4
25.		^{1}S – ^{1}P°	[243]			1	3	240	0.64	0.51	−0.19	C	4
26.	$3p3d$–$3d^2$	^{1}F° – ^{1}G	[299]			7	9	209	0.361	2.49	0.403	C−	5
27.		^{1}P° – ^{1}S	[248]			3	1	540	0.166	0.407	−0.303	C−	5
28.	$3s3p$–$3s4s$	^{3}P° – ^{3}S	65.924	246890	1763800	9	3	2880	0.062	0.122	−0.250	C	6
			66.238	253823	1763800	5	3	1600	0.062	0.068	−0.51	C	6
			65.612	239660	1763800	3	3	980	0.063	0.041	−0.72	C	6
			65.370	233940	1763800	1	3	320	0.062	0.013	−1.21	C	6
29.		^{1}P° – ^{1}S	[69.7]			3	1	1900	0.047	0.032	−0.85	C	6
30.	$3s3p$–$3s4d$	^{1}P° – ^{1}D	59.404	351914	2035300	3	5	3400	0.30	0.18	−0.05	C−	6
31.	$3s3d$–$3s4f$	^{3}D – ^{3}F°											
			70.052	681435	2108900	7	9	8800	0.83	1.3	0.76	C	6
			69.989	679785	2108600	5	7	7900	0.81	0.93	0.61	C−	6
			69.942	678860	2108700	3	5	7400	0.91	0.63	0.44	C	6

Fe xv: Allowed transitions — Continued

No.	Transition Array	Multiplet	λ (Å)	E_i (cm^{-1})	E_k (cm^{-1})	g_i	g_k	A_{ki} (10^8 s^{-1})	f_{ik}	S (at. u.)	log gf	Accuracy	Source
32.		$^1D - {}^1F°$	73.473	762103	2123100	5	7	6200	0.70	0.85	0.54	D	6
33.	$3p^2-3s4f$	$^1D - {}^1F°$	[63.959]	559590	2123100	5	7	1600	0.14	0.15	−0.15	E	6
34.	$3p3d-3p4f$	$^3F° - {}^3G$											
			68.860	949690	2401900	9	11	9200	0.80	1.6	0.86	C	7
35.		$^3D° - {}^3D$											
			70.59			7	7	1700	0.13	0.21	−0.04	D	8
			70.53			7	5	260	0.014	0.023	−1.01	D	8
36.		$^3P° - {}^3D$											
			70.224			1	3	4130	0.92	0.212	−0.038	C	7
37.		$^1F° - {}^1G$	73.199			7	9	8800	0.91	1.5	0.80	C−	7
38.		$^1P° - {}^1D$				3	5	7000				C−	7

^{a}The number in parentheses following the tabulated value indicates the power of ten by which this value has to be multiplied.

Fe xv

Forbidden Transitions

List of tabulated lines

Wavelength (Å)	No.	Wavelength (Å)	No.	Wavelength (Å)	No.	Wavelength (Å)	No.
131.22	21	243.70	20	332.58	12	3675	3
171.91	22	246	14	393.98	11	4522	6
178.70	23	287.55	12	435.18	15	5029	1
189.34	18	292.36	12	470.23	15	5836	3
191.41	18	303.47	19	481.52	16	7058.6	1
196.74	18	305.00	12,19	493.80	8	9928	3
224.29	17	305.86	19	847.67	2	17500	1
226.36	17	307.08	13	978	9	19600	4
227.208	17	307.78	12	999	7	20100	6
227.70	17	312.55	13	1019.5	2	60590	10
233.857	17	321.76	12	1040	5	110000	10
234.76	17	327.03	13	1270	5		

Transition probabilities for forbidden lines involving pairs of levels belonging to the set of configurations $3s^2$, $3s3p$, $3p^2$, and $3s3d$ were computed by Anderson and Anderson[1] using a simplified relativistic self-consistent-field (SCF) approach in which a common set of radial orbitals was generated for all configurations of the $n=3$ complex. They allowed for all electron correlations within the complex. These data are quoted here, but we first converted the transition probabilities to line strengths, which we then reconverted to A-values in order to incorporate more accurate wavelength values. The weakest lines were excluded from this compilation.

Reference

[1]E. K. Anderson and E. M. Anderson, Opt. Spectrosc. (USSR) **55**, 500 (1983).

Fe xv: Forbidden transitions

No.	Transition Array	Multiplet	λ (Å)	E_i (cm^{-1})	E_k (cm^{-1})	g_i	g_k	Type of transition	A_{ki} (s^{-1})	S (at. u.)	Accuracy	Source
1.	$3s3p$–$3s3p$	$^3P° - ^3P°$										
			7058.6	239660	253823	3	5	M1	38.0	2.48	C+	1
			"	"	"	3	5	E2	0.0036	0.19	D−	1
			[17500]	233940	239660	1	3	M1	3.32	1.98	C−	1
			[5029]	233940	253823	1	5	E2	0.0090	0.086	E	1
2.		$^3P° - ^1P°$										
			[1019.5]	253823	351914	5	3	M1	140	0.017	E	1
			[847.67]	233940	351914	1	3	M1	190	0.013	E	1
3.	$3p^2$–$3p^2$	$^3P - ^3P$										
			[5836]	564570	581700	3	5	M1	57	2.1	D	1
			"	"	"	3	5	E2	0.0069	0.14	D−	1
			[9928]	554500	564570	1	3	M1	17.7	1.93	C	1
			[3675]	554500	581700	1	5	E2	0.024	0.048	E	1
4.		$^3P - ^1D$										
			[19600]	554500	559590	1	5	E2	3.7(−6)^a	0.032	E	1
5.		$^3P - ^1S$										
			[1270]			5	1	E2	24	0.048	E	1
			[1040]			3	1	M1	1400	0.060	E	1
6.		$^1D - ^3P$										
			[4522]	559590	581700	5	5	M1	58	0.99	E	1
			"	"	"	5	5	E2	0.025	0.14	E	1
			[20100]	559590	564570	5	3	M1	0.44	0.40	E	1
			"	"	"	5	3	E2	4.6(−6)	0.027	E	1
7.		$^1D - ^1S$	[999]			5	1	E2	270	0.16	E	1
8.	$3p^2$–$3s3d$	$^1D - ^1D$	[493.80]	559590	762103	5	5	E2	160	0.014	E	1
9.		$^1S - ^1D$	[978]			1	5	E2	30	0.081	E	1
10.	$3s3d$–$3s3d$	$^3D - ^3D$										
			[60590]	679785	681435	5	7	M1	0.081	4.7	D+	1
			"	"	"	5	7	E2	2.3(−8)	0.079	E	1
			[110000]	678860	679785	3	5	M1	0.018	4.5	D+	1
			"	"	"	3	5	E2	1.4(−9)	0.069	E	1
11.	$3s^2$–$3s3p$	$^1S - ^3P°$										
			[393.98]	0	253823	1	5	M2	3.39	24.3	C−	1

Fe XV: Forbidden transitions — Continued

No.	Transition Array	Multiplet	λ (Å)	E_i (cm^{-1})	E_k (cm^{-1})	g_i	g_k	Type of transition	A_{ki} (s^{-1})	S (at. u.)	Accu-racy	Source
12.	$3s3p$–$3p^2$	$^3P^\circ$ – ^3P										
			305.00	253823	581700	5	5	M2	18	36	D–	1
			307.78	239660	564570	3	3	M2	9.1	11.4	C–	1
			321.76	253823	564570	5	3	M2	0.012	0.019	E	1
			292.36	239660	581700	3	5	M2	1.8	2.9	D–	1
			[332.58]	253823	554500	5	1	M2	0.378	0.232	C–	1
			[287.55]	233940	581700	1	5	M2	0.12	0.18	D–	1
13.		$^3P^\circ$ – ^1D										
			327.03	253823	559590	5	5	M2	1.6	4.4	E	1
			312.55	239660	559590	3	5	M2	8.4	19	E	1
			[307.08]	233940	559590	1	5	M2	6.8	14	E	1
14.		$^3P^\circ$ – ^1S										
			[246]			5	1	M2	4.2	0.57	D	1
15.		$^1P^\circ$ – ^3P										
			[435.18]	351914	581700	3	5	M2	1.9	22	D–	1
			[470.23]	351914	564570	3	3	M2	0.48	5.0	C–	1
16.		$^1P^\circ$ – ^1D	481.52	351914	559590	3	5	M2	5.6(−4)	0.011	E	1
17.	$3s3p$–$3s3d$	$^3P^\circ$ – ^3D										
			[226.36]	239660	681435	3	7	M2	1.98	1.24	C–	1
			[224.29]	233940	679785	1	5	M2	1.2	0.51	D	1
			233.857	253823	681435	5	7	M2	52	38.2	C–	1
			227.208	239660	679785	3	5	M2	17	7.8	C–	1
			234.76	253823	679785	5	5	M2	7.7	4.12	C–	1
			227.70	239660	678860	3	3	M2	1.6	0.45	D	1
18.		$^3P^\circ$ – ^1D										
			[196.74]	253823	762103	5	5	M2	1.7	0.37	D–	1
			[191.41]	239660	762103	3	5	M2	2.1	0.41	D–	1
			[189.34]	233940	762103	1	5	M2	1.7	0.32	D–	1
19.		$^1P^\circ$ – ^3D										
			[303.47]	351914	681435	3	7	M2	6.4	17.4	C–	1
			[305.00]	351914	679785	3	5	M2	2.04	4.07	C–	1
			[305.86]	351914	678860	3	3	M2	0.13	0.16	D	1
20.		$^1P^\circ$ – ^1D	243.70	351914	762103	3	5	M2	7.1	4.6	D–	1
21.	$3s^2$–$3s3d$	^1S – ^1D	[131.22]	0	762103	1	5	E2	1.6(+6)	0.19	D–	1
22.	$3s^2$–$3p^2$	^1S – ^3P										
			[171.91]	0	581700	1	5	E2	4.3(+4)	0.019	E	1
23.		^1S – ^1D	[178.70]	0	559590	1	5	E2	4.1(+5)	0.22	E	1

^{a}The number in parentheses following the tabulated value indicates the power of ten by which this value has to be multiplied.

Fe XVI

Na Isoelectronic Sequence

Ground State: $1s^2 2s^2 2p^6 3s\ {}^2S_{1/2}$

Ionization Energy: $489.262\ \text{eV} = 3946280\ \text{cm}^{-1}$

Allowed Transitions

List of tabulated lines

Wavelength (Å)	No.	Wavelength (Å)	No.	Wavelength (Å)	No.	Wavelength (Å)	No.
31.041	17	46.725	21	85.027	53	166	62
31.242	17	48.883	20	85.041	53	167.5	43
31.244	17	48.97	20	85.070	53	167.8	43
32.166	4	48.979	20	85.587	37	168.6	43
32.192	4	50.350	2	86.133	37	171.5	70
32.433	15	50.555	2	86.170	37	171.6	70
32.652	15	54.142	9	91.16	36	171.7	70
32.654	15	54.728	9	91.83	36	184.5	74
32.84	14	54.769	9	96.256	46	184.6	74
33.04	14	62.879	8	96.348	46	224	56
34.857	13	63.719	8	96.358	46	225	56
35.106	13	65.746	41	101.6	45	233.2	61
35.112	13	66.081	41	101.7	45	235.1	61
35.333	27	66.089	41	101.8	45	235.3	61
35.368	27	66.263	19	104.0	52	251.058	7
35.369	27	66.368	19	117.2	29	262.967	7
35.71	12	66.392	19	117.7	29	265.007	7
36.01	12	72.317	39	123.4	35	266.7	68
36.749	3	72.385	50	124.5	35	267.0	68
36.803	3	72.438	50	124.6	35	279	60
37.096	25	72.443	50	128.0	65	282	60
37.136	25	72.722	39	128.6	65	304.9	73
37.138	25	72.733	39	139.4	72	305.2	73
39.827	11	74.24	38	139.5	72	305.3	73
40.153	11	74.68	38	144.06	44	311.7	67
40.161	11	76.086	54	144.2	44	312.2	67
40.199	23	76.098	54	144.25	44	314.2	67
40.245	23	76.121	54	146	34	335.407	1
40.246	23	76.330	18	147.0	75	360.798	1
41.095	22	76.502	18	147.1	75	682.6	33
41.137	22	76.796	18	148	34	715.8	33
41.17	22	80.192	48	155.4	63	721.5	33
41.91	10	80.263	48	156.4	63	847.5	28
42.30	10	80.270	48	164	62	908.3	28
46.661	21	80.561	30	165.6	51		
46.718	21	80.723	30	165.7	51		

Line strengths quoted for transitions nl–$n'l'$ $(n,n' = 3,4)$ are the results of the relativistic single-configuration Hartree-Fock calculations of Kim and Cheng.[1] (In the case of the $4d$–$4f$ transition, it is the f-values that are quoted here, since wavelengths were not available with which to make the conversion from line strengths to f- and A-values.)

The lifetimes of the $3p_{1/2,3/2}$ and $3d_{3/2,5/2}$ levels were measured by Buchet $et\ al.$[2] using the beam-foil technique. They included a simulation of cascade effects in their analysis by incorporating theoretical transition-probability data into the fitting procedure. It was found

that the lifetimes of the $3d$ levels which resulted from this approach differed somewhat from those which were determined by a standard multiexponential fit to experimental data. (The effect on the $3p$ levels was well within the uncertainty of their measurements.) Träbert $et\ al.$[3] also measured the lifetimes of the $3d$ levels by the beam-foil method. Although they too included a simulation of repopulation from higher levels, they concluded that for their experimental conditions the incorporation of theoretical data into the cascade analysis resulted in lifetimes for the $3d$ levels which were very similar to those derived by a multiexponential fit to the experimental data alone. The results obtained by the two groups for the 3d levels agree to well within the mutual error estimates, and the reciprocals of the measured lifetimes of the four levels agree very well with the sums of the probabilities of transitions originating from those levels determined theoretically by Kim and Cheng.

Multiplet f-values calculated by Biemont[4] using a fully variational Hartree-Fock approach are quoted for numerous transitions $nl-n'l'$ ($3 \leqslant n \leqslant 5$; $4 \leqslant n' \leqslant 8$; l, $l' = s,p,d,f$). Data for additional transitions (namely, those for which $n > 5$, where n is the principal quantum number of the lower state) can be found in Ref. 4, and transitions involving excited states of higher principal quantum number ($n,n' \leqslant 10$) have been treated by Tull et al.[5] in the frozen-core Hartree-Fock approximation. Whenever wavelengths of individual lines within a multiplet either were available directly or could be deter-

mined from the energy levels, the multiplet strength was distributed among the lines according to LS-coupling rules, except in the case of the $4f-6d$ transition, where the wavelengths of all the lines in the multiplet are identical.

Transitions with small f-values were generally assigned lower accuracy ratings.

References

[1] Y.-K. Kim and K.-T. Cheng, J. Opt. Soc. Am. **68**, 836 (1978).
[2] J. P. Buchet, M. C. Buchet-Poulizac, A. Denis, J. Desesquelles, and M. Druetta, Phys. Rev. A **22**, 2061 (1980).
[3] E. Träbert, K. W. Jones, B. M. Johnson, D. C. Gregory, and T. H. Kruse, Phys. Lett. A **87**, 336 (1982).
[4] E. Biemont, Astron. Astrophys., Suppl. Ser. **31**, 285 (1978).
[5] C. E. Tull, R. P. McEachran, and M. Cohen, At. Data **3**, 169 (1971).

Fe XVI: Allowed transitions

No.	Transition Array	Multiplet	λ (Å)	E_i (cm^{-1})	E_k (cm^{-1})	g_i	g_k	A_{ki} (10^8 s^{-1})	f_{ik}	S (at. u.)	log gf	Accuracy	Source
1.	3s–3p	^{2}S – ^{2}P°	343.46	0	291151	2	6	74.4	0.395	0.893	−0.102	B	1
			335.407	0	298145	2	4	80.1	0.270	0.597	−0.267	B	1
			360.798	0	277163	2	2	63.8	0.125	0.296	−0.603	B	1
2.	3s–4p	^{2}S – ^{2}P°	50.429	0	1983000	2	6	1900	0.217	0.0721	−0.362	B	1
			50.350	0	1985600	2	4	1860	0.141	0.0469	−0.548	B	1
			50.555	0	1977700	2	2	1980	0.0757	0.0252	−0.820	B	1
3.	3s–5p	^{2}S – ^{2}P°	36.766	0	2719900	2	6	1100	0.069	0.017	−0.86	C	4
			36.749	0	2721200	2	4	1100	0.045	0.011	−1.04	C	ls
			36.803	0	2717200	2	2	1200	0.024	0.0057	−1.33	C	ls
4.	3s–6p	^{2}S – ^{2}P°	32.174	0	3108100	2	6	670	0.0314	0.0067	−1.202	C	4
			32.166	0	3108900	2	4	680	0.021	0.0045	−1.37	C	ls
			32.192	0	3106400	2	2	670	0.010	0.0022	−1.68	C	ls
5.	3s–7p	^{2}S – ^{2}P°				2	6		0.0173		−1.461	C	4
6.	3s–8p	^{2}S – ^{2}P°				2	6		0.0106		−1.67	C	4
7.	3p–3d	^{2}P° – ^{2}D	259.01	291151	677243	6	10	170	0.285	1.46	0.234	B	1
			262.967	298145	678421	4	6	163	0.253	0.877	0.006	B	1
			251.058	277163	675477	2	4	156	0.294	0.486	−0.231	B	1
			265.007	298145	675477	4	4	26.5	0.0279	0.0972	−0.953	B	1
8.	3p–4s	^{2}P° – ^{2}S	63.436	291151	1867600	6	2	3230	0.0650	0.0814	−0.409	B	1
			63.719	298145	1867600	4	2	2180	0.0663	0.0556	−0.577	B	1
			62.879	277163	1867600	2	2	1050	0.0623	0.0258	−0.904	B	1
9.	3p–4d	^{2}P° – ^{2}D	54.535	291151	2124900	6	10	4150	0.308	0.332	0.267	B	1
			54.728	298145	2125300	4	6	4160	0.280	0.202	0.050	B	1
			54.142	277163	2124200	2	4	3410	0.300	0.107	−0.222	B	1
			54.769	298145	2124200	4	4	697	0.0313	0.0226	−0.902	B	1

Fe XVI: Allowed transitions — Continued

No.	Transition Array	Multiplet	λ (Å)	E_i (cm^{-1})	E_k (cm^{-1})	g_i	g_k	A_{ki} (10^8 s^{-1})	f_{ik}	S (at. u.)	log gf	Accuracy	Source
10.	3p–5s	^{2}P° – ^{2}S	42.18	291151	2662000	6	2	1390	0.0124	0.0103	−1.128	C	4
			42.30	298145	2662000	4	2	920	0.012	0.0069	−1.30	C	ls
			41.91	277163	2662000	2	2	472	0.0124	0.00343	−1.60	C	ls
11.	3p–5d	^{2}P° – ^{2}D	40.045	291151	2788400	6	10	2400	0.098	0.078	−0.23	C	4
			40.153	298145	2788600	4	6	2500	0.089	0.047	−0.45	C	ls
			39.827	277163	2788100	2	4	2100	0.099	0.026	−0.70	C	ls
			[40.161]	298145	2788100	4	4	410	0.0098	0.0052	−1.41	D	ls
12.	3p–6s	^{2}P° – ^{2}S	35.92	291151	3075000	6	2	740	0.0048	0.0034	−1.54	D	4
			36.01	298145	3075000	4	2	500	0.0049	0.0023	−1.71	D	ls
			35.71	277163	3075000	2	2	240	0.0047	0.0011	−2.03	D	ls
13.	3p–6d	^{2}P° – ^{2}D	35.024	291151	3146400	6	10	1450	0.0445	0.0308	−0.57	C	4
			35.106	298145	3146600	4	6	1440	0.0400	0.0185	−0.80	C	ls
			34.857	277163	3146100	2	4	1230	0.0449	0.0103	−1.047	C	ls
			[35.112]	298145	3146100	4	4	250	0.0045	0.0021	−1.74	D	ls
14.	3p–7s	^{2}P° – ^{2}S	32.96	291151	3325000	6	2	460	0.0025	0.0016	−1.82	D	4
			33.04	298145	3325000	4	2	310	0.0025	0.0011	−2.00	D	ls
			32.84	277163	3325000	2	2	150	0.0025	5.3(−4)^a	−2.31	D	ls
15.	3p–7d	^{2}P° – ^{2}D	32.580	291151	3360600	6	10	920	0.0244	0.0157	−0.83	C	4
			32.652	298145	3360700	4	6	910	0.022	0.0094	−1.06	C	ls
			32.433	277163	3360500	2	4	770	0.024	0.0052	−1.31	C	ls
			[32.654]	298145	3360500	4	4	150	0.0023	0.0010	−2.03	D	ls
16.	3p–8s	^{2}P° – ^{2}S				6	2		0.0015		−2.05	D	4
17.	3p–8d	^{2}P° – ^{2}D	31.176	291151	3498800	6	10	620	0.0150	0.0092	−1.046	C	4
			31.242	298145	3498900	4	6	610	0.013	0.0055	−1.27	C	ls
			31.041	277163	3498700	2	4	520	0.015	0.0031	−1.52	C	ls
			[31.244]	298145	3498700	4	4	100	0.0015	6.1(−4)	−2.23	D	ls
18.	3d–4p	^{2}D – ^{2}P°	76.581	677243	1983000	10	6	752	0.0397	0.100	−0.402	C+	1
			76.502	678421	1985600	6	4	670	0.0392	0.0592	−0.629	B	1
			76.796	675477	1977700	4	2	772	0.0341	0.0345	−0.865	B	1
			[76.330]	675477	1985600	4	4	74	0.0065	0.0065	−1.59	D	1
19.	3d–4f	^{2}D – ^{2}F°	66.326	677243	2184900	10	14	1.01(+4)	0.930	2.03	0.968	B	1
			66.368	678421	2185200	6	8	1.00(+4)	0.885	1.16	0.725	B	1
			66.263	675477	2184600	4	6	9390	0.927	0.809	0.569	B	1
			[66.392]	678421	2184600	6	6	669	0.0442	0.0580	−0.576	B	1
20.	3d–5p	^{2}D – ^{2}P°	48.955	677243	2719900	10	6	300	0.0064	0.010	−1.19	D	4
			48.97	678421	2721200	6	4	260	0.0062	0.0060	−1.43	D	ls
			[48.979]	675477	2717200	4	2	280	0.0051	0.0033	−1.69	D	ls
			[48.883]	675477	2721200	4	4	29	0.0010	6.7(−4)	−2.38	E	ls

J. Phys. Chem. Ref. Data, Vol. 17, Suppl. 4, 1988

Fe XVI: Allowed transitions — Continued

No.	Transition Array	Multiplet	λ (Å)	E_i (cm^{-1})	E_k (cm^{-1})	g_i	g_k	A_{ki} (10^8 s^{-1})	f_{ik}	S (at. u.)	log gf	Accu- racy	Source
21.	3d–5f	^{2}D – ^{2}F°	46.694	677243	2818800	10	14	3710	0.170	0.261	0.230	C	4
			46.718	678421	2818900	6	8	3700	0.161	0.149	−0.014	C	ls
			46.661	675477	2818600	4	6	3460	0.169	0.104	−0.169	C	ls
			[46.725]	678421	2818600	6	6	250	0.0081	0.0075	−1.31	D	ls
22.	3d–6p	^{2}D – ^{2}P°	41.137	677243	3108100	10	6	150	0.0023	0.0031	−1.64	D	4
			41.17	678421	3108900	6	4	140	0.0023	0.0019	−1.85	D	ls
			[41.137]	675477	3106400	4	2	150	0.0018	0.0010	−2.13	D	ls
			[41.095]	675477	3108900	4	4	15	3.9(−4)	2.1(−4)	−2.81	E	ls
23.	3d–6f	^{2}D – ^{2}F°	40.225	677243	3163200	10	14	1900	0.063	0.083	−0.20	C	4
			40.245	678421	3163200	6	8	1800	0.059	0.047	−0.45	C	ls
			40.199	675477	3163100	4	6	1700	0.062	0.033	−0.60	C	ls
			[40.246]	678421	3163100	6	6	120	0.0030	0.0024	−1.74	D	ls
24.	3d–7p	^{2}D – ^{2}P°				10	6		0.0011		−1.96	D	4
25.	3d–7f	^{2}D – ^{2}F°	37.121	677243	3371100	10	14	1070	0.0309	0.0378	−0.51	C	4
			37.138	678421	3371100	6	8	1070	0.0294	0.0216	−0.75	C	ls
			37.096	675477	3371200	4	6	1000	0.0309	0.0151	−0.91	C	ls
			[37.136]	678421	3371200	6	6	73	0.0015	0.0011	−2.05	D	ls
26.	3d–8p	^{2}D – ^{2}P°				10	6		6.4(−4)		−2.19	E	4
27.	3d–8f	^{2}D – ^{2}F°	35.353	677243	3505800	10	14	680	0.0178	0.0207	−0.75	C	4
			35.368	678421	3505800	6	8	680	0.0169	0.0118	−0.99	C	ls
			35.333	675477	3505700	4	6	640	0.018	0.0083	−1.15	C	ls
			[35.369]	678421	3505700	6	6	45	8.4(−4)	5.9(−4)	−2.30	E	ls
28.	4s–4p	^{2}S – ^{2}P°	866.6	1867600	1983000	2	6	16.9	0.571	3.26	0.058	B	1
			[847.5]	1867600	1985600	2	4	18.1	0.391	2.18	−0.107	B	1
			[908.3]	1867600	1977700	2	2	14.6	0.181	1.08	−0.442	B	1
29.	4s–5p	^{2}S – ^{2}P°	117.3	1867600	2719900	2	6	391	0.242	0.187	−0.315	C	4
			[117.2]	1867600	2721200	2	4	393	0.162	0.125	−0.489	C	ls
			[117.7]	1867600	2717200	2	2	390	0.080	0.062	−0.80	C	ls
30.	4s–6p	^{2}S – ^{2}P°	80.613	1867600	3108100	2	6	260	0.076	0.040	−0.82	C	4
			[80.561]	1867600	3108900	2	4	260	0.051	0.027	−0.99	C	ls
			[80.723]	1867600	3106400	2	2	250	0.024	0.013	−1.31	C	ls
31.	4s–7p	^{2}S – ^{2}P°				2	6		0.0353		−1.151	C	4
32.	4s–8p	^{2}S – ^{2}P°				2	6		0.0198		−1.402	C	4
33.	4p–4d	^{2}P° – ^{2}D	704.7	1983000	2124900	6	10	36.4	0.452	6.29	0.433	B	1
			[715.8]	1985600	2125300	4	6	34.8	0.401	3.78	0.205	B	1
			[682.6]	1977700	2124200	2	4	33.3	0.465	2.09	−0.031	B	1
			[721.5]	1985600	2124200	4	4	5.65	0.0441	0.419	−0.753	B	1

Fe XVI: Allowed transitions — Continued

No.	Transition Array	Multiplet	λ (Å)	E_i (cm^{-1})	E_k (cm^{-1})	g_i	g_k	A_{ki} (10^8 s^{-1})	f_{ik}	S (at. u.)	log gf	Accuracy	Source
34.	4p–5s	^{2}P° – ^{2}S	147	1983000	2662000	6	2	990	0.107	0.311	−0.192	C	4
			[148]	1985600	2662000	4	2	650	0.106	0.207	−0.372	C	ls
			[146]	1977700	2662000	2	2	339	0.108	0.104	−0.66	C	ls
35.	4p–5d	^{2}P° – ^{2}D	124.2	1983000	2788400	6	10	700	0.270	0.66	0.210	C	4
			[124.5]	1985600	2788600	4	6	700	0.24	0.40	−0.01	C	ls
			[123.4]	1977700	2788100	2	4	590	0.27	0.22	−0.27	C	ls
			[124.6]	1985600	2788100	4	4	120	0.027	0.044	−0.97	D	ls
36.	4p–6s	^{2}P° – ^{2}S	91.58	1983000	3075000	6	2	500	0.0211	0.0382	−0.90	C	4
			[91.83]	1985600	3075000	4	2	334	0.0211	0.0255	−1.074	C	ls
			[91.16]	1977700	3075000	2	2	170	0.0212	0.0127	−1.373	C	ls
37.	4p–6d	^{2}P° – ^{2}D	85.955	1983000	3146400	6	10	490	0.091	0.15	−0.26	C	4
			[86.133]	1985600	3146600	4	6	480	0.079	0.090	−0.50	C	ls
			[85.587]	1977700	3146100	2	4	400	0.089	0.050	−0.75	C	ls
			[86.170]	1985600	3146100	4	4	79	0.0088	0.010	−1.45	D	ls
38.	4p–7s	^{2}P° – ^{2}S	74.52	1983000	3325000	6	2	300	0.0083	0.012	−1.30	D	4
			[74.68]	1985600	3325000	4	2	190	0.0081	0.0080	−1.49	D	ls
			[74.24]	1977700	3325000	2	2	99	0.0082	0.0040	−1.79	D	ls
39.	4p–7d	^{2}P° – ^{2}D	72.590	1983000	3360600	6	10	328	0.0432	0.062	−0.59	C	4
			[72.722]	1985600	3360700	4	6	320	0.039	0.037	−0.81	C	ls
			[72.317]	1977700	3360500	2	4	280	0.044	0.021	−1.05	C	ls
			[72.733]	1985600	3360500	4	4	54	0.0043	0.0041	−1.77	D	ls
40.	4p–8s	^{2}P° – ^{2}S				6	2		0.0043		−1.59	D	4
41.	4p–8d	^{2}P° – ^{2}D	65.972	1983000	3498800	6	10	225	0.0245	0.0319	−0.83	C	4
			[66.081]	1985600	3498900	4	6	224	0.0219	0.0191	−1.057	C	ls
			[65.746]	1977700	3498700	2	4	189	0.0245	0.0106	−1.310	C	ls
			[66.089]	1985600	3498700	4	4	37	0.0024	0.0021	−2.02	D	ls
42.	4d–4f	^{2}D – ^{2}F°				6	8		0.104		−0.205	C+	1
						4	6		0.111		−0.353	C+	1
						6	6		0.0052		−1.51	D	1
43.	4d–5p	^{2}D – ^{2}P°	168.1	2124900	2719900	10	6	350	0.088	0.49	−0.06	C	4
			[167.8]	2125300	2721200	6	4	310	0.087	0.29	−0.28	C	ls
			[168.6]	2124200	2717200	4	2	340	0.072	0.16	−0.54	C	ls
			[167.5]	2124200	2721200	4	4	36	0.015	0.033	−1.22	D	ls
44.	4d–5f	^{2}D – ^{2}F°	144.1	2124900	2818800	10	14	1700	0.72	3.4	0.86	C	4
			144.25	2125300	2818900	6	8	1600	0.67	1.9	0.60	C	ls
			144.06	2124200	2818600	4	6	1600	0.74	1.4	0.47	C	ls
			[144.2]	2125300	2818600	6	6	110	0.034	0.097	−0.69	D	ls

Fe XVI: Allowed transitions — Continued

No.	Transition Array	Multiplet	λ (Å)	E_i (cm^{-1})	E_k (cm^{-1})	g_i	g_k	A_{ki} (10^8 s^{-1})	f_{ik}	S (at. u.)	log gf	Accuracy	Source
45.	4d–6p	^{2}D – ^{2}P°	101.7	2124900	3108100	10	6	163	0.0152	0.051	−0.82	C	4
			[101.7]	2125300	3108900	6	4	150	0.015	0.031	−1.03	C	ls
			[101.8]	2124200	3106400	4	2	160	0.013	0.017	−1.29	C	ls
			[101.6]	2124200	3108900	4	4	16	0.0025	0.0034	−1.99	D	ls
46.	4d–6f	^{2}D – ^{2}F°	96.311	2124900	3163200	10	14	920	0.179	0.57	0.253	C	4
			[96.348]	2125300	3163200	6	8	930	0.17	0.33	0.02	C	ls
			[96.256]	2124200	3163100	4	6	870	0.18	0.23	−0.14	C	ls
			[96.358]	2125300	3163100	6	6	60	0.0084	0.016	−1.30	D	ls
47.	4d–7p	^{2}D – ^{2}P°				10	6		0.0056		−1.25	D	4
48.	4d–7f	^{2}D – ^{2}F°	80.244	2124900	3371100	10	14	550	0.074	0.20	−0.13	C	4
			[80.270]	2125300	3371100	6	8	540	0.069	0.11	−0.38	C	ls
			[80.192]	2124200	3371200	4	6	520	0.076	0.080	−0.52	C	ls
			[80.263]	2125300	3371200	6	6	37	0.0036	0.0057	−1.67	D	ls
49.	4d–8p	^{2}D – ^{2}P°				10	6		0.0028		−1.55	D	4
50.	4d–8f	^{2}D – ^{2}F°	72.417	2124900	3505800	10	14	353	0.0388	0.093	−0.411	C	4
			[72.438]	2125300	3505800	6	8	350	0.037	0.053	−0.65	C	ls
			[72.385]	2124200	3505700	4	6	330	0.039	0.037	−0.81	C	ls
			[72.443]	2125300	3505700	6	6	24	0.0019	0.0027	−1.95	D	ls
51.	4f–5d	^{2}F° – ^{2}D	165.7	2184900	2788400	14	10	65	0.0192	0.147	−0.57	C	4
			[165.7]	2185200	2788600	8	6	62	0.019	0.084	−0.81	C	ls
			[165.7]	2184600	2788100	6	4	66	0.018	0.059	−0.97	C	ls
			[165.6]	2184600	2788600	6	6	3.1	0.0013	0.0042	−2.11	D	ls
52.	4f–6d	^{2}F° – ^{2}D	[104.0]	2184900	3146400	14	10	28	0.0032	0.015	−1.35	D	4
53.	4f–7d	^{2}F° – ^{2}D	85.056	2184900	3360600	14	10	14	0.0011	0.0043	−1.81	D	4
			[85.070]	2185200	3360700	8	6	14	0.0011	0.0025	−2.05	D	ls
			[85.041]	2184600	3360500	6	4	14	0.0010	0.0017	−2.22	D	ls
			[85.027]	2184600	3360700	6	6	0.66	7.1(−5)	1.2(−4)	−3.37	E	ls
54.	4f–8d	^{2}F° – ^{2}D	76.109	2184900	3498800	14	10	8.9	5.5(−4)	0.0019	−2.11	E	4
			[76.121]	2185200	3498900	8	6	8.4	5.5(−4)	0.0011	−2.36	E	ls
			[76.098]	2184600	3498700	6	4	8.7	5.1(−4)	7.6(−4)	−2.52	E	ls
			[76.086]	2184600	3498900	6	6	0.41	3.6(−5)	5.4(−5)	−3.67	E	ls
55.	5s–5p	^{2}S – ^{2}P°				2	6		0.71		0.15	C	4
56.	5s–6p	^{2}S – ^{2}P°	224	2662000	3108100	2	6	116	0.261	0.385	−0.282	C	4
			[224]	2662000	3108900	2	4	116	0.174	0.257	−0.458	C	ls
			[225]	2662000	3106400	2	2	114	0.086	0.128	−0.76	C	ls
57.	5s–7p	^{2}S – ^{2}P°				2	6		0.083		−0.78	C	4
58.	5s–8p	^{2}S – ^{2}P°				2	6		0.0389		−1.109	C	4
59.	5p–5d	^{2}P° – ^{2}D				6	10		0.59		0.55	C	4

Fe XVI: Allowed transitions — Continued

No.	Transition Array	Multiplet	λ (Å)	E_i (cm^{-1})	E_k (cm^{-1})	g_i	g_k	A_{ki} (10^8 s^{-1})	f_{ik}	S (at. u.)	log gf	Accuracy	Source
60.	5p–6s	^{2}P° – ^{2}S	282	2719900	3075000	6	2	380	0.151	0.84	−0.043	C	4
			[282]	2721200	3075000	4	2	250	0.15	0.56	−0.22	C	ls
			[279]	2717200	3075000	2	2	130	0.15	0.28	−0.52	C	ls
61.	5p–6d	^{2}P° – ^{2}D	234.5	2719900	3146400	6	10	186	0.255	1.18	0.185	C	4
			[235.1]	2721200	3146600	4	6	180	0.23	0.71	−0.04	C	ls
			[233.2]	2717200	3146100	2	4	157	0.256	0.393	−0.291	C	ls
			[235.3]	2721200	3146100	4	4	31	0.025	0.079	−0.99	D	ls
62.	5p–7s	^{2}P° – ^{2}S	165	2719900	3325000	6	2	221	0.0300	0.098	−0.74	C	4
			[166]	2721200	3325000	4	2	140	0.030	0.065	−0.92	C	ls
			[164]	2717200	3325000	2	2	76	0.031	0.033	−1.21	C	ls
63.	5p–7d	^{2}P° – ^{2}D	156.1	2719900	3360600	6	10	140	0.088	0.27	−0.28	C	4
			[156.4]	2721200	3360700	4	6	140	0.078	0.16	−0.51	C	ls
			[155.4]	2717200	3360500	2	4	120	0.088	0.090	−0.75	C	ls
			[156.4]	2721200	3360500	4	4	24	0.0087	0.018	−1.46	D	ls
64.	5p–8s	^{2}P° – ^{2}S				6	2		0.0118		−1.150	C	4
65.	5p–8d	^{2}P° – ^{2}D	128.4	2719900	3498800	6	10	104	0.0428	0.109	−0.59	C	4
			[128.6]	2721200	3498900	4	6	100	0.038	0.065	−0.81	C	ls
			[128.0]	2717200	3498700	2	4	88	0.0431	0.0363	−1.065	C	ls
			[128.6]	2721200	3498700	4	4	17	0.0043	0.0073	−1.76	D	ls
66.	5d–5f	^{2}D – ^{2}F°				10	14		0.199		0.299	C	4
67.	5d–6p	^{2}D – ^{2}P°	312.8	2788400	3108100	10	6	161	0.142	1.46	0.152	C	4
			[312.2]	2788600	3108900	6	4	150	0.14	0.88	−0.07	C	ls
			[314.2]	2788100	3106400	4	2	159	0.118	0.487	−0.327	C	ls
			[311.7]	2788100	3108900	4	4	16	0.024	0.097	−1.02	D	ls
68.	5d–6f	^{2}D – ^{2}F°	266.8	2788400	3163200	10	14	430	0.64	5.6	0.81	C	4
			[267.0]	2788600	3163200	6	8	430	0.61	3.2	0.56	C	ls
			[266.7]	2788100	3163100	4	6	390	0.63	2.2	0.40	C	ls
			[267.0]	2788600	3163100	6	6	28	0.030	0.16	−0.74	D	ls
69.	5d–7p	^{2}D – ^{2}P°				10	6		0.0250		−0.60	C	4
70.	5d–7f	^{2}D – ^{2}F°	171.6	2788400	3371100	10	14	285	0.176	0.99	0.246	C	4
			[171.7]	2788600	3371100	6	8	290	0.17	0.57	0.00	C	ls
			[171.5]	2788100	3371200	4	6	270	0.18	0.40	−0.15	C	ls
			[171.6]	2788600	3371200	6	6	19	0.0083	0.028	−1.30	D	ls
71.	5d–8p	^{2}D – ^{2}P°				10	6		0.0093		−1.03	D	4
72.	5d–8f	^{2}D – ^{2}F°	139.4	2788400	3505800	10	14	190	0.077	0.35	−0.11	C	4
			[139.4]	2788600	3505800	6	8	190	0.073	0.20	−0.36	C	ls
			[139.4]	2788100	3505700	4	6	170	0.076	0.14	−0.52	C	ls
			[139.5]	2788600	3505700	6	6	12	0.0036	0.010	−1.66	D	ls

J. Phys. Chem. Ref. Data, Vol. 17, Suppl. 4, 1988

Fe XVI: Allowed transitions — Continued

No.	Transition Array	Multiplet	λ (Å)	E_i (cm^{-1})	E_k (cm^{-1})	g_i	g_k	A_{ki} (10^8 s^{-1})	f_{ik}	S (at. u.)	log gf	Accuracy	Source
73.	5f–6d	^{2}F° – ^{2}D	305.3	2818800	3146400	14	10	46.7	0.0466	0.66	−0.185	C	4
			[305.2]	2818900	3146600	8	6	45	0.047	0.38	−0.42	C	ls
			[305.3]	2818600	3146100	6	4	46	0.043	0.26	−0.59	C	ls
			[304.9]	2818600	3146600	6	6	2.3	0.0032	0.019	−1.72	D	ls
74.	5f–7d	^{2}F° – ^{2}D	184.6	2818800	3360600	14	10	23	0.0083	0.071	−0.93	D	4
			[184.6]	2818900	3360700	8	6	22	0.0084	0.041	−1.17	D	ls
			[184.5]	2818600	3360500	6	4	23	0.0077	0.028	−1.34	D	ls
			[184.5]	2818600	3360700	6	6	1.1	5.5(−4)	0.0020	−2.48	E	ls
75.	5f–8d	^{2}F° – ^{2}D	147.1	2818800	3498800	14	10	13	0.0030	0.020	−1.38	D	4
			[147.1]	2818900	3498900	8	6	12	0.0028	0.011	−1.64	D	ls
			[147.0]	2818600	3498700	6	4	13	0.0028	0.0080	−1.78	D	ls
			[147.0]	2818600	3498900	6	6	0.61	2.0(−4)	5.7(−4)	−2.93	E	ls

^{a}The number in parentheses following the tabulated value indicates the power of ten by which this value has to be multiplied.

Fe XVI

Forbidden Transitions

List of tabulated lines

Wavelength (Å)	No.	Wavelength (Å)	No.	Wavelength (Å)	No.	Wavelength (Å)	No.
35.860	28	83.886	31	140.8	16	206	47
35.867	28	84.090	31	143	48	207	47
39.348	30	84.324	12	145.5	19	224.3	51
39.670	30	84.360	39	145.6	19	226.2	51
47.052	27	84.918	39	147.40	26	226.3	51
47.077	27	84.926	39	148.04	26	236	58
52.427	29	88.402	6	150.5	7	250.4	61
52.991	29	89.023	6	150.6	7	252.0	61
53.008	29	89.222	6	150.8	7	255.3	13
58.534	3	97.809	8	150.9	7	257.9	13
59.259	3	97.857	8	152.9	52	259.6	13
59.538	3	97.914	8	153.8	52	278.9	14
61.301	36	97.962	8	153.9	52	279.3	14
65.445	41	102.2	11	157.7	10	279.7	14
65.781	41	105	43	157.8	10	283.4	21
66.975	35	108.3	45	157.9	10	283.6	21
66.984	35	108.5	45	174.6	15	283.8	21
68.975	4	108.6	33	174.7	15	284.0	21
69.027	4	118.9	38	174.8	15	290.2	17
69.113	4	119	49	174.9	15	290.3	17
69.166	4	120.0	38	181.0	18	290.4	17
71.762	40	126.8	53	181.1	18	290.5	17
72.176	40	127.5	53	186	42,55	291.8	23
78.186	34	134.5	5	186.4	44	291.9	23
78.217	34	135.9	5	186.6	44	292.0	23
80.887	9	136.7	5	187	55	344.5	56
80.945	9	140.7	16	187.8	44	344.8	56

List of tabulated lines — Continued

Wavelength (Å)	No.	Wavelength (Å)	No.	Wavelength (Å)	No.	Wavelength (Å)	No.
347.5	56	467.1	20	575	63	1020	50
348	54	467.5	20	722.5	24	1030	50
350	54,57	480.5	22	723.6	24	1760	59
377.6	60	480.8	22	724.6	24	1840	59
381.2	60	481.0	22	742.4	25	1850	59
381.4	60	483.3	37	742.9	25	4764.7	1
388.0	32	501.0	37	743.5	25	33960	2
389.7	32	502.5	37	787	46		
466.0	20	559	62	794	46		
466.4	20	562	62	986.2	50		

Electric quadrupole strengths for the $3p_{1/2}$–$3p_{3/2}$ and $3d_{3/2}$–$3d_{5/2}$ transitions, as well as for the $3s$–$3d$ multiplet, were derived from the radial quadrupole integrals calculated by Krueger and Czyzak[1] using a single-configuration Hartree-Fock approach. The strength of the $3s$–$3d$ multiplet was distributed between the two lines of the multiplet according to LS-coupling rules. Quadrupole strengths for numerous multiplets in this sodiumlike ion were determined by Tull *et al.*[2] using the frozen-core Hartree-Fock approximation with no allowance for configuration mixing. Their calculated strength for the $3s$–$3d$ multiplet is in nearly perfect agreement with that of Krueger and Czyzak. LS-coupling rules were applied to obtain strengths of lines within multiplets. The strongest lines for which fairly accurate wavelengths could be derived from experimentally determined energy levels are quoted in this compilation.

The strengths given in Ref. 2 for transitions in which both $\Delta n = 0$ and $\Delta l = 0$ (i.e., transitions between the two levels of a given term) are overstated, and had to be reduced as follows:

$$S(np \; {}^2P^\circ_{1/2} - np \; {}^2P^\circ_{3/2}) = S(\text{Ref. 2}) \times (1/3)$$

$$S(nd \; {}^2D_{3/2} - nd \; {}^2D_{5/2}) = S(\text{Ref. 2}) \times (3/25)$$

$$S(nf \; {}^2F^\circ_{5/2} - nf \; {}^2F^\circ_{7/2}) = S(\text{Ref. 2}) \times (3/49).$$

References

[1]T. K. Krueger and S. J. Czyzak, Mem. R. Astron. Soc. **69**, 145 (1965).
[2]C. E. Tull, M. Jackson, R. P. McEachran, and M. Cohen, J. Quant. Spectrosc. Radiat. Transfer **12**, 893 (1972).

Fe XVI: Forbidden transitions

No.	Transition Array	Multiplet	λ (Å)	E_i (cm^{-1})	E_k (cm^{-1})	g_i	g_k	Type of transition	A_{ki} (s^{-1})	S (at. u.)	Accuracy	Source
1.	$3p$–$3p$	${}^2P^\circ$ – ${}^2P^\circ$										
			[4764.7]	277163	298145	2	4	E2	0.0243	0.142	C	1
2.	$3d$–$3d$	${}^2D$ – ${}^2D$										
			[33960]	675477	678421	4	6	E2	2.7(−7)[a]	0.043	D	1
3.	$3p$–$4p$	${}^2P^\circ$ – ${}^2P^\circ$										
			[59.259]	298145	1985600	4	4	E2	3.4(+7)	0.060	D	2,*ls*
			[59.538]	298145	1977700	4	2	E2	6.7(+7)	0.060	D	2,*ls*
			[58.534]	277163	1985600	2	4	E2	3.7(+7)	0.060	D	2,*ls*
4.	$3d$–$4d$	${}^2D$ – ${}^2D$										
			[69.113]	678421	2125300	6	6	E2	1.78(+7)	0.100	C	2,*ls*
			[69.027]	675477	2124200	4	4	E2	1.6(+7)	0.059	D	2,*ls*
			[69.166]	678421	2124200	6	4	E2	6.6(+6)	0.025	E	2,*ls*
			[68.975]	675477	2125300	4	6	E2	4.5(+6)	0.025	E	2,*ls*

Fe XVI: Forbidden transitions — Continued

No.	Transition Array	Multiplet	λ (Å)	E_i (cm^{-1})	E_k (cm^{-1})	g_i	g_k	Type of transition	A_{ki} (s^{-1})	S (at. u.)	Accuracy	Source
5.	4p–5p	^{2}P° – ^{2}P°										
			[135.9]	1985600	2721200	4	4	E2	5.4(+6)	0.60	C	2,ls
			[136.7]	1985600	2717200	4	2	E2	1.1(+7)	0.60	C	2,ls
			[134.5]	1977700	2721200	2	4	E2	5.7(+6)	0.60	C	2,ls
6.	4p–6p	^{2}P° – ^{2}P°										
			[89.023]	1985600	3108900	4	4	E2	3.5(+6)	0.046	D	2,ls
			[89.222]	1985600	3106400	4	2	E2	6.8(+6)	0.046	D	2,ls
			[88.402]	1977700	3108900	2	4	E2	3.6(+6)	0.046	D	2,ls
7.	4d–5d	^{2}D – ^{2}D										
			[150.8]	2125300	2788600	6	6	E2	3.95(+6)	1.10	C	2,ls
			[150.6]	2124200	2788100	4	4	E2	3.5(+6)	0.64	C	2,ls
			[150.9]	2125300	2788100	6	4	E2	1.48(+6)	0.276	C–	2,ls
			[150.5]	2124200	2788600	4	6	E2	1.00(+6)	0.276	C–	2,ls
8.	4d–6d	^{2}D – ^{2}D										
			[97.914]	2125300	3146600	6	6	E2	2.4(+6)	0.076	D	2,ls
			[97.857]	2124200	3146100	4	4	E2	2.1(+6)	0.044	D	2,ls
			[97.962]	2125300	3146100	6	4	E2	8.8(+5)	0.019	E	2,ls
			[97.809]	2124200	3146600	4	6	E2	5.9(+5)	0.019	E	2,ls
9.	4d–7d	^{2}D – ^{2}D										
			[80.945]	2125300	3360700	6	6	E2	1.5(+6)	0.018	D	2,ls
			[80.887]	2124200	3360500	4	4	E2	1.3(+6)	0.011	D	2,ls
10.	4f–5f	^{2}F° – ^{2}F°										
			[157.8]	2185200	2818900	8	8	E2	2.30(+6)	1.07	C	2,ls
			[157.7]	2184600	2818600	6	6	E2	2.2(+6)	0.77	C	2,ls
			[157.9]	2185200	2818600	8	6	E2	3.68(+5)	0.129	C–	2,ls
			[157.7]	2184600	2818900	6	8	E2	2.78(+5)	0.129	C–	2,ls
11.	4f–6f	^{2}F° – ^{2}F°										
			[102.2]	2185200	3163200	8	8	E2	1.1(+6)	0.061	D	2,ls
			[102.2]	2184600	3163100	6	6	E2	1.1(+6)	0.044	D	2,ls
12.	4f–7f	^{2}F° – ^{2}F°										
			[84.324]	2185200	3371100	8	8	E2	6.4(+5)	0.013	D	2,ls
13.	5p–6p	^{2}P° – ^{2}P°										
			[257.9]	2721200	3108900	4	4	E2	1.26(+6)	3.43	C	2,ls
			[259.6]	2721200	3106400	4	2	E2	2.44(+6)	3.43	C	2,ls
			[255.3]	2717200	3108900	2	4	E2	1.33(+6)	3.43	C	2,ls
14.	5d–6d	^{2}D – ^{2}D										
			[279.3]	2788600	3146600	6	6	E2	1.1(+6)	6.5	C	2,ls
			[279.3]	2788100	3146100	4	4	E2	9.3(+5)	3.78	C	2,ls
			[279.7]	2788600	3146100	6	4	E2	3.97(+5)	1.62	C–	2,ls
			[278.9]	2788100	3146600	4	6	E2	2.69(+5)	1.62	C–	2,ls

Fe XVI: Forbidden transitions — Continued

No.	Transition Array	Multiplet	λ (Å)	E_i (cm^{-1})	E_k (cm^{-1})	g_i	g_k	Type of transition	A_{ki} (s^{-1})	S (at. u.)	Accuracy	Source
15.	$5d$–$7d$	^{2}D – ^{2}D										
			[174.8]	2788600	3360700	6	6	E2	7.3(+5)	0.423	C	2,ls
			[174.7]	2788100	3360500	4	4	E2	6.4(+5)	0.247	C	2,ls
			[174.9]	2788600	3360500	6	4	E2	2.72(+5)	0.106	C–	2,ls
			[174.6]	2788100	3360700	4	6	E2	1.83(+5)	0.106	C–	2,ls
16.	$5d$–$8d$	^{2}D – ^{2}D										
			[140.8]	2788600	3498900	6	6	E2	5.0(+5)	0.098	D	2,ls
			[140.7]	2788100	3498700	4	4	E2	4.3(+5)	0.057	D	2,ls
			[140.8]	2788600	3498700	6	4	E2	1.8(+5)	0.024	E	2,ls
			[140.7]	2788100	3498900	4	6	E2	1.2(+5)	0.024	E	2,ls
17.	$5f$–$6f$	^{3}F° – ^{2}F°										
			[290.4]	2818900	3163200	8	8	E2	8.0(+5)	7.9	C	2,ls
			[290.3]	2818600	3163100	6	6	E2	7.7(+5)	5.7	C	2,ls
			[290.5]	2818900	3163100	8	6	E2	1.3(+5)	0.94	C–	2,ls
			[290.2]	2818600	3163200	6	8	E2	9.6(+4)	0.94	C–	2,ls
18.	$5f$–$7f$	^{2}F° – ^{2}F°										
			[181.1]	2818900	3371100	8	8	E2	4.90(+5)	0.455	C	2,ls
			[181.0]	2818600	3371200	6	6	E2	4.73(+5)	0.328	C	2,ls
			[181.1]	2818900	3371200	8	6	E2	7.9(+4)	0.055	D	2,ls
			[181.0]	2818600	3371100	6	8	E2	5.9(+4)	0.055	D	2,ls
19.	$5f$–$8f$	^{2}F° – ^{2}F°										
			[145.6]	2818900	3505800	8	8	E2	3.1(+5)	0.098	D	2,ls
			[145.5]	2818600	3505700	6	6	E2	3.0(+5)	0.071	D	2,ls
			[145.6]	2818900	3505700	8	6	E2	5.1(+4)	0.012	E	2,ls
			[145.5]	2818600	3505800	6	8	E2	3.9(+4)	0.012	E	2,ls
20.	$6d$–$7d$	^{2}D – ^{2}D										
			[467.1]	3146600	3360700	6	6	E2	3.39(+5)	26.9	C	2,ls
			[466.4]	3146100	3360500	4	4	E2	2.99(+5)	15.7	C	2,ls
			[467.5]	3146600	3360500	6	4	E2	1.3(+5)	6.7	C–	2,ls
			[466.0]	3146100	3360700	4	6	E2	8.5(+4)	6.7	C–	2,ls
21.	$6d$–$8d$	^{2}D – ^{2}D										
			[283.8]	3146600	3498900	6	6	E2	2.55(+5)	1.68	C	2,ls
			[283.6]	3146100	3498700	4	4	E2	2.2(+5)	0.98	C	2,ls
			[284.0]	3146600	3498700	6	4	E2	9.5(+4)	0.419	C–	2,ls
			[283.4]	3146100	3498900	4	6	E2	6.4(+4)	0.419	C–	2,ls
22.	$6f$–$7f$	^{2}F° – ^{2}F°										
			[481.0]	3163200	3371100	8	8	E2	2.86(+5)	35.1	C	2,ls
			[480.5]	3163100	3371200	6	6	E2	2.75(+5)	25.2	C	2,ls
			[480.8]	3163200	3371200	8	6	E2	4.59(+4)	4.21	C–	2,ls
			[480.8]	3163100	3371100	6	8	E2	3.44(+4)	4.21	C–	2,ls
23.	$6f$–$8f$	^{2}F° – ^{2}F°										
			[291.9]	3163200	3505800	8	8	E2	2.00(+5)	2.02	C	2,ls
			[291.9]	3163100	3505700	6	6	E2	1.92(+5)	1.45	C	2,ls
			[292.0]	3163200	3505700	8	6	E2	3.19(+4)	0.242	C–	2,ls
			[291.8]	3163100	3505800	6	8	E2	2.40(+4)	0.242	C–	2,ls

Fe XVI: Forbidden transitions — Continued

No.	Transition Array	Multiplet	λ (Å)	E_i (cm^{-1})	E_k (cm^{-1})	g_i	g_k	Type of transition	A_{ki} (s^{-1})	S (at. u.)	Accu-racy	Source
24.	7d–8d	^{2}D – ^{2}D										
			[723.6]	3360700	3498900	6	6	E2	1.3(+5)	89	C	2,ls
			[723.6]	3360500	3498700	4	4	E2	1.1(+5)	52	C	2,ls
			[724.6]	3360700	3498700	6	4	E2	4.69(+4)	22.3	C−	2,ls
			[722.5]	3360500	3498900	4	6	E2	3.17(+4)	22.3	C−	2,ls
25.	7f–8f	^{2}F° – ^{2}F°										
			[742.4]	3371100	3505800	8	8	E2	1.12(+5)	120	C	2,ls
			[743.5]	3371200	3505700	6	6	E2	1.1(+5)	86	C	2,ls
			[742.9]	3371100	3505700	8	6	E2	1.78(+4)	14.4	C−	2,ls
			[742.9]	3371200	3505800	6	8	E2	1.34(+4)	14.4	C−	2,ls
26.	3s–3d	^{2}S – ^{2}D										
			[147.40]	0	678421	2	6	E2	6.8(+5)	0.170	C	1,ls
			[148.04]	0	675477	2	4	E2	6.7(+5)	0.113	C	1,ls
27.	3s–4d	^{2}S – ^{2}D										
			[47.052]	0	2125300	2	6	E2	1.44(+8)	0.119	C	2,ls
			[47.077]	0	2124200	2	4	E2	1.5(+8)	0.080	D	2,ls
28.	3s–5d	^{2}S – ^{2}D										
			[35.860]	0	2788600	2	6	E2	7.6(+7)	0.016	D	2,ls
			[35.867]	0	2788100	2	4	E2	7.8(+7)	0.011	D	2,ls
29.	3p–4f	^{2}P° – ^{2}F°										
			[52.991]	298145	2185200	4	8	E2	2.6(+8)	0.52	C	2,ls
			[52.427]	277163	2184600	2	6	E2	2.14(+8)	0.303	C	2,ls
			[53.008]	298145	2184600	4	6	E2	5.8(+7)	0.086	D	2,ls
30.	3p–5f	^{2}P° – ^{2}F°										
			[39.670]	298145	2818900	4	8	E2	4.5(+7)	0.021	D	2,ls
			[39.348]	277163	2818600	2	6	E2	3.6(+7)	0.012	D	2,ls
31.	3d–4s	^{2}D – ^{2}S										
			[84.090]	678421	1867600	6	2	E2	1.3(+7)	0.066	D	2,ls
			[83.886]	675477	1867600	4	2	E2	8.9(+6)	0.044	D	2,ls
32.	4s–4d	^{2}S – ^{2}D										
			[388.0]	1867600	2125300	2	6	E2	8.4(+4)	2.64	C	2,ls
			[389.7]	1867600	2124200	2	4	E2	8.2(+4)	1.76	C	2,ls
33.	4s–5d	^{2}S – ^{2}D										
			[108.6]	1867600	2788600	2	6	E2	1.5(+7)	0.81	C	2,ls
			[108.6]	1867600	2788100	2	4	E2	1.5(+7)	0.54	C	2,ls
34.	4s–6d	^{2}S – ^{2}D										
			[78.186]	1867600	3146600	2	6	E2	1.06(+7)	0.111	C	2,ls
			[78.217]	1867600	3146100	2	4	E2	1.1(+7)	0.074	D	2,ls

Fe XVI: Forbidden transitions — Continued

No.	Transition Array	Multiplet	λ (Å)	E_i (cm^{-1})	E_k (cm^{-1})	g_i	g_k	Type of transition	A_{ki} (s^{-1})	S (at. u.)	Accuracy	Source
35.	4s–7d	^{2}S – ^{2}D										
			[66.975]	1867600	3360700	2	6	E2	7.1(+6)	0.034	D	2,ls
			[66.984]	1867600	3360500	2	4	E2	6.9(+6)	0.022	D	2,ls
36.	4s–8d	^{2}S – ^{2}D										
			[61.301]	1867600	3498900	2	6	E2	4.9(+6)	0.015	D	2,ls
37.	4p–4f	^{2}P° – ^{2}F°										
			[501.0]	1985600	2185200	4	8	E2	1.92(+4)	2.88	C	2,ls
			[483.3]	1977700	2184600	2	6	E2	1.78(+4)	1.68	C	2,ls
			[502.5]	1985600	2184600	4	6	E2	4190	0.480	C–	2,ls
38.	4p–5f	^{2}P° – ^{2}F°										
			[120.0]	1985600	2818900	4	8	E2	2.96(+7)	3.51	C	2,ls
			[118.9]	1977700	2818600	2	6	E2	2.42(+7)	2.05	C	2,ls
			[120.0]	1985600	2818600	4	6	E2	6.6(+6)	0.59	C–	2,ls
39.	4p–6f	^{2}P° – ^{2}F°										
			[84.918]	1985600	3163200	4	8	E2	1.26(+7)	0.265	C	2,ls
			[84.360]	1977700	3163100	2	6	E2	1.01(+7)	0.154	C	2,ls
			[84.926]	1985600	3163100	4	6	E2	2.8(+6)	0.044	D	2,ls
40.	4p–7f	^{2}P° – ^{2}F°										
			[72.176]	1985600	3371100	4	8	E2	5.9(+6)	0.055	D	2,ls
			[71.762]	1977700	3371200	2	6	E2	4.7(+6)	0.032	D	2,ls
41.	4p–8f	^{2}P° – ^{2}F°										
			[65.781]	1985600	3505800	4	8	E2	3.2(+6)	0.019	D	2,ls
			[65.445]	1977700	3505700	2	6	E2	2.6(+6)	0.011	D	2,ls
42.	4d–5s	^{2}D – ^{2}S										
			[186]	2125300	2662000	6	2	E2	3.7(+6)	0.98	C–	2,ls
			[186]	2124200	2662000	4	2	E2	2.5(+6)	0.66	C–	2,ls
43.	4d–6s	^{2}D – ^{2}S										
			[105]	2125300	3075000	6	2	E2	2.2(+6)	0.034	D	2,ls
			[105]	2124200	3075000	4	2	E2	1.4(+6)	0.022	D	2,ls
44.	4f–5p	^{2}F° – ^{2}P°										
			[186.6]	2185200	2721200	8	4	E2	9.2(+5)	0.493	C	2,ls
			[187.8]	2184600	2717200	6	2	E2	1.03(+6)	0.287	C	2,ls
			[186.4]	2184600	2721200	6	4	E2	1.5(+5)	0.082	D	2,ls
45.	4f–6p	^{2}F° – ^{2}P°										
			[108.3]	2185200	3108900	8	4	E2	5.6(+5)	0.020	D	2,ls
			[108.5]	2184600	3106400	6	2	E2	6.7(+5)	0.012	D	2,ls
46.	5s–5d	^{2}S – ^{2}D										
			[787]	2662000	2788600	2	6	E2	1.8(+4)	19	D–	2,ls
			[794]	2662000	2788100	2	4	E2	1.7(+4)	13	D–	2,ls

Fe XVI: Forbidden transitions — Continued

No.	Transition Array	Multiplet	λ (Å)	E_i (cm^{-1})	E_k (cm^{-1})	g_i	g_k	Type of transition	A_{ki} (s^{-1})	S (at. u.)	Accuracy	Source
47.	5s–6d	^{2}S – ^{2}D										
			[206]	2662000	3146600	2	6	E2	2.77(+6)	3.67	C–	2,ls
			[207]	2662000	3146100	2	4	E2	2.71(+6)	2.45	C–	2,ls
48.	5s–7d	^{2}S – ^{2}D										
			[143]	2662000	3360700	2	6	E2	2.29(+6)	0.489	C	2,ls
			[143]	2662000	3360500	2	4	E2	2.29(+6)	0.326	C	2,ls
49.	5s–8d	^{2}S – ^{2}D										
			[119]	2662000	3498900	2	6	E2	1.70(+6)	0.145	C	2,ls
			[119]	2662000	3498700	2	4	E2	1.7(+6)	0.097	D	2,ls
50.	5p–5f	^{2}P° – ^{2}F°										
			[1020]	2721200	2818900	4	8	E2	5200	27.4	C	2,ls
			[986.2]	2717200	2818600	2	6	E2	4800	16.0	C	2,ls
			[1030]	2721200	2818600	4	6	E2	1100	4.57	C–	2,ls
51.	5p–6f	^{2}P° – ^{2}F°										
			[226.2]	2721200	3163200	4	8	E2	5.3(+6)	15.0	C	2,ls
			[224.3]	2717200	3163100	2	6	E2	4.3(+6)	8.8	C	2,ls
			[226.3]	2721200	3163100	4	6	E2	1.18(+6)	2.50	C–	2,ls
52.	5p–7f	^{2}P° – ^{2}F°										
			[153.9]	2721200	3371100	4	8	E2	3.26(+6)	1.34	C	2,ls
			[152.9]	2717200	3371200	2	6	E2	2.6(+6)	0.78	C	2,ls
			[153.8]	2721200	3371200	4	6	E2	7.3(+5)	0.224	C–	2,ls
53.	5p–8f	^{2}P° – ^{2}F°										
			[127.5]	2721200	3505800	4	8	E2	1.95(+6)	0.313	C	2,ls
			[126.8]	2717200	3505700	2	6	E2	1.56(+6)	0.183	C	2,ls
			[127.5]	2721200	3505700	4	6	E2	4.3(+5)	0.052	D	2,ls
54.	5d–6s	^{2}D – ^{2}S										
			[350]	2788600	3075000	6	2	E2	1.1(+6)	6.9	D+	2,ls
			[348]	2788100	3075000	4	2	E2	7.6(+5)	4.6	D+	2,ls
55.	5d–7s	^{2}D – ^{2}S										
			[187]	2788600	3325000	6	2	E2	7.6(+5)	0.207	C–	2,ls
			[186]	2788100	3325000	4	2	E2	5.2(+5)	0.138	C–	2,ls
56.	5f–6p	^{2}F° – ^{2}P°										
			[344.8]	2818900	3108900	8	4	E2	4.5(+5)	5.2	C	2,ls
			[347.5]	2818600	3106400	6	2	E2	5.0(+5)	3.03	C	2,ls
			[344.5]	2818600	3108900	6	4	E2	7.5(+4)	0.87	C–	2,ls
57.	6s–7d	^{2}S – ^{2}D										
			[350]	3075000	3360700	2	6	E2	6.9(+5)	13	D+	2,ls
			[350]	3075000	3360500	2	4	E2	6.9(+5)	8.6	D+	2,ls

Fe XVI: Forbidden transitions — Continued

No.	Transition Array	Multiplet	λ (Å)	E_i (cm^{-1})	E_k (cm^{-1})	g_i	g_k	Type of transition	A_{ki} (s^{-1})	S (at. u.)	Accuracy	Source
58.	$6s$–$8d$	^{2}S – ^{2}D										
			[236]	3075000	3498900	2	6	E2	6.5(+5)	1.7	D+	2,ls
			[236]	3075000	3498700	2	4	E2	6.3(+5)	1.1	D+	2,ls
59.	$6p$–$6f$	^{2}P° – ^{2}F°										
			[1840]	3108900	3163200	4	8	E2	1440	145	C–	2,ls
			[1760]	3106400	3163100	2	6	E2	1400	84	C–	2,ls
			[1850]	3108900	3163100	4	6	E2	310	24	D	2,ls
60.	$6p$–$7f$	^{2}P° – ^{2}F°										
			[381.4]	3108900	3371100	4	8	E2	1.30(+6)	49.9	C	2,ls
			[377.6]	3106400	3371200	2	6	E2	1.06(+6)	29.1	C	2,ls
			[381.2]	3108900	3371200	4	6	E2	2.9(+5)	8.3	C–	2,ls
61.	$6p$–$8f$	^{2}P° – ^{2}F°										
			[252.0]	3108900	3505800	4	8	E2	9.8(+5)	4.73	C	2,ls
			[250.4]	3106400	3505700	2	6	E2	7.8(+5)	2.76	C	2,ls
			[252.0]	3108900	3505700	4	6	E2	2.2(+5)	0.79	C–	2,ls
62.	$6d$–$7s$	^{2}D – ^{2}S										
			[562]	3146600	3325000	6	2	E2	4.8(+5)	32	D+	2,ls
			[559]	3146100	3325000	4	2	E2	3.4(+5)	22	D+	2,ls
63.	$7s$–$8d$	^{2}S – ^{2}D										
			[575]	3325000	3498900	2	6	E2	1.7(+5)	38	D+	2,ls
			[575]	3325000	3498700	2	4	E2	1.7(+5)	25	D+	2,ls

[a]The number in parentheses following the tabulated value indicates the power of ten by which this value has to be multiplied.

Fe XVII

Ne Isoelectronic Sequence

Ground State: $1s^2 2s^2 2p^6 \ {}^1S_0$

Ionization Energy: 1262.2 eV = 10180000 cm^{-1}

Allowed Transitions

List of tabulated lines

Wavelength (Å)	No.	Wavelength (Å)	No.	Wavelength (Å)	No.	Wavelength (Å)	No.
11.023	11	17.054	12	51.177	40	99.50	3
11.043	10	41.37	44	52.748	42	100	7
12.123	21	46.7	25	52.9	37	254.48	30
12.264	20	47.48	27	55.528	41	263.1	30
12.322	19	47.6	26	55.96	35	269.88	30
12.526	18	47.8	26	57.32	36	283.8	34
12.681	17	47.85	26	58.76	43	284.01	32
13.823	9	48.47	29	87.30	2	284.3	31
13.891	8	48.85	28	90.375	5	350.58	23
15.015	16	49.427	38	90.77	1	366.8	33
15.262	15	49.5	38	91	5	409.91	22
15.450	14	49.7	38	92	6	705.2	24
16.777	13	50.26	39	95.29	4		

For resonance transitions to $J = 1$ levels of the $2p^5 3s$ and $2p^5 3d$ configurations, we quote f-values which were calculated by Shorer[1] using the relativistic random phase approximation (RRPA). These calculations allowed for mixing between configurations of type $2p^5 ns$ and $2p^5 nd$, as well as correlation effects due to configurations having a vacancy in the $1s$ or $2s$ subshell. Shorer showed by numerical comparison the effects of including various configurations in his basis, thus providing an illustrative example of the rather drastic changes due to configuration interaction that can result in the f-values of transitions in heavy ions.

A-values for numerous transitions involving an electron jump of the type $2s$–np ($n = 2$–4), $2p$–ns, $2p$–nd, $3s$–np, $3p$–nd ($n = 3,4$), or $3p$–$4s$ were calculated by Loulergue and Nussbaumer[2] using scaled Thomas-Fermi wavefunctions. The following configurations were included in their basis: $2s^2 2p^6$, $2s^2 2p^5 nl$, and $2s 2p^6 nl$ (for $n = 3$: $l = s, p, d$; for $n = 4$: $l = s, p, d, f$). Their results are quoted here, but, in cases where better wavelength data were available, their transition probabilities were first converted to line strengths, which were then reconverted to f- and A-values by using the more accurate wavelengths. Data for resonance lines were not modified, as the calculated wavelengths of Ref. 2 for these lines are fairly accurate.

Transition probabilities for a few lines for which Loulergue and Nussbaumer did not report results were taken from the work of Pokleba and Safronova,[3] who used

wavefunctions calculated by a charge-expansion perturbation theory approach with allowance for mixing of configurations in which a single $2s$ or $2p$ electron is excited to an $n = 3$ orbital but with no inclusion of configurations in which an electron occupies the $n = 4$ shell. As with the data of Ref. 2, the results of Ref. 3 were modified to incorporate more accurate wavelengths.

Oscillator strengths for three lines not treated in any of the abovementioned sources are the results of the Hartree-XR (Hartree-Fock with relativistic effects and statistical allowance for exchange) calculations of Fawcett et al.[4] Additional data reported by them could not be used, as it was impossible to determine the appropriate LS-coupling designations of the levels involved in the transitions, which were designated in $J_1 l$ coupling in Ref. 4.

Transitions involving levels of the $2p^5 3p$ and $2p^5 3d$ configurations which are indicated by Jupen and Litzen[5] to be of low to moderate purity in LS coupling are excluded here, as are very weak lines. Transitions involving the corresponding levels in the $2p^5 4l$ configurations are excluded as well, as no percentage composition data were available for these levels. The pattern of levels within the $2s 2p^6 3d$ configuration resulting from the calculations of Loulergue and Nussbaumer is entirely different from that determined by Vainshtein and Safronova,[6] whose energy levels were apparently used by Pokleba and Safronova in their transition probability calculations. We have thus excluded transitions out of these levels from our tabulation.

References

[1]P. Shorer, Phys. Rev. A **20**, 642 (1979).

[2]M. Loulergue and H. Nussbaumer, Astron. Astrophys. **45**, 125 (1975).

[3]A. K. Pokleba and U. I. Safronova, Preprint No. 11, Akad. Nauk SSSR, Ot. Ob. Fiz. Astron., Inst. Spektrosk. (Moscow, 1981).

[4]B. C. Fawcett, G. E. Bromage, and R. W. Hayes, Mon. Not. R. Astron. Soc. **186**, 113 (1979).

[5]C. Jupen and U. Litzen, Phys. Scr. **30**, 112 (1984).

[6]L. A. Vainshtein and U. I. Safronova, *Spektroskopicheskie Konstanty Atomov*, 5–122 (Ed. V. B. Belyanin, Akad. Nauk SSSR, Ot. Ob. Fiz. Astron., Nauch. Sov. Spektrosk., Moscow, 1977).

Fe XVII: Allowed transitions

No.	Transition Array	Multiplet	λ (Å)	E_i (cm^{-1})	E_k (cm^{-1})	g_i	g_k	A_{ki} (10^8 s^{-1})	f_{ik}	S (at. u.)	log gf	Accuracy	Source
1.	$2s^22p^5(^2P^o_{3/2})3s-$ $2s2p^63s$	$(^3/_2,^1/_2)^o - {}^3S$											
			90.77	5848400	6966000	5	3	990	0.074	0.11	−0.43	D	2
2.		$(^3/_2,^1/_2)^o - {}^1S$											
			87.30	5863700	7009000	3	1	670	0.026	0.022	−1.12	D	2
3.	$2s^22p^5(^2P^o_{1/2})3s-$ $2s2p^63s$	$(^1/_2,^1/_2)^o - {}^3S$											
			[99.50]	5960500	6966000	3	3	250	0.037	0.036	−0.96	D	2
4.		$(^1/_2,^1/_2)^o - {}^1S$											
			95.29	5960500	7009000	3	1	420	0.019	0.018	−1.24	D	2
5.	$2s^22p^53p-$ $2s2p^63p$	$^3S -{}^3P^o$											
			[90.375]	6092400	7198900	3	3	140	0.017	0.015	−1.30	E	2
			[91]			3	1	460	0.019	0.017	−1.25	D	2
6.		$^3D - {}^3P^o$											
			[92]			7	5	780	0.071	0.15	−0.31	D	2
7.	$2s^22p^54p-$ $2s2p^64p$	$^3S - {}^3P^o$											
			[100]			3	3	110	0.016	0.016	−1.31	E	2
			[100]			3	1	460	0.023	0.023	−1.16	E	2
8.	$2s^22p^6-2s2p^63p$	$^1S - {}^3P^o$											
			13.891	0	7198900	1	3	3400	0.030	0.0013	−1.53	E	2
9.		$^1S - {}^1P^o$	13.823	0	7234300	1	3	3.3(+4)^a	0.28	0.013	−0.55	D	2
10.	$2s^22p^6-2s2p^64p$	$^1S - {}^3P^o$											
			11.043	0	9055500	1	3	2900	0.016	5.8(−4)	−1.80	E	2
11.		$^1S - {}^1P^o$	11.023	0	9071900	1	3	2.1(+4)	0.11	0.0042	−0.94	D	2
12.	$2p^6-$ $2p^5(^2P^o_{3/2})3s$	$^1S - (^3/_2,^1/_2)^o$											
			17.054	0	5863700	1	3	9330	0.122	0.00685	−0.914	C+	1
13.	$2p^6-$ $2p^5(^2P^o_{1/2})3s$	$^1S - (^1/_2,^1/_2)^o$											
			16.777	0	5960500	1	3	8290	0.105	0.00580	−0.979	C+	1

Fe XVII: Allowed transitions — Continued

No.	Transition Array	Multiplet	λ (Å)	E_i (cm^{-1})	E_k (cm^{-1})	g_i	g_k	A_{ki} (10^8 s^{-1})	f_{ik}	S (at. u.)	log gf	Accuracy	Source
14.	$2p^6$–$2p^53d$	^{1}S – ^{3}P°											
			15.450	0	6472500	1	3	900	0.0097	4.9(−4)	−2.01	E	1
15.		^{1}S – ^{3}D°											
			15.262	0	6552200	1	3	6.0(+4)	0.63	0.032	−0.20	D	1
16.		^{1}S – ^{1}P°	15.015	0	6660000	1	3	2.28(+5)	2.31	0.114	0.364	C+	1
17.	$2p^6$– $2p^5(^2P^\circ_{3/2})4s$	^{1}S – $(^3/_2,^1/_2)°$											
			12.681	0	7885800	1	3	3500	0.025	0.0011	−1.60	D	2
18.	$2p^6$– $2p^5(^2P^\circ_{1/2})4s$	^{1}S – $(^1/_2,^1/_2)°$											
			12.526	0	7983400	1	3	3000	0.021	8.7(−4)	−1.67	D	2
19.	$2p^6$–$2p^54d$	^{1}S – ^{3}P°											
			12.322	0	8115600	1	3	530	0.0036	1.5(−4)	−2.44	E	2
20.		^{1}S – ^{3}D°											
			12.264	0	8153900	1	3	5.9(+4)	0.40	0.016	−0.40	D	2
21.		^{1}S – ^{1}P°	12.123	0	8248800	1	3	8.0(+4)	0.53	0.021	−0.28	D	2
22.	$2p^5(^2P^\circ_{3/2})3s$– $2p^53p$	$(^3/_2,^1/_2)°$ – ^{3}S											
			409.91	5848400	6092400	5	3	33	0.050	0.34	−0.60	D	2
23.		$(^3/_2,^1/_2)°$ – ^{3}D											
			350.58	5848400	6133600	5	7	64	0.16	0.95	−0.08	D	2
24.	$2p^5(^2P^\circ_{1/2})3s$– $2p^53p$	$(^1/_2,^1/_2)°$ – ^{3}S											
			[705.2]	5950600	6092400	1	3	0.11	0.0025	0.0059	−2.59	E	3
25.	$2p^5(^2P^\circ_{3/2})3s$– $2p^54p$	$(^3/_2,^1/_2)°$ – ^{3}D											
			[46.7]			5	7	2600	0.12	0.092	−0.22	D	4
26.	$2s2p^63s$– $2s2p^64p$	^{3}S – ^{3}P°	47.7			3	9	2500	0.25	0.12	−0.12	D	2
			[47.6]			3	5	2600	0.15	0.069	−0.35	D	2
			[47.85]	6966000	9055500	3	3	2400	0.083	0.039	−0.61	D	2
			[47.8]			3	1	2700	0.031	0.015	−1.03	D	2
27.		^{3}S – ^{1}P°											
			[47.48]	6966000	9071900	3	3	360	0.012	0.0057	−1.44	E	2
28.		^{1}S – ^{3}P°											
			[48.85]	7009000	9055500	1	3	400	0.043	0.0069	−1.37	E	2

Fe XVII: Allowed transitions — Continued

No.	Transition Array	Multiplet	λ (Å)	E_i (cm^{-1})	E_k (cm^{-1})	g_i	g_k	A_{ki} (10^8 s^{-1})	f_{ik}	S (at. u.)	log gf	Accu-racy	Source
29.		^{1}S – ^{1}P°	[48.47]	7009000	9071900	1	3	2400	0.26	0.041	−0.59	D	2
30.	$2p^53p$–$2p^53d$	^{3}S – ^{3}P°	258.9	6092400	6478600	3	9	77	0.23	0.59	−0.16	E	2
			254.48	6092400	6485400	3	5	54	0.088	0.22	−0.58	E	2
			[263.1]	6092400	6472500	3	3	96	0.10	0.26	−0.52	D	2
			269.88	6092400	6462900	3	1	110	0.041	0.11	−0.91	D	2
31.		^{3}D – ^{3}P°	[284.3]	6133600	6485400	7	5	4.6	0.0040	0.026	−1.56	E	3
32.		^{3}D – ^{3}F°	284.01	6133600	6485700	7	9	110	0.17	1.1	0.07	D	2
33.		^{3}P – ^{3}P°	[366.8]	6199900	6472500	1	3	1.6	0.0099	0.012	−2.00	E	3
34.		^{3}P – ^{3}D°	[283.8]	6199900	6552200	1	3	47	0.17	0.16	−0.77	D	3
35.	$2p^53p$– $2p^5(^2P^°_{3/2})4s$	^{3}S – ($^3/_2$,$^1/_2$)°	[55.96]	6092400	7879000	3	5	670	0.052	0.029	−0.80	D	2
36.		^{3}D – ($^3/_2$,$^1/_2$)°	57.32	6133600	7879000	7	5	1700	0.060	0.079	−0.38	D	2
37.	$2p^53p$– $2p^5(^2P^°_{1/2})4s$	^{3}S – ($^1/_2$,$^1/_2$)°	[52.9]			3	1	74	0.0010	5.4(−4)	−2.51	E	2
38.	$2p^53p$–$2p^54d$	^{3}S – ^{3}P°	49.5			3	9	2400	0.27	0.13	−0.10	E	2
			[49.5]			3	5	2000	0.12	0.060	−0.43	E	2
			[49.427]	6092400	8115600	3	3	4000	0.15	0.071	−0.36	D	2
			[49.7]			3	1	510	0.0063	0.0031	−1.72	E	2
39.		^{3}D – ^{3}F°	50.26	6133600	8124000	7	9	6000	0.29	0.34	0.31	D	2
40.		^{3}P – ^{3}D°	[51.177]	6199900	8153900	1	3	2400	0.28	0.048	−0.55	D	2
41.		^{1}S – ^{3}D°	[55.528]	6353000	8153900	1	3	790	0.11	0.020	−0.96	E	2
42.		^{1}S – ^{1}P°	[52.748]	6353000	8248800	1	3	2700	0.33	0.058	−0.48	D	2
43.	$2p^53d$–$2p^54f$	^{3}F° – ^{3}G	58.76	6485700	8188000	9	11	1.2(+4)	0.78	1.4	0.85	D	4
44.	$2p^53d$–$2p^55f$	^{3}F° – ^{3}G	41.37	6485700	8903000	9	11	4800	0.15	0.18	0.13	D	4

^aThe number in parentheses following the tabulated value indicates the power of ten by which this value has to be multiplied.

Fe XVII

Forbidden Transitions

A-values were calculated by Bhatia et al.[1] for numerous forbidden transitions within the $2p^5 3s$, $2p^5 3p$, and $2p^5 3d$ configurations, as well as for lines of the $2p^5 3s$–$2p^5 3d$ and $2p^6$–$2p^5 3p$ arrays. Their calculations employed scaled Thomas-Fermi wavefunctions with limited allowance for configuration interaction. A number of these data are quoted here, but the A-values were first converted to line strengths, which were then reconverted to transition probabilities by using more accurate wavelengths. Bhatia et al. did not indicate which of their results were due to magnetic dipole, and which were due to electric quadrupole, radiation. In those cases where it was impossible to make a definitive determination of the type of radiation solely on the basis of selection rules, the A-value could not be converted to a line strength. It appears, however, that the A-values quoted for the three transitions within the $2p^5 3s$ configuration for which the type of transition is not indicated in our tabulation are due to magnetic dipole radiation.

An A-value was reported in Ref. 1 for the magnetic quadrupole transition from the ground state to the $J=2$ level of the $2p\,3s$ configuration. We chose, however, to quote the result of Loulergue and Nussbaumer,[2] who used a method similar to that of Bhatia et al. but allowed for more extensive configuration mixing. The A-value tabulated for the magnetic quadrupole resonance transition to one of the $J=2$ levels of the $2p^5 3d$ configuration is the result of the relativistic Dirac-Fock calculations of Fielder et al.,[3] who did not allow for configuration interaction.

Transitions involving levels of the $2p^5 3p$ and $2p^5 3d$ configurations which are indicated by Jupen and Litzen[4] to be of low to moderate purity in LS coupling are excluded here, as are very weak lines.

References

[1] A. K. Bhatia, U. Feldman, and J. F. Seely, At. Data Nucl. Data Tables **32**, 435 (1985).

[2] M. Loulergue and H. Nussbaumer, Astron. Astrophys. **45**, 125 (1975).

[3] W. Fielder, Jr., D. L. Lin, and D. Ton-That, Phys. Rev. A **19**, 741 (1979).

[4] C. Jupen and U. Litzen, Phys. Scr. **30**, 112 (1984).

Fe XVII: Forbidden transitions

No.	Transition Array	Multiplet	λ (Å)	E_i (cm^{-1})	E_k (cm^{-1})	g_i	g_k	Type of transition	A_{ki} (s^{-1})	S (at. u.)	Accuracy	Source
1.	$2p^5(^2P^\circ_{3/2})3s$– $2p^5(^2P^\circ_{3/2})3s$	$(^3/_2,^1/_2)^\circ$ – $(^3/_2,^1/_2)^\circ$										
			[6530]	5848400	5863700	5	3		36.9		C−	1
2.	$2p^5(^2P^\circ_{3/2})3s$– $2p^5(^2P^\circ_{1/2})3s$	$(^3/_2,^1/_2)^\circ$ – $(^1/_2,^1/_2)^\circ$										
			[892.1]	5848400	5960500	5	3		1.67(+4)[a]		C−	1
			[1150]	5863700	5950600	3	1	M1	1.6(+4)	0.91	C−	1
			[1030]	5863700	5960500	3	3		2900		D+	1
3.	$2p^5 3d$–$2p^5 3d$	$^3P^\circ$ – $^3P^\circ$										
			[4440]	6462900	6485400	1	5	E2	0.0035	0.018	E	1
4.	$2p^6$– $2p^5(^2P^\circ_{3/2})3s$	^1S – $(^3/_2,^1/_2)^\circ$										
			17.100	0	5848400	1	5	M2	2.0(+5)	0.22	D+	2
5.	$2p^6$–$2p^5 3d$	^1S – $^3P^\circ$										
			[15.419]	0	6485400	1	5	M2	6.3(+6)	4.1	D	3
6.	$2p^5(^2P^\circ_{3/2})3s$– $2p^5 3d$	$(^3/_2,^1/_2)^\circ$ – $^3P^\circ$										
			[160.2]	5848400	6472500	5	3		3.7(+5)		E	1
			[160.8]	5863700	6485400	3	5		8.3(+4)		E	1
			[162.7]	5848400	6462900	5	1	E2	3.5(+5)	0.024	E	1

Fe xvii: Forbidden transitions — Continued

No.	Transition Array	Multiplet	λ (Å)	E_i (cm^{-1})	E_k (cm^{-1})	g_i	g_k	Type of transition	A_{ki} (s^{-1})	S (at. u.)	Accu-racy	Source
7.		$(^3/_2,^1/_2)^\circ - {}^3F^\circ$	[156.9]	5848400	6485700	5	9	E2	4.1(+5)	0.21	D−	1

^{a}The number in parentheses following the tabulated value indicates the power of ten by which this value has to be multiplied.

Fe xviii

F Isoelectronic Sequence

Ground State: $1s^2 2s^2 2p^5\ {}^2P^\circ_{3/2}$

Ionization Energy: $1362\ eV = 10985000\ cm^{-1}$

Allowed Transitions

List of tabulated lines

Wavelength (Å)	No.	Wavelength (Å)	No.	Wavelength (Å)	No.	Wavelength (Å)	No.
13.919	10	14.255	7	15.258	5	15.870	2
13.954	10	14.361	9	15.491	5	16.024	3
14.121	10	14.419	8	15.623	4	16.073	2
14.150	9	14.467	7	15.764	3	93.93	1
14.209	8	14.49	6	15.869	4	103.95	1

Oscillator strengths for lines of the multiplet $2s^2 2p^5\ {}^2P^\circ - 2s 2p^6\ {}^2S$ are the results of the Dirac-Fock calculations of Cheng et al.,[1] which included a perturbative treatment of the Breit interaction and the Lamb shift.

For lines of the arrays $2p^5 - 2p^4 3s$ and $2p^5 - 2p^4 3d$, we quote the f-values calculated by Fawcett[2] using Cowan's Hartree-Fock-Relativistic (HFR) method and incorporating scaling of energy parameters on the basis of a least-squares fit to observed energies. Fawcett's calculations included fairly extensive allowance for configuration mixing in both odd- and even-parity states. Transitions involving levels which are indicated by Fawcett to be of low to moderate purity in LS coupling are excluded from this compilation, as are lines characterized by very small f-values.

The ratio of A-values for the two resonance lines out of the $2s 2p^6\ {}^2S_{1/2}$ level as given in Ref. 1 is in reasonably good agreement with the result of Stratton et al.[3] derived from relative-intensity measurements.

The lifetime of the $2s 2p^6\ {}^2S_{1/2}$ level has been measured by Buchet et. al.[4] with the beam-foil technique and was found to be 12.2 ns with an estimated error of ±0.8 ns. This value is about 50% larger than the lifetime of 8.1 ns that may be derived from the tabulated theoretical data, which we estimate as being of "C+" accuracy. Thus the two results disagree by about 15% outside their respective estimated error ranges.

References

[1]K. T. Cheng, Y.-K. Kim, and J. P. Desclaux, At. Data Nucl. Data Tables **24**, 111 (1979).

[2]B. C. Fawcett, At. Data Nucl. Data Tables **31**, 495 (1984).

[3]B. C. Stratton, H. W. Moos, S. Suckewer, U. Feldman, J. F. Seely, and A. K. Bhatia, Phys. Rev. A **31**, 2534 (1985).

[4]J. P. Buchet, M. C. Buchet-Poulizac, A. Denis, J. Desesquelles, and M. Druetta, Phys. Rev. A **22**, 2061 (1980).

Fe XVIII: Allowed transitions

No.	Transition Array	Multiplet	λ (Å)	E_i (cm^{-1})	E_k (cm^{-1})	g_i	g_k	A_{ki} (10^8 s^{-1})	f_{ik}	S (at. u.)	log gf	Accuracy	Source
1.	$2s^2 2p^5 - 2s\,2p^6$	^{2}P° − ^{2}S	97.051	34190	1064580	6	2	1240	0.0584	0.112	−0.455	C+	1
			93.93	0	1064580	4	2	913	0.0604	0.0747	−0.617	C+	1
			103.95	102580	1064580	2	2	331	0.0537	0.0368	−0.969	C+	1
2.	$2p^5 - 2p^4(^3$P$)3s$	^{2}P° − ^{4}P											
			16.073	0	6221600	4	6	910	0.0053	0.0011	−1.67	E	2
			15.870	0	6301200	4	2	2000	0.0038	7.9(−4)ᵃ	−1.82	E	2
3.		^{2}P° − ^{2}P											
			16.024	102580	6343600	2	2	1.5(+4)	0.059	0.0062	−0.93	D	2
			15.764	0	6343600	4	2	1.4(+4)	0.026	0.0054	−0.98	D	2
4.	$2p^5 - 2p^4(^1$D$)3s$	^{2}P° − ^{2}D											
			15.623	0	6400800	4	6	1.1(+4)	0.062	0.013	−0.61	D	2
			15.869	102580	6404200	2	4	1.3(+4)	0.10	0.010	−0.70	D	2
5.	$2p^5 - 2p^4(^1$S$)3s$	^{2}P° − ^{2}S	15.328	34190	6558000	6	2	1.4(+4)	0.017	0.0050	−1.00	E	2
			15.258	0	6558000	4	2	2800	0.0048	9.6(−4)	−1.72	E	2
			15.491	102580	6558000	2	2	1.1(+4)	0.039	0.0040	−1.11	D	2
6.	$2p^5 - 2p^4(^3$P$)3d$	^{2}P° − ^{4}F											
			[14.49]			4	4	7600	0.024	0.0046	−1.02	E	2
7.	$2p^5 - 2p^4(^1$D$)3d$	^{2}P° − ^{2}S	14.325	34190	7014900	6	2	1.8(+5)	0.19	0.053	0.05	D	2
			14.255	0	7014900	4	2	1.6(+5)	0.24	0.045	−0.02	D	2
			14.467	102580	7014900	2	2	2.7(+4)	0.084	0.0080	−0.77	D	2
8.		^{2}P° − ^{2}P											
			[14.209]	0	7037900	4	4	1.9(+5)	0.59	0.11	0.37	E	2
			14.419	102580	7037900	2	4	3.2(+4)	0.20	0.019	−0.40	E	2
9.		^{2}P° − ^{2}D											
			14.361	102580	7067100	2	4	1.5(+5)	0.92	0.087	0.26	E	2
			14.150	0	7067100	4	4	4.3(+4)	0.13	0.024	−0.28	E	2
10.	$2p^5 - 2p^4(^1$S$)3d$	^{2}P° − ^{2}D	14.007	34190	7173600	6	10	6.9(+4)	0.34	0.093	0.30	E	2
			13.954	0	7166400	4	6	1.1(+4)	0.050	0.0092	−0.70	D	2
			14.121	102580	7184300	2	4	1.5(+5)	0.89	0.083	0.25	D	2
			[13.919]	0	7184300	4	4	960	0.0028	5.1(−4)	−1.95	E	2

ᵃThe number in parentheses following the tabulated value indicates the power of ten by which this value has to be multiplied.

Fe XVIII

Forbidden Transitions

Line strengths for the magnetic dipole and electric quadrupole contributions to the transition between the two levels of the $2p^5$ configuration are the results of the Dirac-Fock calculations of Cheng et al.[1] These relativistic calculations included a perturbative treatment of the Breit interaction and the Lamb shift. The strength of the electric quadrupole transition as defined in Ref. 1 was multiplied by the factor $2/3$ which is needed to bring this value into conformance with the definition of quadrupole strengths used in the NBS tables.

Reference

[1] K. T. Cheng, Y.-K. Kim, and J. P. Desclaux, At. Data Nucl. Data Tables **24**, 111 (1979).

Fe XVIII: Forbidden transitions

No.	Transition Array	Multiplet	λ (Å)	E_i (cm^{-1})	E_k (cm^{-1})	g_i	g_k	Type of transition	A_{ki} (s^{-1})	S (at. u.)	Accuracy	Source
1.	$2p^5$–$2p^5$	$^2P°$ – $^2P°$										
			974.86	0	102580	4	2	M1	1.94(+4)[a]	1.33	B	1
			"	"	"	4	2	E2	1.9	0.0020	D	1

[a]The number in parentheses following the tabulated value indicates the power of ten by which this value has to be multiplied.

Fe XIX

O Isoelectronic Sequence

Ground State: $1s^2 2s^2 2p^4\ ^3P_2$

Ionization Energy: $1469\ \text{eV} = 11850000\ \text{cm}^{-1}$

Allowed Transitions

List of tabulated lines

Wavelength (Å)	No.	Wavelength (Å)	No.	Wavelength (Å)	No.	Wavelength (Å)	No.
13.271	29	13.796	20	14.929	11	91.02	4
13.38	27	13.83	21	14.966	10	101.55	1
13.413	31	13.836	25	14.995	14	106.12	6
13.426	24	13.934	20	15.015	18	106.33	1
13.47	28	13.94	19	15.040	13	108.37	1
13.520	22	13.961	20	15.111	13	109.97	1
13.56	27	14.534	15	15.138	10	111.70	1
13.67	21	14.604	15	15.172	10	115.42	8
13.68	23	14.625	12	15.2	9	120.00	1
13.69	30	14.668	11,15,16	15.4	9	132.63	3
13.700	32	14.671	17	78.90	2	151.61	5
13.71	30	14.735	11	83.89	2		
13.738	26	14.806	16	84.89	2		
13.789	20	14.833	12	87.02	7		

The tabulated oscillator strengths for transitions of the arrays $2s^2 2p^4 - 2s 2p^5$ and $2s 2p^5 - 2p^6$ are the results of the multiconfiguration Dirac-Fock (MCDF) calculations of Cheng et al.[1] These relativistic calculations included a perturbative treatment of the Breit interaction and the Lamb shift. Allowance for configuration mixing was limited to the $n = 2$ complex. The results should be quite accurate, except in the case of weak lines. (The $2s^2 2p^4 \, ^1D_2 - 2s 2p^5 \, ^3P_1^o$ transition has been omitted from this tabulation, because its f-value as reported in Ref. 1 is extremely small, and thus very uncertain.)

Transition probabilities for lines of the $2s^2 2p^4 - 2s 2p^5$ array were calculated by Froese Fischer and Saha[2] using the multiconfiguration Hartree-Fock (MCHF) method with Breit-Pauli corrections. Their basis set included many configurations outside the $n = 2$ complex, but relativistic effects were not treated to the same degree as in Ref. 1. Line strengths derived from these two sources are in reasonably good agreement, particularly for the stronger transitions.

A few experimental data are available for this ion. The lifetime of the $2s 2p^5 \, ^3P_2^o$ level was measured by Buchet et al.[3] using the beam-foil technique. The reciprocal of the sum of the probabilities of all downward transitions from this level, derived from the f-values presented in Ref. 1 and assuming the accuracy estimates given in this compilation, lies within the quoted uncertainty of the experimentally determined lifetime. Stratton et al.[4] measured ratios of transition probabilities for two pairs of transitions, one of these pairs originating from the $2s 2p^5 \, ^3P_2^o$

level and the other from the $2s 2p^5 \, ^3P_1^o$ level. The former agrees fairly well with the theoretical data of Cheng et al.; the latter is nearly a factor of two larger than theory, but it is claimed to be rather uncertain.

A-values for lines of the $2p^4 \, ^3P - 2p^3(^4S^o)3s \, ^5S^o$ multiplet are taken from the scaled Thomas-Fermi approach of Kastner et al.[5] with configuration interaction and relativistic effects. For all other lines of the $2p^4 - 2p^3 3s$ array and for lines of the $2p^4 - 2p^3 3d$ array, we quote the f-values calculated by Fawcett[6] using Cowan's Hartree-Fock-Relativistic (HFR) method and incorporating scaling of energy parameters on the basis of a least-squares fit to observed energies. Fawcett's calculations included fairly extensive allowance for configuration mixing in both odd- and even-parity states. The weakest lines were not reported and thus are not tabulated here. Transitions involving levels which are indicated by Fawcett to be of low to moderate purity in LS coupling are excluded from this compilation.

References

[1] K. T. Cheng, Y.-K. Kim, and J. P. Desclaux, At. Data Nucl. Data Tables 24, 111 (1979).
[2] C. Froese Fischer and H. P. Saha, J. Phys. B 17, 943 (1984).
[3] J. P. Buchet, M. C. Buchet-Poulizac, A. Denis, J. Desesquelles, and M. Druetta, Phys. Rev. A 22, 2061 (1980).
[4] B. C. Stratton, H. W. Moos, S. Suckewer, U. Feldman, J. F. Seely, and A. K. Bhatia, Phys. Rev. A 31, 2534 (1985).
[5] S. O. Kastner, A. K. Bhatia, and L. Cohen, Phys. Scr. 15, 259 (1977).
[6] B. C. Fawcett, At. Data Nucl. Data Tables 34, 215 (1986).

Fe XIX: Allowed transitions

No.	Transition Array	Multiplet	λ (Å)	E_i (cm^{-1})	E_k (cm^{-1})	g_i	g_k	A_{ki} (10^8 s^{-1})	f_{ik}	S (at. u.)	log gf	Accuracy	Source
1.	$2s^2 2p^4 - 2s 2p^5$	$^3P - ^3P^o$	109.04	38180	955310	9	9	540	0.096	0.31	−0.06	C	1
			108.37	0	922760	5	5	390	0.068	0.12	−0.47	C	1
			111.70	89430	984690	3	3	126	0.0235	0.0259	−1.152	C	1
			101.55	0	984690	5	3	317	0.0294	0.0491	−0.83	C	1
			106.33	89430	1029900	3	1	610	0.0342	0.0359	−0.99	C	1
			120.00	89430	922760	3	5	104	0.0374	0.0443	−0.95	C	1
			109.97	75350	984690	1	3	160	0.087	0.031	−1.06	C	1
2.		$^3P - ^1P^o$											
			78.90	0	1267430	5	3	130	0.0071	0.0092	−1.45	E	1
			84.89	89430	1267430	3	3	9.3	0.0010	8.4(−4)[a]	−2.52	E	1
			83.89	75350	1267430	1	3	16	0.0050	0.0014	−2.30	E	1
3.		$^1D - ^3P^o$											
			132.63	168770	922760	5	5	22	0.0059	0.013	−1.53	E	1
4.		$^1D - ^1P^o$	91.02	168770	1267430	5	3	1490	0.111	0.166	−0.256	C	1
5.		$^1S - ^3P^o$											
			[151.61]	325100	984690	1	3	7.9	0.0082	0.0041	−2.09	E	1
6.		$^1S - ^1P^o$	106.12	325100	1267430	1	3	110	0.054	0.019	−1.27	C	1

Fe XIX: Allowed transitions — Continued

No.	Transition Array	Multiplet	λ (Å)	E_i (cm^{-1})	E_k (cm^{-1})	g_i	g_k	A_{ki} (10^8 s^{-1})	f_{ik}	S (at. u.)	log gf	Accuracy	Source
7.	$2s2p^5$–$2p^6$	^{3}P° – ^{1}S											
			87.02	984690	2133830	3	1	120	0.0045	0.0039	−1.87	E	1
8.		^{1}P° – ^{1}S	115.42	1267430	2133830	3	1	1610	0.107	0.122	−0.493	C	1
9.	$2p^4$–$2p^3(^4$S°$)3s$	^{3}P – ^{5}S°											
			[15.2]			5	5	450	0.0016	3.9(−4)	−2.11	E	5
			[15.4]			3	5	17	1.0(−4)	1.5(−5)	−3.52	E	5
10.	$2p^4$–$2p^3(^4$S°$)3d$	^{3}P – ^{3}S°	*15.052*	*38180*	*6681800*	9	3	3.8(+4)	0.043	0.019	−0.42	C−	6
			14.966	0	6681800	5	3	2.5(+4)	0.051	0.013	−0.59	C−	6
			15.172	89430	6681800	3	3	6700	0.023	0.0034	−1.16	C−	6
			15.138	75350	6681800	1	3	5100	0.053	0.0026	−1.28	C−	6
11.	$2p^4$ – $2p^3(^2$D°$)3d$	^{3}P – ^{3}D°											
			14.668	0	6817600	5	7	1.1(+4)	0.051	0.012	−0.59	C	6
			14.929	89430	6786600	3	5	2500	0.014	0.0021	−1.38	D	6
			14.735	0	6786600	5	5	9800	0.032	0.0078	−0.80	D	6
			14.929	89430	6787800	3	3	1.2(+4)	0.041	0.0060	−0.41	D	6
12.		^{3}P – ^{1}D°											
			[14.625]	0	6837700	5	5	1400	0.0044	0.0011	−1.66	E	6
			14.663	89430	6837700	3	5	2700	0.0015	0.0022	−1.35	E	6
13.	$2p^4$ – $2p^3(^2$D°$)3s$	^{1}D – ^{3}D°											
			[15.589]	168770	6817600	5	7	1100	0.0050	0.0011	−1.66	E	6
			15.663	168770	6837700	5	5	1300	0.0046	0.0022	−1.35	E	6
14.		^{1}D – ^{1}D°	14.995	168770	6837700	5	5	2.2(+4)	0.074	0.018	−0.43	D	6
15.	$2p^4$ – $2p^3(^2$P°$)3s$	^{1}D – ^{3}P°											
			16.668	89430	6907000	3	1	1.1(+4)	0.012	0.0017	−1.44	C	6
			16.534	89430	6969800	3	5	6800	0.036	0.0052	−0.97	D	6
			[16.604]	75350	6922800	1	3	7500	0.072	0.0035	−1.14	D	6
16.		^{1}D – ^{3}P°											
			14.668	168770	6969800	5	3	1800	0.022	0.0053	−0.96	E	6
			14.806	168770	6969800	5	3	5600	0.011	0.0027	−1.26	E	6
17.		^{1}D – ^{1}P°	[14.671]	168770	6985100	5	3	1.1(+4)	0.021	0.0051	−0.91	D	6
18.		^{3}S – ^{1}P°	15.015	325100	6985100	1	3	1.4(+4)	0.14	0.0069	−0.85	D	6
19.	$2p^4$ – $2p^3(^4$S°$)3d$	^{3}P – ^{5}D°											
			[13.94]			5	5	2600	0.0076	0.0017	−1.42	E	6
			[13.94]			5	3	2700	0.0011	0.0011	−1.62	E	6
20.		^{3}P – ^{3}D°											
			13.796	0	7248500	5	1	7.0(+4)	0.028	0.0064	0.15	D	6
			13.934	75350	7252100	1	5	4.51(+4)	0.394	0.0181	−0.405	C−	6
			[13.961]	89430	7252100	3	3	2.0(+4)	0.058	0.0080	−0.76	C−	6
			[13.789]	0	7252100	5	3	2800	0.0048	0.0011	−1.62	D	6

J. Phys. Chem. Ref. Data, Vol. 17, Suppl. 4, 1988

Fe XIX: Allowed transitions — Continued

No.	Transition Array	Multiplet	λ (Å)	E_i (cm^{-1})	E_k (cm^{-1})	g_i	g_k	A_{ki} (10^8 s^{-1})	f_{ik}	S (at. u.)	log gf	Accuracy	Source
21.	$2p^4 - 2p^3(^2D°)3d$	$^3P - {}^3F°$											
			[13.83]			3	5	5000	0.024	0.0033	−1.14	E	6
			[13.83]			5	5	1.4(+4)	0.039	0.0088	−0.71	E	6
22.	$2p^4 - 2p^3(^2D°)3d$	$^3P - {}^3D°$											
			13.520	0	7396400	5	7	2.0(+5)	0.76	0.17	0.58	D	6
23.		$^3P - {}^3P°$											
			[13.68]			3	1	8.0(+4)	0.075	0.010	−0.65	D	6
24.		$^3P - {}^1F°$											
			13.426	0	7747900	5	7	4.8(+4)	0.18	0.040	−0.05	E	6
25.		$^1D - {}^3D°$											
			[13.836]	168770	7396400	5	7	3700	0.015	0.0034	−1.12	E	6
26.	$2p^4 - 2p^3(^2D°)3d$	$^1D - {}^1F°$	13.738	168770	7447900	5	7	1.0(+4)	0.40	0.090	0.30	D	6
27.	$2p^4 - 2p^3(^2P°)3d$	$^3P - {}^3F°$											
			[13.38]			5	7	3200	0.012	0.0026	−1.22	E	6
			[13.56]			3	5	1.0(+4)	0.046	0.0062	−0.86	E	6
28.		$^3P - {}^3P°$											
			[13.47]			3	1	1.5(+5)	0.14	0.019	−0.38	D	6
29.		$^3P - {}^1P°$											
			[13.271]	89430	7624400	3	3	8700	0.023	0.0030	−1.16	E	6
30.		$^1D - {}^3F°$											
			[13.69]			5	7	2.3(+4)	0.091	0.021	−0.34	E	6
			[13.71]			5	5	2.2(+4)	0.063	0.014	−0.50	E	6
31.		$^1D - {}^1P°$	[13.413]	168770	7624400	5	3	1.3(+4)	0.021	0.0046	−0.98	D	6
32.	$2p^4 - 2p^3(^2D°)3d$	$^1S - {}^1P°$	13.700	325100	7624400	1	3	2.7(+5)	2.3	0.10	0.36	D	6

[a]The number in parenthesis following the tabulated value indicates the power of ten by which this value has to be multiplied.

Fe XIX

Forbidden Transitions

Line strengths tabulated for magnetic dipole and electric quadrupole transitions within the $2p^4$ configuration are the results of the multiconfiguration Dirac-Fock (MCDF) calculations of Cheng et al.[1] These relativistic calculations included a perturbative treatment of the Breit interaction and the Lamb shift. Allowance for configuration mixing was limited to the $n=2$ complex. Strengths of electric quadrupole transitions as defined in Ref. 1 were multiplied by the factor $2/3$ which is needed to bring these values into conformance with the definition of quadrupole strengths used in the NBS tables.

Transition probabilities for these same lines were calculated by Froese Fischer and Saha[2] using the multiconfiguration Hartree-Fock (MCHF) method with Breit-Pauli corrections. Their basis included many configurations outside the $n=2$ complex, but relativistic effects were not treated to the same degree as in Ref. 1. Line strengths derived from these data are in quite good agreement with the data of Cheng et al. For this ion of the oxygen isoelectronic sequence, correlation effects due to mixing with configurations outside the complex were found by Froese Fischer and Saha to be rather small, as shown by a comparison of the results of their calculations employing an extensive basis to those derived by the same technique but limited to configurations within the $n=2$ complex.

A-values tabulated for forbidden transitions within the $2s2p^5$ configuration, and for transitions of the $2s^22p^4-2p^6$ array, were calculated by Loulergue et al.[3] using scaled Thomas-Fermi wavefunctions.

The weakest lines are excluded from this compilation, as their transition probabilities are considered to be very uncertain. (This applies to all lines of the $2s^22p^4-2p^6$ array.)

References

[1] K. T. Cheng, Y.-K Kim, and J. P. Desclaux, At. Data Nucl. Data Tables 24, 111 (1979).
[2] C. Froese Fischer and H. P. Saha, Phys. Rev. A 28, 3169 (1983).
[3] M. Loulergue, H. E. Mason, H. Nussbaumer, and P. J. Storey, Astron. Astrophys. 150, 246 (1985).

Fe XIX: Forbidden transitions

No.	Transition Array	Multiplet	λ (Å)	E_i (cm^{-1})	E_k (cm^{-1})	g_i	g_k	Type of transition	A_{ki} (s^{-1})	S (at. u.)	Accuracy	Source
1.	$2p^4-2p^4$	$^3P - ^3P$										
			1118.06	0	89430	5	3	M1	1.45(+4)[a]	2.25	C	1
			"	"	"	5	3	E2	0.61	0.0019	E	1
			[7100]	75350	89430	1	3	M1	40	1.6	C	1
			[1327]	0	75350	5	1	E2	0.49	0.0012	E	1
2.		$^3P - ^1D$										
			592.234	0	168770	5	5	M1	1.7(+4)	0.67	D	1
			"	"	"	5	5	E2	6.0	0.0013	E	1
			[1260]	89430	168770	3	5	M1	670	0.25	D	1
3.		$^3P - ^1S$										
			424.26	89430	325100	3	1	M1	1.5(+5)	0.42	D	1
4.		$^1D - ^1S$	[639.67]	168770	325100	5	1	E2	49	0:0031	E	1
5.	$2s2p^5-2s2p^5$	$^3P° - ^3P°$										
			[161.5]	922760	984690	5	3	M1	5200	2.4	C	3
			[2211]	984690	1029900	3	1	M1	4820	1.93	C	3
6.		$^3P° - ^1P°$										
			[290.13]	922760	1267430	5	3	M1	2.9(+4)	0.079	D−	3
			[353.68]	984690	1267430	3	3	M1	9400	0.046	D−	3
			[421.00]	1029900	1267430	1	3	M1	7700	0.064	D−	3

[a]The number in parentheses following the tabulated value indicates the power of ten by which this value has to be multiplied.

Fe xx

N Isoelectronic Sequence

Ground State: $1s^2 2s^2 2p^3\ {}^4S^\circ_{3/2}$

Ionization Energy: $1582.0\ \text{eV} = 12708000\ \text{cm}^{-1}$

Allowed Transitions

List of tabulated lines

Wavelength (Å)	No.	Wavelength (Å)	No.	Wavelength (Å)	No.	Wavelength (Å)	No.
12.51	37,38	13.13	35	14.14	20	111.60	14
12.59	36	13.14	30	14.18	18	113.34	6
12.60	45	13.22	29	14.23	20	114.72	11
12.61	45	13.23	35	14.26	20	115.36	15
12.66	45	13.24	35	80.51	4	115.42	6
12.67	45	13.28	29	80.59	13	118.66	1
12.69	28	13.35	34	82.035	13	121.83	1
12.70	28	13.46	21,34	83.23	13	122.00	16
12.73	40,41	13.49	33	83.24	8	127.86	10
12.77	27	13.70	17,22	83.69	3	131.70	15
12.78	40,46	13.71	24	88.24	13	132.85	1
12.79	40	13.72	22	89.976	13	136.06	10
12.82	39	13.77	22	90.60	8	138.49	16
12.88	39	13.78	17	92.63	12	139.08	10
12.89	46	13.79	22	93.78	8	140.44	16
12.90	26,46	13.83	24	94.64	7	141.95	5
12.92	43	13.90	17	95.95	2	146.51	5
12.93	32,44	13.91	19	98.09	14	155.04	5
12.98	43	13.92	23	98.38	12	162.74	5,16
12.99	32	13.98	19	101.83	12	171.63	9
13.00	32	13.99	19	106.98	11	173.33	5
13.01	42	14.04	18	108.71	6	200.95	9
13.03	25,43,44	14.05	18,23	108.83	12	232.80	9
13.07	31	14.06	23	109.66	14		
13.12	42	14.13	18	110.63	6		

The tabulated oscillator strengths for transitions of the arrays $2s^2 2p^3$–$2s 2p^4$ and $2s 2p^4$–$2p^5$ are the results of the multiconfiguration Dirac-Fock (MCDF) calculations of Cheng et al.[1] These relativistic calculations included a perturbative treatment of the Breit interaction and the Lamb shift. The results should be quite accurate, except in the case of weak lines. (A few very weak lines have been omitted from this tabulation.)

The scaled Thomas-Fermi approach with configuration interaction and relativistic effects was used by Mason and Bhatia[2] to compute oscillator strengths for lines of the arrays $2p^3$–$2p^2 3s$ and $2p^3$–$2p^2 3d$. Their results for the stronger lines are quoted here, but transitions involving levels which are indicated to be less than 60% pure in LS coupling are omitted.

In addition, all transitions involving the $2p^2({}^3P)3d\ {}^2P_{1/2}$ and ${}^4P_{1/2}$ levels are omitted, since there is an inconsis-

tency in Ref. 2 between the designations of these levels and their dominant eigenvector components.

Oscillator strengths for a few lines of the array $2s 2p^4$–$2s 2p^3 3d$ are available from another source,[3] but they have not been tabulated here since no indication of the percentage compositions of the levels is provided.

References

[1] K. T. Cheng, Y.-K. Kim, and J. P. Desclaux, At. Data Nucl. Data Tables **24**, 111 (1979).

[2] H. E. Mason and A. K. Bhatia, Astron. Astrophys. Suppl. Ser. **52**, 181 (1983).

[3] G. E. Bromage, R. D. Cowan, B. C. Fawcett, H. Gordon, M. G. Hobby, N. J. Peacock, and A. Ridgeley, United Kingdom Atomic Energy Authority Report CLM-R170 (August 1977).

Fe xx: Allowed transitions

No.	Transition Array	Multiplet	λ (Å)	E_i (cm^{-1})	E_k (cm^{-1})	g_i	g_k	A_{ki} (10^8 s^{-1})	f_{ik}	S (at. u.)	log gf	Accuracy	Source
1.	$2s^2 2p^3 - 2s 2p^4$	$^4S° - ^4P$	126.51	0	790430	4	12	160	0.115	0.192	−0.336	C	1
			132.85	0	752730	4	6	130	0.052	0.091	−0.68	C	1
			121.83	0	820820	4	4	186	0.0413	0.066	−0.78	C	1
			118.66	0	842740	4	2	209	0.0221	0.0345	−1.054	C	1
2.		$^4S° - ^2D$											
			95.95	0	1042210	4	4	19	0.0026	0.0033	−1.98	E	1
3.		$^4S° - ^2S$											
			83.69	0	1194850	4	2	19	0.0010	0.0011	−2.40	E	1
4.		$^4S° - ^2P$											
			80.51	0	1242080	4	4	46	0.0045	0.0048	−1.74	E	1
5.		$^2D° - ^4P$											
			[173.33]	175810	752730	6	6	2.7	0.0012	0.0041	−2.14	E	1
			[146.51]	138270	820820	4	4	1.3	4.2(−4)[a]	8.1(−4)	−2.77	E	1
			[155.04]	175810	820820	6	4	0.46	1.1(−4)	3.4(−4)	−3.18	E	1
			[141.95]	138270	842740	4	2	3.4	5.1(−4)	9.5(−4)	−2.69	E	1
			[162.74]	138270	752730	4	6	6.4	0.0038	0.0081	−1.82	E	1
6.		$^2D° - ^2D$	112.24	160790	1051760	10	10	360	0.068	0.25	−0.17	C−	1
			113.34	175810	1058130	6	6	330	0.063	0.14	−0.42	C	1
			110.63	138270	1042210	4	4	430	0.078	0.11	−0.51	C	1
			[115.42]	175810	1042210	6	4	0.43	5.7(−5)	1.3(−4)	−3.47	E	1
			[108.71]	138270	1058130	4	6	0.27	7.1(−5)	1.0(−4)	−3.55	E	1
7.		$^2D° - ^2S$											
			94.64	138270	1194850	4	2	450	0.030	0.037	−0.92	E	1
8.		$^2D° - ^2P$	89.781	160790	1274610	10	6	930	0.068	0.20	−0.17	C	1
			93.78	175810	1242080	6	4	1000	0.089	0.16	−0.27	C	1
			83.24	138270	1339680	4	2	291	0.0151	0.0166	−1.219	C	1
			90.60	138270	1242080	4	4	147	0.0181	0.0216	−1.140	C	1
9.		$^2P° - ^4P$											
			[232.80]	323180	752730	4	6	0.27	3.3(−4)	0.0010	−2.88	E	1
			[200.95]	323180	820820	4	4	1.8	0.0011	0.0029	−2.36	E	1
			[171.63]	260090	842740	2	2	2.5	0.0011	0.0012	−2.66	E	1
10.		$^2P° - ^2D$	133.40	302150	1051760	6	10	52	0.023	0.061	−0.86	C−	1
			136.06	323180	1058130	4	6	60	0.0250	0.0448	−1.000	C	1
			127.86	260090	1042210	2	4	29.8	0.0146	0.0123	−1.53	C	1
			[139.08]	323180	1042210	4	4	6.9	0.0020	0.0037	−2.10	D	1
11.		$^2P° - ^2S$	112.02	302150	1194850	6	2	360	0.023	0.050	−0.87	C−	1
			114.72	323180	1194850	4	2	30	0.0030	0.0045	−1.92	D	1
			106.98	260090	1194850	2	2	370	0.064	0.045	−0.89	C	1

Fe xx: Allowed transitions — Continued

No.	Transition Array	Multiplet	λ (Å)	E_i (cm^{-1})	E_k (cm^{-1})	g_i	g_k	A_{ki} (10^8 s^{-1})	f_{ik}	S (at. u.)	log gf	Accuracy	Source
12.		$^2P°$ – ^2P	*102.83*	*302150*	*1274610*	6	6	425	0.067	0.137	−0.393	C−	1
			108.83	323180	1242080	4	4	94	0.0167	0.0239	−1.175	C	1
			92.63	260090	1339680	2	2	44	0.0057	0.0035	−1.94	D	1
			98.38	323180	1339680	4	2	960	0.070	0.091	−0.55	C	1
			101.83	260090	1242080	2	4	91	0.0284	0.0190	−1.246	C	1
13.	$2s2p^4$–$2p^5$	^4P – $^2P°$											
			83.23	752730	1954150	6	4	30	0.0021	0.0035	−1.90	E	1
			80.59	820820	2061730	4	2	2.7	1.3(−4)	1.4(−4)	−3.28	E	1
			88.24	820820	1954150	4	4	16	0.0019	0.0022	−2.12	E	1
			[82.035]	842740	2061730	2	2	9.6	9.7(−4)	5.2(−4)	−2.71	E	1
			[89.976]	842740	1954150	2	4	5.4	0.0013	7.7(−4)	−2.59	E	1
14.		^2D – $^2P°$	*106.58*	*1051760*	*1990010*	10	6	590	0.060	0.21	−0.22	C	1
			111.60	1058130	1954150	6	4	430	0.054	0.12	−0.49	C	1
			98.09	1042210	2061730	4	2	462	0.0333	0.0430	−0.88	C	1
			109.66	1042210	1954150	4	4	176	0.0317	0.0458	−0.90	C	1
15.		^2S – $^2P°$	*125.76*	*1194850*	*1990010*	2	6	75	0.053	0.0442	−0.97	C−	1
			131.70	1194850	1954150	2	4	90	0.0469	0.0407	−1.028	C	1
			[115.36]	1194850	2061730	2	2	23	0.0046	0.0035	−2.04	D	1
16.		^2P – $^2P°$	*139.78*	*1274610*	*1990010*	6	6	420	0.12	0.34	−0.13	C	1
			140.44	1242080	1954150	4	4	310	0.092	0.17	−0.43	C	1
			138.49	1339680	2061730	2	2	320	0.093	0.085	−0.73	C	1
			122.00	1242080	2061730	4	2	370	0.0413	0.066	−0.78	C	1
			[162.74]	1339680	1954150	2	4	17.9	0.0142	0.0152	−1.55	C	1
17.	$2p^3$–$2p^2(^3P)3s$	$^4S°$ – ^4P	*13.76*			4	12	1.1(+4)	0.094	0.017	−0.43	D	2
			[13.70]			4	6	1.1(+4)	0.045	0.0081	−0.74	D	2
			[13.78]			4	4	1.0(+4)	0.029	0.0053	−0.94	D	2
			[13.90]			4	2	1.2(+4)	0.017	0.0031	−1.17	D	2
18.		$^2D°$ – ^4P											
			[14.04]			6	6	2300	0.0068	0.0019	−1.39	E	2
			[14.05]			4	4	880	0.0026	4.8(−4)	−1.98	E	2
			[14.13]			6	4	800	0.0016	4.5(−4)	−2.02	E	2
			[14.18]			4	2	1700	0.0026	4.9(−4)	−1.98	E	2
19.		$^2D°$ – ^2P	*13.98*			10	6	2.1(+4)	0.037	0.017	−0.43	E	2
			[13.98]			6	4	1.6(+4)	0.032	0.0088	−0.72	D	2
			[13.99]			4	2	2.2(+4)	0.033	0.0061	−0.88	D	2
			[13.91]			4	4	3300	0.0096	0.0018	−1.42	E	2
20.		$^2P°$ – ^2P											
			[14.26]			4	4	2500	0.0075	0.0014	−1.52	E	2
			[14.23]			2	2	6300	0.019	0.0018	−1.42	D	2
			[14.14]			2	4	5500	0.033	0.0031	−1.18	D	2
21.	$2p^3$–$2p^2(^1D)3s$	$^4S°$ – ^2D											
			[13.46]			4	6	390	0.0016	2.8(−4)	−2.19	E	2

Fe xx: Allowed transitions — Continued

No.	Transition Array	Multiplet	λ (Å)	E_i (cm⁻¹)	E_k (cm⁻¹)	g_i	g_k	A_{ki} (10⁸ s⁻¹)	f_{ik}	S (at. u.)	log gf	Accuracy	Source
22.		^{2}D° – ^{2}D	*13.75*			10	10	1.2(+4)	0.035	0.016	−0.45	E	2
			[13.79]			6	6	1.2(+4)	0.034	0.0093	−0.69	D	2
			[13.70]			4	4	4300	0.012	0.0022	−1.32	D	2
			[13.77]			6	4	1500	0.0028	7.6(−4)	−1.77	E	2
			[13.72]			4	6	5000	0.021	0.0038	−1.08	D	2
23.		^{2}P° – ^{2}D	*14.01*			6	10	8800	0.043	0.012	−0.58	E	2
			[14.06]			4	6	2500	0.011	0.0020	−1.36	D	2
			[13.92]			2	4	930	0.0054	4.9(−4)	−1.97	E	2
			[14.05]			4	4	1.7(+4)	0.049	0.0091	−0.71	D	2
24.	$2p^3$–$2p^2$(^{1}S)$3s$	^{2}P° – ^{2}S	*13.79*			6	2	1.9(+4)	0.018	0.0050	−0.96	D	2
			[13.83]			4	2	9800	0.014	0.0025	−1.25	D	2
			[13.71]			2	2	9900	0.028	0.0025	−1.25	D	2
25.	$2p^3$–$2p^2$(^{3}P)$3d$	^{4}S° – ^{4}F											
			[13.03]			4	4	6700	0.017	0.0029	−1.17	E	2
26.		^{4}S° – ^{4}D											
			[12.90]			4	2	6200	0.0077	0.0013	−1.51	E	2
27.		^{4}S° – ^{4}P											
			[12.77]			4	4	2.1(+5)	0.51	0.086	0.31	D	2
28.		^{4}S° – ^{2}D											
			[12.69]			4	6	1.2(+4)	0.042	0.0070	−0.77	E	2
			[12.70]			4	4	2100	0.0051	8.5(−4)	−1.69	E	2
29.		^{2}D° – ^{4}F											
			[13.22]			6	8	3100	0.011	0.0029	−1.18	E	2
			[13.28]			4	4	6100	0.016	0.0028	−1.19	E	2
30.		^{2}D° – ^{4}D											
			[13.14]			4	2	2400	0.0031	5.4(−4)	−1.91	E	2
31.		^{2}D° – ^{4}P											
			[13.07]			6	4	8200	0.014	0.0036	−1.08	E	2
32.		^{2}D° – ^{2}D											
			[12.99]			6	6	5.1(+4)	0.13	0.033	−0.11	D	2
			[13.00]			6	4	1.1(+4)	0.018	0.0046	−0.97	D	2
			[12.93]			4	6	1.6(+5)	0.62	0.11	0.39	D	2
33.		^{2}P° – ^{4}F											
			[13.49]			2	4	240	0.0013	1.2(−4)	−2.59	E	2
34.		^{2}P° – ^{4}D											
			[13.35]			2	2	4500	0.012	0.0011	−1.62	E	2
			[13.46]			4	2	2000	0.0027	4.8(−4)	−1.97	E	2

Fe xx: Allowed transitions — Continued

No.	Transition Array	Multiplet	λ (Å)	E_i (cm^{-1})	E_k (cm^{-1})	g_i	g_k	A_{ki} (10^8 s^{-1})	f_{ik}	S (at. u.)	log gf	Accuracy	Source
35.		^{2}P° – ^{2}D	*13.20*			6	10	4.1(+4)	0.18	0.047	0.03	E	2
			[13.23]			4	6	1600	0.0064	0.0011	−1.59	E	2
			[13.13]			2	4	8.9(+4)	0.46	0.040	−0.04	D	2
			[13.24]			4	4	1.2(+4)	0.032	0.0056	−0.89	D	2
36.	$2p^3$–$2p^2(^1$D)$3d$	^{4}S° – ^{2}D											
			[12.59]			4	4	1300	0.0031	5.1(−4)	−1.91	E	2
37.		^{4}S° – ^{2}P											
			[12.51]			4	4	1200	0.0028	4.6(−4)	−1.95	E	2
38.		^{4}S° – ^{2}S											
			[12.51]			4	2	2000	0.0023	3.8(−4)	−2.04	E	2
39.		^{2}D° – ^{2}D											
			[12.82]			4	4	1.1(+5)	0.28	0.047	0.05	D	2
			[12.88]			6	4	2.7(+4)	0.045	0.011	−0.57	D	2
40.		^{2}D° – ^{2}P	*12.78*			10	6	3.6(+4)	0.052	0.022	−0.28	E	2
			[12.79]			6	4	1.7(+4)	0.027	0.0068	−0.79	D	2
			[12.78]			4	2	6.9(+4)	0.085	0.014	−0.47	D	2
			[12.73]			4	4	3300	0.0080	0.0013	−1.49	E	2
41.		^{2}D° – ^{2}S											
			[12.73]			4	2	4.0(+4)	0.048	0.0080	−0.72	E	2
42.		^{2}P° – ^{2}D											
			[13.01]			2	4	3.0(+4)	0.15	0.013	−0.52	D	2
			[13.12]			4	4	4300	0.011	0.0019	−1.36	D	2
43.		^{2}P° – ^{2}P											
			[13.03]			4	4	1.4(+5)	0.36	0.062	0.16	D	2
			[12.98]			2	2	6.7(+4)	0.17	0.015	−0.47	D	2
			[12.92]			2	4	1.7(+4)	0.084	0.0071	−0.77	D	2
44.		^{2}P° – ^{2}S	*13.00*			6	2	1.0(+5)	0.086	0.022	−0.29	D	2
			[13.03]			4	2	8.6(+4)	0.11	0.019	−0.36	D	2
			[12.93]			2	2	1.2(+4)	0.030	0.0026	−1.22	D	2
45.	$2p^3$–$2p^2(^1$S)$3d$	^{2}D° – ^{2}D	*12.64*			10	10	7400	0.018	0.0074	−0.75	E	2
			[12.67]			6	6	1.0(+4)	0.025	0.0063	−0.82	D	2
			[12.60]			4	4	1100	0.0027	4.5(−4)	−1.97	E	2
			[12.66]			6	4	940	0.0015	3.8(−4)	−2.05	E	2
			[12.61]			4	6	420	0.0015	2.5(−4)	−2.22	E	2
46.		^{2}P° – ^{2}D	*12.86*			6	10	1.6(+5)	0.67	0.17	0.60	D	2
			[12.90]			4	6	1.4(+5)	0.54	0.092	0.33	D	2
			[12.78]			2	4	1.4(+5)	0.68	0.057	0.13	D	2
			[12.89]			4	4	4.4(+4)	0.11	0.019	−0.36	D	2

[a]The number in parentheses following the tabulated value indicates the power of ten by which this value has to be multiplied.

Fe xx

Forbidden Transitions

Line strengths tabulated for magnetic dipole and electric quadrupole transitions within the $2p^3$ configuration are the results of the multiconfiguration Dirac-Fock (MCDF) calculations of Cheng et al.[1] These relativistic calculations included a perturbative treatment of the Breit interaction and the Lamb shift. Allowance for configuration mixing was limited to the $n=2$ complex. Strengths of electric quadrupole transitions as defined in Ref. 1 were multiplied by the factor $^2/_3$ which is needed to bring these values into conformance with the definition of quadrupole strengths used in the NBS tables. The weakest lines are excluded from this compilation, as their strengths are considered to be very uncertain.

Transition probabilities for these same lines were calculated by Godefroid and Froese Fischer[2] using the multiconfiguration Hartree-Fock (MCHF) method with Breit-Pauli corrections. Their basis included many configurations outside the $n=2$ complex, but relativistic effects were not treated to the same degree as in Ref. 1. Line strengths derived from these data are in rather good agreement with the data of Cheng et al.

A-values for the M1 and E2 components of the single transition within the $2p^5$ configuration were obtained by applying Z-expansion formulas published by Oboladze and Safronova.[3] Their values for the magnetic dipole contribution to this line are in very good agreement with the results of the scaled Thomas-Fermi calculations of Bhatia et al.[4] and Bhatia[5] for nitrogenlike Ti and Mn, respectively. It is not clear whether Oboladze and Safronova incorporated configuration interaction into their calculations. Thus the A-value for the E2 contribution should be considered rather uncertain.

References

[1]K. T. Cheng, Y.-K. Kim, and J. P. Desclaux, At. Data Nucl. Data Tables **24**, 111 (1979).

[2]M. Godefroid and C. Froese Fischer, J. Phys. B **17**, 681 (1984).

[3]N. S. Oboladze and U. I. Safronova, Opt. Spectrosc. (USSR) **48**, 469 (1980).

[4]A. K. Bhatia, U. Feldman, and G. A. Doschek, J. Appl. Phys. **51**, 1464 (1980).

[5]A. K. Bhatia, J. Appl. Phys. **53**, 59 (1982).

Fe xx: Forbidden transitions

No.	Transition Array	Multiplet	λ (Å)	E_i (cm^{-1})	E_k (cm^{-1})	g_i	g_k	Type of transition	A_{ki} (s^{-1})	S (at. u.)	Accuracy	Source
1.	$2p^3$–$2p^3$	$^4S°$ – $^2D°$										
			567.76	0	175810	4	6	M1	1300	0.052	D−	1
			[723.22]	0	138270	4	4	M1	1.6(+4)^a	0.92	D	1
2.		$^4S°$ – $^2P°$										
			309.26	0	323180	4	4	M1	3.0(+4)	0.13	D	1
			[384.48]	0	260090	4	2	M1	3.3(+4)	0.14	D	1
3.		$^2D°$ – $^2D°$										
			2665.1	138270	175810	4	6	M1	418	1.76	C	1
			"	"	"	4	6	E2	0.0027	0.0013	E	1
4.		$^2D°$ – $^2P°$										
			[1187]	175810	260090	6	2	E2	0.75	0.0021	E	1
			679.3	175810	323180	6	4	M1	1.3(+4)	0.59	D	1
			"	"	"	6	4	E2	15	0.0051	E	1
			[820.88]	138270	260090	4	2	M1	6100	0.25	D	1
			"	"	"	4	2	E2	5.2	0.0023	E	1
			541.35	138270	323180	4	4	M1	4.7(+4)	1.1	D	1
5.		$^2P°$ – $^2P°$										
			[1585]	260090	323180	2	4	M1	1600	0.94	C−	1

Fe xx: Forbidden transitions — Continued

No.	Transition Array	Multiplet	λ (Å)	E_i (cm^{-1})	E_k (cm^{-1})	g_i	g_k	Type of transition	A_{ki} (s^{-1})	S (at. u.)	Accu-racy	Source
6.	$2p^5$–$2p^5$	^{2}P° – ^{2}P°										
			[929.54]	1954150	2061730	4	2	M1	2.24(+4)	1.33	C+	3
			"	"	"	4	2	E2	2.2	0.0018	E	3

[a]The number in parentheses following the tabulated value indicates the power of ten by which this value has to be multiplied.

Fe XXI

C Isoelectronic Sequence

Ground State: $1s^2 2s^2 2p^2\ ^3P_0$

Ionization Energy: 1689 eV = 13620000 cm^{-1}

Allowed Transitions

List of tabulated lines

Wavelength (Å)	No.	Wavelength (Å)	No.	Wavelength (Å)	No.	Wavelength (Å)	No.
8.47	72	9.85	51	13.43	31	123.33	19
8.53	70,71	12.02	38	13.65	33	123.83	3
8.56	69,71	12.10	37	79.781	16	125.29	19
8.57	70	12.13	38	84.26	6	127.04	28
8.61	69	12.18	39	86.26	16	128.73	2
8.64	68	12.19	38	87.462	6	138.11	14
8.65	68,76	12.21	37	89.054	21	138.61	19
8.66	74,75	12.25	36	91.28	4	142.05	19
8.72	66	12.28	37	94.999	5	142.16	2
8.74	73,78	12.30	36	96.166	18	142.27	2
8.81	67	12.36	36	97.88	4	143.19	8
8.83	77	12.37	44	98.36	11	144.79	25
9.34	56	12.38	43	98.69	18	145.65	2
9.41	56	12.43	36	99.08	5	146.86	8
9.42	55	12.46	35	102.22	4	151.50	2
9.44	54	12.47	35,42	103.77	17	151.63	2
9.45	53	12.49	41	103.83	17	155.06	22
9.46	55	12.53	35	104.29	17	156.21	22
9.47	53,54	12.57	47	105.02	23	162.89	24
9.52	53	12.66	46	108.12	3	178.59	7
9.54	52	12.67	40	112.06	20	179.67	27
9.56	52,61,62	12.73	40	112.47	15	180.71	13
9.58	52,58,60	12.79	30	113.30	10	181.57	22
9.59	48,59	12.82	45	114.30	20	187.48	7
9.62	48	12.91	30	115.01	17	187.67	7
9.63	48	12.95	29	115.08	17	192.39	24
9.67	65	12.99	30	115.15	3	208.42	26
9.68	57	13.00	29	117.42	9	242.06	1
9.69	48,64	13.03	29	117.51	3	246.74	12
9.70	57	13.13	29	118.69	3	259.29	26
9.73	48	13.14	29	118.71	17	270.47	1
9.74	50	13.20	29,32	121.21	3		
9.75	49	13.25	31	121.36	19		
9.79	63	13.41	34	122.61	19		

The tabulated oscillator strengths for transitions of the arrays $2s^2 2p^2$–$2s 2p^3$ and $2s 2p^3$–$2p^4$ are the results of the multiconfiguration Dirac-Fock (MCDF) calculations of Cheng et al.[1] These relativistic calculations included a perturbative treatment of the Breit interaction and the Lamb shift. Allowance for configuration mixing was limited to the $n=2$ complex. The results should be quite accurate, except in the case of weak lines. (A few very weak lines have been omitted from this tabulation.)

Transition probabilities for lines of the $2s^2 2p^2$–$2s 2p^3$ array were calculated by Froese Fischer and Saha[2] using the multiconfiguration Hartree-Fock (MCHF) method with Breit-Pauli corrections. Their basis included many configurations outside the $n=2$ complex, but relativistic effects were not treated to the same degree as in Ref. 1. Line strengths derived from these two sources are in reasonably good agreement, particularly for the stronger transitions.

Stratton et al.[3] measured the ratio of A-values for two lines out of the $2s 2p^3\,^3S_1^\circ$ level. Their result is significantly smaller than the corresponding ratio derived from the theoretical data of Cheng et al.

The gf-values calculated by Mason et al.[4] using the scaled Thomas-Fermi approach with configuration inter-action and relativistic effects are quoted for the arrays $2p^2$–$2pns$ and $2p^2$–$2pnd$ ($n=3$–5). The weakest lines have been omitted, as have those which involve levels which are indicated in Ref. 2 to be strongly mixed in LS coupling.

Oscillator strength data are available for a number of transitions of the $2s 2p^3$–$2s 2p^2 3d$ array,[5] but they are not tabulated here since no indication of the percentage compositions of the levels is provided.

References

[1] K. T. Cheng, Y.–K. Kim, and J. P. Desclaux, At. Data Nucl. Data Tables 24, 111 (1979).
[2] C. Froese Fischer and H. P. Saha, Phys. Scr. 32, 181 (1985).
[3] B. C. Stratton, H. W. Moos, S. Suckewer, U. Feldman, J. F. Seely, and A. K. Bhatia, Phys. Rev. A 31, 2534 (1985).
[4] H. E. Mason, G. A. Doschek, U. Feldman, and A. K. Bhatia, Astron. Astrophys. 73, 74 (1979).
[5] G. E. Bromage, R. D. Cowan, B. C. Fawcett, H. Gordon, M. G. Hobby, N. J. Peacock, and A. Ridgeley, United Kingdom Atomic Energy Authority Report CLM-R170 (August 1977).

Fe XXI: Allowed transitions

No.	Transition Array	Multiplet	λ (Å)	E_i (cm^{-1})	E_k (cm^{-1})	g_i	g_k	A_{ki} (10^8 s^{-1})	f_{ik}	S (at. u.)	log gf	Accuracy	Source
1.	$2s^2 2p^2$–$2s 2p^3$	^{3}P – ^{5}S°											
			[270.47]	117300	487030	5	5	0.35	3.8(−4)^a	0.0017	−2.72	E	1
			[242.06]	73910	487030	3	5	0.36	5.3(−4)	0.0013	−2.80	E	1
2.		^{3}P – ^{3}D°	142.89	89800	789630	9	15	88	0.045	0.19	−0.39	D	1
			145.65	117300	803900	5	7	66	0.0295	0.071	−0.83	C	1
			142.16	73910	777350	3	5	100	0.051	0.072	−0.82	C	1
			128.73	0	776810	1	3	120	0.093	0.039	−1.03	C	1
			[151.50]	117300	777350	5	5	0.13	4.4(−5)	1.1(−4)	−3.66	E	1
			142.27	73910	776810	3	3	7.9	0.0024	0.0034	−2.14	D	1
			[151.63]	117300	776810	5	3	0.73	1.5(−4)	3.7(−4)	−3.12	E	1
3.		^{3}P – ^{3}P°	118.51	89800	933640	9	9	237	0.0498	0.175	−0.348	C−	1
			121.21	117300	942330	5	5	217	0.0479	0.096	−0.62	C	1
			117.51	73910	924880	3	3	171	0.0354	0.0411	−0.97	C	1
			[123.83]	117300	924880	5	3	32	0.0044	0.0090	−1.66	D	1
			118.69	73910	916460	3	1	241	0.0170	0.0199	−1.292	C	1
			115.15	73910	942330	3	5	3.6	0.0012	0.0014	−2.44	D	1
			108.12	0	924880	1	3	42.8	0.0225	0.0080	−1.65	C	1

Fe XXI: Allowed transitions — Continued

No.	Transition Array	Multiplet	λ (Å)	E_i (cm^{-1})	E_k (cm^{-1})	g_i	g_k	A_{ki} (10^8 s^{-1})	f_{ik}	S (at. u.)	log gf	Accuracy	Source
4.		$^3P - {}^3S°$	*99.427*	*89800*	1095560	9	3	1000	0.051	0.15	−0.34	C	1
			102.22	117300	1095560	5	3	640	0.060	0.10	−0.52	C	1
			97.88	73910	1095560	3	3	264	0.0379	0.0366	−0.94	C	1
			91.28	0	1095560	1	3	99	0.0370	0.0111	−1.432	C	1
5.		$^3P - {}^1D°$											
			99.08	117300	1126550	5	5	88	0.013	0.021	−1.19	E	1
			[94.999]	73910	1126550	3	5	4.2	9.5(−4)	8.9(−4)	−2.55	E	1
6.		$^3P - {}^1P°$											
			[87.462]	117300	1260650	5	3	2.3	1.6(−4)	2.3(−4)	−3.10	E	1
			84.26	73910	1260650	3	3	53	0.0056	0.0047	−1.77	E	1
7.		$^1D - {}^3D°$											
			[178.59]	243950	803900	5	7	10	0.0070	0.021	−1.46	E	1
			[187.48]	243950	777350	5	5	0.38	2.0(−4)	6.2(−4)	−3.00	E	1
			[187.67]	243950	776810	5	3	2.0	6.4(−4)	0.0020	−2.49	E	1
8.		$^1D - {}^3P°$											
			[143.19]	243950	942330	5	5	2.4	7.5(−4)	0.0018	−2.43	E	1
			[146.86]	243950	924880	5	3	3.0	5.9(−4)	0.0014	−2.53	E	1
9.		$^1D - {}^3S°$	[117.42]	243950	1095560	5	3	3.0	3.7(−4)	7.2(−4)	−2.73	E	1
10.		$^1D - {}^1D°$	113.30	243950	1126550	5	5	480	0.092	0.17	−0.34	C	1
11.		$^1D - {}^1P°$	98.36	243950	1260650	5	3	710	0.062	0.10	−0.51	C	1
12.		$^1S - {}^3D°$											
			[246.74]	371520	776810	1	3	0.55	0.0015	0.0012	−2.82	E	1
13.		$^1S - {}^3P°$											
			[180.71]	371520	924880	1	3	1.6	0.0024	0.0014	−2.62	E	1
14.		$^1S - {}^3S°$	[138.11]	371520	1095560	1	3	6.9	0.0059	0.0027	−2.23	E	1
15.		$^1S - {}^1P°$	112.47	371520	1260650	1	3	183	0.104	0.0385	−0.98	C	1
16.	$2s2p^3 - 2p^4$	$^5S° - {}^3P$											
			86.26	487030	1646320	5	5	14	0.0016	0.0023	−2.10	E	1
			[79.781]	487030	1740460	5	3	2.8	1.6(−4)	2.1(−4)	−3.10	E	1

Fe XXI: Allowed transitions — Continued

No.	Transition Array	Multiplet	λ (Å)	E_i (cm^{-1})	E_k (cm^{-1})	g_i	g_k	A_{ki} (10^8 s^{-1})	f_{ik}	S (at. u.)	log gf	Accuracy	Source
17.		^{3}D° – ^{3}P	*111.36*	*789630*	*1687630*	15	9	460	0.051	0.282	−0.114	C	1
			118.71	803900	1646320	7	5	309	0.0467	0.128	−0.486	C	1
			103.83	777350	1740460	5	3	227	0.0220	0.0376	−0.96	C	1
			104.29	776810	1735720	3	1	373	0.0203	0.0209	−1.215	C	1
			115.08	777350	1646320	5	5	147	0.0292	0.055	−0.84	C	1
			103.77	776810	1740460	3	3	156	0.0252	0.0258	−1.121	C	1
			115.01	776810	1646320	3	5	37.8	0.0125	0.0142	−1.426	C	1
18.		^{3}D° – ^{1}D											
			98.69	803900	1817220	7	5	59	0.0062	0.014	−1.36	E	1
			[96.166]	777350	1817220	5	5	7.2	0.0010	0.0016	−2.30	E	1
19.		^{3}P° – ^{3}P	*132.63*	*933640*	*1687630*	9	9	140	0.036	0.14	−0.49	D	1
			142.05	942330	1646320	5	5	36.7	0.0111	0.0260	−1.256	C	1
			[122.61]	924880	1740460	3	3	1.5	3.4(−4)	4.1(−4)	−2.99	E	1
			125.29	942330	1740460	5	3	177	0.0250	0.052	−0.90	C	1
			123.33	924880	1735720	3	1	204	0.0155	0.0189	−1.333	C	1
			138.61	924880	1646320	3	5	38.3	0.0184	0.0252	−1.258	C	1
			121.36	916460	1740460	1	3	51	0.0341	0.0136	−1.467	C	1
20.		^{3}P° – ^{1}D											
			114.30	942330	1817220	5	5	25	0.0049	0.0092	−1.61	E	1
			[112.06]	924880	1817220	3	5	13	0.0041	0.0045	−1.91	E	1
21.		^{3}P° – ^{1}S											
			[89.054]	924880	2047800	3	1	45	0.0018	0.0016	−2.27	E	1
22.		^{3}S° – ^{3}P	*168.90*	*1095560*	*1687630*	3	9	100	0.13	0.22	−0.40	C	1
			181.57	1095560	1646320	3	5	68	0.056	0.10	−0.77	C	1
			155.06	1095560	1740460	3	3	140	0.052	0.080	−0.81	C	1
			156.21	1095560	1735720	3	1	193	0.0235	0.0363	−1.152	C	1
23.		^{3}S° – ^{1}S	[105.02]	1095560	2047800	3	1	58	0.0032	0.0033	−2.02	E	1
24.		^{1}D° – ^{3}P											
			[192.39]	1126550	1646320	5	5	8.5	0.0047	0.015	−1.63	E	1
			[162.89]	1126550	1740460	5	3	3.6	8.5(−4)	0.0023	−2.37	E	1
25.		^{1}D° – ^{1}D	144.79	1126550	1817220	5	5	356	0.112	0.267	−0.252	C	1
26.		^{1}P° – ^{3}P											
			[259.29]	1260650	1646320	3	5	1.1	0.0019	0.0049	−2.24	E	1
			[208.42]	1260650	1740460	3	3	7.4	0.0048	0.0099	−1.84	E	1

Fe xxi: Allowed transitions — Continued

No.	Transition Array	Multiplet	λ (Å)	E_i (cm^{-1})	E_k (cm^{-1})	g_i	g_k	A_{ki} (10^8 s^{-1})	f_{ik}	S (at. u.)	log gf	Accuracy	Source
27.		^{1}P° – ^{1}D	[179.67]	1260650	1817220	3	5	50	0.0400	0.071	−0.92	C	1
28.		^{1}P° – ^{1}S	127.04	1260650	2047800	3	1	840	0.068	0.085	−0.69	C	1
29.	$2p^2$–$2p3s$	^{3}P – ^{3}P°	13.06			9	9	2.0(+4)	0.052	0.020	−0.33	D	4
			[13.03]			5	5	1.3(+4)	0.033	0.0071	−0.78	D	4
			[13.13]			3	3	3900	0.010	0.0013	−1.52	D	4
			[13.20]			5	3	1.2(+4)	0.019	0.0041	−1.02	D	4
			[13.14]			3	1	2.0(+4)	0.017	0.0022	−1.29	D	4
			[12.95]			3	5	6200	0.026	0.0033	−1.11	D	4
			[13.00]			1	3	7200	0.055	0.0024	−1.26	D	4
30.		^{3}P – ^{1}P°											
			[12.99]			5	3	1100	0.0016	3.4(−4)	−2.10	E	4
			[12.91]			3	3	1200	0.0030	3.8(−4)	−2.05	E	4
			[12.79]			1	3	140	0.0010	4.2(−5)	−3.00	E	4
31.		^{1}D – ^{3}P°											
			[13.25]			5	5	3400	0.0089	0.0019	−1.35	E	4
			[13.43]			5	3	920	0.0015	3.3(−4)	−2.12	E	4
32.		^{1}D – ^{1}P°	[13.20]			5	3	2.3(+4)	0.036	0.0078	−0.74	D	4
33.		^{1}S – ^{3}P°											
			[13.65]			1	3	260	0.0022	9.9(−5)	−2.66	E	4
34.		^{1}S – ^{1}P°	[13.41]			1	3	7300	0.059	0.0026	−1.23	E	4
35.	$2p^2$–$2p3d$	^{3}P – ^{3}F°											
			[12.47]			5	7	5.8(+4)	0.19	0.039	−0.02	E	4
			[12.46]			3	5	1500	0.0058	7.1(−4)	−1.76	E	4
			[12.53]			5	5	1.5(+4)	0.035	0.0072	−0.76	E	4
36.		^{3}P – ^{3}D°											
			[12.30]			5	7	2.1(+5)	0.67	0.14	0.53	D	4
			[12.25]			1	3	2.1(+5)	1.4	0.056	0.15	D	4
			[12.36]			3	3	3.6(+4)	0.083	0.010	−0.60	D	4
			[12.43]			5	3	2100	0.0029	5.9(−4)	−1.84	E	4
37.		^{3}P – ^{3}P°											
			[12.21]			3	3	1.2(+5)	0.26	0.031	−0.11	D	4
			[12.28]			5	3	5.2(+4)	0.071	0.014	−0.45	D	4
			[12.21]			3	1	1.5(+5)	0.11	0.013	−0.48	D	4
			[12.10]			1	3	230	0.0015	6.0(−5)	−2.82	E	4
38.		^{3}P – ^{1}P°											
			[12.19]			5	3	6400	0.0086	0.0017	−1.37	E	4
			[12.13]			3	3	1.8(+4)	0.040	0.0048	−0.92	E	4
			[12.02]			1	3	1.3(+4)	0.083	0.0033	−1.08	E	4
39.		^{3}P – ^{1}F°											
			[12.18]			5	7	2.2(+4)	0.067	0.013	−0.47	E	4

Fe XXI: Allowed transitions — Continued

No.	Transition Array	Multiplet	λ (Å)	E_i (cm^{-1})	E_k (cm^{-1})	g_i	g_k	A_{ki} (10^8 s^{-1})	f_{ik}	S (at. u.)	log gf	Accuracy	Source
40.		$^1D - \,^3F°$											
			[12.67]			5	7	2400	0.0080	0.0017	−1.40	E	4
			[12.73]			5	5	8200	0.020	0.0042	−1.00	E	4
41.		$^1D - \,^3D°$											
			[12.49]			5	7	1.3(+4)	0.041	0.0084	−0.69	E	4
42.		$^1D - \,^3P°$											
			[12.47]			5	3	1.3(+4)	0.018	0.0037	−1.05	E	4
43.		$^1D - \,^1P°$	[12.38]			5	3	6900	0.0095	0.0019	−1.32	E	4
44.		$^1D - \,^1F°$	[12.37]			5	7	3.1(+5)	1.0	0.20	0.70	D	4
45.		$^1S - \,^3D°$											
			[12.82]			1	3	1200	0.0088	3.7(−4)	−2.06	E	4
46.		$^1S - \,^3P°$											
			[12.66]			1	3	980	0.0071	3.0(−4)	−2.15	E	4
47.		$^1S - \,^1P°$	[12.57]			1	3	7.2(+4)	0.51	0.021	−0.29	E	4
48.	$2p^2 - 2p\,4s$	$^3P - \,^3P°$											
			[9.63]			5	5	3000	0.0042	6.7(−4)	−1.68	E	4
			[9.73]			5	3	2300	0.0020	3.2(−4)	−2.00	E	4
			[9.69]			3	1	3600	0.0017	1.6(−4)	−2.29	E	4
			[9.59]			3	5	1700	0.0040	3.8(−4)	−1.92	E	4
			[9.62]			1	3	1200	0.0049	1.6(−4)	−2.31	E	4
49.		$^1D - \,^3P°$											
			[9.75]			5	5	770	0.0011	1.8(−4)	−2.26	E	4
50.		$^1D - \,^1P°$	[9.74]			5	3	5300	0.0045	7.2(−4)	−1.65	E	4
51.		$^1S - \,^1P°$	[9.85]			1	3	2200	0.0095	3.1(−4)	−2.02	E	4
52.	$2p^2 - 2p\,4d$	$^3P - \,^3F°$											
			[9.56]			5	7	3.2(+4)	0.061	0.0096	−0.52	E	4
			[9.54]			3	5	750	0.0017	1.6(−4)	−2.29	E	4
			[9.58]			5	5	5200	0.0071	0.0011	−1.45	E	4
53.		$^3P - \,^3D°$											
			[9.47]			5	7	4.9(+4)	0.093	0.014	−0.33	D	4
			[9.45]			1	3	5.2(+4)	0.21	0.0065	−0.68	D	4
			[9.52]			3	3	8100	0.011	0.0010	−1.48	D	4
54.		$^3P - \,^1D°$											
			[9.47]			5	5	6100	0.0082	0.0013	−1.39	E	4
			[9.44]			3	5	1.7(+4)	0.037	0.0034	−0.95	E	4

Fe XXI: Allowed transitions — Continued

No.	Transition Array	Multiplet	λ (Å)	E_i (cm^{-1})	E_k (cm^{-1})	g_i	g_k	A_{ki} (10^8 s^{-1})	f_{ik}	S (at. u.)	log gf	Accuracy	Source
55.		^{3}P – ^{3}P°											
			[9.42]			3	3	3.3(+4)	0.044	0.0041	−0.88	D	4
			[9.46]			5	3	1.5(+4)	0.012	0.0019	−1.22	D	4
			[9.42]			3	1	4.3(+4)	0.019	0.0018	−1.24	D	4
56.		^{3}P – ^{1}P°											
			[9.41]			3	3	1300	0.0017	1.6(−4)	−2.29	E	4
			[9.34]			1	3	2200	0.0086	2.6(−4)	−2.07	E	4
57.		^{1}D – ^{3}F°											
			[9.68]			5	7	4000	0.0079	0.0013	−1.40	E	4
			[9.70]			5	5	1900	0.0027	4.3(−4)	−1.87	E	4
58.		^{1}D – ^{3}D°											
			[9.58]			5	7	1700	0.0033	5.2(−4)	−1.78	E	4
59.		^{1}D – ^{1}D°	[9.59]			5	5	1.0(+4)	0.014	0.0022	−1.15	D	4
60.		^{1}D – ^{3}P°											
			[9.58]			5	3	3900	0.0032	5.0(−4)	−1.80	E	4
61.		^{1}D – ^{1}F°	[9.56]			5	7	8.9(+4)	0.17	0.027	−0.07	D	4
62.		^{1}D – ^{1}P°	[9.56]			5	3	2300	0.0019	3.0(−4)	−2.02	E	4
63.		^{1}S – ^{3}D°											
			[9.79]			1	3	2200	0.0094	3.0(−4)	−2.03	E	4
64.		^{1}S – ^{3}P°											
			[9.69]			1	3	500	0.0021	6.7(−5)	−2.68	E	4
65.		^{1}S – ^{1}P°	[9.67]			1	3	5.7(+4)	0.24	0.0076	−0.62	D	4
66.	$2p^2$–$2p\,5s$	^{1}D – ^{1}P°	[8.72]			5	3	2000	0.0014	2.0(−4)	−2.15	E	4
67.		^{1}S – ^{1}P°	[8.81]			1	3	890	0.0031	9.0(−5)	−2.51	E	4
68.	$2p^2$–$2p\,5d$	^{3}P – ^{3}F°											
			[8.64]			5	7	1.5(+4)	0.024	0.0034	−0.92	E	4
			[8.65]			5	5	2500	0.0028	4.0(−4)	−1.85	E	4
69.		^{3}P – ^{3}D°											
			[8.56]			5	7	2.0(+4)	0.030	0.0042	−0.82	D	4
			[8.56]			1	3	2.1(+4)	0.070	0.0020	−1.15	D	4
			[8.61]			3	3	3200	0.0036	3.1(−4)	−1.97	E	4
70.		^{3}P – ^{1}D°											
			[8.57]			5	5	2800	0.0031	4.4(−4)	−1.81	E	4
			[8.53]			3	5	6100	0.011	9.3(−4)	−1.48	E	4

Fe XXI: Allowed transitions — Continued

No.	Transition Array	Multiplet	λ (Å)	E_i (cm⁻¹)	E_k (cm⁻¹)	g_i	g_k	A_{ki} (10⁸ s⁻¹)	f_{ik}	S (at. u.)	log gf	Accuracy	Source
71.		$^3P - {}^3P°$											
			[8.53]			3	3	1.5(+4)	0.016	0.0013	−1.32	D	4
			[8.56]			5	3	6500	0.0043	6.1(−4)	−1.67	E	4
			[8.53]			3	1	1.8(+4)	0.0066	5.6(−4)	−1.70	E	4
72.		$^3P - {}^1P°$											
			[8.47]			1	3	1400	0.0046	1.3(−4)	−2.34	E	4
73.		$^1D - {}^3F°$											
			[8.74]			5	7	2700	0.0044	6.3(−4)	−1.66	E	4
74.		$^1D - {}^1D°$	[8.66]			5	5	4400	0.0049	7.0(−4)	−1.61	E	4
75.		$^1D - {}^3P°$											
			[8.66]			5	3	1600	0.0011	1.6(−4)	−2.26	E	4
76.		$^1D - {}^1F°$	[8.65]			5	7	3.9(+4)	0.061	0.0087	−0.52	D	4
77.		$^1S - {}^3D°$											
			[8.83]			1	3	1600	0.0055	1.6(−4)	−2.26	E	4
78.		$^1S - {}^1P°$	[8.74]			1	3	2.5(+4)	0.085	0.0024	−1.07	D	4

[a]The number in parentheses following the tabulated value indicates the power of ten by which this value has to be multiplied.

Fe XXI

Forbidden Transitions

Line strengths tabulated for magnetic dipole and electric quadrupole transitions within the $2p^2$ configuration are the results of the multiconfiguration Dirac-Fock (MCDF) calculations of Cheng et al.[1] These relativistic calculations included a perturbative treatment of the Breit interaction and the Lamb shift. Allowance for configuration mixing was limited to the $n=2$ complex. Strengths of electric quadrupole transitions as defined in Ref. 1 were multiplied by the factor $^2/_3$ which is needed to bring these values into conformance with the definition of quadrupole strengths used in the NBS tables. The weakest lines are excluded from this compilation, as their strengths are considered to be very uncertain.

Transition probabilities for these same lines were calculated by Froese Fischer and Saha[2] using the multiconfiguration Hartree-Fock (MCHF) method with Breit-Pauli corrections. Their basis included many configurations outside the $n=2$ complex, but relativistic effects were not treated to the same degree as in Ref. 1. Line strengths derived from these data are in good agreement with the data of Cheng et al.

References

[1]K. T. Cheng, Y.-K. Kim, and J. P. Desclaux, At. Data Nucl. Data Tables **24**, 111 (1979).
[2]C. Froese Fischer and H. P. Saha, Phys. Scr. **32**, 181 (1985).

Fe XXI:　Forbidden transitions

No.	Transition Array	Multiplet	λ (Å)	E_i (cm^{-1})	E_k (cm^{-1})	g_i	g_k	Type of transition	A_{ki} (s^{-1})	S (at. u.)	Accuracy	Source
1.	$2p^2$–$2p^2$	^3P – ^3P										
			2298.0	73910	117300	3	5	M1	840	1.90	C	1
			"	"	"	3	5	E2	0.0068	0.0013	E	1
			1354.1	0	73910	1	3	M1	6500	1.79	C	1
			[852.51]	0	117300	1	5	E2	0.75	0.0010	E	1
2.		^3P – ^1D										
			786.1	117300	243950	5	5	M1	1.6(+4)[a]	1.4	C	1
			"	"	"	5	5	E2	2.5	0.0022	E	1
			585.8	73910	243950	3	5	M1	1.6(+4)	0.59	D	1
3.		^3P – ^1S										
			[336.01]	73910	371520	3	1	M1	1.4(+5)	0.20	D	1
4.		^1D – ^1S	[783.88]	243950	371520	5	1	E2	15	0.0027	E	1

[a]The number in parentheses following the tabulated value indicates the power of ten by which this value has to be multiplied.

Fe XXII

B Isoelectronic Sequence

Ground State:　$1s^2 2s^2 2p \ ^2P^\circ_{1/2}$

Ionization Energy:　1799 eV = 14510000 cm^{-1}

Allowed Transitions

List of tabulated lines

Wavelength (Å)	No.	Wavelength (Å)	No.	Wavelength (Å)	No.	Wavelength (Å)	No.
8.71	20	11.72	17	89.781	7	151.54	9
8.74	20	11.748	24,25,30	100.78	3	153.96	16
8.75	19	11.763	22	100.93	6	155.92	2
8.81	19,20	11.789	25,30	102.23	4	157.03	9
8.951	36	11.797	25	103.54	6	157.37	13
8.960	41	11.823	24	106.91	6	161.74	2
8.977	39	11.837	25	109.53	6	169.08	16
8.992	38	11.886	24	112.21	10	173.21	13
9.002	36	11.898	31	113.31	6	184.19	12
9.006	38	11.922	22	114.41	3	192.29	8
9.050	36	11.929	22	115.19	10	230.10	12
9.064	35	11.976	23	116.28	4	239.11	15
9.065	37	12.027	34	117.17	3	246.54	1
9.163	42	12.045	27,32	117.52	5	247.44	12
9.182	35	12.053	27	120.03	10	248.11	11
9.183	41	12.077	31,33	125.71	5	252.76	1
9.215	40	12.095	27	129.17	13	291.69	1
9.241	39	12.193	26,28	134.65	5	347.96	1
11.42	18	12.29	21	135.78	2	359.26	14
11.48	18	12.325	29	136.01	3	378.39	11
11.56	17	12.48	21	139.64	13		
11.58	18	85.717	7	144.85	9		
11.62	17	85.911	7	149.87	9		

The tabulated oscillator strengths for transitions of the arrays $2s^2 2p-2s 2p^2$ and $2s 2p^2-2p^3$ are the results of the multiconfiguration Dirac-Fock (MCDF) calculations of Cheng et al.[1] These relativistic calculations included a perturbative treatment of the Breit interaction and the Lamb shift. The results should be quite accurate, except in the case of weak lines. (A few very weak lines have been omitted from this tabulation.)

There exist a number of additional reliable sources of data[2-5] for $2s-2p$ transitions in Fe XXII in which the theoretical method used allowed for relativistic effects and for more extensive configuration interaction than the approach described in Ref. 1. The results agree very well with those of Cheng et al.—with the exception of the weak lines, for which the values reported in all of these works are subject to rather large uncertainties. In view of this outcome, and given the fact that the only authors who reported results for all electric-dipole transitions of the arrays $2s^2 2p-2s 2p^2$ and $2s 2p^2-2p^3$ were Cheng et al., we have chosen to tabulate their results exclusively.

According to several sources (see, e.g., Refs. 1, 2, 4, and 5), the lower of the two levels $2s 2p^2 \, ^2P_{1/2}$ and $^2S_{1/2}$ is mostly of ^2P character, having "crossed" the $^2S_{1/2}$ level at about V XIX or Cr XX. We have thus labeled these two levels accordingly, in contrast to their labeling by Cheng et al., which is consistent with their ordering at the neutral end of the B sequence.

Oscillator strengths determined by Mason and Storey[6] with the scaled Thomas-Fermi approach including allowance for relativistic effects and considerable configuration interaction are quoted for transitions of the arrays $2p-ns$, $2p-nd$, and $2s^2 2p-2s 2p np$ ($n = 3,4$). Transitions involving doublet levels of the configurations $2s 2p 3p$ and $2s 2p 4p$ are excluded from our tabulation, since the parent terms are not indicated in Ref. 6. Also excluded are very weak transitions, as well as those which involve $J = ^3/_2$ levels whose term designations are reported by Mason and Storey to be ambiguous.

The f-values which were calculated by Bromage et al.[7] using the Hartree-XR (Hartree-Fock with statistical allowance for exchange and relativistic effects) method are quoted for transitions of the type $2s 2p^2-2s 2p nd$ ($n = 3,4$). They have reported their results for only the strongest transitions of these arrays.

References

[1] K. T. Cheng, Y.-K. Kim, and J. P. Desclaux, At. Data Nucl. Data Tables 24, 111 (1979).
[2] R. Glass, J. Phys. B 13, 15 (1980).
[3] R. Glass, J. Phys. B 13, 899 (1980).
[4] M. Vajed-Samii, D. Ton-That, and L. Armstrong, Jr., Phys. Rev. A 23, 3034 (1981).
[5] M. Dankwort and E. Trefftz, Astron. Astrophys. 65, 93 (1978).
[6] H. E. Mason and P. J. Storey, Mon. Not. R. Astron. Soc. 191, 631 (1980).
[7] G. E. Bromage, R. D. Cowan, B. C. Fawcett, and A. Ridgeley, J. Opt. Soc. Am. 68, 48 (1978).

Fe XXII: Allowed transitions

No.	Transition Array	Multiplet	λ (Å)	E_i (cm^{-1})	E_k (cm^{-1})	g_i	g_k	A_{ki} (10^8 s^{-1})	f_{ik}	S (at. u.)	log gf	Accuracy	Source
1.	$2s^2 2p-2s 2p^2$	$^2P° - {}^4P$											
			[252.76]	118230	513870	4	6	0.70	0.0010	0.0033	−2.40	E	1
			[291.69]	118230	461060	4	4	0.086	1.1(−4)ᵃ	4.2(−4)	−3.36	E	1
			[246.54]	0	405620	2	2	0.87	7.9(−4)	0.0013	−2.80	E	1
			[347.96]	118230	405620	4	2	0.14	1.3(−4)	6.0(−4)	−3.28	E	1
2.		$^2P° - {}^2D$	148.91	78820	750350	6	10	80	0.044	0.13	−0.58	D	1
			155.92	118230	759590	4	6	62	0.0340	0.070	−0.87	C	1
			135.78	0	736490	2	4	110	0.062	0.055	−0.91	C	1
			[161.74]	118230	736490	4	4	0.38	1.5(−4)	3.2(−4)	−3.22	E	1
3.		$^2P° - {}^2P$	115.32	78820	946000	6	6	440	0.088	0.20	−0.28	C	1
			114.41	118230	992260	4	4	450	0.088	0.13	−0.45	C	1
			117.17	0	853490	2	2	390	0.080	0.062	−0.80	C	1
			136.01	118230	853490	4	2	0.12	1.6(−5)	2.9(−5)	−4.19	E	1
			100.78	0	992260	2	4	62	0.0189	0.0125	−1.423	C	1
4.		$^2P° - {}^2S$	111.19	78820	978190	6	2	430	0.026	0.058	−0.80	C−	1
			116.28	118230	978190	4	2	353	0.0358	0.055	−0.84	C	1
			102.23	0	978190	2	2	27	0.0042	0.0028	−2.08	D	1

Fe XXII: Allowed transitions — Continued

No.	Transition Array	Multiplet	λ (Å)	E_i (cm^{-1})	E_k (cm^{-1})	g_i	g_k	A_{ki} (10^8 s^{-1})	f_{ik}	S (at. u.)	log gf	Accu-racy	Source
5.	$2s\,2p^2$–$2p^3$	^{4}P – ^{4}S°	*128.48*	*478230*	1256540	12	4	449	0.0370	0.188	−0.352	C	1
			134.65	513870	1256540	6	4	196	0.0356	0.095	−0.67	C	1
			125.71	461060	1256540	4	4	152	0.0360	0.060	−0.84	C	1
			117.52	405620	1256540	2	4	103	0.0428	0.0331	−1.068	C	1
6.		^{4}P – ^{2}D°											
			109.53	513870	1426850	6	6	19	0.0035	0.0076	−1.68	E	1
			[106.91]	461060	1396400	4	4	23	0.0039	0.0055	−1.81	E	1
			[113.31]	513870	1396400	6	4	2.2	2.8(−4)	6.3(−4)	−2.77	E	1
			[103.54]	461060	1426850	4	6	0.50	1.2(−4)	1.6(−4)	−3.32	E	1
			[100.93]	405620	1396400	2	4	0.33	1.0(−4)	6.6(−5)	−3.70	E	1
7.		^{4}P – ^{2}P°											
			[89.781]	513870	1627690	6	4	1.5	1.2(−4)	2.1(−4)	−3.14	E	1
			[85.717]	461060	1627690	4	4	2.9	3.2(−4)	3.6(−4)	−2.89	E	1
			[85.911]	405620	1569610	2	2	2.3	2.5(−4)	1.4(−4)	−3.30	E	1
8.		^{2}D – ^{4}S°											
			[192.29]	736490	1256540	4	4	1.3	7.4(−4)	0.0019	−2.53	E	1
9.		^{2}D – ^{2}D°	*150.53*	*750350*	*1414670*	10	10	149	0.050	0.250	−0.297	C	1
			149.87	759590	1426850	6	6	128	0.0432	0.128	−0.59	C	1
			151.54	736490	1396400	4	4	76	0.0260	0.052	−0.98	C	1
			157.03	759590	1396400	6	4	50	0.0124	0.0385	−1.128	C	1
			144.85	736490	1426850	4	6	35.4	0.0167	0.0319	−1.175	C	1
10.		^{2}D – ^{2}P°	*116.55*	*750350*	*1608330*	10	6	230	0.0281	0.108	−0.55	C−	1
			115.19	759590	1627690	6	4	143	0.0189	0.0430	−0.95	C	1
			120.03	736490	1569610	4	2	296	0.0320	0.051	−0.89	C	1
			112.21	736490	1627690	4	4	51	0.0097	0.014	−1.41	D	1
11.		^{2}P – ^{4}S°											
			[378.39]	992260	1256540	4	4	0.31	6.7(−4)	0.0033	−2.57	E	1
			[248.11]	853490	1256540	2	4	1.2	0.0022	0.0036	−2.36	E	1
12.		^{2}P – ^{2}D°	*213.37*	*946000*	*1414670*	6	10	42	0.047	0.20	−0.55	D	1
			[230.10]	992260	1426850	4	6	33.8	0.0402	0.122	−0.79	C	1
			[184.19]	853490	1396400	2	4	66	0.067	0.081	−0.87	C	1
			[247.44]	992260	1396400	4	4	0.44	4.0(−4)	0.0013	−2.80	E	1
13.		^{2}P – ^{2}P°	*150.98*	*946000*	*1608330*	6	6	190	0.064	0.19	−0.42	C−	1
			157.37	992260	1627690	4	4	200	0.075	0.16	−0.52	C	1
			139.64	853490	1569610	2	2	26	0.0075	0.0069	−1.82	D	1
			173.21	992260	1569610	4	2	22	0.0050	0.011	−1.70	D	1
			129.17	853490	1627690	2	4	37.4	0.0187	0.0159	−1.427	C	1
14.		^{2}S – ^{4}S°	[359.26]	978190	1256540	2	4	0.12	4.8(−4)	0.0011	−3.02	E	1
15.		^{2}S – ^{2}D°											
			[239.11]	978190	1396400	2	4	8.5	0.0146	0.0230	−1.53	C	1

Fe XXII: Allowed transitions — Continued

No.	Transition Array	Multiplet	λ (Å)	E_i (cm^{-1})	E_k (cm^{-1})	g_i	g_k	A_{ki} (10^8 s^{-1})	f_{ik}	S (at. u.)	log gf	Accuracy	Source
16.		^{2}S – ^{2}P°	*158.69*	978190	*1608330*	2	6	58	0.066	0.069	−0.88	C	1
			153.96	978190	1627690	2	4	17.4	0.0124	0.0126	−1.61	C	1
			169.08	978190	1569610	2	2	120	0.050	0.056	−1.00	C	1
17.	$2s^22p-$ $2s2p(^3$P°$)3p$	^{2}P° – ^{4}D											
			[11.62]			4	6	490	0.0015	2.3(−4)	−2.22	E	6
			[11.56]			2	2	1.4(+4)	0.028	0.0021	−1.25	D	6
			[11.72]			4	2	2100	0.0022	3.4(−4)	−2.06	E	6
18.		^{2}P° – ^{4}P											
			[11.48]			4	6	2.0(+4)	0.058	0.0088	−0.63	D	6
			[11.42]			2	2	2300	0.0045	3.4(−4)	−2.05	E	6
			[11.58]			4	2	1300	0.0013	2.0(−4)	−2.28	E	6
19.	$2s^22p-$ $2s2p(^3$P°$)4p$	^{2}P° – ^{4}D											
			[8.81]			4	6	1500	0.0026	3.0(−4)	−1.98	E	6
			[8.75]			2	2	7000	0.0080	4.6(−4)	−1.80	E	6
20.		^{2}P° – ^{4}P											
			[8.74]			4	6	6400	0.011	0.0013	−1.36	D	6
			[8.71]			2	2	2900	0.0033	1.9(−4)	−2.18	E	6
			[8.81]			4	2	2100	0.0012	1.4(−4)	−2.32	E	6
21.	$2p-3s$	^{2}P° – ^{2}S	*12.42*			6	2	2.5(+4)	0.019	0.0047	−0.94	D	6
			[12.48]			4	2	1.7(+4)	0.020	0.0033	−1.10	D	6
			[12.29]			2	2	7500	0.017	0.0014	−1.47	D	6
22.	$2p-3d$	^{2}P° – ^{2}D	*11.869*	*78820*	*8504100*	6	10	1.8(+5)	0.64	0.15	0.58	D	6
			11.922	118230	8506100	4	6	1.8(+5)	0.59	0.093	0.37	D	6
			11.763	0	8501200	2	4	1.6(+5)	0.65	0.050	0.11	D	6
			[11.929]	118230	8501200	4	4	3.0(+4)	0.065	0.010	−0.59	D	6
23.	$2s2p^2-$ $2s2p(^3$P°$)3d$	^{4}P – ^{4}F°											
			11.976	513870	8863900	6	8	5.9(+4)	0.17	0.040	0.01	D	7
24.		^{4}P – ^{4}P°											
			11.748	461060	8972000	4	4	1.2(+5)	0.25	0.039	0.00	D	7
			11.823	513870	8972000	6	4	7.9(+4)	0.11	0.026	−0.18	D	7
			11.748	461060	8973000	4	2	1.8(+5)	0.19	0.029	−0.12	D	7
			11.886	461060	8874400	4	6	1.3(+5)	0.42	0.066	0.23	D	7
25.		^{4}P – ^{4}D°											
			11.837	513870	8962000	6	8	2.3(+5)	0.65	0.15	0.59	D	7
			11.748	461060	8973000	4	6	4.8(+4)	0.15	0.023	−0.22	D	7
			11.797	405620	8882000	2	4	1.7(+5)	0.70	0.054	0.15	D	7
			11.837	513870	8973000	6	6	1.7(+5)	0.35	0.082	0.32	D	7
			11.789	405620	8888000	2	2	2.6(+5)	0.55	0.043	0.04	D	7

Fe XXII: Allowed transitions — Continued

No.	Transition Array	Multiplet	λ (Å)	E_i (cm^{-1})	E_k (cm^{-1})	g_i	g_k	A_{ki} (10^8 s^{-1})	f_{ik}	S (at. u.)	log gf	Accuracy	Source
26.		^{2}D – ^{2}D°											
			12.193	736490	8937900	4	6	9.9(+4)	0.33	0.053	0.12	D	7
27.		^{2}D – ^{2}F°	12.049	750350	9049500	10	14	2.0(+5)	0.61	0.24	0.78	D	7
			12.045	759590	9061800	6	8	2.4(+5)	0.71	0.17	0.63	D	7
			12.053	736490	9033200	4	6	6.1(+4)	0.20	0.032	−0.10	D	7
			12.095	759590	9033200	6	6	7.8(+4)	0.17	0.041	0.01	D	7
28.		^{2}P – ^{2}P°											
			12.193	853490	9054900	2	4	7.2(+4)	0.32	0.026	−0.19	D	7
29.		^{2}S – ^{2}P°											
			12.325	978190	9092000	2	2	1.5(+5)	0.35	0.028	−0.15	D	7
30.	$2s\,2p^2$– $2s\,2p(^1$P°$)3d$	^{2}D – ^{2}F°											
			11.789	759590	9242100	6	8	1.2(+5)	0.32	0.075	0.28	D	7
			11.748	736490	9248600	4	6	1.6(+5)	0.49	0.076	0.29	D	7
31.		^{2}P – ^{2}D°											
			12.077	992260	9273000	4	6	2.4(+5)	0.78	0.12	0.49	D	7
			11.898	853490	9258300	2	4	8.2(+4)	0.35	0.027	−0.15	D	7
32.		^{2}P – ^{2}P°											
			12.045	992260	9292800	4	4	9.7(+4)	0.21	0.033	−0.08	D	7
33.		^{2}S – ^{2}D°											
			12.077	978190	9258300	2	4	1.0(+5)	0.44	0.035	−0.06	D	7
34.		^{2}S – ^{2}P°											
			12.027	978190	9292800	2	4	6.9(+4)	0.30	0.024	−0.22	D	7
35.	$2p$–$4s$	^{2}P° – ^{2}S	9.142	78820	11030000	6	2	7400	0.0031	5.6(−4)	−1.73	E	6
			9.182	118230	11030000	4	2	5500	0.0035	4.2(−4)	−1.85	E	6
			9.064	0	11030000	2	2	1900	0.0023	1.4(−4)	−2.34	E	6
36.	$2p$–$4d$	^{2}P° – ^{2}D	8.985	78820	11210000	6	10	5.6(+4)	0.11	0.020	−0.17	D	6
			9.002	118230	11230000	4	6	5.5(+4)	0.10	0.012	−0.40	D	6
			8.951	0	11170000	2	4	4.6(+4)	0.11	0.0065	−0.66	D	6
			[9.050]	118230	11170000	4	4	9800	0.012	0.0014	−1.32	D	6
37.	$2s\,2p^2$– $2s\,2p(^3$P°$)4d$	^{4}P – ^{4}F°											
			9.065	461060	11490000	4	6	3.5(+4)	0.065	0.0078	−0.59	D	7
38.		^{4}P – ^{4}D°											
			9.006	513870	11600000	6	8	5.7(+4)	0.093	0.017	−0.25	D	7
			8.992	405620	11530000	2	4	4.9(+4)	0.12	0.0071	−0.62	D	7
			9.006	513870	11600000	6	6	5.3(+4)	0.065	0.012	−0.41	D	7

Fe XXII: Allowed transitions — Continued

No.	Transition Array	Multiplet	λ (Å)	E_i (cm^{-1})	E_k (cm^{-1})	g_i	g_k	A_{ki} (10^8 s^{-1})	f_{ik}	S (at. u.)	log gf	Accuracy	Source
39.		^{2}D – ^{2}F°											
			8.977	759590	11900000	6	8	2.5(+4)	0.040	0.0071	−0.62	D	7
			9.241	736490	11600000	4	6	5.1(+4)	0.098	0.012	−0.41	D	7
40.		^{2}D – ^{2}D°											
			9.215	759590	11610000	6	6	2.7(+4)	0.035	0.0064	−0.68	D	7
41.	$2s\,2p^2-$ $2s\,2p(^1$P°$)4d$	^{2}D – ^{2}F°											
			9.183	759590	11600000	6	8	8.3(+4)	0.14	0.025	−0.08	D	7
			8.960	736490	11900000	4	6	3.8(+4)	0.068	0.0080	−0.57	D	7
42.		^{2}P – ^{2}D°											
			9.163	992260	11900000	4	6	6.9(+4)	0.13	0.016	−0.28	D	7

[a]The number in parentheses following the tabulated value indicates the power of ten by which this value has to be multiplied.

Fe XXII

Forbidden Transitions

The line strengths tabulated for the single magnetic dipole and single electric quadrupole transition within the $2s^2 2p$ ground state configuration are the results of the multiconfiguration Dirac-Fock (MCDF) calculations of Cheng et al.[1] These relativistic calculations include a perturbative treatment of the Breit interaction and the Lamb shift. Allowance for configuration mixing is limited to the $n=2$ complex. The strength of the electric quadrupole transition as defined in Ref. 1 was multiplied by the factor $^2/_3$ in order to bring this value into conformance with the definition of the quadrupole strength used in the NBS tables.

Transition probabilities for the same lines were calculated by Froese Fischer and Saha[2] using the multiconfig-uration Hartree-Fock (MCHF) method with Breit-Pauli corrections. Their orbital basis includes many configurations outside the $n=2$ complex, but relativistic effects were not treated to the same degree as in Ref. 1. The line strengths for both the M1 and E2 transitions are in very good agreement with the data of Cheng et al.[1]

References

[1]K. T. Cheng, Y.-K. Kim, and J. P. Desclaux, At. Data Nucl. Data Tables **24**, 111 (1979).

[2]C. Froese Fischer and H. P. Saha, Phys. Rev. A **28**, 3169 (1983).

Fe XXII: Forbidden transitions

No.	Transition Array	Multiplet	λ (Å)	E_i (cm^{-1})	E_k (cm^{-1})	g_i	g_k	Type of transition	A_{ki} (s^{-1})	S (at. u.)	Accuracy	Source
1.	2p–2p	$^2P° - ^2P°$										
			845.55	0	118270	2	4	M1	1.48(+4)[a]	1.33	B	1
			"	"	"	2	4	E2	1.39	0.00143	C	1

[a]The number in parentheses following the tabulated value indicates the power of ten by which this value has to be multiplied.

Fe XXIII

Be Isoelectronic Sequence

Ground State: $1s^2 2s^2\ ^1S_0$

Ionization Energy: 1958.6 eV = 15797000 cm^{-1}

Allowed Transitions

List of tabulated lines

Wavelength (Å)	No.	Wavelength (Å)	No.	Wavelength (Å)	No.	Wavelength (Å)	No.
7.445	36	8.557	29	11.247	15	11.846	48
7.472	35	8.559	30	11.255	15	11.857	49
7.680	66	8.578	29	11.298	42	11.873	37
7.733	66	8.614	60	11.325	42	11.882	39
7.778	69	8.618	60	11.338	42	11.897	39
7.826	69	8.640	54	11.341	47	11.898	53
7.849	70	8.664	64	11.411	42	11.984	39
7.854	68	8.669	63	11.429	47	12.048	39
7.883	67	8.672	63	11.433	47	12.095	39
8.197	28	8.710	54	11.445	42	12.098	40
8.218	26	8.722	56	11.458	42	12.176	38
8.260	28	8.752	65	11.460	47	12.427	41
8.272	25	8.756	56	11.482	13	12.654	20
8.273	24,27	8.764	56,62	11.485	45	13.014	21
8.277	25	8.814	61	11.491	14,47	121.20	4
8.281	26	8.822	56	11.519	47	132.84	2
8.290	25	8.853	56,58	11.520	45	136.53	4
8.291	24	8.858	57	11.524	45	144.36	3
8.292	24	8.910	55	11.544	45	147.24	3
8.293	25	9.028	59	11.593	52	149.22	7
8.307	23	10.793	12	11.596	51	154.27	3
8.312	24	10.902	12	11.613	46	166.74	3
8.320	22	10.910	12	11.615	45	173.31	3
8.437	34	10.927	11	11.674	46	180.10	3
8.456	33	10.934	11	11.675	45	221.33	6
8.470	32	10.979	10	11.691	44	263.76	1
8.474	30	10.981	11	11.698	50	313.62	5
8.479	31	11.018	9,11	11.702	37	364.48	5
8.529	60	11.047	11	11.737	43	491.98	5
8.550	60	11.086	17	11.744	37		
8.552	60	11.165	16	11.835	39		

Oscillator strengths for transitions of the arrays $2s^2$–$2s2p$ and $2s2p$–$2p^2$ are taken from the multiconfiguration Dirac-Fock (MCDF) calculations of Cheng et al.[1] These relativistic calculations include the configuration interaction most relevant for the states of these configurations, as well as a perturbative treatment of the Breit interaction and the Lamb shift. The results should be quite accurate, except for the weakest intercombination lines. (The $2s2p$ $^3P_1^\circ$–$2p^2$ $^1S_0$ transition has been omitted from this tabulation, since its f-value is considerably smaller than those of the other lines of the array.)

A number of sources of reliable data, from relativistic calculations, are available for the $2s$–$2p$ transitions. However, with the exception of some of the weaker lines, they all agree very well with the results of Cheng et al.[1] The latter are quoted exclusively here since they provide data from a single set of comprehensive calculations, all done at a uniform and reasonably accurate level of approximation, for the valence shell $2s$–$2p$ transitions for all ions of the isoelectronic sequence.

Beam-foil lifetime measurements are also available from the work of Buchet et al.[2] for some of these transitions, and they all agree well with the values quoted here. A preliminary result derived from the beam-foil lifetime experiment by Dietrich et al.[3] for the $2s^2$ $^1S_0$–$2s2p$ $^3P_1^\circ$ transition deviates considerably from the theoretical value of Cheng et al.[1] which we have selected, although the experimental result with its stated error limits lies within the theoretical uncertainty given here. Nussbaumer and Storey[4] have calculated A-values for all downward transitions from levels of the configurations $2\ell'n\ell$ ($\ell'=s, p$; $n=3,4$; $\ell=s, p, d$), as well as $2s4f$ and $2p4f$, in order to construct simulated decay curves for the $2s2p$ $^3P_1^\circ$ and $^1P_1^\circ$ levels. They conclude that the suspected blending of $2s^2$ $^1S_0$–$2s2p$ $^3P_1^\circ$ with the second-order line of $2s^2$ $^1S_0$–$2s2p$ $^1P_1^\circ$ in the experiment of Dietrich et al. should not have been a significant factor in determining the lifetime, and that there must have been additional problems in the experiment.

The f-values for the $2s^2$–$2s3p$, $2s2p$–$2p3p$, $2s2p$–$2s3s$, $2p^2$–$2p3s$, $2s2p$–$2s3d$, and $2p^2$–$2p3d$ arrays of transitions are taken from the work of Fawcett,[5] who used Cowan's version of the relativistic Hartree-Fock method with intermediate coupling and configuration interaction. This work provides a comprehensive set of data for the entire isoelectronic sequence, calculated at a uniform level of approximation. Some of these transitions, for some ions

of this sequence, have also been calculated by Bhatia et al.[6] using the program SUPERSTRUCTURE, which includes configuration interaction and intermediate coupling. Where they overlap, these two sets of calculations agree to within the uncertainties assigned here. Transitions involving the $J=1$ levels of $2p3p$ ^3S and ^3P have been omitted because of erratic behavior of the f-values along the sequence.

Data for the $2s$–$4p$ and $2p$–$4s$ transitions, as well as for lines of the $2s2p$–$2s4d$ array, have been taken from the comprehensive tabulation of Fawcett et al.,[7] which reports f-values calculated at the same level of approximation as described above. Additional transitions, not available in Ref. 5, are taken from the calculations of Nussbaumer,[8] which are based on the Thomas-Fermi-Dirac model including configuration interaction. The f-values for transitions of the $2p^2$–$2p4d$ array, as well as transitions where the upper state involves principal quantum number 5, have been taken from the Hartree-XR calculations of Bromage et al.[9]

Weak transitions for cases where there is severe mixing of the configurations or LS terms have been omitted. Some multiplet f-values for transitions involving the outer electron alone, $2s3s$–$2s3p$ and $2s3p$–$2s3d$, have been interpolated from systematic trends along the isoelectronic sequence and assigned a low accuracy.

References

[1] K. T. Cheng, Y.-K. Kim, and J. P. Desclaux, At. Data Nucl. Data Tables **24**, 111 (1979).

[2] J. P. Buchet, M. C. Buchet-Poulizac, A. Denis, J. Desesquelles, M. Druetta, J. P. Grandin, M. Huet, X. Husson, and D. Lecler, Phys. Rev. A **30**, 309 (1984).

[3] D. D. Dietrich, J. A. Leavitt, S. Bashkin, J. G. Conway, H. Gould, D. MacDonald, R. Marrus, B. M. Johnson, and D. J. Pegg, Phys. Rev. A **18**, 208 (1978).

[4] H. Nussbaumer and P. J. Storey, J. Phys. B **12**, 1647 (1979).

[5] B. C. Fawcett, At. Data Nucl. Data Tables **30**, 1 (1984); **33**, 479 (1985).

[6] A. K. Bhatia, U. Feldman, and J. F. Seeley, At. Data Nucl. Data Tables **35**, 449 (1986).

[7] B. C. Fawcett, C. Jordan, J. R. Lemen, and K. J. H. Phillips, Rutherford Appleton Laboratory Report RAL–86–094 (1986).

[8] H. Nussbaumer, Astron. Astrophys. **16**, 77 (1972).

[9] G. E. Bromage, R. D. Cowan, B. C. Fawcett and A. Ridgeley, J. Opt. Soc. Am. **68**, 48 (1978).

Fe XXIII: Allowed transitions

No.	Transition Array	Multiplet	λ (Å)	E_i (cm^{-1})	E_k (cm^{-1})	g_i	g_k	A_{ki} (10^8 s^{-1})	f_{ik}	S (at. u.)	log gf	Accuracy	Source
1.	$2s^2$–$2s2p$	^{1}S – ^{3}P°											
			263.76	0	379130	1	3	0.48	0.0015	0.0013	−2.82	D	1
2.		^{1}S – ^{1}P°	132.84	0	752840	1	3	195	0.155	0.0678	−0.810	B	1
3.	$2s2p$–$2p^2$	^{3}P° – ^{3}P	*162.12*	*427160*	*1044000*	9	9	133	0.0523	0.251	−0.328	B	1
			166.74	471780	1071700	5	5	75.8	0.0316	0.0867	−0.801	B	1
			154.27	379130	1027200	3	3	41.8	0.0149	0.0227	−1.350	B	1
			180.10	471780	1027200	5	3	44.6	0.0130	0.0385	−1.187	B	1
			173.31	379130	956100	3	1	123	0.0185	0.0317	−1.256	B	1
			144.36	379130	1071700	3	5	54.3	0.0283	0.0403	−1.071	B	1
			147.24	348180	1027200	1	3	65.9	0.0643	0.0312	−1.192	B	1
4.		^{3}P° – ^{1}D											
			136.53	471780	1204200	5	5	48.3	0.0135	0.0303	−1.171	C	1
			[121.20]	379130	1204200	3	5	4.4	0.0016	0.0019	−2.31	D	1
5.		^{1}P° – ^{3}P											
			[313.62]	752840	1071700	3	5	3.7	0.0090	0.028	−1.57	D	1
			[364.48]	752840	1027200	3	3	0.075	1.5(−4)[a]	5.4(−4)	−3.35	E	1
			[491.98]	752840	956100	3	1	0.23	2.8(−4)	0.0014	−3.08	E	1
6.		^{1}P° – ^{1}D	221.33	752840	1204200	3	5	46.1	0.0564	0.123	−0.772	B	1
7.		^{1}P° – ^{1}S	149.22	752840	1423000	3	1	327	0.0364	0.0536	−0.962	B	1
8.	$2s3d$–$2p3d$	^{3}D – ^{3}F°											
						7	9	24				C+	4
9.	$2s^2$–$2s3p$	^{1}S – ^{3}P°											
			11.018	0	9076000	1	3	4.9(+4)	0.27	0.0098	−0.57	C−	5
10.		^{1}S – ^{1}P°	10.979	0	9107000	1	3	7.9(+4)	0.43	0.016	−0.37	C−	5
11.	$2s2p$–$2p3p$	^{3}P° – ^{3}D											
			10.927	471780	9624000	5	7	6.0(+4)	0.15	0.027	−0.12	C−	5
			10.934	379130	9524000	3	5	5.4(+4)	0.16	0.017	−0.32	C−	5
			[10.981]	348180	9455000	1	3	1.5(+4)	0.084	0.0030	−1.08	D	5
			[11.047]	471780	9524000	5	5	1900	0.0034	6.2(−4)	−1.77	D	5
			[11.018]	379130	9455000	3	3	2.7(+4)	0.050	0.0054	−0.82	D	5
12.		^{3}P° – ^{3}P											
			[10.902]	471780	9644000	5	5	5.3(+4)	0.094	0.017	−0.33	C−	5
			[10.910]	379130	[9545000]	3	1	6.7(+4)	0.040	0.0043	−0.92	D	5
			[10.793]	379130	9644000	3	5	2200	0.0063	6.7(−4)	−1.72	D	5
13.		^{1}P° – ^{1}P	[11.482]	752840	[9462000]	3	3	1.7(+4)	0.033	0.0037	−1.00	D	5
14.		^{1}P° – ^{3}D											
			[11.491]	752840	9455000	3	3	1.5(+4)	0.029	0.0033	−1.06	D	5

Fe XXIII: Allowed transitions — Continued

No.	Transition Array	Multiplet	λ (Å)	E_i (cm^{-1})	E_k (cm^{-1})	g_i	g_k	A_{ki} (10^8 s^{-1})	f_{ik}	S (at. u.)	log gf	Accuracy	Source
15.		^{1}P° – ^{3}P											
			[11.247]	752840	9644000	3	5	2.1(+4)	0.067	0.0074	−0.70	D	5
			[11.255]	752840	[9638000]	3	3	3.7(+4)	0.070	0.0078	−0.68	C−	5
16.		^{1}P° – ^{1}D	[11.165]	752840	9709000	3	5	6.7(+4)	0.21	0.023	−0.20	C−	5
17.		^{1}P° – ^{1}S	[11.086]	752840	[9773000]	3	1	6.5(+4)	0.040	0.0044	−0.92	D	5
18.	$2p^2$–$2s3p$	^{3}P – ^{3}P°											
						5	5	400				C	4
						3	1	380				C	4
						3	5	300				C	4
19.		^{1}D – ^{3}P°											
						5	5	79				D	4
20.		^{1}D – ^{1}P°	[12.654]	1204200	9107000	5	3	1700	0.0024	5.1(−4)	−1.91	D	8
21.		^{1}S – ^{1}P°	[13.014]	1423000	9107000	1	3	1400	0.011	4.6(−4)	−1.97	D	8
22.	$2s^2$–$2s4p$	^{1}S – ^{3}P°											
			[8.320]	0	[12020000]	1	3	1.3(+4)	0.039	0.0011	−1.41	D	7
23.		^{1}S – ^{1}P°	[8.307]	0	[12038000]	1	3	4.8(+4)	0.15	0.0041	−0.82	D	7
24.	$2s2p$–$2p4p$	^{3}P° – ^{3}D											
			[8.273]	471780	12560000	5	7	2.9(+4)	0.042	0.0057	−0.68	C−	7
			[8.292]	379130	[12439000]	3	5	2.6(+4)	0.044	0.0036	−0.88	D	7
			[8.291]	348180	[12410000]	1	3	9100	0.028	7.6(−4)	−1.55	D	7
			[8.312]	379130	[12410000]	3	3	1.5(+4)	0.016	0.0013	−1.32	D	7
25.		^{3}P° – ^{3}P											
			[8.277]	471780	[12554000]	5	5	2.2(+4)	0.023	0.0031	−0.94	D	7
			[8.293]	379130	[12437000]	3	3	8100	0.0084	6.9(−4)	−1.60	D	7
			[8.290]	379130	[12442000]	3	1	2.6(+4)	0.0090	7.4(−4)	−1.57	D	7
			[8.272]	348180	[12437000]	1	3	2.0(+4)	0.062	0.0017	−1.21	D	7
26.		^{3}P° – ^{1}P											
			[8.281]	471780	[12547000]	5	3	8300	0.0051	7.0(−4)	−1.59	D	7
			[8.218]	379130	[12547000]	3	3	4600	0.0047	3.8(−4)	−1.85	D	7
27.		^{3}P° – ^{3}S											
			[8.273]	471780	[12559000]	5	3	2.4(+4)	0.015	0.0020	−1.12	D	7
28.		^{3}P° – ^{1}D											
			[8.260]	471780	[12579000]	5	5	7200	0.0074	0.0010	−1.43	D	7
			[8.197]	379130	[12579000]	3	5	2800	0.0047	3.8(−4)	−1.85	D	7
29.		^{1}P° – ^{3}D											
			[8.557]	752840	[12439000]	3	5	4700	0.0086	7.3(−4)	−1.59	D	7
			[8.578]	752840	[12410000]	3	3	6600	0.0073	6.2(−4)	−1.66	D	7

Fe XXIII: Allowed transitions — Continued

No.	Transition Array	Multiplet	λ (Å)	E_i (cm^{-1})	E_k (cm^{-1})	g_i	g_k	A_{ki} (10^8 s^{-1})	f_{ik}	S (at. u.)	log gf	Accuracy	Source
30.		$^1P°$ – ^3P											
			[8.474]	752840	[12554000]	3	5	8400	0.015	0.0013	−1.35	D	7
			[8.559]	752840	[12437000]	3	3	4300	0.0047	4.0(−4)	−1.85	D	7
31.		$^1P°$ – ^1P	[8.479]	752840	[12547000]	3	3	2.2(+4)	0.024	0.0020	−1.14	D	7
32.		$^1P°$ – ^3S	[8.470]	752840	[12559000]	3	3	8400	0.0090	7.5(−4)	−1.57	D	7
33.		$^1P°$ – ^1D	[8.456]	752840	[12579000]	3	5	2.2(+4)	0.040	0.0033	−0.92	D	7
34.		$^1P°$ – ^1S	[8.437]	752840	[12606000]	3	1	2.1(+4)	0.0073	6.1(−4)	−1.66	D	7
35.	$2s^2$–$2s\,5p$	^1S – $^1P°$	7.472	0	13383000	1	3	2.5(+4)	0.063	0.0015	−1.20	D	9
36.	$2s\,2p$–$2p\,5p$	$^3P°$ – ^3D											
			7.445	471780	13904000	5	7	1.5(+4)	0.018	0.0022	−1.05	D	9
37.	$2s\,2p$–$2s\,3s$	$^3P°$ – ^3S	*11.811*	*427160*	8894000	9	3	3.7(+4)	0.026	0.0091	−0.63	D	5
			[11.873]	471780	8894000	5	3	2.1(+4)	0.026	0.0051	−0.89	D	5
			[11.744]	379130	8894000	3	3	1.3(+4)	0.026	0.0030	−1.10	D	5
			[11.702]	348180	8894000	1	3	4400	0.027	0.0010	−1.57	D	5
38.		$^1P°$ – ^1S	[12.176]	752840	[8966000]	3	1	1.5(+4)	0.011	0.0013	−1.48	D	5
39.	$2p^2$–$2p\,3s$	^3P – $^3P°$	*11.936*	*1044000*	[*9422000*]	9	9	2.5(+4)	0.054	0.019	−0.32	D	5
			[11.897]	1071700	[9477000]	5	5	1.6(+4)	0.034	0.0067	−0.77	D	5
			[11.984]	1027200	[9372000]	3	3	4600	0.010	0.0012	−1.52	D	5
			[12.048]	1071700	[9372000]	5	3	1.1(+4)	0.015	0.0030	−1.12	D	5
			[12.095]	1027200	9295000	3	1	2.3(+4)	0.017	0.0020	−1.29	D	5
			[11.835]	1027200	[9477000]	3	5	9400	0.033	0.0039	−1.00	D	5
			[11.882]	956100	[9372000]	1	3	8800	0.056	0.0022	−1.25	D	5
40.		^1D – $^1P°$	[12.098]	1204200	9470000	5	3	2.1(+4)	0.028	0.0056	−0.85	D	5
41.		^1S – $^1P°$	[12.427]	1423000	9470000	1	3	7900	0.055	0.0023	−1.26	D	5
42.	$2s\,2p$–$2s\,3d$	$^3P°$ – ^3D	*11.388*	*427160*	*9208000*	9	15	2.2(+5)	0.71	0.24	0.81	C−	5
			[11.441]	471780	9212000	5	7	2.2(+5)	0.60	0.11	0.48	C−	5
			[11.325]	379130	9209000	3	5	1.7(+5)	0.55	0.062	0.22	C−	5
			[11.298]	348180	9199000	1	3	1.3(+5)	0.74	0.028	−0.13	C−	5
			[11.445]	471780	9209000	5	5	5.6(+4)	0.11	0.021	−0.26	C−	5
			[11.338]	379130	9199000	3	3	9.3(+4)	0.18	0.020	−0.27	C−	5
			[11.458]	471780	9199000	5	3	6100	0.0072	0.0014	−1.44	D	5
43.		$^1P°$ – ^1D	[11.737]	752840	9273000	3	5	1.8(+5)	0.61	0.071	0.26	C−	5
44.	$2p^2$–$2p\,3d$	^3P – $^3F°$											
			[11.691]	1071700	9625000	5	7	7.7(+4)	0.220	0.0423	0.041	C−	5
45.		^3P – $^3D°$	*11.523*	*1044000*	[*9722000*]	9	15	2.1(+5)	0.70	0.24	0.80	C−	5
			[11.524]	1071700	9749000	5	7	2.3(+5)	0.63	0.12	0.50	C−	5
			[11.485]	1027200	[9734000]	3	5	1.40(+5)	0.463	0.053	0.143	C−	5
			[11.520]	956100	9637000	1	3	2.16(+5)	1.29	0.0489	0.111	C−	5
			[11.544]	1071700	[9734000]	5	5	1.5(+4)	0.030	0.0057	−0.82	D	5
			[11.615]	1027200	9637000	3	3	4.4(+4)	0.090	0.010	−0.57	C−	5
			[11.675]	1071700	9637000	5	3	2100	0.0026	5.0(−4)	−1.89	D	5

Fe XXIII: Allowed transitions — Continued

No.	Transition Array	Multiplet	λ (Å)	E_i (cm^{-1})	E_k (cm^{-1})	g_i	g_k	A_{ki} (10^8 s^{-1})	f_{ik}	S (at. u.)	log gf	Accuracy	Source
46.		^{3}P – ^{1}D°											
			[11.674]	1071700	9638000	5	5	2.3(+4)	0.048	0.0092	−0.62	C−	5
			[11.613]	1027200	9638000	3	5	1.0(+5)	0.34	0.039	0.01	D	5
47.		^{3}P – ^{3}P°	*11.469*	*1044000*	*[9763000]*	9	9	1.5(+5)	0.29	0.10	0.42	C−	5
			[11.519]	1071700	9753000	5	5	1.16(+5)	0.230	0.0436	0.061	C−	5
			[11.433]	1027200	[9774000]	3	3	1.2(+5)	0.24	0.027	−0.14	C−	5
			[11.491]	1071700	[9774000]	5	3	5.9(+4)	0.070	0.013	−0.46	C−	5
			[11.429]	1027200	[9777000]	3	1	1.7(+5)	0.11	0.012	−0.48	C−	5
			[11.460]	1027200	9753000	3	5	1.9(+4)	0.063	0.0071	−0.72	D	5
			[11.341]	956100	[9774000]	1	3	360	0.0021	7.8(−5)	−2.68	D	5
48.		^{1}D – ^{3}F°											
			[11.846]	1204200	[9646000]	5	5	8100	0.017	0.0033	−1.07	D	5
49.		^{1}D – ^{1}D°	[11.857]	1204200	9638000	5	5	2.3(+4)	0.048	0.0094	−0.62	C−	5
50.		^{1}D – ^{3}P°											
			[11.698]	1204200	9753000	5	5	7.3(+4)	0.15	0.029	−0.12	C−	5
51.		^{1}D – ^{1}P°	[11.596]	1204200	9828000	5	3	1.2(+4)	0.015	0.0029	−1.12	D	5
52.		^{1}D – ^{1}F°	[11.593]	1204200	9830000	5	7	3.58(+5)	1.01	0.193	0.70	C−	5
53.		^{1}S – ^{1}P°	[11.898]	1423000	9828000	1	3	2.03(+5)	1.29	0.051	0.111	C−	5
54.	$2s\,2p$–$2s\,4s$	^{3}P° – ^{3}S											
			[8.710]	471780	[11953000]	5	3	9700	0.0066	9.5(−4)	−1.48	D	7
			[8.640]	379130	[11953000]	3	3	5700	0.0064	5.5(−4)	−1.72	D	7
55.		^{1}P° – ^{1}S	[8.910]	752840	[11976000]	3	1	9300	0.0037	3.3(−4)	−1.95	D	7
56.	$2p^2$–$2p\,4s$	^{3}P – ^{3}P°											
			[8.756]	1071700	[12493000]	5	5	2100	0.0024	3.5(−4)	−1.92	D	7
			[8.853]	1071700	[12367000]	5	3	7200	0.0051	7.4(−4)	−1.59	D	7
			[8.822]	1027200	[12362000]	3	1	1.2(+4)	0.0047	4.1(−4)	−1.85	D	7
			[8.722]	1027200	[12493000]	3	5	2300	0.0043	3.7(−4)	−1.89	D	7
			[8.764]	956100	[12367000]	1	3	4300	0.015	4.3(−4)	−1.82	D	7
57.		^{1}D – ^{3}P°											
			[8.858]	1204200	[12493000]	5	5	4900	0.0058	8.5(−4)	−1.54	D	7
58.		^{1}D – ^{1}P°	[8.853]	1204200	[12500000]	5	3	1.3(+4)	0.0089	0.0013	−1.35	D	7
59.		^{1}S – ^{1}P°	[9.028]	1423000	[12500000]	1	3	4900	0.018	5.3(−4)	−1.74	D	7
60.	$2s\,2p$–$2s\,4d$	^{3}P° – ^{3}D											
			[8.614]	471780	12081000	5	7	7.7(+4)	0.12	0.017	−0.22	C−	7
			[8.550]	379130	12075000	3	5	6.0(+4)	0.11	0.0093	−0.48	C−	7
			[8.529]	348180	12073000	1	3	4.3(+4)	0.14	0.0039	−0.85	D	7
			[8.618]	471780	12075000	5	5	1.9(+4)	0.021	0.0030	−0.98	D	7
			[8.552]	379130	12073000	3	3	3.2(+4)	0.035	0.0030	−0.98	D	7
61.		^{1}P° – ^{1}D	[8.814]	752840	12098000	3	5	6.2(+4)	0.12	0.010	−0.44	C−	7

Fe XXIII: Allowed transitions — Continued

No.	Transition Array	Multiplet	λ (Å)	E_i (cm^{-1})	E_k (cm^{-1})	g_i	g_k	A_{ki} (10^8 s^{-1})	f_{ik}	S (at. u.)	log gf	Accuracy	Source
62.	$2p^2$-$2p\,4d$	^{3}P – ^{3}F°											
			8.764	1071700	12484000	5	7	4.6(+4)	0.074	0.011	−0.43	D	9
63.		^{3}P – ^{3}D°											
			8.669	1071700	12603000	5	7	6.1(+4)	0.096	0.014	−0.32	D	9
			[8.672]	956100	12488000	1	3	6.8(+4)	0.23	0.0066	−0.64	D	9
64.		^{3}P – ^{3}P°											
			8.664	1027200	12615000	3	3	4.4(+4)	0.050	0.0043	−0.82	D	9
65.		^{1}D – ^{1}F°	8.752	1204200	12631000	5	7	1.2(+5)	0.19	0.027	−0.02	D	9
66.	$2s\,2p$-$2s\,5d$	^{3}P° – ^{3}D											
			7.733	471780	13404000	5	7	3.0(+4)	0.038	0.0048	−0.72	D	9
			7.680	379130	13400000	3	5	2.5(+4)	0.037	0.0028	−0.95	D	9
			7.680	348180	13369000	1	3	1.8(+4)	0.047	0.0012	−1.33	D	9
67.		^{1}P° – ^{1}D	7.883	752840	13438000	3	5	2.8(+4)	0.043	0.0033	−0.89	D	9
68.	$2p^2$-$2p\,5d$	^{3}P – ^{3}F°											
			7.854	1071700	13804000	5	7	2.3(+4)	0.030	0.0039	−0.82	D	9
69.		^{3}P – ^{3}D°											
			7.778	1071700	13929000	5	7	2.5(+4)	0.032	0.0041	−0.80	D	9
			7.826	1027200	13805000	3	5	2.6(+4)	0.040	0.0031	−0.92	D	9
70.		^{1}D – ^{1}F°	7.849	1204200	13945000	5	7	4.9(+4)	0.064	0.0083	−0.49	D	9
71.	$2s\,3s$-$2s\,3p$	^{3}S – ^{3}P°				3	9		0.12		−0.44	D	interp.
72.		^{1}S – ^{1}P°				1	3		0.050		−1.30	E	interp.
73.	$2s\,3p$-$2s\,3d$	^{3}P° – ^{3}D				9	15		0.027		−0.61	E	interp.
74.		^{1}P° – ^{1}D				3	5		0.047		−0.85	E	interp.

[a]The number in parentheses following the tabulated value indicates the power of ten by which this value has to be multiplied.

Fe XXIII

Forbidden Transitions

Transition probabilities for magnetic dipole and electric quadrupole transitions within the $2s\,2p$ and $2p^2$ configurations and for lines of the $2s^2$-$2p^2$ transition array, as well as for magnetic quadrupole transitions of the arrays $2s^2$-$2s\,2p$ and $2s\,2p$-$2p^2$, were calculated by Glass,[1,2,3] using relativistic intermediate-coupling wavefunctions.

Glass carried out a rather extensive treatment of configuration interaction and achieved very good agreement between calculated and experimentally obtained transition energies, usually between 1 and 2%. We have tabulated the transition probability data which he obtained with experimentally derived wavelengths.

Of his results for electric quadrupole transitions, we have tabulated data for only the strongest lines. The calculated A-values for the M2 transitions and for the remaining E2 transitions are extremely small, and are therefore not listed.

References

[1]R. Glass, Astrophys. Space Sci. **91**, 417 (1983).
[2]R. Glass, Astrophys. Space Sci. **92**, 307 (1983).
[3]R. Glass, Astrophys. Space Sci. **87**, 41 (1983).

Fe XXIII: Forbidden transitions

No.	Transition Array	Multiplet	λ (Å)	E_i (cm^{-1})	E_k (cm^{-1})	g_i	g_k	Type of transition	A_{ki} (s^{-1})	S (at. u.)	Accuracy	Source
1.	$2s^2$–$2p^2$	^1S – ^3P										
			[93.31]	0	1071700	1	5	E2	2400	5.1($-$5)[a]	E	2
			[97.35]	0	1027200	1	3	M1	1.3($+$4)	0.0013	E	1
2.	$2s2p$–$2s2p$	$^3P°$ – $^3P°$										
			1079.3	379130	471780	3	5	M1	9980	2.33	C+	1
			[3230.1]	348180	379130	1	3	M1	473	1.77	C+	1
3.		$^3P°$ – $^1P°$										
			[355.80]	471780	752840	5	3	M1	1.2($+$4)	0.060	D	1
			[267.59]	379130	752840	3	3	M1	1.2($+$4)	0.026	D	1
			"	"	"	3	3	E2	47	1.2($-$4)	E	2
			[247.12]	348180	752840	1	3	M1	2.9($+$4)	0.049	D	1
4.	$2p^2$–$2p^2$	^3P – ^3P										
			[2246.5]	1027200	1071700	3	5	M1	910	1.9	C	1
			[1406.5]	956100	1027200	1	3	M1	7200	2.2	C	1
5.		^3P – ^1D										
			[754.7]	1071700	1204200	5	5	M1	1.9($+$4)	1.51	C	1
			[565.0]	1027200	1204200	3	5	M1	2.1($+$4)	0.70	D+	1
6.		^3P – ^1S										
			[252.7]	1027200	1423000	3	1	M1	2.3($+$5)	0.14	D	1
7.		^1D – ^1S										
			[457.0]	1204200	1423000	5	1	E2	180	0.0021	E	2

[a]The number in parentheses following the tabulated value indicates the power of ten by which this value has to be multipled.

Fe XXIV

Li Isoelectronic Sequence

Ground State: $1s^2 2s \, ^2S_{1/2}$

Ionization Energy: 2045.8 eV = 16500000 cm^{-1}

Allowed Transitions

List of tabulated lines

Wavelength (Å)	No.	Wavelength (Å)	No.	Wavelength (Å)	No.	Wavelength (Å)	No.
1.8523	8	1.876	1	8.316	18	30.7	22
1.8552	3	1.8767	5	8.369	17	30.9	22
1.8563	8	1.893	4	10.619	10	31.5	26
1.8572	3	1.898	4	10.663	10	31.9	26
1.858	7	6.58	14	11.030	16	37.0	36
1.8604	2	6.749	21	11.171	16	37.3	36
1.8614	7	6.787	13	11.187	16	44.2	32
1.8626	6	6.808	21	11.262	15	44.8	35
1.8627	7	6.972	20	11.422	15	45.2	35
1.8637	2	7.033	20	17.1	29	67.6	31
1.8655	6	7.169	12	17.3	29	68.5	34
1.8672	7	7.370	19	18.3	24	69.4	34
1.8678	6	7.438	19	18.7	28	192.04	9
1.8700	5	7.983	11	18.8	28	255.10	9
1.8721	5	7.993	11	21.4	23		
1.8730	1	8.231	18	21.8	27		
1.8739	5	8.280	17	22.0	27		

Transition probabilities for the inner-shell transitions to doubly excited $n = 2$ states are the results of the multiconfiguration Dirac-Fock (MCDF) calculations of Hata and Grant.[1] Their results are in good agreement with those of the multiconfiguration scaled Thomas-Fermi calculations with intermediate coupling by Bely-Dubau et al.[2]

Oscillator strengths for lines of the principal ($2s-2p$) resonance multiplet are the results of the MCDF calculations of Cheng et al.,[3] which include a perturbative treatment of the Breit interaction and the Lamb shift. Other sources of reliable theoretical data for these $2s-2p$ transitions are the Hartree-Fock line strength calculations of Weiss[4] with relativistic corrections and the MCDF approach of Armstrong et al.[5]

Lifetimes of the $2p$ levels have been determined by Dietrich et al.[6] using the beam-foil technique. The associated oscillator strengths for the $2s-2p$ transitions are in excellent agreement with the results mentioned above.

The results of the relativistic Hartree-Fock calculations of Kim and Desclaux[7] were averaged with the results of Armstrong et al.[5] for the $2s-3p$ transitions. The data of Ref. 5 are quoted for the lines of the $2p-3d$ multiplet too.

The results of the scaled Thomas-Fermi calculations of Hayes[8] are tabulated for the $2p-3s$ transitions. He used the Breit-Pauli approximation to account for relativistic effects. The Hartree-Fock results of Doschek et al.[9] that included configuration interaction and relativistic corrections are quoted for transitions of the type $2l-4l'$. The $2p-5d$ f-values are the results of the Hartree-Fock calculations with statistical exchange (HX) of Burkhalter et al.[10]

The f-value for the $3d-4f$ transition was taken from a study of systematic trends along isoelectronic sequences by Smith and Wiese.[11] The tabulated data for the remaining transitions were taken from the theoretical analysis of Martin and Wiese,[12] which was based on a generalized study of systematic trends for several spectral series of the lithium isoelectronic sequence. For these transitions, no relativistic calculations were available. However, the relativistic calculations of Younger and Weiss[13] for the hydrogen isoelectronic sequence provide a means of assessing the magnitude of relativistic corrections since the Li sequence is very similar in structure to the H sequence. For those transitions for which relativistic effects were estimated to be significant (specifically, whenever the ratio of the weighted relativistic hydrogenic f-values gf_{ik} of any two lines within a multiplet was found to deviate from the corresponding LS-coupling line-strength ratio by more than 5% for the appropriate value of the nuclear charge Z), the f-values were excluded from the compilation. A more detailed discussion of this comparison is given in Ref. 12.

Transition probability data are available for numerous transitions involving doubly excited states with the spectator electron occupying the $n = 3$ shell, or higher.[1,14,15] These have not been tabulated, however, since they be-

J. Phys. Chem. Ref. Data, Vol. 17, Suppl. 4, 1988

long to, or are very close to belonging to, the unresolved satellites of the helium-like ion.

References

[1] J. Hata and I. P. Grant, Mon. Not R. Astron. Soc. **211**, 549 (1984).

[2] F. Bely-Dubau, A. H. Gabriel, and S. Volonte, Mon. Not. R. Astron. Soc. **186**, 405 (1979).

[3] K. T. Cheng, Y.-K. Kim, and J. P. Desclaux, At. Data Nucl. Data Tables **24**, 111 (1979).

[4] A. W. Weiss, J. Quant. Spectrosc. Radiat. Transfer **18**, 481 (1977).

[5] L. Armstrong, Jr., W. R. Fielder, and D. L. Lin, Phys. Rev. A **14**, 1114 (1976).

[6] D. D. Dietrich, J. A. Leavitt, S. Bashkin, J. G. Conway, H. Gould, D. MacDonald, R. Marrus, B. M. Johnson, and D. J. Pegg, Phys. Rev. A **18**, 208 (1978).

[7] Y.-K. Kim and J. P. Desclaux, Phys. Rev. Lett. **36**, 139 (1976) and private communication.

[8] M. A. Hayes, Mon. Not. R. Astron. Soc. **189**, 55P (1979).

[9] G. A. Doschek, J. F. Meekins, and R. D. Cowan, Astrophys. J. **177**, 261 (1972).

[10] P. G. Burkhalter, C. M. Dozier, C. Stallings, and R. D. Cowan, J. Appl. Phys. **49**, 1092 (1978).

[11] M. W. Smith and W. L. Wiese, Astrophys. J. Suppl. Ser. **23**, No. 196, 103 (1971).

[12] G. A. Martin and W. L. Wiese, J. Phys. Chem. Ref. Data **5**, 537 (1976).

[13] S. M. Younger and A. W. Weiss, J. Res. Nat. Bur. Stand., Sect. A **79**, 629 (1975).

[14] L. A. Vainshtein and U. I. Safronova, At. Data Nucl. Data Tables **25**, 311 (1980).

[15] F. Bely-Dubau, A. H. Gabriel, and S. Volonte, Mon. Not. R. Astron. Soc. **189**, 801 (1979).

Fe XXIV: Allowed transitions

No.	Transition Array	Multiplet	λ (Å)	E_i (cm^{-1})	E_k (cm^{-1})	g_i	g_k	A_{ki} (10^8 s^{-1})	f_{ik}	S (at. u.)	log gf	Accuracy	Source
1.	$1s^2 2s-$ $1s(^2S)2s2p(^3P°)$	$^2S - ^4P°$											
			1.8730	0	53390000	2	4	1.5(+5)^a	0.016	1.9(−4)	−1.50	D+	1
			[1.874]			2	2	4.2(+4)	0.0022	2.7(−5)	−2.35	D+	1
2.		$^2S - ^2P°$	*1.8615*	0	*53720000*	2	6	6.70(+5)	0.104	0.00128	−0.680	C	1
			1.8604	0	53752000	2	4	4.4(+4)	0.0046	5.6(−5)	−2.04	D	1
			1.8637	0	53657000	2	2	1.91(+6)	0.0995	0.00122	−0.701	C	1
3.	$1s^2 2s-$ $1s(^2S)2s2p(^1P°)$	$^2S - ^2P°$	*1.8559*	0	*53883000*	2	6	4.23(+6)	0.656	0.00801	0.118	C	1
			1.8552	0	53903000	2	4	4.82(+6)	0.497	0.00608	−0.002	C	1
			1.8572	0	53844000	2	2	3.06(+6)	0.158	0.00193	−0.500	C	1
4.	$1s^2 2p - 1s 2s^2$	$^2P° - ^2S$	*1.895*			6	2	1.9(+5)	0.0035	1.3(−4)	−1.68	D+	1
			[1.897]			4	2	9.8(+4)	0.0026	6.6(−5)	−1.98	D+	1
			[1.891]			2	2	9.7(+4)	0.0052	6.5(−5)	−1.98	D+	1
5.	$1s^2 2p - 1s 2p^2$	$^2P° - ^4P$											
			1.8721	520720	53937000	4	6	3.2(+5)	0.025	6.2(−4)	−1.00	D	1
			1.8700	392000	53877000	2	4	1000	1.0(−4)	1.3(−6)	−3.68	E	1
			1.8739	520720	53877000	4	4	8.3(+4)	0.0044	1.1(−4)	−1.76	D	1
			1.8721	392000	53807000	2	2	2.0(+5)	0.011	1.3(−4)	−1.68	D	1
			1.8767	520720	53807000	4	2	2500	6.6(−5)	1.6(−6)	−3.58	E	1
6.		$^2P° - ^2D$	*1.8648*	*477810*	*54104000*	6	10	2.6(+6)	0.23	0.0084	0.14	C	1
			1.8655	520720	54126000	4	6	2.14(+6)	0.167	0.00411	−0.174	C	1
			1.8626	392000	54070000	2	4	3.16(+6)	0.329	0.00403	−0.182	C	1
			1.8678	520720	54070000	4	4	3.5(+5)	0.018	4.5(−4)	−1.14	D	1
7.		$^2P° - ^2P$	*1.8618*	*477810*	*54188000*	6	6	6.59(+6)	0.343	0.0126	0.313	C	1
			1.8614	520720	54244000	4	4	6.24(+6)	0.324	0.00795	0.113	C	1
			[1.8627]	392000	54077000	2	2	5.47(+6)	0.285	0.00349	−0.245	C	1
			1.8672	520720	54077000	4	2	1.63(+6)	0.0426	0.00105	−0.767	C	1
			1.858	392000	54244000	2	4	1.2(+5)	0.012	1.5(−4)	−1.60	D	1

Fe xxiv: Allowed transitions — Continued

No.	Transition Array	Multiplet	λ (Å)	E_i (cm^{-1})	E_k (cm^{-1})	g_i	g_k	A_{ki} (10^8 s^{-1})	f_{ik}	S (at. u.)	log gf	Accuracy	Source
8.		^{2}P° – ^{2}S	1.8550	477810	54385000	6	2	2.59(+6)	0.0450	0.00163	−0.574	C	1
			1.8563	520720	54385000	4	2	2.43(+6)	0.0641	0.00157	−0.591	C	1
			1.8523	392000	54385000	2	2	1.0(+5)	0.0051	6.3(−5)	−1.99	D	1
9.	2s–2p	^{2}S – ^{2}P°	209.29	0	477810	2	6	33.2	0.0654	0.0901	−0.883	B+	3
			192.04	0	520720	2	4	43.2	0.0478	0.0604	−1.020	B+	3
			255.10	0	392000	2	2	18.1	0.0177	0.0297	−1.451	B+	3
10.	2s–3p	^{2}S – ^{2}P°	10.634	0	9404100	2	6	7.36(+4)	0.374	0.0262	−0.126	B+	5,7
			10.619	0	9417100	2	4	7.28(+4)	0.246	0.0172	−0.308	B+	5,7
			10.663	0	9378200	2	2	7.51(+4)	0.128	0.00899	−0.592	B+	5,7
11.	2s–4p	^{2}S – ^{2}P°	7.987	0	12520000	2	6	3.4(+4)	0.097	0.0051	−0.71	C+	9
			7.983	0	12530000	2	4	3.43(+4)	0.0655	0.00344	−0.883	C+	9
			7.993	0	12510000	2	2	3.4(+4)	0.033	0.0017	−1.18	C+	9
12.	2s–5p	^{2}S – ^{2}P°	7.169	0	13950000	2	6	1.7(+4)	0.040	0.0019	−1.10	C+	12
13.	2s–6p	^{2}S – ^{2}P°	6.787	0	14730000	2	6	1.02(+4)	0.0212	9.47(−4)	−1.373	C+	12
14.	2s–7p	^{2}S – ^{2}P°	[6.58]			2	6	6420	0.0125	5.4(−4)	−1.602	C+	12
15.	2p–3s	^{2}P° – ^{2}S	11.370	477810	9272500	6	2	2.6(+4)	0.017	0.0038	−0.99	D	8
			11.422	520720	9272500	4	2	1.80(+4)	0.0176	0.00265	−1.152	C	8
			11.262	392000	9272500	2	2	7900	0.015	0.0011	−1.52	D	8
16.	2p–3d	^{2}P° – ^{2}D	11.124	477810	9467100	6	10	2.19(+5)	0.678	0.149	0.609	B	5
			11.171	520720	9472500	4	6	2.18(+5)	0.611	0.0899	0.388	B	5
			11.030	392000	9459000	2	4	1.84(+5)	0.670	0.0487	0.127	B	5
			11.187	520720	9459000	4	4	3.6(+4)	0.068	0.010	−0.57	B	5
17.	2p–4s	^{2}P° – ^{2}S	8.339	477810	12470000	6	2	1.0(+4)	0.0036	6.0(−4)	−1.66	D	9
			[8.369]	520720	12470000	4	2	6900	0.0036	4.0(−4)	−1.84	D	9
			[8.280]	392000	12470000	2	2	3600	0.0037	2.0(−4)	−2.13	D	9
18.	2p–4d	^{2}P° – ^{2}D	8.284	477810	12550000	6	10	7.16(+4)	0.123	0.0201	−0.133	C+	9
			8.316	520720	12550000	4	6	7.07(+4)	0.110	0.0120	−0.357	C+	9
			8.231	392000	12550000	2	4	6.10(+4)	0.124	0.00672	−0.606	C+	9
			8.316	520720	12550000	4	4	1.18(+4)	0.0122	0.00134	−1.312	C	9
19.	2p–5d	^{2}P° – ^{2}D	7.412	477810	13970000	6	10	3.3(+4)	0.045	0.0066	−0.57	C−	10
			7.438	520720	13970000	4	6	3.26(+4)	0.0405	0.00397	−0.79	C	10
			7.370	392000	13970000	2	4	2.8(+4)	0.046	0.0022	−1.04	C	10
			7.438	520720	13970000	4	4	5400	0.0045	4.4(−4)	−1.74	D	10
20.	2p–6d	^{2}P° – ^{2}D	7.012	477810	14740000	6	10	1.79(+4)	0.0220	0.00305	−0.879	C+	12
			7.033	520720	14740000	4	6	1.78(+4)	0.0198	0.00183	−1.102	C+	ls
			6.972	392000	14740000	2	4	1.52(+4)	0.0222	0.00102	−1.352	C+	ls
			7.033	520720	14740000	4	4	2900	0.0022	2.0(−4)	−2.06	D	ls

Fe XXIV: Allowed transitions — Continued

No.	Transition Array	Multiplet	λ (Å)	E_i (cm^{-1})	E_k (cm^{-1})	g_i	g_k	A_{ki} (10^8 s^{-1})	f_{ik}	S (at. u.)	log gf	Accuracy	Source
21.	2p–7d	^{2}P° – ^{2}D	6.788	477810	15210000	6	10	1.09(+4)	0.0126	0.00169	−1.121	C+	12
			6.808	520720	15210000	4	6	1.08(+4)	0.0113	0.00101	−1.346	C+	ls
			[6.749]	392000	15210000	2	4	9280	0.0127	5.6(−4)	−1.596	C+	ls
			6.808	520720	15210000	4	4	1800	0.0012	1.1(−4)	−2.31	D	ls
22.	3s–4p	^{2}S – ^{2}P°	30.8	9272500	12520000	2	6	1.1(+4)	0.45	0.091	−0.05	C	12
			[30.7]	9272500	12530000	2	4	1.1(+4)	0.30	0.061	−0.22	C	ls
			[30.9]	9272500	12510000	2	2	1.0(+4)	0.15	0.030	−0.53	C	ls
23.	3s–5p	^{2}S – ^{2}P°	[21.4]	9272500	13950000	2	6	5200	0.108	0.0152	−0.67	C	12
24.	3s–6p	^{2}S – ^{2}P°	[18.3]	92725000	14730000	2	6	3200	0.048	0.0058	−1.02	C	12
25.	3s–7p	^{2}S – ^{2}P°				2	6		0.0250		−1.301	C	12
26.	3p–4d	^{2}P° – ^{2}D	31.8	9404100	12550000	6	10	2.4(+4)	0.60	0.38	0.56	B	12
			[31.9]	9417100	12550000	4	6	2.4(+4)	0.55	0.23	0.34	B	ls
			[31.5]	9378200	12550000	2	4	2.1(+4)	0.63	0.13	0.10	B	ls
			[31.9]	9417100	12550000	4	4	3900	0.060	0.025	−0.62	C+	ls
27.	3p–5d	^{2}P° – ^{2}D	21.9	9404100	13970000	6	10	1.15(+4)	0.138	0.0597	−0.082	C+	12
			[22.0]	9417100	13970000	4	6	1.14(+4)	0.124	0.0358	−0.306	C+	ls
			[21.8]	9378200	13970000	2	4	9730	0.139	0.0199	−0.557	C+	ls
			[22.0]	9417100	13970000	4	4	1900	0.014	0.0040	−1.26	D	ls
28.	3p–6d	^{2}P° – ^{2}D	18.7	9404100	14740000	6	10	6390	0.0558	0.0206	−0.475	C+	12
			[18.8]	9417100	14740000	4	6	6300	0.0501	0.0124	−0.698	C+	ls
			[18.7]	9378200	14740000	2	4	5320	0.0558	0.00687	−0.952	C+	ls
			[18.8]	9417100	14740000	4	4	1100	0.0057	0.0014	−1.65	D	ls
29.	3p–7d	^{2}P° – ^{2}D	17.2	9404100	15210000	6	10	3910	0.0289	0.00982	−0.761	C+	12
			[17.3]	9417100	15210000	4	6	3840	0.0259	0.00589	−0.985	C+	ls
			[17.1]	9378200	15210000	2	4	3310	0.0290	0.00327	−1.236	C+	ls
			[17.3]	9417100	15210000	4	4	640	0.0029	6.5(−4)	−1.94	D	ls
30.	3d–4f	^{2}D – ^{2}F°				10	14		1.00	1.000		B	11
31.	4s–5p	^{2}S – ^{2}P°	[67.6]	12470000	13950000	2	6	2330	0.478	0.213	−0.020	C	12
32.	4s–6p	^{2}S – ^{2}P°	[44.2]	12470000	14730000	2	6	1460	0.128	0.0373	−0.59	C	12
33.	4s–7p	^{2}S – ^{2}P°				2	6		0.056		−0.95	C	12
34.	4p–5d	^{2}P° – ^{2}D	69.0	12520000	13970000	6	10	4920	0.585	0.797	0.545	C+	12
			[69.4]	12530000	13970000	4	6	4830	0.523	0.478	0.321	C+	ls
			[68.5]	12510000	13970000	2	4	4190	0.590	0.266	0.072	C+	ls
			[69.4]	12530000	13970000	4	4	800	0.058	0.053	−0.63	D	ls
35.	4p–6d	^{2}P° – ^{2}D	45.0	12520000	14740000	6	10	2810	0.142	0.126	−0.70	C+	12
			[45.2]	12530000	14740000	4	6	2760	0.127	0.0756	−0.294	C+	ls
			[44.8]	12510000	14740000	2	4	2370	0.142	0.0420	−0.545	C+	ls
			[45.2]	12530000	14740000	4	4	460	0.014	0.0084	−1.25	D	ls

J. Phys. Chem. Ref. Data, Vol. 17, Suppl. 4, 1988

Fe XXIV: Allowed transitions — Continued

No.	Transition Array	Multiplet	λ (Å)	E_i (cm^{-1})	E_k (cm^{-1})	g_i	g_k	A_{ki} (10^8 s^{-1})	f_{ik}	S (at. u.)	log gf	Accu-racy	Source
36.	4p–7d	^{2}P° – ^{2}D	37.2	12520000	15210000	6	10	1780	0.0617	0.0453	−0.432	C+	12
			[37.3]	12530000	15210000	4	6	1770	0.0554	0.0272	−0.655	C+	ls
			[37.0]	12510000	15210000	2	4	1510	0.0620	0.0151	−0.907	C+	ls
			[37.3]	12530000	15210000	4	4	290	0.0061	0.0030	−1.61	D	ls

[a]The number in parentheses following the tabulated value indicates the power of ten by which this value has to be multiplied.

Fe XXIV

Forbidden Transitions

The single magnetic dipole transition within the $1s^2 2p$ configuration has the line strength of 1.33 in the absence of relativistic effects in the wavefunctions.[1] It is estimated that these effects are negligible, since comprehensive relativistic calculations by Cheng et al.[2] for the analogous transition in the $1s^2 2s^2 2p$ configuration of the boron sequence show that such relativistic corrections are negligible until much more highly charged ions.

The listed transition probability data are also expected to be quite accurate since the energy levels are derived from experimental data.

An electric quadrupole transition at the same wavelength is estimated to be of negligible strength, as calculated by Bhatia[3] for this transition in the case of Mn XXIII. (He obtains a ratio of about 10^{-3} for the ratio of E2 to M1 line strengths).

References

[1]W. L. Wiese, M. W. Smith, and B. M. Miles, "Atomic Transition Probabilities", Vol. II, NSRDS–NBS 22, U.S. Govt. Print. Office, Washington, DC 1969.

[2]K. T. Cheng, Y.–K. Kim, and J. P. Desclaux, At. Data Nucl. Data Tables 24, 111 (1979).

[3]A. K. Bhatia, private communication (1986).

Fe XXIV: Forbidden transitions

No.	Transition Array	Multiplet	λ (Å)	E_i (cm^{-1})	E_k (cm^{-1})	g_i	g_k	Type of transition	A_{ki} (s^{-1})	S (at. u.)	Accu-racy	Source
1.	2p–2p	^{2}P° – ^{2}P°										
			[776.88]	392000	520720	2	4	M1	1.91(+4)[a]	1.33	B	interp.

[a]The number in parentheses following the tabulated value indicates the power of ten by which this value has to be multiplied.

Fe xxv

He Isoelectronic Sequence

Ground State: $1s^2\ {}^1S_0$

Ionization Energy: 8828.1 eV $= 71203000\ cm^{-1}$

Allowed Transitions

List of tabulated lines

Wavelength (Å)	No.	Wavelength (Å)	No.	Wavelength (Å)	No.	Wavelength (Å)	No.
1.4607	19	1.792	9	6.9468	38	28.950	41
1.4611	18	1.793	3	7.4924	25	29.253	42
1.4945	17	1.794	9	7.6191	26	29.795	46
1.4952	16	1.797	11	7.6527	33	30.224	47
1.5730	15	1.798	11	7.7930	34	62.846	57
1.5749	14	1.800	5	10.038	23	63.295	58
1.778	4	1.802	7	10.221	24	64.608	60
1.782	13	1.810	8	10.371	29	65.342	61
1.787	6,10	1.8502	2	10.586	30	194.9	21
1.788	3,9	1.8593	1	19.934	43	272.6	20
1.789	9	6.7073	27	20.139	44	384.3	22
1.790	9	6.8157	28	20.272	50	398.9	20
1.791	3,12	6.8288	37	20.527	51	426.6	20

Oscillator strengths for transitions of the $1s^2$–$1s2p$ array are taken from the results of Drake,[1] who incorporated accurate nonrelativistic matrix elements and Dirac hydrogenic matrix elements into a Z-expansion technique in order to provide f-values which would accurately reflect correlation effects for low-Z ions and relativistic effects for high-Z ions of the helium isoelectronic sequence. The f-values for the $1s^2\ {}^1S$—$1snp\ {}^3P°$ ($n=3$—5) transitions were interpolated from results of the relativistic random phase approximation (RRPA) calculations of Johnson and Lin.[2] For other s-p and p-s transitions, we tabulate the published RRPA data of Lin et al.[3,4] A lifetime experiment by Buchet et al.[5] agrees well with theory, within the accuracy limits assigned here, for the $1s2s\ {}^3S_1$ – $1s2p\ {}^3P°_2$ transition.

The charge expansion results of Laughlin[6] are given for various p-d and d-p transitions, as well as transitions between $4d$ and $4f$ levels. For those multiplets involving no change in principal quantum number ($3p$-$3d$, $4p$-$4d$, $4d$-$4f$) the f-values should be considered rather uncertain, since they are sensitive to energy differences. Oscillator strengths for the $2p$-$3d$ transitions, and for $1s3p\ {}^3P°$ – $1s3d\ {}^3D$, were interpolated from the variational calculations of Weiss.[7] Both of these calculations indicate that, unlike the triplets, the $nd\ {}^1D$ energy levels ($n=3,4$) lie below the $np\ {}^1P°$ levels, and the $4f\ {}^1F°$ lies below the $4d\ {}^1D$.

Brown and Cortez[8] have provided f-values for numerous d-f and f-d transitions for the isoelectronic sequence by fitting Z-expansion formulas to the results of variational calculations for the low-Z ions. Their results for transitions between the lower-lying D and F° terms are tabulated here.

Transition probabilities for the stronger transitions involving the doubly excited $n=2$ states are taken from the calculations of Dubau et al.,[9] who used the program SUPERSTRUCTURE and included configuration interaction and intermediate coupling. Numerous data are also available for transitions involving doubly excited states where the spectator electron has principal quantum number $n=3$ or higher.[9,10] However, these data are not tabulated here since most of the transitions are very close to belonging to the unresolved satellites of the H-like ions, if they do not in fact do so.

References

[1] G. W. F. Drake, Phys. Rev. A **19**, 1387 (1979).

[2] W. R. Johnson and C. D. Lin, Phys. Rev. A **14**, 565 (1976).

[3] C. D. Lin, W. R. Johnson, and A. Dalgarno, Astrophys. J. **217**, 1011 (1977).

[4] C. D. Lin, W. R. Johnson, and A. Dalgarno, Phys. Rev. A **15**, 154 (1977).

[5] J. P. Buchet, M. C. Buchet-Poulizac, J. Denis, J. Desesquelles, M. Druetta, J. P. Grandin, M. Huet, X. Husson, and D. Lecler, Phys. Rev. A **30**, 309 (1984).

[6] C. J. Laughlin, J. Phys. B **6**, 1942 (1973).

[7] A. W. Weiss, J. Res. Nat. Bur. Stand., Sect A **71**, 163 (1967).

[8] R. T. Brown and J.-L. M. Cortez, Astrophys. J. **176**, 267 (1972).

[9] J. Dubau, A. H. Gabriel, M. Loulergue, L. Steenman-Clark, and S. Volonté, Mon. Not. R. Astron. Soc. **195**, 705 (1981).

[10] L. A. Vainshtein and U. I. Safronova, At. Data Nucl. Data Tables **25**, 311 (1980).

Fe xxv: Allowed transitions

No.	Transition Array	Multiplet	λ (Å)	E_i (cm⁻¹)	E_k (cm⁻¹)	g_i	g_k	A_{ki} (10⁸ s⁻¹)	f_{ik}	S (at. u.)	log gf	Accuracy	Source
1.	$1s^2$–$1s2p$	^{1}S – ^{3}P°											
			[1.8593]	0	[53785000]	1	3	4.42(+5)ᵃ	0.0687	4.21(−4)	−1.163	B	1
2.		^{1}S – ^{1}P°	[1.8502]	0	[54047400]	1	3	4.57(+6)	0.703	0.00428	−0.153	B	1
3.	$1s2s$–$2s2p$	^{3}S – ^{3}P°	1.790	[53534300]	[109400000]	3	9	2.6(+6)	0.38	0.0067	0.06	C	9
			[1.788]	[53534300]	[109450000]	3	5	2.68(+6)	0.214	0.00378	−0.192	C	9
			[1.791]	[53534300]	[109350000]	3	3	2.59(+6)	0.125	0.00220	−0.428	C	9
			[1.793]	[53534300]	[109300000]	3	1	2.67(+6)	0.0429	7.6(−4)	−0.89	C	9
4.		^{3}S – ^{1}P°											
			[1.778]	[53534300]	[109760000]	3	3	8.7(+4)	0.0041	7.2(−5)	−1.91	D	9
5.		^{1}S – ^{3}P°											
			[1.800]	[53787200]	[109350000]	1	3	8.6(+4)	0.013	7.4(−5)	−1.90	D	9
6.		^{1}S – ^{1}P°	[1.787]	[53787200]	[109760000]	1	3	2.57(+6)	0.369	0.00217	−0.433	C	9
7.	$1s2p$–$2s^2$	^{3}P° – ^{1}S											
			[1.802]	[53785000]	[109290000]	3	1	4.1(+5)	0.0067	1.2(−4)	−1.70	D	9
8.		^{1}P° – ^{1}S	[1.810]	[54047400]	[109290000]	3	1	5.9(+5)	0.0097	1.7(−4)	−1.54	D	9
9.	$1s2p$–$2p^2$	^{3}P° – ^{3}P	1.791	[53847700]	[109670000]	9	9	4.74(+6)	0.228	0.0121	0.312	C	9
			[1.792]	[53901100]	[109700000]	5	5	2.81(+6)	0.135	0.00399	−0.170	C	9
			[1.790]	[53785000]	[109650000]	3	3	1.23(+6)	0.059	0.00104	−0.75	C	9
			[1.794]	[53901100]	[109650000]	5	3	2.22(+6)	0.064	0.00190	−0.493	C	9
			[1.792]	[53785000]	[109590000]	3	1	4.92(+6)	0.079	0.00140	−0.63	C	9
			[1.788]	[53785000]	[109700000]	3	5	1.63(+6)	0.130	0.00230	−0.408	C	9
			[1.789]	[53768700]	[109650000]	1	3	1.78(+6)	0.256	0.00151	−0.59	C	9
10.		^{3}P° – ^{1}D											
			[1.787]	[53901100]	[101410000]	5	5	1.19(+6)	0.057	0.00168	−0.55	C	9
11.		^{1}P° – ^{3}P											
			[1.797]	[54047400]	[109700000]	3	5	8.8(+5)	0.071	0.0013	−0.67	D	9
			[1.798]	[54047400]	[109650000]	3	3	1.0(+5)	0.0048	8.6(−5)	−1.84	D	9
12.		^{1}P° – ^{1}D	[1.791]	[54047400]	[109870000]	3	5	4.10(+6)	0.329	0.0058	−0.006	C	9
13.		^{1}P° – ^{1}S	[1.782]	[54047400]	[110160000]	3	1	4.69(+6)	0.074	0.00131	−0.65	C	9
14.	$1s^2$–$1s3p$	^{1}S – ^{3}P°											
			[1.5749]	0	[63496000]	1	3	1.5(+5)	0.017	8.8(−5)	−1.77	E	interp.
15.		^{1}S – ^{1}P°	[1.5730]	0	[63570800]	1	3	1.24(+6)	0.138	7.15(−4)	−0.860	B	3
16.	$1s^2$–$1s4p$	^{1}S – ^{3}P°											
			[1.4952]	0	[66881100]	1	3	6.0(+4)	0.0060	3.0(−5)	−2.22	E	interp.
17.		^{1}S – ^{1}P°	[1.4945]	0	[66912100]	1	3	5.05(+5)	0.0507	2.49(−4)	−1.295	B	3

Fe xxv: Allowed transitions — Continued

No.	Transition Array	Multiplet	λ (Å)	E_i (cm^{-1})	E_k (cm^{-1})	g_i	g_k	A_{ki} (10^8 s^{-1})	f_{ik}	S (at. u.)	log gf	Accu-racy	Source
18.	$1s^2$–$1s5p$	^{1}S – ^{3}P°											
			[1.4611]	0	[68443500]	1	3	3.1(+4)	0.0030	1.4(−5)	−2.52	E	*interp.*
19.		^{1}S – ^{1}P°	[1.4607]	0	[68459300]	1	3	2.54(+5)	0.0244	1.17(−4)	−1.613	B	3
20.	$1s2s$–$1s2p$	^{3}S – ^{3}P°	*319.1*	[53534300]	[*53847700*]	3	9	8.94	0.0409	0.129	−0.911	B	4
			[272.6]	[53534300]	[53901100]	3	5	14.7	0.0273	0.0735	−1.087	B	4
			[398.9]	[53534300]	[53785000]	3	3	4.31	0.0103	0.0405	−1.511	B	4
			[426.6]	[53534300]	[53768700]	3	1	3.82	0.00347	0.0146	−1.982	B	4
21.		^{3}S – ^{1}P°											
			[194.9]	[53534300]	[54047400]	3	3	3.46	0.00197	0.00379	−2.228	B	4
22.		^{1}S – ^{1}P°	[384.3]	[53787200]	[54047400]	1	3	4.96	0.0329	0.0417	−1.482	B	4
23.	$1s2s$–$1s3p$	^{3}S – ^{3}P°											
			[10.038]	[53534300]	[63496000]	3	3	8.08(+4)	0.122	0.0121	−0.437	B	3
24.		^{1}S – ^{1}P°	[10.221]	[53787200]	[63570800]	1	3	7.75(+4)	0.364	0.0122	−0.439	B	3
25.	$1s2s$–$1s4p$	^{3}S – ^{3}P°											
			[7.4924]	[53534300]	[66881100]	3	3	3.6(+4)	0.030	0.0022	−1.05	B	3
26.		^{1}S – ^{1}P°	[7.6191]	[53787200]	[66912100]	1	3	3.4(+4)	0.088	0.0022	−1.06	B	3
27.	$1s2s$–$1s5p$	^{3}S – ^{3}P°											
			[6.7073]	[53534300]	[68443500]	3	3	1.8(+4)	0.012	7.9(−4)	−1.44	B	3
28.		^{1}S – ^{1}P°	[6.8157]	[53787200]	[68459300]	1	3	1.7(+4)	0.036	8.1(−4)	−1.44	B	3
29.	$1s2p$–$1s3s$	^{3}P° – ^{3}S											
			[10.371]	[53785000]	[63426900]	3	3	8700	0.014	0.0014	−1.38	B	3
30.		^{1}P° – ^{1}S	[10.586]	[54047400]	[63493700]	3	1	2.5(+4)	0.014	0.0015	−1.38	B	3
31.	$1s2p$–$1s3d$	^{3}P° – ^{3}D				9	15		0.69		0.79	C+	*interp.*
32.		^{1}P° – ^{1}D				3	5		0.70		0.32	C+	*interp.*
33.	$1s2p$–$1s4s$	^{3}P° – ^{3}S											
			[7.6527]	[53785000]	[66852300]	3	3	3500	0.0031	2.3(−4)	−2.03	C	3
34.		^{1}P° – ^{1}S	[7.7930]	[54047400]	[66879400]	3	1	1.0(+4)	0.0031	2.4(−4)	−2.03	C	3
35.	$1s2p$–$1s4d$	^{3}P° – ^{3}D				9	15		0.12		0.03	C	6
36.		^{1}P° – ^{1}D				3	5		0.12		−0.44	C	6
37.	$1s2p$–$1s5s$	^{3}P° – ^{3}S											
			[6.8288]	[53785000]	[68428900]	3	3	1700	0.0012	8.1(−5)	−2.44	C	3

Fe xxv: Allowed transitions — Continued

No.	Transition Array	Multiplet	λ (Å)	E_i (cm^{-1})	E_k (cm^{-1})	g_i	g_k	A_{ki} (10^8 s^{-1})	f_{ik}	S (at. u.)	log gf	Accuracy	Source
38.		^{1}P° – ^{1}S	[6.9468]	[54047400]	[68442500]	3	1	5000	0.0012	8.2($-$5)	$-$2.44	C	3
39.	1s3s–1s3p	^{3}S – ^{3}P°											
						3	3		0.016		$-$1.32	C	3
40.		^{1}S – ^{1}P°				1	3		0.056		$-$1.25	C	3
41.	1s3s–1s4p	^{3}S – ^{3}P°											
			[28.950]	[63426900]	[66881100]	3	3	1.07($+$4)	0.135	0.0386	$-$0.393	B	3
42.		^{1}S – ^{1}P°	[29.253]	[63493700]	[66912100]	1	3	1.04($+$4)	0.400	0.0385	$-$0.398	B	3
43.	1s3s–1s5p	^{3}S – ^{3}P°											
			[19.934]	[63426900]	[68443500]	3	3	5700	0.034	0.0067	$-$0.99	B	3
44.		^{1}S – ^{1}P°	[20.139]	[63493700]	[68459300]	1	3	5650	0.103	0.00683	$-$0.987	B	3
45.	1s3p–1s3d	^{3}P° – ^{3}D				9	15		0.012		$-$0.97	D	*interp.*
46.	1s3p–1s4s	^{3}P° – ^{3}S											
			[29.795]	[63496000]	[66852300]	3	3	2500	0.033	0.0097	$-$1.00	B	3
47.		^{1}P° – ^{1}S	[30.224]	[63570800]	[66879400]	3	1	7400	0.034	0.010	$-$0.99	B	3
48.	1s3p–1s4d	^{3}P° – ^{3}D				9	15		0.60		0.73	C	6
49.		^{1}P° – ^{1}D				3	5		0.62		0.27	C	6
50.	1s3p–1s5s	^{3}P° – ^{3}S											
			[20.272]	[63496000]	[68428900]	3	3	1200	0.0073	0.0015	$-$1.66	D	3
51.		^{1}P° – ^{1}S	[20.527]	[63570800]	[68442500]	3	1	3700	0.0077	0.0016	$-$1.64	C	3
52.	1s3d–1s3p	^{1}D – ^{1}P°				5	3		0.0020		$-$2.00	E	6
53.	1s3d–1s4p	^{3}D – ^{3}P°				15	9		0.012		$-$0.74	C	6
54.		^{1}D – ^{1}P°				5	3		0.011		$-$1.26	C	6
55.	1s4s–1s4p	^{3}S – ^{3}P°											
						3	3		0.023		$-$1.16	E	3
56.		^{1}S – ^{1}P°				1	3		0.078		$-$1.11	D	3
57.	1s4s–1s5p	^{3}S – ^{3}P°											
			[62.846]	[66852300]	[68443500]	3	3	2530	0.150	0.0931	$-$0.347	B	3
58.		^{1}S – ^{1}P°	[63.295]	[66879400]	[68459300]	1	3	2480	0.446	0.0929	$-$0.351	B	3
59.	1s4p–1s4d	^{3}P° – ^{3}D				9	15		0.019		$-$0.77	D	6
60.	1s4p–1s5s	^{3}P° – ^{3}S											
			[64.608]	[66881100]	[68428900]	3	3	850	0.053	0.034	$-$0.80	B	3

Fe xxv: Allowed transitions — Continued

No.	Transition Array	Multiplet	λ (Å)	E_i (cm^{-1})	E_k (cm^{-1})	g_i	g_k	A_{ki} (10^8 s^{-1})	f_{ik}	S (at. u.)	log gf	Accu- racy	Source
61.		^{1}P$^\circ$ – ^{1}S	[65.342]	[66912100]	[68442500]	3	1	2600	0.055	0.035	−0.78	B	3
62.	1s4d–1s4p	^{1}D – ^{1}P$^\circ$				5	3		0.0031		−1.81	E	6
63.	1s4d–1s4f	^{3}D – ^{3}F$^\circ$				15	21		7.8(−4)		−1.93	E	6
64.	1s4d–1s5f	^{3}D – ^{3}F$^\circ$				15	21		0.89		1.13	B	8
65.		^{1}D – ^{1}F$^\circ$				5	7		0.89		0.65	B	8
66.	1s4f–1s4d	^{1}F$^\circ$ – ^{1}D				7	5		4.2(−4)		−2.53	E	6
67.	1s4f–1s5d	^{3}F$^\circ$ – ^{3}D				21	15		0.0089		−0.73	C	8
68.		^{1}F$^\circ$ – ^{1}D				7	5		0.0089		1.21	C	8
69.	1s5s–1s5p	^{3}S – ^{3}P$^\circ$											
						3	3		0.029		−1.06	E	3
70.		^{1}S – ^{1}P$^\circ$				1	3		0.099		−1.00	E	3

^{a}The number in parentheses following the tabulated value indicates the power of ten by which this value has to be multiplied.

Fe xxv

Forbidden Transitions

The results of multi-configuration Dirac-Fock calculations by Hata and Grant[1] have been selected for this tabulation. Their work includes both a very detailed consideration of configuration interaction—with configurational wavefunction sets containing as many as 51 interacting states—as well as a fully relativistic treatment based on the Dirac Hamiltonian. Their calculated wavelengths are in very close agreement with experiment, and the agreement between the experimental lifetime data[2,3] and the theoretical result is excellent, with differences in the few percent range. A comprehensive comparison table containing all experimental data on these He-sequence transitions is given in the introduction to the forbidden lines of Ti xxi.

References

[1] J. Hata and I. P. Grant, Mon. Not. R. Astr. Soc. **211**, 549 (1984).
[2] H. Gould, R. Marrus, and P. J. Mohr, Phys. Rev. Lett. **33**, 676 (1974).
[3] J. P. Buchet, M. C. Buchet-Poulizac, A. Denis, J. Desesquelles, M. Druetta, J. P. Grandin, M. Huet, X. Husson, and D. Lecler, Phys. Rev. **A 30**, 309 (1984).

Fe xxv: Forbidden transitions

No.	Transition Array	Multiplet	λ (Å)	E_i (cm^{-1})	E_k (cm^{-1})	g_i	g_k	Type of transition	A_{ki} (s^{-1})	S (at. u.)	Accu- racy	Source
1.	1s^2–1s2s	^{1}S – ^{3}S	[1.8682]	0	[53527090]	1	3	M1	2.12(+8)^a	1.54(−4)	B	1
2.	1s^2–1s2p	^{1}S – ^{3}P$^\circ$										
			[1.8554]	0	53895550	1	5	M2	6.64(+9)	0.110	B	1

^{a}The number in parentheses following the tabulated value indicates the power of ten by which this value has to be multiplied.

Fe XXVI

H Isoelectronic Sequence

Ground State: $1s\ ^2S_{1/2}$

Ionization Energy: $9277.76\ \text{eV} = 74829600\ \text{cm}^{-1}$

Allowed Transitions

Electric dipole transition probability data for this hydrogen-like ion can be obtained directly, in a non-relativistic approximation, from the data for neutral hydrogen.[1] The oscillator strength is independent of Z along the entire isoelectronic sequence and is therefore identical to the value for the hydrogen atom. Line strengths scale as Z^{-2} and transition probabilities scale as Z^4, i.e.,

$$S_Z = Z^{-2} S_H, \qquad A_Z = Z^4 A_H.$$

For higher nuclear charges in this sequence, relativistic corrections will cause these values to deviate increasingly from the non-relativistic ones. The first effect of relativity will be to alter the transition energies, or wavelengths, from the non-relativistic, even though the line strength itself is still well approximated by the non-relativistic value. In this case, experimental energies should be used in the standard conversion formulas, given in the general introduction to this volume, to calculate the most accurate values of f and A. It should be noted that the relativistic removal of the j-degeneracy introduces dipole transitions which do not occur in the non-relativistic theory, e.g., $2s_{1/2} - 2p_{3/2}$.

For very high Z, it is necessary to use the four-component Dirac spinors rather than two-component Schroedinger functions in theoretical calculations, and this introduces relativistic corrections to the line strengths themselves. Several recent systematic studies of the problem[2,3] indicate that these corrections are not large for stages of ionization in the range 20–30. Corrections for $Z = 30$ are usually no larger than 5–10% and generally substantially less than 5%. If an accuracy greater than this is required, the reader is referred to these papers[2,3] for a more detailed error analysis.

References

[1] W. L. Wiese, M. W. Smith, and B. M. Glennon, Atomic Transition Probabilities – Hydrogen through Neon (A Critical Data Compilation), Vol. I, 157 pp., Nat. Stand. Ref. Data Ser., Nat. Bur. Stand. (U.S.), 4 (May 1966).

[2] S. M. Younger and A. W. Weiss, J. Res. Nat. Bur. Stand., Sect. A **79**, 629 (1975).

[3] S. J. Rose, Rutherford Appleton Laboratory Report RL–82–114 (December 1982).

Cobalt

Co I

Ground State: $1s^2 2s^2 2p^6 3s^2 3p^6 3d^7 4s^2\ ^4F_{9/2}$

Ionization Energy: $7.86\,\text{eV} = 63400\,\text{cm}^{-1}$

Allowed Transitions

List of tabulated lines

Wavelength (Å)	No.	Wavelength (Å)	No.	Wavelength (Å)	No.	Wavelength (Å)	No.
2287.80	47	3013.59	8	3385.22	32	3550.59	2
2295.22	22	3017.55	13	3388.16	33	3552.72	4
2309.03	21	3034.43	11	3395.37	34	3560.89	31
2323.13	21	3042.48	8	3405.12	33	3564.95	29
2325.53	23	3044.00	9	3409.17	36	3569.37	60
2335.98	24	3048.89	9	3409.65	35	3574.97	31
2338.66	21	3054.13	5	3412.34	34	3575.36	2
2353.36	21	3054.72	10	3412.63	4	3584.80	4
2355.48	21	3061.82	12	3413.52	3	3585.15	31
2358.18	24	3062.20	11	3414.74	35	3587.19	60
2365.06	18	3064.37	10	3415.52	3	3594.87	2
2371.85	22	3071.96	11	3417.15	37	3602.08	2
2384.86	17	3072.34	13	3417.80	29	3605.37	30
2392.03	18	3082.61	8	3431.58	4	3608.31	30
2402.06	17	3086.78	9	3433.05	33	3618.01	61
2407.25	18	3089.60	8	3442.92	4	3624.96	31
2412.76	20	3098.19	8	3443.64	32	3627.81	29
2414.46	18	3105.93	38	3449.17	32	3631.39	2
2415.29	18	3110.82	13	3449.44	32	3647.66	2
2424.93	17	3118.25	12	3453.51	32	3652.54	2
2429.23	19	3118.64	11	3455.24	4	3656.96	31
2432.21	17	3121.42	7	3456.92	3	3677.98	30
2435.82	16	3121.57	9	3460.72	60	3704.06	60
2436.66	17	3127.25	38	3462.80	37	3745.49	56
2439.04	17	3129.01	39	3465.79	3	3808.10	27
2460.80	17	3132.22	5	3474.02	2,36	3811.07	54
2467.69	17	3136.73	6	3474.53	35	3841.46	58
2470.27	45	3137.33	8	3483.41	33	3842.05	57
2476.64	43	3139.95	7	3483.80	3	3845.47	56
2504.52	42	3147.06	8	3489.40	61	3850.95	27
2511.02	43	3149.31	7	3490.74	30	3861.16	57
2521.36	15	3158.29	11	3491.32	4	3873.12	28
2528.97	15	3158.77	8	3495.68	32	3873.95	28
2530.13	46	3186.35	6	3496.68	29	3881.87	28
2535.96	15	3189.75	7	3502.28	31	3884.60	55
2536.50	44	3191.30	5	3502.63	4	3885.28	54
2544.25	15	3193.16	38	3506.32	31	3894.07	56
2562.12	15	3198.66	38	3509.84	32	3894.98	28
2567.34	15	3199.32	7	3510.43	4	3906.29	27
2574.35	15	3203.03	7	3512.64	31	3909.93	1
2685.34	41	3219.15	6	3513.48	3	3922.76	59
2695.85	41	3223.15	39	3518.34	61	3933.92	27
2764.19	40	3227.75	6	3520.08	2	3935.96	55
2815.56	40	3237.03	5	3521.58	30	3940.89	28
2862.60	14	3250.00	38	3523.42	31	3941.73	27
2886.44	14	3281.59	6	3526.85	2	3945.33	52
2928.81	10	3337.17	34	3529.03	3	3952.33	26
2987.17	12	3354.37	37	3529.82	32	3952.92	51
2989.59	10	3367.11	32	3533.36	3	3957.93	28
3000.55	10	3370.32	35	3542.98	29	3965.01	54

List of tabulated lines — Continued

Wavelength (Å)	No.	Wavelength (Å)	No.	Wavelength (Å)	No.	Wavelength (Å)	No.
3965.24	53	4553.34	93	5476.47	97	6490.34	75
3974.73	28	4574.94	50	5477.09	99	6508.79	108
3978.65	27	4580.14	49	5483.35	64	6563.40	74
3979.52	1	4619.33	49	5483.96	99	6623.36	106
3987.12	26	4685.86	48	5530.78	63	6632.44	84
3991.68	27	4699.18	49	5590.74	81	6638.40	106
3994.54	27	4781.43	73	5647.23	85	6644.03	96
3995.31	54	4782.56	48	5688.59	81	6652.29	106
3997.90	58	4920.27	73	5878.10	69	6678.82	68
3998.55	57	4932.88	72	5883.42	81	6771.04	68
4020.90	26	5034.06	82	5890.49	77	6784.85	111
4045.39	54	5034.97	95	5915.55	77	6808.94	108
4058.18	26	5094.96	83	5935.39	70	6814.95	68
4066.37	53	5113.22	82	5984.09	62	6872.32	68
4076.12	26	5146.16	65	5991.89	81	7016.60	68
4082.59	26	5146.75	95	6005.03	62	7052.87	68
4092.39	52	5149.80	64	6058.23	100	7054.04	90
4110.53	52	5165.16	64	6082.43	94	7106.37	109
4118.77	51	5176.09	83	6093.14	62	7154.69	79
4121.32	51	5212.70	95	6116.99	62	7354.58	67
4132.16	53	5230.21	64	6129.12	80	7388.69	89
4150.43	26	5235.19	78	6181.01	104	7417.38	79
4180.70	91	5247.92	64	6189.01	62	7590.57	79
4223.77	91	5265.79	95	6230.94	62	7610.24	86
4240.79	92	5266.51	78	6246.39	94	7712.66	86
4259.87	91	5280.63	97	6249.50	76	7743.27	101
4270.43	52	5287.57	99	6282.64	62	7838.12	103
4309.44	91	5301.04	64	6352.75	106	8093.93	105
4359.43	92	5331.46	64	6386.69	102	8372.79	107
4436.43	66	5332.65	95	6417.86	84	8835.21	110
4450.79	92	5333.15	63	6429.91	75	8926.21	88
4484.51	49	5352.05	97	6430.34	104	9356.98	87
4490.31	93	5369.59	64	6444.68	96		
4526.52	25	5381.11	71	6451.14	104		
4550.47	93	5469.31	71	6455.00	98		

For this spectrum, we have chosen the work of Cardon et al.,[1] who determined relative oscillator strengths for 362 lines. These relative data were converted to an absolute scale by using radiative lifetimes measured by Figger et al.[2] and by Marek and Vogt,[3] who employed selective tunable dye laser excitation.

Cardon et al. measured 159 lines in absorption by the hook (anomalous dispersion) method, and 314 additional lines in emission with a hollow cathode discharge. These emission lines were recorded with the 1 m Fourier transform spectrometer at Kitt Peak National Observatory. The improved resolution of this instrument allowed the measurement of f-values for some previously unresolved lines. Oscillator strengths for 95 lines were measured both in emission and absorption and subjected to the "bowtie" procedure, which fits the combined emission and absorption data to a set of optimally consistent rela-

tive f-values. This technique was recently developed by Cardon and co-workers and has been described in an earlier paper.[4]

In this compilation, we have tabulated data for 338 Co I lines. We have generally omitted blended lines as well as lines where only upper limits to the log gf-values, instead of specific values, were determined.

References

[1]B. L. Cardon, P. L. Smith, J. M. Scalo, L. Testerman, and W. Whaling, Astrophys. J. **260**, 395 (1982).

[2]H. Figger, J. Heldt, K. Siomos, and H. Walther, Astron. Astrophys. **43**, 389 (1975).

[3]J. Marek and K. Vogt, Z. Phys. A **280**, 235 (1977).

[4]B. L. Cardon, P. L. Smith, and W. Whaling, Phys. Rev. A **20**, 2411 (1979).

Co I: Allowed transitions

No.	Multiplet	λ (Å)	E_i (cm^{-1})	E_k (cm^{-1})	g_i	g_k	A_{ki} (10^8 s^{-1})	f_{ik}	S (at. u.)	log gf	Accuracy	Source
1.	$a\ ^4F - z\ ^6G°$ (3)											
		3909.93	0.0	25569	10	12	0.0011	3.0(−4)[a]	0.039	−2.52	C	1
		3979.52	816.0	25938	8	10	8.8(−4)	2.6(−4)	0.027	−2.68	D	1
2.	$a\ ^4F - z\ ^4F°$ (4)	*3565.8*	*793.1*	*28829*	28	28	0.14	0.026	8.7	−0.13	C	1
		3526.85	0.0	28346	10	10	0.13	0.024	2.8	−0.62	C	1
		3575.36	816.0	28777	8	8	0.096	0.018	1.7	−0.83	C	1
		3594.87	1406.8	29216	6	6	0.092	0.018	1.3	−0.97	C	1
		3602.08	1809.3	29563	4	4	0.10	0.019	0.92	−1.11	C	1
		3474.02	0.0	28777	10	8	0.034	0.0049	0.56	−1.31	C	1
		3520.08	816.0	29216	8	6	0.046	0.0064	0.59	−1.29	C	1
		3550.59	1406.8	29563	6	4	0.040	0.0050	0.35	−1.52	C	1
		3631.39	816.0	28346	8	10	0.0052	0.0013	0.12	−1.99	C	1
		3652.54	1406.8	28777	6	8	0.0086	0.0023	0.17	−1.86	C	1
		3647.66	1809.3	29216	4	6	0.010	0.0031	0.15	−1.91	C	1
3.	$a\ ^4F - z\ ^4G°$ (5)											
		3465.79	0.0	28845	10	12	0.092	0.020	2.3	−0.70	C	1
		3513.48	816.0	29270	8	10	0.078	0.018	1.7	−0.84	C	1
		3529.03	1406.8	29735	6	8	0.088	0.022	1.5	−0.88	D	1
		3533.36	1809.3	30103	4	6	0.091	0.026	1.2	−0.99	D	1
		3415.52	0.0	29270	10	10	2.6(−4)	4.5(−5)	0.0050	−3.35	C	1
		3456.92	816.0	29735	8	8	0.0018	3.2(−4)	0.029	−2.59	D	1
		3483.80	1406.8	30103	6	6	0.0025	4.6(−4)	0.032	−2.56	D	1
		3413.52	816.0	30103	8	6	2.3(−4)	3.1(−5)	0.0028	−3.61	D	1
4.	$a\ ^4F - z\ ^4D°$ (6)	*3438.6*	*793.1*	*29866*	28	20	0.17	0.021	6.7	−0.23	C	1
		3412.63	0.0	29295	10	8	0.12	0.017	1.9	−0.78	C+	1
		3431.58	816.0	29949	8	6	0.11	0.014	1.3	−0.94	C	1
		3442.92	1406.8	30444	6	4	0.12	0.014	0.96	−1.07	C	1
		3455.24	1809.3	30743	4	2	0.19	0.017	0.77	−1.17	C	1
		3510.43	816.0	29295	8	8	0.038	0.0070	0.65	−1.25	C+	1
		3502.63	1406.8	29949	6	6	0.052	0.0096	0.66	−1.24	C	1
		3491.32	1809.3	30444	4	4	0.050	0.0091	0.42	−1.44	C	1
		3584.80	1406.8	29295	6	8	0.0020	5.2(−4)	0.036	−2.51	C+	1
		3552.72	1809.3	29949	4	6	0.0021	5.9(−4)	0.027	−2.63	C	1
5.	$a\ ^4F - z\ ^2G°$ (7)											
		3132.22	816.0	32733	8	8	7.4(−4)	1.1(−4)	0.0090	−3.06	D	1
		3054.13	0.0	32733	10	8	2.2(−4)	2.5(−5)	0.0025	−3.60	E	1
		3237.03	816.0	31700	8	10	0.0071	0.0014	0.12	−1.95	C	1
		3191.30	1406.8	32733	6	8	8.2(−4)	1.7(−4)	0.011	−3.00	D	1
6.	$a\ ^4F - z\ ^2F°$ (8)											
		3136.73	0.0	31871	10	8	0.0023	2.7(−4)	0.028	−2.57	C	1
		3219.15	816.0	31871	8	8	0.0048	7.5(−4)	0.064	−2.22	C+	1
		3186.35	1406.8	32782	6	6	0.0019	2.8(−4)	0.018	−2.77	C	1
		3281.59	1406.8	31871	6	8	4.3(−4)	9.2(−5)	0.0059	−3.26	D	1
		3227.75	1809.3	32782	4	6	7.7(−4)	1.8(−4)	0.0077	−3.14	D	1

Co I: Allowed transitions — Continued

No.	Multiplet	λ (Å)	E_i (cm^{-1})	E_k (cm^{-1})	g_i	g_k	A_{ki} (10^8 s^{-1})	f_{ik}	S (at. u.)	log gf	Accuracy	Source
7.	$a\ ^4F - y\ ^4D°$ (9)											
		3121.42	0.0	32028	10	8	0.022	0.0026	0.26	−1.59	C+	1
		3139.95	816.0	32655	8	6	0.025	0.0027	0.23	−1.66	C	1
		3149.31	1406.8	33151	6	4	0.028	0.0028	0.17	−1.78	C	1
		3203.03	816.0	32028	8	8	0.0010	1.5(−4)	0.013	−2.91	C	1
		3199.32	1406.8	32655	6	6	0.0017	2.6(−4)	0.017	−2.80	C	1
		3189.75	1809.3	33151	4	4	0.0031	4.7(−4)	0.020	−2.73	C	1
8.	$a\ ^4F - y\ ^4G°$ (10)											
		3082.61	0.0	32431	10	12	0.027	0.0046	0.46	−1.34	C+	1
		3158.77	816.0	32465	8	10	0.022	0.0041	0.34	−1.48	C+	1
		3147.06	1406.8	33173	6	8	0.045	0.0090	0.56	−1.27	C+	1
		3137.33	1809.3	33674	4	6	0.047	0.010	0.43	−1.38	C+	1
		3089.60	816.0	33173	8	8	0.024	0.0034	0.27	−1.57	C+	1
		3098.19	1406.8	33674	6	6	0.022	0.0032	0.19	−1.72	C+	1
		3013.59	0.0	33173	10	8	0.014	0.0015	0.15	−1.81	C+	1
		3042.48	816.0	33674	8	6	0.019	0.0019	0.16	−1.81	C+	1
9.	$a\ ^4F - y\ ^4F°$ (11)											
		3044.00	0.0	32842	10	10	0.19	0.026	2.6	−0.59	C+	1
		3086.78	1809.3	34196	4	4	0.19	0.027	1.1	−0.96	C+	1
		3048.89	1406.8	34196	6	4	0.075	0.0069	0.42	−1.38	C+	1
		3121.57	816.0	32842	8	10	0.010	0.0018	0.15	−1.83	C+	1
10.	$a\ ^4F - y\ ^2G°$ (13)											
		2989.59	0.0	33440	10	10	0.038	0.0051	0.50	−1.29	C+	1
		3000.55	816.0	34134	8	8	0.0070	9.5(−4)	0.075	−2.12	C	1
		2928.81	0.0	34134	10	8	0.0025	2.6(−4)	0.025	−2.59	C	1
		3064.37	816.0	33440	8	10	0.0055	9.7(−4)	0.078	−2.11	C+	1
		3054.72	1406.8	34134	6	8	0.0016	3.0(−4)	0.018	−2.74	C	1
11.	$a\ ^4F - z\ ^2D°$ (12)											
		3062.20	816.0	33463	8	6	0.0040	4.2(−4)	0.034	−2.47	C	1
		3034.43	1406.8	34352	6	4	0.014	0.0013	0.079	−2.10	C	1
		3118.64	1406.8	33463	6	6	2.6(−4)	3.7(−5)	0.0023	−3.65	D	1
		3071.96	1809.3	34352	4	4	0.011	0.0015	0.062	−2.21	C	1
		3158.29	1809.3	33463	4	6	3.9(−5)	8.7(−6)	3.6(−4)	−4.46	E	1
12.	$a\ ^4F - (\ °)^b$											
		3118.25	1406.8	33467	6	8	0.0023	4.5(−4)	0.028	−2.57	C+	1
		3061.82	816.0	33467	8	8	0.16	0.022	1.8	−0.75	C+	1
		2987.17	0.0	33467	10	8	0.049	0.0052	0.52	−1.28	B	1
13.	$a\ ^4F - (\ °)^b$											
		3110.82	1809.3	33946	4	6	0.0026	5.7(−4)	0.023	−2.64	C+	1
		3072.34	1406.8	33946	6	6	0.15	0.021	1.3	−0.90	C+	1
		3017.55	816.0	33946	8	6	0.069	0.0070	0.56	−1.25	C+	1

Co I: Allowed transitions — Continued

No.	Multiplet	λ (Å)	E_i (cm^{-1})	E_k (cm^{-1})	g_i	g_k	A_{ki} (10^8 s^{-1})	f_{ik}	S (at. u.)	log gf	Accuracy	Source
14.	$a\ ^4F - y\ ^2F°$ (uv 1)											
		2886.44	816.0	35451	8	8	0.016	0.0020	0.15	−1.79	C	1
		2862.60	1406.8	36330	6	6	0.0066	8.2(−4)	0.046	−2.31	C	1
15.	$a\ ^4F - x\ ^4D°$ (uv 3)											
		2521.36	0.0	39649	10	8	3.0	0.23	19	0.36	D	1
		2528.97	816.0	40346	8	6	2.8	0.20	14	0.21	D	1
		2535.96	1406.8	40828	6	4	1.9	0.12	6.2	−0.13	C	1
		2544.25	1809.3	41102	4	2	3.0	0.15	4.9	−0.23	D	1
		2574.35	816.0	39649	8	8	0.17	0.017	1.2	−0.86	C	1
		2567.34	1406.8	40346	6	6	0.30	0.030	1.5	−0.75	C	1
		2562.12	1809.3	40828	4	4	0.39	0.039	1.3	−0.81	C	1
16.	$a\ ^4F - ^6P°$											
		2435.82	0.0	41041	10	8	0.019	0.0013	0.11	−1.87	D	1
17.	$a\ ^4F - x\ ^4F°$ (uv 5)											
		2424.93	0.0	41226	10	10	3.2	0.28	22	0.45	C	1
		2432.21	816.0	41918	8	8	2.6	0.23	15	0.26	D	1
		2436.66	1406.8	42434	6	6	2.6	0.24	11	0.15	D	1
		2439.04	1809.3	42797	4	4	2.7	0.24	7.7	−0.02	C	1
		2384.86	0.0	41918	10	8	0.24	0.017	1.3	−0.78	C	1
		2402.06	816.0	42434	8	6	0.51	0.033	2.1	−0.58	C	1
		2467.69	1406.8	41918	6	8	0.070	0.0085	0.42	−1.29	D	1
		2460.80	1809.3	42434	4	6	0.12	0.016	0.52	−1.19	D	1
18.	$a\ ^4F - x\ ^4G°$ (uv 6)											
		2407.25	0.0	41529	10	12	3.6	0.38	30	0.58	C	1
		2414.46	1406.8	42811	6	8	3.4	0.39	19	0.37	C	1
		2415.29	1809.3	43200	4	6	3.6	0.48	15	0.28	C+	1
		2365.06	0.0	42269	10	10	0.13	0.011	0.83	−0.97	C	1
		2392.03	1406.8	43200	6	6	0.40	0.034	1.6	−0.69	D	1
19.	$a\ ^4F - z\ ^4P°$ (uv 7)											
		2429.23	816.0	41969	8	6	0.047	0.0031	0.20	−1.60	D	1
20.	$a\ ^4F - w\ ^4D°$ (uv 10)											
		2412.76	1809.3	43243	4	6	0.65	0.085	2.7	−0.47	C	1
21.	$a\ ^4F - w\ ^4F°$ (uv 11)											
		2309.03	0.0	43295	10	10	0.56	0.045	3.4	−0.35	C+	1
		2323.13	816.0	43848	8	8	0.50	0.040	2.5	−0.49	C	1
		2338.66	1809.3	44556	4	4	0.77	0.063	1.9	−0.60	D	1
		2353.36	816.0	43295	8	10	0.15	0.015	0.95	−0.91	D	1
		2355.48	1406.8	43848	6	8	0.13	0.015	0.69	−1.05	D	1

Co I: Allowed transitions — Continued

No.	Multiplet	λ (Å)	E_i (cm^{-1})	E_k (cm^{-1})	g_i	g_k	A_{ki} (10^8 s^{-1})	f_{ik}	S (at. u.)	log gf	Accuracy	Source
22.	$a\ ^4F - x\ ^2F°$ (uv 12)											
		2295.22	0.0	43555	10	8	0.22	0.014	1.0	−0.86	C	1
		2371.85	1406.8	43555	6	8	0.073	0.0082	0.38	−1.31	D	1
23.	$a\ ^4F - w\ ^4G°$ (uv 14)											
		2325.53	1406.8	44394	6	8	0.11	0.012	0.55	−1.14	D	1
24.	$a\ ^4F - (\ °)^b$											
		2335.98	1406.8	44202	6	6	0.51	0.042	1.9	−0.60	C	1
		2358.18	1809.3	44202	4	6	0.14	0.018	0.56	−1.14	D	1
25.	$b\ ^4F - z\ ^6G°$											
		4526.52	3482.8	25569	10	12	8.8(−6)	3.2(−6)	4.8(−4)	−4.49	D	1
26.	$b\ ^4F - z\ ^4F°$ (16)											
		4020.90	3482.8	28346	10	10	0.0035	8.5(−4)	0.11	−2.07	C	1
		4058.18	4142.7	28777	8	8	0.0015	3.8(−4)	0.040	−2.52	C	1
		4076.12	4690.2	29216	6	6	7.3(−4)	1.8(−4)	0.015	−2.96	C	1
		4082.59	5075.8	29563	4	4	5.9(−4)	1.5(−4)	0.0079	−3.23	C	1
		3952.33	3482.8	28777	10	8	0.0022	4.1(−4)	0.053	−2.39	C	1
		3987.12	4142.7	29216	8	6	0.0012	2.1(−4)	0.022	−2.77	C	1
		4150.43	4690.2	28777	6	8	5.3(−4)	1.8(−4)	0.015	−2.96	C	1
27.	$b\ ^4F - z\ ^4G°$ (17)											
		3941.73	3482.8	28845	10	12	0.0033	9.3(−4)	0.12	−2.03	C	1
		3978.65	4142.7	29270	8	10	0.0020	6.0(−4)	0.063	−2.32	C	1
		3991.68	4690.2	29735	6	8	0.0015	4.7(−4)	0.037	−2.55	D	1
		3994.54	5075.8	30103	4	6	0.0014	4.9(−4)	0.026	−2.71	D	1
		3906.29	4142.7	29735	8	8	0.0029	6.7(−4)	0.069	−2.27	D	1
		3933.92	4690.2	30103	6	6	0.0012	2.9(−4)	0.023	−2.76	D	1
		3808.10	3482.8	29735	10	8	0.0015	2.6(−4)	0.033	−2.58	D	1
		3850.95	4142.7	30103	8	6	0.0022	3.6(−4)	0.037	−2.54	D	1
28.	$b\ ^4F - z\ ^4D°$ (18)											
		3873.12	3482.8	29295	10	8	0.12	0.022	2.8	−0.66	C+	1
		3873.95	4142.7	29949	8	6	0.10	0.017	1.7	−0.87	C	1
		3881.87	4690.2	30444	6	4	0.082	0.012	0.95	−1.13	C	1
		3894.98	5075.8	30743	4	2	0.088	0.010	0.51	−1.40	C	1
		3974.73	4142.7	29295	8	8	0.0025	6.0(−4)	0.063	−2.32	C+	1
		3957.93	4690.2	29949	6	6	0.0062	0.0015	0.11	−2.06	C	1
		3940.89	5075.8	30444	4	4	0.0083	0.0019	0.10	−2.11	C	1
29.	$b\ ^4F - z\ ^2G°$ (19)											
		3542.98	3482.8	31700	10	10	6.1(−5)	1.1(−5)	0.0013	−3.94	E	1
		3496.68	4142.7	32733	8	8	0.034	0.0063	0.58	−1.30	C	1
		3417.80	3482.8	32733	10	8	0.0039	5.5(−4)	0.062	−2.26	C	1
		3627.81	4142.7	31700	8	10	0.047	0.012	1.1	−1.03	C	1
		3564.95	4690.2	32733	6	8	0.070	0.018	1.3	−0.97	C	1

Co I: Allowed transitions — Continued

No.	Multiplet	λ (Å)	E_i (cm^{-1})	E_k (cm^{-1})	g_i	g_k	A_{ki} (10^8 s^{-1})	f_{ik}	S (at. u.)	log gf	Accuracy	Source
30.	$b\ ^4F - z\ ^2F°$ (20)											
		3521.58	3482.8	31871	10	8	0.18	0.026	3.0	−0.58	C+	1
		3490.74	4142.7	32782	8	6	0.016	0.0022	0.20	−1.75	D	1
		3605.37	4142.7	31871	8	8	0.038	0.0074	0.70	−1.23	C+	1
		3677.98	4690.2	31871	6	8	4.3(−4)	1.2(−4)	0.0084	−3.16	D	1
		3608.31	5075.8	32782	4	6	7.1(−4)	2.1(−4)	0.0099	−3.08	D	1
31.	$b\ ^4F - y\ ^4D°$ (21)	*3517.0*	*4157.6*	*32583*	28	20	1.0	0.13	43	0.57	C+	1
		3502.28	3482.8	32028	10	8	0.80	0.12	14	0.07	C+	1
		3506.32	4142.7	32655	8	6	0.82	0.11	11	−0.04	C+	1
		3512.64	4690.2	33151	6	4	1.0	0.12	8.6	−0.13	C	1
		3523.42	5075.8	33449	4	2	0.98	0.091	4.2	−0.44	D	1
		3585.15	4142.7	32028	8	8	0.071	0.014	1.3	−0.96	C+	1
		3574.97	4690.2	32655	6	6	0.15	0.028	2.0	−0.77	C+	1
		3560.89	5075.8	33151	4	4	0.23	0.043	2.0	−0.76	C	1
		3656.96	4690.2	32028	6	8	0.0025	6.6(−4)	0.048	−2.40	C	1
		3624.96	5075.8	32655	4	6	0.0052	0.0015	0.074	−2.21	C+	1
32.	$b\ ^4F - y\ ^4G°$ (22)	*3488.7*	*4157.6*	*32813*	28	36	0.99	0.23	75	0.81	C+	1
		3453.51	3482.8	32431	10	12	1.1	0.24	27	0.38	C+	1
		3529.82	4142.7	32465	8	10	0.46	0.11	9.9	−0.07	C+	1
		3509.84	4690.2	33173	6	8	0.32	0.080	5.5	−0.32	C+	1
		3495.68	5075.8	33674	4	6	0.49	0.13	6.2	−0.27	C+	1
		3449.44	3482.8	32465	10	10	0.18	0.032	3.6	−0.50	C+	1
		3443.64	4142.7	33173	8	8	0.69	0.12	11	−0.01	C+	1
		3449.17	4690.2	33674	6	6	0.76	0.14	9.2	−0.09	C+	1
		3367.11	3482.8	33173	10	8	0.060	0.0081	0.90	−1.09	C+	1
		3385.22	4142.7	33674	8	6	0.11	0.014	1.3	−0.95	C+	1
33.	$b\ ^4F - y\ ^4F°$ (23)											
		3405.12	3482.8	32842	10	10	1.0	0.18	20	0.25	C+	1
		3433.05	5075.8	34196	4	4	1.0	0.19	8.4	−0.13	C+	1
		3388.16	4690.2	34196	6	4	0.24	0.027	1.8	−0.79	C+	1
		3483.41	4142.7	32842	8	10	0.055	0.013	1.1	1.00	C+	1
34.	$b\ ^4F - y\ ^2G°$ (25)											
		3337.17	3482.8	33440	10	10	0.0034	5.6(−4)	0.062	−2.25	C	1
		3412.34	4142.7	33440	8	10	0.61	0.13	12	0.03	C+	1
		3395.37	4690.2	34134	6	8	0.29	0.066	4.5	−0.40	C	1
35.	$b\ ^4F - z\ ^2D°$ (24)											
		3409.65	4142.7	33463	8	6	8.0(−4)	1.0(−4)	0.0093	−3.08	D	1
		3370.32	4690.2	34352	6	4	0.021	0.0024	0.16	−1.84	C	1
		3474.53	4690.2	33463	6	6	0.0038	6.8(−4)	0.047	−2.39	C	1
		3414.74	5075.8	34352	4	4	0.088	0.015	0.69	−1.21	C	1
36.	$b\ ^4F - (\ °)^b$											
		3474.02	4690.2	33467	6	8	0.56	0.14	9.3	−0.09	B	1
		3409.17	4142.7	33467	8	8	0.42	0.074	6.6	−0.23	B	1

Co I: Allowed transitions — Continued

No.	Multiplet	λ (Å)	E_i (cm^{-1})	E_k (cm^{-1})	g_i	g_k	A_{ki} (10^8 s^{-1})	f_{ik}	S (at. u.)	log gf	Accuracy	Source
37.	$b\ ^4F - (\ °)^b$											
		3462.80	5075.8	33946	4	6	0.79	0.21	9.7	−0.07	C+	1
		3417.15	4690.2	33946	6	6	0.32	0.056	3.8	−0.47	C+	1
		3354.37	4142.7	33946	8	6	0.11	0.014	1.2	−0.95	C+	1
38.	$b\ ^4F - y\ ^2F°$ (26)											
		3127.25	3482.8	35451	10	8	0.0050	5.9(−4)	0.061	−2.23	C	1
		3105.93	4142.7	36330	8	6	0.0012	1.3(−4)	0.010	−3.00	D	1
		3193.16	4142.7	35451	8	8	0.0033	5.0(−4)	0.042	−2.40	C	1
		3250.00	4690.2	35451	6	8	0.0099	0.0021	0.13	−1.90	C	1
		3198.66	5075.8	36330	4	6	0.0038	8.7(−4)	0.037	−2.46	D	1
39.	$b\ ^4F - y\ ^2D°$											
		3129.01	4142.7	36092	8	6	0.0020	2.2(−4)	0.018	−2.76	D	1
		3223.15	5075.8	36092	4	6	0.0017	4.1(−4)	0.017	−2.79	E	1
40.	$b\ ^4F - x\ ^4D°$ (uv 52)											
		2764.19	3482.8	39649	10	8	0.043	0.0040	0.36	−1.40	C	1
		2815.56	4142.7	39649	8	8	0.032	0.0038	0.28	−1.52	C	1
41.	$b\ ^4F - x\ ^4F°$ (uv 53)											
		2695.85	4142.7	41226	8	10	0.045	0.0061	0.43	−1.31	C	1
		2685.34	4690.2	41918	6	8	0.055	0.0080	0.42	−1.32	D	1
42.	$b\ ^4F - w\ ^4D°$ (uv 55)											
		2504.52	3482.8	43399	10	8	0.18	0.014	1.1	−0.86	D	1
43.	$b\ ^4F - w\ ^4F°$ (uv 56)											
		2511.02	3482.8	43295	10	10	0.92	0.087	7.2	−0.06	C	1
		2476.64	3482.8	43848	10	8	0.22	0.016	1.3	−0.80	D	1
44.	$b\ ^4F - x\ ^2F°$											
		2536.50	4142.7	43555	8	8	0.30	0.029	1.9	−0.64	D	1
45.	$b\ ^4F - w\ ^4G°$ (uv 57)											
		2470.27	3482.8	43952?	10	12	0.15	0.017	1.3	−0.78	D	1
46.	$b\ ^4F - (\ °)^b$											
		2530.13	4690.2	44202	6	6	0.071	0.0068	0.34	−1.39	D	1
47.	$b\ ^4F - 6°$ (uv 64)											
		2287.80	4142.7	47839	8	8	0.86	0.067	4.0	−0.27	D	1

Co I: Allowed transitions — Continued

No.	Multiplet	λ (Å)	E_i (cm^{-1})	E_k (cm^{-1})	g_i	g_k	A_{ki} (10^8 s^{-1})	f_{ik}	S (at. u.)	log gf	Accuracy	Source
48.	$a\ ^2F - z\ ^4F°$											
		4782.56	7442.4	28346	8	10	2.5(−5)	1.1(−5)	0.0013	−4.07	C	1
		4685.86	7442.4	28777	8	8	5.6(−5)	1.8(−5)	0.0023	−3.83	C	1
49.	$a\ ^2F - z\ ^4G°$ (27)											
		4580.14	7442.4	29270	8	10	2.7(−4)	1.1(−4)	0.013	−3.07	C	1
		4699.18	8460.8	29735	6	8	7.4(−5)	3.2(−5)	0.0030	−3.71	D	1
		4484.51	7442.4	29735	8	8	1.4(−4)	4.1(−5)	0.0049	−3.48	D	1
		4619.33	8460.8	30103	6	6	2.9(−5)	9.4(−6)	8.6(−4)	−4.25	E	1
50.	$a\ ^2F - z\ ^4D°$											
		4574.94	7442.4	29295	8	8	3.2(−5)	9.9(−6)	0.0012	−4.10	D	1
51.	$a\ ^2F - z\ ^2G°$ (28)	*4117.4*	*7878.9*	*32159*	14	18	0.18	0.058	11	−0.09	C	1
		4121.32	7442.4	31700	8	10	0.19	0.060	6.5	−0.32	C	1
		4118.77	8460.8	32733	6	8	0.16	0.054	4.4	−0.49	C	1
		3952.92	7442.4	32733	8	8	0.016	0.0039	0.40	−1.51	C	1
52.	$a\ ^2F - z\ ^2F°$	*4100.2*	*7878.9*	*32261*	14	14	0.063	0.016	3.0	−0.65	D	1
		4092.39	7442.4	31871	8	8	0.057	0.014	1.5	−0.94	C+	1
		4110.53	8460.8	32782	6	6	0.055	0.014	1.1	−1.08	D	1
		3945.33	7442.4	32782	8	6	0.022	0.0039	0.40	−1.51	D	1
		4270.43	8460.8	31871	6	8	1.0(−4)	3.6(−5)	0.0031	−3.66	D	1
53.	$a\ ^2F - y\ ^4D°$ (30)											
		4066.37	7442.4	32028	8	8	0.011	0.0027	0.29	−1.66	C	1
		4132.16	8460.8	32655	6	6	9.9(−4)	2.5(−4)	0.021	−2.82	C	1
		3965.24	7442.4	32655	8	6	2.6(−4)	4.6(−5)	0.0049	−3.43	E	1
54.	$a\ ^2F - y\ ^4G°$ (31)											
		3995.31	7442.4	32465	8	10	0.25	0.075	7.9	−0.22	C+	1
		4045.39	8460.8	33173	6	8	0.024	0.0080	0.64	−1.32	C+	1
		3885.28	7442.4	33173	8	8	0.0030	6.9(−4)	0.070	−2.26	C	1
		3965.01	8460.8	33674	6	6	5.2(−4)	1.2(−4)	0.0097	−3.13	E	1
		3811.07	7442.4	33674	8	6	0.0020	3.3(−4)	0.033	−2.58	C	1
55.	$a\ ^2F - y\ ^4F°$ (32)											
		3935.96	7442.4	32842	8	10	0.062	0.018	1.9	−0.84	C+	1
		3884.60	8460.8	34196	6	4	0.016	0.0024	0.18	−1.84	C+	1
56.	$a\ ^2F - y\ ^2G°$ (34)	*3864.5*	*7878.9*	*33748*	14	18	0.60	0.17	31	0.39	C	1
		3845.47	7442.4	33440	8	10	0.46	0.13	13	0.01	C+	1
		3894.07	8460.8	34134	6	8	0.69	0.21	16	0.10	C	1
		3745.49	7442.4	34134	8	8	0.075	0.016	1.6	−0.90	C	1

Co ɪ: Allowed transitions — Continued

No.	Multiplet	λ (Å)	E_i (cm^{-1})	E_k (cm^{-1})	g_i	g_k	A_{ki} (10^8 s^{-1})	f_{ik}	S (at. u.)	log gf	Accuracy	Source
57.	$a\ ^2F - z\ ^2D°$ (33)	3853.9	7878.9	33819	14	10	0.13	0.021	3.7	−0.54	C	1
		3842.05	7442.4	33463	8	6	0.13	0.021	2.1	−0.77	C	1
		3861.16	8460.8	34352	6	4	0.14	0.021	1.6	−0.89	C	1
		3998.55	8460.8	33463	6	6	4.2(−4)	1.0(−4)	0.0079	−3.22	D	1
58.	$a\ ^2F - (\ °)^b$											
		3997.90	8460.8	33467	6	8	0.070	0.022	1.8	−0.87	C+	1
		3841.46	7442.4	33467	8	8	0.0042	9.3(−4)	0.094	−2.13	C	1
59.	$a\ ^2F - (\ °)^b$											
		3922.76	8460.8	33946	6	6	0.0057	0.0013	0.10	−2.10	C+	1
60.	$a\ ^2F - y\ ^2F°$ (35)	3576.9	7878.9	35828	14	14	1.6	0.30	50	0.63	C	1
		3569.37	7442.4	35451	8	8	1.5	0.29	28	0.37	C	1
		3587.19	8460.8	36330	6	6	1.4	0.26	19	0.20	C	1
		3460.72	7442.4	36330	8	6	0.0031	4.2(−4)	0.039	−2.47	D	1
		3704.06	8460.8	35451	6	8	0.12	0.034	2.5	−0.69	C	1
61.	$a\ ^2F - y\ ^2D°$	3504.6	7878.9	36405	14	10	1.4	0.19	30	0.42	C	1
		3489.40	7442.4	36092	8	6	1.3	0.18	16	0.15	C	1
		3518.34	8460.8	36875	6	4	1.6	0.20	14	0.07	C	1
		3618.01	8460.8	36092	6	6	0.0020	3.9(−4)	0.028	−2.63	D	1
62.	$a\ ^4P - z\ ^4D°$ (37)											
		6282.64	14036	29949	4	6	0.0019	0.0017	0.14	−2.16	C	1
		6230.94	14399	30444	2	4	0.0014	0.0016	0.066	−2.49	C	1
		6189.01	13796	29949	6	6	0.0010	5.9(−4)	0.072	−2.45	C	1
		6093.14	14036	30444	4	4	0.0016	9.1(−4)	0.073	−2.44	C	1
		6116.99	14399	30743	2	2	0.0029	0.0016	0.065	−2.49	D	1
		6005.03	13796	30444	6	4	2.2(−4)	8.0(−5)	0.0095	−3.32	D	1
		5984.09	14036	30743	4	2	4.7(−4)	1.3(−4)	0.0099	−3.30	D	1
63.	$a\ ^4P - z\ ^2F°$ (38)											
		5530.78	13796	31871	6	8	0.0024	0.0015	00.16	−2.06	C+	1
		5333.15	14036	32782	4	6	1.1(−4)	6.7(−5)	0.0047	−3.57	D	1
64.	$a\ ^4P - y\ ^4D°$	5373.1	13977	32583	12	20	0.012	0.0085	1.8	−0.99	C	1
		5483.35	13796	32028	6	8	0.0090	0.0054	0.58	−1.49	C	1
		5369.59	14036	32655	4	6	0.0086	0.0056	0.40	−1.65	C+	1
		5331.46	14399	33151	2	4	0.0064	0.0055	0.19	−1.96	C	1
		5301.04	13796	32655	6	6	0.0040	0.0017	0.17	−2.00	C	1
		5230.21	14036	33151	4	4	0.0088	0.0036	0.25	−1.84	C	1
		5247.92	14399	33449	2	2	0.010	0.0043	0.15	−2.07	D	1
		5165.16	13796	33151	6	4	8.2(−4)	2.2(−4)	0.022	−2.88	C	1
		5149.80	14036	33449	4	2	0.0022	4.3(−4)	0.029	−2.76	D	1
65.	$a\ ^4P - z\ ^2D°$											
		5146.16	14036	33463	4	6	9.2(−5)	5.5(−5)	0.0037	−3.66	D	1

Co I: Allowed transitions — Continued

No.	Multiplet	λ (Å)	E_i (cm^{-1})	E_k (cm^{-1})	g_i	g_k	A_{ki} (10^8 s^{-1})	f_{ik}	S (at. u.)	log gf	Accuracy	Source
66.	$a\ ^4P - y\ ^2F°$											
		4436.43	13796	36330	6	6	9.8(−4)	2.9(−4)	0.025	−2.76	D	1
67.	$b\ ^4P - z\ ^4F°$ (53)											
		7354.58	15184	28777	6	8	3,3(−4)	3.6(−4)	0.052	−2.67	C	1
68.	$b\ ^4P - z\ ^4D°$ (54)											
		7052.87	15744	29949	4	6	0.0054	0.0060	0.56	−1.62	C	1
		7016.60	16196	30444	2	4	0.0046	0.0067	0.31	−1.87	C	1
		6771.04	15184	29949	6	6	0.0026	0.0018	0.24	−1.97	C	1
		6814.95	15744	30444	4	4	0.0045	0.0031	0.28	−1.90	C	1
		6872.32	16196	30743	2	2	0.010	0.0071	0.32	−1.85	D	1
		6678.82	15744	30743	4	2	0.0016	5.2(−4)	0.046	−2.68	D	1
69.	$b\ ^4P - z\ ^2F°$											
		5878.10	15744	32782	4	6	2.2(−4)	1.7(−4)	0.013	−3.16	D	1
70.	$b\ ^4P - y\ ^4D°$ (55)											
		5935.39	15184	32028	6	8	4.9(−4)	3.5(−4)	0.041	−2.68	D	1
71.	$b\ ^4P - z\ ^2D°$ (56)											
		5469.31	15184	33463	6	6	0.0011	4.9(−4)	0.053	−2.53	C	1
		5381.11	15744	34352	4	4	0.0025	0.0011	0.076	−2.37	C	1
72.	$b\ ^4P - y\ ^2F°$											
		4932.88	15184	35451	6	8	2.4(−4)	1.2(−4)	0.011	−3.15	E	1
73.	$b\ ^4P - y\ ^2F°$ (57)											
		4781.43	15184	36092	6	6	0.0034	0.0012	0.11	−2.15	C	1
		4920.27	15744	36092	4	6	0.0012	6.4(−4)	0.042	−2.59	C	1
74.	$a\ ^2G - z\ ^2G°$ (80)											
		6563.40	16468	31700	10	10	0.0021	0.0013	0.29	−1.87	C	1
75.	$a\ ^2G - z\ ^2F°$ (81)											
		6490.34	16468	31871	10	8	6.0(−4)	3.0(−4)	0.065	−2.52	C	1
		6429.91	17234	32782	8	6	0.0010	4.9(−4)	0.082	−2.41	D	1
76.	$a\ ^2G - y\ ^4G°$											
		6249.50	16468	32465	10	10	6.6(−4)	3.9(−4)	0.080	−2.41	C	1

Co I: Allowed transitions — Continued

No.	Multiplet	λ (Å)	E_i (cm^{-1})	E_k (cm^{-1})	g_i	g_k	A_{ki} (10^8 s^{-1})	f_{ik}	S (at. u.)	log gf	Accuracy	Source
77.	$a\ ^2G - y\ ^2G°$ (82)											
		5890.49	16468	33440	10	10	0.0013	6.9(−4)	0.13	−2.16	C	1
		5915.55	17234	34134	8	8	0.0024	0.0013	0.19	−2.00	C	1
78.	$a\ ^2G - y\ ^2F°$ (83)											
		5266.51	16468	35451	10	8	0.017	0.0058	1.0	−1.24	C	1
		5235.19	17234	36330	8	6	0.014	0.0042	0.58	−1.47	C	1
79.	$a\ ^2D - z\ ^4D°$ (89)											
		7417.38	16471	29949	4	6	0.0017	0.0021	0.21	−2.07	C	1
		7590.57	16778	29949	6	6	3.4(−4)	2.9(−4)	0.043	−2.76	E	1
		7154.69	16471	30444	4	4	0.0012	9.5(−4)	0.090	−2.42	C	1
80.	$a\ ^2D - z\ ^2F°$											
		6129.12	16471	32782	4	6	3.6(−4)	3.0(−4)	0.024	−2.92	D	1
81.	$a\ ^2D - z\ ^2D°$	*5824.5*	*16655*	*33819*	10	10	0.0058	0.0030	0.57	−1.53	C	1
		5991.89	16778	33463	6	6	0.0044	0.0024	0.28	−1.85	C	1
		5590.74	16471	34352	4	4	0.0072	0.0034	0.25	−1.87	C	1
		5688.59	16778	34352	6	4	8.4(−4)	2.7(−4)	0.030	−2.79	C	1
		5883.42	16471	33463	4	6	1.0(−4)	7.9(−5)	0.0061	−3.50	D	1
82.	$a\ ^2D - y\ ^2F°$ (91)											
		5034.06	16471	36330	4	6	2.4(−4)	1.4(−4)	0.0091	−3.26	E	1
		5113.22	16778	36330	6	6	0.0024	9.6(−4)	0.097	−2.24	C	1
83.	$a\ ^2D - y\ ^2D°$ (92)											
		5176.09	16778	36092	6	6	0.0079	0.0032	0.32	−1.72	C	1
		5094.96	16471	36092	4	6	0.0044	0.0026	0.17	−1.99	C	1
84.	$a\ ^2P - z\ ^2D°$ (111)											
		6632.44	18390	33463	4	6	0.0025	0.0025	0.22	−2.00	C	1
		6417.86	18775	34352	2	4	6.3(−4)	7.7(−4)	0.033	−2.81	C	1
85.	$a\ ^2P - y\ ^2D°$ (112)											
		5647.23	18390	36092	4	6	0.0096	0.0069	0.51	−1.56	C	1
86.	$b\ ^2P - z\ ^2D°$ (126)											
		7712.66	20501	33463	4	6	0.0050	0.0067	0.68	−1.57	C	1
		7610.24	21216	34352	2	4	0.0044	0.0076	0.38	−1.82	D	1
87.	$a\ ^2H - y\ ^4G°$											
		9356.98	21780	32465	12	10	0.0012	0.0013	0.48	−1.81	D	1

Co I: Allowed transitions — Continued

No.	Multiplet	λ (Å)	E_i (cm^{-1})	E_k (cm^{-1})	g_i	g_k	A_{ki} (10^8 s^{-1})	f_{ik}	S (at. u.)	log gf	Accuracy	Source
88.	$b\ ^2D - z\ ^2D°$											
		8926.21	23153	34352	4	4	0.0020	0.0023	0.27	−2.03	D	1
89.	$b\ ^2D - y\ ^2F°$ (139)											
		7388.69	21920	35451	6	8	0.0034	0.0037	0.54	−1.65	D	1
90.	$b\ ^2D - y\ ^2D°$ (140)											
		7054.04	21920	36092	6	6	0.0066	0.0047	0.69	−1.53	C	1
91.	$z\ ^6F° - f\ ^4F$											
		4180.70	23612	47524	12	10	0.0016	3.4(−4)	0.056	−2.39	C+	1
		4223.77	23856	47524	10	10	0.0012	3.1(−4)	0.043	−2.51	C	1
		4309.44	24326	47524	8	10	0.0077	0.0027	0.30	−1.67	B	1
		4259.87	24733	48202	6	8	0.0053	0.0019	0.16	−1.94	C+	1
92.	$z\ ^6D° - f\ ^4F$											
		4359.43	25269	48202	8	8	0.0059	0.0017	0.19	−1.87	B	1
		4240.79	24628	48202	10	8	9.9(−4)	2.1(−4)	0.030	−2.67	D	1
		4450.79	25740	48202	6	8	0.0019	7.4(−4)	0.065	−2.35	C+	1
93.	$z\ ^6G° - f\ ^4F$											
		4553.34	25569	47524	12	10	0.0052	0.0014	0.24	−1.79	B	1
		4490.31	25938	48202	10	8	0.0021	5.1(−4)	0.076	−2.29	C+	1
		4550.47	26232	48202	8	8	0.0020	6.1(−4)	0.073	−2.31	C	1
94.	$z\ ^4F° - e\ ^4F$ (169)											
		6082.43	28346	44782	10	10	0.054	0.030	6.0	−0.52	B	1
		6246.39	28777	44782	8	10	0.0011	8.3(−4)	0.14	−2.18	C	1
95.	$z\ ^4F° - f\ ^4F$ (170)											
		5212.70	28346	47524	10	10	0.19	0.078	13	−0.11	B	1
		5146.75	28777	48202	8	8	0.15	0.060	8.1	−0.32	B	1
		5034.97	28346	48202	10	8	0.0050	0.0015	0.25	−1.82	B	1
		5332.65	28777	47524	8	10	0.038	0.020	2.8	−0.79	B	1
		5265.79	29216	48202	6	8	0.050	0.028	2.9	−0.78	B	1
96.	$z\ ^4G° - e\ ^4F$											
		6444.68	29270	44782	10	10	0.0062	0.0039	0.83	−1.41	C+	1
		6644.03	29735	44782	8	10	0.0011	9.5(−4)	0.17	−2.12	C	1
97.	$z\ ^4G° - f\ ^4F$ (172)											
		5352.05	28845	47524	12	10	0.27	0.096	20	0.06	B	1
		5280.63	29270	48202	10	8	0.28	0.093	16	−0.03	D	1
		5476.47	29270	47524	10	10	0.0056	0.0025	0.45	−1.60	B	1

Co I: Allowed transitions — Continued

No.	Multiplet	λ (Å)	E_i (cm^{-1})	E_k (cm^{-1})	g_i	g_k	A_{ki} (10^8 s^{-1})	f_{ik}	S (at. u.)	log gf	Accuracy	Source
98.	z ^{4}D° – e ^{4}F (174)											
		6455.00	29295	44782	8	10	0.090	0.070	12	−0.25	B	1
99.	z ^{4}D° – f ^{4}F (175)											
		5483.96	29295	47524	8	10	0.073	0.041	6.0	−0.48	B	1
		5477.09	29949	48202	6	8	0.068	0.041	4.4	−0.61	B	1
		5287.57	29295	48202	8	8	0.030	0.013	1.7	−1.00	B	1
100.	z ^{2}G° – f ^{4}F											
		6058.23	31700	48202	10	8	0.0069	0.0030	0.60	−1.52	C+	1
101.	z ^{2}F° – e ^{4}F (183)											
		7743.27	31871	44782	8	10	0.0099	0.011	2.3	−1.05	C	1
102.	z ^{2}F° – f ^{4}F											
		6386.69	31871	47524	8	10	0.0043	0.0033	0.55	−1.58	C+	1
103.	y ^{4}D° – e ^{4}F											
		7838.12	32028	44782	8	10	0.054	0.063	13	−0.30	B	1
104.	y ^{4}D° – f ^{4}F											
		6451.14	32028	47524	8	10	0.027	0.021	3.5	−0.78	B	1
		6430.34	32655	48202	6	8	0.027	0.022	2.8	−0.88	B	1
		6181.01	32028	48202	8	8	0.013	0.0077	1.3	−1.21	B	1
105.	y ^{4}G° – e ^{4}F (189)											
		8093.93	32431	44782	12	10	0.20	0.16	52	0.29	B	1
106.	y ^{4}G° – f ^{4}F											
		6623.36	32431	47524	12	10	0.0039	0.0021	0.56	−1.59	C+	1
		6352.75	32465	48202	10	8	0.0048	0.0023	0.49	−1.63	D	1
		6638.40	32465	47524	10	10	8.3(−4)	5.5(−4)	0.12	−2.26	E	1
		6652.29	33173	48202	8	8	0.0024	0.0016	0.28	−1.90	C	1
107.	y ^{4}F° – e ^{4}F (193)											
		8372.79	32842	44782	10	10	0.087	0.091	25	−0.04	B	1
108.	y ^{4}F° – f ^{4}F											
		6808.94	32842	47524	10	10	0.0059	0.0041	0.91	−1.39	C+	1
		6508.79	32842	48202	10	8	0.0054	0.0028	0.59	−1.56	C+	1
109.	y ^{2}G° – f ^{4}F											
		7106.37	34134	48202	8	8	0.0011	8.1(−4)	0.15	−2.19	E	1

Co I: Allowed transitions — Continued

No.	Multiplet	λ (Å)	E_i (cm^{-1})	E_k (cm^{-1})	g_i	g_k	A_{ki} (10^8 s^{-1})	f_{ik}	S (at. u.)	log gf	Accuracy	Source
110.	(°)^b – e ^{4}F											
		8835.21	33467	44782	8	10	0.0043	0.0063	1.5	−1.30	C	1
111.	(°)^b – f ^{4}F											
		6784.85	33467	48202	8	8	0.0064	0.0044	0.79	−1.45	C+	1

^{a}The number in parentheses following the tabulated value indicates the power of ten by which this value has to be multiplied.
^{b}The *LS*-coupling designation of this term was not provided in the NBS energy level computation (J. Phys. Chem. Ref. Data **10**, 1097 (1981)), so we have accordingly omitted it from this work.

Co II

Fe Isoelectronic Sequence

Ground State: $1s^2 2s^2 2p^6 3s^2 3p^6 3d^8$ ^{3}F$_4$

Ionization Energy: 17.083 eV = 137795 cm^{-1}

Allowed transitions

List of tabulated lines

Wavelength (Å)	No.	Wavelength (Å)	No.	Wavelength (Å)	No.	Wavelength (Å)	No.
2283.52	3	2326.47	2	2389.54	1	2694.68	5
2286.15	3	2330.36	2	2393.9	2	2810.86	4
2293.38	3	2344.28	2	2404.17	1	2825.25	4
2301.4	3	2353.41	2	2414.07	1	2834.95	4
2307.85	3	2363.8	2	2417.66	1	3352.79	7
2311.61	3	2375.19	2	2428.3	1	3415.77	7
2314.05	3	2378.62	1	2436.98	1	3621.22	6
2314.97	3	2383.45	1	2449.16	1		
2324.3	2	2388.92	1	2663.53	5		

For this spectrum, we have chosen the work of Salih *et al.*,[1] who determined absolute transition probabilities by combining lifetime measurements with branching ratios. These authors measured radiative lifetimes by using time-resolved laser fluorescence spectroscopy. Branching ratios were obtained from a hollow cathode discharge and were recorded with the one meter Fourier transform spectrometer at Kitt Peak National Observatory.

For the strong branches, the data are estimated to be accurate within 10 percent, while for the weaker lines, the uncertainties are greater. For a few of these weak lines, Salih *et al*. have tabulated transition probabilities consisting of only one significant figure. We have not included such lines in this compilation.

Reference
[1]S. Salih, J. E. Lawler, and W. Whaling, Phys. Rev. A **31**, 744 (1985).

Co II: Allowed transitions

No.	Multiplet	λ (Å)	E_i (cm^{-1})	E_k (cm^{-1})	g_i	g_k	A_{ki} (10^8 s^{-1})	f_{ik}	S (at. u.)	log gf	Accuracy	Source
1.	$a\ ^5F - z\ ^5F°$ (uv 7)											
		2388.92	3351	45198	11	11	2.8	0.24	21	0.42	B	1
		2417.66	4029	45379	9	9	0.85	0.074	5.3	−0.17	B	1
		2414.07	4561	45972	7	7	0.72	0.063	3.5	−0.36	B	1
		2404.17	5205	46787	3	3	1.5	0.13	3.1	−0.41	C	1
		2378.62	3351	45379	11	9	1.9	0.13	11	0.16	B	1
		2383.45	4029	45972	9	7	1.8	0.12	8.4	0.03	B	1
		2389.54	4950	46787	5	3	1.5	0.077	3.0	−0.41	C	1
		2428.30	4029	45198	9	11	0.074	0.0080	0.58	−1.14	B	1
		2449.16	4561	45379	7	9	0.071	0.0082	0.46	−1.24	B	1
		2436.98	4950	45972	5	7	0.14	0.017	0.70	−1.06	B	1
2.	$a\ ^5F - z\ ^5D°$ (uv 8)											
		2326.47	3351	46321	11	9	0.79	0.052	4.4	−0.24	B	1
		2324.30	4029	47039	9	7	0.78	0.049	3.4	−0.35	B	1
		2330.36	4950	47849	5	3	1.32	0.0645	2.47	−0.492	B	1
		2363.80	4029	46321	9	9	2.1	0.18	12	0.20	B	1
		2353.41	4561	47039	7	7	1.9	0.16	8.6	0.04	B	1
		2344.28	5205	47849	3	3	1.5	0.12	2.9	−0.43	B	1
		2393.90	4561	46321	7	9	0.10	0.011	0.61	−1.11	C+	1
		2375.19	4950	47039	5	7	0.13	0.015	0.60	−1.11	B	1
3.	$a\ ^5F - z\ ^5G°$ (uv 9)											
		2286.15	3351	47079	11	13	3.3	0.31	25	0.53	B	1
		2307.85	4029	47346	9	11	2.6	0.25	17	0.36	B	1
		2311.61	4561	47808	7	9	2.8	0.29	15	0.31	B	1
		2314.05	4950	48151	5	7	2.8	0.31	12	0.20	B	1
		2314.97	5205	48389	3	5	2.7	0.36	8.3	0.04	B	1
		2283.52	4029	47808	9	9	0.20	0.016	1.1	−0.85	C+	1
		2293.38	4561	48151	7	7	0.33	0.026	1.4	−0.74	B	1
		2301.40	4950	48389	5	5	0.38	0.030	1.1	−0.82	B	1
4.	$b\ ^3F - z\ ^5F°$											
		2825.25	9813	45198	9	11	0.017	0.0025	0.21	−1.65	C	1
		2810.86	9813	45379	9	9	0.014	0.0017	0.14	−1.83	C	1
		2834.95	10708	45972	7	7	0.010	0.0012	0.079	−2.07	C	1
5.	$b\ ^3F - z\ ^5G°$ (uv 13)											
		2663.53	9813	47346	9	11	0.53	0.069	5.4	−0.21	C	1
		2694.68	10708	47808	7	9	0.30	0.042	2.6	−0.53	C	1
6.	$a\ ^5P - z\ ^5F°$ (1)											
		3621.22	17772	45379	7	9	0.026	0.0066	0.55	−1.34	C	1
7.	$a\ ^5P - z\ ^5D°$ (2)											
		3415.77	17772	47039	7	7	0.015	0.0026	0.21	−1.74	E	1
		3352.79	18032	47849	5	3	0.029	0.0029	0.16	−1.83	E	1

Co III

Mn Isoelectronic Sequence

Ground State: $1s^2 2s^2 2p^6 3s^2 3p^6 3d^7\ {}^4F_{9/2}$

Ionization Energy: 33.50 eV = 270200 cm^{-1}

Forbidden Transitions

List of tabulated lines

Wavelength (Å)	No.	Wavelength (Å)	No.	Wavelength (Å)	No.	Wavelength (Å)	No.
1733.1	8	2917.7	24	5165.0	4	11063	11
1758.8	8	2926.1	27	5192.5	16	11347	11
1773.6	8	2987.7	27	5246.8	4	11864	11
1777.9	8	3030.6	27	5333.0	4	12722	11
1791.1	8	4265.8	5	5454.1	4	13098	11
1792.9	8	4272.8	6	5626.3	3	15485	15
1806.4	8	4335.1	6	5839.4	20	17410	15
2352.4	13	4387.2	6	5887.8	3	17641	15
2365.0	13	4399.9	5	5906.0	3	18200	10
2378.9	13	4424.7	5	5942.0	20	19568	10
2386.7	13	4468.8	6	6126.7	3	20015	10
2391.8	13	4499.3	6	6194.8	3	20963	10
2413.9	13	4520.5	12	6209.2	20	22799	10
2455.1	17	4547.4	5	6576.1	2	24737	19
2503.6	17	4567.2	12	6849.3	23	30136	19
2533.7	17	4569.1	5	6853.6	2	34913	19
2665.8	21	4581.7	12	6961.4	2	84874	25
2678.9	7	4626.3	6	6961.5	2	118800	1
2699.8	21	4629.7	12	7012.3	26	126660	14
2700.3	7	4713.5	12	7152.6	2	138080	18
2718.2	21	4717.2	6	7160.7	26	139980	22
2740.7	7	4835.0	29	7169.2	2	163920	1
2753.7	21	4905.1	29	7201.7	23	211180	30
2763.1	7	4915.5	16	7270.2	2	240260	1
2787.3	7	4948.3	29	7358.3	23	260410	9
2810.5	7	4988.0	16	7372.1	2	338320	28
2820.0	7	5021.7	29	7497.1	2	442160	9
2843.7	7	5114.0	16	7643.8	26		
2886.1	27	5134.7	4	7820.5	26		

For this spectrum, we have chosen the work of Hansen et al.,[1] who calculated M1 and E2 transition probabilities for transitions within the $3d^7$ ground configuration. These authors used a single configuration approximation, which should be fairly reliable, since the ground configuration is fairly well separated from other configurations of the same parity. Also, the authors determined eigenvector components by a parametric fitting of theoretical energy expressions to observed energy levels. Finally, Hartree-Fock calculations were used to determine s_q, the radial electric quadrupole integral, which is needed in the calculation of E2 transition probabilities.

Reference

[1] J. E. Hansen, A. J. J. Raassen, and P. H. M. Uylings, Astrophys. J. **277**, 435 (1984).

Co III: Forbidden transitions

No.	Transition Array	Multiplet	λ (Å)	E_i (cm^{-1})	E_k (cm^{-1})	g_i	g_k	Type of transition	A_{ki} (s^{-1})	S (at. u.)	Accuracy	Source
1.	$3d^7$–$3d^7$	${}^4F$ – ${}^4F$										
			[118800]	0.0	841.5	10	8	M1	0.020	9.9	C+	1
			[163920]	841.5	1451.4	8	6	M1	0.013	13	C+	1
			[240260]	1451.4	1867.5	6	4	M1	0.0047	9.7	C+	1

Co III: Forbidden transitions — Continued

No.	Transition Array	Multiplet	λ (Å)	E_i (cm^{-1})	E_k (cm^{-1})	g_i	g_k	Type of transition	A_{ki} (s^{-1})	S (at. u.)	Accuracy	Source
2.		$^4F - {}^4P$										
			[6576.1]	0.0	15202	10	6	E2	0.048	2.1	E	1
			[6853.6]	841.5	15428	8	4	E2	0.027	0.97	E	1
			[6961.4]	1451.4	15812	6	2	E2	0.020	0.39	E	1
			[6961.5]	841.5	15202	8	6	M1	0.0015	1.1(−4)[a]	E	1
			"	"	"	8	6	E2	0.012	0.70	E	1
			[7152.6]	1451.4	15428	6	4	M1	2.8(−4)	1.5(−5)	E	1
			"	"	"	6	4	E2	0.016	0.71	E	1
			[7169.2]	1867.5	15812	4	2	E2	0.026	0.59	E	1
			[7270.2]	1451.4	15202	6	6	M1	5.8(−4)	5.0(−5)	E	1
			"	"	"	6	6	E2	0.0021	0.15	E	1
			[7372.1]	1867.5	15428	4	4	M1	1.4(−5)	8.3(−7)	E	1
			"	"	"	4	4	E2	0.0044	0.23	E	1
			[7497.1]	1867.5	15202	4	6	M1	2.0(−4)	1.9(−5)	E	1
			"	"	"	4	6	E2	2.0(−4)	0.017	E	1
3.		$^4F - {}^2G$										
			[5887.8]	0.0	16980	10	10	M1	0.40	0.030	E	1
			[5906.0]	841.5	17769	8	8	M1	0.15	0.0092	E	1
			[5626.3]	0.0	17769	10	8	M1	0.014	7.4(−4)	E	1
			[6194.8]	841.5	16980	8	10	M1	0.12	0.011	E	1
			[6126.7]	1451.4	17769	6	8	M1	0.11	0.0075	E	1
4.		$^4F - {}^2P$										
			[5165.0]	841.5	20197	8	4	E2	0.0051	0.045	E	1
			[5134.7]	1451.4	20921	6	2	E2	0.0022	0.0093	E	1
			[5333.0]	1451.4	20197	6	4	M1	0.062	0.0014	E	1
			"	"	"	6	4	E2	0.0023	0.024	E	1
			[5246.8]	1867.5	20921	4	2	M1	0.0012	1.3(−5)	E	1
			"	"	"	4	2	E2	0.0012	0.0057	E	1
			[5454.1]	1867.5	20197	4	4	M1	0.044	0.0011	E	1
			"	"	"	4	4	E2	5.7(−4)	0.0066	E	1
5.		$^4F - {}^2H$										
			[4569.1]	841.5	22722	8	12	E2	9.5(−6)	1.4(−4)	E	1
			[4547.4]	1451.4	23436	6	10	E2	9.5(−6)	1.1(−4)	E	1
			[4399.9]	0.0	22722	10	12	M1	4.9(−4)	1.9(−5)	E	1
			"	"	"	10	12	E2	1.3(−4)	0.0015	E	1
			[4424.7]	841.5	23436	8	10	M1	0.0022	7.1(−5)	E	1
			"	"	"	8	10	E2	3.8(−5)	3.8(−4)	E	1
			[4265.8]	0.0	23436	10	10	M1	0.0043	1.2(−4)	E	1
			"	"	"	10	10	E2	1.8(−5)	1.5(−4)	E	1
6.		$^4F - {}^2D2$										
			[4335.1]	0.0	23061	10	6	E2	5.6(−4)	0.0031	E	1
			[4272.8]	841.5	24239	8	4	E2	5.8(−4)	0.0020	E	1
			[4499.3]	841.5	23061	8	6	M1	0.75	0.015	E	1
			"	"	"	8	6	E2	2.1(−4)	0.0014	E	1
			[4387.2]	1451.4	24239	6	4	M1	0.73	0.0091	E	1
			"	"	"	6	4	E2	4.1(−4)	0.0016	E	1
			[4626.3]	1451.4	23061	6	6	M1	0.081	0.0018	E	1
			"	"	"	6	6	E2	1.2(−5)	9.1(−5)	E	1
			[4468.8]	1867.5	24239	4	4	M1	0.39	0.0052	E	1
			"	"	"	4	4	E2	1.7(−5)	7.2(−5)	E	1
			[4717.2]	1867.5	23061	4	6	M1	0.035	8.2(−4)	E	1
			"	"	"	4	6	E2	1.4(−6)	1.2(−5)	E	1

Co III: Forbidden transitions — Continued

No.	Transition Array	Multiplet	λ (Å)	E_i (cm^{-1})	E_k (cm^{-1})	g_i	g_k	Type of transition	A_{ki} (s^{-1})	S (at. u.)	Accuracy	Source
7.		$^4F - {}^2F$										
			[2700.3]	0.0	37022	10	6	E2	5.0(−4)	2.6(−4)	E	1
			[2678.9]	0.0	37318	10	8	M1	0.12	6.8(−4)	D	1
			"	"	"	10	8	E2	0.0035	0.0023	E	1
			[2763.1]	841.5	37022	8	6	M1	0.031	1.5(−4)	D	1
			"	"	"	8	6	E2	5.4(−4)	3.1(−4)	E	1
			[2740.7]	841.5	37318	8	8	M1	0.023	1.4(−4)	D	1
			"	"	"	8	8	E2	4.1(−4)	3.0(−4)	E	1
			[2810.5]	1451.4	37022	6	6	M1	0.024	1.2(−4)	D	1
			"	"	"	6	6	E2	7.2(−4)	4.5(−4)	E	1
			[2787.3]	1451.4	37318	6	8	M1	0.051	3.3(−4)	D	1
			"	"	"	6	8	E2	0.0012	9.6(−4)	E	1
			[2843.7]	1867.5	37022	4	6	M1	0.12	6.1(−4)	D	1
			"	"	"	4	6	E2	0.0027	0.0018	E	1
			[2820.0]	1867.5	37318	4	8	E2	2.8(−4)	2.4(−4)	E	1
8.		$^4F - {}^2D1$										
			[1733.1]	0.0	57699	10	6	E2	0.039	0.0022	E	1
			[1773.6]	841.5	57226	8	4	E2	0.0076	3.2(−4)	E	1
			[1758.8]	841.5	57699	8	6	M1	0.11	1.3(−4)	E	1
			"	"	"	8	6	E2	0.0029	1.7(−4)	E	1
			[1792.9]	1451.4	57226	6	4	M1	0.13	1.1(−4)	E	1
			"	"	"	6	4	E2	1.8(−5)	7.9(−7)	E	1
			[1777.9]	1451.4	57699	6	6	M1	0.017	2.1(−5)	E	1
			"	"	"	6	6	E2	0.0020	1.3(−4)	E	1
			[1806.4]	1867.5	57226	4	4	M1	0.085	7.4(−5)	E	1
			"	"	"	4	4	E2	0.0015	6.9(−5)	E	1
			[1791.1]	1867.5	57699	4	6	M1	0.0077	9.8(−6)	E	1
			"	"	"	4	6	E2	7.8(−4)	5.1(−5)	E	1
9.		$^4P - {}^4P$										
			[442160]	15202	15428	6	4	M1	2.7(−4)	3.5	C+	1
			[260410]	15428	15812	4	2	M1	0.0025	3.3	C+	1
10.		$^4P - {}^2P$										
			[20015]	15202	20197	6	4	M1	0.15	0.18	D	1
			[18200]	15428	20921	4	2	M1	1.9(−4)	8.5(−5)	D	1
			[20963]	15428	20197	4	4	M1	0.080	0.11	D	1
			[19568]	15812	20921	2	2	M1	0.20	0.11	D	1
			[22799]	15812	20197	2	4	M1	0.033	0.058	D	1
11.		$^4P - {}^2D2$										
			[12722]	15202	23061	6	6	M1	0.047	0.022	E	1
			[11347]	15428	24239	4	4	M1	0.014	0.0030	E	1
			[11063]	15202	24239	6	4	M1	0.0026	5.2(−4)	E	1
			[13098]	15428	23061	4	6	M1	0.024	0.012	E	1
			[11864]	15812	24239	2	4	M1	0.0099	0.0025	E	1

Co III: Forbidden transitions — Continued

No.	Transition Array	Multiplet	λ (Å)	E_i (cm^{-1})	E_k (cm^{-1})	g_i	g_k	Type of transition	A_{ki} (s^{-1})	S (at. u.)	Accuracy	Source
12.		$^4P - {}^2F$										
			[4567.2]	15428	37318	4	8	E2	0.0019	0.018	E	1
			[4713.5]	15812	37022	2	6	E2	5.4(−4)	0.0045	E	1
			[4520.5]	15202	37318	6	8	M1	1.8(−4)	4.9(−6)	E	1
			"	"	"	6	8	E2	1.6(−4)	0.0014	E	1
			[4629.7]	15428	37022	4	6	M1	6.1(−5)	1.3(−6)	E	1
			"	"	"	4	6	E2	8.7(−4)	0.0066	E	1
			[4581.7]	15202	37022	6	6	M1	0.0016	3.4(−5)	E	1
			"	"	"	6	6	E2	3.6(−5)	2.6(−4)	E	1
13.		$^4P - {}^2D1$										
			[2352.4]	15202	57699	6	6	M1	0.86	0.0025	E	1
			"	"	"	6	6	E2	7.4(−4)	1.9(−4)	E	1
			[2391.8]	15428	57226	4	4	M1	0.32	6.5(−4)	E	1
			"	"	"	4	4	E2	0.064	0.012	E	1
			[2378.9]	15202	57226	6	4	M1	0.11	2.2(−4)	E	1
			"	"	"	6	4	E2	1.5(−4)	2.7(−5)	E	1
			[2365.0]	15428	57699	4	6	M1	0.14	4.1(−4)	E	1
			"	"	"	4	6	E2	0.12	0.032	E	1
			[2413.9]	15812	57226	2	4	M1	0.083	1.7(−4)	E	1
			"	"	"	2	4	E2	0.031	0.0060	E	1
			[2386.7]	15812	57699	2	6	E2	0.010	0.0028	E	1
14.		$^2G - {}^2G$										
			[126660]	16980	17769	10	8	M1	0.0072	4.3	C+	1
15.		$^2G - {}^2H$										
			[17410]	16980	22722	10	12	M1	0.042	0.099	E	1
			[17641]	17769	23436	8	10	M1	0.039	0.079	E	1
			[15485]	16980	23436	10	10	M1	0.13	0.18	E	1
16.		$^2G - {}^2F$										
			[4988.0]	16980	37022	10	6	E2	0.024	0.26	E	1
			[4915.5]	16980	37318	10	8	M1	0.094	0.0033	E	1
			"	"	"	10	8	E2	0.071	0.97	E	1
			[5192.5]	17769	37022	8	6	M1	0.078	0.0024	E	1
			"	"	"	8	6	E2	0.061	0.82	E	1
			[5114.0]	17769	37318	8	8	M1	0.15	0.0060	E	1
			"	"	"	8	8	E2	0.0058	0.097	E	1
17.		$^2G - {}^2D1$										
			[2455.1]	16980	57699	10	6	E2	6.4	2.0	E	1
			[2533.7]	17769	57226	8	4	E2	6.4	1.6	E	1
			[2503.6]	17769	57699	8	6	M1	0.0014	4.9(−6)	E	1
			"	"	"	8	6	E2	0.46	0.16	E	1
18.		$^2P - {}^2P$										
			[138080]	20197	20921	4	2	M1	0.0064	1.2	C+	1
19.		$^2P - {}^2D2$										
			[34913]	20197	23061	4	6	M1	0.018	0.17	E	1
			[30136]	20921	24239	2	4	M1	0.027	0.11	E	1
			[24737]	20197	24239	4	4	M1	0.15	0.34	E	1

Co III: Forbidden transitions — Continued

No.	Transition Array	Multiplet	λ (Å)	E_i (cm^{-1})	E_k (cm^{-1})	g_i	g_k	Type of transition	A_{ki} (s^{-1})	S (at. u.)	Accuracy	Source
20.		^{2}P – ^{2}F										
			[5839.4]	20197	37318	4	8	E2	0.0088	0.28	E	1
			[6209.2]	20921	37022	2	6	E2	0.0080	0.26	E	1
			[5942.0]	20197	37022	4	6	M1	0.0012	5.6(−5)	E	1
			"	"	"	4	6	E2	0.013	0.34	E	1
21.		^{2}P – ^{2}D1										
			[2718.2]	20921	57699	2	6	E2	0.45	0.24	E	1
			[2665.8]	20197	57699	4	6	M1	0.063	2.7(−4)	E	1
			"	"	"	4	6	E2	1.6	0.77	E	1
			[2753.7]	20921	57226	2	4	M1	0.0014	4.3(−6)	E	1
			"	"	"	2	4	E2	0.91	0.34	E	1
			[2699.8]	20197	57226	4	4	M1	0.013	3.8(−5)	E	1
			"	"	"	4	4	E2	0.55	0.19	E	1
22.		^{2}H – ^{2}H										
			[139980]	22722	23436	12	10	M1	0.0053	5.4	C+	1
23.		^{2}H – ^{2}F										
			[6849.3]	22722	37318	12	8	E2	0.044	3.2	E	1
			[7358.3]	23436	37022	10	6	E2	0.030	2.3	E	1
			[7201.7]	23436	37318	10	8	M1	5.4(−4)	6.0(−5)	E	1
			"	"	"	10	8	E2	0.0032	0.30	E	1
24.		^{2}H – ^{2}D1										
			[2917.7]	23436	57699	10	6	E2	0.086	0.065	E	1
25.		^{2}D2 – ^{2}D2										
			[84874]	23061	24239	6	4	M1	0.025	2.3	C+	1
26.		^{2}D2 – ^{2}F										
			[7643.8]	24239	37318	4	8	E2	0.0029	0.36	E	1
			[7012.3]	23061	37318	6	8	M1	0.0095	9.7(−4)	E	1
			"	"	"	6	8	E2	0.018	1.5	E	1
			[7820.5]	24239	37022	4	6	M1	0.0059	6.3(−4)	E	1
			"	"	"	4	6	E2	0.0079	0.83	E	1
			[7160.7]	23061	37022	6	6	M1	0.020	0.0016	E	1
			"	"	"	6	6	E2	0.0036	0.24	E	1
27.		^{2}D2 – ^{2}D1										
			[2886.1]	23061	57699	6	6	M1	0.0029	1.6(−5)	D	1
			"	"	"	6	6	E2	0.40	0.29	E	1
			[3030.6]	24239	57226	4	4	M1	2.8(−4)	1.2(−6)	D	1
			"	"	"	4	4	E2	0.55	0.33	E	1
			[2926.1]	23061	57226	6	4	M1	0.51	0.0019	D	1
			"	"	"	6	4	E2	0.11	0.056	E	1
			[2987.7]	24239	57699	4	6	M1	0.27	0.0016	D	1
			"	"	"	4	6	E2	3.7(−4)	3.1(−4)	E	1
28.		^{2}F – ^{2}F										
			[338320]	37022	37318	6	8	M1	3.0(−4)	3.4	C+	1

Co III: Forbidden transitions — Continued

No.	Transition Array	Multiplet	λ (Å)	E_i (cm^{-1})	E_k (cm^{-1})	g_i	g_k	Type of transition	A_{ki} (s^{-1})	S (at. u.)	Accuracy	Source
29.		^{2}F – ^{2}D1										
			[5021.7]	37318	57226	8	4	E2	0.039	0.30	E	1
			[4905.1]	37318	57699	8	6	M1	0.20	0.0053	E	1
			"	"	"	8	6	E2	0.31	3.1	E	1
			[4948.3]	37022	57226	6	4	M1	0.20	0.0036	E	1
			"	"	"	6	4	E2	0.30	2.1	E	1
			[4835.0]	37022	57699	6	6	M1	0.37	0.0093	E	1
			"	"	"	6	6	E2	0.048	0.45	E	1
30.		^{2}D1 – ^{2}D1										
			[211180]	57226	57699	4	6	M1	0.0011	2.3	C+	1

^{a}The number in parentheses following the tabulated value indicates the power of ten by which this value has to be multiplied.

Co VIII

Ca Isoelectronic Sequence

Ground State: $1s^2 2s^2 2p^6 3s^2 3p^6 3d^2 \; ^3F_2$

Ionization Energy: 157.8 eV = 1273000 cm^{-1}

Allowed Transitions

List of tabulated lines

Wavelength (Å)	No.	Wavelength (Å)	No.	Wavelength (Å)	No.	Wavelength (Å)	No.
122.273	45	125.071	51	153.005	19	165.191	13
122.320	45	125.155	46,51	153.926	12	166.256	6
122.472	45	125.268	49	156.958	21	167.016	20
122.488	45	125.340	50	157.266	23	167.152	6
122.577	45	125.350	49	157.416	23	167.738	16
122.956	44	125.566	49	157.687	23	168.084	16
123.022	44	125.821	54	157.773	24	168.921	22
123.045	44	127.916	53	157.984	23	169.051	16
123.173	44	128.397	52	158.066	23	169.196	16
123.239	44	132.756	55	158.783	5	169.537	16
123.307	44	149.718	25	161.479	18	169.711	16
123.489	43	150.701	26	161.733	18	169.819	14
123.753	43	150.958	9	161.917	18	170.169	2
124.649	48	151.944	19	162.095	18	170.589	8
124.795	51	152.200	19	162.337	18	171.107	15
124.830	51	152.534	19	162.57	18	171.460	7
124.871	47	152.597	17,19	162.708	18	171.522	11
124.878	51	152.896	17	164.721	13	172.402	7

List of tabulated lines — Continued

Wavelength (Å)	No.	Wavelength (Å)	No.	Wavelength (Å)	No.	Wavelength (Å)	No.
172.767	7	182.355	29	185.041	40	189.040	3
172.776	10	182.686	29	185.461	40	189.472	32
173.373	10	183.167	29	185.835	34	190.342	42
173.561	10	183.266	29	187.092	33	190.574	31,37
173.742	10	183.686	28	187.375	39	191.262	30
179.068	4	183.939	35	187.909	38	191.645	36
179.147	1	184.203	28	188.054	38	191.757	36
179.731	1	184.265	28	188.165	38	192.332	36
179.949	1	184.356	27	188.241	38	192.619	41
180.422	1	184.850	27	188.345	38		
181.786	29	184.861	40	188.674	38		

For this spectrum, we have chosen the data of Fawcett, Ridgeley, and Ekberg.[1] These authors experimentally observed and classified about 140 Co VIII lines. For 120 lines in the $3d^2$–$3d\,4p$, $3d^2$–$3d\,4f$, and $3p^6 3d^2$–$3p^5 3d^3$ transition arrays, Fawcett $et\ al.$ calculated oscillator strengths by the Hartree-XR method (self-consistent-field calculations with exchange, configuration interaction, and relativistic effects). In general, these data should be accurate to within fifty percent. We estimate that the f-values for intercombination (spin-forbidden) lines and weak lines (log $gf < -2.0$) are not as reliable, so that in these cases we have assigned accuracies of "E."

Reference

[1]B. C. Fawcett, A. Ridgeley, and J. O. Ekberg, Phys. Scr. **21**, 155 (1980).

Co VIII: Allowed transitions

No.	Transition Array	Multiplet	λ (Å)	E_i (cm⁻¹)	E_k (cm⁻¹)	g_i	g_k	A_{ki} (10^8 s⁻¹)	f_{ik}	S (at. u.)	log gf	Accuracy	Source
1.	$3p^6 3d^2$– $3p^5(^2P°)3d^3(^2H)$	^3F – $^3G°$											
			179.147	3144	561346	9	11	860	0.51	2.7	0.66	D	1
			179.731	1430	557817	7	9	780	0.48	2.0	0.53	D	1
			179.949	0	555699	5	7	480	0.32	0.96	0.21	D	1
			180.422	1430	555699	7	7	13	0.0064	0.027	−1.35	D	1
2.		^3F – $^1H°$											
			170.169	3144	590805	9	11	40	0.021	0.11	−0.72	E	1
3.		^1G – $^3G°$											
			189.040	32360	561346	9	11	19	0.012	0.068	−0.96	E	1
4.		^1G – $^1H°$	179.068	32360	590805	9	11	660	0.39	2.0	0.54	D	1
5.		^1G – $^1G°$	158.783	32360	662151	9	9	3700	1.4	6.6	1.10	D	1
6.	$3p^6 3d^2$– $3p^5(^2P°)3d^3(^2F)$	^3F – $^3D°$											
			167.152	1430	599641	7	5	190	0.058	0.22	−0.39	D	1
			166.256	1430	602844	7	7	31	0.013	0.049	−1.05	D	1
7.		^1D – $^3D°$											
			171.460	19624	602844	5	7	660	0.41	1.2	0.31	E	1
			172.402	19624	599641	5	5	380	0.17	0.48	−0.07	E	1
			172.767	19624	598440	5	3	4.1	0.0011	0.0031	−2.26	E	1

Co VIII: Allowed transitions — Continued

No.	Transition Array	Multiplet	λ (Å)	E_i (cm^{-1})	E_k (cm^{-1})	g_i	g_k	A_{ki} (10^8 s^{-1})	f_{ik}	S (at. u.)	log gf	Accuracy	Source
8.		^1D – $^1D°$	170.589	19624	605841	5	5	1200	0.50	1.4	0.40	D	1
9.		^1D – $^1F°$	150.958	19624	682051	5	7	2600	1.3	3.1	0.80	D	1
10.		^3P – $^3D°$											
			172.776	24055	602844	5	7	390	0.25	0.70	0.09	D	1
			173.373	22839	599641	3	5	560	0.42	0.72	0.10	D	1
			173.561	22304	598440	1	3	500	0.68	0.39	−0.17	D	1
			173.742	24055	599641	5	5	88	0.040	0.11	−0.70	D	1
11.		^3P – $^1D°$											
			171.522	22839	605841	3	5	150	0.11	0.19	−0.48	E	1
12.		^1G – $^1F°$	153.926	32360	682051	9	7	3300	0.90	4.1	0.91	D	1
13.	$3p^63d^2$– $3p^5(^2P°)3d^3(^2G)$	^3F – $^1F°$											
			165.191	3144	608501	9	7	64	0.020	0.099	−0.74	E	1
			164.721	1430	608501	7	7	56	0.023	0.086	−0.80	E	1
14.		^1D – $^1F°$	169.819	19624	608501	5	7	630	0.38	1.1	0.28	D	1
15.		^3P – $^1F°$											
			171.107	24055	608501	5	7	610	0.37	1.0	0.27	E	1
16.	$3p^63d^2$– $3p^5(^2P°)3d^3(^4P)$	^3P – $^3P°$	*168.61*	*23455*	*616526*	9	9	1400	0.60	3.0	0.73	D	1
			168.084	24055	619010	5	5	1200	0.49	1.4	0.39	D	1
			169.196	22839	613869	3	3	450	0.19	0.32	−0.24	D	1
			169.537	24055	613869	5	3	430	0.11	0.31	−0.25	D	1
			169.711	22839	612076	3	1	1400	0.21	0.34	−0.21	D	1
			167.738	22839	619010	3	5	260	0.18	0.30	−0.26	D	1
			169.051	22304	613869	1	3	450	0.58	0.32	−0.24	D	1
17.		^3P – $^3S°$											
			152.896	24055	678094	5	3	2500	0.53	1.3	0.42	D	1
			152.597	22839	678094	3	3	1700	0.58	0.87	0.24	D	1
18.	$3p^63d^2$– $3p^5(^2P°)3d^3(^4F)$	^3F – $^3F°$	*162.08*	*1824*	*618817*	21	21	2500	0.98	11	1.31	D	1
			161.917	3144	620737	9	9	2500	0.99	4.8	0.95	D	1
			162.095	1430	618348	7	7	2200	0.88	3.3	0.79	D	1
			162.337	0	616019	5	5	2200	0.87	2.3	0.64	D	1
			162.57	3144	618348	9	7	100	0.031	0.15	−0.55	D	1
			162.708	1430	616019	7	5	140	0.038	0.14	−0.57	D	1
			161.479	1430	620737	7	9	160	0.080	0.30	−0.25	D	1
			161.733	0	618348	5	7	190	0.10	0.28	−0.28	D	1
19.		^3F – $^3D°$											
			153.005	3144	656715	9	7	3200	0.86	3.9	0.89	D	1
			152.534	1430	657020	7	5	3000	0.75	2.6	0.72	D	1
			151.944	0	658136	5	3	2800	0.59	1.5	0.47	D	1
			152.597	1430	656715	7	7	180	0.062	0.22	−0.36	D	1
			152.200	0	657020	5	5	280	0.098	0.25	−0.31	D	1

Co VIII: Allowed transitions — Continued

No.	Transition Array	Multiplet	λ (Å)	E_i (cm^{-1})	E_k (cm^{-1})	g_i	g_k	A_{ki} (10^8 s^{-1})	f_{ik}	S (at. u.)	log gf	Accuracy	Source
20.		^{1}D – ^{3}F°											
			167.016	19624	618348	5	7	62	0.036	0.10	−0.74	E	1
21.		^{1}D – ^{3}D°											
			156.958	19624	656715	5	7	70	0.036	0.094	−0.74	E	1
22.		^{3}P – ^{3}F°											
			168.921	24055	616019	5	5	13	0.0058	0.016	−1.54	D	1
23.		^{3}P – ^{3}D°											
			158.066	24055	656715	5	7	780	0.41	1.1	0.31	D	1
			157.687	22839	657020	3	5	720	0.45	0.70	0.13	D	1
			157.266	22304	658136	1	3	460	0.51	0.27	−0.29	D	1
			157.984	24055	657020	5	5	170	0.062	0.16	−0.51	D	1
			157.416	22839	658136	3	3	310	0.12	0.18	−0.46	D	1
24.	$3p^6 3d^2$– $3p^5(^2P°)3d^3(^2P)$	^{1}D – ^{1}P°	157.773	19624	653446	5	3	870	0.20	0.51	−0.01	D	1
25.	$3p^6 3d^2$– $3p^5(^2P°)3d^3(^2D)$	^{1}D – ^{1}P°	149.718	19624	687584	5	3	2500	0.50	1.2	0.40	D	1
26.		^{3}P – ^{1}P°											
			150.701	24055	687584	5	3	380	0.078	0.19	−0.41	E	1
27.	$3d^2$–$3d\,4p$	^{3}F – ^{1}D°											
			184.850	1430	542430	7	5	39	0.014	0.061	−1.00	E	1
			184.356	0	542430	5	5	99	0.050	0.15	−0.60	E	1
28.		^{3}F – ^{3}D°											
			184.265	3144	545834	9	7	350	0.14	0.76	0.10	D	1
			184.203	1430	544314	7	5	350	0.13	0.54	−0.05	D	1
			184.265	0	542701	5	3	470	0.14	0.44	−0.14	D	1
			183.686	1430	545834	7	7	80	0.040	0.17	−0.55	D	1
29.		^{3}F – ^{3}F°											
			182.355	3144	551524	9	9	170	0.082	0.45	−0.13	D	1
			182.686	1430	548799	7	7	77	0.038	0.16	−0.57	D	1
			182.686	0	547400	5	5	84	0.042	0.13	−0.68	D	1
			183.266	3144	548799	9	7	73	0.029	0.16	−0.59	D	1
			183.167	1430	547400	7	5	37	0.013	0.056	−1.03	D	1
			181.786	1430	551524	7	9	6.5	0.0041	0.017	−1.54	D	1
30.		^{1}D – ^{1}D°	191.262	19624	542430	5	5	150	0.083	0.26	−0.38	D	1
31.		^{1}D – ^{3}D°											
			190.574	19624	544314	5	5	37	0.020	0.063	−1.00	E	1
32.		^{1}D – ^{3}F°											
			189.472	19624	547400	5	5	74	0.040	0.12	−0.70	E	1

Co VIII: Allowed transitions — Continued

No.	Transition Array	Multiplet	λ (Å)	E_i (cm^{-1})	E_k (cm^{-1})	g_i	g_k	A_{ki} (10^8 s^{-1})	f_{ik}	S (at. u.)	log gf	Accu-racy	Source
33.		$^1D - {}^3P°$											
			187.092	19624	554082	5	3	52	0.016	0.050	−1.09	E	1
34.		$^1D - {}^1F°$	185.835	19624	557736	5	7	8.9	0.0065	0.020	−1.49	D	1
35.		$^1D - {}^1P°$	183.939	19624	563271	5	3	200	0.062	0.19	−0.51	D	1
36.		$^3P - {}^3D°$											
			191.645	24055	545834	5	7	54	0.042	0.13	−0.68	D	1
			191.757	22839	544314	3	5	62	0.057	0.11	−0.77	D	1
			192.332	22839	542701	3	3	20	0.011	0.021	−1.47	D	1
37.		$^3P - {}^3F°$											
			190.574	24055	548799	5	7	7.1	0.0054	0.017	−1.57	D	1
38.		$^3P - {}^3P°$	*188.27*	*23455*	*554614*	9	9	310	0.16	0.91	0.17	D	1
			188.345	24055	554998	5	5	260	0.14	0.43	−0.16	D	1
			188.241	22839	554082	3	3	99	0.053	0.098	−0.80	D	1
			188.674	24055	554082	5	3	99	0.032	0.098	−0.80	D	1
			188.165	22839	554287	3	1	360	0.064	0.12	−0.72	D	1
			187.909	22839	554998	3	5	41	0.037	0.068	−0.96	D	1
			188.054	22304	554082	1	3	100	0.16	0.098	−0.80	D	1
39.		$^3P - {}^1F°$											
			187.375	24055	557736	5	7	6.5	0.0048	0.015	−1.62	E	1
40.		$^3P - {}^1P°$											
			185.461	24055	563271	5	3	55	0.017	0.052	−1.07	E	1
			185.041	22839	563271	3	3	2.0	0.0010	0.0019	−2.51	E	1
			184.861	22304	563271	1	3	9.8	0.015	0.0092	−1.82	E	1
41.		$^1G - {}^3F°$											
			192.619	32360	551524	9	9	0.44	2.4(−4)[a]	0.0014	−2.66	E	1
42.		$^1G - {}^1F°$	190.342	32360	557736	9	7	470	0.20	1.1	0.25	D	1
43.	$3d^2 - 3d4f$	$^3F - {}^1G°$											
			123.753	3144	811205	9	9	110	0.024	0.089	−0.66	E	1
			123.489	1430	811205	7	9	100	0.030	0.085	−0.68	E	1
44.		$^3F - {}^3F°$											
			123.307	3144	814130	9	9	1100	0.25	0.93	0.36	D	1
			123.173	1430	813298	7	7	1100	0.25	0.70	0.24	D	1
			123.022	0	812862	5	5	1300	0.30	0.60	0.17	D	1
			123.239	1430	812862	7	5	130	0.022	0.061	−0.82	D	1
			123.045	1430	814130	7	9	210	0.061	0.17	−0.37	D	1
			122.956	0	813298	5	7	290	0.094	0.19	−0.33	D	1

Co VIII: Allowed transitions — Continued

No.	Transition Array	Multiplet	λ (Å)	E_i (cm^{-1})	E_k (cm^{-1})	g_i	g_k	A_{ki} (10^8 s^{-1})	f_{ik}	S (at. u.)	log gf	Accuracy	Source
45.		^{3}F – ^{3}G°											
			122.472	3144	819657	9	11	3100	0.84	3.1	0.88	D	1
			122.320	1430	818958	7	9	2700	0.77	2.2	0.73	D	1
			122.273	0	817839	5	7	2500	0.78	1.6	0.59	D	1
			122.577	3144	818958	9	9	360	0.080	0.29	−0.14	D	1
			122.488	1430	817839	7	7	340	0.077	0.22	−0.27	D	1
46.		^{1}D – ^{1}D°	125.155	19624	818633	5	5	1700	0.39	0.80	0.29	D	1
47.		^{1}D – ^{3}D°											
			124.871	19624	820450	5	7	1600	0.54	1.1	0.43	E	1
48.		^{1}D – ^{1}F°	124.649	19624	821881	5	7	720	0.23	0.48	0.07	D	1
49.		^{3}P – ^{3}D°											
			125.566	24055	820450	5	7	550	0.18	0.38	−0.04	D	1
			125.350	22839	820605	3	5	2100	0.82	1.0	0.39	D	1
			125.268	22304	820599	1	3	1700	1.2	0.50	0.08	D	1
			125.350	22839	820599	3	3	470	0.11	0.14	−0.48	D	1
50.		^{3}P – ^{1}F°											
			125.340	24055	821881	5	7	1700	0.56	1.2	0.45	E	1
51.		^{3}P – ^{3}P°											
			125.155	24055	823064	5	5	1900	0.44	0.90	0.34	D	1
			124.878	22839	823613	3	3	920	0.22	0.27	−0.19	D	1
			125.071	24055	823613	5	3	730	0.10	0.21	−0.29	D	1
			124.830	22839	823928	3	1	1800	0.14	0.18	−0.37	D	1
			124.795	22304	823613	1	3	130	0.093	0.038	−1.03	D	1
52.		^{1}G – ^{1}G°	128.397	32360	811205	9	9	840	0.21	0.79	0.27	D	1
53.		^{1}G – ^{3}F°											
			127.916	32360	814130	9	9	120	0.030	0.11	−0.57	E	1
54.		^{1}G – ^{1}H°	125.821	32360	827140	9	11	3100	0.90	3.4	0.91	D	1
55.		^{1}S – ^{1}P°	132.756	74247	827508	1	3	1800	1.4	0.62	0.15	D	1

[a]The number in parentheses following the tabulated value indicates the power of ten by which this value has to be multiplied.

Co VIII

Forbidden Transitions

List of tabulated lines

Wavelength (Å)	No.	Wavelength (Å)	No.	Wavelength (Å)	No.	Wavelength (Å)	No.
1346.9	5	3421.8	4	4780.8	3	22560	6
1830.7	8	4156.0	3	5094.4	2	31100	6
1945.2	11	4377.2	3	5494.8	2	58330	1
1992.3	11	4418.6	3	6066.3	2	69910	1
3089.3	4	4482.2	3	7849.6	7	82210	9
3232.2	4	4669.6	3	12040	10	186900	9

For this ion, we selected the work of Warner and Kirkpatrick,[1] who used a single-configuration approximation and calculated radial integrals with scaled Thomas-Fermi wavefunctions. We have tabulated M1 and E2 transition probabilities for lines within the $3d^2$ configuration. Warner and Kirkpatrick also calculated electric quadrupole A-values for transitions within the $3d^2$–$3d\,4s$ transition array. We have omitted these lines, however, since accurate experimental energy levels within the $3d\,4s$ configuration were unavailable. For

lines within the $3d^2\,^3F$ and $3d^2\,^3P$ terms, we have recalculated Warner and Kirkpatrick's A-values by using observed energy-level data instead of theoretically derived values.

Reference

[1]B. Warner and R. C. Kirkpatrick, Mon. Not. R. Astron. Soc. **144**, 397 (1969).

Co VIII: Forbidden transitions

No.	Transition Array	Multiplet	λ (Å)	E_i (cm^{-1})	E_k (cm^{-1})	g_i	g_k	Type of transition	A_{ki} (s^{-1})	S (at. u.)	Accuracy	Source
1.	$3d^2$–$3d^2$	$^3F - ^3F$										
			[58330]	1430	3144	7	9	M1	0.10	6.8	C	1n
			[69910]	0	1430	5	7	M1	0.081	7.2	C	1n
2.		$^3F - ^1D$ (1F)										
			[6066.3]	3144	19624	9	5	E2	0.0021	0.051	E	1
			[5494.8]	1430	19624	7	5	M1	0.94	0.029	E	1
			"	"	"	7	5	E2	6.5(−4)^a	0.0097	E	1
			[5094.4]	0	19624	5	5	M1	0.62	0.015	E	1
			"	"	"	5	5	E2	4.7(−4)	0.0048	E	1
3.		$^3F - ^3P$ (2F)										
			[4780.8]	3144	24055	9	5	E2	0.047	0.35	E	1
			[4669.6]	1430	22839	7	3	E2	0.051	0.20	E	1
			[4482.2]	0	22304	5	1	E2	0.096	0.10	E	1
			[4418.6]	1430	24055	7	5	M1	0.13	0.0021	E	1
			"	"	"	7	5	E2	0.019	0.095	E	1
			[4377.2]	0	22839	5	3	M1	0.0034	3.2(−5)	E	1
			"	"	"	5	3	E2	0.035	0.10	E	1
			[4156.0]	0	24055	5	5	M1	0.042	5.6(−4)	E	1
			"	"	"	5	5	E2	0.0031	0.011	E	1

Co VIII: Forbidden transitions — Continued

No.	Transition Array	Multiplet	λ (Å)	E_i (cm^{-1})	E_k (cm^{-1})	g_i	g_k	Type of transition	A_{ki} (s^{-1})	S (at. u.)	Accu-racy	Source
4.		$^3F - {}^1G$										
			[3421.8]	3144	32360	9	9	M1	0.74	0.0099	E	1
			"	"	"	9	9	E2	2.6(−4)	6.5(−4)	E	1
			[3232.2]	1430	32360	7	9	M1	0.53	0.0060	E	1
			[3089.3]	0	32360	5	9	E2	5.8(−4)	8.7(−4)	E	1
5.		$^3F - {}^1S$										
			[1346.9]	0	74247	5	1	E2	0.16	4.2(−4)	E	1
6.		$^1D - {}^3P$										
			[22560]	19624	24055	5	5	M1	0.26	0.55	E	1
			[31100]	19624	22839	5	3	M1	0.054	0.18	E	1
7.		$^1D - {}^1G$	[7849.6]	19624	32360	5	9	E2	8.9(−4)	0.14	E	1
8.		$^1D - {}^1S$	[1830.7]	19624	74247	5	1	E2	18	0.22	E	1
9.		$^3P - {}^3P$										
			[82210]	22839	24055	3	5	M1	0.0190	1.96	C	1n
			[186900]	22304	22839	1	3	M1	0.00270	1.96	C	1n
10.		$^3P - {}^1G$										
			[12040]	24055	32360	5	9	E2	1.8(−5)	0.024	E	1
11.		$^3P - {}^1S$										
			[1992.3]	24055	74247	5	1	E2	1.6	0.030	E	1
			[1945.2]	22839	74247	3	1	M1	12	0.0033	E	1

aThe number in parentheses following the tabulated value indicates the power of ten by which this value has to be multiplied.

Co IX

K Isoelectronic Sequence

Ground State: $1s^2 2s^2 2p^6 3s^2 3p^6 3d\ ^2D_{3/2}$

Ionization Energy: $186.13\ \text{eV} = 1501300\ \text{cm}^{-1}$

Allowed Transitions

For this spectrum, we have chosen the data of Tiwary,[1,2] who calculated absolute multiplet oscillator strengths for the $3p^6 3d$–$3p^5 3d^2$ and $3p^6 3d$–$3p^5 3d\,4s$ tran-sition arrays by using configuration interaction wave-functions We then converted these multiplet strengths to individual line strengths according to LS-coupling rules.

For the isoelectronic ions of Cr VI, Mn VII, and Fe VIII, within the $3p^63d–3p^53d4s$ array, the LS-coupling line strengths generally agree quite well with the intermediate-coupling calculations of Cowan.[3] There are, however, two multiplets for these ions—$3p^63d\,^2D$ – $3p^53d(^3D°)4s\,^2D°$ and $3p^63d\,^2D$ –$3p^53d(^1F°)4s\,^2F°$—which exhibit gross disagreement (particularly for weak lines) between f-values derived from LS-coupling and intermediate-coupling calculations. Therefore, we have

omitted these multiplets from this compilation. Also, we have tabulated data only for lines which have been experimentally observed.

References

[1] S. N. Tiwary, Chem. Phys. Lett. **96**, 333 (1983).
[2] S. N. Tiwary, Astrophys. J. **272**, 781 (1983).
[3] R. D. Cowan, Astrophys. J. **147**, 377 (1967).

Co IX: Allowed transitions

No.	Transition Array	Multiplet	λ (Å)	E_i (cm^{-1})	E_k (cm^{-1})	g_i	g_k	A_{ki} (10^8 s^{-1})	f_{ik}	S (at. u.)	log gf	Accuracy	Source
1.	$3p^63d–$ $3p^5(^2P°)3d^2(^1D)$	^2D – $^2F°$	204.11	1471	491392	10	14	41	0.036	0.24	−0.44	D−	1
			207.180	2451	485123	6	8	40	0.034	0.14	−0.69	D−	ls
			200.100	0	499750	4	6	40	0.036	0.096	−0.84	D−	ls
			201.086	2451	499750	6	6	2.9	0.0017	0.0069	−1.98	E	ls
2.	$3p^63d–$ $3p^5(^2P°)3d^2(^3F)$	^2D – $^2F°$	171.35	1471	585063	10	14	1200	0.72	4.1	0.86	D−	1
			170.695	2451	588291	6	8	1200	0.68	2.3	0.61	D−	ls
			172.190	0	580759	4	6	1100	0.71	1.6	0.45	D−	ls
			172.917	2451	580759	6	6	78	0.035	0.12	−0.68	E	ls
3.		^2D – $^2D°$	155.38	1471	645069	10	10	3600	1.3	6.6	1.11	D−	1
			155.669	2451	644843	6	6	3300	1.2	3.7	0.86	D−	ls
			154.942	0	645408	4	4	3300	1.2	2.4	0.67	D−	ls
			155.530	2451	645408	6	4	350	0.085	0.26	−0.29	E	ls
			155.076	0	644843	4	6	240	0.13	0.26	−0.29	E	ls
4.	$3p^63d–$ $3p^5(^2P°)3d^2(^3P)$	^2D – $^2P°$	153.43	1471	653217	10	6	2600	0.56	2.8	0.75	D−	1
			153.308	2451	654735	6	4	2400	0.56	1.7	0.53	D−	ls
			153.803	0	650182	4	2	2600	0.46	0.93	0.26	D−	ls
			152.733	0	654735	4	4	270	0.094	0.19	−0.42	E	ls
5.	$3p^63d–$ $3p^53d(^3P°)4s$	^2D – $^2P°$	101.19	1471	989707	10	6	520	0.048	0.16	−0.32	D−	2
			101.107	2451	991510	6	4	470	0.048	0.096	−0.54	D−	ls
			101.410	0	986100	4	2	510	0.040	0.053	−0.80	D−	ls
			100.856	0	991510	4	4	54	0.0083	0.011	−1.48	E	ls
6.	$3p^63d–$ $3p^53d(^3F°)4s$	^2D – $^2F°$	99.55	1471	1005996	10	14	530	0.11	0.36	0.04	D−	2
			99.921	2451	1003240	6	8	530	0.11	0.21	−0.19	D−	ls
			99.042	0	1009670	4	6	490	0.11	0.14	−0.37	D−	ls
			99.284	2451	1009670	6	6	35	0.0051	0.010	−1.51	E	ls

Co x

Ar Isoelectronic Sequence

Ground State: $1s^2 2s^2 2p^6 3s^2 3p^6 \; {}^1S_0$

Ionization Energy: $275.4 \, \text{eV} = 2221000 \; \text{cm}^{-1}$

Allowed Transitions

The line strength for the $3p^6$–$3p^5 3d$ resonance transition of this argon-like ion was interpolated from the superposition-of-configurations (SOC) calculations of Weiss[1] for neighboring ions, which are expected to be fairly accurate. The remainder of the oscillator strengths were interpolated from the Dirac-Hartree-Fock data of Lin et al.,[2] who included correlation only in the lower state.

References

[1] A. W. Weiss, private communication.
[2] D. L. Lin, W. Fielder, Jr., and L. Armstrong, Jr., Phys. Rev. A **16**, 589 (1977).

Co x: Allowed transitions

No.	Transition Array	Multiplet	λ (Å)	E_i (cm^{-1})	E_k (cm^{-1})	g_i	g_k	A_{ki} (10^8 s^{-1})	f_{ik}	S (at. u.)	log gf	Accuracy	Source
1.	$3p^6$–$3p^5 3d$	^1S – $^1P°$	158.87	0	629450	1	3	2200	2.5	1.3	0.40	C	interp.
2.	$3p^6$– $3p^5(^2P°_{3/2})4s$	^1S – $(^3/_2, ^1/_2)°$											
			90.47	0	1105000	1	3	430	0.16	0.048	−0.80	D	interp.
3.	$3p^6$– $3p^5(^2P°_{1/2})4s$	^1S – $(^1/_2, ^1/_2)°$											
			88.99	0	1124000	1	3	650	0.23	0.067	−0.64	D	interp.
4.	$3p^6$– $3p^5(^2P°_{3/2})4d$	^1S – $^2[^3/_2]°$											
			72.45	0	1380000	1	3	1700	0.39	0.093	−0.41	D	interp.
5.	$3p^6$– $3p^5(^2P°_{1/2})4d$	^1S – $^2[^3/_2]°$											
			71.48	0	1399000	1	3	870	0.20	0.047	−0.70	D	interp.

Co XI

Cl Isoelectronic Sequence

Ground State: $1s^2 2s^2 2p^6 3s^2 3p^5\ ^2P^o_{3/2}$

Ionization Energy: 305 eV = 2460000 cm^{-1}

Allowed Transitions

Line strengths for transitions of the arrays $3s^2 3p^5$–$3s\,3p^6$ and $3p^5$–$3p^4 3d$ are the results of the multiconfiguration Dirac-Fock (MCDF) calculations of Huang et al.[1] These relativistic calculations include a perturbative treatment of the Breit interaction and the Lamb shift. Configuration mixing was limited to some configurations within the $n=3$ complex. Those configurations which were assumed to lie far above $3p^5$ or $3p^4 3d$ in energy were excluded, as were all configurations outside the complex.

According to the semi-empirical HX (Hartree-Fock with statistical allowance for exchange) calculations of Bromage et al.[2] for Fe X, some levels of the $3p^4 3d$ configuration are strongly mixed in the LS basis, and in a few cases the LS designations given in Ref. 2 differed from those of Huang et al. The level designations used in this compilation are in accord with the theoretical results of Refs. 1 and 2 for Fe X. Percentage compositions published by Bromage[3] for the levels of the $3p^4 3d$ configuration in V VII and Ni XII indicate that the designations for the iron ion are appropriate for the neighboring ions of the chlorine isoelectronic sequence. Transitions involving highly mixed levels have been excluded, as have the very weak transitions.

The calculated wavelengths of Huang et al. differ appreciably from the observed ones found in the literature. Thus the available experimentally determined wavelengths were used in making the conversion from line strengths to f- and A-values. (Otherwise, the calculated wavelengths of Huang et al. were used, but they provide only a rough idea of the spectral-line positions.) Bromage et al. indicate that it was necessary to scale down some configuration-interaction parameters by a greater amount than usual in order to fit their calculated energy levels for Fe X to the experimental data. This could be an indication that neglecting to take configuration interaction into account on a larger scale yields significant errors in the energy levels and/or f-values.

References

[1] K.-N. Huang, Y.-K. Kim, K. T. Cheng, and J. P. Desclaux, At. Data Nucl. Data Tables 28, 355 (1983).
[2] G. E. Bromage, R. D. Cowan, and B. C. Fawcett, Phys. Scr. 15, 177 (1977).
[3] G. E. Bromage, Astron. Astrophys., Suppl. Ser. 41, 79 (1980).

Co XI: Allowed transitions

No.	Transition Array	Multiplet	λ (Å)	E_i (cm^{-1})	E_k (cm^{-1})	g_i	g_k	A_{ki} (10^8 s^{-1})	f_{ik}	S (at. u.)	log gf	Accuracy	Source
1.	$3s^2 3p^5$–$3s\,3p^6$	$^2P^o$ – ^2S	325.54	6450	313630	6	2	64	0.0339	0.218	−0.69	C−	1
			318.85	0	313630	4	2	45.0	0.0343	0.144	−0.86	C−	1
			339.81	19350	313630	2	2	19	0.033	0.074	−1.18	C−	1
2.	$3p^5$–$3p^4(^3P)3d$	$^2P^o$ – ^4F											
			[207]			4	6	0.42	4.0(−4)^a	0.0011	−2.79	E	1
			[215]			2	4	0.12	1.7(−4)	2.4(−4)	−3.47	E	1
3.		$^2P^o$ – ^4P											
			[211]			2	4	0.092	1.2(−4)	1.7(−4)	−3.61	E	1
			[203]			4	4	0.37	2.3(−4)	6.1(−4)	−3.04	E	1
			[214]			2	2	0.84	5.7(−4)	8.1(−4)	−2.94	E	1
			[205]			4	2	3.6	0.0011	0.0031	−2.34	E	1

Co XI: Allowed transitions —Continued

No.	Transition Array	Multiplet	λ (Å)	E_i (cm^{-1})	E_k (cm^{-1})	g_i	g_k	A_{ki} (10^8 s^{-1})	f_{ik}	S (at. u.)	log gf	Accuracy	Source
4.		^{2}P° – ^{2}F											
			[195]			4	6	0.17	1.4(−4)	3.7(−4)	−3.24	E	1
5.		^{2}P° – ^{2}D	162.51	6450	621790	6	10	1940	1.28	4.11	0.89	C−	1
			162.57	0	615120	4	6	1940	1.15	2.47	0.66	C	1
			163.32	19350	631790	2	4	1860	1.49	1.60	0.474	C	1
			158.28	0	631790	4	4	47	0.018	0.037	−1.15	D	1
6.	$3p^5$–$3p^4(^1$D$)3d$	^{2}P° – ^{2}F											
			[184]			4	6	1.8	0.0014	0.0034	−2.25	E	1
7.		^{2}P° – ^{2}S	173.59	6450	582510	6	2	1700	0.26	0.89	0.19	C−	1
			171.67	0	582510	4	2	1300	0.29	0.65	0.06	C−	1
			177.59	19350	582510	2	2	438	0.207	0.242	−0.383	C−	1
8.	$3p^5$–$3p^4(^1$S$)3d$	^{2}P° – ^{2}D											
			[178]			2	4	11	0.010	0.012	−1.69	D	1
			[172]			4	4	3.6	0.0016	0.0036	−2.20	E	1

^{a}The number in parentheses following the tabulated value indicates the power of ten by which this value has to be multiplied.

Co XI

Forbidden Transitions

Line strengths for the magnetic dipole and electric quadrupole contributions to the transition between the two levels of the $3p^5$ configuration are the results of the multiconfiguration Dirac-Fock (MCDF) calculations of Huang *et al.*[1] These relativistic calculations included a perturbative treatment of the Breit interaction and the Lamb shift. Allowance for mixing among odd-parity configurations was limited to the set $3s^23p^5$, $3s3p^53d$, $3p^53d^2$, and $3s^23p^33d^2$. The strength of the electric quadrupole transition as defined in Ref. 1 was multiplied by the factor $^2/_3$ which is needed to bring this value into conformance with the definition of quadrupole strengths used in the NBS tables.

Reference

[1]K.-N. Huang, Y.-K. Kim, K. T. Cheng, and J. P. Desclaux, At. Data Nucl. Data Tables **28**, 355 (1983).

Co XI: Forbidden transitions

No.	Transition Array	Multiplet	λ (Å)	E_i (cm^{-1})	E_k (cm^{-1})	g_i	g_k	Type of transition	A_{ki} (s^{-1})	S (at. u.)	Accuracy	Source
1.	$3p^5$–$3p^5$	$^2P°$ – $^2P°$										
			[5167]	0	19350	4	2	M1	130	1.33	B	1
			"	"	"	4	2	E2	0.036	0.16	D–	1

Co XII

S Isoelectronic Sequence

Ground State: $1s^2 2s^2 2p^6 3s^2 3p^4$ $^3P_2$

Ionization Energy: 336 eV = 2710000 cm^{-1}

Allowed Transitions

Oscillator strengths for a few transitions of the arrays $3s^2 3p^4$–$3s\,3p^5$ and $3p^4$–$3p^3 3d$ were interpolated from the results of Mason[1] and Bromage et al.[2] for Fe XI and those of Bromage[3] for Ni XIII. The term designations used here are in accord with the results of Refs. 2 and 3.

References

[1] H. E. Mason, Mon. Not. R. Astron. Soc. **170**, 651 (1975).
[2] G. E. Bromage, R. D. Cowan, and B. C. Fawcett, Phys. Scr. **15**, 177 (1977).
[3] G. E. Bromage, Astron. Astrophys., Suppl. Ser. **41**, 79 (1980).

Co XII: Allowed transitions

No.	Transition Array	Multiplet	λ (Å)	E_i (cm^{-1})	E_k (cm^{-1})	g_i	g_k	A_{ki} (10^8 s^{-1})	f_{ik}	S (at. u.)	log gf	Accuracy	Source
1.	$3s^2 3p^4$–$3s\,3p^5$	^3P – $^3P°$											
			326.12	0	306640	5	5	23	0.036	0.19	−0.74	E	interp.
2.		^1D – $^1P°$	286.64			5	3	88	0.065	0.31	−0.49	D	interp.
3.	$3p^4$–$3p^3(^2D°)3d$	^3P – $^3P°$											
			175.44	0	570000	5	5	1200	0.55	1.6	0.44	D–	interp.
						3	1		0.24		−0.14	D	interp.
4.		^1D – $^1D°$	172.41			5	5	1300	0.58	1.6	0.46	D	interp.
5.		^1D – $^1F°$	169.91			5	7	1750	1.06	2.96	0.72	C–	interp.
6.	$3p^4$–$3p^3(^2P°)3d$	^1S – $^1P°$	172.33			1	3	1540	2.06	1.17	0.314	C–	interp.

Co XII

Forbidden Transitions

Transition probabilities for magnetic dipole and electric quadrupole lines within the $3p^4$ configuration are the results of the scaled Thomas-Fermi calculations of Mendoza and Zeippen.[1] They included a number of correlation configurations in their basis set and introduced Breit-Pauli relativistic corrections as a perturbation to the nonrelativistic Hamiltonian.

Reference

[1]C. Mendoza and C. J. Zeippen, Mon. Not. R. Astron. Soc. **202**, 981 (1983).

Co XII: Forbidden transitions

No.	Transition Array	Multiplet	λ (Å)	E_i (cm^{-1})	E_k (cm^{-1})	g_i	g_k	Type of transition	A_{ki} (s^{-1})	S (at. u.)	Accu-racy	Source
1.	$3p^4$–$3p^4$	^{3}P – ^{3}P										
			[6319]	0	15820	5	3	M1	84.0	2.36	C+	1
			"	"	"	5	3	E2	0.0082	0.15	D–	1
			[79300]	15820	17080	3	1	M1	0.097	1.8	C–	1
			[5853]	0	17080	5	1	E2	0.019	0.078	E	1
2.		^{3}P – ^{1}D										
			[2370]			5	5	M1	160	0.39	E	1
			"	"	"	5	5	E2	0.25	0.056	E	1
			[3800]			3	5	M1	13	0.13	E	1
			"	"	"	3	5	E2	0.0036	0.0085	E	1
			[4000]			1	5	E2	0.0022	0.0067	E	1
3.		^{3}P – ^{1}S										
			[1120]			5	1	E2	2.3	0.0024	E	1
			[1370]			3	1	M1	1600	0.15	E	1
4.		^{1}D – ^{1}S	[2140]			5	1	E2	9.7	0.26	D–	1

Co XIII

P Isoelectronic Sequence

Ground State: $1s^2 2s^2 2p^6 3s^2 3p^3\ {}^4S^{\circ}_{3/2}$

Ionization Energy: 379 eV = 3057000 cm^{-1}

Allowed Transitions

List of tabulated lines

Wavelength (Å)	No.	Wavelength (Å)	No.	Wavelength (Å)	No.	Wavelength (Å)	No.
163	12	183.65	14	203	13	310.67	2
168	15	185.39	17	205	4,11	313.91	2
170	15	188.42	17	208	11	315	2
174.82	9	189	16	209	11	320.40	1
177	14	190	8	227	7	325.70	1
179	14	192	16	228	7	338.80	1
179.59	5	193	8	230	7	348	3
180.87	5	194	8	239	6	355	3
182.09	14	196	8	253	10	358	3
182.52	5	202	4	306	2		

Line strengths for transitions of the arrays $3s^2 3p^3$–$3s 3p^4$ and $3p^3$–$3p^2 3d$ are the results of the multiconfiguration Dirac-Fock (MCDF) calculations of Huang.[1] These relativistic calculations included a perturbative treatment of the Breit interaction and the Lamb shift. Allowance for configuration mixing was limited to configurations within the $n = 3$ complex having no more than two electrons in the $3d$ subshell.

Huang published neither an energy-level diagram nor percentage compositions for levels of the $3s^2 3p^3$, $3s 3p^4$, and $3s^2 3p^2 3d$ configurations in Co XIII. We have used the percentages given by Bromage et al.[2] for Fe XII, and by Bromage[3] for V IX and Ni XIV, as a guide to naming the levels; their values resulted from Hartree-Fock calculations with relativistic effects and statistical allowance for exchange (HXR), and incorporated correlation effects

due to a few configurations within the $n = 3$ complex. Whenever a term designation of a level in Fe XII, as given in Ref. 1, is different from that indicated in Ref. 2, all transitions involving the corresponding level in Co XIII are omitted from this compilation.

Transitions involving levels which are indicated to be of low purity in LS coupling are omitted here. Lines which are characterized by very small f-values are assigned lower accuracy ratings; the weakest lines have been excluded.

References

[1]K.-N. Huang, At. Data Nucl. Data Tables **30**, 313 (1984).
[2]G. E. Bromage, R. D. Cowan, and B. C. Fawcett, Mon. Not. R. Astron. Soc. **183**, 19 (1978).
[3]G. E. Bromage, Astron. Astrophys., Suppl. Ser. **41**, 79 (1980).

Co XIII: Allowed transitions

No.	Transition Array	Multiplet	λ (Å)	E_i (cm^{-1})	E_k (cm^{-1})	g_i	g_k	A_{ki} (10^8 s^{-1})	f_{ik}	S (at. u.)	log gf	Accuracy	Source
1.	$3s^2 3p^3$–$3s 3p^4$	$^4S^{\circ}$ – ^4P	331.19	0	301940	4	12	20	0.096	0.42	−0.41	D	1
			338.80	0	295160	4	6	18	0.047	0.21	−0.73	D	1
			325.70	0	307030	4	4	21	0.033	0.14	−0.88	D	1
			320.40	0	312110	4	2	21	0.016	0.069	−1.18	D	1

Co XIII: Allowed transitions — Continued

No.	Transition Array	Multiplet	λ (Å)	E_i (cm^{-1})	E_k (cm^{-1})	g_i	g_k	A_{ki} (10^8 s^{-1})	f_{ik}	S (at. u.)	log gf	Accuracy	Source
2.		^{2}D° – ^{2}D	*312.61*			10	10	36	0.052	0.54	−0.28	E	1
			313.91			6	6	33	0.048	0.30	−0.54	D	1
			310.67			4	4	41	0.059	0.24	−0.63	D	1
			[315]			6	4	0.32	3.2(−4)^a	0.0020	−2.71	E	1
			[306]			4	6	0.18	3.7(−4)	0.0015	−2.83	E	1
3.		^{2}P° – ^{2}D	*353*			6	10	5.5	0.017	0.12	−0.99	E	1
			[355]			4	6	6.8	0.019	0.090	−1.11	D	1
			[348]			2	4	3.4	0.012	0.028	−1.61	D	1
			[358]			4	4	0.31	5.9(−4)	0.0028	−2.62	E	1
4.	3p^3–3p^2(^{3}P)3d	^{4}S° – ^{4}D											
			[202]			4	6	4.0	0.0036	0.0097	−1.84	E	1
			[205]			4	4	3.8	0.0024	0.0064	−2.02	E	1
5.		^{4}S° – ^{4}P	*181.48*	0	*551040*	4	12	900	1.3	3.2	0.73	D	1
			182.52	0	547890	4	6	890	0.67	1.6	0.43	D	1
			180.87	0	552880	4	4	940	0.46	1.1	0.27	D	1
			179.59	0	556820	4	2	940	0.23	0.54	−0.04	D	1
6.		^{2}D° – ^{4}F											
			[239]			4	4	2.2	0.0019	0.0060	−2.12	E	1
7.		^{2}D° – ^{4}D											
			[227]			6	6	1.3	0.0010	0.0046	−2.21	E	1
			[230]			6	4	4.6	0.0024	0.011	−1.84	E	1
			[228]			4	2	5.6	0.0022	0.0066	−2.06	E	1
8.		^{2}D° – ^{4}P											
			[196]			6	6	16	0.0090	0.035	−1.27	E	1
			[194]			6	4	5.3	0.0020	0.0077	−1.92	E	1
			[190]			4	2	47	0.013	0.032	−1.29	E	1
			[193]			4	6	7.0	0.0059	0.015	−1.63	E	1
9.		^{2}D° – ^{2}F											
			174.82			6	8	1100	0.70	2.4	0.62	E	1
10.		^{2}P° – ^{4}D											
			[253]			4	2	2.9	0.0014	0.0046	−2.26	E	1
11.		^{2}P° – ^{4}P											
			[209]			4	4	2.8	0.0018	0.0050	−2.14	E	1
			[205]			2	2	7.2	0.0045	0.0061	−2.04	E	1
			[208]			4	2	7.8	0.0025	0.0069	−2.00	E	1
12.	3p^3–3p^2(^{1}D)3d	^{4}S° – ^{2}D											
			[163]			4	6	2.3	0.0014	0.0029	−2.27	E	1
13.		^{2}D° – ^{2}G											
			[203]			6	8	3.6	0.0030	0.012	−1.75	E	1

Co XIII: Allowed transitions — Continued

No.	Transition Array	Multiplet	λ (Å)	E_i (cm^{-1})	E_k (cm^{-1})	g_i	g_k	A_{ki} (10^8 s^{-1})	f_{ik}	S (at. u.)	log gf	Accuracy	Source
14.		^{2}D° – ^{2}D	183.02			10	10	590	0.30	1.8	0.48	D	1
			183.65			6	6	490	0.25	0.89	0.17	D	1
			182.09			4	4	650	0.33	0.78	0.11	D	1
			[179]			6	4	130	0.042	0.15	−0.59	D	1
			[177]			4	6	18	0.012	0.029	−1.30	D	1
15.		^{2}D° – ^{2}P											
			[170]			6	4	9.6	0.0028	0.0093	−1.78	E	1
			[168]			4	4	19	0.0081	0.018	−1.49	E	1
16.		^{2}P° – ^{2}D	191			6	10	81	0.074	0.28	−0.35	E	1
			[192]			4	6	76	0.063	0.16	−0.60	D	1
			[189]			2	4	83	0.088	0.11	−0.75	D	1
			[192]			4	4	5.7	0.0032	0.0080	−1.90	E	1
17.		^{2}P° – ^{2}P											
			188.42			4	4	540	0.29	0.71	0.06	E	1
			185.39			2	4	190	0.20	0.24	−0.41	E	1

^{a}The number in parentheses following the tabulated value indicates the power of ten by which this value has to be multiplied.

Co XIII

Forbidden Transitions

Line strengths for magnetic dipole and electric quadrupole transitions within the $3p^3$ configuration are the results of the multiconfiguration Dirac-Fock (MCDF) calculations of Huang.[1] These relativistic calculations included a perturbative treatment of the Breit interaction and the Lamb shift. Allowance for configuration mixing was limited to configurations within the $n = 3$ complex having no more than two electrons in the $3d$ subshell. Strengths of electric quadrupole transitions as defined in Ref. 1 were multiplied by the factor $^2/_3$ which is needed to bring these values into conformance with the definition of quadrupole strengths used in the NBS tables. We have excluded from this compilation the electric quadrupole contributions to the ^{4}S$^\circ_{3/2}$ – ^{2}P$^\circ_{3/2}$ and ^{4}S$^\circ_{3/2}$ – ^{2}P$^\circ_{1/2}$ transitions, since their strengths are very small and thus subject to considerable uncertainty.

Data for these same transitions calculated by Mendoza and Zeippen[2] with the scaled Thomas-Fermi approach with allowance for correlation are generally in very good agreement with the results of Ref. 1. These latter calculations treated relativistic effects by introducing Breit-Pauli corrections as a perturbation to the nonrelativistic Hamiltonian.

References

[1]K.-N. Huang, At. Data Nucl. Data Tables 30, 313 (1984).
[2]C. Mendoza and C. J. Zeippen, Mon. Not. R. Astron. Soc. 198, 127 (1982).

Co XIII: Forbidden transitions

No.	Transition Array	Multiplet	λ (Å)	E_i (cm^{-1})	E_k (cm^{-1})	g_i	g_k	Type of transition	A_{ki} (s^{-1})	S (at. u.)	Accuracy	Source
1.	$3p^3$–$3p^3$	$^4S°$ – $^2D°$										
			[2010]			4	6	M1	3.8	0.0068	E	1
			"			4	6	E2	0.15	0.018	E	1
			[2290]			4	4	M1	90	0.16	C	1
			"			4	4	E2	0.056	0.0084	E	1
2.		$^4S°$ – $^2P°$										
			[1130]			4	4	M1	470	0.10	C	1
			[1260]			4	2	M1	280	0.042	D	1
3.		$^2D°$ – $^2D°$										
			[17000]			4	6	M1	1.88	2.05	C+	1
			"			4	6	E2	1.5(−5)[a]	0.074	E	1
4.		$^2D°$ – $^2P°$										
			[3350]			6	2	E2	0.34	0.17	D−	1
			[2600]			6	4	M1	130	0.34	C	1
			"			6	4	E2	1.7	0.48	D−	1
			[2790]			4	2	M1	110	0.17	C	1
			"			4	2	E2	1.0	0.21	D−	1
			[2250]			4	4	M1	350	0.59	C	1
			"			4	4	E2	0.87	0.12	D−	1
5.		$^2P°$ – $^2P°$										
			[11500]			2	4	M1	4.97	1.12	C+	1
			"			2	4	E2	6.5(−5)	0.031	E	1

[a]The number in parentheses following the tabulated value indicates the power of ten by which this value has to be multiplied.

Co XIV

Si Isoelectronic Sequence

Ground State: $1s^2 2s^2 2p^6 3s^2 3p^2\ {}^3P_0$

Ionization Energy: $411\ \mathrm{eV} = 3315000\ \mathrm{cm}^{-1}$

Allowed Transitions

List of tabulated lines

Wavelength (Å)	No.	Wavelength (Å)	No.	Wavelength (Å)	No.	Wavelength (Å)	No.
56.115	14	190.82	7	278	2	333	1
68.807	15	195.66	13	288	2	334.21	1
73.402	16	200	11	289	2	342.21	1
160	9	220	5	289.26	2	346	1
170	8	221	5	297	2	346.50	1
184.41	12	225	5	298.42	2	387	3
187	6	244	10	321	1	393	4

Line strengths for transitions of the arrays $3s^2 3p^2$–$3s3p^3$ and $3p^2$–$3p3d$ are the results of the multiconfiguration Dirac-Fock (MCDF) calculations of Huang.[1] These relativistic calculations included a perturbative treatment of the Breit interaction and the Lamb shift. Allowance for configuration mixing included all configurations within the $n = 3$ complex.

Huang published neither an energy-level diagram nor percentage compositions for levels of the $3s^2 3p^2$, $3s3p^3$, and $3s^2 3p3d$ configurations in Co XIV. We have used the percentages given by Bromage et al.[2] for Fe XIII, and by Bromage[3] for V X and Ni XV, as a guide to naming the levels; their values resulted from Hartree-Fock calculations with relativistic effects and statistical allowance for exchange (HXR), and incorporated correlation effects due to a partial set of configurations within the $n = 3$ complex. Whenever the term designation of a level in Fe XIII, as given in Ref. 1, is different from that indicated in Ref. 2, all transitions involving the corresponding level in Co XIV are omitted from this compilation.

A few f-values for transitions to configurations in which one electron occupies the $n = 4$ shell were interpolated from the results of Kastner et al.[4] for Fe XIII and Zn XVII, which were computed by a multiconfiguration scaled Thomas-Fermi approach.

Transitions involving levels which are indicated to be of low purity in LS coupling are omitted here. Lines which are characterized by very small f-values are assigned lower accuracy ratings; the weakest lines have been excluded.

References

[1] K.-N. Huang, At. Data Nucl. Data Tables **32**, 503 (1985).
[2] G. E. Bromage, R. D. Cowan, and B. C. Fawcett, Mon. Not. R. Astron. Soc. **183**, 19 (1978).
[3] G. E. Bromage, Astron. Astrophys., Suppl. Ser. **41**, 79 (1980).
[4] S. O. Kastner, M. Swartz, A. K. Bhatia, and J. Lapides, J. Opt. Soc. Am. **68**, 1558 (1978).

Co XIV: Allowed transitions

No.	Transition Array	Multiplet	λ (Å)	E_i (cm^{-1})	E_k (cm^{-1})	g_i	g_k	A_{ki} (10^8 s^{-1})	f_{ik}	S (at. u.)	log gf	Accuracy	Source
1.	$3s^2 3p^2$–$3s3p^3$	${}^3P$ – ${}^3D°$	*337*			9	15	16	0.045	0.45	−0.39	E	1
			342.21	22640	314860	5	7	14	0.034	0.19	−0.77	D	1
			334.21	12030	311240	3	5	17	0.048	0.16	−0.84	D	1
			[321]			1	3	16	0.072	0.076	−1.14	D	1
			[346.50]	22640	311240	5	5	0.35	6.3(−4)^a	0.0036	−2.50	E	1
			[333]			3	3	3.5	0.0058	0.019	−1.76	D−	1
			[346]			5	3	0.15	1.6(−4)	9.3(−4)	−3.09	E	1

Co XIV: Allowed transitions — Continued

No.	Transition Array	Multiplet	λ (Å)	E_i (cm^{-1})	E_k (cm^{-1})	g_i	g_k	A_{ki} (10^8 s^{-1})	f_{ik}	S (at. u.)	log gf	Accuracy	Source
2.		^{3}P – ^{3}P°	*292*			9	9	43	0.055	0.48	−0.30	D	1
			298.42	22640	357740	5	5	37	0.049	0.24	−0.61	D	1
			[288]			3	3	23	0.028	0.081	−1.07	D	1
			[297]			5	3	9.5	0.0076	0.037	−1.42	D−	1
			[289]			3	1	46	0.019	0.055	−1.24	C−	1
			[289.26]	12030	357740	3	5	4.0	0.0084	0.024	−1.60	D−	1
			[278]			1	3	14	0.049	0.045	−1.31	D	1
3.		^{1}D – ^{3}D°											
			[387]			5	7	1.2	0.0038	0.024	−1.72	E	1
4.		^{1}S – ^{3}P°											
			[393]			1	3	0.30	0.0021	0.0027	−2.68	E	1
5.	$3p^2$–$3p3d$	^{3}P – ^{3}F°											
			[221]			5	7	4.3	0.0044	0.016	−1.66	E	1
			[220]			3	5	1.4	0.0017	0.0038	−2.28	E	1
			[225]			5	5	2.0	0.0015	0.0056	−2.12	E	1
6.		^{3}P – ^{3}P°											
			[187]			3	1	530	0.092	0.17	−0.56	D	1
7.		^{3}P – ^{3}D°											
			190.82	22640	546690	5	7	710	0.54	1.7	0.43	D	1
8.		^{3}P – ^{1}F°											
			[170]			5	7	45	0.027	0.076	−0.87	E	1
9.		^{3}P – ^{1}P°											
			[160]			1	3	5.8	0.0066	0.0035	−2.18	E	1
10.		^{1}D – ^{3}F°											
			[244]			5	5	4.5	0.0040	0.016	−1.70	E	1
11.		^{1}D – ^{3}D°											
			[200]			5	7	32	0.027	0.089	−0.87	E	1
12.		^{1}D – ^{1}F°	184.41			5	7	720	0.51	1.56	0.410	C	1
13.		^{1}S – ^{1}P°	195.66			1	3	600	1.0	0.67	0.02	D	1
14.	$3p^2$–$3p4d$	^{1}D – ^{1}F°	56.115			5	7	5100	0.34	0.31	0.23	D	*interp.*
15.	$3p3d$–$3p4f$	^{3}F° – ^{3}G											
			68.807			9	11	8200	0.71	1.4	0.81	D	*interp.*
16.		^{1}F° – ^{1}G	73.402			7	9	7300	0.76	1.3	0.73	E	*interp.*

[a]The number in parentheses following the tabulated value indicates the power of ten by which this value has to be multiplied.

Co XIV

Forbidden Transitions

List of tabulated lines

Wavelength (Å)	No.	Wavelength (Å)	No.	Wavelength (Å)	No.	Wavelength (Å)	No.
340	12	770	6	2160	4	8310	1
420	10	780	6	2200	8	9422	1
431.35	11	1120	3	2300	8	25000	7
570	9	1200	5	2320	2	27600	7
580	9	1280	3	2331	8		
610	9	2100	8	3100	2		
640	9	2150	8	4416	1		

Line strengths for magnetic dipole and electric quadrupole transitions are the results of the multiconfiguration Dirac-Fock (MCDF) calculations of Huang.[1] These relativistic calculations included a perturbative treatment of the Breit interaction and the Lamb shift. Allowance for configuration interaction encompassed all configurations within the $n = 3$ complex. Huang calculated line strengths for transitions within the $3p^2$ configuration, as well as for transitions between pairs of odd-parity levels whose lower level is one of the four lowest-lying odd-parity levels in the $n = 3$ complex. Transitions involving odd-parity levels which are indicated by Bromage et al.[2] (for Fe XIII) or Bromage[3] (for V X and Ni XV) to be of low purity in LS coupling in Fe-group species are omitted here, as are lines whose

strengths are very small. The strength of the magnetic dipole contribution to the $3s\,3p^3$ $^3D_1^o$ - $3s\,3p^3$ $^3D_2^o$ transition is excluded from the tabulation, because its wavelength uncertainty is unacceptably large. Strengths of electric quadrupole transitions as reported in Ref. 1 were multiplied by the factor $^2/_3$ which is needed to bring these values into conformance with the definition of quadrupole strengths used in the NBS tables.

References

[1] K.-N. Huang, At. Data Nucl. Data Tables **32**, 503 (1985).
[2] G. E. Bromage, R. D. Cowan, and B. C. Fawcett, Mon. Not. R. Astron. Soc. **183**, 19 (1978).
[3] G. E. Bromage, Astron. Astrophys., Suppl. Ser. **41**, 79 (1980).

Co XIV: Forbidden transitions

No.	Transition Array	Multiplet	λ (Å)	E_i (cm^{-1})	E_k (cm^{-1})	g_i	g_k	Type of transition	A_{ki} (s^{-1})	S (at. u.)	Accuracy	Source
1.	$3p^2$–$3p^2$	^3P – ^3P										
			[9422]	12030	22640	3	5	M1	14.4	2.24	C+	1
			"	"	"	3	5	E2	5.4(−4)^a	0.12	D−	1
			[8310]	0	12030	1	3	M1	29.9	1.91	C+	1
			[4416]	0	22640	1	5	E2	0.014	0.070	E	1
2.		^3P – ^1D										
			[3100]			5	5	M1	120	0.69	E	1
			"	"	"	5	5	E2	0.11	0.091	E	1
			[2320]			3	5	M1	110	0.26	E	1
			"	"	"	3	5	E2	0.070	0.014	E	1
3.		^3P – ^1S										
			[1280]			5	1	E2	5.9	0.012	E	1
			[1120]			3	1	M1	1600	0.084	E	1
4.		^1D – ^1S	[2160]			5	1	E2	8.6	0.24	D−	1

Co XIV: Forbidden transitions — Continued

No.	Transition Array	Multiplet	λ (Å)	E_i (cm^{-1})	E_k (cm^{-1})	g_i	g_k	Type of transition	A_{ki} (s^{-1})	S (at. u.)	Accu-racy	Source
5.	$3s\,3p^3$–$3s\,3p^3$	$^5S°$ – $^3D°$										
			[1200]			5	7	M1	2.9	0.0013	E	1
			"	"	"	5	7	E2	0.66	0.0068	E	1
			[1200]			5	5	M1	81	0.026	E	1
			"	"	"	5	5	E2	0.61	0.0045	E	1
			[1200]			5	3	M1	26	0.0050	E	1
			"	"	"	5	3	E2	0.25	0.0011	E	1
6.		$^5S°$ – $^3P°$										
			[770]			5	5	M1	930	0.079	E	1
			[780]			5	3	M1	530	0.028	E	1
7.		$^3D°$ – $^3D°$										
			[27600]	311240	314860	5	7	M1	0.79	4.3	D+	1
			"	"	"	5	7	E2	6.4(−7)	0.043	E	1
			[25000]			3	7	E2	2.9(−7)	0.012	E	1
8.		$^3D°$ – $^3P°$										
			[2300]			7	3	E2	2.2	0.25	D−	1
			[2200]			5	1	E2	5.5	0.17	D−	1
			[2331]	314860	357740	7	5	M1	130	0.30	E	1
			"	"	"	7	5	E2	2.1	0.44	D−	1
			[2100]			5	3	E2	0.52	0.038	E	1
			[2200]			3	1	M1	200	0.078	E	1
			[2150]	311240	357740	5	5	M1	110	0.21	E	1
			"	"	"	5	5	E2	1.8	0.24	D−	1
			[2100]	"	"	3	3	M1	220	0.23	E	1
			"	"	"	3	3	E2	2.9	0.21	D−	1
			[2100]	"	"	3	5	M1	34	0.058	E	1
			"	"	"	3	5	E2	0.74	0.090	E	1
9.	$3s\,3p^3$–$3s^2\,3p\,3d$	$^3D°$ – $^3F°$										
			[570]			5	9	E2	17	0.0054	E	1
			[610]			3	7	E2	6.8	0.0024	E	1
			[580]			7	9	M1	920	0.060	E	1
			[640]			5	5	M1	31	0.0015	E	1
10.		$^3D°$ – $^3P°$										
			[420]			5	1	E2	410	0.0032	E	1
11.		$^3D°$ – $^3D°$										
			[431.35]	314860	546690	7	7	M1	72	0.0015	E	1
12.		$^3D°$ – $^1F°$										
			[340]			7	7	M1	110	0.0011	E	1

[a]The number in parentheses following the tabulated value indicates the power of ten by which the value has to be multiplied.

Co xv

Al Isoelectronic Sequence

Ground State: $1s^2 2s^2 2p^6 3s^2 3p \ ^2P^{\circ}_{1/2}$

Ionization Energy: $444 \text{ eV} = 3580000 \text{ cm}^{-1}$

Allowed Transitions

List of tabulated lines

Wavelength (Å)	No.	Wavelength (Å)	No.	Wavelength (Å)	No.	Wavelength (Å)	No.
157	25	205.85	20	249	18	278	16
159	33	206.94	20	250	19	309.86	2
168	36	207	29,40	251	18	320	9
169	35,36	208	22,29,31,41	252	6	325	15
170	35	216	32,39	253.34	4	329	11
175	34	217	7	255.88	3	330.24	2
176	34	220	32	259	6	332	8
177	24,34	221	32	263	17	333.58	2
178	30	222	7	265	17	334	11
181	30	225	21,32	266	10	336	8
191	38	227	21	267	17	343	15
192	37	228	7,21	268	10	346	15
193	22,38	230	28	269	10,16	368	13
196	22,23	231	28	270.43	6	374	13
197	23	234.41	4	271.84	3	381	13
197.54	20	239.42	4	273	26	419	1
201	23,29	241	27	274	26	445	14
203	22,40	247	18,19	275	5	472	12
204	31,41	247.76	4	276	26	493	12
205	40,41	248	19	277	16		

Line strengths for transitions of the arrays $3s^2 3p$–$3s 3p^2$, $3s 3p^2$–$3p^3$, $3s^2 3d$–$3s 3p 3d$, $3s^2 3p$–$3s^2 3d$, and $3s 3p^2$–$3s 3p 3d$ are the results of the multiconfiguration Dirac-Fock (MCDF) calculations of Huang.[1] These relativistic calculations included a perturbative treatment of the Breit interaction. Allowance for configuration mixing included all configurations within the $n = 3$ complex.

Huang published neither an energy-level diagram nor percentage compositions for levels of the $3s^2 3p$, $3s 3p^2$, $3s^2 3d$, $3p^3$, and $3s 3p 3d$ configurations in Co xv. We have used the percentages given by Fawcett[2] for the adjacent Al-like ions as a guide to naming the levels; the latter's values resulted from Hartree-Fock calculations with relativistic effects and statistical allowance for exchange (HXR), and incorporated correlation effects due to all configurations within the $n = 3$ complex.

Transitions involving levels which are indicated to be of low purity in LS coupling in one or both adjacent Al-like ions are omitted here. Lines which are characterized by very small f-values are assigned lower accuracy ratings; the weakest lines have been excluded. A few wavelengths computed by Huang for transitions in Fe xiv differ significantly from those which resulted from the fitting and scaling procedure applied by Fawcett[2]; lines for which the wavelengths are in serious disagreement have been omitted in our tabulation for Co xv.

References

[1] K.-N. Huang, At. Data Nucl. Data Tables **34**, 1 (1986) and private communication.

[2] B. C. Fawcett, At. Data Nucl. Data Tables **28**, 557 (1983).

Co xv: Allowed transitions

No.	Transition Array	Multiplet	λ (Å)	E_i (cm^{-1})	E_k (cm^{-1})	g_i	g_k	A_{ki} (10^8 s^{-1})	f_{ik}	S (at. u.)	log gf	Accuracy	Source
1.	$3s^2 3p$–$3s 3p^2$	$^2P°$ – ^4P											
			[419]			4	6	0.37	0.0015	0.0081	−2.23	E	1
2.		$^2P°$ – ^2D	323.36	15300	324550	6	10	23	0.059	0.38	−0.45	E	1
			330.24	22950	325760	4	6	21	0.051	0.22	−0.69	D	1
			309.86	0	322730	2	4	27	0.078	0.16	−0.80	D	1
			[333.58]	22950	322730	4	4	0.60	0.0010	0.0044	−2.40	E	1
3.		$^2P°$ – ^2S	266.30	15300	390810	6	2	200	0.072	0.38	−0.36	E	1
			[271.84]	22950	390810	4	2	7.6	0.0042	0.015	−1.78	E	1
			255.88	0	390810	2	2	220	0.21	0.36	−0.37	E	1
4.		$^2P°$ – ^2P	244.90	15300	423630	6	6	440	0.39	1.9	0.37	E	1
			247.76	22950	426600	4	4	370	0.340	1.11	0.134	C−	1
			239.42	0	417680	2	2	130	0.11	0.18	−0.64	E	1
			253.34	22950	417680	4	2	240	0.11	0.38	−0.34	E	1
			234.41	0	426600	2	4	85	0.140	0.216	−0.55	C−	1
5.	$3s 3p^2$–$3p^3$	^4P – $^2D°$											
			[275]			6	6	1.3	0.0015	0.0079	−2.06	E	1
6.		^4P – $^4S°$	263			12	4	390	0.13	1.4	0.21	D	1
			270.43			6	4	170	0.13	0.67	−0.12	D	1
			[259]			4	4	130	0.13	0.46	−0.27	D	1
			[252]			2	4	73	0.14	0.23	−0.56	D	1
7.		^4P – $^2P°$											
			[228]			6	4	2.6	0.0014	0.0061	−2.09	E	1
			[222]			4	4	6.5	0.0048	0.014	−1.72	E	1
			[217]			2	4	3.0	0.0043	0.0061	−2.07	E	1
8.		^2D – $^2D°$											
			[336]			6	6	33	0.056	0.37	−0.48	E	1
			[332]			4	6	3.5	0.0087	0.038	−1.46	E	1
9.		^2D – $^4S°$											
			[320]			4	4	1.9	0.0028	0.012	−1.94	E	1
10.		^2D – $^2P°$	268			10	6	130	0.085	0.75	−0.07	D	1
			[268]			6	4	110	0.079	0.42	−0.32	D	1
			[269]			4	2	150	0.079	0.28	−0.50	D	1
			[266]			4	4	14	0.015	0.052	−1.23	D	1
11.		^2S – $^2P°$	331			2	6	14	0.069	0.15	−0.86	E	1
			[329]			2	4	21	0.069	0.15	−0.86	E	1
			[334]			2	2	0.57	9.5(−4)[a]	0.0021	−2.72	E	1
12.		^2P – $^4S°$											
			[493]			4	4	0.85	0.0031	0.020	−1.91	E	1
			[472]			2	4	0.20	0.0014	0.0042	−2.57	E	1

Co xv: Allowed transitions — Continued

No.	Transition Array	Multiplet	λ (Å)	E_i (cm^{-1})	E_k (cm^{-1})	g_i	g_k	A_{ki} (10^8 s^{-1})	f_{ik}	S (at. u.)	log gf	Accuracy	Source
13.		^{2}P – ^{2}P°											
			[374]			4	4	30	0.063	0.31	−0.60	D	1
			[368]			2	2	39	0.078	0.19	−0.80	E	1
			[381]			4	2	7.3	0.0080	0.040	−1.50	E	1
14.	$3s^2 3d$ – $3s 3p(^3$P°$)3d$	^{2}D – ^{4}P°											
			[445]			6	6	0.37	0.0011	0.0097	−2.18	E	1
15.		^{2}D – ^{2}F°	*333*			10	14	22	0.051	0.56	−0.29	E	1
			[325]			6	8	24	0.051	0.33	−0.51	E	1
			[343]			4	6	15	0.040	0.18	−0.80	E	1
			[346]			6	6	4.3	0.0078	0.053	−1.33	E	1
16.		^{2}D – ^{2}P°	*275*			10	6	10	0.0070	0.063	−1.16	E	1
			[278]			6	4	5.7	0.0044	0.024	−1.58	E	1
			[269]			4	2	2.1	0.0012	0.0041	−2.33	E	1
			[277]			4	4	8.3	0.0096	0.035	−1.42	E	1
17.	$3s^2 3d$ – $3s 3p(^1$P°$)3d$	^{2}D – ^{2}F°	*265*			10	14	300	0.45	3.9	0.65	E	1
			[267]			6	8	290	0.42	2.2	0.40	E	1
			[263]			4	6	300	0.46	1.6	0.27	E	1
			[265]			6	6	9.6	0.010	0.053	−1.22	E	1
18.		^{2}D – ^{2}D°	*249*			10	10	250	0.23	1.9	0.37	E	1
			[249]			6	6	240	0.22	1.1	0.13	E	1
			[249]			4	4	110	0.10	0.33	−0.40	E	1
			[251]			6	4	140	0.087	0.43	−0.28	E	1
			[247]			4	6	2.7	0.0037	0.012	−1.83	E	1
19.		^{2}D – ^{2}P°	*249*			10	6	350	0.20	1.6	0.29	D	1
			[248]			6	4	210	0.13	0.63	−0.11	D	1
			[250]			4	2	360	0.17	0.55	−0.18	D	1
			[247]			4	4	150	0.14	0.45	−0.26	D	1
20.	$3p$–$3d$	^{2}P° – ^{2}D	*203.07*	*15300*	*507740*	6	10	460	0.47	1.9	0.45	D	1
			205.85	22950	508740	4	6	430	0.41	1.1	0.21	D	1
			197.54	0	506230	2	4	390	0.45	0.59	−0.04	D	1
			206.94	22950	506230	4	4	91	0.059	0.16	−0.63	D	1
21.	$3s 3p^2$ – $3s 3p(^3$P°$)3d$	^{4}P – ^{4}F°											
			[227]			6	8	3.9	0.0040	0.018	−1.62	E	1
			[225]			4	6	2.2	0.0025	0.0073	−2.01	E	1
			[228]			4	4	1.3	0.0010	0.0030	−2.40	E	1
22.		^{4}P – ^{4}P°											
			[208]			6	6	33	0.021	0.087	−0.90	E	1
			[193]			2	2	1.0	5.7(−4)	7.2(−4)	−2.95	E	1
			[196]			4	2	320	0.093	0.24	−0.43	D	1
			[203]			4	6	270	0.25	0.68	0.01	E	1

Co XV: Allowed transitions — Continued

No.	Transition Array	Multiplet	λ (Å)	E_i (cm^{-1})	E_k (cm^{-1})	g_i	g_k	A_{ki} (10^8 s^{-1})	f_{ik}	S (at. u.)	log gf	Accuracy	Source
23.		^{4}P – ^{4}D°											
			[201]			6	8	449	0.363	1.44	0.338	C–	1
			[196]			4	6	120	0.10	0.27	−0.38	D	1
			[201]			6	6	300	0.18	0.71	0.03	D	1
			[197]			2	2	450	0.26	0.34	−0.28	D	1
			[201]			4	2	7.4	0.0022	0.0059	−2.05	E	1
24.		^{4}P – ^{2}F°											
			[177]			6	8	6.4	0.0040	0.014	−1.62	E	1
25.		^{4}P – ^{2}P°											
			[157]			2	4	4.6	0.0034	0.0035	−2.17	E	1
26.		^{2}D – ^{4}F°											
			[273]			6	6	1.1	0.0012	0.0064	−2.15	E	1
			[274]			4	4	1.4	0.0016	0.0058	−2.19	E	1
			[276]			6	4	1.4	0.0010	0.0057	−2.20	E	1
27.		^{2}D – ^{4}P°											
			[241]			6	6	12	0.010	0.048	−1.22	E	1
28.		^{2}D – ^{4}D°											
			[231]			6	8	4.9	0.0053	0.024	−1.50	E	1
			[230]			4	6	1.1	0.0014	0.0041	−2.27	E	1
29.		^{2}D – ^{2}F°	204			10	14	190	0.16	1.1	0.21	E	1
			[201]			6	8	180	0.15	0.59	−0.05	E	1
			[207]			4	6	140	0.14	0.38	−0.25	E	1
			[208]			6	6	35	0.023	0.093	−0.87	E	1
30.		^{2}D – ^{2}P°											
			[178]			4	2	1.7	3.9(−4)	9.2(−4)	−2.80	E	1
			[181]			4	4	0.25	1.2(−4)	2.9(−4)	−3.31	E	1
31.		^{2}S – ^{2}P°	207			2	6	350	0.68	0.93	0.14	E	1
			[208]			2	4	460	0.60	0.82	0.08	E	1
			[204]			2	2	130	0.082	0.11	−0.79	E	1
32.		^{2}P – ^{2}P°	222			6	6	200	0.15	0.65	−0.05	E	1
			[225]			4	4	110	0.084	0.25	−0.47	D	1
			[216]			2	2	340	0.24	0.34	−0.32	E	1
			[220]			4	2	50	0.018	0.053	−1.14	D	1
			[221]			2	4	2.4	0.0035	0.0051	−2.15	E	1
33.	$3s3p^2$– $3s3p(^1$P°$)3d$	^{4}P – ^{2}F°											
			[159]			6	8	5.0	0.0025	0.0079	−1.82	E	1

Co xv: Allowed transitions — Continued

No.	Transition Array	Multiplet	λ (Å)	E_i (cm^{-1})	E_k (cm^{-1})	g_i	g_k	A_{ki} (10^8 s^{-1})	f_{ik}	S (at. u.)	log gf	Accuracy	Source
34.		^{2}D – ^{2}F°	*176*			10	14	320	0.21	1.2	0.32	E	1
			[177]			6	8	310	0.19	0.67	0.06	E	1
			[175]			4	6	300	0.20	0.47	−0.09	E	1
			[176]			6	6	15	0.0072	0.025	−1.37	E	1
35.		^{2}D – ^{2}D°											
			[169]			4	4	1.6	6.7(−4)	0.0015	−2.57	E	1
			[170]			6	4	2.3	6.6(−4)	0.0022	−2.41	E	1
36.		^{2}D – ^{2}P°	*168*			10	6	5.5	0.0014	0.0077	−1.86	E	1
			[168]			6	4	3.6	0.0010	0.0034	−2.21	E	1
			[169]			4	2	1.4	3.1(−4)	6.8(−4)	−2.91	E	1
			[168]			4	4	3.8	0.0016	0.0036	−2.19	E	1
37.		^{2}S – ^{2}D°											
			[192]			2	4	12	0.013	0.017	−1.57	E	1
38.		^{2}S – ^{2}P°	*192*			2	6	120	0.21	0.26	−0.39	E	1
			[191]			2	4	73	0.080	0.10	−0.80	E	1
			[193]			2	2	230	0.13	0.16	−0.60	E	1
39.		^{2}P – ^{2}F°											
			[216]			4	6	5.4	0.0056	0.016	−1.65	E	1
40.		^{2}P – ^{2}D°	*204*			6	10	640	0.67	2.7	0.60	E	1
			[205]			4	6	670	0.63	1.7	0.40	E	1
			[203]			2	4	540	0.67	0.90	0.13	E	1
			[207]			4	4	41	0.026	0.071	−0.98	E	1
41.		^{2}P – ^{2}P°											
			[205]			4	4	340	0.21	0.58	−0.07	D	1
			[204]			2	2	58	0.036	0.049	−1.14	E	1
			[208]			4	2	110	0.034	0.094	−0.86	C−	1

^aThe number in parentheses following the tabulated value indicates the power of ten by which this value has to be multiplied.

Co xv

Forbidden Transitions

Line strengths for magnetic dipole and electric quadrupole transitions within the $3s^2 3p$ $^2P°$ and $3s 3p^2$ ^4P terms are the results of the multiconfiguration Dirac-Fock (MCDF) calculations of Huang.[1] These relativistic calculations included a perturbative treatment of the Breit interaction and the Lamb shift. Allowance for configuration mixing included all configurations within the $n = 3$ complex. Strengths of electric quadrupole transitions as reported in Ref. 1 were multiplied by the factor $^2/_3$ which is needed to bring these values into conformance with the definition of quadrupole strengths used in the NBS tables.

Reference

[1]K.-N. Huang, At. Data Nucl. Data Tables **34**, 1 (1986).

Co xv: Forbidden transitions

No.	Transition Array	Multiplet	λ (Å)	E_i (cm^{-1})	E_k (cm^{-1})	g_i	g_k	Type of transition	A_{ki} (s^{-1})	S (at. u.)	Accuracy	Source
1.	$3p$–$3p$	$^2P°$ – $^2P°$										
			[4356]	0	22950	2	4	M1	109	1.33	C+	1
			"	"	"	2	4	E2	0.032	0.12	D–	1
2.	$3s 3p^2$–$3s 3p^2$	^4P – ^4P										
			[8680]			4	6	M1	24.3	3.54	C	1
			"			4	6	E2	9.7(−4)^a	0.17	D–	1
			[10500]			2	4	M1	19.0	3.27	C	1
			"			2	4	E2	4.9(−5)	0.015	E	1
			[4750]			2	6	E2	0.015	0.13	D–	1

^{a}The number in parentheses following the tabulated value indicates the power of ten by which this value has to be multiplied.

Co xvi

Mg Isoelectronic Sequence

Ground State: $1s^2 2s^2 2p^6 3s^2$ $^1S_0$

Ionization Energy: 511.96 eV = 4129200 cm^{-1}

Allowed Transitions

List of tabulated lines

Wavelength (Å)	No.	Wavelength (Å)	No.	Wavelength (Å)	No.	Wavelength (Å)	No.
47.489	13	212.800	15	229.074	16	287.96	3
61.025	22	213.396	15	265.74	2	293	4
62.131	23	219.947	15	271.38	3	302.69	3
64.537	24	220.980	15	281.88	3	308	4
210.249	15	221.62	15	284.42	3	389	1

Oscillator strengths were interpolated from the results of theoretical calculations reported by various researchers for the neighboring magnesium-like ions Fe XV and Ni XVII. Data for the three transitions $3s^2\,^1S_0 - 3snp\,^1P_1^\circ$ ($n = 3-5$) were reported by Shorer et al.,[1] who applied the relativistic random phase approximation (RRPA) with allowance for correlation within the context of a frozen core. The source of f-values for most transitions of the arrays $3s3p-3p^2$, $3s3d-3p3d$, $3s3p-3s3d$, and $3p^2-3p3d$ is the work of Fawcett,[2] who performed Hartree-Fock calculations which included relativistic effects and statistical allowance for exchange (HXR); he incorporated correlation effects due to all configurations in the $n = 3$ complex.

Kastner et al.[3] calculated A-values for a number of lines of the array $3p3d-3p4f$ in Fe XV and Zn XIX by application of a multiconfiguration scaled Thomas-Fermi (STF) approach. These transition probabilities were converted to oscillator strengths, from which f-values for a few transitions of this array in Co XVI were interpolated.

A-values for the three intercombination lines tabulated here were calculated for Co XVI by Kastner and Bhatia[4] using a scaled STF approach that allowed for correlation due to all configurations in the $n = 3$ complex.

Transitions involving levels which are indicated in Ref. 2 or 3 to be of low purity in LS coupling in neighboring Mg-like ions are omitted here. Lines which are characterized by very small f-values are assigned lower accuracy ratings.

References

[1]P. Shorer, C. D. Lin, and W. R. Johnson, Phys. Rev. A **16**, 1109 (1977).

[2]B. C. Fawcett, At. Data Nucl. Data Tables **28**, 579 (1983).

[3]S. O. Kastner, M. Swartz, A. K. Bhatia, and J. Lapides, J. Opt. Soc. Am. **68**, 1558 (1978).

[4]S. O. Kastner and A. K. Bhatia, J. Opt. Soc. Am. **69**, 1391 (1979).

Co XVI: Allowed transitions

No.	Transition Array	Multiplet	λ (Å)	E_i (cm^{-1})	E_k (cm^{-1})	g_i	g_k	A_{ki} (10^8 s^{-1})	f_{ik}	S (at. u.)	log gf	Accuracy	Source
1.	$3s^2-3s3p$	$^1S - ^3P^\circ$											
			[389]			1	3	0.54	0.0037	0.0047	−2.43	E	4
2.		$^1S - ^1P^\circ$	265.74	0	376310	1	3	251	0.796	0.696	−0.099	C+	interp.
3.	$3s3p-3p^2$	$^3P^\circ - ^3P$											
			284.42			5	5	130	0.16	0.75	−0.10	D	interp.
			[287.96]			3	3	54	0.067	0.19	−0.70	C−	interp.
			302.69			5	3	79	0.065	0.32	−0.49	C	interp.
						3	1		0.087		−0.58	C	interp.
			271.38			3	5	46	0.085	0.23	−0.59	D−	interp.
			281.88			1	3	78	0.28	0.26	−0.55	C	interp.
4.		$^3P^\circ - ^1D$											
			[308]			5	5	25	0.036	0.18	−0.75	E	4
			[293]			3	5	14	0.030	0.087	−1.05	E	4
5.		$^1P^\circ - ^1D$				3	5		0.085		−0.59	E	interp.
6.		$^1P^\circ - ^1S$				3	1		0.10		−0.52	C	interp.
7.	$3s3d-3p3d$	$^3D - ^3F^\circ$											
						7	9		0.155		0.035	C	interp.
						5	7		0.13		−0.19	C−	interp.
						3	5		0.12		−0.44	D−	interp.
						7	7		0.023		−0.79	C	interp.
						5	5		0.021		−0.98	D	interp.
						7	5		1.7(−4)^a		−2.92	E	interp.
8.		$^3D - ^3D^\circ$											
						7	7		0.11		−0.11	C−	interp.
						3	3		0.032		−1.02	E	interp.
						5	3		0.062		−0.51	E	interp.
						5	7		0.041		−0.69	C	interp.

Co XVI: Allowed transitions — Continued

No.	Transition Array	Multiplet	λ (Å)	E_i (cm^{-1})	E_k (cm^{-1})	g_i	g_k	A_{ki} (10^8 s^{-1})	f_{ik}	S (at. u.)	log gf	Accuracy	Source
9.		^{3}D – ^{3}P°											
						5	3		0.026		−0.89	E	interp.
						3	1		0.052		−0.81	C	interp.
						3	3		0.11		−0.48	E	interp.
10.		^{1}D – ^{1}D°				5	5		0.030		−0.82	D−	interp.
11.		^{1}D – ^{1}F°				5	7		0.41		0.31	D	interp.
12.		^{1}D – ^{1}P°				5	3		0.11		−0.26	D	interp.
13.	$3s^2$–$3s4p$	^{1}S – ^{1}P°	47.489	0	2105800	1	3	3760	0.381	0.060	−0.419	C	interp.
14.	$3s^2$–$3s5p$	^{1}S – ^{1}P°				1	3		0.117		−0.93	C	interp.
15.	$3s3p$–$3s3d$	^{3}P° – ^{3}D	216.56			9	15	246	0.288	1.85	0.414	C−	interp.
			219.947			5	7	235	0.239	0.87	0.077	C−	interp.
			212.800			3	5	190	0.22	0.46	−0.18	C−	interp.
			210.249			1	3	150	0.29	0.20	−0.54	C−	interp.
			220.980			5	5	59	0.043	0.16	−0.67	C−	interp.
			213.396			3	3	110	0.073	0.15	−0.66	C	interp.
			[221.62]			5	3	6.3	0.0028	0.010	−1.85	D−	interp.
16.		^{1}P° – ^{1}D	229.074	376310	812850	3	5	450	0.59	1.3	0.25	D	interp.
17.	$3p^2$–$3p3d$	^{3}P – ^{3}D°											
						5	7		0.26		0.11	D−	interp.
						1	3		0.60		−0.22	E	interp.
						3	3		0.035		−0.98	E	interp.
						5	3		0.0014		−2.15	E	interp.
18.		^{3}P – ^{3}P°											
						3	3		0.11		−0.48	E	interp.
						5	3		0.032		−0.80	E	interp.
						3	1		0.053		−0.80	C−	interp.
19.		^{1}D – ^{1}D°				5	5		0.13		−0.19	E	interp.
20.		^{1}D – ^{1}P°				5	3		0.0013		−2.19	E	interp.
21.		^{1}S – ^{1}P°				1	3		0.60		−0.22	C−	interp.
22.	$3p3d$–$3p4f$	^{3}F° – ^{3}G											
			61.025			9	11	1.2(+4)	0.80	1.4	0.86	C	interp.
23.		^{3}P° – ^{3}D											
			62.131			1	3	5300	0.92	0.19	−0.04	C	interp.
24.		^{1}F° – ^{1}G	64.537			7	9	1.1(+4)	0.90	1.3	0.80	C−	interp.

[a]The number in parentheses following the tabulated value indicates the power of ten by which this value has to be multiplied.

Co XVII

Na Isoelectronic Sequence

Ground State: $1s^2 2s^2 2p^6 3s\ ^2S_{1/2}$

Ionization Energy: 546.58 eV = 4408500 cm^{-1}

Allowed Transitions

List of tabulated lines

Wavelength (Å)	No.	Wavelength (Å)	No.	Wavelength (Å)	No.	Wavelength (Å)	No.
28.868	4	43.348	20	85.412	46	152.5	70
28.874	4	43.367	20	85.529	46	163.2	74
28.960	15	45.319	2	90.066	45	163.3	74
29.171	15	45.527	2	90.123	45	201.5	56
29.175	15	48.564	9	90.196	45	201.8	56
31.142	13	49.133	9	92.047	52	208.5	61
31.386	13	49.171	9	92.081	52	210.4	61
31.390	13	58.842	19	92.098	52	210.6	61
32.910	25	58.948	19	105.1	29	234.95	7
32.950	25	58.969	19	105.6	29	236.8	68
32.951	25	64.470	39	110.4	35	237.2	68
32.995	3	64.872	39	111.6	35	247.56	7
33.046	3	64.893	39	111.7	35	249.84	7
35.617	11	67.277	18	127.96	44	269.8	73
35.660	23	67.445	18	128.2	44	270.0	73
35.707	23	67.737	18	128.20	44	270.1	73
35.932	11	71.169	48	129.7	34	276.4	67
35.943	11	71.250	48	131.4	34	276.9	67
36.446	22	71.256	48	138.6	63	277.0	67
36.455	22	72.228	30	139.4	63	312.54	1
36.495	22	72.265	30	139.5	63	339.50	1
37.410	10	75.279	53	146.3	51	636.1	33
37.768	10	75.307	53	146.5	51	672.5	33
41.404	21	75.313	53	147.7	43	679.8	33
41.462	21	76.383	37	148.0	43	786.8	28
41.468	21	76.953	37	148.7	43	854.7	28
43.279	20	76.976	37	152.3	70		

Strengths of the lines of the 3s–3p and 3p–3d transitions were taken from Edlén's interpolation formulae.[1] These were based on the results of Weiss' Hartree-Fock calculations,[2] in which ratios of relativistic Dirac to nonrelativistic line strengths in hydrogenic ions were applied as scaling factors to the nonrelativistic Hartree-Fock line strengths in the corresponding sodiumlike species. Oscillator strengths for the 4p–4d transitions were derived by Gruzdev and Sherstyuk[3] using the relativistic variant of their effective orbital quantum number method, which utilizes a Coulomb potential in conjunction with a semiempirical orbital quantum number which is determined from experimental energy levels. Strengths of the lines of the 3s–4p and 3p–4d transitions, as well as f-values of the 3d–4f transitions, were interpolated from the results of the relativistic single-configuration Hartree-Fock calculations of Kim and

Cheng[4] for Fe XVI and, depending on the transition, either Kr XXVI or Mo XXXII.

Multiplet f-values calculated by Tull et al.[5] using the frozen-core Hartree-Fock approach are quoted for numerous transitions $nl-n'l'$ $(3 \leqslant n \leqslant 5;\ 4 \leqslant n' \leqslant 8;\ l, l'=s,p,d,f)$. Data for additional transitions (namely, those for which $n,n' \leqslant 10$, where n,n' are the principal quantum numbers of the lower and upper states, respectively) can be found in Ref. 5. Whenever wavelengths of individual lines within a multiplet either were available directly or could be determined from the energy levels, the multiplet strength was distributed among the lines according to LS-coupling rules. The strength of the $3p\ ^2P° - 4s\ ^2S$ multiplet was not distributed between the two lines in the multiplet, however, since the calculations of Kim and Cheng indicate that in the corresponding transition in neighboring sodiumlike ions (Fe XVI and

Mo XXXII) the ratio of the two line strengths deviates somewhat from the value that would be obtained in the case of pure *LS* coupling.

Transitions with small *f*-values were generally assigned lower accuracy ratings.

References

[1]B. Edlén, Phys. Scr. **17**, 565 (1978).
[2]A. W. Weiss, J. Quant. Spectrosc. Radiat. Transfer **18**, 481 (1977).
[3]P. F. Gruzdev and A. I. Sherstyuk, Opt. Spectrosc. (USSR) **46**, 353 (1979).
[4]Y.-K. Kim and K.-T. Cheng, J. Opt. Soc. Am. **68**, 836 (1978).
[5]C. E. Tull, R. P. McEachran, and M. Cohen, At. Data **3**, 169 (1971).

Co XVII: Allowed transitions

No.	Transition Array	Multiplet	λ (Å)	E_i (cm^{-1})	E_k (cm^{-1})	g_i	g_k	A_{ki} (10^8 s^{-1})	f_{ik}	S (at. u.)	log gf	Accuracy	Source
1.	3s–3p	^{2}S – ^{2}P°	321.04	0	311490	2	6	81.9	0.380	0.803	−0.119	B	1
			312.54	0	319960	2	4	89.3	0.261	0.538	−0.282	B	1
			339.50	0	294550	2	2	68.6	0.119	0.265	−0.625	B	1
2.	3s–4p	^{2}S – ^{2}P°	45.389	0	2203200	2	6	2400	0.22	0.067	−0.35	C	interp.
			45.319	0	2206600	2	4	2370	0.146	0.0435	−0.54	C	interp.
			45.527	0	2196500	2	2	2530	0.0787	0.0236	−0.803	C+	interp.
3.	3s–5p	^{2}S – ^{2}P°	33.012	0	3029200	2	6	1500	0.072	0.016	−0.84	C	5
			32.995	0	3030800	2	4	1600	0.051	0.011	−0.99	C	ls
			33.046	0	3026100	2	2	1500	0.024	0.0053	−1.31	C	ls
4.	3s–6p	^{2}S – ^{2}P°	28.870	0	3463800	2	6	870	0.0326	0.0062	−1.186	C	5
			28.868	0	3464000	2	4	860	0.022	0.0041	−1.37	C	ls
			28.874	0	3463300	2	2	880	0.011	0.0021	−1.66	C	ls
5.	3s–7p	^{2}S – ^{2}P°				2	6		0.0179		−1.446	C	5
6.	3s–8p	^{2}S – ^{2}P°				2	6		0.0110		−1.66	C	5
7.	3p–3d	^{2}P° – ^{2}D	243.36	311490	722410	6	10	181	0.268	1.29	0.207	B	1
			247.56	319960	723900	4	6	172	0.238	0.775	−0.022	B	1
			234.95	294550	720170	2	4	168	0.279	0.431	−0.254	B	1
			249.84	319960	720170	4	4	27.8	0.0260	0.0856	−0.983	B	1
8.	3p–4s	^{2}P° – ^{2}S	56.561	311490	2079500	6	2	3800	0.061	0.068	−0.44	C−	5
9.	3p–4d	^{2}P° – ^{2}D	48.943	311490	2354700	6	10	5300	0.319	0.308	0.281	C	interp.
			49.133	319960	2355300	4	6	5400	0.291	0.188	0.065	C	interp.
			48.564	294550	2353700	2	4	4400	0.31	0.099	−0.21	C	interp.
			49.171	319960	2353700	4	4	900	0.0326	0.0211	−0.88	C	interp.
10.	3p–5s	^{2}P° – ^{2}S	37.648	311490	2967700	6	2	1710	0.0121	0.0090	−1.139	C	5
			37.768	319960	2967700	4	2	1100	0.012	0.0060	−1.32	C	ls
			[37.410]	294550	2967700	2	2	580	0.012	0.0030	−1.61	C	ls
11.	3p–5d	^{2}P° – ^{2}D	35.827	311490	3102700	6	10	3150	0.101	0.071	−0.218	C	5
			35.932	319960	3103000	4	6	3100	0.091	0.043	−0.44	C	ls
			35.617	294550	3102200	2	4	2700	0.10	0.024	−0.69	C	ls
			[35.943]	319960	3102200	4	4	510	0.0099	0.0047	−1.40	D	ls
12.	3p–6s	^{2}P° – ^{2}S				6	2		0.0047		−1.55	D	5

Co XVII: Allowed transitions — Continued

No.	Transition Array	Multiplet	λ (Å)	E_i (cm^{-1})	E_k (cm^{-1})	g_i	g_k	A_{ki} (10^8 s^{-1})	f_{ik}	S (at. u.)	log gf	Accuracy	Source
13.	$3p$–$6d$	^{2}P$^\circ$ – ^{2}D	*31.305*	*311490*	*3505900*	6	10	1870	0.0458	0.0283	−0.56	C	5
			31.386	319960	3506100	4	6	1860	0.0411	0.0170	−0.78	C	*ls*
			31.142	294550	3505700	2	4	1600	0.046	0.0094	−1.04	C	*ls*
			[31.390]	319960	3505700	4	4	310	0.0046	0.0019	−1.74	D	*ls*
14.	$3p$–$7s$	^{2}P$^\circ$ – ^{2}S				6	2		0.0024		−1.84	D	5
15.	$3p$–$7d$	^{2}P$^\circ$ – ^{2}D	*29.100*	*311490*	*3747900*	6	10	1190	0.0251	0.0144	−0.82	C	5
			29.171	319960	3748100	4	6	1200	0.022	0.0086	−1.05	C	*ls*
			28.960	294550	3747600	2	4	1000	0.0252	0.00480	−1.298	C	*ls*
			[29.175]	319960	3747600	4	4	200	0.0025	9.6(−4)^a	−2.00	D	*ls*
16.	$3p$–$8s$	^{2}P$^\circ$ – ^{2}S				6	2		0.0014		−2.08	D	5
17.	$3p$–$8d$	^{2}P$^\circ$ – ^{2}D				6	10		0.0154		−1.034	C	5
18.	$3d$–$4p$	^{2}D – ^{2}P$^\circ$	*67.531*	*722410*	*2203200*	10	6	890	0.0367	0.082	−0.435	C−	5
			[67.445]	723900	2206600	6	4	810	0.037	0.049	−0.66	C−	*ls*
			[67.737]	720170	2196500	4	2	880	0.030	0.027	−0.92	C−	*ls*
			[67.277]	720170	2206600	4	4	91	0.0062	0.0055	−1.60	D	*ls*
19.	$3d$–$4f$	^{2}D – ^{2}F$^\circ$	*58.907*	*722410*	*2420000*	10	14	1.3(+4)	0.93	1.8	0.97	C	*interp.*
			58.948	723900	2420300	6	8	1.3(+4)	0.89	1.0	0.73	C	*interp.*
			58.842	720170	2419700	4	6	1.2(+4)	0.93	0.72	0.57	C	*interp.*
			[58.969]	723900	2419700	6	6	850	0.0442	0.051	−0.58	C	*interp.*
20.	$3d$–$5p$	^{2}D – ^{2}P$^\circ$	*43.350*	*722410*	*3029200*	10	6	350	0.0060	0.0086	−1.22	D	5
			[43.348]	723900	3030800	6	4	320	0.0061	0.0052	−1.44	D	*ls*
			[43.367]	720170	3026100	4	2	360	0.0051	0.0029	−1.69	D	*ls*
			[43.279]	720170	3030800	4	4	36	0.0010	5.7(−4)	−2.40	E	*ls*
21.	$3d$–$5f$	^{2}D – ^{2}F$^\circ$	*41.439*	*722410*	*3135600*	10	14	4740	0.171	0.233	0.233	C	5
			41.462	723900	3135700	6	8	4730	0.162	0.133	−0.011	C	*ls*
			41.404	720170	3135400	4	6	4400	0.17	0.093	−0.17	C	*ls*
			[41.468]	723900	3135400	6	6	320	0.0082	0.0067	−1.31	D	*ls*
22.	$3d$–$6p$	^{2}D – ^{2}P$^\circ$	*36.478*	*722410*	*3463800*	10	6	180	0.0021	0.0025	−1.68	D	5
			[36.495]	723900	3464000	6	4	160	0.0021	0.0015	−1.90	D	*ls*
			[36.455]	720170	3463300	4	2	170	0.0017	8.3(−4)	−2.16	D	*ls*
			[36.446]	720170	3464000	4	4	18	3.5(−4)	1.7(−4)	−2.85	E	*ls*
23.	$3d$–$6f$	^{2}D – ^{2}F$^\circ$	*35.688*	*722410*	*3524500*	10	14	2400	0.063	0.074	−0.20	C	5
			35.707	723900	3524500	6	8	2300	0.060	0.042	−0.45	C	*ls*
			35.660	720170	3524500	4	6	2200	0.064	0.030	−0.59	C	*ls*
			[35.707]	723900	3524500	6	6	160	0.0030	0.0021	−1.75	D	*ls*
24.	$3d$–$7p$	^{2}D – ^{2}P$^\circ$				10	6		0.0010		−2.00	D	5
25.	$3d$–$7f$	^{2}D – ^{2}F$^\circ$	*32.935*	*722410*	*3758700*	10	14	1360	0.0309	0.0335	−0.51	C	5
			32.951	723900	3758700	6	8	1350	0.0293	0.0191	−0.75	C	*ls*
			32.910	720170	3758800	4	6	1270	0.0309	0.0134	−0.91	C	*ls*
			[32.950]	723900	3758800	6	6	91	0.0015	9.6(−4)	−2.05	D	*ls*

Co XVII: Allowed transitions — Continued

No.	Transition Array	Multiplet	λ (Å)	E_i (cm⁻¹)	E_k (cm⁻¹)	g_i	g_k	A_{ki} (10^8 s⁻¹)	f_{ik}	S (at. u.)	log gf	Accuracy	Source
26.	$3d$–$8p$	^{2}D – ^{2}P°				10	6		6.0(−4)		−2.22	E	5
27.	$3d$–$8f$	^{2}D – ^{2}F°				10	14		0.0178		−0.75	C	5
28.	$4s$–$4p$	^{2}S – ^{2}P°	808.4	2079500	2203200	2	6	18	0.52	2.8	0.02	C	5
			[786.8]	2079500	2206600	2	4	20	0.37	1.9	−0.13	C	ls
			[854.7]	2079500	2196500	2	2	15	0.17	0.93	−0.48	C	ls
29.	$4s$–$5p$	^{2}S – ^{2}P°	105.3	2079500	3029200	2	6	510	0.254	0.176	−0.294	C	5
			[105.1]	2079500	3030800	2	4	510	0.169	0.117	−0.471	C	ls
			[105.6]	2079500	3026100	2	2	510	0.085	0.059	−0.77	C	ls
30.	$4s$–$6p$	^{2}S – ^{2}P°	72.239	2079500	3463800	2	6	340	0.079	0.038	−0.80	C	5
			[72.228]	2079500	3464000	2	4	340	0.053	0.025	−0.98	C	ls
			[72.265]	2079500	3463300	2	2	350	0.027	0.013	−1.26	C	ls
31.	$4s$–$7p$	^{2}S – ^{2}P°				2	6		0.0366		−1.135	C	5
32.	$4s$–$8p$	^{2}S – ^{2}P°				2	6		0.0205		−1.387	C	5
33.	$4p$–$4d$	^{2}P° – ^{2}D	660.1	2203200	2354700	6	10	37	0.40	5.2	0.38	C	3
			[672.5]	2206600	2355300	4	6	34	0.35	3.1	0.15	C	3
			[636.1]	2196500	2353700	2	4	34	0.41	1.7	−0.09	C	3
			[679.8]	2206600	2353700	4	4	5.6	0.039	0.35	−0.81	C	3
34.	$4p$–$5s$	^{2}P° – ^{2}S	130.8	2203200	2967700	6	2	1200	0.103	0.266	−0.209	C	5
			[131.4]	2206600	2967700	4	2	790	0.102	0.177	−0.388	C	ls
			[129.7]	2196500	2967700	2	2	410	0.10	0.089	−0.68	C	ls
35.	$4p$–$5d$	^{2}P° – ^{2}D	111.2	2203200	3102700	6	10	930	0.287	0.63	0.236	C	5
			[111.6]	2206600	3103000	4	6	920	0.26	0.38	0.01	C	ls
			[110.4]	2196500	3102200	2	4	790	0.29	0.21	−0.24	C	ls
			[111.7]	2206600	3102200	4	4	150	0.029	0.042	−0.94	D	ls
36.	$4p$–$6s$	^{2}P° – ^{2}S				6	2		0.0206		−0.91	C	5
37.	$4p$–$6d$	^{2}P° – ^{2}D	76.764	2203200	3505900	6	10	650	0.095	0.14	−0.24	C	5
			[76.953]	2206600	3506100	4	6	620	0.083	0.084	−0.48	C	ls
			[76.383]	2196500	3505700	2	4	530	0.093	0.047	−0.73	C	ls
			[76.976]	2206600	3505700	4	4	100	0.0092	0.0093	−1.44	D	ls
38.	$4p$–$7s$	^{2}P° – ^{2}S				6	2		0.0081		−1.31	D	5
39.	$4p$–$7d$	^{2}P° – ^{2}D	64.737	2203200	3747900	6	10	428	0.0448	0.057	−0.57	C	5
			[64.872]	2206600	3748100	4	6	420	0.040	0.034	−0.80	C	ls
			[64.470]	2196500	3747600	2	4	360	0.045	0.019	−1.05	C	ls
			[64.893]	2206600	3747600	4	4	70	0.0044	0.0038	−1.75	D	ls
40.	$4p$–$8s$	^{2}P° – ^{2}S				6	2		0.0042		−1.60	D	5
41.	$4p$–$8d$	^{2}P° – ^{2}D				6	10		0.0253		−0.82	C	5
42.	$4d$–$4f$	^{2}D – ^{2}F°				10	14		0.106		0.025	C	5

Co XVII: Allowed transitions — Continued

No.	Transition Array	Multiplet	λ (Å)	E_i (cm^{-1})	E_k (cm^{-1})	g_i	g_k	A_{ki} (10^8 s^{-1})	f_{ik}	S (at. u.)	log gf	Accuracy	Source
43.	4d–5p	^{2}D – ^{2}P°	148.3	2354700	3029200	10	6	420	0.083	0.41	−0.08	C	5
			[148.0]	2355300	3030800	6	4	390	0.086	0.25	−0.29	C	ls
			[148.7]	2353700	3026100	4	2	430	0.071	0.14	−0.54	C	ls
			[147.7]	2353700	3030800	4	4	42	0.014	0.027	−1.26	D	ls
44.	4d–5f	^{2}D – ^{2}F°	128.1	2354700	3135600	10	14	2100	0.73	3.1	0.86	C	5
			128.20	2355300	3135700	6	8	2200	0.71	1.8	0.63	C	ls
			127.96	2353700	3135400	4	6	1900	0.71	1.2	0.45	C	ls
			[128.2]	2355300	3135400	6	6	140	0.035	0.089	−0.68	D	ls
45.	4d–6p	^{2}D – ^{2}P°	90.163	2354700	3463800	10	6	197	0.0144	0.0427	−0.84	C	5
			[90.196]	2355300	3464000	6	4	177	0.0144	0.0256	−1.064	C	ls
			[90.123]	2353700	3463300	4	2	197	0.0120	0.0142	−1.320	C	ls
			[90.066]	2353700	3464000	4	4	19	0.0024	0.0028	−2.02	D	ls
46.	4d–6f	^{2}D – ^{2}F°	85.485	2354700	3524500	10	14	1170	0.180	0.51	0.255	C	5
			[85.529]	2355300	3524500	6	8	1200	0.17	0.29	0.01	C	ls
			[85.412]	2353700	3524500	4	6	1100	0.18	0.20	−0.15	C	ls
			[85.529]	2355300	3524500	6	6	81	0.0089	0.015	−1.27	D	ls
47.	4d–7p	^{2}D – ^{2}P°				10	6		0.0053		−1.28	D	5
48.	4d–7f	^{2}D – ^{2}F°	71.225	2354700	3758700	10	14	690	0.074	0.17	−0.13	C	5
			[71.256]	2355300	3758700	6	8	680	0.069	0.097	−0.38	C	ls
			[71.169]	2353700	3758800	4	6	640	0.073	0.068	−0.54	C	ls
			[71.250]	2355300	3758800	6	6	46	0.0035	0.0049	−1.68	D	ls
49.	4d–8p	^{2}D – ^{2}P°				10	6		0.0026		−1.59	D	5
50.	4d–8f	^{2}D – ^{2}F°				10	14		0.0390		−0.409	C	5
51.	4f–5d	^{2}F° – ^{2}D	146.5	2420000	3102700	14	10	80	0.0185	0.125	−0.59	C	5
			[146.5]	2420300	3103000	8	6	76	0.018	0.071	−0.83	C	ls
			[146.5]	2419700	3102200	6	4	81	0.017	0.050	−0.98	C	ls
			[146.3]	2419700	3103000	6	6	3.9	0.0012	0.0036	−2.13	D	ls
52.	4f–6d	^{2}F° – ^{2}D	92.090	2420000	3505900	14	10	34	0.0031	0.013	−1.36	D	5
			[92.098]	2420300	3506100	8	6	32	0.0031	0.0074	−1.61	D	ls
			[92.081]	2419700	3505700	6	4	34	0.0029	0.0052	−1.77	D	ls
			[92.047]	2419700	3506100	6	6	1.6	2.0(−4)	3.7(−4)	−2.91	E	ls
53.	4f–7d	^{2}F° – ^{2}D	75.307	2420000	3747900	14	10	18	0.0011	0.0038	−1.81	D	5
			[75.313]	2420300	3748100	8	6	17	0.0011	0.0022	−2.05	D	ls
			[75.307]	2419700	3747600	6	4	18	0.0010	0.0015	−2.22	D	ls
			[75.279]	2419700	3748100	6	6	0.87	7.4(−5)	1.1(−4)	−3.35	E	ls
54.	4f–8d	^{2}F° – ^{2}D				14	10		5.2(−4)		−2.14	E	5
55.	5s–5p	^{2}S – ^{2}P°				2	6		0.67		0.13	C	5
56.	5s–6p	^{2}S – ^{2}P°	201.6	2967700	3463800	2	6	150	0.275	0.365	−0.260	C	5
			[201.5]	2967700	3464000	2	4	150	0.183	0.243	−0.436	C	ls
			[201.8]	2967700	3463300	2	2	150	0.092	0.122	−0.74	C	ls

Co XVII: Allowed transitions — Continued

No.	Transition Array	Multiplet	λ (Å)	E_i (cm^{-1})	E_k (cm^{-1})	g_i	g_k	A_{ki} (10^8 s^{-1})	f_{ik}	S (at. u.)	log gf	Accuracy	Source
57.	5s–7p	^{2}S – ^{2}P°				2	6		0.086		−0.76	C	5
58.	5s–8p	^{2}S – ^{2}P°				2	6		0.0403		−1.094	C	5
59.	5p–5d	^{2}P° – ^{2}D				6	10		0.55		0.52	C	5
60.	5p–6s	^{2}P° – ^{2}S				6	2		0.147		−0.055	C	5
61.	5p–6d	^{2}P° – ^{2}D	209.8	3029200	3505900	6	10	247	0.272	1.13	0.213	C	5
			[210.4]	3030800	3506100	4	6	250	0.25	0.68	−0.01	C	ls
			[208.5]	3026100	3505700	2	4	211	0.275	0.377	−0.260	C	ls
			[210.6]	3030800	3505700	4	4	41	0.027	0.075	−0.97	D	ls
62.	5p–7s	^{2}P° – ^{2}S				6	2		0.0293		−0.75	C	5
63.	5p–7d	^{2}P° – ^{2}D	139.1	3029200	3747900	6	10	190	0.092	0.25	−0.26	C	5
			[139.4]	3030800	3748100	4	6	190	0.082	0.15	−0.49	C	ls
			[138.6]	3026100	3747600	2	4	160	0.091	0.083	−0.74	C	ls
			[139.5]	3030800	3747600	4	4	32	0.0093	0.017	−1.43	D	ls
64.	5p–8s	^{2}P° – ^{2}S				6	2		0.0116		−1.157	C	5
65.	5p–8d	^{2}P° – ^{2}D				6	10		0.0445		−0.57	C	5
66.	5d–5f	^{2}D – ^{2}F°				10	14		0.188		0.274	C	5
67.	5d–6p	^{2}D – ^{2}P°	276.9	3102700	3463800	10	6	194	0.134	1.22	0.127	C	5
			[277.0]	3103000	3464000	6	4	170	0.13	0.73	−0.10	C	ls
			[276.9]	3102200	3463300	4	2	194	0.112	0.407	−0.350	C	ls
			[276.4]	3102200	3464000	4	4	19	0.022	0.081	−1.05	D	ls
68.	5d–6f	^{2}D – ^{2}F°	237.1	3102700	3524500	10	14	550	0.65	5.1	0.81	C	5
			[237.2]	3103000	3524500	6	8	550	0.62	2.9	0.57	C	ls
			[236.8]	3102200	3524500	4	6	510	0.64	2.0	0.41	C	ls
			[237.2]	3103000	3524500	6	6	38	0.032	0.15	−0.72	D	ls
69.	5d–7p	^{2}D – ^{2}P°				10	6		0.0239		−0.62	C	5
70.	5d–7f	^{2}D – ^{2}F°	152.4	3102700	3758700	10	14	365	0.178	0.89	0.250	C	5
			[152.5]	3103000	3758700	6	8	360	0.17	0.51	0.01	C	ls
			[152.3]	3102200	3758800	4	6	340	0.18	0.36	−0.14	C	ls
			[152.5]	3103000	3758800	6	6	24	0.0083	0.025	−1.30	D	ls
71.	5d–8p	^{2}D – ^{2}P°				10	6		0.0090		−1.05	D	5
72.	5d–8f	^{2}D – ^{2}F°				10	14		0.078		−0.11	C	5
73.	5f–6d	^{2}F° – ^{2}D	270.1	3135600	3505900	14	10	58	0.0453	0.56	−0.198	C	5
			[270.0]	3135700	3506100	8	6	55	0.045	0.32	−0.44	C	ls
			[270.1]	3135400	3505700	6	4	57	0.041	0.22	−0.61	C	ls
			[269.8]	3135400	3506100	6	6	2.8	0.0030	0.016	−1.74	D	ls

Co XVII: Allowed transitions — Continued

No.	Transition Array	Multiplet	λ (Å)	E_i (cm^{-1})	E_k (cm^{-1})	g_i	g_k	A_{ki} (10^8 s^{-1})	f_{ik}	S (at. u.)	log gf	Accuracy	Source
74.	5f–7d	^{2}F° – ^{2}D	163.3	3135600	3747900	14	10	28	0.0081	0.061	−0.95	D	5
			[163.3]	3135700	3748100	8	6	27	0.0081	0.035	−1.19	D	ls
			[163.3]	3135400	3747600	6	4	28	0.0074	0.024	−1.35	D	ls
			[163.2]	3135400	3748100	6	6	1.3	5.3(−4)	0.0017	−2.50	E	ls
75.	5f–8d	^{2}F° – ^{2}D				14	10		0.0029		−1.39	D	5

^{a}The number in parentheses following the tabulated value indicates the power of ten by which this value has to be multiplied.

Co XVII

Forbidden Transitions

List of tabulated lines

Wavelength (Å)	No.	Wavelength (Å)	No.	Wavelength (Å)	No.	Wavelength (Å)	No.
32.227	22	79.529	6	139.7	10	304.6	43
35.515	24	79.573	6	139.8	10	305.0	43
42.457	21	86.775	8	154.8	15	338.4	45
42.486	21	86.806	8	154.9	15	339.2	45
47.057	23	86.896	8	155.0	15	339.3	45
47.612	23	86.926	8	155.1	15	362.6	26
47.626	23	90.514	11	160.4	17	364.7	26
52.301	3	90.563	11	160.5	17	412.5	18
53.005	3	95.813	36	162.9	34	413.2	18
53.291	3	97.704	27	163.3	34	413.4	18
59.930	29	97.780	27	163.6	35	414.1	18
59.948	29	106.5	31	163.8	35	426.8	19
61.158	4	107.6	31	164.9	35	427.0	19
61.218	4	107.7	31	185.7	38	448.0	30
61.297	4	119.9	5	185.9	38	467.9	30
61.357	4	121.3	5	200.6	41	469.3	30
64.008	33	122.0	5	202.6	41	739.1	37
64.429	33	128.1	39	228.4	13	743.5	37
70.097	28	128.2	39	230.8	13	914.9	40
70.116	28	133.5	7	231.2	13	953.3	40
71.798	9	133.6	7	247.6	14	956.0	40
73.567	25	133.7	7	247.8	14	1630	44
73.768	25	133.9	7	248.1	14	1650	44
74.716	12	136.5	42	248.3	14	3934	1
75.301	32	137.4	42	257.0	16	26800	2
75.878	32	138.14	20	257.2	16		
78.895	6	138.86	20	304.3	43		

Electric quadrupole strengths for numerous multiplets in this sodiumlike ion were determined by Tull *et al.*[1] using the frozen-core Hartree-Fock approach with no allowance for configuration mixing. *LS*-coupling rules were applied to obtain strengths of lines within multiplets. The strongest lines for which fairly accurate wavelengths could be derived from experimentally determined energy levels are quoted in this compilation.

The strengths given in Ref. 1 for transitions in which both $\Delta n = 0$ and $\Delta l = 0$ (i.e., transitions between the two levels of a given term) are overstated, and had to be reduced as follows:

$$S(np\ ^2P^\circ_{1/2} - np\ ^2P^\circ_{3/2}) = S(\text{Ref. 1}) \times (1/3)$$
$$S(nd\ ^2D_{3/2} - nd\ ^2D_{5/2}) = S(\text{Ref. 1}) \times (3/25)$$
$$S(nf\ ^2F^\circ_{5/2} - nf\ ^2F^\circ_{7/2}) = S(\text{Ref. 1}) \times (3/49).$$

Reference

[1]C. E. Tull, M. Jackson, R. P. McEachran, and M. Cohen, J. Quant. Spectrosc. Radiat. Transfer **12**, 893 (1972).

Co XVII: Forbidden transitions

No.	Transition Array	Multiplet	λ (Å)	E_i (cm^{-1})	E_k (cm^{-1})	g_i	g_k	Type of transition	A_{ki} (s^{-1})	S (at. u.)	Accuracy	Source
1.	3p–3p	^{2}P° – ^{2}P°										
			[3934]	294550	319960	2	4	E2	0.051	0.115	C	1
2.	3d–3d	^{2}D – ^{2}D										
			[26800]	720170	723900	4	6	E2	6.9(−7)^a	0.034	D−	1
3.	3p–4p	^{2}P° – ^{2}P°										
			[53.005]	319960	2206600	4	4	E2	4.8(+7)	0.048	D	1,*ls*
			[53.291]	319960	2196500	4	2	E2	9.4(+7)	0.048	D	1,*ls*
			[52.301]	294550	2206600	2	4	E2	5.2(+7)	0.048	D	1,*ls*
4.	3d–4d	^{2}D – ^{2}D										
			[61.297]	723900	2355300	6	6	E2	2.6(+7)	0.079	D	1,*ls*
			[61.218]	720170	2353700	4	4	E2	2.2(+7)	0.046	D	1,*ls*
			[61.357]	723900	2353700	6	4	E2	9.7(+6)	0.020	E	1,*ls*
			[61.158]	720170	2355300	4	6	E2	6.5(+6)	0.020	E	1,*ls*
5.	4p–5p	^{2}P° – ^{2}P°										
			[121.3]	2206600	3030800	4	4	E2	7.6(+6)	0.477	C	1,*ls*
			[122.0]	2206600	3026100	4	2	E2	1.48(+7)	0.477	C	1,*ls*
			[119.9]	2196500	3030800	2	4	E2	8.1(+6)	0.477	C	1,*ls*
6.	4p–6p	^{2}P° – ^{2}P°										
			[79.529]	2206600	3464000	4	4	E2	4.9(+6)	0.037	D	1,*ls*
			[79.573]	2206600	3463300	4	2	E2	9.7(+6)	0.037	D	1,*ls*
			[78.895]	2196500	3464000	2	4	E2	5.1(+6)	0.037	D	1,*ls*
7.	4d–5d	^{2}D – ^{2}D										
			[133.7]	2355300	3103000	6	6	E2	5.7(+6)	0.87	C	1,*ls*
			[133.6]	2353700	3102200	4	4	E2	5.0(+6)	0.51	C	1,*ls*
			[133.9]	2355300	3102200	6	4	E2	2.12(+6)	0.217	C−	1,*ls*
			[133.5]	2353700	3103000	4	6	E2	1.43(+6)	0.217	C−	1,*ls*
8.	4d–6d	^{2}D – ^{2}D										
			[86.896]	2355300	3506100	6	6	E2	3.4(+6)	0.060	D	1,*ls*
			[86.806]	2353700	3505700	4	4	E2	3.0(+6)	0.035	D	1,*ls*
			[86.926]	2355300	3505700	6	4	E2	1.3(+6)	0.015	E	1,*ls*
			[86.775]	2353700	3506100	4	6	E2	8.5(+5)	0.015	E	1,*ls*

Co XVII: Forbidden transitions — Continued

No.	Transition Array	Multiplet	λ (Å)	E_i (cm^{-1})	E_k (cm^{-1})	g_i	g_k	Type of transition	A_{ki} (s^{-1})	S (at. u.)	Accuracy	Source
9.	4d–7d	^{2}D – ^{2}D										
			[71.798]	2355300	3748100	6	6	E2	2.1(+6)	0.014	D	1,*ls*
10.	4f–5f	^{2}F° – ^{2}F°										
			[139.8]	2420300	3135700	8	8	E2	3.3(+6)	0.84	C	1,*ls*
			[139.7]	2419700	3135400	6	6	E2	3.2(+6)	0.61	C	1,*ls*
			[139.8]	2420300	3135400	8	6	E2	5.3(+5)	0.101	C−	1,*ls*
			[139.7]	2419700	3135700	6	8	E2	3.99(+5)	0.101	C−	1,*ls*
11.	4f–6f	^{2}F° – ^{2}F°										
			[90.563]	2420300	3524500	8	8	E2	1.6(+6)	0.047	D	1,*ls*
			[90.514]	2419700	3524500	6	6	E2	1.6(+6)	0.034	D	1,*ls*
12.	4f–7f	^{2}F° – ^{2}F°										
			[74.716]	2420300	3758700	8	8	E2	9.0(+5)	0.010	D	1,*ls*
13.	5p–6p	^{2}P° – ^{2}P°										
			[230.8]	3030800	3464000	4	4	E2	1.76(+6)	2.74	C	1,*ls*
			[231.2]	3030800	3463300	4	2	E2	3.48(+6)	2.74	C	1,*ls*
			[228.4]	3026100	3464000	2	4	E2	1.85(+6)	2.74	C	1,*ls*
14.	5d–6d	^{2}D – ^{2}D										
			[248.1]	3103000	3506100	6	6	E2	1.5(+6)	5.1	C	1,*ls*
			[247.8]	3102200	3505700	4	4	E2	1.33(+6)	2.97	C	1,*ls*
			[248.3]	3103000	3505700	6	4	E2	5.7(+5)	1.27	C−	1,*ls*
			[247.6]	3102200	3506100	4	6	E2	3.82(+5)	1.27	C−	1,*ls*
15.	5d–7d	^{2}D – ^{2}D										
			[155.0]	3103000	3748100	6	6	E2	1.04(+6)	0.333	C	1,*ls*
			[154.9]	3102200	3747600	4	4	E2	9.1(+5)	0.194	C	1,*ls*
			[155.1]	3103000	3747600	6	4	E2	3.9(+5)	0.083	D	1,*ls*
			[154.8]	3102200	3748100	4	6	E2	2.6(+5)	0.083	D	1,*ls*
16.	5f–6f	^{2}F° – ^{2}F°										
			[257.2]	3135700	3524500	8	8	E2	1.1(+6)	6.1	C	1,*ls*
			[257.0]	3135400	3524500	6	6	E2	1.10(+6)	4.41	C	1,*ls*
			[257.2]	3135700	3524500	8	6	E2	1.8(+5)	0.73	C−	1,*ls*
			[257.0]	3135400	3524500	6	8	E2	1.4(+5)	0.73	C−	1,*ls*
17.	5f–7f	^{2}F° – ^{2}F°										
			[160.5]	3135700	3758700	8	8	E2	7.0(+5)	0.357	C	1,*ls*
			[160.4]	3135400	3758800	6	6	E2	6.8(+5)	0.257	C	1,*ls*
			[160.5]	3135700	3758800	8	6	E2	1.1(+5)	0.043	D	1,*ls*
			[160.4]	3135400	3758700	6	8	E2	8.5(+4)	0.043	D	1,*ls*
18.	6d–7d	^{2}D – ^{2}D										
			[413.2]	3506100	3748100	6	6	E2	4.90(+5)	21.1	C	1,*ls*
			[413.4]	3505700	3747600	4	4	E2	4.28(+5)	12.3	C	1,*ls*
			[414.1]	3506100	3747600	6	4	E2	1.8(+5)	5.3	C−	1,*ls*
			[412.5]	3505700	3748100	4	6	E2	1.2(+5)	5.3	C−	1,*ls*

Co XVII: Forbidden transitions — Continued

No.	Transition Array	Multiplet	λ (Å)	E_i (cm^{-1})	E_k (cm^{-1})	g_i	g_k	Type of transition	A_{ki} (s^{-1})	S (at. u.)	Accuracy	Source
19.	$6f$–$7f$	$^2F^\circ$ – $^2F^\circ$										
			[427.0]	3524500	3758700	8	8	E2	4.07(+5)	27.5	C	1,ls
			[426.8]	3524500	3758800	6	6	E2	3.91(+5)	19.8	C	1,ls
			[426.8]	3524500	3758800	8	6	E2	6.5(+4)	3.30	C–	1,ls
			[427.0]	3524500	3758700	6	8	E2	4.88(+4)	3.30	C–	1,ls
20.	$3s$–$3d$	^2S – ^2D										
			[138.14]	0	723900	2	6	E2	7.6(+5)	0.136	C	1,ls
			[138.86]	0	720170	2	4	E2	7.4(+5)	0.091	D	1,ls
21.	$3s$–$4d$	^2S – ^2D										
			[42.457]	0	2355300	2	6	E2	2.05(+8)	0.101	C	1,ls
			[42.486]	0	2353700	2	4	E2	2.1(+8)	0.068	D	1,ls
22.	$3s$–$5d$	^2S – ^2D										
			[32.227]	0	3103000	2	6	E2	1.0(+8)	0.013	D	1,ls
23.	$3p$–$4f$	$^2P^\circ$ – $^2F^\circ$										
			[47.612]	319960	2420300	4	8	E2	3.61(+8)	0.421	C	1,ls
			[47.057]	294550	2419700	2	6	E2	2.97(+8)	0.245	C	1,ls
			[47.626]	319960	2419700	4	6	E2	8.0(+7)	0.070	D	1,ls
24.	$3p$–$5f$	$^2P^\circ$ – $^2F^\circ$										
			[35.515]	319960	3135700	4	8	E2	5.9(+7)	0.016	D	1,ls
25.	$3d$–$4s$	^2D – ^2S										
			[73.768]	723900	2079500	6	2	E2	1.9(+7)	0.050	D	1,ls
			[73.567]	720170	2079500	4	2	E2	1.3(+7)	0.033	D	1,ls
26.	$4s$–$4d$	^2S – ^2D										
			[362.6]	2079500	2355300	2	6	E2	9.4(+4)	2.11	C	1,ls
			[364.7]	2079500	2353700	2	4	E2	9.1(+4)	1.40	C	1,ls
27.	$4s$–$5d$	^2S – ^2D										
			[97.704]	2079500	3103000	2	6	E2	2.2(+7)	0.69	C	1,ls
			[97.780]	2079500	3102200	2	4	E2	2.16(+7)	0.460	C	1,ls
28.	$4s$–$6d$	^2S – ^2D										
			[70.097]	2079500	3506100	2	6	E2	1.5(+7)	0.091	D	1,ls
			[70.116]	2079500	3505700	2	4	E2	1.5(+7)	0.060	D	1,ls
29.	$4s$–$7d$	^2S – ^2D										
			[59.930]	2079500	3748100	2	6	E2	9.8(+6)	0.027	D	1,ls
			[59.948]	2079500	3747600	2	4	E2	9.8(+6)	0.018	D	1,ls
30.	$4p$–$4f$	$^2P^\circ$ – $^2F^\circ$										
			[467.9]	2206600	2420300	4	8	E2	2.12(+4)	2.26	C	1,ls
			[448.0]	2196500	2419700	2	6	E2	2.05(+4)	1.32	C	1,ls
			[469.3]	2206600	2419700	4	6	E2	4640	0.377	C–	1,ls

Co XVII: Forbidden transitions — Continued

No.	Transition Array	Multiplet	λ (Å)	E_i (cm^{-1})	E_k (cm^{-1})	g_i	g_k	Type of transition	A_{ki} (s^{-1})	S (at. u.)	Accuracy	Source
31.	4p–5f	$^2P°$ – $^2F°$										
			[107.6]	2206600	3135700	4	8	E2	4.18(+7)	2.87	C	1,ls
			[106.5]	2196500	3135400	2	6	E2	3.41(+7)	1.67	C	1,ls
			[107.7]	2206600	3135400	4	6	E2	9.2(+6)	0.478	C–	1,ls
32.	4p–6f	$^2P°$ – $^2F°$										
			[75.878]	2206600	3524500	4	8	E2	1.72(+7)	0.206	C	1,ls
			[75.301]	2196500	3524500	2	6	E2	1.39(+7)	0.120	C	1,ls
			[75.878]	2206600	3524500	4	6	E2	3.8(+6)	0.034	D	1,ls
33.	4p–7f	$^2P°$ – $^2F°$										
			[64.429]	2206600	3758700	4	8	E2	7.9(+6)	0.042	D	1,ls
			[64.008]	2196500	3758800	2	6	E2	6.3(+6)	0.024	D	1,ls
34.	4d–5s	^2D – ^2S										
			[163.3]	2355300	2967700	6	2	E2	5.4(+6)	0.74	C	1,ls
			[162.9]	2353700	2967700	4	2	E2	3.63(+6)	0.496	C	1,ls
35.	4f–5p	$^2F°$ – $^2P°$										
			[163.8]	2420300	3030800	8	4	E2	1.32(+6)	0.370	C	1,ls
			[164.9]	2419700	3026100	6	2	E2	1.49(+6)	0.216	C	1,ls
			[163.6]	2419700	3030800	6	4	E2	2.2(+5)	0.062	D	1,ls
36.	4f–6p	$^2F°$ – $^2P°$										
			[95.813]	2420300	3464000	8	4	E2	8.3(+5)	0.016	D	l,ls
37.	5s–5d	^2S – ^2D										
			[739.1]	2967700	3103000	2	6	E2	1.90(+4)	15.0	C	1,ls
			[743.5]	2967700	3102200	2	4	E2	1.85(+4)	10.0	C	1,ls
38.	5s–6d	^2S – ^2D										
			[185.7]	2967700	3506100	2	6	E2	3.97(+6)	3.13	C	1,ls
			[185.9]	2967700	3505700	2	4	E2	3.93(+6)	2.08	C	1,ls
39.	5s–7d	^2S – ^2D										
			[128.1]	2967700	3748100	2	6	E2	3.25(+6)	0.400	C	1,ls
			[128.2]	2967700	3747600	2	4	E2	3.24(+6)	0.267	C	1,ls
40.	5p–5f	$^2P°$ – $^2F°$										
			[953.3]	3030800	3135700	4	8	E2	5800	21.6	C	1,ls
			[914.9]	3026100	3135400	2	6	E2	5500	12.6	C	1,ls
			[956.0]	3030800	3135400	4	6	E2	1260	3.60	C–	1,ls
41.	5p–6f	$^2P°$ – $^2F°$										
			[202.6]	3030800	3524500	4	8	E2	7.6(+6)	12.3	C	1,ls
			[200.6]	3026100	3524500	2	6	E2	6.2(+6)	7.2	C	1,ls
			[202.6]	3030800	3524500	4	6	E2	1.69(+6)	2.06	C–	1,ls

Co XVII: Forbidden transitions — Continued

No.	Transition Array	Multiplet	λ (Å)	E_i (cm^{-1})	E_k (cm^{-1})	g_i	g_k	Type of transition	A_{ki} (s^{-1})	S (at. u.)	Accu-racy	Source
42.	5p–7f	^{2}P° – ^{2}F°										
			[137.4]	3030800	3758700	4	8	E2	4.55(+6)	1.06	C	1,ls
			[136.5]	3026100	3758800	2	6	E2	3.7(+6)	0.62	C	1,ls
			[137.4]	3030800	3758800	4	6	E2	1.01(+6)	0.176	C−	1,ls
43.	5f–6p	^{2}F° – ^{2}P°										
			[304.6]	3135700	3464000	8	4	E2	6.3(+5)	3.91	C	1,ls
			[305.0]	3135400	3463300	6	2	E2	7.3(+5)	2.28	C	1,ls
			[304.3]	3135400	3464000	6	4	E2	1.0(+5)	0.65	C−	1,ls
44.	6p–6f	^{2}P° – ^{2}F°										
			[1650]	3464000	3524500	4	8	E2	1960	114	C−	1,ls
			[1630]	3463300	3524500	2	6	E2	1600	66	C−	1,ls
			[1650]	3464000	3524500	4	6	E2	430	19	D+	1,ls
45.	6p–7f	^{2}P° – ^{2}F°										
			[339.3]	3464000	3758700	4	8	E2	1.92(+6)	41.1	C	1,ls
			[338.4]	3463300	3758800	2	6	E2	1.51(+6)	24.0	C	1,ls
			[339.2]	3464000	3758800	4	6	E2	4.2(+5)	6.8	C−	1,ls

[a]The number in parentheses following the tabulated value indicates the power of ten by which this value has to be multiplied.

Co XVIII

Ne Isoelectronic Sequence

Ground State: $1s^2 2s^2 2p^6\ ^1S_0$

Ionization Energy: 1397.2 eV = 11269000 cm^{-1}

Allowed Transitions

List of tabulated lines

Wavelength (Å)	No.	Wavelength (Å)	No.	Wavelength (Å)	No.	Wavelength (Å)	No.
10.030	8	13.634	13	87	5	271	24,25
10.975	18	13.868	12	88	2	272	27
11.108	17	14.041	11	90	1	326	20
11.15	16	15.169	10	103	3	362	26
11.321	15	15.437	9	241	23	382	19
11.486	14	45.35	31	251	23	711	21
12.606	7	85	4	256	23		

For resonance transitions to $J = 1$ levels of the $2p^5 3d$ configuration, we quote A-values which were calculated by Vainshtein and Safronova[1] using a charge-expansion perturbation theory approach with allowance for mixing of the $2p^5 3s$, $2p^5 3d$, and $2s 2p^6 3p$ configurations. Their results for Fe XVII are in rather good agreement with those of Shorer,[2] who used the relativistic random phase approximation (RRPA) with allowance for mixing between configurations of type $2p^5 ns$ and $2p^5 nd$, as well as correlation effects due to configurations having a vacancy in the $1s$ or $2s$ subshell. The results of Ref. 1 for resonance transitions to levels of the $2p^5 3s$ configuration in Fe XVII differed by about a factor of two from those of Ref. 2, so for the $2p^6 - 2p^5 3s$ transitions we interpolated f-values from the data of Shorer for Fe XVII and the results of Loulergue and Nussbaumer[3] for Ni XIX. The latter were calculated by a scaled Thomas-Fermi approach with a basis that consisted of the following configurations: $2s^2 2p^6$, $2s^2 2p^5 nl$, and $2s 2p^6 nl$ (for $n = 3$: $l = s, p, d$; for $n = 4$: $l = s, p, d, f$).

A-values quoted here for a number of transitions involving an electron jump of the type $2s - 2p$, $3s - 3p$, or $3p - 3d$ were taken from the work of Pokleba and Safronova,[4] who used wavefunctions calculated by a charge-expansion perturbation theory approach with allowance for mixing of configurations in which a single $2s$ or $2p$ electron is excited to an $n = 3$ orbital but with no inclusion of configurations in which an electron occupies the $n = 4$ shell. Transitions involving levels of the $2p^5 3p$ and $2p^5 3d$ configurations which are indicated by Jupen and Litzen[5] to be of low to moderate purity in LS coupling in Fe XVII are excluded here, as are very weak lines. The pattern of levels within the $2s 2p^6 3d$ configuration in the isoelectronic ions Fe XVII and Ni XIX resulting from the calculations of Loulergue and Nussbaumer is entirely different from that determined by Vainshtein and Safronova, whose energy levels were apparently used by Pokleba and Safronova in their transition probability calculations. We have thus excluded transitions out of these levels from our tabulation.

Oscillator strengths for a number of additional lines were interpolated from the data of Loulergue and Nussbaumer.

References

[1] L. A. Vainshtein and U. I. Safronova, *Spektroskopicheskie Konstanty Atomov*, 5–122 (Ed. V. B. Belyanin, Akad. Nauk SSSR, Ot. Ob. Fiz. Astron., Nauch. Sov. Spektrosk., Moscow, 1977).
[2] P. Shorer, Phys. Rev. A **20**, 642 (1979).
[3] M. Loulergue and H. Nussbaumer, Astron. Astrophys. **45**, 125 (1975).
[4] A. K. Pokleba and U. I. Safronova, Preprint No. 11, Akad. Nauk SSSR, Ot. Ob. Fiz. Astron., Inst. Spektrosk. (Moscow, 1981).
[5] C. Jupen and U. Litzen, Phys. Scr. **30**, 112 (1984).

Co XVIII: Allowed transitions

No.	Transition Array	Multiplet	λ (Å)	E_i (cm^{-1})	E_k (cm^{-1})	g_i	g_k	A_{ki} (10^8 s^{-1})	f_{ik}	S (at. u.)	log gf	Accuracy	Source
1.	$2s^2 2p^5(^2P^\circ_{3/2})3s -$ $2s 2p^6 3s$	$(^3/_2,^1/_2)^\circ - {}^3S$											
			[90]			5	3	820	0.060	0.089	-0.52	E	4
2.		$(^3/_2,^1/_2)^\circ - {}^1S$											
			[88]			3	1	650	0.025	0.022	-1.12	E	*interp.*
3.	$2s^2 2p^5(^2P^\circ_{1/2})3s -$ $2s 2p^6 3s$	$(^1/_2,^1/_2)^\circ - {}^3S$											
			[103]			3	3	190	0.030	0.031	-1.04	E	4
4.	$2s^2 2p^5 3p -$ $2s 2p^6 3p$	${}^3S - {}^3P^\circ$											
			[85]			3	3	200	0.022	0.018	-1.19	E	4
			[85]			3	1	630	0.023	0.019	-1.17	D	4
5.		${}^3D - {}^3P^\circ$											
			[87]			7	5	720	0.058	0.12	-0.39	D	4
6.	$2s^2 2p^5 4p -$ $2s 2p^6 4p$	${}^3S - {}^3P^\circ$											
						3	3		0.015		-1.35	E	*interp.*
						3	1		0.024		-1.14	E	*interp.*

Co XVIII: Allowed transitions — Continued

No.	Transition Array	Multiplet	λ (Å)	E_i (cm^{-1})	E_k (cm^{-1})	g_i	g_k	A_{ki} (10^8 s^{-1})	f_{ik}	S (at. u.)	log gf	Accuracy	Source
7.	$2s^2 2p^6 - 2s\,2p^6 3p$	$^1S - {}^1P°$	12.606	0	7932700	1	3	4.1(+ 4)^a	0.29	0.012	−0.54	D	*interp.*
8.	$2s^2 2p^6 - 2s\,2p^6 4p$	$^1S - {}^1P°$	10.030	0	9970100	1	3	2.7(+4)	0.12	0.0040	−0.92	D	*interp.*
9.	$2p^6-$ $2p^5({}^2P_{3/2}^°)3s$	$^1S - ({}^3\!/_2,{}^1\!/_2)°$											
			15.437	0	6477900	1	3	1.11(+4)	0.119	0.0060	−0.92	C−	*interp.*
10.	$2p^6-$ $2p^5({}^2P_{1/2}^°)3s$	$^1S - ({}^1\!/_2,{}^1\!/_2)°$											
			15.169	0	6592400	1	3	1.01(+4)	0.105	0.0052	−0.98	C−	*interp.*
11.	$2p^6-2p^5 3d$	$^1S - {}^3P°$	14.041	0	7122000	1	3	1300	0.012	5.3(−4)	−1.94	E	1
12.		$^1S - {}^3D°$	13.868	0	7210800	1	3	8.1(+4)	0.70	0.032	−0.15	D	1
13.		$^1S - {}^1P°$	13.634	0	7334600	1	3	2.87(+5)	2.40	0.108	0.380	C−	1
14.	$2p^6-$ $2p^5({}^2P_{3/2}^°)4s$	$^1S - ({}^3\!/_2,{}^1\!/_2)°$											
			11.486	0	8706300	1	3	4200	0.025	9.5(−4)	−1.60	D	*interp.*
15.	$2p^6-$ $2p^5({}^2P_{1/2}^°)4s$	$^1S - ({}^1\!/_2,{}^1\!/_2)°$											
			11.321	0	8833100	1	3	3800	0.022	8.2(−4)	−1.66	D	*interp.*
16.	$2p^6-2p^5 4d$	$^1S - {}^3P°$	[11.15]			1	3	610	0.0034	1.2(−4)	−2.47	E	*interp.*
17.		$^1S - {}^3D°$	11.108	0	9002500	1	3	7.6(+4)	0.42	0.015	−0.38	D−	*interp.*
18.		$^1S - {}^1P°$	10.975	0	9111600	1	3	9.4(+4)	0.51	0.018	−0.29	D−	*interp.*
19.	$2p^5({}^2P_{3/2}^°)3s-$ $2p^5 3p$	$({}^3\!/_2,{}^1\!/_2)° - {}^3S$	[382]			5	3	37	0.049	0.31	−0.61	D	4
20.		$({}^3\!/_2,{}^1\!/_2)° - {}^3D$	[326]			5	7	64	0.14	0.77	−0.15	D	4
21.	$2p^5({}^2P_{1/2}^°)3s-$ $2p^5 3p$	$({}^1\!/_2,{}^1\!/_2)° - {}^3S$	[711]			1	3	0.086	0.0020	0.0046	−2.71	E	4

Co XVIII: Allowed transitions — Continued

No.	Transition Array	Multiplet	λ (Å)	E_i (cm⁻¹)	E_k (cm⁻¹)	g_i	g_k	A_{ki} (10⁸ s⁻¹)	f_{ik}	S (at. u.)	log gf	Accuracy	Source
22.	$2s\,2p^6 3s -$ $2s\,2p^6 4p$	³S – ³P°											
						3	5		0.15		−0.35	D	interp.
						3	3		0.083		−0.60	D	interp.
						3	1		0.032		−1.02	D−	interp.
23.	$2p^5 3p - 2p^5 3d$	³S – ³P°	*246*			3	9	73	0.20	0.48	−0.23	E	4
			[241]			3	5	51	0.074	0.18	−0.65	E	4
			[251]			3	3	90	0.085	0.21	−0.59	D	4
			[256]			3	1	110	0.036	0.091	−0.97	D	4
24.		³D – ³P°											
			[271]			7	5	4.9	0.0039	0.024	−1.57	E	4
25.		³D – ³F°											
			[271]			7	9	100	0.14	0.88	−0.00	D	4
26.		³P – ³P°											
			[362]			1	3	1.4	0.0083	0.0098	−2.08	E	4
27.		³P – ³D°											
			[272]			1	3	50	0.17	0.15	−0.78	D	4
28.	$2p^5 3p -$ $2p^5(^2P^°_{3/2})4s$	³S – (³/₂,¹/₂)°											
						3	5		0.051		−0.82	D	interp.
29.		³D – (³/₂,¹/₂)°											
						7	5		0.060		−0.38	D	interp.
30.	$2p^5 3p - 2p^5 4d$	³S – ³P°											
						3	5		0.12		−0.44	E	interp.
						3	3		0.16		−0.32	D−	interp.
31.		³D – ³F°											
			[45.35]			7	9	7600	0.30	0.31	0.32	D−	interp.
32.		³P – ³D°											
						1	3		0.30		−0.52	E	interp.
33.		¹S – ³D°											
						1	3		0.10		−1.00	E	interp.
34.		¹S – ¹P°				1	3		0.34		−0.47	D	interp.

ᵃThe number in parentheses following the tabulated value indicates the power of ten by which this value has to be multiplied.

Co XIX

F Isoelectronic Sequence

Ground State: $1s^2 2s^2 2p^5\ {}^2P^\circ_{3/2}$

Ionization Energy: 1504.6 eV = 12135000 cm^{-1}

Allowed Transitions

Oscillator strengths for lines of the multiplet $2s^2 2p^5\ {}^2P^\circ - 2s 2p^6\ {}^2S$ are the results of the Dirac-Fock calculations of Cheng *et al.*,[1] which included a perturbative treatment of the Breit interaction and the Lamb shift.

For a few lines of the arrays $2p^5-2p^4 3s$ and $2p^5-2p^4 3d$, we interpolated *f*-values from the Hartree-Fock-Relativistic (HFR) results of Fawcett[2] for the isoelectronic ions Fe XVIII and Ni XX.

References

[1]K. T. Cheng, Y.-K. Kim, and J. P. Desclaux, At. Data Nucl. Data Tables **24**, 111 (1979).

[2]B. C. Fawcett, At. Data Nucl. Data Tables **31**, 495 (1984).

Co XIX: Allowed transitions

No.	Transition Array	Multiplet	λ (Å)	E_i (cm^{-1})	E_k (cm^{-1})	g_i	g_k	A_{ki} (10^8 s^{-1})	f_{ik}	S (at. u.)	log gf	Accuracy	Source
1.	$2s^2 2p^5 - 2s 2p^6$	${}^2P^\circ - {}^2S$	*91.66*	*41000*	1132000	6	2	1340	0.0563	0.102	−0.471	C+	1
			88.35	0	1132000	4	2	1000	0.0586	0.0682	−0.630	C+	1
			99.02	122000	1132000	2	2	350	0.0514	0.0335	−0.988	C+	1
2.	$2p^5 - 2p^4({}^3P)3s$	${}^2P^\circ - {}^2P$											
			14.553	122000	6993000	2	2	1.8(+4)^a	0.056	0.0054	−0.95	D−	*interp.*
			14.300	0	6993000	4	2	1.6(+4)	0.024	0.0045	−1.02	E	*interp.*
3.	$2p^5 - 2p^4({}^1D)3s$	${}^2P^\circ - {}^2D$											
			14.178	0	7053200	4	6	1.3(+4)	0.061	0.011	−0.61	D	*interp.*
			14.418	122000	7058000	2	4	1.6(+4)	0.099	0.0094	−0.70	D	*interp.*
4.	$2p^5 - 2p^4({}^1S)3s$	${}^2P^\circ - {}^2S$											
			14.074	122000	7227000	2	2	1.3(+4)	0.040	0.0037	−1.10	D−	*interp.*
5.	$2p^5 - 2p^4({}^1D)3d$	${}^2P^\circ - {}^2S$	*13.05*	*41000*	7703600	6	2	2.2(+5)	0.19	0.048	0.05	E	*interp.*
			12.981	0	7703600	4	2	1.9(+5)	0.24	0.041	−0.02	D	*interp.*
			13.188	122000	7703600	2	2	3.0(+4)	0.079	0.0069	−0.80	E	*interp.*

^{a}The number in parentheses following the tabulated value indicates the power of ten by which this value has to be multiplied.

Co XIX

Forbidden Transitions

Line strengths for the magnetic dipole and electric quadrupole contributions to the transition between the two levels of the $2p^5$ configuration are the results of the Dirac-Fock calculations of Cheng *et al.*[1] These relativistic calculations included a perturbative treatment of the Breit interaction and the Lamb shift. The strength of the electric quadrupole transition as defined in Ref. 1 was multiplied by the factor $^2/_3$ which is needed to bring this value into conformance with the definition of quadrupole strengths used in the NBS tables.

Reference

[1]K. T. Cheng, Y.-K. Kim, and J. P. Desclaux, At. Data Nucl. Data Tables **24**, 111 (1979).

Co XIX:　Forbidden transitions

No.	Transition Array	Multiplet	λ (Å)	E_i (cm^{-1})	E_k (cm^{-1})	g_i	g_k	Type of transition	A_{ki} (s^{-1})	S (at. u.)	Accuracy	Source
1.	$2p^5$-$2p^5$	$^2P°$ – $^2P°$										
			[820]	0	122000	4	2	M1	3.25(+4)^a	1.33	C+	1
			"	"	"	4	2	E2	3.9	0.0017	D–	1

^{a}The number in parentheses following the tabulated value indicates the power of ten by which this value has to be multiplied.

Co XX

O Isoelectronic Sequence

Ground State:　$1s^2 2s^2 2p^4\ {}^3P_2$

Ionization Energy:　$1603\ \text{eV} = 12930000\ \text{cm}^{-1}$

Allowed Transitions

List of tabulated lines

Wavelength (Å)	No.	Wavelength (Å)	No.	Wavelength (Å)	No.	Wavelength (Å)	No.
12.348	16	13.372	13	80.51	2	105.72	1
12.551	18	13.496	14	82.48	7	109.14	8
12.606	15	13.517	11	86.19	4	114.40	1
13.172	9	13.643	10	94.94	1	126.22	3
13.240	13	13.661	12	99.89	1	144.91	5
13.307	13	13.786	9	101.39	6		
13.317	11	74.10	2	101.88	1		
13.356	10	79.01	2	103.16	1		

The tabulated oscillator strengths for transitions of the arrays $2s^2 2p^4 - 2s 2p^5$ and $2s 2p^5 - 2p^6$ are the results of the multiconfiguration Dirac-Fock (MCDF) calculations of Cheng et al.[1] These relativistic calculations included a perturbative treatment of the Breit interaction and the Lamb shift. Allowance for configuration mixing was limited to the $n = 2$ complex. The results should be quite accurate, except in the case of weak lines. (The $2s^2 2p^4\ {}^1D_2 - 2s 2p^5\ {}^3P_1^\circ$ transition has been omitted from this tabulation, because its f-value as reported in Ref. 1 is extremely small, and thus very uncertain.)

Transition probabilities for lines of the $2s^2 2p^4 - 2s 2p^5$ array were calculated by Froese Fischer and Saha[2] using the multiconfiguration Hartree-Fock (MCHF) method with Breit-Pauli corrections. Their basis set included many configurations outside the $n = 2$ complex, but rela-

tivistic effects were not treated to the same degree as in Ref. 1. Line strengths derived from these two sources are in reasonably good agreement, particularly for the stronger transitions.

For a few lines of the arrays $2p^4 - 2p^3 3s$ and $2p^4 - 2p^3 3d$, we interpolated f-values from the Hartree-Fock-Relativistic (HFR) results of Fawcett[3] for the isoelectronic ions Fe XIX and Ni XXI.

References

[1] K. T. Cheng, Y.-K. Kim, and J. P. Desclaux, At. Data Nucl. Data Tables **24**, 111 (1979).
[2] C. Froese Fischer and H. P. Saha, J. Phys. B **17**, 943 (1984).
[3] B. C. Fawcett, At. Data Nucl. Data Tables **34**, 215 (1986).

Co XX: Allowed transitions

No.	Transition Array	Multiplet	λ (Å)	E_i (cm^{-1})	E_k (cm^{-1})	g_i	g_k	A_{ki} (10^8 s^{-1})	f_{ik}	S (at. u.)	log gf	Accuracy	Source
1.	$2s^2 2p^4 - 2s 2p^5$	${}^3P - {}^3P^\circ$	102.62	45130	1019570	9	9	580	0.092	0.28	−0.08	C	1
			101.88	0	981550	5	5	420	0.065	0.11	−0.49	C	1
			105.72	107410	1053300	3	3	134	0.0225	0.0235	−1.171	C	1
			94.94	0	1053300	5	3	364	0.0295	0.0461	−0.83	C	1
			99.89	107410	1108510	3	1	670	0.0333	0.0329	−1.000	C	1
			114.40	107410	981550	3	5	109	0.0358	0.0404	−0.97	C	1
			103.16	83930	1053300	1	3	170	0.083	0.028	−1.08	C	1
2.		${}^3P - {}^1P^\circ$											
			74.10	0	1349530	5	3	150	0.0074	0.0090	−1.43	E	1
			80.51	107410	1349530	3	3	13	0.0013	0.0010	−2.41	E	1
			79.01	83930	1349530	1	3	19	0.0054	0.0014	−2.27	E	1
3.		${}^1D - {}^3P^\circ$											
			126.22	189300	981550	5	5	27	0.0065	0.014	−1.49	E	1
4.		${}^1D - {}^1P^\circ$	86.19	189300	1349530	5	3	1590	0.106	0.150	−0.276	C	1
5.		${}^1S - {}^3P^\circ$											
			[144.91]	363240	1053300	1	3	9.4	0.0089	0.0042	−2.05	E	1
6.		${}^1S - {}^1P^\circ$	101.39	363240	1349530	1	3	110	0.052	0.017	−1.28	C	1
7.	$2s 2p^5 - 2p^6$	${}^3P^\circ - {}^1S$											
			82.48	1053300	2265780	3	1	160	0.0056	0.0046	−1.77	E	1
8.		${}^1P^\circ - {}^1S$	109.14	1349530	2265780	3	1	1730	0.103	0.111	−0.51	C	1
9.	$2p^4 - 2p^3({}^2S^\circ)3s$	${}^3P - {}^3S^\circ$											
			13.172	89430	7337600	5	3	2.9(+4)^a	0.048	0.011	−0.62	D−	interp.
			13.786	83930	7337600	1	3	6200	0.053	0.0024	−1.28	D	interp.
10.	$2p^4 - 2p^3({}^2D^\circ)3s$	${}^3P - {}^3D^\circ$											
			13.356	0	7487300	5	7	1.3(+4)	0.047	0.010	−0.63	D+	interp.
			13.643	107410	7442000	3	5	1.5(+4)	0.042	0.0057	−0.90	D−	interp.

Co xx: Allowed transitions — Continued

No.	Transition Array	Multiplet	λ (Å)	E_i (cm⁻¹)	E_k (cm⁻¹)	g_i	g_k	A_{ki} (10⁸ s⁻¹)	f_{ik}	S (at. u.)	log gf	Accuracy	Source
11.		³P – ¹D°											
			[13.317]	0	7509400	5	5	1700	0.0045	9.9(+4)	−1.65	E	interp.
			13.517	107410	7509400	3	5	3100	0.014	0.0019	−1.38	E	interp.
12.		¹D – ¹D°	13.661	189300	7509400	5	5	2.5(+4)	0.070	0.016	−0.46	E	interp.
13.	2p⁴ – 2p³(²P°)3s	³P – ³P°											
			13.372	107410	7585700	3	1	1.2(+4)	0.011	0.0015	−1.48	C−	interp.
			13.240	104710	7660300	3	5	7800	0.034	0.0044	−0.99	E	interp.
			[13.307]	83930	7598900	1	3	8700	0.069	0.0030	−1.16	E	interp.
14.		¹D – ³P°											
			13.496	189300	7598900	5	3	6700	0.011	0.0024	−1.26	E	interp.
15.	2p⁴ – 2p³(⁴S°)3d	³P – ³D°											
			12.606	0	7932700	5	7	9.3(+4)	0.31	0.064	0.19	E	interp.
						5	3		0.0049		−1.61	D−	interp.
16.	2p⁴ – 2p³(²D°)3d	³P – ³P°											
			12.348	0	8098500	5	7	2.4(+5)	0.76	0.15	0.58	E	interp.
17.		³P – ³P°											
						3	1		0.075		−0.65	D−	interp.
18.	2p⁴ – 2p³(²D°)3d	¹D – ¹F°	12.551	189300	8156800	5	7	1.2(+5)	0.39	0.081	0.29	E	interp.
19.	2p⁴ – 2p³(²P°)3d	³P – ³P°											
						3	1		0.14		−0.38	D−	interp.
20.		¹D – ¹P°				5	3		0.020		−1.00	D	interp.
21.		¹S – ¹P°				1	3		2.4		0.38	D	interp.

ᵃThe number in parenthesis following the tabulated value indicates the power of ten by which this value has to be multiplied.

Co xx

Forbidden Transitions

Line strengths tabulated for magnetic dipole and electric quadrupole transitions within the $2p^4$ configuration are the results of the multiconfiguration Dirac-Fock (MCDF) calculations of Cheng et al.[1] These relativistic calculations included a perturbative treatment of the Breit interaction and the Lamb shift. Allowance for configuration mixing was limited to the $n=2$ complex. Strengths of electric quadrupole transitions as defined in

Ref. 1 were multiplied by the factor $^2/_3$ which is needed to bring these values into conformance with the definition of quadrupole strengths used in the NBS tables.

Transition probabilities for these same lines were calculated by Froese Fischer and Saha[2] using the multiconfiguration Hartree-Fock (MCHF) method with Breit-Pauli corrections. Their basis included many configurations outside the $n = 2$ complex, but relativistic effects were not treated to the same degree as in Ref. 1. Line strengths derived from these data are in quite good agreement with the data of Cheng *et al*. For this ion of the oxygen isoelectronic sequence, correlation effects due to mixing with configurations outside the complex were found by Froese Fischer and Saha to be rather small, as shown by a comparison of the results of their calculations employing an extensive basis to those derived by the same technique but limited to configurations within the $n = 2$ complex.

The weakest lines are excluded from this compilation, as their transition probabilities are considered to be very uncertain.

References

[1]K. T. Cheng, Y.-K Kim, and J. P. Desclaux, At. Data Nucl. Data Tables **24**, 111 (1979).

[2]C. Froese Fischer and H. P. Saha, Phys. Rev. A **28**, 3169 (1983).

Co xx: Forbidden transitions

No.	Transition Array	Multiplet	λ (Å)	E_i (cm^{-1})	E_k (cm^{-1})	g_i	g_k	Type of transition	A_{ki} (s^{-1})	S (at. u.)	Accuracy	Source
1.	$2p^4$–$2p^4$	$^3P - ^3P$										
			[931.01]	0	107410	5	3	M1	2.46(+4)[a]	2.21	C	1
			"	"	"	5	3	E2	1.3	0.0016	E	1
			[4258]	83930	107410	1	3	M1	170	1.5	C	1
			[1192]	0	83930	5	1	E2	0.77	0.0011	E	1
2.		$^3P - ^1D$										
			[528.26]	0	189300	5	5	M1	2.7(+4)	0.75	D	1
			"	"	"	5	5	E2	9.8	0.0012	E	1
			[1221]	107410	189300	3	5	M1	830	0.28	D	1
3.		$^3P - ^1S$										
			[390.88]	107410	363240	3	1	M1	2.3(+5)	0.50	D	1
4.		$^1D - ^1S$	[574.91]	189300	363240	5	1	E2	67	0.0025	E	1

[a]The number in parentheses following the tabulated value indicates the power of ten by which this value has to be multiplied.

Co XXI

N Isoelectronic Sequence

Ground State: $1s^2 2s^2 2p^3 \, {}^4S^\circ_{3/2}$

Ionization Energy: 1735 eV = 13990000 cm^{-1}

Allowed Transitions

List of tabulated lines

Wavelength (Å)	No.	Wavelength (Å)	No.	Wavelength (Å)	No.	Wavelength (Å)	No.
75.87	4	89.25	7	106.76	6	132.24	5
75.90	13	90.31	2	107.57	15	133.06	16
77.291	13	91.76	14	110.08	11	133.64	10
77.69	8	93.00	12	110.71	1	136.53	5
78.71	13	96.36	12	113.70	1	145.35	5
78.90	3	101.30	11	113.76	16	153.38	5
84.03	13	101.93	6	120.91	10	157.40	16
85.40	8	103.93	14	124.67	15	160.51	9
85.741	13	104.14	6	125.15	1	164.61	5
86.66	12	104.27	12	130.02	10	192.13	9
88.77	8	106.23	14	131.09	16	227.27	9

The tabulated oscillator strengths for transitions of the arrays $2s^2 2p^3$–$2s 2p^4$ and $2s 2p^4$–$2p^5$ are the results of the multiconfiguration Dirac-Fock (MCDF) calculations of Cheng et al.[1] These relativistic calculations included a perturbative treatment of the Breit interaction and the Lamb shift. The results should be quite accurate, except in the case of weak lines. (A few very weak lines have been omitted from this tabulation.)

Reference

[1]K. T. Cheng, Y.-K. Kim, and J. P. Desclaux, At. Data Nucl. Data Tables **24**, 111 (1979).

Co XXI: Allowed transitions

No.	Transition Array	Multiplet	λ (Å)	E_i (cm^{-1})	E_k (cm^{-1})	g_i	g_k	A_{ki} (10^8 s^{-1})	f_{ik}	S (at. u.)	log gf	Accuracy	Source
1.	$2s^2 2p^3$–$2s 2p^4$	$^4S^\circ$ – ^4P	*118.59*	0	*843230*	4	12	175	0.111	0.173	−0.353	C	1
			125.15	0	799040	4	6	139	0.0491	0.081	−0.71	C	1
			113.70	0	879510	4	4	208	0.0404	0.060	−0.79	C	1
			110.71	0	903260	4	2	236	0.0217	0.0316	−1.061	C	1
2.		$^4S^\circ$ – ^2D											
			90.31	0	1107330	4	4	30	0.0037	0.0044	−1.83	E	1
3.		$^4S^\circ$ – ^2S											
			78.90	0	1267470	4	2	28	0.0013	0.0014	−2.28	E	1
4.		$^4S^\circ$ – ^2P											
			75.87	0	1318040	4	4	60	0.0052	0.0052	−1.68	E	1

Co XXI: Allowed transitions — Continued

No.	Transition Array	Multiplet	λ (Å)	E_i (cm^{-1})	E_k (cm^{-1})	g_i	g_k	A_{ki} (10^8 s^{-1})	f_{ik}	S (at. u.)	log gf	Accuracy	Source
5.		^{2}D° – ^{4}P											
			[164.61]	191530	799040	6	6	3.4	0.0014	0.0046	−2.08	E	1
			[136.53]	147080	879510	4	4	2.0	5.6(−4)^a	0.0010	−2.65	E	1
			[145.35]	191530	879510	6	4	0.71	1.5(−4)	4.3(−4)	−3.05	E	1
			[132.24]	147080	903260	4	2	5.1	6.7(−4)	0.0012	−2.57	E	1
			[153.38]	147080	799040	4	6	9.5	0.0050	0.010	−1.70	E	1
6.		^{2}D° – ^{2}D											
			106.76	191530	1128170	6	6	360	0.061	0.13	−0.44	C	1
			104.14	147080	1107330	4	4	470	0.076	0.10	−0.52	C	1
			[101.93]	147080	1128170	4	6	0.64	1.5(−4)	2.0(−4)	−3.22	E	1
7.		^{2}D° – ^{2}S											
			89.25	147080	1267470	4	2	500	0.030	0.035	−0.92	E	1
8.		^{2}D° – ^{2}P	*84.529*	*173750*	*1356780*	10	6	1000	0.065	0.18	−0.19	C	1
			88.77	191530	1318040	6	4	1100	0.086	0.15	−0.29	C	1
			77.69	147080	1434250	4	2	285	0.0129	0.0132	−1.287	C	1
			85.40	147080	1318040	4	4	137	0.0150	0.0169	−1.222	C	1
9.		^{2}P° – ^{4}P											
			[227.27]	359030	799040	4	6	0.28	3.3(−4)	9.9(−4)	−2.88	E	1
			[192.13]	359030	879510	4	4	2.3	0.0013	0.0033	−2.28	E	1
			[160.51]	280260	903260	2	2	3.6	0.0014	0.0015	−2.55	E	1
10.		^{2}P° – ^{2}D	*127.06*	*332770*	*1119830*	6	10	54	0.022	0.055	−0.88	C−	1
			130.02	359030	1128170	4	6	63	0.0241	0.0413	−1.016	C	1
			120.91	280260	1107330	2	4	30.3	0.0133	0.0106	−1.58	C	1
			[133.64]	359030	1107330	4	4	6.7	0.0018	0.0032	−2.14	D	1
11.		^{2}P° – ^{2}S	*106.99*	*332770*	*1267470*	6	2	360	0.021	0.044	−0.90	C−	1
			110.08	359030	1267470	4	2	22	0.0020	0.0029	−2.10	D	1
			101.30	280260	1267470	2	2	400	0.062	0.041	−0.91	C	1
12.		^{2}P° – ^{2}P	*97.655*	*332770*	*1356780*	6	6	464	0.066	0.128	−0.400	C−	1
			104.27	359030	1318040	4	4	98	0.0159	0.0218	−1.197	C	1
			86.66	280260	1434250	2	2	42	0.0047	0.0027	−2.03	D	1
			93.00	359030	1434250	4	2	1100	0.069	0.085	−0.56	C	1
			96.36	280260	1318040	2	4	104	0.0290	0.0184	−1.237	C	1
13.	$2s\,2p^4$–$2p^5$	^{4}P – ^{2}P°											
			78.71	799040	2069560	6	4	40	0.0025	0.0039	−1.82	E	1
			75.90	879510	2197080	4	2	3.9	1.7(−4)	1.7(−4)	−3.17	E	1
			84.03	879510	2069560	4	4	24	0.0025	0.0028	−2.00	E	1
			[77.291]	903260	2197080	2	2	12	0.0011	5.6(−4)	−2.66	E	1
			[85.741]	903260	2069560	2	4	7.7	0.0017	9.6(−4)	−2.47	E	1
14.		^{2}D – ^{2}P°	*100.78*	*1119830*	*2112070*	10	6	630	0.057	0.19	−0.24	C	1
			106.23	1128170	2069560	6	4	460	0.052	0.11	−0.51	C	1
			91.76	1107330	2197080	4	2	477	0.0301	0.0364	−0.92	C	1
			103.93	1107330	2069560	4	4	208	0.0337	0.0461	−0.87	C	1

Co xxi: Allowed transitions — Continued

No.	Transition Array	Multiplet	λ (Å)	E_i (cm^{-1})	E_k (cm^{-1})	g_i	g_k	A_{ki} (10^8 s^{-1})	f_{ik}	S (at. u.)	log gf	Accuracy	Source
15.		^{2}S – ^{2}P°	*118.40*	1267470	*2112070*	2	6	86	0.055	0.0425	−0.96	C−	1
			124.67	1267470	2069560	2	4	100	0.0467	0.0383	−1.030	C	1
			[107.57]	1267470	2197080	2	2	34	0.0059	0.0042	−1.93	D	1
16.		^{2}P – ^{2}P°	*132.40*	*1356780*	*2112070*	6	6	440	0.11	0.30	−0.16	C	1
			133.06	1318040	2069560	4	4	320	0.086	0.15	−0.46	C	1
			131.09	1434250	2197080	2	2	350	0.089	0.077	−0.75	C	1
			113.76	1318040	2197080	4	2	437	0.0424	0.064	−0.77	C	1
			[157.40]	1434250	2069560	2	4	16.6	0.0123	0.0127	−1.61	C	1

[a]The number in parentheses following the tabulated value indicates the power of ten by which this value has to be multiplied.

Co xxi

Forbidden Transitions

Line strengths tabulated for magnetic dipole and electric quadrupole transitions within the $2p^3$ configuration are the results of the multiconfiguration Dirac-Fock (MCDF) calculations of Cheng et al.[1] These relativistic calculations included a perturbative treatment of the Breit interaction and the Lamb shift. Allowance for configuration mixing was limited to the $n=2$ complex. Strengths of electric quadrupole transitions as defined in Ref. 1 were multiplied by the factor $^2/_3$ which is needed to bring these values into conformance with the definition of quadrupole strengths used in the NBS tables. The weakest lines are excluded from this compilation, as their strengths are considered to be very uncertain.

A-values for the M1 and E2 components of the single transition within the $2p^5$ configuration were obtained by applying Z-expansion formulas published by Oboladze and Safronova.[2] Their values for the magnetic dipole contribution to this line are in very good agreement with the results of the scaled Thomas-Fermi calculations of Bhatia et al.[3] and Bhatia[4] for nitrogenlike Ti and Mn, respectively. It is not clear whether Oboladze and Safronova incorporated configuration interaction into their calculations. Thus the A-value for the E2 contribution should be considered rather uncertain.

References

[1]K. T. Cheng, Y.-K. Kim, and J. P. Desclaux, At. Data Nucl. Data Tables **24**, 111 (1979).

[2]N. S. Oboladze and U. I. Safronova, Opt. Spectrosc. (USSR) **48**, 469 (1980).

[3]A. K. Bhatia, U. Feldman, and G. A. Doschek, J. Appl. Phys. **51**, 1464 (1980).

[4]A. K. Bhatia, J. Appl. Phys. **53**, 59 (1982).

Co XXI: Forbidden transitions

No.	Transition Array	Multiplet	λ (Å)	E_i (cm^{-1})	E_k (cm^{-1})	g_i	g_k	Type of transition	A_{ki} (s^{-1})	S (at. u.)	Accuracy	Source
1.	$2p^3$–$2p^3$	$^4S°$ – $^2D°$										
			[522.11]	0	191530	4	6	M1	2400	0.077	D−	1
			[679.90]	0	147080	4	4	M1	2.6(+4)^a	1.2	D	1
2.		$^4S°$ – $^2P°$										
			[278.53]	0	359030	4	4	M1	3.7(+4)	0.12	D	1
			[356.81]	0	280260	4	2	M1	5.0(+4)	0.17	D	1
3.		$^2D°$ – $^2D°$										
			[2249]	147080	191530	4	6	M1	670	1.70	C	1
			"	"	"	4	6	E2	0.0054	0.0011	E	1
4.		$^2D°$ – $^2P°$										
			[1127]	191530	280260	6	2	E2	0.83	0.0018	E	1
			[597.01]	191530	359030	6	4	M1	2.0(+4)	0.62	D	1
			"	"	"	6	4	E2	23	0.0042	E	1
			[750.86]	147080	280260	4	2	M1	7600	0.24	D	1
			"	"	"	4	2	E2	6.7	0.0019	E	1
			[471.81]	147080	359030	4	4	M1	7.7(+4)	1.2	D	1
5.		$^2P°$ – $^2P°$										
			[1270]	280260	359030	2	4	M1	3000	0.92	C−	1
6.	$2p^5$–$2p^5$	$^2P°$ – $^2P°$										
			[784.19]	2069560	2197080	4	2	M1	3.72(+4)	1.33	C+	2
			"	"	"	4	2	E2	4.3	0.0015	E	2

^{a}The number in parentheses following the tabulated value indicates the power of ten by which this value has to be multiplied.

Co XXII

C Isoelectronic Sequence

Ground State: $1s^2 2s^2 2p^2\ ^3P_0$

Ionization Energy: $1846\ \text{eV} = 14890000\ \text{cm}^{-1}$

Allowed Transitions

List of tabulated lines

Wavelength (Å)	No.	Wavelength (Å)	No.	Wavelength (Å)	No.	Wavelength (Å)	No.
75.125	16	98.07	17	116.22	3	143.87	22
78.98	6	100.14	3	116.97	19	146.40	22
82.058	6	105.69	20	118.31	19	153.05	24
82.09	16	106.23	15	119.55	28	170.09	27
83.702	21	107.49	10	119.92	2	170.18	7
85.43	4	107.58	3	132.46	14	171.49	13
89.173	5	108.16	20	132.63	19	171.79	22
90.211	18	108.84	17	134.13	2	180.37	7
92.61	4	109.14	17	134.57	2	181.17	7
93.00	18	110.14	3	135.42	8	185.03	24
93.02	11	111.47	3	136.49	25	196.59	26
93.12	5	112.54	9	136.56	19	225.39	1
96.88	4	113.24	17	136.75	2	239.05	12
96.93	17	113.37	3	139.52	8	252.40	1
97.16	17	113.93	19	143.25	2	252.70	26
97.764	23	115.35	19	143.76	2		

The tabulated oscillator strengths for transitions of the arrays $2s^2 2p^2 - 2s 2p^3$ and $2s 2p^3 - 2p^4$ are the results of the multiconfiguration Dirac-Fock (MCDF) calculations of Cheng et al.[1] These relativistic calculations included a perturbative treatment of the Breit interaction and the Lamb shift. Allowance for configuration mixing was limited to the $n=2$ complex. The results should be quite accurate, except in the case of weak lines. (A few very weak lines have been omitted from this tabulation.)

Transition probabilities for lines of the $2s^2 2p^2 - 2s 2p^3$ array were calculated by Froese Fischer and Saha[2] using the multiconfiguration Hartree-Fock (MCHF) method with Breit-Pauli corrections. Their basis included many configurations outside the $n=2$ complex, but relativistic effects were not treated to the same degree as in Ref. 1. Line strengths derived from these two sources are in reasonably good agreement, particularly for the stronger transitions.

References

[1] K. T. Cheng, Y.-K. Kim, and J. P. Desclaux, At. Data Nucl. Data Tables 24, 111 (1979).
[2] C. Froese Fischer and H. P. Saha, Phys. Scr. 32, 181 (1985).

Co XXII: Allowed transitions

No.	Transition Array	Multiplet	λ (Å)	E_i (cm^{-1})	E_k (cm^{-1})	g_i	g_k	A_{ki} (10^8 s^{-1})	f_{ik}	S (at. u.)	log gf	Accuracy	Source
1.	$2s^2 2p^2 - 2s 2p^3$	$^3P - {}^5S°$											
			[252.40]	138240	534430	5	5	0.48	4.6(−4)ᵃ	0.0019	−2.64	E	1
			[225.39]	90750	534430	3	5	0.56	7.1(−4)	0.0016	−2.67	E	1

Co XXII: Allowed transitions — Continued

No.	Transition Array	Multiplet	λ (Å)	E_i (cm^{-1})	E_k (cm^{-1})	g_i	g_k	A_{ki} (10^8 s^{-1})	f_{ik}	S (at. u.)	log gf	Accuracy	Source
2.		$^3P - {}^3D°$	*134.36*	*107050*	*851320*	9	15	95	0.043	0.17	−0.42	D	1
			136.75	138240	869520	5	7	70	0.0275	0.062	−0.86	C	1
			134.13	90750	836320	3	5	110	0.0496	0.066	−0.83	C	1
			119.92	0	833860	1	3	150	0.096	0.038	−1.02	C	1
			[143.25]	138240	836320	5	5	0.49	1.5(−4)	3.5(−4)	−3.12	E	1
			134.57	90750	833860	3	3	6.3	0.0017	0.0023	−2.29	D	1
			[143.76]	138240	833860	5	3	1.2	2.2(−4)	5.2(−4)	−2.96	E	1
3.		$^3P - {}^3P°$	*110.81*	*107050*	*1009480*	9	9	260	0.049	0.16	−0.36	D	1
			113.37	138240	1020310	5	5	249	0.0480	0.090	−0.62	C	1
			110.14	90750	998650	3	3	203	0.0369	0.0401	−0.96	C	1
			116.22	138240	998650	5	3	30	0.0037	0.0071	−1.73	D	1
			111.47	90750	987840	3	1	266	0.0165	0.0182	−1.305	C	1
			107.58	90750	1020310	3	5	2.3	6.7(−4)	7.1(−4)	−2.70	E	1
			100.14	0	998650	1	3	43.9	0.0198	0.0065	−1.70	C	1
4.		$^3P - {}^3S°$	*94.034*	*107050*	*1170490*	9	3	1100	0.0488	0.136	−0.357	C	1
			96.88	138240	1170490	5	3	700	0.059	0.094	−0.53	C	1
			92.61	90750	1170490	3	3	272	0.0350	0.0320	−0.98	C	1
			85.43	0	1170490	1	3	103	0.0339	0.0095	−1.470	C	1
5.		$^3P - {}^1D°$											
			93.12	138240	1212170	5	5	110	0.014	0.021	−1.15	E	1
			[89.173]	90750	1212170	3	5	5.5	0.0011	9.7(−4)	−2.48	E	1
6.		$^3P - {}^1P°$											
			[82.058]	138240	1356890	5	3	3.6	2.2(−4)	3.0(−4)	−2.96	E	1
			78.98	90750	1356890	3	3	62	0.0058	0.0045	−1.76	E	1
7.		$^1D - {}^3D°$											
			[170.18]	281890	869520	5	7	13	0.0079	0.022	−1.40	E	1
			[180.37]	281890	836320	5	5	0.39	1.9(−4)	5.6(−4)	−3.02	E	1
			[181.17]	281890	833860	5	3	2.4	7.2(−4)	0.0021	−2.44	E	1
8.		$^1D - {}^3P°$											
			[135.42]	281890	1020310	5	5	2.3	6.4(−4)	0.0014	−2.49	E	1
			[139.52]	281890	998650	5	3	2.9	5.0(−4)	0.0011	−2.60	E	1
9.		$^1D - {}^3S°$	[112.54]	281890	1170490	5	3	4.7	5.4(−4)	0.0010	−2.57	E	1
10.		$^1D - {}^1D°$	107.49	281890	1212170	5	5	500	0.087	0.15	−0.36	C	1
11.		$^1D - {}^1P°$	93.02	281890	1356890	5	3	770	0.060	0.092	−0.52	C	1
12.		$^1S - {}^3D°$											
			[239.05]	415540	833860	1	3	0.58	0.0015	0.0012	−2.82	E	1
13.		$^1S - {}^3P°$											
			[171.49]	415540	998650	1	3	2.0	0.0026	0.0015	−2.59	E	1
14.		$^1S - {}^3S°$	[132.46]	415540	1170490	1	3	8.5	0.0067	0.0029	−2.17	E	1

Co XXII: Allowed transitions — Continued

No.	Transition Array	Multiplet	λ (Å)	E_i (cm^{-1})	E_k (cm^{-1})	g_i	g_k	A_{ki} (10^8 s^{-1})	f_{ik}	S (at. u.)	log gf	Accuracy	Source
15.		^1S – $^1P°$	106.23	415540	1356890	1	3	197	0.100	0.0350	−1.000	C	1
16.	$2s\,2p^3$–$2p^4$	$^5S°$ – ^3P											
			82.09	534430	1752610	5	5	21	0.0021	0.0028	−1.98	E	1
			[75.125]	534430	1865550	5	3	3.3	1.7(−4)	2.1(−4)	−3.07	E	1
17.		$^3D°$ – ^3P	105.25	851320	1801470	15	9	494	0.0493	0.256	−0.131	C	1
			113.24	869520	1752610	7	5	321	0.0441	0.115	−0.51	C	1
			97.16	836320	1865550	5	3	238	0.0202	0.0323	−1.000	C	1
			98.07	833860	1853540	3	1	404	0.0194	0.0188	−1.235	C	1
			109.14	836320	1752610	5	5	162	0.0290	0.052	−0.84	C	1
			96.93	833860	1865550	3	3	167	0.0235	0.0225	−1.152	C	1
			108.84	833860	1752610	3	5	47.6	0.0141	0.0152	−1.374	C	1
18.		$^3D°$ – ^1D											
			93.00	869520	1944830	7	5	73	0.0068	0.015	−1.32	E	1
			[90.211]	836320	1944830	5	5	8.2	0.0010	0.0015	−2.30	E	1
19.		$^3P°$ – ^3P	126.26	1009480	1801470	9	9	150	0.035	0.13	−0.50	D	1
			[136.56]	1020310	1752610	5	5	37.9	0.0106	0.0238	−1.276	C	1
			[115.35]	998650	1865550	3	3	4.1	8.1(−4)	9.2(−4)	−2.61	E	1
			118.31	1020310	1865550	5	3	198	0.0249	0.0485	−0.90	C	1
			116.97	998650	1853540	3	1	216	0.0148	0.0171	−1.353	C	1
			132.63	998650	1752610	3	5	41.2	0.0181	0.0237	−1.265	C	1
			113.93	987840	1865550	1	3	57	0.0334	0.0125	−1.476	C	1
20.		$^3P°$ – ^1D											
			108.16	1020310	1944830	5	5	38	0.0067	0.012	−1.47	E	1
			[105.69]	998650	1944830	3	5	18	0.0049	0.0051	−1.83	E	1
21.		$^3P°$ – ^1S											
			[83.702]	998650	2193360	3	1	60	0.0021	0.0017	−2.20	E	1
22.		$^3S°$ – ^3P	158.48	1170490	1801470	3	9	109	0.123	0.193	−0.432	C	1
			171.79	1170490	1752610	3	5	69	0.051	0.087	−0.82	C	1
			143.87	1170490	1865550	3	3	160	0.051	0.072	−0.82	C	1
			146.40	1170490	1853540	3	1	221	0.0237	0.0343	−1.148	C	1
23.		$^3S°$ – ^1S	[97.764]	1170490	2193360	3	1	65	0.0031	0.0030	−2.03	E	1
24.		$^1D°$ – ^3P											
			[185.03]	1212170	1752610	5	5	9.4	0.0048	0.015	−1.62	E	1
			[153.05]	1212170	1865550	5	3	5.7	0.0012	0.0030	−2.22	E	1
25.		$^1D°$ – ^1D	136.49	1212170	1944830	5	5	380	0.106	0.238	−0.276	C	1
26.		$^1P°$ – ^3P											
			[252.70]	1356890	1752610	3	5	1.2	0.0019	0.0047	−2.24	E	1
			[196.59]	1356890	1865550	3	3	9.0	0.0052	0.010	−1.81	E	1
27.		$^1P°$ – ^1D	170.09	1356890	1944830	3	5	54	0.0389	0.065	−0.93	C	1
28.		$^1P°$ – ^1S	119.55	1356890	2193360	3	1	910	0.065	0.077	−0.71	C	1

[a]The number in parentheses following the tabulated value indicates the power of ten by which this value has to be multiplied.

Co XXII

Forbidden Transitions

Line strengths tabulated for magnetic dipole and electric quadrupole transitions within the $2p^2$ configuration are the results of the multiconfiguration Dirac-Fock (MCDF) calculations of Cheng et al.[1] These relativistic calculations included a perturbative treatment of the Breit interaction and the Lamb shift. Allowance for configuration mixing was limited to the $n=2$ complex. Strengths of electric quadrupole transitions as defined in Ref. 1 were multiplied by the factor $2/3$ which is needed to bring these values into conformance with the definition of quadrupole strengths used in the NBS tables. The weakest lines are excluded from this compilation, as their strengths are considered to be very uncertain.

Transition probabilities for these same lines were calculated by Froese Fischer and Saha[2] using the multiconfiguration Hartree-Fock (MCHF) method with Breit-Pauli corrections. Their basis included many configurations outside the $n=2$ complex, but relativistic effects were not treated to the same degree as in Ref. 1. Line strengths derived from these data are in good agreement with the data of Cheng et al.

References

[1]K. T. Cheng, Y.-K. Kim, and J. P. Desclaux, At. Data Nucl. Data Tables **24**, 111 (1979).
[2]C. Froese Fischer and H. P. Saha, Phys. Scr. **32**, 181 (1985).

Co XXII: Forbidden transitions

No.	Transition Array	Multiplet	λ (Å)	E_i (cm^{-1})	E_k (cm^{-1})	g_i	g_k	Type of transition	A_{ki} (s^{-1})	S (at. u.)	Accuracy	Source
1.	$2p^2$–$2p^2$	^3P – ^3P										
			[2105]	90750	138240	3	5	M1	1050	1.81	C	1
			"	"	"	3	5	E2	0.0089	0.0011	E	1
			[1102]	0	90750	1	3	M1	1.19(+4)[a]	1.77	C	1
2.		^3P – ^1D										
			[696.14]	138240	281890	5	5	M1	2.4(+4)	1.5	C	1
			"	"	"	5	5	E2	4.3	0.0021	E	1
			[523.18]	90750	281890	3	5	M1	2.6(+4)	0.68	D	1
3.		^3P– ^1S										
			[307.89]	90750	415540	3	1	M1	2.1(+5)	0.23	D	1
4.		^1D – ^1S	[748.22]	281890	415540	5	1	E2	16	0.0022	E	1

[a]The number in parentheses following the tabulated value indicates the power of ten by which this value has to be multiplied.

Co XXIII

B Isoelectronic Sequence

Ground State: $1s^2 2s^2 2p\ ^2P^\circ_{1/2}$

Ionization Energy: 1962 eV = 15820000 cm^{-1}

Allowed Transitions

List of tabulated lines

Wavelength (Å)	No.	Wavelength (Å)	No.	Wavelength (Å)	No.	Wavelength (Å)	No.
80.223	7	108.03	3	142.05	9	229.61	11
80.611	7	109.70	4	143.30	9	229.65	15
84.313	7	110.23	5	143.89	16	231.71	1
93.90	3	110.71	3	146.86	13	237.30	12
94.808	6	113.17	10	147.09	2	237.82	1
95.16	4	118.68	13	149.88	9	277.77	1
97.672	6	119.12	5	154.04	2	342.15	1
101.32	6	126.82	2	160.97	16	347.35	14
103.80	6	128.37	5	164.70	13	365.18	11
104.45	10	130.06	13	171.50	12		
107.91	10	130.90	3	181.72	8		
107.93	6	136.12	9	218.24	12		

The tabulated oscillator strengths for transitions of the arrays $2s^2 2p$–$2s\,2p^2$ and $2s\,2p^2$–$2p^3$ are the results of the multiconfiguration Dirac-Fock (MCDF) calculations of Cheng et al.[1] These relativistic calculations included a perturbative treatment of the Breit interaction and the Lamb shift. The results should be quite accurate, except in the case of weak lines. (A few very weak lines have been omitted from this tabulation.)

According to several sources (see, e.g., introduction to Fe XXII), the lower of the two levels $2s\,2p^2\ ^2P_{1/2}$ and $^2S_{1/2}$

is mostly of ^2P character, having "crossed" the $^2S_{1/2}$ level at about V XIX or Cr XX. We have thus labeled these two levels accordingly, in contrast to their labeling by Cheng. et al., which is consistent with their ordering at the neutral end of the B sequence.

Reference

[1]K. T. Cheng, Y.-K. Kim, and J. P. Desclaux, At. Data Nucl. Data Tables **24**, 111 (1979).

Co XXIII: Allowed transitions

No.	Transition Array	Multiplet	λ (Å)	E_i (cm^{-1})	E_k (cm^{-1})	g_i	g_k	A_{ki} (10^8 s^{-1})	f_{ik}	S (at. u.)	log gf	Accuracy	Source
1.	$2s^2 2p$–$2s\,2p^2$	$^2P^\circ$ – ^4P											
			[237.82]	139310	559800	4	6	1.0	0.0013	0.0041	−2.28	E	1
			[277.77]	139310	499320	4	4	0.10	1.2(−4)^a	4.4(−4)	−3.32	E	1
			[231.71]	0	431580	2	2	1.2	9.8(−4)	0.0015	−2.71	E	1
			[342.15]	139310	431580	4	2	0.17	1.5(−4)	6.8(−4)	−3.22	E	1
2.		$^2P^\circ$ – ^2D	140.05	92870	806890	6	10	85	0.0416	0.115	−0.60	C−	1
			147.09	139310	819160	4	6	67	0.0327	0.063	−0.88	C	1
			126.82	0	788490	2	4	130	0.062	0.052	−0.91	C	1
			[154.04]	139310	788490	4	4	0.12	4.3(−5)	8.7(−5)	−3.76	E	1

Co XXIII: Allowed transitions — Continued

No.	Transition Array	Multiplet	λ (Å)	E_i (cm^{-1})	E_k (cm^{-1})	g_i	g_k	A_{ki} (10^8 s^{-1})	f_{ik}	S (at. u.)	log gf	Accuracy	Source
3.		^{2}P° – ^{2}P	108.91	92870	1011040	6	6	500	0.088	0.19	−0.28	C−	1
			108.03	139310	1064940	4	4	490	0.086	0.12	−0.46	C	1
			110.71	0	903250	2	2	430	0.079	0.058	−0.80	C	1
			[130.90]	139310	903250	4	2	0.77	9.9(−5)	1.7(−4)	−3.40	E	1
			93.90	0	1064940	2	4	65	0.0171	0.0106	−1.466	C	1
4.		^{2}P° – ^{2}S	104.38	92870	1050890	6	2	460	0.025	0.052	−0.82	C−	1
			109.70	139310	1050890	4	2	384	0.0346	0.050	−0.86	C	1
			95.16	0	1050890	2	2	24	0.0032	0.0020	−2.19	D	1
5.	2s2p^2–2p^3	^{4}P – ^{4}S°	121.88	518270	1338780	12	4	476	0.0353	0.170	−0.373	C	1
			128.37	559800	1338780	6	4	206	0.0340	0.086	−0.69	C	1
			119.12	499320	1338780	4	4	159	0.0338	0.053	−0.87	C	1
			110.23	431580	1338780	2	4	116	0.0422	0.0306	−1.074	C	1
6.		^{4}P – ^{2}D°											
			103.80	559800	1523160	6	6	27	0.0044	0.0090	−1.58	E	1
			[101.32]	499320	1486340	4	4	34	0.0052	0.0069	−1.68	E	1
			[107.93]	559800	1486340	6	4	4.0	4.6(−4)	9.8(−4)	−2.56	E	1
			[97.672]	499320	1523160	4	6	0.65	1.4(−4)	1.8(−4)	−3.25	E	1
			[94.808]	431580	1486340	2	4	0.78	2.1(−4)	1.3(−4)	−3.38	E	1
7.		^{4}P – ^{2}P°											
			[84.313]	559800	1745850	6	4	1.7	1.2(−4)	2.0(−4)	−3.14	E	1
			[80.223]	499320	1745850	4	4	3.5	3.4(−4)	3.6(−4)	−2.87	E	1
			[80.611]	431580	1672110	2	2	2.8	2.7(−4)	1.4(−4)	−3.27	E	1
8.		^{2}D – ^{4}S°											
			[181.72]	788490	1338780	4	4	2.2	0.0011	0.0026	−2.36	E	1
9.		^{2}D – ^{2}D°	142.54	806890	1508430	10	10	161	0.0490	0.230	−0.310	C	1
			142.05	819160	1523160	6	6	136	0.0412	0.116	−0.61	C	1
			143.30	788490	1486340	4	4	78	0.0240	0.0453	−1.018	C	1
			149.88	819160	1486340	6	4	55	0.0124	0.0367	−1.128	C	1
			136.12	788490	1523160	4	6	42.7	0.0178	0.0319	−1.148	C	1
10.		^{2}D – ^{2}P°	109.36	806890	1721270	10	6	250	0.027	0.098	−0.57	C−	1
			107.91	819160	1745850	6	4	152	0.0177	0.0377	−0.97	C	1
			113.17	788490	1672110	4	2	329	0.0316	0.0471	−0.90	C	1
			104.45	788490	1745850	4	4	56	0.0091	0.013	−1.44	D	1
11.		^{2}P – ^{4}S°											
			[365.18]	1064940	1338780	4	4	0.37	7.4(−4)	0.0036	−2.53	E	1
			[229.61]	903250	1338780	2	4	2.0	0.0032	0.0048	−2.19	E	1
12.		^{2}P – ^{2}D°	201.05	1011040	1508430	6	10	47	0.048	0.19	−0.54	D	1
			[218.24]	1064940	1523160	4	6	35.6	0.0381	0.109	−0.82	C	1
			[171.50]	903250	1486340	2	4	77	0.068	0.077	−0.87	C	1
			[237.30]	1064940	1486340	4	4	0.45	3.8(−4)	0.0012	−2.82	E	1

Co XXIII: Allowed transitions — Continued

No.	Transition Array	Multiplet	λ (Å)	E_i (cm^{-1})	E_k (cm^{-1})	g_i	g_k	A_{ki} (10^8 s^{-1})	f_{ik}	S (at. u.)	log gf	Accuracy	Source
13.		^{2}P – ^{2}P°	140.80	1011040	1721270	6	6	210	0.061	0.17	−0.44	C−	1
			146.86	1064940	1745850	4	4	230	0.074	0.14	−0.53	C	1
			[130.06]	903250	1672110	2	2	34	0.0087	0.0075	−1.76	D	1
			[164.70]	1064940	1672110	4	2	22	0.0044	0.0095	−1.75	D	1
			118.68	903250	1745850	2	4	36.9	0.0156	0.0122	−1.51	C	1
14.		^{2}S – ^{4}S°	[347.35]	1050890	1338780	2	4	0.15	5.5(−4)	0.0013	−2.96	E	1
15.		^{2}S – ^{2}D°											
			[229.65]	1050890	1486340	2	4	7.3	0.0116	0.0175	−1.63	C	1
16.		^{2}S – ^{2}P°	149.17	1050890	1721270	2	6	64	0.064	0.063	−0.89	C	1
			143.89	1050890	1745850	2	4	22.6	0.0140	0.0133	−1.55	C	1
			160.97	1050890	1672110	2	2	122	0.0475	0.050	−1.022	C	1

^{a}The number in parentheses following the tabulated value indicates the power of ten by which this value has to be multiplied.

Co XXIII

Forbidden Transitions

The line strengths tabulated for the single magnetic dipole and single electric quadrupole transition within the $2s^2 2p$ ground state configuration are the results of the multiconfiguration Dirac-Fock (MCDF) calculations of Cheng et al.[1] These relativistic calculations include a perturbative treatment of the Breit interaction and the Lamb shift. Allowance for configuration mixing is limited to the $n=2$ complex. The strength of the electric quadrupole transition as defined in Ref. 1 was multiplied by the factor $^2/_3$ in order to bring this value into conformance with the definition of the quadrupole strength used in the NBS tables.

Transition probabilities for the same lines were calculated by Froese Fischer and Saha[2] using the multiconfiguration Hartree-Fock (MCHF) method with Breit-Pauli corrections. Their orbital basis includes many configurations outside the $n=2$ complex, but relativistic effects were not treated to the same degree as in Ref. 1. The line strengths for both the M1 and E2 transitions, derived from these data by interpolation between appropriately spaced ions of the B sequence, are in very good agreement with the data of Cheng et al.[1]

References

[1]K. T. Cheng, Y.-K. Kim, and J. P. Desclaux, At. Data Nucl. Data Tables **24**, 111 (1979).
[2]C. Froese Fischer and H. P. Saha, Phys. Rev. A **28**, 3169 (1983).

Co XXIII: Forbidden transitions

No.	Transition Array	Multiplet	λ (Å)	E_i (cm^{-1})	E_k (cm^{-1})	g_i	g_k	Type of transition	A_{ki} (s^{-1})	S (at. u.)	Accuracy	Source
1.	$2p$–$2p$	^{2}P° – ^{2}P°										
			[717.9]	0	139290	2	4	M1	2.42(+4)^a	1.33	B	1
			"	"	"	2	4	E2	2.67	0.00121	C	1

^{a}The number in parentheses following the tabulated value indicates the power of ten by which this value has to be multiplied.

Co xxiv

Be Isoelectronic Sequence

Ground State: $1s^2 2s^2 \, ^1S_0$

Ionization Energy: $2119 \, \text{eV} = 17090000 \, \text{cm}^{-1}$

Allowed Transitions

List of tabulated lines

Wavelength (Å)	No.	Wavelength (Å)	No.	Wavelength (Å)	No.	Wavelength (Å)	No.
9.97	11	10.54	29	10.80	25	144.83	3
10.07	10	10.55	24	10.81	19,31	159.00	3
10.08	11	10.57	24	10.93	35	165.75	3
10.12	8,10	10.58	24,28	10.94	19	172.42	3
10.16	10	10.59	13,27,29	11.14	21,22	204.10	6
10.18	10	10.64	28	11.43	23	250.80	1
10.26	15	10.67	33,34	112.32	4	294.90	5
10.39	14	10.68	27	125.15	2	344.65	5
10.41	24	10.71	27	128.24	4	492.61	5
10.44	29	10.74	27	135.24	3		
10.45	24	10.76	19,27	137.73	3		
10.50	12	10.77	32	139.80	7		

Oscillator strengths for transitions of the arrays $2s^2$–$2s2p$ and $2s2p$–$2p^2$ are taken from the multiconfiguration Dirac-Fock (MCDF) calculations of Cheng et al.[1] These relativistic calculations include the configuration interaction most relevant for the states of these configurations, as well as a perturbative treatment of the Breit interaction and the Lamb shift. The results should be quite accurate, except for the weakest intercombination lines. (The $^3P_1^o$ – $^1S_0$ transition of the $2s2p$–$2p^2$ array has been omitted here, since the f-value is considerably smaller than those of the other lines of this array.)

A number of sources of reliable data, from other relativistic calculations, are available for the $2s$–$2p$ transitions. However, with the exception of some of the weaker lines, they all agree well with the results of Cheng et al.[1] The latter are quoted exclusively here since they provide data from a single set of comprehensive calculations, all done at a uniform and reasonably accurate level of approximation, for the valence shell $2s$–$2p$ transitions for all ions of the isoelectronic sequence.

The f-values for the $2s^2$–$2s3p$, $2s2p$–$2p3p$, $2s2p$–$2s3s$, $2p^2$–$2p3s$, $2s2p$–$2s3d$, and $2p^2$–$2p3d$ arrays of transitions have been obtained by interpolating the isoelectronic sequence data of Fawcett,[2] who used Cowan's version of the relativistic Hartree-Fock method with intermediate coupling and configuration interaction. Calculated data were available for the neighboring ions of the sequence,

Fe xxiii and Ni xxv, so the interpolation should be quite reliable. Some of these transitions, for some ions of this sequence, have also been calculated by Bhatia et al.[3] using the program SUPERSTRUCTURE, which includes configuration interaction and intermediate coupling, too. Where they overlap, these two sets of calculations agree to within the uncertainties assigned here. Transitions involving the $J=1$ levels of $2p3p$ ^3S and ^3P have been omitted because of erratic behavior of the f-values along the sequence.

Oscillator strengths for the transition array $2s^2$–$2s4p$ have been interpolated from the relativistic random phase approximation (RRPA) calculations along the isoelectronic sequence by Lin and Johnson.[4]

A few multiplet f-values for transitions involving the outer electron alone, $2s3s$–$2s3p$ and $2s3p$–$2s3d$, have been interpolated along the isoelectronic sequence and assigned a low accuracy.

References

[1] K. T. Cheng, Y.-K. Kim, and J. P. Desclaux, At. Data Nucl. Data Tables **24**, 111 (1979).

[2] B. C. Fawcett, At. Data Nucl. Data Tables **30**, 1 (1984); **33**, 479 (1985).

[3] A. K. Bhatia, U. Feldman, and J. F. Seely, At. Data Nucl. Data Tables **35**, 449 (1986).

[4] C. D. Lin and W. R. Johnson, Phys. Rev. A **15**, 1046 (1977).

Co XXIV: Allowed transitions

No.	Transition Array	Multiplet	λ (Å)	E_i (cm⁻¹)	E_k (cm⁻¹)	g_i	g_k	A_{ki} (10⁸ s⁻¹)	f_{ik}	S (at. u.)	log gf	Accuracy	Source
1.	$2s^2$–$2s2p$	¹S – ³P°											
			[250.80]	0	398720	1	3	0.64	0.0018	0.0015	−2.74	D	1
2.		¹S – ¹P°	[125.15]	0	799040	1	3	216	0.152	0.0626	−0.818	B	1
3.	$2s2p$–$2p^2$	³P° – ³P	*153.72*	*456150*	*1106700*	9	9	141	0.0498	0.227	−0.348	B	1
			[159.00]	509210	1138140	5	5	76.5	0.0290	0.0759	−0.839	B	1
			[144.83]	398720	1089190	3	3	46.1	0.0145	0.0207	−1.362	B	1
			[172.42]	509210	1089190	5	3	46.4	0.0124	0.0352	−1.208	B	1
			[165.75]	398720	1002040	3	1	130	0.0179	0.0293	−1.270	B	1
			[135.24]	398720	1138140	3	5	61.5	0.0281	0.0375	−1.074	B	1
			[137.73]	363130	1089190	1	3	74.2	0.0633	0.0287	−1.199	B	1
4.		³P° – ¹D											
			[128.24]	509210	1289000	5	5	61	0.0151	0.0319	−1.122	C	1
			[112.32]	398720	1289000	3	5	5.7	0.0018	0.0020	−2.27	D	1
5.		¹P° – ³P											
			[294.90]	799040	1138140	3	5	4.6	0.010	0.029	−1.52	D	1
			[344.65]	799040	1089190	3	3	0.10	1.8(−4)ᵃ	6.1(−4)	−3.27	E	1
			[492.61]	799040	1002040	3	1	0.25	3.0(−4)	0.0015	−3.05	E	1
6.		¹P° – ¹D	[204.10]	799040	1289000	3	5	52.1	0.0542	0.109	−0.789	B	1
7.		¹P° – ¹S	[139.80]	799040	1514350	3	1	360	0.0352	0.0486	−0.976	B	1
8.	$2s^2$–$2s3p$	¹S – ³P°											
			[10.12]	0	9886000	1	3	5.9(+4)	0.27	0.0090	−0.57	C−	*interp.*
9.		¹S – ¹P°				1	3		0.44		−0.36	C−	*interp.*
10.	$2s2p$–$2p3p$	³P° – ³D											
			[10.07]	509210	10444000	5	7	7.0(+4)	0.15	0.025	−0.12	C−	*interp.*
			[10.07]	398720	10333000	3	5	6.3(+4)	0.16	0.016	−0.32	C−	*interp.*
			[10.12]	363130	10245000	1	3	1.8(+4)	0.084	0.0028	−1.08	D	*interp.*
			[10.18]	509210	10333000	5	5	1700	0.0027	4.5(−4)	−1.87	E	*interp.*
			[10.16]	398720	10245000	3	3	3.4(+4)	0.052	0.0052	−0.81	E	*interp.*
11.		³P° – ³P											
			[10.08]	509210	10425000	5	5	6.0(+4)	0.092	0.015	−0.34	C−	*interp.*
						3	1		0.040		−0.92	D	*interp.*
			[9.97]	398720	10425000	3	5	1900	0.0048	4.7(−4)	−1.84	D	*interp.*
12.		¹P° – ¹P	[10.50]	799040	10320000	3	3	1.8(+4)	0.029	0.0030	−1.06	D	*interp.*
13.		¹P° – ³D											
			[10.59]	799040	10245000	3	3	1.6(+4)	0.027	0.0028	−1.09	D	*interp.*
14.		¹P° – ³P											
			[10.39]	799040	10425000	3	5	2.7(+4)	0.072	0.0074	−0.67	D	*interp.*
			[10.39]	799040	10425000	3	3	4.3(+4)	0.070	0.0072	−0.68	C−	*interp.*

Co XXIV: Allowed trasitions — Continued

No.	Transition Array	Multiplet	λ (Å)	E_i (cm⁻¹)	E_k (cm⁻¹)	g_i	g_k	A_{ki} (10⁸ s⁻¹)	f_{ik}	S (at. u.)	log gf	Accuracy	Source
15.		¹P° – ¹D	[10.26]	799040	10541000	3	5	7.6(+4)	0.20	0.020	−0.22	C−	interp.
16.		¹P° – ¹S				3	1		0.040		−0.92	D	interp.
17.	2s²-2s4p	¹S – ³P°											
						1	3		0.029		−1.54	D	interp.
18.		¹S – ¹P°				1	3		0.15		−0.82	D	interp.
19.	2s2p-2s3s	³P° – ³S	10.87	456150	9653000	9	3	4.5(+4)	0.026	0.0085	−0.62	D	interp.
			[10.94]	509210	9653000	5	3	2.4(+4)	0.026	0.0047	−0.89	D	interp.
			[10.81]	398720	9653000	3	3	1.5(+4)	0.026	0.0028	−1.11	D	interp.
			[10.76]	363130	9653000	1	3	5200	0.027	9.6(−4)	−1.57	D	interp.
20.		¹P° – ¹S				3	1		0.011		−1.48	D	interp.
21.	2p²-2p3s	³P – ³P°											
						5	5		0.033		−0.78	D	interp.
						3	3		0.0096		−1.54	D	interp.
						5	3		0.015		−1.12	D	interp.
			[11.14]	1089190	10065000	3	1	2.6(+4)	0.016	0.0018	−1.32	D	interp.
						3	5		0.035		−0.98	D	interp.
						1	3		0.055		−1.26	D	interp.
22.		¹D – ¹P°	[11.14]	1289000	10264000	5	3	2.4(+4)	0.027	0.0050	−0.87	D	interp.
23.		¹S – ¹P°	[11.43]	1514350	10264000	1	3	9200	0.054	0.0020	−1.27	D	interp.
24.	2s2p-2s3d	³P° – ³D	10.50	456150	9977000	9	15	2.6(+5)	0.71	0.22	0.80	C−	interp.
			[10.55]	509210	9986000	5	7	2.6(+5)	0.60	0.10	0.48	C−	interp.
			[10.45]	398720	9971000	3	5	2.0(+5)	0.55	0.057	0.22	C−	interp.
			[10.41]	363130	9965000	1	3	1.5(+5)	0.75	0.026	−0.12	C−	interp.
			[10.57]	509210	9971000	5	5	6.6(+4)	0.11	0.019	−0.26	C−	interp.
			[10.45]	398720	9965000	3	3	1.1(+5)	0.18	0.019	−0.27	C−	interp.
			[10.58]	509210	9965000	5	3	7200	0.0072	0.0013	−1.44	C−	interp.
25.		¹P° – ¹D	[10.80]	799040	10058000	3	5	2.1(+5)	0.61	0.065	0.26	C−	interp.
26.	2p²-2p3d	³P – ³F°											
						5	7		0.25		0.10	C−	interp.
27.		³P – ³D°											
						5	7		0.61		0.48	C−	interp.
			[10.71]	1089190	10430000	3	5	1.67(+5)	0.478	0.051	0.157	C−	interp.
			[10.59]	1002040	10449000	1	3	2.58(+5)	1.30	0.0453	0.114	C−	interp.
			[10.76]	1138140	10430000	5	5	1.8(+4)	0.031	0.0055	−0.81	D	interp.
			[10.68]	1089190	10449000	3	3	5.1(+4)	0.087	0.0092	−0.58	C−	interp.
			[10.74]	1138140	10449000	5	3	2900	0.0030	5.3(−4)	−1.82	D	interp.
28.		³P – ¹D°											
			[10.64]	1138140	10539000	5	5	3.3(+4)	0.056	0.0098	−0.55	C−	interp.
			[10.58]	1089190	10539000	3	5	1.1(+5)	0.32	0.033	−0.02	D	interp.

Co XXIV: Allowed transitions — Continued

No.	Transition Array	Multiplet	λ (Å)	E_i (cm⁻¹)	E_k (cm⁻¹)	g_i	g_k	A_{ki} (10⁸ s⁻¹)	f_{ik}	S (at. u.)	log gf	Accuracy	Source
29.		³P – ³P°	*10.56*	*1106700*	10578000	9	9	1.8(+5)	0.29	0.092	0.42	C−	*interp.*
			[10.59]	1138140	10578000	5	5	1.25(+5)	0.210	0.0366	0.021	C−	*interp.*
			[10.54]	1089190	10578000	3	3	1.5(+5)	0.25	0.026	−0.12	C−	*interp.*
			[10.59]	1138140	10578000	5	3	6.6(+4)	0.067	0.012	−0.47	C−	*interp.*
			[10.54]	1089190	10578000	3	1	2.0(+5)	0.11	0.011	−0.48	C−	*interp.*
			[10.54]	1089190	10578000	3	5	2.3(+4)	0.063	0.0066	−0.72	D	*interp.*
			[10.44]	1002040	10578000	1	3	310	0.0015	5.2(−5)	−2.82	E	*interp.*
30.		¹D – ³F°				5	5		0.014		−1.15	D	*interp.*
31.		¹D – ¹D°	[10.81]	1289000	10539000	5	5	2.6(+4)	0.045	0.0080	−0.65	C−	*interp.*
32.		¹D – ³P°	[10.77]	1289000	10578000	5	5	9.8(+4)	0.17	0.030	−0.07	C−	*interp.*
33.		¹D – ¹P°	[10.67]	1289000	10661000	5	3	1.5(+4)	0.015	0.0026	−1.12	D	*interp.*
34.		¹D – ¹F°	[10.67]	1289000	10658000	5	7	4.18(+5)	1.00	0.176	0.70	C−	*interp.*
35.		¹S – ¹P°	[10.93]	1514350	10661000	1	3	2.38(+5)	1.28	0.0461	0.107	C−	*interp.*
36.	2s3s–2s3p	³S – ³P°				3	9		0.12		−0.44	D	*interp.*
37.		¹S – ¹P°				1	3		0.050		−1.30	E	*interp.*
38.	2s3p–2s3d	³P° – ³D				9	15		0.026		−0.63	E	*interp.*
39.		¹P° – ¹D				3	5		0.045		−0.87	E	*interp.*

ªThe number in parentheses following the tabulated value indicates the power of ten by which this value has to be multiplied.

Co XXIV

Forbidden Transitions

Transition probabilities for magnetic dipole transitions within the $2s2p$ and $2p^2$ configurations were calculated by Oboladze and Safronova[1] using a nuclear charge expansion method. For other ions of the Be sequence, where their results could be compared with other recent calculations, the agreement is typically within 20%.

The transition probability for the one electric quadrupole transition listed, which is relatively strong compared to other E2 transitions, has been interpolated from the data of Anderson and Anderson[2] and Glass[3,4] for neighboring ions of the Be sequence. This A-value exhibits a smooth nuclear charge dependence.

References

[1] N. S. Oboladze and U. I. Safronova, Opt. Spectrosc. (USSR) **48**, 469 (1980).
[2] E. K. Anderson and E. M. Anderson, Opt. Spectrosc. (USSR) **52**, 478 (1982).
[3] R. Glass, Z. Phys. A **320**, 545 (1985).
[4] R. Glass, Astrophys. Space Sci. **92**, 307 (1983).

Co XXIV: Forbidden transitions

No.	Transition Array	Multiplet	λ (Å)	E_i (cm^{-1})	E_k (cm^{-1})	g_i	g_k	Type of transition	A_{ki} (s^{-1})	S (at. u.)	Accuracy	Source
1.	$2s2p$–$2s2p$	$^3P°$ – $^3P°$										
			[905.1]	398720	509210	3	5	M1	1.66(+4)[a]	2.28	C+	1
			[2809.0]	363130	398720	1	3	M1	718	1.77	C+	1
2.		$^3P°$ – $^1P°$										
			[345.03]	509210	799040	5	3	M1	1.6(+4)	0.073	D	1
			[249.80]	398720	799040	3	3	M1	2.6(+4)	0.045	D	1
			"	"	"	3	3	E2	68	1.2(−4)	D−	interp.
			[229.41]	363130	799040	1	3	M1	4.4(+4)	0.059	D	1
3.	$2p^2$–$2p^2$	^3P – ^3P										
			[2042.2]	1089190	1138140	3	5	M1	1060	1.67	C	1
			[1147.5]	1002040	1089190	1	3	M1	1.02(+4)	1.71	C	1
4.		^3P – ^1D										
			[662.87]	1138140	1289000	5	5	M1	2.79(+4)	1.51	C	1
			[500.48]	1089190	1289000	3	5	M1	3.1(+4)	0.72	D+	1
5.		^3P – ^1S										
			[235.21]	1089190	1514350	3	1	M1	2.9(+5)	0.14	D	1

[a]The number in parentheses following the tabulated value indicates the power of ten by which this value has to be multiplied.

Co XXV

Li Isoelectronic Sequence

Ground State: $1s^2 2s\ ^2S_{1/2}$

Ionization Energy: 2219.0 eV = 17897000 cm^{-1}

Allowed Transitions

Transition probabilities for the strongest inner-shell transitions to doubly excited $n = 2$ states are taken from the multiconfiguration Dirac-Fock (MCDF) calculations of Hata and Grant.[1] Their results are in good agreement with the Z-expansion perturbation calculations of Vainshtein and Safronova,[2] who included relativistic corrections at the level of the Pauli approximation.

Oscillator strengths for lines of the principal ($2s$–$2p$) resonance multiplet are the results of the MCDF calcula-

tions of Cheng et al.,[3] which include a perturbative treatment of the Breit interaction and the Lamb shift.

The results of the scaled Thomas-Fermi calculations of Hayes,[4] which included relativistic effects, and the Hartree-XR (Hartree-Fock with statistical exchange and relativistic effects) calculations of Fawcett et al.[5] for the $2p$–$3s$ transitions in Fe XXIV and Ni XXVI, respectively, were used to derive interpolated f-values for these transitions in Co XXV. Similarly, the results of the MCDF calculations of Armstrong *et al.*[6] for Fe XXIV and those of Fawcett *et al.*[5] for Ni XXVI were used to interpolate *f*-values for the 2p–3d transitions in Co XXV.

The *f*-value for the 3d–4f transition was taken from a study of systematic trends along isoelectronic sequences by Smith and Wiese.[7] The tabulated data for many additional transitions were taken from the theoretical analysis of Martin and Wiese,[8] which was based on a generalized study of systematic trends for several spectral series of the lithium isoelectronic sequence.

Results of the relativistic Hartree-Fock calculations of Kim and Desclaux[9] for several ions of the Li sequence were incorporated into the data of Ref. 8 for the 2s–3p transitions. For all other transitions for which the results of Ref. 8 are quoted here, no relativistic calculations were available. However, the relativistic calculations of Younger and Weiss[10] for the hydrogen isoelectronic sequence provide a means of assessing the magnitude of relativistic corrections since the Li sequence is very similar in structure to the H sequence. For those transitions for which relativistic effects were estimated to be significant (specifically, whenever the ratio of the weighted relativistic hydrogenic *f*-values gf_{ik} of any two lines within a multiplet was found to deviate from the corresponding *LS*-coupling linestrength ratio by more than 5% for the appropriate value of the nuclear charge *Z*), the *f*-values were excluded from the compilation. A more detailed discussion of this comparison is given in Ref. 8.

Transition probability data are available for numerous transitions involving doubly excited states with spectator electron occupying the $n = 3$ shell, or higher.[11] These have not been tabulated, however, since they belong to, or are very close to belonging to, the unresolved satellites of the helium-like ion.

References

[1] J. Hata and I. P. Grant, Mon. Not. R. Astron Soc. **211**, 549 (1984).
[2] L. A. Vainshtein and U. I. Safronova, At. Data Nucl. Data Tables **21**, 49 (1978).
[3] K. T. Cheng, Y.-.K Kim, and J. P. Desclaux, At. Data Nucl. Data Tables **24**, 111 (1979).
[4] M. A. Hayes, Mon. Not. R. Astron. Soc. **189**, 55P (1979).
[5] B. C. Fawcett, A. Ridgeley, and T. P. Hughes, Mon. Not. R. Astron. Soc. **188**, 365 (1979).
[6] L. Armstrong, Jr., W. R. Fielder, and D. L. Lin, Phys. Rev. A **14**, 1114 (1976).
[7] M. W. Smith and W. L. Wiese, Astrophys. J. Suppl. Ser. **23**, No. 196, 103 (1971).
[8] G. A. Martin and W. L. Wiese, J. Phys. Chem. Ref. Data **5**, 537 (1976).
[9] Y.-K. Kim and J. P. Desclaux, Phys. Rev. Lett. **36**, 139 (1976) and private communication.
[10] S. M. Younger and A. W. Weiss, J. Res. Nat. Bur. Stand., Sect. A **79**, 629 (1975).
[11] L. A. Vainshtein and U. I. Safronova, At. Data Nucl. DataTables **25**, 311 (1980).

Co XXV: Allowed transitions

No.	Transition Array	Multiplet	λ (Å)	E_i (cm^{-1})	E_k (cm^{-1})	g_i	g_k	A_{ki} (10^8 s^{-1})	f_{ik}	S (at. u.)	log gf	Accuracy	Source
1.	$1s^22s$– $1s(^2S)2s2p(^3P°)$	^2S – $^2P°$	*1.723*			2	6	7.3(+5)^a	0.097	0.0011	−0.71	D	1
			[1.721]			2	4	5500	4.9(−4)	5.6(−6)	−3.01	D	1
			[1.724]			2	2	2.2(+6)	0.096	0.0011	−0.72	D	1
2.	$1s^22s$– $1s(^2S)2s2p(^1P°)$	^2S – $^2P°$											
			[1.718]			2	2	3.6(+6)	0.16	0.0018	−0.50	D	1
3.	$1s^22p$–$1s2p^2$	$^2P°$ – ^2D											
			[1.725]			4	6	2.4(+6)	0.16	0.0036	−0.19	D	1
			[1.723]			2	4	3.8(+6)	0.34	0.0038	−0.17	D	1
4.		$^2P°$ – ^2P											
			[1.722]			4	4	7.1(+6)	0.31	0.0071	0.10	D	1
			[1.723]			2	2	6.3(+6)	0.28	0.0032	−0.25	D	1
			[1.727]			4	2	1.9(+6)	0.042	9.7(−4)	−0.77	D	1
5.		$^2P°$ – ^2S											
			[1.717]			4	2	2.9(+6)	0.064	0.0014	−0.59	D	1

Co xxv: Allowed transitions — Continued

No.	Transition Array	Multiplet	λ (Å)	E_i (cm^{-1})	E_k (cm^{-1})	g_i	g_k	A_{ki} (10^8 s^{-1})	f_{ik}	S (at. u.)	log gf	Accuracy	Source
6.	2s–2p	^{2}S – ^{2}P°	*195*			2	6	37.7	0.0645	0.0828	−0.889	B+	3
			[178]			2	4	49.9	0.0474	0.0556	−1.023	B+	3
			[243]			2	2	19.2	0.0170	0.0272	−1.469	B+	3
7.	2s–3p	^{2}S – ^{2}P°	*9.81*			2	6	8.69(+4)	0.376	0.0243	−0.124	B+	8
			[9.80]			2	4	8.58(+4)	0.247	0.0159	−0.306	B+	8
			9.839	0	10160000	2	2	8.89(+4)	0.129	0.00836	−0.588	B+	8
8.	2s–4p	^{2}S – ^{2}P°				2	6		0.100		−0.699	C+	8
9.	2s–5p	^{2}S – ^{2}P°				2	6		0.040		−1.10	C+	8
10.	2s–6p	^{2}S – ^{2}P°				2	6		0.0213		−1.371	C+	8
11.	2s–7p	^{2}S – ^{2}P°				2	6		0.0125		−1.602	C+	8
12.	2p–3s	^{2}P° – ^{2}S	*10.49*			6	2	3.2(+4)	0.018	0.0037	−0.97	D	*interp.*
			[10.54]			4	2	2.2(+4)	0.018	0.0025	−1.14	C	*interp.*
			[10.38]			2	2	1.1(+4)	0.017	0.0012	−1.47	D	*interp.*
13.	2p–3d	^{2}P° – ^{2}D	*10.24*			6	10	2.59(+5)	0.68	0.137	0.61	C	*interp.*
			10.286			4	6	2.56(+5)	0.608	0.0824	0.386	C+	*interp.*
			10.151			2	4	2.19(+5)	0.678	0.0453	0.132	C+	*interp.*
			[10.32]			4	4	4.3(+4)	0.068	0.0092	−0.57	C	*interp.*
14.	2p–4d	^{2}P° – ^{2}D				6	10		0.12		−0.14	C+	8
15.	2p–5d	^{2}P° – ^{2}D				6	10		0.0450		−0.569	C+	8
16.	2p–6d	^{2}P° – ^{2}D				6	10		0.0220		−0.879	C+	8
17.	2p–7d	^{2}P° – ^{2}D				6	10		0.0126		−1.121	C+	8
18.	3s–4p	^{2}S – ^{2}P°				2	6		0.45		−0.05	C	8
19.	3s–5p	^{2}S – ^{2}P°				2	6		0.108		−0.67	C	8
20.	3s–6p	^{2}S – ^{2}P°				2	6		0.048		−1.02	C	8
21.	3s–7p	^{2}S – ^{2}P°				2	6		0.0250		−1.301	C	8
22.	3p–4d	^{2}P° – ^{2}D				6	10		0.60		0.56	B	8
23.	3p–5d	^{2}P° – ^{2}D				6	10		0.138		−0.082	C+	8
24.	3p–6d	^{2}P° – ^{2}D				6	10		0.0558		−0.475	C+	8
25.	3p–7d	^{2}P° – ^{2}D				6	10		0.0289		−0.761	C+	8
26.	3d–4f	^{2}D – ^{2}F°				10	14		1.00		1.000	B	7
27.	4s–5p	^{2}S – ^{2}P°				2	6		0.481		−0.017	C	8
28.	4s–6p	^{2}S – ^{2}P°				2	6		0.129		−0.59	C	8
29.	4s–7p	^{2}S – ^{2}P°				2	6		0.056		−0.95	C	8
30.	4p–5d	^{2}P° – ^{2}D				6	10		0.585		0.545	C+	8

Co xxv: Allowed transitions — Continued

No.	Transition Array	Multiplet	λ (Å)	E_i (cm^{-1})	E_k (cm^{-1})	g_i	g_k	A_{ki} (10^8 s^{-1})	f_{ik}	S (at. u.)	log gf	Accuracy	Source
31.	4p–6d	^{2}P° – ^{2}D				6	10		0.142		−0.070	C+	8
32.	4p–7d	^{2}P° – ^{2}D				6	10		0.0617		−0.432	C+	8

[a]The number in parentheses following the tabulated value indicates the power of ten by which this value has to be multiplied.

Co xxv

Forbidden Transitions

The single magnetic dipole transition within the $1s^2 2p$ configuration has the line strength of 1.33 in the absence of relativistic effects in the wavefunctions.[1] It is estimated that these effects are negligible, since comprehensive relativistic calculations by Cheng *et al.*[2] for the analogous transition in the $1s^2 2s^2 2p$ configuration of the boron sequence show that such relativistic corrections are negligible until much more highly charged ions.

The listed transition probability data are also expected to be quite accurate since the energy levels are derived from experimental data.

An electric quadrupole transition at the same wavelength is estimated to be of negligible strength, as calculated by Bhatia[3] for this transition in the case of Mn XXIII. (He obtains a ratio of about 10^{-3} for the ratio of E2 to M1 line strengths).

References

[1]W. L. Wiese, M. W. Smith, and B. M. Miles, "Atomic Transition Probabilities", Vol. II, NSRDS–NBS 22, U.S. Govt. Print. Office, Washington, DC 1969.
[2]K. T. Cheng, Y.-K. Kim, and J. P. Desclaux, At. Data Nucl. Data Tables **24**, 111 (1979).
[3]A. K. Bhatia, private communication (1986).

Co xxv: Forbidden transitions

No.	Transition Array	Multiplet	λ (Å)	E_i (cm^{-1})	E_k (cm^{-1})	g_i	g_k	Type of transition	A_{ki} (s^{-1})	S (at. u.)	Accuracy	Source
1.	2p–2p	^{2}P° – ^{2}P°										
			[659.37]	[409520]	[561180]	2	4	M1	3.13(+4)[a]	1.33	B	*interp.*

[a]The number in parentheses following the tabulated value indicates the power of ten by which this value has to be multiplied.

Co XXVI

He Isoelectronic Sequence

Ground State: $1s^2 \, {}^1S_0$

Ionization Energy: 9544.1 eV $= 76979000 \, \text{cm}^{-1}$

Allowed Transitions

List of tabulated lines

Wavelength (Å)	No.	Wavelength (Å)	No.	Wavelength (Å)	No.	Wavelength (Å)	No.
1.3512	19	1.660	3,9	6.4208	38	26.767	41
1.3515	18	1.661	3,9	6.9269	25	27.024	42
1.3824	17	1.662	9	7.0393	26	27.524	46
1.3831	16	1.664	11	7.0708	33	27.941	47
1.4552	15	1.666	11	7.2032	34	58.106	57
1.4569	14	1.667	5	9.2825	23	58.473	58
1.648	4	1.669	7,11	9.4422	24	59.684	60
1.651	13	1.677	8	9.5814	29	60.408	61
1.655	6,10	1.7118	2	9.7862	30	181.2	21
1.656	3,9	1.7203	1	18.427	43	249.3	20
1.657	9	6.2007	27	18.607	44	348.3	22
1.658	9	6.2972	28	18.730	50	379.5	20
1.659	12	6.3098	37	18.973	51	406.8	20

Oscillator strengths for transitions of the $1s^2$–$1s2p$ array are taken from the results of Drake,[1] who incorporated accurate nonrelativistic matrix elements and Dirac hydrogenic matrix elements into a Z-expansion technique in order to provide f-values which would accurately reflect correlation effects for low-Z ions and relativistic effects for high-Z ions of the helium isoelectronic sequence. The f-values for the $1s^2 \, {}^1S$ – $1snp \, {}^3P°$ ($n = 3$–5) transitions were interpolated from results of the relativistic random phase approximation (RRPA) calculations of Johnson and Lin.[2] Data for other s–p and p–s transitions were interpolated from the RRPA results of Lin *et al.*,[3] with the exception of the $2s$–$2p$ transitions, where we tabulate the actual published RRPA A-values of these same authors.[4]

The charge expansion results of Laughlin[5] are given for various p–d and d–p transitions, as well as transitions between $4d$ and $4f$ levels. For those multiplets involving no change in principal quantum number ($3p$–$3d$, $4p$–$4d$, $4d$–$4f$) the f-values should be considered rather uncertain, since they are sensitive to energy differences. Oscillator strengths for the $2p$–$3d$ transitions, and for $1s3p \, {}^3P°$ – $1s3d \, {}^3D$, were interpolated from the variational calculations of Weiss.[6] Both of these calculations indicate that, unlike the triplets, the $nd \, {}^1D$ energy levels ($n = 3,4$) lie below the $np \, {}^1P°$ levels, and the $4f \, {}^1F°$ lies below the $4d \, {}^1D$.

Brown and Cortez[7] have provided f-values for numerous d–f and f–d transitions for the isoelectronic sequence

by fitting Z-expansion formulas to the results of variational calculations for the low-Z ions. Their results for transitions between the lower-lying D and F° terms are tabulated here.

Transition probabilities for the stronger transitions involving the doubly excited $n = 2$ states are taken from the comprehensive, charge expansion perturbation theory calculations of Vainshtein and Safronova.[8] Numerous data are also available for transitions involving doubly excited states where the spectator electron has principal quantum number $n = 3$.[9] However, these data are not tabulated here since most of the transitions are very close to belonging to the unresolved satellites of the H-like ions, if they do not in fact do so.

References

[1] G. W. F. Drake, Phys. Rev. A **19**, 1387 (1979).

[2] W. R. Johnson and C. D. Lin, Phys. Rev. A **14**, 565 (1976).

[3] C. D. Lin, W. R. Johnson, and A. Dalgarno, Astrophys. J. **217**, 1011 (1977).

[4] C. D. Lin, W. R. Johnson, and A. Dalgarno, Phys. Rev. A **15**, 154 (1977).

[5] C. J. Laughlin, J. Phys. B **6**, 1942 (1973).

[6] A. W. Weiss, J. Res. Nat. Bur. Stand., Sect. A **71**, 163 (1967).

[7] R. T. Brown and J.-L. M. Cortez, Astrophys. J. **176**, 267 (1972).

[8] L. A. Vainshtein and U. I. Safronova, At. Data Nucl. Data Tables **21**, 49 (1978).

[9] L. A. Vainshtein and U. I. Safronova, At. Data Nucl. Data Tables **25**, 311 (1980).

Co XXVI: Allowed transitions

No.	Transition Array	Multiplet	λ (Å)	E_i (cm^{-1})	E_k (cm^{-1})	g_i	g_k	A_{ki} (10^8 s^{-1})	f_{ik}	S (at. u.)	log gf	Accuracy	Source
1.	$1s^2$–$1s2p$	^{1}S – ^{3}P°											
			[1.7203]	0	[58128200]	1	3	5.89(+5)^a	0.0784	4.44(−4)	−1.106	B	1
2.		^{1}S – ^{1}P°	[1.7118]	0	[58416500]	1	3	5.26(+6)	0.693	0.00391	−0.159	B	1
3.	$1s2s$–$2s2p$	^{3}S – ^{3}P°	*1.658*	[57864700]	[*118180000*]	3	9	3.3(+6)	0.41	0.0067	0.09	C	8
			[1.656]	[57864700]	[118240000]	3	5	3.3(+6)	0.23	0.0037	−0.17	C	8
			[1.660]	[57864700]	[118110000]	3	3	3.2(+6)	0.13	0.0022	−0.40	C	8
			[1.661]	[57864700]	[118060000]	3	1	3.3(+6)	0.045	7.5(−4)	−0.86	C	8
4.		^{3}S – ^{1}P°											
			[1.648]	[57864700]	[118540000]	3	3	1.5(+5)	0.0061	9.9(−5)	−1.74	D	8
5.		^{1}S – ^{3}P°											
			[1.667]	[58129400]	[118110000]	1	3	1.5(+5)	0.019	1.0(−4)	−1.73	D	8
6.		^{1}S – ^{1}P°	[1.655]	[58129400]	[118540000]	1	3	3.2(+6)	0.39	0.0021	−0.40	C	8
7.	$1s2p$–$2s^2$	^{3}P° – ^{1}S											
			[1.669]	[58128200]	[118050000]	3	1	6.4(+5)	0.0089	1.5(−4)	−1.57	D	8
8.		^{1}P° – ^{1}S	[1.677]	[58416500]	[118050000]	3	1	6.2(+5)	0.0087	1.4(−4)	−1.58	D	8
9.	$1s2p$–$2p^2$	^{3}P° – ^{3}P	*1.659*	[*58202700*]	[*118470000*]	9	9	5.8(+6)	0.240	0.0118	0.335	C	8
			[1.660]	[58265900]	[118510000]	5	5	3.2(+6)	0.13	0.0036	−0.18	C	8
			[1.658]	[58128200]	[118440000]	3	3	1.5(+6)	0.062	0.0010	−0.73	C	8
			[1.662]	[58265900]	[118440000]	5	3	2.8(+6)	0.070	0.0019	−0.46	C	8
			[1.661]	[58128200]	[118330000]	3	1	6.0(+6)	0.083	0.0014	−0.61	C	8
			[1.656]	[58128200]	[118510000]	3	5	2.1(+6)	0.14	0.0024	−0.36	C	8
			[1.657]	[58110500]	[118440000]	1	3	2.2(+6)	0.27	0.0015	−0.57	C	8
10.		^{3}P° – ^{1}D											
			[1.655]	[58265900]	[118670000]	5	5	1.8(+6)	0.074	0.0020	−0.43	C	8
11.		^{1}P° – ^{3}P											
			[1.664]	[58416500]	[118510000]	3	5	1.3(+6)	0.090	0.0015	−0.57	C	8
			[1.666]	[58416500]	[118440000]	3	3	1.6(+5)	0.0067	1.1(−4)	−1.70	D	8
			[1.669]	[58416500]	[118330000]	3	1	1.0(+5)	0.0014	2.3(−5)	−2.38	D	8
12.		^{1}P° – ^{1}D	[1.659]	[58416500]	[118670000]	3	5	4.9(+6)	0.34	0.0055	0.00	C	8
13.		^{1}P° – ^{1}S	[1.651]	[58416500]	[118980000]	3	1	5.9(+6)	0.080	0.0013	−0.62	C	8
14.	$1s^2$–$1s3p$	^{1}S – ^{3}P°											
			[1.4569]	0	[68637700]	1	3	1.9(+5)	0.018	8.6(−5)	−1.74	E	*interp.*
15.		^{1}S – ^{1}P°	[1.4552]	0	[68720200]	1	3	1.42(+6)	0.135	6.47(−4)	−0.870	B	*interp.*
16.	$1s^2$–$1s4p$	^{1}S – ^{3}P°											
			[1.3831]	0	[72301100]	1	3	7.7(+4)	0.0066	3.0(−5)	−2.18	E	*interp.*
17.		^{1}S – ^{1}P°	[1.3824]	0	[72335400]	1	3	5.68(+5)	0.0488	2.22(−4)	−1.312	B	*interp.*

Co XXVI: Allowed transitions — Continued

No.	Transition Array	Multiplet	λ (Å)	E_i (cm^{-1})	E_k (cm^{-1})	g_i	g_k	A_{ki} (10^8 s^{-1})	f_{ik}	S (at. u.)	log gf	Accuracy	Source
18.	$1s^2-1s\,5p$	^{1}S – ^{3}P°											
			[1.3515]	0	[73991900]	1	3	4.0(+4)	0.0033	1.5(−5)	−2.48	E	*interp.*
19.		^{1}S – ^{1}P°	[1.3512]	0	[74009400]	1	3	2.92(+5)	0.0240	1.07(−4)	−1.620	B	*interp.*
20.	$1s\,2s-1s\,2p$	^{3}S – ^{3}P°	*295.9*	[57864700]	*[58202700]*	3	9	10.3	0.0407	0.119	−0.913	B	4
			[249.3]	[57864700]	[58265900]	3	5	17.8	0.0276	0.0681	−1.081	B	4
			[379.5]	[57864700]	[58128200]	3	3	4.56	0.00985	0.0369	−1.530	B	4
			[406.8]	[57864700]	[58110500]	3	1	4.08	0.00337	0.0136	−1.995	B	4
21.		^{3}S – ^{1}P°											
			[181.2]	[57864700]	[58416500]	3	3	4.56	0.00224	0.00402	−2.172	B	4
22.		^{1}S – ^{1}P°	[348.3]	[58129400]	[58416500]	1	3	6.10	0.0333	0.0382	−1.478	B	4
23.	$1s\,2s-1s\,3p$	^{3}S – ^{3}P°											
			[9.2825]	[57864700]	[68637700]	3	3	9.37(+4)	0.121	0.0111	−0.440	B	*interp.*
24.		^{1}S – ^{1}P°	[9.4422]	[58129400]	[68720200]	1	3	8.93(+4)	0.358	0.0111	−0.446	B	*interp.*
25.	$1s\,2s-1s\,4p$	^{3}S – ^{3}P°											
			[6.9269]	[57864700]	[72301100]	3	3	4.3(+4)	0.031	0.0021	−1.03	B	*interp.*
26.		^{1}S – ^{1}P°	[7.0393]	[58129400]	[72335400]	1	3	3.9(+4)	0.086	0.0020	−1.07	B	*interp.*
27.	$1s\,2s-1s\,5p$	^{3}S – ^{3}P°											
			[6.2007]	[57864700]	[73991900]	3	3	2.1(+4)	0.012	7.3(−4)	−1.44	B	*interp.*
28.		^{1}S – ^{1}P°	[6.2972]	[58129400]	[74009400]	1	3	2.0(+4)	0.035	7.3(−4)	−1.46	B	*interp.*
29.	$1s\,2p-1s\,3s$	^{3}P° – ^{3}S											
			[9.5814]	[58128200]	[68565100]	3	3	1.0(+4)	0.014	0.0013	−1.38	B	*interp.*
30.		^{1}P° – ^{1}S	[9.7862]	[58416500]	[68635000]	3	1	2.9(+4)	0.014	0.0014	−1.38	B	*interp.*
31.	$1s\,2p-1s\,3d$	^{3}P° – ^{3}D				9	15		0.69		0.79	C+	*interp.*
32.		^{1}P° – ^{1}D				3	5		0.70		0.32	C+	*interp.*
33.	$1s\,2p-1s\,4s$	^{3}P° – ^{3}S											
			[7.0708]	[58128200]	[72270900]	3	3	4000	0.0030	2.1(−4)	−2.05	C	*interp.*
34.		^{1}P° – ^{1}S	[7.2032]	[58416500]	[72299200]	3	1	1.2(+4)	0.0031	2.2(−4)	−2.03	C	*interp.*
35.	$1s\,2p-1s\,4d$	^{3}P° – ^{3}D				9	15		0.12		0.03	C	5
36.		^{1}P° – ^{1}D				3	5		0.12		−0.44	C	5
37.	$1s\,2p-1s\,5s$	^{3}P° – ^{3}S											
			[6.3098]	[58128200]	[73976600]	3	3	2000	0.0012	7.5(−5)	−2.44	C	*interp.*
38.		^{1}P° – ^{1}S	[6.4208]	[58416500]	[73990800]	3	1	5800	0.0012	7.6(−5)	−2.44	C	*interp.*

Co XXVI:　Allowed transitions — Continued

No.	Transition Array	Multiplet	λ (Å)	E_i (cm^{-1})	E_k (cm^{-1})	g_i	g_k	A_{ki} (10^8 s^{-1})	f_{ik}	S (at. u.)	log gf	Accuracy	Source
39.	$1s\,3s$–$1s\,3p$	^{3}S – ^{3}P°											
						3	3		0.016		−1.32	C	*interp.*
40.		^{1}S – ^{1}P°				1	3		0.057		−1.24	C	*interp.*
41.	$1s\,3s$–$1s\,4p$	^{3}S – ^{3}P°											
			[26.767]	[68565100]	[72301100]	3	3	1.24(+4)	0.133	0.0352	−0.399	B	*interp.*
42.		^{1}S – ^{1}P°	[27.024]	[68635000]	[72335400]	1	3	1.20(+4)	0.393	0.0350	−0.406	B	*interp.*
43.	$1s\,3s$–$1s\,5p$	^{3}S – ^{3}P°											
			[18.427]	[68565100]	[73991900]	3	3	6700	0.034	0.0062	−0.99	B	*interp.*
44.		^{1}S – ^{1}P°	[18.607]	[68635000]	[74009400]	1	3	6490	0.101	0.00619	−0.996	B	*interp.*
45.	$1s\,3p$–$1s\,3d$	^{3}P° – ^{3}D				9	15		0.011		−1.00	D	*interp.*
46.	$1s\,3p$–$1s\,4s$	^{3}P° – ^{3}S											
			[27.524]	[68637700]	[72270900]	3	3	2800	0.032	0.0087	−1.02	B	*interp.*
47.		^{1}P° – ^{1}S	[27.941]	[68720200]	[72299200]	3	1	8500	0.033	0.0091	−1.00	B	*interp.*
48.	$1s\,3p$–$1s\,4d$	^{3}P° – ^{3}D				9	15		0.61		0.74	C	5
49.		^{1}P° – ^{1}D				3	5		0.62		0.27	C	5
50.	$1s\,3p$–$1s\,5s$	^{3}P° – ^{3}S											
			[18.730]	[68637700]	[73976600]	3	3	1300	0.0071	0.0013	−1.67	C	*interp.*
51.		^{1}P° – ^{1}S	[18.973]	[68720200]	[73990800]	3	1	4200	0.0075	0.0014	−1.65	C	*interp.*
52.	$1s\,3d$–$1s\,3p$	^{1}D – ^{1}P°				5	3		0.0020		−2.00	E	5
53.	$1s\,3d$–$1s\,4p$	^{3}D – ^{3}P°				15	9		0.012		−0.74	C	5
54.		^{1}D – ^{1}P°				5	3		0.011		−1.26	C	5
55.	$1s\,4s$–$1s\,4p$	^{3}S – ^{3}P°											
						3	3		0.025		−1.12	E	*interp.*
56.		^{1}S – ^{1}P°				1	3		0.069		−1.16	D	*interp.*
57.	$1s\,4s$–$1s\,5p$	^{3}S – ^{3}P°											
			[58.106]	[72270900]	[73991900]	3	3	2960	0.150	0.0861	−0.347	B	*interp.*
58.		^{1}S – ^{1}P°	[58.473]	[72299200]	[74009400]	1	3	2850	0.438	0.0843	−0.359	B	*interp.*
59.	$1s\,4p$–$1s\,4d$	^{3}P° – ^{3}D				9	15		0.019		−0.77	D	5
60.	$1s\,4p$–$1s\,5s$	^{3}P° – ^{3}S											
			[59.684]	[72301100]	[73976600]	3	3	970	0.052	0.031	−0.81	B	*interp.*
61.		^{1}P° – ^{1}S	[60.408]	[72335400]	[73990800]	3	1	3000	0.054	0.032	−0.79	B	*interp.*

Co XXVI: Allowed transitions — Continued

No.	Transition Array	Multiplet	λ (Å)	E_i (cm^{-1})	E_k (cm^{-1})	g_i	g_k	A_{ki} (10^8 s^{-1})	f_{ik}	S (at. u.)	log gf	Accu-racy	Source
62.	1s4d–1s4p	^{1}D – ^{1}P°				5	3		0.0030		−1.82	E	5
63.	1s4d–1s4f	^{3}D – ^{3}F°				15	21		7.6(−4)		−1.94	E	5
64.	1s4d–1s5f	^{3}D – ^{3}F°				15	21		0.89		1.13	B	7
65.		^{1}D – ^{1}F°				5	7		0.89		0.65	B	7
66.	1s4f–1s4d	^{1}F° – ^{1}D				7	5		4.0(−4)		−2.55	E	5
67.	1s4f–1s5d	^{3}F° – ^{3}D				21	15		0.0089		−0.73	C	7
68.		^{1}F° – ^{1}D				7	5		0.0089		−1.21	C	7
69.	1s5s–1s5p	^{3}S – ^{3}P°											
						3	3		0.027		−1.09	E	*interp.*
70.		^{1}S – ^{1}P°				1	3		0.10		−1.00	E	*interp.*

aThe number in parentheses following the tabulated value indicates the power of ten by which this value has to be multiplied.

Co XXVI

Forbidden Transitions

The results of multi-configuration Dirac-Fock calculations by Hata and Grant[1] have been selected for this tabulation. Their work includes both a very detailed consideration of configuration interaction—with configurational wavefunction sets containing as many as 51 interacting states—as well as a fully relativistic treatment based on the Dirac Hamiltonian. Their calculated wavelengths are in very close agreement with experimental values. For the ions Ti XXI, V XXII and Fe XXV, where accurate experimental lifetime data are available, the agreement between these and the theoretical results of Hata and Grant[1] is excellent, with differences not exceeding a few percent (see the comparison table in the introduction to the forbidden lines of Ti XXI).

Reference

[1] J. Hata and I. P. Grant, Mon. Not. R. Astr. Soc. **211**, 549 (1984).

Co XXVI: Forbidden transitions

No.	Transition Array	Multiplet	λ (Å)	E_i (cm^{-1})	E_k (cm^{-1})	g_i	g_k	Type of transition	A_{ki} (s^{-1})	S (at. u.)	Accu-racy	Source
1.	1s^2–1s2s	^{1}S – ^{3}S	[1.7284]	0	[57856690]	1	3	M1	3.12(+8)^a	1.79(−4)	B	1
2.	1s^2–1s2p	^{1}S – ^{3}P°										
			[1.7164]	0	[58260720]	1	5	M2	9.05(+9)	0.102	B	1

aThe number in parentheses following the tabulated value indicates the power of ten by which this value has to be multiplied.

Co XXVII

H Isoelectronic Sequence

Ground State: $1s\ ^2S_{1/2}$

Ionization Energy: $10012.20\ \text{eV} = 80753200\ \text{cm}^{-1}$

Allowed Transitions

Electric dipole transition probability data for this hydrogen-like ion can be obtained directly, in a non-relativistic approximation, from the data for neutral hydrogen.[1] The oscillator strength is independent of Z along the entire isoelectronic sequence and is therefore identical to the value for the hydrogen atom. Line strengths scale as Z^{-2} and transition probabilities scale as Z^4, i.e.,

$$S_Z = Z^{-2}S_H, \qquad A_Z = Z^4 A_H.$$

For higher nuclear charges in this sequence, relativistic corrections will cause these values to deviate increasingly from the non-relativistic ones. The first effect of relativity will be to alter the transition energies, or wavelengths, from the non-relativistic, even though the line strength itself is still well approximated by the non-relativistic value. In this case, experimental energies should be used in the standard conversion formulas, given in the general introduction to this volume, to calculate the most accurate values of f and A. It should be noted that the relativistic removal of the j-degeneracy introduces dipole transitions which do not occur in the non-relativistic theory, e.g., $2s_{1/2} - 2p_{3/2}$.

For very high Z, it is necessary to use the four-component Dirac spinors rather than two-component Schroedinger functions in theoretical calculations, and this introduces relativistic corrections to the line strengths themselves. Several recent systematic studies of the problem[2,3] indicate that these corrections are not large for stages of ionization in the range 20–30. Corrections for $Z = 30$ are usually no larger than 5–10% and generally substantially less than 5%. If an accuracy greater than this is required, the reader is referred to these papers[2,3] for a more detailed error analysis.

References

[1]W. L. Wiese, M. W. Smith, and B. M. Glennon, Atomic Transition Probabilities – Hydrogen through Neon (A Critical Data Compilation), Vol. I, 157 pp., Nat. Stand. Ref. Data Ser., Nat. Bur. Stand. (U.S.), 4 (May 1966).

[2]S. M. Younger and A. W. Weiss, J. Res. Nat. Bur. Stand., Sect. A **79**, 629 (1975).

[3]S. J. Rose, Rutherford Appleton Laboratory Report RL–82–114 (December 1982).

Nickel

Ni I

Ground State: $1s^2 2s^2 2p^6 3s^2 3p^6 3d^8 4s^2 \, ^3F_4$

Ionization Energy: 7.6375 eV = 61600 cm^{-1}

Allowed Transitions

List of tabulated lines

Wavelength (Å)	No.	Wavelength (Å)	No.	Wavelength (Å)	No.	Wavelength (Å)	No.
1963.85	68	2182.38	28	2907.46	16	3320.26	10
1968.90	34	2183.91	87	2914.01	15	3322.31	79
1976.87	68	2190.22	57	2943.91	47	3337.01	37
1981.61	68	2191.21	86	2981.65	47	3351.06	3
1990.25	68	2197.35	57	2984.13	13	3361.55	39
1994.29	33	2201.59	85	2991.11	15	3362.80	44
2000.49	65	2212.15	27	2992.59	45	3365.76	77
2001.83	66	2220.71	58	2994.46	48	3366.16	9
2007.01	65	2221.94	27	3002.48	47	3367.88	41
2007.69	34	2230.96	57	3003.62	47	3369.56	6
2014.25	68	2244.46	53	3012.00	81	3371.99	8
2025.40	32	2251.48	53	3019.14	11	3374.22	37
2026.62	32	2253.57	52	3031.87	11	3380.57	78
2029.29	66	2254.81	25	3037.93	45	3380.89	8
2033.56	60	2258.15	56	3045.01	13	3391.04	5
2041.16	65	2259.56	56	3050.82	45	3392.98	40
2047.35	63	2261.42	24	3054.31	45	3409.58	5
2050.84	67	2266.35	52	3057.64	47	3413.48	5
2052.04	30	2267.55	21	3064.62	47	3413.94	37
2052.45	30	2271.95	55	3080.75	47	3414.76	39
2053.91	29	2274.66	21	3097.12	11	3420.73	10
2055.50	32	2287.32	53	3099.11	14	3423.71	41
2059.92	61	2288.40	52	3101.56	45	3433.56	39
2060.20	61	2289.98	23	3101.88	80	3437.28	3
2062.37	32	2293.11	56	3105.46	13	3446.26	41
2063.42	65	2300.77	54	3107.71	13	3452.88	37
2064.39	61	2302.97	56	3114.12	46	3458.46	39
2069.52	64	2307.35	55	3129.30	13	3461.66	37
2082.87	32	2312.34	20	3134.11	45	3467.50	3
2085.37	62	2313.98	21	3145.12	8	3469.48	9
2085.57	90	2317.16	26	3145.71	11	3472.55	40
2088.98	61	2320.03	20	3159.52	11	3483.77	7
2089.09	31	2321.38	21	3165.51	42	3485.88	37
2095.13	90	2324.65	25	3184.37	11	3492.96	38
2105.85	64	2325.79	20	3195.57	13	3498.19	2
2109.79	30	2329.96	26	3197.11	46	3500.85	7
2111.73	29	2345.54	19	3200.42	44	3502.60	3
2114.43	88	2346.63	22	3221.65	9	3507.69	3
2121.40	59	2347.51	18	3225.02	79	3510.33	38
2124.80	89	2348.73	51	3226.98	8	3513.93	37
2125.62	28	2419.31	19	3232.93	8	3515.05	39
2128.41	31	2476.88	17	3234.65	42	3519.76	5
2129.96	58	2540.02	83	3235.75	11	3523.07	74
2147.80	58	2553.37	17	3243.05	43	3523.44	36
2151.93	29	2561.42	17	3248.46	42	3524.54	38
2152.23	59	2696.48	82	3249.44	12	3527.98	6
2157.83	57	2746.74	50	3250.74	79	3548.19	41
2158.31	57	2798.65	50	3271.11	44	3551.53	5
2161.04	58	2805.08	15	3282.69	8	3561.75	2
2166.15	58	2821.29	49	3286.94	39	3566.37	76
2173.54	84	2834.55	16	3310.20	77	3571.86	5
2174.48	57	2865.50	50	3315.66	43	3577.24	3

List of tabulated lines — Continued

Wavelength (Å)	No.	Wavelength (Å)	No.	Wavelength (Å)	No.	Wavelength (Å)	No.
3587.93	36	4729.28	193	5094.42	163	6105.78	206
3597.70	38	4731.81	162	5099.95	154	6108.12	94
3602.28	3	4732.47	193	5102.97	97	6111.06	189
3609.31	36	4740.17	123	5115.40	167	6119.78	200
3610.46	38	4752.43	138	5129.37	153	6128.99	91
3612.74	7	4756.52	122	5137.08	98	6130.17	203
3619.39	75	4758.42	172	5155.14	179	6133.95	188
3620.02	3	4762.63	115	5155.76	182	6175.42	196
3624.73	2	4773.41	164	5157.99	128	6176.81	186
3641.64	6	4786.29	99	5176.57	181	6177.26	103
3664.09	4	4786.54	122	5179.14	175	6186.74	188
3669.24	2	4791.00	115	5187.82	161	6198.65	206
3670.43	4	4793.47	159	5197.17	176	6204.64	185
3674.06	35	4807.00	156	5220.31	131	6223.99	186
3674.15	71	4808.86	160	5262.83	130	6230.09	187
3688.41	5	4812.00	136	5265.71	142	6256.36	92
3722.49	38	4814.62	122	5347.71	147	6271.76	117
3736.81	69	4815.92	137	5353.42	114	6272.65	200
3739.23	2	4817.82	208	5371.33	214	6314.66	111
3749.05	1	4821.14	208	5388.35	114	6316.61	203
3775.57	73	4829.03	137	5392.37	205	6322.17	204
3783.53	69	4831.18	128	5424.65	114	6327.60	93
3792.34	2	4838.64	212	5435.87	114	6364.60	111
3793.61	4	4843.17	99	5452.85	143	6370.38	134
3807.14	72	4843.51	193	5453.26	190	6375.22	210
3831.69	70	4845.17	132	5462.49	171	6378.26	201
3832.87	1	4852.56	136	5468.10	171	6384.67	202
3858.30	71	4855.41	136	5475.57	161	6414.60	200
3885.87	1	4900.97	122	5476.91	104	6421.52	209
3946.19	1	4904.41	135	5494.89	190	6482.80	110
3973.55	70	4912.03	128	5499.41	166	6532.89	109
4009.98	150	4913.97	138	5504.12	165	6586.33	109
4027.67	184	4918.36	167	5514.80	170	6598.59	204
4035.96	150	4925.58	142	5537.11	169	6635.15	213
4093.62	1	4935.83	167	5553.69	112	6643.64	92
4200.46	120	4937.34	131	5578.72	96	6661.39	202
4295.88	168	4945.46	147	5587.87	114	6767.77	102
4331.65	101	4946.04	149	5589.38	178	6772.36	134
4382.87	183	4953.20	128	5592.28	112	6842.07	133
4401.54	118	4967.55	142	5593.74	179	6850.48	152
4410.52	119	4976.16	129	5625.33	199	6914.56	106
4431.03	194	4976.35	97	5637.12	197	6973.52	216
4437.57	157	4976.71	208	5638.82	177	7001.54	109
4450.13	168	4980.17	129	5641.88	192	7030.06	133
4462.46	118	4995.65	147	5642.66	177	7034.42	121
4470.48	118	4996.85	146	5643.10	211	7062.97	109
4513.00	162	4998.23	128	5664.02	215	7110.91	108
4519.99	100	5000.34	147	5682.20	191	7122.24	133
4551.24	195	5003.75	99	5695.00	198	7197.02	106
4553.18	139	5010.96	146	5709.55	95	7220.79	220
4567.42	126	5012.46	128	5711.91	112	7261.93	106
4600.37	122	5017.58	128	5748.34	94	7286.56	127
4604.99	122	5032.75	180	5749.28	166	7291.45	107
4606.23	124	5035.37	144	5754.68	113	7327.67	141
4633.03	123	5039.37	145	5760.85	190	7381.94	219
4648.66	122	5042.20	137	5805.23	192	7385.24	116
4666.99	148	5048.85	174	5847.00	93	7401.13	218
4675.64	132	5079.96	105	5892.88	113	7414.51	106
4686.22	122	5080.53	144	5996.74	204	7422.30	140
4698.39	193	5081.11	173	6007.31	91	7481.49	217
4701.34	125	5082.35	136	6025.73	207	7488.73	158
4701.54	193	5084.08	155	6039.31	203	7714.32	106
4714.42	122	5085.49	136	6053.68	201	7727.66	151
4715.78	122	5088.96	155	6086.29	204	7788.94	106

For this spectrum, we provide data for 465 spectral lines and have utilized four data sources.[1-4] Huber and Sandeman[1] measured relative oscillator strengths by using the anomalous dispersion (hook) method and obtained data for a few additional, weak resonance lines by employing the absorption technique. They normalized their relative *f*-values to an absolute scale by using lifetime measurements of Becker *et al.*,[5] who used the zero-field level-crossing (Hanle) technique. Doerr and Kock[2] performed emission as well as hook measurements to determine relative oscillator strengths. The emission work was done with a hollow cathode discharge and a Fourier transform spectrometer for the data acquisition. The hook measurements were made in a high-temperature furnace. The absolute scale of Ref. 2 is based on lifetime data of Becker *et al.*[6] who employed selective laser excitation. Another source is the work of Lennard *et al.*,[3] who measured branching ratios in emission with a hollow cathode discharge and then normalized these data to beam-foil lifetimes, which they also determined. In addition, we utilized the data of Kostyk,[4] who derived log *gf*-values from solar spectra.

Our accuracy ratings for the data of Ref. 1 directly reflect the authors' own uncertainty estimates. The authors considered uncertainties in the relative values due primarily to measurement errors in the distance between line and hook, and also took into account the number of measurements taken, plus the uncertainty in the absolute scale as given by Ref. 5. The accuracies of the absolute log *gf*-values of Ref. 1 vary between 20 and 85 percent, but they are generally in the 25–30% range.

In their paper, Doerr and Kock separate their data into three distinct sets. Their "basic set" is emission *A*-values (or branching ratios) which they have normalized directly to the lifetimes of Ref. 6. We view these data as being the most reliable presently available and have assigned accuracies of "C+." The second set contains hook measurements for lines originating from lower levels of the basic set. The final set comprises emission lines starting from the same upper level as lines measured by the hook method. Because of the additional normalizations required for the second and third sets of data, they are not as reliable as the first, and we have reduced the accuracies accordingly. Refs. 1 and 2 have 105 lines in common, and for 91 of these, the *f*-values agree within ±50 percent. The situation for strong lines

(log *gf* > −1.0) is even better — data for 85 percent of these lines agree within ±25 percent.

The third data source included in this compilation is the experiment by Lennard *et al.*[3] Their beam-foil lifetimes are, on the average, about 20 percent longer than those of Ref. 5. When Refs. 2 and 3 are compared, considerable scatter is present for weak lines, and the *f*-values of Lennard *et al.* are much lower (by about 50 percent) than those of Doerr and Kock. Because of likely cascade problems in the beam-foil measurements, we have renormalized the data of Ref. 3 to the lifetimes of Ref. 5 whenever possible. Lennard et al. also provided transition probabilities for lines arising from high-lying upper levels not measured by the authors of Refs. 5 or 6. Thus, in these cases, we had to rely on the somewhat less accurate beam-foil lifetimes of Ref. 3 for the absolute scale, and we have lowered the accuracy ratings accordingly.

Another source we utilized is the paper by Kostyk.[4] This author determined log *gf*-values for 175 lines from the central intensities of solar Fe I lines. We compared Refs. 2 and 4 and found that the *f*-values for 11 of 15 overlapping lines agree within ± 52 percent. There are no apparent systematic trends or shifts in scale. However, because of saturation effects in the sun, Kostyk's *f*-values for stronger lines (log *gf* > −1.5) are probably not as reliable. We have therefore reduced the accuracies for these lines to "D−."

In this compilation, we have given first priority to either Ref. 1 or 2, depending largely on the authors' own error estimates. For lines not covered by either Ref. 1 or 2, we included the normalized (or unnormalized) data of Lennard *et al.* Finally, for the remaining (generally weak) lines, we tabulated the log *gf*-values of Kostyk.

References

[1] M. C. E. Huber and R. J. Sandeman, Astron. Astrophys. **86**, 95 (1980).

[2] A. Doerr and M. Kock, J. Quant. Spectrosc. Radiat. Transfer **33**, 307 (1985).

[3] W. N. Lennard, W. Whaling, J. M. Scalo, and L. Testerman, Astrophys. J. **197**, 517 (1975).

[4] R. I. Kostyk, Astrometriya Astrofiz. **46**, 58 (1982).

[5] U. Becker, L. H. Gobel, and W. D. Klotz, Astron. Astrophys. **33**, 241 (1974).

[6] U. Becker, H. Kerkhoff, M. Schmidt, and P. Zimmermann, J. Quant. Spectrosc. Radiat. Transfer **25**, 339 (1981).

Ni I: Allowed transitions

No.	Multiplet	λ (Å)	E_i (cm^{-1})	E_k (cm^{-1})	g_i	g_k	A_{ki} (10^8 s^{-1})	f_{ik}	S (at. u.)	log gf	Accuracy	Source
1.	$a\ ^3F - z\ ^5D°$ (1)											
		3946.19	1332.2	26666	7	7	2.1(−6)[a]	4.8(−7)	4.4(−5)	−5.47	C−	1
		3749.05	0.0	26666	9	7	2.7(−4)	4.4(−5)	0.0049	−3.40	C−	1
		3832.87	1332.2	27415	7	5	2.0(−4)	3.1(−5)	0.0028	−3.66	C−	1
		3885.87	2216.5	27944	5	3	4.1(−5)	5.5(−6)	3.5(−4)	−4.56	C	1
		4093.62	1332.2	25754	7	9	9.0(−7)	2.9(−7)	2.8(−5)	−5.69	C	1
2.	$a\ ^3F - z\ ^5G°$ (2)											
		3624.73	0.0	27580	9	11	0.0026	6.2(−4)	0.067	−2.25	C	2
		3739.23	1332.2	28068	7	9	0.0024	6.4(−4)	0.055	−2.35	E	1
		3792.34	2216.5	28578	5	7	3.8(−4)	1.2(−4)	0.0072	−3.24	D+	2
		3561.75	0.0	28068	9	9	0.0029	5.6(−4)	0.059	−2.30	C	2
		3669.24	1332.2	28578	7	7	0.0020	4.0(−4)	0.034	−2.55	D+	2
		3498.19	0.0	28578	9	7	1.1(−5)	1.6(−6)	1.7(−4)	−4.84	C−	1
3.	$a\ ^3F - z\ ^5F°$ (3)											
		3502.60	0.0	28542	9	11	0.0015	3.3(−4)	0.034	−2.53	C	2
		3602.28	1332.2	29084	7	9	0.0044	0.0011	0.092	−2.11	C	2
		3620.02	2216.5	29833	5	7	4.3(−5)	1.2(−5)	7.0(−4)	−4.23	D+	2
		3437.28	0.0	29084	9	9	0.044	0.0079	0.80	−1.15	C	1,2
		3507.69	1332.2	29833	7	7	0.0024	4.4(−4)	0.036	−2.51	C	2
		3577.24	2216.5	30163	5	5	1.6(−4)	3.0(−5)	0.0018	−3.82	D+	2
		3351.06	0.0	29833	9	7	2.8(−5)	3.7(−6)	3.7(−4)	−4.48	D+	2
		3467.50	1332.2	30163	7	5	0.012	0.0015	0.12	−1.98	C	1,2
4.	$a\ ^3F - z\ ^3P°$ (4)											
		3670.43	1332.2	28569	7	5	0.0061	8.8(−4)	0.075	−2.21	D+	2
		3664.09	2216.5	29501	5	3	0.020	0.0024	0.15	−1.92	D+	2
		3793.61	2216.5	28569	5	5	0.0018	4.0(−4)	0.025	−2.70	D+	2
5.	$a\ ^3F - z\ ^3F°$ (5)	*3480.0*	*971.8*	*29699*	21	21	0.066	0.012	2.9	−0.60	C	1,2,3n
		3391.04	0.0	29481	9	9	0.066	0.011	1.1	−0.99	C	1,2
		3571.86	1332.2	29321	7	7	0.052	0.0099	0.81	−1.16	C+	1
		3519.76	2216.5	30619	5	5	0.041	0.0076	0.44	−1.42	C	1
		3409.58	0.0	29321	9	7	0.0037	5.0(−4)	0.050	−2.35	C	1,2
		3413.48	1332.2	30619	7	5	0.038	0.0047	0.37	−1.48	C	1
		3551.53	1332.2	29481	7	9	0.0016	3.9(−4)	0.032	−2.56	E	3n
		3688.41	2216.5	29321	5	7	0.0045	0.0013	0.078	−2.19	D+	2
6.	$a\ ^3F - (\ °)$[b]											
		3369.56	0.0	29669	9	7	0.18	0.024	2.4	−0.66	C	1,2
		3527.98	1332.2	29669	7	7	0.0042	7.9(−4)	0.064	−2.26	C	2
		3641.64	2216.5	29669	5	7	8.4(−5)	2.3(−5)	0.0014	−3.93	D+	2
7.	$a\ ^3F - z\ ^3D°$ (6)											
		3500.85	1332.2	29889	7	5	0.046	0.0061	0.49	−1.37	C	1,2
		3483.77	2216.5	30913	5	3	0.14	0.016	0.89	−1.11	C	1
		3612.74	2216.5	29889	5	5	0.042	0.0081	0.48	−1.39	C	1

Ni I: Allowed transitions — Continued

No.	Multiplet	λ (Å)	E_i (cm^{-1})	E_k (cm^{-1})	g_i	g_k	A_{ki} (10^8 s^{-1})	f_{ik}	S (at. u.)	log gf	Accuracy	Source
8.	$a\,^3F - z\,^3G°$ (7)	*3310.9*	*971.8*	*31166*	21	27	0.050	0.010	2.4	−0.66	C	1,2
		3232.93	0.0	30923	9	11	0.073	0.014	1.3	−0.90	C	1,2
		3371.99	1332.2	30980	7	9	0.026	0.0057	0.44	−1.40	C	1
		3380.89	2216.5	31786	5	7	0.038	0.0091	0.51	−1.34	C	1
		3226.98	0.0	30980	9	9	0.0023	3.5(−4)	0.034	−2.50	D	1
		3282.69	1332.2	31786	7	7	0.0060	9.7(−4)	0.073	−2.17	C	2
		3145.12	0.0	31786	9	7	2.1(−4)	2.4(−5)	0.0023	−3.66	D+	2
9.	$a\,^3F - z\,^1F°$ (8)											
		3221.65	0.0	31031	9	7	0.016	0.0019	0.18	−1.76	C	1,2
		3366.16	1332.2	31031	7	7	0.040	0.0068	0.53	−1.32	D+	2
		3469.48	2216.5	31031	5	7	0.013	0.0032	0.18	−1.80	C	1
10.	$a\,^3F - z\,^1D°$ (9)											
		3320.26	1332.2	31442	7	5	0.049	0.0058	0.45	−1.39	C	1,2
		3420.73	2216.5	31442	5	5	9.9(−4)	1.7(−4)	0.0098	−3.06	D+	2
11.	$a\,^3F - ^3F°$ (11)	*3104.8*	*971.8*	*33171*	21	21	0.0390	0.00564	1.21	−0.927	C+	2
		3031.87	0.0	32973	9	9	0.017	0.0024	0.21	−1.67	C+	2
		3145.71	1332.2	33112	7	7	0.0080	0.0012	0.086	−2.08	C+	2
		3184.37	2216.5	33611	5	5	0.0071	0.0011	0.056	−2.27	C+	2
		3019.14	0.0	33112	9	7	0.064	0.0069	0.61	−1.21	C+	2
		3097.12	1332.2	33611	7	5	0.033	0.0033	0.24	−1.63	C+	2
		3159.52	1332.2	32973	7	9	3.0(−4)	5.7(−5)	0.0041	−3.40	C+	2
		3235.75	2216.5	33112	5	7	6.4(−4)	1.4(−4)	0.0075	−3.15	C+	2
12.	$a\,^3F - z\,^1P°$ (10)											
		3249.44	2216.5	32982	5	3	0.0041	3.9(−4)	0.021	−2.71	C+	2
13.	$a\,^3F - y\,^3D°$ (12)	*3035.8*	*971.8*	*33903*	21	15	0.0603	0.00596	1.25	−0.903	C	2
		2984.13	0.0	33501	9	7	0.066	0.0069	0.61	−1.21	C+	2
		3045.01	1332.2	34163	7	5	0.028	0.0028	0.20	−1.71	C	2
		3105.46	2216.5	34409	5	3	0.071	0.0062	0.32	−1.51	D+	2
		3107.71	1332.2	33501	7	7	0.0013	1.9(−4)	0.014	−2.87	C+	2
		3129.30	2216.5	34163	5	5	0.0053	7.8(−4)	0.040	−2.41	D+	2
		3195.57	2216.5	33501	5	7	0.0058	0.0012	0.065	−2.21	C+	2
14.	$a\,^3F - z\,^1G°$ (13)											
		3099.11	1332.2	33590	7	9	0.021	0.0038	0.27	−1.57	C+	2
15.	$a\,^3F - y\,^1F°$ (uv 1)											
		2805.08	0.0	35639	9	7	0.0088	8.0(−4)	0.067	−2.14	C	2
		2914.01	1332.2	35639	7	7	0.0058	7.3(−4)	0.049	−2.29	C	2
		2991.11	2216.5	35639	5	7	0.0012	2.3(−4)	0.011	−2.94	D+	2

Ni I: Allowed transitions — Continued

No.	Multiplet	λ (Å)	E_i (cm^{-1})	E_k (cm^{-1})	g_i	g_k	A_{ki} (10^8 s^{-1})	f_{ik}	S (at. u.)	log gf	Accu- racy	Source
16.	$a\ ^3F - y\ ^1D°$ (uv 2)											
		2834.55	1332.2	36601	7	5	0.0066	5.7(−4)	0.037	−2.40	D+	2
		2907.46	2216.5	36601	5	5	0.020	0.0025	0.12	−1.90	D+	2
17.	$a\ ^3F - ^5P°$											
		2476.88	0.0	40361	9	7	0.026	0.0018	0.14	−1.78	C	1
		2553.37	1332.2	40484	7	5	0.0060	4.2(−4)	0.025	−2.53	D−	1
		2561.42	1332.2	40361	7	7	0.0037	3.7(−4)	0.022	−2.59	D−	1
18.	$a\ ^3F - ^3F°$											
		2347.51	0.0	42585	9	9	0.22	0.018	1.3	−0.78	D−	1
19.	$a\ ^3F - ^3D°$											
		2345.54	0.0	42621	9	7	2.2	0.14	9.9	0.11	C	1
		2419.31	1332.2	42654	7	5	0.20	0.012	0.69	−1.06	E	1
20.	$a\ ^3F - ^3G°$											
		2320.03	0.0	43090	9	11	6.9	0.69	47	0.79	C	1
		2325.79	1332.2	44315	7	9	3.5	0.37	20	0.41	C	1
		2312.34	1332.2	44565	7	7	5.5	0.44	24	0.49	C	1
21.	$a\ ^3F - ^3F°$											
		2274.66	1332.2	45281	7	7	0.052	0.0040	0.21	−1.55	D−	1
		2313.98	2216.5	45419	5	5	5.0	0.40	15	0.30	E	1
		2267.55	1332.2	45419	7	5	0.080	0.0044	0.23	−1.51	D−	1
		2321.38	2216.5	45281	5	7	5.6	0.63	24	0.50	E	1
22.	$a\ ^3F - ^3P°$											
		2346.63	1332.2	43933	7	5	0.55	0.033	1.8	−0.64	D−	1
23.	$a\ ^3F - (\ °)^b$											
		2289.98	0.0	43655	9	7	2.1	0.13	8.7	0.06	C	1
24.	$a\ ^3F - ^5D°$											
		2261.42	0.0	44206	9	7	0.091	0.0054	0.36	−1.31	C	1
25.	$a\ ^3F - (\ °)^b$											
		2254.81	0.0	44336	9	9	0.096	0.0073	0.49	−1.18	C	1
		2324.65	1332.2	44336	7	9	0.18	0.019	1.0	−0.88	D−	1
26.	$a\ ^3F - ^3D°$											
		2317.16	1332.2	44475	7	5	3.8	0.22	12	0.18	C	1
		2329.96	2216.5	45122	5	3	5.3	0.26	9.9	0.11	C	1
27.	$a\ ^3F - x\ ^3P°$ (uv 15)											
		2212.15	1332.2	46523	7	5	0.058	0.0031	0.16	−1.67	D−	1
		2221.94	2216.5	47208	5	3	0.22	0.0096	0.35	−1.32	D	1

Ni I: Allowed transitions — Continued

No.	Multiplet	λ (Å)	E_i (cm^{-1})	E_k (cm^{-1})	g_i	g_k	A_{ki} (10^8 s^{-1})	f_{ik}	S (at. u.)	log gf	Accuracy	Source
28.	$a\ ^3F - v\ ^3D°$ (uv 16)											
		2125.62	0.0	47030	9	7	0.051	0.0027	0.17	−1.62	C	1
		2182.38	1332.2	47139	7	5	0.13	0.0068	0.34	−1.32	C	1
29.	$a\ ^3F - {}^1F°$											
		2053.91	0.0	48672	9	7	0.0075	3.7(−4)	0.022	−2.48	D	1
		2111.73	1332.2	48672	7	7	0.065	0.0043	0.21	−1.52	C	1
		2151.93	2216.5	48672	5	7	0.032	0.0031	0.11	−1.81	C	1
30.	$a\ ^3F - {}^3F°$											
		2052.04	0.0	48715	9	9	0.097	0.0061	0.37	−1.26	C	1
		2052.45	1332.2	50039	7	5	0.032	0.0014	0.068	−2.00	C	1
		2109.79	1332.2	48715	7	9	0.015	0.0013	0.063	−2.04	C	1
31.	$a\ ^3F - {}^1D°$											
		2089.09	1332.2	49185	7	5	0.097	0.0045	0.22	−1.50	D	1
		2128.41	2216.5	49185	5	5	0.056	0.0038	0.13	−1.72	C	1
32.	$a\ ^3F - {}^3D°$											
		2026.62	0.0	49328	9	7	0.24	0.012	0.70	−0.98	C	1
		2025.40	1332.2	50689	7	5	0.23	0.010	0.47	−1.15	D−	1
		2055.50	2216.5	50851	5	3	0.33	0.013	0.43	−1.20	C	1
		2082.87	1332.2	49328	7	7	0.085	0.0056	0.27	−1.41	C	1
		2062.37	2216.5	50689	5	5	0.046	0.0030	0.10	−1.83	E	1
33.	$a\ ^3F - (\ °)^b$											
		1994.29	0.0	50143	9	7	0.057	0.0027	0.16	−1.62	C	1
34.	$a\ ^3F - u\ ^3F°$ (uv 23)											
		1968.90	0.0	50790	9	9	0.045	0.0026	0.15	−1.63	C	1
		2007.69	1332.2	51125	7	7	0.090	0.0054	0.25	−1.42	C	1
35.	$a\ ^3D - z\ ^5D°$ (15)											
		3674.06	204.8	27415	7	5	0.0027	3.8(−4)	0.033	−2.57	D−	1
36.	$a\ ^3D - z\ ^5G°$ (16)											
		3587.93	204.8	28068	7	9	0.0026	6.5(−4)	0.054	−2.34	C	1,2
		3609.31	879.8	28578	5	7	0.0059	0.0016	0.097	−2.09	C	1,2
		3523.44	204.8	28578	7	7	0.0027	5.1(−4)	0.041	−2.45	D+	2
37.	$a\ ^3D - z\ ^5F°$ (17)											
		3461.66	204.8	29084	7	9	0.27	0.062	5.0	−0.36	C+	1
		3452.88	879.8	29833	5	7	0.098	0.025	1.4	−0.91	C+	1
		3513.93	1713.1	30163	3	5	0.011	0.0033	0.11	−2.01	C−	1
		3374.22	204.8	29833	7	7	0.015	0.0025	0.19	−1.76	C	1,2
		3413.94	879.8	30163	5	5	0.022	0.0038	0.21	−1.72	D+	2
		3485.88	1713.1	30392	3	3	0.013	0.0024	0.081	−2.15	C	1,2
		3337.01	204.8	30163	7	5	8.3(−5)	9.9(−6)	7.6(−4)	−4.16	D+	2

Ni I: Allowed transitions — Continued

No.	Multiplet	λ (Å)	E_i (cm^{-1})	E_k (cm^{-1})	g_i	g_k	A_{ki} (10^8 s^{-1})	f_{ik}	S (at. u.)	log gf	Accuracy	Source
38.	$a\ ^3D - z\ ^3P°$ (18)	*3529.0*	*731.5*	*29060*	15	9	1.1	0.13	22	0.28	C+	1,2
		3524.54	204.8	28569	7	5	1.0	0.13	11	−0.03	C	1,2
		3492.96	879.8	29501	5	3	0.98	0.11	6.2	−0.27	C+	1
		3510.33	1713.1	30192	3	1	1.2	0.071	2.5	−0.67	C+	1
		3610.46	879.8	28569	5	5	0.072	0.014	0.84	−1.15	C+	1
		3597.70	1713.1	29501	3	3	0.14	0.027	0.96	−1.09	C	1,2
		3722.49	1713.1	28569	3	5	0.0080	0.0028	0.10	−2.08	D+	2
39.	$a\ ^3D - z\ ^3F°$ (19)	*3451.2*	*731.5*	*29699*	15	21	0.596	0.149	25.4	0.349	C	1,2
		3414.76	204.8	29481	7	9	0.55	0.12	9.8	−0.06	C	1
		3515.05	879.8	29321	5	7	0.42	0.11	6.4	−0.26	C	1,2
		3458.46	1713.1	30619	3	5	0.61	0.18	6.3	−0.26	C+	1
		3433.56	204.8	29321	7	7	0.17	0.031	2.4	−0.67	C+	1
		3361.55	879.8	30619	5	5	0.048	0.0081	0.45	−1.39	C	1,2
		3286.94	204.8	30619	7	5	0.0047	5.4(−4)	0.041	−2.42	C	1,2
40.	$a\ ^3D - (\ °)^b$											
		3392.98	204.8	29669	7	7	0.24	0.041	3.2	−0.54	C+	1
		3472.55	879.8	29669	5	7	0.12	0.031	1.8	−0.81	C+	1
41.	$a\ ^3D - z\ ^3D°$ (20)											
		3446.26	879.8	29889	5	5	0.44	0.078	4.4	−0.41	C+	1
		3423.71	1713.1	30913	3	3	0.33	0.058	2.0	−0.76	C	1,2
		3367.88	204.8	29889	7	5	0.0019	2.3(−4)	0.018	−2.79	D+	2
		3548.19	1713.1	29889	3	5	0.029	0.0090	0.31	−1.57	D+	2
42.	$a\ ^3D - z\ ^3G°$ (21)											
		3248.46	204.8	30980	7	9	0.0047	9.7(−4)	0.072	−2.17	C	1
		3234.65	879.8	31786	5	7	0.020	0.0045	0.24	−1.65	C	1,2
		3165.51	204.8	31786	7	7	4.3(−4)	6.5(−5)	0.0048	−3.34	D+	2
43.	$a\ ^3D - z\ ^1F°$ (22)											
		3243.05	204.8	31031	7	7	0.048	0.0075	0.56	−1.28	C	1,2
		3315.66	879.8	31031	5	7	0.053	0.012	0.67	−1.21	C+	1
44.	$a\ ^3D - z\ ^1D°$ (23)											
		3200.42	204.8	31442	7	5	0.0028	3.1(−4)	0.023	−2.66	C	2
		3271.11	879.8	31442	5	5	0.0086	0.0014	0.075	−2.16	C	1,2
		3362.80	1713.1	31442	3	5	0.0021	5.9(−4)	0.020	−2.75	D+	2
45.	$a\ ^3D - \ ^3F°$	*3081.8*	*731.5*	*33171*	15	21	0.841	0.168	25.5	0.400	C+	2
		3050.82	204.8	32973	7	9	0.60	0.11	7.6	−0.12	C+	2
		3101.56	879.8	33112	5	7	0.63	0.13	6.4	−0.20	C+	2
		3134.11	1713.1	33611	3	5	0.73	0.18	5.5	−0.27	C+	2
		3037.93	204.8	33112	7	7	0.28	0.039	2.8	−0.56	C+	2
		3054.31	879.8	33611	5	5	0.40	0.056	2.8	−0.55	C+	2
		2992.59	204.8	33611	7	5	0.054	0.0052	0.36	−1.44	C+	2

Ni I: Allowed transitions — Continued

No.	Multiplet	λ (Å)	E_i (cm^{-1})	E_k (cm^{-1})	g_i	g_k	A_{ki} (10^8 s^{-1})	f_{ik}	S (at. u.)	log gf	Accuracy	Source
46.	$a\ ^3D - z\ ^1P°$ (24)											
		3114.12	879.8	32982	5	3	0.058	0.0050	0.26	−1.60	C+	2
		3197.11	1713.1	32982	3	3	0.029	0.0044	0.14	−1.88	C+	2
47.	$a\ ^3D - y\ ^3D°$ (uv 24)	*3013.8*	*731.5*	*33903*	15	15	0.982	0.134	19.9	0.302	C−	1,2
		3002.48	204.8	33501	7	7	0.80	0.11	7.5	−0.12	C+	2
		3003.62	879.8	34163	5	5	0.69	0.094	4.6	−0.33	D−	1
		3057.64	1713.1	34409	3	3	1.0	0.14	4.3	−0.37	D−	1
		2943.91	204.8	34163	7	5	0.11	0.0099	0.67	−1.16	C	1
		2981.65	879.8	34409	5	3	0.28	0.022	1.1	−0.95	C	2
		3064.62	879.8	33501	5	7	0.11	0.021	1.1	−0.98	C+	2
		3080.75	1713.1	34163	3	5	0.087	0.021	0.63	−1.21	C	1,2
48.	$a\ ^3D - z\ ^1G°$ (27)											
		2994.46	204.8	33590	7	9	0.087	0.015	1.0	−0.98	C+	2
49.	$a\ ^3D - y\ ^1F°$ (uv 25)											
		2821.29	204.8	35639	7	7	0.049	0.0058	0.38	−1.39	C	1,2
50.	$a\ ^3D - y\ ^1D°$ (uv 26)											
		2746.74	204.8	36601	7	5	0.017	0.0013	0.084	−2.03	C	1,2
		2798.65	879.8	36601	5	5	0.058	0.0068	0.31	−1.47	C	1,2
		2865.50	1713.1	36601	3	5	0.018	0.0037	0.11	−1.95	C	2
51.	$a\ ^3D - ^3F°$											
		2348.73	204.8	42768	7	7	0.22	0.018	0.97	−0.90	E	1
52.	$a\ ^3D - ^3G°$											
		2266.35	204.8	44315	7	9	0.023	0.0023	0.12	−1.80	D−	1
		2288.40	879.8	44565	5	7	0.081	0.0089	0.34	−1.35	D−	1
		2253.57	204.8	44565	7	7	0.19	0.015	0.76	−0.99	C	1
53.	$a\ ^3D - ^3F°$											
		2251.48	879.8	45281	5	7	0.040	0.0043	0.16	−1.67	D	1
		2287.32	1713.1	45419	3	5	0.18	0.024	0.55	−1.14	D−	1
		2244.46	879.8	45419	5	5	0.38	0.029	1.1	−0.84	C	1
54.	$a\ ^3D - (\ °)^b$											
		2300.77	204.8	43655	7	7	0.75	0.060	3.2	−0.38	C	1
55.	$a\ ^3D - ^5D°$											
		2307.35	879.8	44206	5	7	0.16	0.017	0.66	−1.06	C	1
		2271.95	204.8	44206	7	7	0.050	0.0038	0.20	−1.57	C	1

Ni I: Allowed transitions — Continued

No.	Multiplet	λ (Å)	E_i (cm^{-1})	E_k (cm^{-1})	g_i	g_k	A_{ki} (10^8 s^{-1})	f_{ik}	S (at. u.)	log gf	Accuracy	Source
56.	$a\ ^3D - ^3D°$											
		2293.11	879.8	44475	5	5	0.38	0.030	1.1	−0.82	C	1
		2302.97	1713.1	45122	3	3	0.45	0.036	0.81	−0.97	C	1
		2258.15	204.8	44475	7	5	0.17	0.0094	0.49	−1.18	C	1
		2259.56	879.8	45122	5	3	0.20	0.0091	0.34	−1.34	C	1
57.	$a\ ^3D - x\ ^3P°$ (uv 36)	*2166.2*	*731.5*	*46881*	15	9	1.1	0.046	4.9	−0.16	C−	1
		2158.31	204.8	46523	7	5	0.69	0.034	1.7	−0.62	C	1
		2157.83	879.8	47208	5	3	0.41	0.017	0.60	−1.07	C	1
		2174.48	1713.1	47687	3	1	0.89	0.021	0.45	−1.20	E	1
		2190.22	879.8	46523	5	5	0.30	0.021	0.77	−0.97	C	1
		2197.35	1713.1	47208	3	3	0.78	0.057	1.2	−0.77	C	1
		2230.96	1713.1	46523	3	5	0.052	0.0065	0.14	−1.71	E	1
58.	$a\ ^3D - v\ ^3D°$ (uv 37)											
		2161.04	879.8	47139	5	5	0.13	0.0089	0.32	−1.35	C	1
		2129.96	204.8	47139	7	5	0.042	0.0020	0.099	−1.85	C	1
		2147.80	879.8	47425	5	3	0.47	0.020	0.69	−1.01	C	1
		2166.15	879.8	47030	5	7	0.066	0.0065	0.23	−1.49	C	1
		2220.71	1713.1	47139	3	5	0.082	0.010	0.22	−1.52	C	1
59.	$a\ ^3D - ^5S°$											
		2121.40	204.8	47329	7	5	0.28	0.014	0.67	−1.02	C	1
		2152.23	879.8	47329	5	5	0.032	0.0022	0.078	−1.96	C	1
60.	$a\ ^3D - ^3F°$											
		2033.56	879.8	50039	5	5	0.030	0.0019	0.062	−2.03	D−	1
61.	$a\ ^3D - w\ ^3P°$ (uv 40)											
		2059.92	204.8	48735	7	5	0.21	0.0097	0.46	−1.17	D−	1
		2060.20	879.8	49403	5	3	0.23	0.0087	0.30	−1.36	E	1
		2064.39	1713.1	50139	3	1	0.40	0.0086	0.17	−1.59	D−	1
		2088.98	879.8	48735	5	5	0.042	0.0028	0.095	−1.86	D−	1
62.	$a\ ^3D - ^1P°$											
		2085.37	879.8	48818	5	3	0.077	0.0030	0.10	−1.82	C	1
63.	$a\ ^3D - ^1D°$											
		2047.35	204.8	49033	7	5	0.18	0.0082	0.39	−1.24	C	1
64.	$a\ ^3D - ^1D°$											
		2069.52	879.8	49185	5	5	0.11	0.0071	0.24	−1.45	D−	1
		2105.85	1713.1	49185	3	5	0.030	0.0033	0.069	−2.00	C	1
65.	$a\ ^3D - ^3D°$											
		2007.01	879.8	50689	5	5	0.17	0.010	0.35	−1.28	C	1
		2000.49	879.8	50851	5	3	0.054	0.0020	0.064	−2.01	C	1
		2063.42	879.8	49328	5	7	0.050	0.0045	0.15	−1.65	D−	1
		2041.16	1713.1	50689	3	5	0.032	0.0033	0.067	−2.00	D−	1

Ni I: Allowed transitions — Continued

No.	Multiplet	λ (Å)	E_i (cm^{-1})	E_k (cm^{-1})	g_i	g_k	A_{ki} (10^8 s^{-1})	f_{ik}	S (at. u.)	log gf	Accuracy	Source
66.	$a\,^3D - (\ ^\circ)^b$											
		2001.83	204.8	50143	7	7	0.073	0.0044	0.20	−1.51	C	1
		2029.29	879.8	50143	5	7	0.023	0.0020	0.065	−2.01	C	1
67.	$a\,^3D - x\,^1P^\circ$ (uv 45)											
		2050.84	1713.1	50458	3	3	0.076	0.0048	0.098	−1.84	E	1
68.	$a\,^3D - u\,^3F^\circ$ (uv 47)											
		1976.87	204.8	50790	7	9	1.1	0.080	3.7	−0.25	C	1
		1990.25	879.8	51125	5	7	0.83	0.069	2.3	−0.46	C	1
		2014.25	1713.1	51344	3	5	0.93	0.094	1.9	−0.55	C	1
		1963.85	204.8	51125	7	7	0.11	0.0064	0.29	−1.35	C	1
		1981.61	879.8	51344	5	5	0.13	0.0078	0.25	−1.41	C	1
69.	$a\,^1D - z\,^5F^\circ$ (30)											
		3783.53	3409.9	29833	5	7	0.033	0.0098	0.61	−1.31	C	1
		3736.81	3409.9	30163	5	5	0.014	0.0029	0.18	−1.84	E	1
70.	$a\,^1D - z\,^3P^\circ$ (31)											
		3973.55	3409.9	28569	5	5	0.0038	8.9(−4)	0.058	−2.35	D+	2
		3831.69	3409.9	29501	5	3	0.015	0.0020	0.13	−2.00	C	1
71.	$a\,^1D - z\,^3F^\circ$ (32)											
		3858.30	3409.9	29321	5	7	0.069	0.021	1.4	−0.97	C	1
		3674.15	3409.9	30619	5	5	0.020	0.0041	0.25	−1.69	D+	2
72.	$a\,^1D - (\ ^\circ)^b$											
		3807.14	3409.9	29669	5	7	0.043	0.013	0.83	−1.18	C	1
73.	$a\,^1D - z\,^3D^\circ$ (33)											
		3775.57	3409.9	29889	5	5	0.042	0.0089	0.56	−1.35	C	1
74.	$a\,^1D - z\,^3G^\circ$ (34)											
		3523.07	3409.9	31786	5	7	4.4(−4)	1.2(−4)	0.0067	−3.24	D+	2
75.	$a\,^1D - z\,^1F^\circ$ (35)	3619.39	3409.9	31031	5	7	0.66	0.18	11.0	−0.04	C	1,2
76.	$a\,^1D - z\,^1D^\circ$ (36)	3566.37	3409.9	31442	5	5	0.56	0.11	6.3	−0.27	C	1
77.	$a\,^1D - \,^3F^\circ$											
		3365.76	3409.9	33112	5	7	0.054	0.013	0.72	−1.19	C+	2
		3310.20	3409.9	33611	5	5	0.0012	2.0(−4)	0.011	−3.01	C+	2

Ni I: Allowed transitions — Continued

No.	Multiplet	λ (Å)	E_i (cm^{-1})	E_k (cm^{-1})	g_i	g_k	A_{ki} (10^8 s^{-1})	f_{ik}	S (at. u.)	log gf	Accuracy	Source
78.	a ^{1}D – z ^{1}P° (37)	3380.57	3409.9	32982	5	3	1.3	0.14	7.5	−0.17	C+	2
79.	a ^{1}D – y ^{3}D° (39)											
		3322.31	3409.9	33501	5	7	0.058	0.014	0.74	−1.17	C+	2
		3250.74	3409.9	34163	5	5	0.022	0.0035	0.19	−1.76	C	2
		3225.02	3409.9	34409	5	3	0.093	0.0087	0.46	−1.36	C	1,2
80.	a ^{1}D – y ^{1}F° (40)	3101.88	3409.9	35639	5	7	0.49	0.098	5.0	−0.31	C	1
81.	a ^{1}D – y ^{1}D° (41)	3012.00	3409.9	36601	5	5	1.3	0.18	9.0	−0.04	C	1,2
82.	a ^{1}D – ^{5}P°											
		2696.48	3409.9	40484	5	5	0.014	0.0015	0.067	−2.12	D−	1
83.	a ^{1}D – ^{3}F°											
		2540.02	3409.9	42768	5	7	0.026	0.0035	0.15	−1.76	D−	1
84.	a ^{1}D – w ^{3}P° (uv 59)											
		2173.54	3409.9	49403	5	3	0.15	0.0065	0.23	−1.49	E	1
85.	a ^{1}D – ^{1}P°	2201.59	3409.9	48818	5	3	0.73	0.032	1.1	−0.80	C	1
86.	a ^{1}D – ^{1}D°	2191.21	3409.9	49033	5	5	0.046	0.0033	0.12	−1.78	E	1
87.	a ^{1}D – ^{1}D°	2183.91	3409.9	49185	5	5	0.12	0.0089	0.32	−1.35	D	1
88.	a ^{1}D – ^{3}D°											
		2114.43	3409.9	50689	5	5	0.097	0.0065	0.23	−1.49	C	1
89.	a ^{1}D – x ^{1}P° (uv 63)	2124.80	3409.9	50458	5	3	0.38	0.016	0.54	−1.11	C	1
90.	a ^{1}D – u ^{3}F° (uv 65)											
		2095.13	3409.9	51125	5	7	0.11	0.010	0.35	−1.30	E	1
		2085.57	3409.9	51344	5	5	2.6	0.17	5.8	−0.07	D−	1
91.	b ^{1}D – z ^{5}F° (42)											
		6128.99	13521	29833	5	7	1.2(−4)	9.4(−5)	0.0094	−3.33	D+	2
		6007.31	13521	30163	5	5	1.7(−4)	9.4(−5)	0.0093	−3.33	D	4
92.	b ^{1}D – z ^{3}P° (43)											
		6643.64	13521	28569	5	5	0.0015	0.0010	0.11	−2.30	D	3n
		6256.36	13521	29501	5	3	0.0019	6.6(−4)	0.068	−2.48	D	3n

Ni I: Allowed transitions — Continued

No.	Multiplet	λ (Å)	E_i (cm^{-1})	E_k (cm^{-1})	g_i	g_k	A_{ki} (10^8 s^{-1})	f_{ik}	S (at. u.)	log gf	Accuracy	Source
93.	$b\ ^1D - z\ ^3F°$ (44)											
		6327.60	13521	29321	5	7	1.7(−4)	1.4(−4)	0.015	−3.15	E	3n
		5847.00	13521	30619	5	5	2.4(−4)	1.2(−4)	0.012	−3.21	D+	2
94.	$b\ ^1D - z\ ^3D°$ (45)											
		6108.12	13521	29889	5	5	0.0013	7.1(−4)	0.071	−2.45	D+	2
		5748.34	13521	30913	5	3	3.7(−4)	1.1(−4)	0.010	−3.26	D	4
95.	$b\ ^1D - z\ ^1F°$ (46)	5709.55	13521	31031	5	7	0.0020	0.0014	0.13	−2.17	D+	2
96.	$b\ ^1D - z\ ^1D°$ (47)	5578.72	13521	31442	5	5	9.8(−4)	4.6(−4)	0.042	−2.64	D+	2
97.	$b\ ^1D - ^3F°$											
		5102.97	13521	33112	5	7	8.8(−4)	4.8(−4)	0.040	−2.62	C+	2
		4976.35	13521	33611	5	5	4.3(−4)	1.6(−4)	0.013	−3.10	C+	2
98.	$b\ ^1D - z\ ^1P°$ (48)	5137.08	13521	32982	5	3	0.0086	0.0020	0.17	−1.99	C+	2
99.	$b\ ^1D - y\ ^3D°$ (50)											
		5003.75	13521	33501	5	7	6.0(−4)	3.2(−4)	0.026	−2.80	C+	2
		4843.17	13521	34163	5	5	2.9(−4)	1.0(−4)	0.0080	−3.30	D+	2
		4786.29	13521	34409	5	3	7.4(−4)	1.5(−4)	0.012	−3.12	D+	2
100.	$b\ ^1D - y\ ^1F°$ (51)	4519.99	13521	35639	5	7	6.1(−4)	2.6(−4)	0.020	−2.88	D+	2
101.	$b\ ^1D - y\ ^1D°$ (52)	4331.65	13521	36601	5	5	0.0056	0.0016	0.11	−2.10	D+	2
102.	$a\ ^1S - z\ ^3P°$ (57)											
		6767.77	14729	29501	1	3	0.0033	0.0068	0.15	−2.17	D	3n
103.	$a\ ^1S - z\ ^3D°$ (58)											
		6177.26	14729	30913	1	3	1.8(−4)	3.2(−4)	0.0064	−3.50	D	4
104.	$a\ ^1S - z\ ^1P°$ (59)	5476.91	14729	32982	1	3	0.095	0.13	2.3	−0.89	C+	2
105.	$a\ ^1S - y\ ^3D°$ (60)											
		5079.96	14729	34409	1	3	0.0015	0.0018	0.030	−2.75	D+	2

Ni I: Allowed transitions — Continued

No.	Multiplet	λ (Å)	E_i (cm^{-1})	E_k (cm^{-1})	g_i	g_k	A_{ki} (10^8 s^{-1})	f_{ik}	S (at. u.)	log gf	Accuracy	Source
106.	$a\ ^3P - z\ ^3P°$ (62)	*7481.3*	*15697*	*29060*	9	9	0.0029	0.0024	0.54	−1.66	D−	3n,4
		7714.32	15610	28569	5	5	0.0014	0.0013	0.16	−2.20	D	3n
		7261.93	15734	29501	3	3	8.4(−4)	6.7(−4)	0.048	−2.70	E	3n
		7197.02	15610	29501	5	3	9.0(−4)	4.2(−4)	0.050	−2.68	E	3n
		6914.56	15734	30192	3	1	0.0075	0.0018	0.12	−2.27	D	4
		7788.94	15734	28569	3	5	8.4(−4)	0.0013	0.097	−2.42	D	3n
		7414.51	16017	29501	1	3	0.0011	0.0027	0.066	−2.57	E	3n
107.	$a\ ^3P - z\ ^3F°$ (63)											
		7291.45	15610	29321	5	7	3.0(−4)	3.3(−4)	0.040	−2.78	E	3n
108.	$a\ ^3P - (\ °)^b$											
		7110.91	15610	29669	5	7	2.0(−4)	2.1(−4)	0.025	−2.98	D	4
109.	$a\ ^3P - z\ ^3D°$ (64)											
		7062.97	15734	29889	3	5	8.5(−5)	1.1(−4)	0.0074	−3.50	E	3n
		7001.54	15610	29889	5	5	6.0(−5)	4.4(−5)	0.0050	−3.66	E	3n
		6586.33	15734	30913	3	3	7.9(−4)	5.2(−4)	0.034	−2.81	D	4
		6532.89	15610	30913	5	3	2.1(−4)	8.1(−5)	0.0088	−3.39	D	4
110.	$a\ ^3P - z\ ^1F°$ (66)											
		6482.80	15610	31031	5	7	5.3(−4)	4.7(−4)	0.050	−2.63	E	3n
111.	$a\ ^3P - z\ ^1D°$ (67)											
		6314.66	15610	31442	5	5	0.0057	0.0034	0.35	−1.77	D	3n
		6364.60	15734	31442	3	5	3.5(−5)	3.6(−5)	0.0022	−3.97	E	3n
112.	$a\ ^3P - \ ^3F°$											
		5711.91	15610	33112	5	7	0.0016	0.0011	0.10	−2.27	C+	2
		5592.28	15734	33611	3	5	0.0011	9.0(−4)	0.050	−2.57	C+	2
		5553.69	15610	33611	5	5	2.5(−4)	1.2(−4)	0.011	−3.23	C+	2
113.	$a\ ^3P - z\ ^1P°$ (68)											
		5754.68	15610	32982	5	3	0.0031	9.4(−4)	0.089	−2.33	C+	2
		5892.88	16017	32982	1	3	0.0029	0.0045	0.087	−2.35	C+	2
114.	$a\ ^3P - y\ ^3D°$ (70)											
		5587.87	15610	33501	5	7	0.0022	0.0014	0.13	−2.14	C+	2
		5424.65	15734	34163	3	5	7.7(−4)	5.7(−4)	0.030	−2.77	D+	2
		5435.87	16017	34409	1	3	0.0019	0.0026	0.046	−2.59	D+	2
		5388.35	15610	34163	5	5	1.3(−4)	5.5(−5)	0.0049	−3.56	D+	2
		5353.42	15734	34409	3	3	0.0012	5.3(−4)	0.028	−2.80	D+	2

Ni I: Allowed transitions — Continued

No.	Multiplet	λ (Å)	E_i (cm^{-1})	E_k (cm^{-1})	g_i	g_k	A_{ki} (10^8 s^{-1})	f_{ik}	S (at. u.)	log gf	Accuracy	Source
115.	$a\ ^3P - y\ ^1D°$ (71)											
		4762.63	15610	36601	5	5	0.0022	7.6(−4)	0.060	−2.42	D+	2
		4791.00	15734	36601	3	5	2.3(−4)	1.3(−4)	0.0063	−3.40	D+	2
116.	$a\ ^1G - y\ ^1F°$ (84)	7385.24	22102	35639	9	7	0.0019	0.0012	0.26	−1.97	D	4
117.	$z\ ^5D° - e\ ^3D$											
		6271.76	26666	42606	7	7	5.8(−4)	3.4(−4)	0.050	−2.62	D	4
118.	$z\ ^5D° - e\ ^5F$ (86)											
		4401.54	25754	48467	9	11	0.38	0.13	17	0.08	D	3
		4470.48	27415	49778	5	7	0.19	0.080	5.9	−0.40	D−	4
		4462.46	27944	50346	3	5	0.17	0.084	3.7	−0.60	D−	4
119.	$z\ ^5D° - e\ ^3F$ (88)											
		4410.52	26666	49333	7	9	0.032	0.012	1.2	−1.08	D−	4
120.	$z\ ^5D° - f\ ^3F$ (89)											
		4200.46	26666	50466	7	9	0.018	0.0062	0.60	−1.36	D−	4
121.	$z\ ^5G° - e\ ^3D$ (97)											
		7034.42	28578	42790	7	5	0.0026	0.0014	0.23	−2.01	D	4
122.	$z\ ^5G° - e\ ^5F$ (98)											
		4714.42	27261	48467	13	11	0.46	0.13	26	0.23	D	3
		4648.66	27580	49086	11	9	0.24	0.063	11	−0.16	D−	4
		4604.99	28068	49778	9	7	0.23	0.057	7.8	−0.29	D−	4
		4600.37	29013	50745	5	3	0.26	0.049	3.7	−0.61	D−	4
		4786.54	27580	48467	11	11	0.18	0.061	11	−0.17	D	3
		4756.52	28068	49086	9	9	0.15	0.051	7.2	−0.34	D−	4
		4715.78	28578	49778	7	7	0.20	0.065	7.1	−0.34	D−	4
		4686.22	29013	50346	5	5	0.14	0.046	3.5	−0.64	D−	4
		4900.97	28068	48467	9	11	0.0050	0.0022	0.32	−1.70	E	3
		4814.62	29013	49778	5	7	0.0086	0.0042	0.33	−1.68	D	4
123.	$z\ ^5G° - e\ ^3G$ (99)											
		4740.17	28068	49159	9	11	0.0034	0.0014	0.20	−1.90	E	3
		4633.03	27580	49159	11	11	7.8(−4)	2.5(−4)	0.042	−2.56	E	3
124.	$z\ ^5G° - f\ ^3D$ (100)											
		4606.23	29013	50717	5	3	0.10	0.020	1.5	−1.00	D−	4

Ni I: Allowed transitions — Continued

No.	Multiplet	λ (Å)	E_i (cm^{-1})	E_k (cm^{-1})	g_i	g_k	A_{ki} (10^8 s^{-1})	f_{ik}	S (at. u.)	log gf	Accuracy	Source
125.	$z\ ^5G° - e\ ^3F$ (101)											
		4701.34	28068	49333	9	9	0.020	0.0067	0.93	−1.22	D−	4
126.	$z\ ^5G° - f\ ^3F$ (102)											
		4567.42	28578	50466	7	9	0.0014	5.6(−4)	0.058	−2.41	D	4
127.	$z\ ^5F° - e\ ^3D$ (109)											
		7286.56	30392	44112	3	3	0.0025	0.0020	0.14	−2.22	D	4
128.	$z\ ^5F° - e\ ^5F$ (111)											
		5017.58	28542	48467	11	11	0.20	0.076	14	−0.08	D	3
		4998.23	29084	49086	9	9	0.049	0.018	2.7	−0.78	D−	4
		5012.46	29833	49778	7	7	0.11	0.041	4.8	−0.54	D−	4
		4953.20	30163	50346	5	5	0.12	0.043	3.5	−0.67	D−	4
		4912.03	30392	50745	3	3	0.15	0.053	2.6	−0.80	D−	4
		4831.18	29084	49778	9	7	0.16	0.042	6.0	−0.42	D−	4
		5157.99	29084	48467	9	11	0.0059	0.0029	0.44	−1.59	E	3
129.	$z\ ^5F° - e\ ^3G$ (112)											
		4976.16	29084	49175	9	9	0.013	0.0050	0.73	−1.35	D−	4
		4980.17	29084	49159	9	11	0.19	0.086	13	−0.11	D	3
130.	$z\ ^5F° - e\ ^3P$											
		5262.83	30163	49159	5	5	5.7(−4)	2.3(−4)	0.020	−2.93	E	3
131.	$z\ ^5F° - e\ ^3F$ (114)											
		4937.34	29084	49333	9	9	0.12	0.045	6.6	−0.39	D−	4
		5220.31	30163	49314	5	7	0.017	0.0098	0.84	−1.31	D−	4
132.	$z\ ^5F° - f\ ^3F$ (115)											
		4675.64	29084	50466	9	9	0.0062	0.0020	0.28	−1.74	D	4
		4845.17	29833	50466	7	9	0.0016	7.2(−4)	0.080	−2.30	D	4
133.	$z\ ^3P° - e\ ^3D$ (126)											
		7122.24	28569	42606	5	7	0.21	0.22	26	0.04	D−	4
		7030.06	28569	42790	5	5	0.0050	0.0037	0.43	−1.73	D	4
		6842.07	29501	44112	3	3	0.016	0.011	0.75	−1.48	D−	4
134.	$z\ ^3P° - e\ ^1D$ (127)											
		6370.38	28569	44263	5	5	0.0038	0.0023	0.24	−1.94	D	4
		6772.36	29501	44263	3	5	0.030	0.035	2.3	−0.98	D−	4

Ni I: Allowed transitions — Continued

No.	Multiplet	λ (Å)	E_i (cm^{-1})	E_k (cm^{-1})	g_i	g_k	A_{ki} (10^8 s^{-1})	f_{ik}	S (at. u.)	log gf	Accuracy	Source
135.	$z\,^3P^\circ - e\,^3S$ (129)											
		4904.41	28569	48953	5	3	0.62	0.14	11	-0.17	D$-$	4
136.	$z\,^3P^\circ - e\,^3P$ (130)											
		4855.41	28569	49159	5	5	0.57	0.20	16	0.00	D	3
		5082.35	29501	49171	3	3	0.25	0.096	4.8	-0.54	D$-$	4
		4852.56	28569	49171	5	3	0.080	0.017	1.4	-1.07	D$-$	4
		4812.00	29501	50276	3	1	0.095	0.011	0.52	-1.48	D$-$	4
		5085.49	29501	49159	3	5	0.017	0.011	0.55	-1.48	E	3
137.	$z\,^3P^\circ - f\,^3D$ (131)											
		4829.03	28569	49272	5	7	0.19	0.094	7.4	-0.33	D$-$	4
		5042.20	29501	49328	3	5	0.14	0.088	4.4	-0.58	D$-$	4
		4815.92	28569	49328	5	5	0.0093	0.0032	0.26	-1.79	D	4
138.	$z\,^3P^\circ - e\,^1P$ (132)											
		4752.43	29501	50537	3	3	0.20	0.067	3.1	-0.70	D$-$	4
		4913.97	30192	50537	1	3	0.22	0.23	3.8	-0.63	D$-$	4
139.	$z\,^3P^\circ - e\,^1S$ (135)											
		4553.18	29501	51457	3	1	0.070	0.0073	0.33	-1.66	D	4
140.	$z\,^3F^\circ - e\,^3D$ (139)											
		7422.30	29321	42790	7	5	0.18	0.10	18	-0.14	D$-$	4
141.	$z\,^3F^\circ - e\,^1D$ (140)											
		7327.67	30619	44263	5	5	0.0042	0.0034	0.41	-1.77	D	4
142.	$z\,^3F^\circ - e\,^5F$ (141)											
		5265.71	29481	48467	9	11	0.0037	0.0019	0.29	-1.77	E	3
		4925.58	29481	49778	9	7	0.059	0.017	2.5	-0.82	D$-$	4
		4967.55	30619	50745	5	3	0.024	0.0054	0.44	-1.57	D	4
143.	$z\,^3F^\circ - e\,^3S$											
		5452.85	30619	48953	5	3	0.016	0.0044	0.39	-1.66	D	4
144.	$z\,^3F^\circ - e\,^3G$ (143)											
		5080.53	29481	49159	9	11	0.32	0.15	23	0.13	D	3
		5035.37	29321	49175	7	9	0.57	0.28	32	0.29	D$-$	4
145.	$z\,^3F^\circ - e\,^3P$ (142)											
		5039.37	29321	49159	7	5	0.037	0.010	1.2	-1.15	E	3

Ni I: Allowed transitions — Continued

No.	Multiplet	λ (Å)	E_i (cm^{-1})	E_k (cm^{-1})	g_i	g_k	A_{ki} (10^8 s^{-1})	f_{ik}	S (at. u.)	log gf	Accu-racy	Source
146.	$z\ ^3F^\circ - f\ ^3D$ (144)											
		4996.85	29321	49328	7	5	0.056	0.015	1.7	−0.98	D−	4
		5010.96	29321	49272	7	7	0.051	0.019	2.2	−0.87	D−	4
147.	$z\ ^3F^\circ - e\ ^3F$ (145)											
		5000.34	29321	49314	7	7	0.14	0.053	6.1	−0.43	D−	4
		4945.46	30619	50834	5	5	0.083	0.030	2.5	−0.82	D−	4
		4995.65	29321	49333	7	9	0.0078	0.0038	0.43	−1.58	D	4
		5347.71	30619	49314	5	7	0.0030	0.0018	0.16	−2.04	D	4
148.	$z\ ^3F^\circ - f\ ^3F$ (146)											
		4666.99	30619	52041	5	5	0.063	0.020	1.6	−0.99	D−	4
149.	$z\ ^3F^\circ - e\ ^1F$ (148)											
		4946.04	30619	50832	5	7	0.020	0.010	0.84	−1.29	D−	4
150.	$z\ ^3F^\circ - g\ ^3F$ (150)											
		4009.98	29321	54251	7	7	0.0086	0.0021	0.19	−1.84	E	3
		4035.96	29481	54251	9	7	0.0048	9.0(−4)	0.11	−2.09	E	3
151.	$(\ ^\circ)^b - e\ ^3D$											
		7727.66	29669	42606	7	7	0.11	0.097	17	−0.17	D−	4
152.	$(\ ^\circ)^b - e\ ^1D$											
		6850.48	29669	44263	7	5	0.0023	0.0011	0.18	−2.10	D	4
153.	$(\ ^\circ)^b - e\ ^3P$											
		5129.37	29669	49159	7	5	0.12	0.033	4.0	−0.63	D	3
154.	$(\ ^\circ)^b - f\ ^3D$											
		5099.95	29669	49272	7	7	0.29	0.11	13	−0.10	D−	4
155.	$(\ ^\circ)^b - e\ ^3F$											
		5084.08	29669	49333	7	9	0.31	0.15	18	0.03	D−	4
		5088.96	29669	49314	7	7	0.017	0.0065	0.77	−1.34	D−	4
156.	$(\ ^\circ)^b - f\ ^3F$											
		4807.00	29669	50466	7	9	0.073	0.033	3.6	−0.64	D−	4
157.	$(\ ^\circ)^b - g\ ^3D$											
		4437.57	29669	52197	7	7	0.028	0.0082	0.84	−1.24	D−	4
158.	$z\ ^3D^\circ - e\ ^1D$ (157)											
		7488.73	30913	44263	3	5	0.0012	0.0017	0.13	−2.29	D	4

Ni I: Allowed transitions — Continued

No.	Multiplet	λ (Å)	E_i (cm^{-1})	E_k (cm^{-1})	g_i	g_k	A_{ki} (10^8 s^{-1})	f_{ik}	S (at. u.)	log gf	Accuracy	Source
159.	$z\ ^3$D° – $e\ ^5$F (158)											
		4793.47	29889	50745	5	3	0.0062	0.0013	0.10	−2.19	D	4
160.	$z\ ^3$D° – $e\ ^3$G (160)											
		4808.86	29889	50678	5	7	0.016	0.0078	0.62	−1.41	D−	4
161.	$z\ ^3$D° – $e\ ^3$P (159)											
		5187.82	29889	49159	5	5	0.0061	0.0025	0.21	−1.91	E	3
		5475.57	30913	49171	3	3	0.0059	0.0026	0.14	−2.10	D	4
162.	$z\ ^3$D° – $f\ ^3$F (163)											
		4731.81	30913	52041	3	5	0.084	0.047	2.2	−0.85	D−	4
		4513.00	29889	52041	5	5	0.022	0.0068	0.50	−1.47	D−	4
163.	$z\ ^3$D° – $e\ ^1$P (164)											
		5094.42	30913	50537	3	3	0.071	0.028	1.4	−1.08	D−	4
164.	$z\ ^3$D° – $e\ ^1$F (167)											
		4773.41	29889	50832	5	7	0.012	0.0059	0.46	−1.53	D	4
165.	$z\ ^3$G° – $e\ ^5$F (175)											
		5504.12	30923	49086	11	9	0.0044	0.0016	0.32	−1.75	D	4
166.	$z\ ^3$G° – $e\ ^3$G (176)											
		5499.41	30980	49159	9	11	6.2(−4)	3.4(−4)	0.056	−2.51	E	3
		5749.28	31786	49175	7	9	0.0023	0.0015	0.19	−1.99	D	4
167.	$z\ ^3$G° – $f\ ^3$F (177)											
		5115.40	30923	50466	11	9	0.22	0.071	13	−0.11	D−	4
		4918.36	30980	51306	9	7	0.23	0.064	9.3	−0.24	D−	4
		4935.83	31786	52041	7	5	0.24	0.064	7.3	−0.35	D−	4
168.	$z\ ^3$G° – $g\ ^3$F (178)											
		4295.88	30980	54251	9	7	0.17	0.037	4.7	−0.48	D	3
		4450.13	31786	54251	7	7	0.0082	0.0024	0.25	−1.77	E	3
169.	$z\ ^1$F° – $e\ ^5$F (188)											
		5537.11	31031	49086	7	9	0.0015	9.0(−4)	0.12	−2.20	D	4

Ni I: Allowed transitions — Continued

No.	Multiplet	λ (Å)	E_i (cm^{-1})	E_k (cm^{-1})	g_i	g_k	A_{ki} (10^8 s^{-1})	f_{ik}	S (at. u.)	log gf	Accu-racy	Source
170.	z ^{1}F° $- e$ ^{3}P (189)											
		5514.80	31031	49159	7	5	0.0045	0.0015	0.19	−1.99	E	3
171.	z ^{1}F° $- e$ ^{3}F (192)											
		5462.49	31031	49333	7	9	0.029	0.017	2.1	−0.93	D−	4
		5468.10	31031	49314	7	7	0.0078	0.0035	0.44	−1.61	D	4
172.	z ^{1}F° $- f$ ^{3}F (193)											
		4758.42	31031	52041	7	5	0.0046	0.0011	0.12	−2.11	D	4
173.	z ^{1}F° $- e$ ^{1}G (194)	5081.11	31031	50706	7	9	0.57	0.29	33	0.30	D−	4
174.	z ^{1}F° $- e$ ^{1}F (195)	5048.85	31031	50832	7	7	0.16	0.060	6.9	−0.38	D−	4
175.	z ^{1}D° $- e$ ^{5}F (202)											
		5179.14	31442	50745	5	3	0.028	0.0068	0.58	−1.47	D−	4
176.	z ^{1}D° $- e$ ^{3}G (204)											
		5197.17	31442	50678	5	7	0.023	0.013	1.1	−1.19	D−	4
177.	z ^{1}D° $- e$ ^{3}P (203)											
		5642.66	31442	49159	5	5	0.0040	0.0019	0.18	−2.02	E	3
		5638.82	31442	49171	5	3	0.013	0.0038	0.35	−1.72	D	4
178.	z ^{1}D° $- f$ ^{3}D (205)											
		5589.38	31442	49328	5	5	0.031	0.014	1.3	−1.14	D−	4
179.	z ^{1}D° $- e$ ^{3}F (206)											
		5593.74	31442	49314	5	7	0.044	0.029	2.7	−0.84	D−	4
		5155.14	31442	50834	5	5	0.11	0.045	3.8	−0.65	D−	4
180.	z ^{1}D° $- f$ ^{3}F (207)											
		5032.75	31442	51306	5	7	0.020	0.011	0.89	−1.27	D−	4
181.	z ^{1}D° $- f$ ^{1}D (209)	5176.57	31442	50754	5	5	0.18	0.073	6.2	−0.44	D−	4
182.	z ^{1}D° $- e$ ^{1}F (210)	5155.76	31442	50832	5	7	0.29	0.16	14	−0.09	D-	4
183.	z ^{1}D° $- g$ ^{3}F											
		4382.87	31442	54251	5	7	0.015	0.0060	0.44	−1.52	E	3

Ni I: Allowed transitions — Continued

No.	Multiplet	λ (Å)	E_i (cm^{-1})	E_k (cm^{-1})	g_i	g_k	A_{ki} (10^8 s^{-1})	f_{ik}	S (at. u.)	log gf	Accuracy	Source
184.	$z\,^1D° - g\,^1F$	4027.67	31442	56263	5	7	0.13	0.046	3.0	−0.64	D−	4
185.	$^3F° - e\,^5F$											
		6204.64	32973	49086	9	9	0.014	0.0082	1.5	−1.13	D−	4
186.	$^3F° - e\,^3G$											
		6176.81	32973	49159	9	11	0.047	0.033	6.0	−0.53	D	3
		6223.99	33112	49175	7	9	0.020	0.015	2.1	−0.99	D−	4
187.	$^3F° - e\,^3P$											
		6230.09	33112	49159	7	5	0.019	0.0079	1.1	−1.26	E	3
188.	$^3F° - f\,^3D$											
		6133.95	32973	49272	9	7	0.0037	0.0016	0.30	−1.83	D	4
		6186.74	33112	49272	7	7	0.027	0.016	2.2	−0.96	D−	4
189.	$^3F° - e\,^3F$											
		6111.06	32973	49333	9	9	0.027	0.015	2.7	−0.87	D−	4
190.	$^3F° - f\,^3F$											
		5494.89	33112	51306	7	7	0.022	0.0099	1.3	−1.16	D−	4
		5453.26	32973	51306	9	7	0.010	0.0036	0.58	−1.49	D−	4
		5760.85	33112	50466	7	9	0.035	0.023	3.0	−0.80	D−	4
191.	$^3F° - e\,^1G$											
		5682.20	33112	50706	7	9	0.078	0.048	6.3	−0.47	D−	4
192.	$^3F° - e\,^1F$											
		5641.88	33112	50832	7	7	0.025	0.012	1.6	−1.07	D−	4
		5805.23	33611	50832	5	7	0.065	0.046	4.4	−0.64	D−	4
193.	$^3F° - g\,^3F$											
		4701.54	32973	54237	9	9	0.14	0.045	6.3	−0.39	D−	4
		4729.28	33112	54251	7	7	0.027	0.0090	0.98	−1.20	E	3
		4698.39	32973	54251	9	7	0.062	0.016	2.2	−0.84	D	3
		4732.47	33112	54237	7	9	0.093	0.040	4.4	−0.55	D−	4
		4843.51	33611	54251	5	7	0.044	0.021	1.7	−0.97	D	3
194.	$^3F° - f\,^3G$											
		4431.03	33611	56173	5	7	0.041	0.017	1.2	−1.07	D−	4
195.	$^3F° - f\,^1F$											
		4551.24	33611	55577	5	7	0.061	0.026	2.0	−0.88	D−	4
196.	$z\,^1P° - e\,^3P$ (217)											
		6175.42	32982	49171	3	3	0.17	0.098	6.0	−0.53	D−	4

J. Phys. Chem. Ref. Data, Vol. 17, Suppl. 4, 1988

Ni I: Allowed transitions — Continued

No.	Multiplet	λ (Å)	E_i (cm^{-1})	E_k (cm^{-1})	g_i	g_k	A_{ki} (10^8 s^{-1})	f_{ik}	S (at. u.)	log gf	Accu-racy	Source
197.	z ^{1}P° – f ^{3}D (218)											
		5637.12	32982	50717	3	3	0.11	0.050	2.8	−0.82	D−	4
198.	z ^{1}P° – e ^{1}P (220)	5695.00	32982	50537	3	3	0.17	0.082	4.6	−0.61	D−	4
199.	z ^{1}P° – f ^{1}D (221)	5625.33	32982	50754	3	5	0.084	0.067	3.7	−0.70	D−	4
200.	y ^{3}D° – e ^{5}F (244)											
		6414.60	33501	49086	7	9	0.011	0.0086	1.3	−1.22	D−	4
		6272.65	34409	50346	3	5	0.0059	0.0058	0.36	−1.76	D	4
		6119.78	34409	50745	3	3	0.027	0.015	0.90	−1.35	D−	4
201.	y ^{3}D° – e ^{3}G (247)											
		6378.26	33501	49175	7	9	0.023	0.018	2.7	−0.89	D−	4
		6053.68	34163	50678	5	7	0.022	0.017	1.7	−1.07	D−	4
202.	y ^{3}D° – e ^{3}P (246)											
		6384.67	33501	49159	7	5	0.024	0.011	1.6	−1.13	E	3
		6661.39	34163	49171	5	3	0.013	0.0054	0.59	−1.57	D	4
203.	y ^{3}D° – f ^{3}D (248)											
		6130.17	34409	50717	3	3	0.065	0.037	2.2	−0.96	D−	4
		6316.61	33501	49328	7	5	0.0042	0.0018	0.26	−1.90	D	4
		6039.31	34163	50717	5	3	0.0057	0.0019	0.19	−2.03	D	4
204.	y ^{3}D° – e ^{3}F (249)											
		6598.59	34163	49314	5	7	0.023	0.021	2.3	−0.98	D−	4
		6086.29	34409	50834	3	5	0.11	0.098	5.9	−0.53	D−	4
		6322.17	33501	49314	7	7	0.016	0.0097	1.4	−1.17	D−	4
		5996.74	34163	50834	5	5	0.032	0.017	1.7	−1.06	D−	4
205.	y ^{3}D° – f ^{3}F (250)											
		5392.37	33501	52041	7	5	0.022	0.0068	0.85	−1.32	D−	4
206.	y ^{3}D° – e ^{1}P											
		6105.78	34163	50537	5	3	0.0058	0.0020	0.20	−2.01	D	4
		6198.65	34409	50537	3	3	0.0046	0.0026	0.16	−2.10	D	4
207.	y ^{3}D° – f ^{1}D (251)											
		6025.73	34163	50754	5	5	0.0064	0.0035	0.34	−1.76	D	4

Ni I: Allowed transitions — Continued

No.	Multiplet	λ (Å)	E_i (cm^{-1})	E_k (cm^{-1})	g_i	g_k	A_{ki} (10^8 s^{-1})	f_{ik}	S (at. u.)	log gf	Accuracy	Source
208.	$y\ ^3D° - g\ ^3F$ (254)											
		4821.14	33501	54237	7	9	0.045	0.020	2.2	−0.85	D−	4
		4976.71	34163	54251	5	7	0.016	0.0083	0.68	−1.38	E	3
		4817.82	33501	54251	7	7	0.070	0.024	2.7	−0.77	D	3
209.	$z\ ^1G° - e\ ^3G$ (258)											
		6421.52	33590	49159	9	11	0.0074	0.0056	1.1	−1.30	E	3
210.	$z\ ^1G° - f\ ^3D$											
		6375.22	33590	49272	9	7	0.0020	9.7(−4)	0.18	−2.06	D	4
211.	$z\ ^1G° - f\ ^3F$ (259)											
		5643.10	33590	51306	9	7	0.017	0.0064	1.1	−1.24	D−	4
212.	$z\ ^1G° - g\ ^3F$ (260)											
		4838.64	33590	54251	9	7	0.22	0.060	8.6	−0.27	D	3
213.	$y\ ^1F° - e\ ^1G$ (264)	6635.15	35639	50706	7	9	0.025	0.022	3.3	−0.82	D−	4
214.	$y\ ^1F° - g\ ^3F$											
		5371.33	35639	54251	7	7	0.16	0.070	8.7	−0.31	D	3
215.	$y\ ^1D° - g\ ^3F$ (272)											
		5664.02	36601	54251	5	7	0.11	0.074	6.9	−0.43	D	3
216.	$^3F° - i\ ^3D$											
		6973.52	42768	57104	7	7	0.025	0.018	3.0	−0.89	D−	4
217.	$^3G° - e\ ^3H$											
		7481.49	44315	57678	9	11	0.070	0.072	16	−0.19	D−	4
218.	$^3F° - i\ ^3F$											
		7401.13	43259	56767	9	9	0.089	0.073	16	−0.18	D−	4
219.	$^3F° - g\ ^3G$											
		7381.94	43259	56802	9	11	0.097	0.097	21	−0.06	D−	4
220.	$^3F° - i\ ^3D$											
		7220.79	43259	57104	9	7	0.058	0.035	7.5	−0.50	D−	4

[a] The number in parentheses following the tabulated value indicates the power of ten by which this value has to be multiplied.

[b] The LS-coupling designation of this term was not provided in the NBS energy level compilation (J. Sugar and C. Corliss, J. Phys. Chem. Ref. Data **14**, Suppl. 2 (1985)), so we have accordingly omitted it from this work.

Ni I

Forbidden Transitions

List of tabulated lines

Wavelength (Å)	No.	Wavelength (Å)	No.	Wavelength (Å)	No.	Wavelength (Å)	No.
1977.2	13	6437.70	11	7464.39	4	31191	8
2030.2	6	6489.61	11	7507.44	9	39513	8
2080.6	18	6604.30	11	7908.30	9	45182	20
2635.2	22	6730.25	11	7929.70	16	47868	20
2788.8	24	6787.00	11	7989.9	3	58917	8
2798.5	24	6941.63	4	8111.97	16	75046	1
4523.16	5	7002.02	4	8194.57	16	82794	19
4565.4	12	7130.24	11	8201.77	2	113040	1
4710.7	12	7193.97	11	8466.38	9	119980	7
4813.27	5	7218.7	10	8832.31	15	148100	7
5027.34	5	7243.99	4	8843.42	2	352890	23
5348.3	17	7393.71	2	9887.18	14	805210	23
6404.46	4	7395.79	4	11650	21		

For this spectrum, we have utilized the work by Garstang,[1] who calculated M1 and E2 transition probabilities for lines arising from the three lowest configurations: $3d^8 4s^2$, $3d^9 4s$, and $3d^{10}$. Garstang included limited configuration interaction in his calculations. As is usually the case, the data for electric quadrupole transitions are expected to be less accurate than those for magnetic dipole transitions.

Reference

[1]R. H. Garstang, J. Res. Nat. Bur. Stand., Sect. A **68**, 61 (1964).

Ni I: Forbidden transitions

No.	Multiplet	λ (Å)	E_i (cm^{-1})	E_k (cm^{-1})	g_i	g_k	Type of transition	A_{ki} (s^{-1})	S (at. u.)	Accuracy	Source
1.	$a\,^3F - a\,^3F$										
		[75046]	0.0	1332.2	9	7	M1	0.062	6.8	C+	1
		[113040]	1332.2	2216.5	7	5	M1	0.025	6.7	C+	1
2.	$a\,^3F - b\,^1D$ (1F)										
		7393.71	0.0	13521	9	5	E2	0.0056	0.37	E	1
		8201.77	1332.2	13521	7	5	M1	0.39	0.040	D	1
		"	"	"	7	5	E2	7.6(−4)^a	0.084	E	1
		8843.42	2216.5	13521	5	5	M1	0.17	0.022	D	1
		"	"	"	5	5	E2	2.0(−4)	0.032	E	1
3.	$a\,^3F - a\,^1S$										
		[7989.9]	2216.5	14729	5	1	E2	1.8(−4)	0.0035	E	1

Ni I: Forbidden transitions — Continued

No.	Multiplet	λ (Å)	E_i (cm^{-1})	E_k (cm^{-1})	g_i	g_k	Type of transition	A_{ki} (s^{-1})	S (at. u.)	Accuracy	Source
4.	$a\ ^3F - a\ ^3P$ (2F)										
		6404.46	0.0	15610	9	5	E2	0.032	1.0	E	1
		6941.63	1332.2	15734	7	3	E2	0.025	0.72	E	1
		7243.99	2216.5	16017	5	1	E2	0.031	0.37	E	1
		7002.02	1332.2	15610	7	5	M1	0.15	0.0095	D	1
		"	"	"	7	5	E2	0.0056	0.28	E	1
		7395.79	2216.5	15734	5	3	M1	0.0022	9.9(−5)	D	1
		"	"	"	5	3	E2	0.0090	0.36	E	1
		7464.39	2216.5	15610	5	5	M1	0.039	0.0030	D	1
		"	"	"	5	5	E2	4.0(−4)	0.028	E	1
5.	$a\ ^3F - a\ ^1G$ (3F)										
		4523.16	0.0	22102	9	9	M1	0.32	0.0099	D	1
		"	"	"	9	9	E2	2.2(−4)	0.0022	E	1
		4813.27	1332.2	22102	7	9	M1	0.16	0.0060	D	1
		"	"	"	7	9	E2	3.8(−6)	5.3(−5)	E	1
		5027.34	2216.5	22102	5	9	E2	3.0(−4)	0.0052	E	1
6.	$a\ ^3F - e\ ^1S$										
		[2030.2]	2216.5	51457	5	1	E2	0.17	0.0034	E	1n
7.	$a\ ^3D - a\ ^3D$										
		[148100]	204.8	879.8	7	5	M1	0.0070	4.2	C+	1
		[119980]	879.8	1713.1	5	3	M1	0.021	4.0	C+	1
8.	$a\ ^3D - a\ ^1D$										
		[31191]	204.8	3409.9	7	5	M1	0.078	0.44	D	1
		[39513]	879.8	3409.9	5	5	M1	0.0062	0.071	D	1
		[58917]	1713.1	3409.9	3	5	M1	0.011	0.42	D	1
9.	$a\ ^3D - b\ ^1D$ (4F)										
		7507.44	204.8	13521	7	5	E2	0.014	0.99	E	1
		7908.30	879.8	13521	5	5	E2	0.017	1.6	E	1
		8466.38	1713.1	13521	3	5	E2	4.5(−4)	0.058	E	1
10.	$a\ ^3D - a\ ^1S$										
		[7218.7]	879.8	14729	5	1	E2	0.068	0.79	E	1
11.	$a\ ^3D - a\ ^3P$ (5F)										
		6437.70	204.8	15734	7	3	E2	0.092	1.8	E	1
		6604.30	879.8	16017	5	1	E2	0.19	1.4	E	1
		6489.61	204.8	15610	7	5	E2	0.074	2.5	E	1
		6730.25	879.8	15734	5	3	E2	0.012	0.30	E	1
		6787.00	879.8	15610	5	5	E2	0.018	0.77	E	1
		7130.24	1713.1	15734	3	3	E2	0.053	1.7	E	1
		7193.97	1713.1	15610	3	5	E2	0.0093	0.53	E	1
12.	$a\ ^3D - a\ ^1G$										
		[4710.7]	879.8	22102	5	9	E2	0.080	0.99	E	1
		[4565.4]	204.8	22102	7	9	E2	7.9(−4)	0.0084	E	1

Ni I: Forbidden transitions — Continued

No.	Multiplet	λ (Å)	E_i (cm^{-1})	E_k (cm^{-1})	g_i	g_k	Type of transition	A_{ki} (s^{-1})	S (at. u.)	Accuracy	Source
13.	$a\ ^3D - e\ ^1S$										
		[1977.2]	879.8	51457	5	1	E2	8.3	0.15	E	1n
14.	$a\ ^1D - b\ ^1D$ (6F)	9887.18	3409.9	13521	5	5	E2	0.012	3.4	E	1
15.	$a\ ^1D - a\ ^1S$ (7F)	8832.31	3409.9	14729	5	1	E2	0.31	9.9	E	1
16.	$a\ ^1D - a\ ^3P$ (8F)										
		8194.57	3409.9	15610	5	5	E2	0.024	2.6	E	1
		8111.97	3409.9	15734	5	3	E2	4.7(−4)	0.029	E	1
		7929.70	3409.9	16017	5	1	E2	1.1(−5)	2.1(−4)	E	1
17.	$a\ ^1D - a\ ^1G$	[5348.3]	3409.9	22102	5	9	E2	0.44	10	E	1
18.	$a\ ^1D - e\ ^1S$	[2080.6]	3409.9	51457	5	1	E2	82	1.9	E	1n
19.	$b\ ^1D - a\ ^1S$	[82794]	13521	14729	5	1	E2	2.9(−4)	670	E	1
20.	$b\ ^1D - a\ ^3P$										
		[47868]	13521	15610	5	5	M1	0.072	1.5	D	1
		[45182]	13521	15734	5	3	M1	0.063	0.65	D	1
21.	$b\ ^1D - a\ ^1G$	[11650]	13521	22102	5	9	E2	4.1(−4)	0.47	E	1
22.	$b\ ^1D - e\ ^1S$	[2635.2]	13521	51457	5	1	E2	9.4	0.71	E	1n
23.	$a\ ^3P - a\ ^3P$										
		[805210]	15610	15734	5	3	M1	3.2(−5)	1.9	C+	1
		[352890]	15734	16017	3	1	M1	0.0012	2.0	C+	1
24.	$a\ ^3P - e\ ^1S$										
		[2788.8]	15610	51457	5	1	E2	2.9	0.29	E	1n
		[2798.5]	15734	51457	3	1	M1	5.3	0.0043	D	1n

[a]The number in parentheses following the tabulated value indicates the power of ten by which this value has to be multiplied.

Ni II

Co Isoelectronic Sequence

Ground State: $1s^2 2s^2 2p^6 3s^2 3d^9 \, ^2D_{5/2}$

Ionization Energy: $18.16898 \text{ eV} = 146\,541.56 \text{ cm}^{-1}$

Allowed Transitions

List of tabulated lines

Wavelength (Å)	No.	Wavelength (Å)	No.	Wavelength (Å)	No.	Wavelength (Å)	No.
1751.92	1	2174.67	5	2264.46	3	2375.42	12
2034.05	6	2175.15	4	2270.21	3	2387.76	10
2053.30	6	2184.61	4	2278.77	13	2394.52	11
2080.85	7	2188.05	3	2287.09	13	2410.74	9
2090.10	6	2201.41	4	2296.55	12	2412.27	2
2093.56	6	2206.72	4	2297.14	2	2413.04	10
2125.12	5	2210.38	4	2297.49	2	2416.13	11
2125.91	4	2216.48	3	2298.27	12	2433.56	10
2128.58	6	2220.40	14	2303.00	2	2437.89	10
2138.58	4	2222.96	3	2316.04	2	2510.87	9
2158.74	4	2224.36	12	2326.45	2	2545.90	9
2161.22	5	2224.86	3	2334.58	11	2630.27	8
2165.55	4	2226.33	3	2356.40	13		
2169.10	4	2253.85	3	2367.39	2		

For this spectrum, we have chosen the *f*-value data of Bell *et al.*,[1] as normalized by Lawler and Salih.[2] Bell *et al.* determined relative oscillator strengths in emission with a wall-stabilized arc operated in argon and small admixtures of nickel carbonyl. All observations were performed photoelectrically, and digital data processing techniques were employed. Since the measured lines are located in the near ultraviolet, the intensity calibrations presented a special problem, which was solved by utilizing the continuous emission of a hydrogen arc operated at well-diagnosed plasma conditions. Lawler and Salih measured radiative lifetimes of twelve levels of Ni II by the laser-induced fluorescence technique. They then used these lifetimes to convert the data of Ref. 1 to an absolute scale.

The *f*-values tabulated here are consistently 60 percent lower than those published in our earlier NBS compilation,[4] which utilized the data of Bell *et al.* on an absolute basis. The most probable reason for this difference lies in the method with which Bell *et al.* determined their absolute scale. In the absence of reliable lifetime data, they utilized local thermodynamic equilibrium (LTE) relations to tie their Ni II *f*-value scale to that of Ni I, which they had determined on an absolute scale. Recent LTE studies in stabilized arcs showed however that at the electron densities of their experiment, LTE conditions are not yet achieved, so that this normalization procedure is not applicable. Therefore, in this com-pilation, we have tabulated the data and accuracy estimates of Lawler and Salih, whenever available. For lines not covered by Ref. 2, we normalized the data of Ref. 1 (which should be quite reliable on a *relative* scale). This normalization consisted of subtracting 0.21 from the original log *gf*-values of Ref. 1.

For two lines, we did not use Ref. 1 as the original data source. For the 2326.45 Å line, Lawler and Salih found that the branching ratio of Ref. 1 was inconsistent with other measured and calculated data. We therefore used Lawler and Salih's recommended transition probability for this line. The log *gf*-value for the 1751.92 Å line was determined from the semi-empirical scaled Thomas-Fermi-Dirac calculations of Kurucz and Peytremann.[3] For this line, the branching ratio of Ref. 3 was adjusted to the lifetime of the $3d^8(^3F)4p \, ^2F^{\circ}_{7/2}$ level, as measured by Lawler and Salih.

References

[1]G. D. Bell, D. R. Paquette, and W. L. Wiese, Astrophys. J. **143**, 559 (1966).

[2]J. E. Lawler and S. Salih, Phys. Rev. A **35**, 5046 (1987).

[3]R. L. Kurucz and E. Peytremann, Smithsonian Astrophysical Observatory Special Report 362 (1975).

[4]J. R. Fuhr, G. A. Martin, W. L. Wiese, and S. M. Younger, J. Phys. Chem. Ref. Data **10**, 305 (1981).

Ni II: Allowed transitions

No.	Transition Array	Multiplet	λ (Å)	E_i (cm^{-1})	E_k (cm^{-1})	g_i	g_k	A_{ki} (10^8 s^{-1})	f_{ik}	S (at. u.)	log gf	Accu-racy	Source
1.	$3d^9-$ $3d^8(^3F)4p$	^{2}D - ^{2}F°											
			1751.92	0	57081	6	8	0.48	0.029	1.0	−0.75	C	3n
2.	$3d^8(^3F)4s-$ $3d^8(^3F)4p$	^{4}F - ^{4}D°											
			2316.04	8394	51558	10	8	2.88	0.185	14.1	0.268	B	1n
			2303.00	9330	52738	8	6	2.9	0.17	11	0.14	C	1n
			2297.14	10116	53635	6	4	2.70	0.142	6.46	−0.068	B	1n
			2297.49	10664	54176	4	2	3.0	0.12	3.6	−0.32	B	1n
			2367.39	9330	51558	8	8	0.074	0.0062	0.39	−1.30	B	1n
			2326.45	10664	53635	4	4	0.33	0.027	0.82	−0.97	D	2
			2412.27	10116	51558	6	8	0.0024	2.8(−4)^a	0.013	−2.78	C	1n
3.		^{4}F - ^{4}G°											
			2216.48	8394	53496	10	12	3.4	0.30	22	0.48	B	1n
			2270.21	9330	53365	8	10	1.56	0.151	9.01	0.081	B	1n
			2264.46	10116	54263	6	8	1.43	0.147	6.56	−0.056	B	1n
			2253.85	10664	55019	4	6	1.98	0.226	6.71	−0.043	C+	1n
			2222.96	8394	53365	10	10	0.98	0.073	5.3	−0.14	B	1n
			2224.86	9330	54263	8	8	1.55	0.115	6.74	−0.036	B	1n
			2226.33	10116	55019	6	6	1.3	0.097	4.3	−0.24	B	1n
			2188.05	9330	55019	8	6	0.057	0.0031	0.18	−1.61	C+	1n
4.		^{4}F - ^{4}F°	*2171.3*	*9355*	*55395*	28	28	3.14	0.222	44.4	0.793	C+	1n
			2165.55	8394	54557	10	10	2.4	0.17	12	0.23	C	1n
			2169.10	9330	55418	8	8	1.58	0.111	6.37	−0.050	B	1n
			2175.15	10116	56075	6	6	1.77	0.126	5.39	−0.123	B	1n
			2184.61	10664	56424	4	4	2.90	0.207	5.97	−0.081	B	1n
			2125.91	8394	55418	10	8	0.050	0.0027	0.19	−1.57	B	1n
			2138.58	9330	56075	8	6	0.177	0.00910	0.513	−1.138	C+	1n
			2158.74	10116	56424	6	4	0.35	0.016	0.70	−1.01	C	1n
			2210.38	9330	54557	8	10	0.39	0.036	2.1	−0.54	C	1n
			2206.72	10116	55418	6	8	1.66	0.162	7.04	−0.013	B	1n
			2201.41	10664	56075	4	6	1.3	0.14	4.1	−0.25	B	1n
5.		^{4}F - ^{2}G°											
			2125.12	9330	56371	8	8	0.064	0.0043	0.24	−1.46	C	1n
			2174.67	9330	55300	8	10	1.43	0.127	7.26	0.006	B	1n
			2161.22	10116	56371	6	8	0.20	0.019	0.80	−0.95	C	1n
6.		^{4}F - ^{2}F°											
			2053.30	8394	57081	10	8	0.025	0.0013	0.086	−1.90	B	1n
			2034.05	9330	58493	8	6	0.023	0.0011	0.057	−2.07	C	1n
			2093.56	9330	57081	8	8	0.065	0.0043	0.24	−1.47	C+	1n
			2128.58	10116	57081	6	8	0.248	0.0225	0.944	−0.870	B	1n
			2090.10	10664	58493	4	6	0.070	0.0069	0.19	−1.56	C	1n
7.		^{4}F - ^{2}D°											
			2080.85	10664	58706	4	4	0.080	0.0052	0.14	−1.68	C	1n
8.		^{2}F - ^{4}D°											
			2630.27	13550	51558	8	8	0.0068	7.1(−4)	0.049	−2.25	C	1n

Ni II: Allowed transitions — Continued

No.	Transition Array	Multiplet	λ (Å)	E_i (cm^{-1})	E_k (cm^{-1})	g_i	g_k	A_{ki} (10^8 s^{-1})	f_{ik}	S (at. u.)	log gf	Accuracy	Source
9.		^{2}F – ^{4}G°											
			2510.87	13550	53365	8	10	0.58	0.069	4.5	−0.26	B	1n
			2545.90	14996	54263	6	8	0.156	0.0202	1.02	−0.916	B	1n
			2410.74	13550	55019	8	6	0.010	6.5(−4)	0.042	−2.28	D	1n
10.		^{2}F – ^{4}F°											
			2437.89	13550	54557	8	10	0.54	0.060	3.9	−0.32	C	1n
			2387.76	13550	55418	8	8	0.159	0.0136	0.855	−0.964	B	1n
			2433.56	14996	56075	6	6	0.073	0.0065	0.31	−1.41	B	1n
			2413.04	14996	56424	6	4	0.083	0.0048	0.23	−1.54	B	1n
11.		^{2}F – ^{2}G°	*2402.8*	*14170*	*55776*	14	18	2.23	0.248	27.5	0.541	C+	1n
			2394.52	13550	55300	8	10	1.70	0.183	11.5	0.165	B	1n
			2416.13	14996	56371	6	8	2.1	0.25	12	0.17	C	1n
			2334.58	13550	56371	8	8	0.80	0.065	4.0	−0.28	C	1n
12.		^{2}F – ^{2}F°	*2297.3*	*14170*	*57686*	14	14	3.94	0.312	33.0	0.640	C+	1n
			2296.55	13550	57081	8	8	1.98	0.157	9.47	0.098	B	1n
			2298.27	14996	58493	6	6	2.8	0.22	10	0.12	C	1n
			2224.36	13550	58493	8	6	0.32	0.018	1.0	−0.85	C	1n
			2375.42	14996	57081	6	8	0.66	0.074	3.5	−0.35	B	1n
13.		^{2}F – ^{2}D°	*2284.3*	*14170*	*57934*	14	10	2.97	0.166	17.5	0.367	C	1n
			2278.77	13550	57420	8	6	2.8	0.16	9.8	0.12	C	1n
			2287.09	14996	58706	6	4	2.8	0.15	6.6	−0.06	C	1n
			2356.40	14996	57420	6	6	0.28	0.023	1.1	−0.85	C	1n
14.	$3d^8(^3$P$)4s$ – $3d^8(^1$D$)4p$	^{4}P - ^{2}F°											
			2220.40	23108	68131	6	8	2.3	0.23	9.9	0.13	C	1n

^{a}The number in parentheses following the tabulated value indicates the power of ten by which this value has to be multiplied.

J. Phys. Chem. Ref. Data, Vol. 17, Suppl. 4, 1988

Ni II

Forbidden Transitions

List of tabulated lines

Wavelength (Å)	No.	Wavelength (Å)	No.	Wavelength (Å)	No.	Wavelength (Å)	No.
1020.9	7	5132.6	6	7413.33	22	17646	2
3074.11	26	5269.16	12	7612.7	9	18953	18
3076.1	26	5274.27	6	7694.6	5	19388	2
3223.1	26	5275.83	12	8033.5	5	20487	2
3378.55	24	5281.46	6	8303.23	22	20805	14
3439.29	24	5431.39	6	8703.9	10	21014	14
3559.86	24	5703.64	12	8896.1	7	23079	2
3627.35	24	5711.46	12	9377.33	21	23343	14
3993.1	25	6007.1	9	9757.4	10	23606	14
4025.3	23	6365.1	9	9885.74	21	23688	2
4033.0	23	6441.3	8	9956.6	10	24779	18
4143.17	11	6467.3	5	10460	7	29106	2
4147.30	11	6668.16	22	10618	16	51850	15
4201.2	25	6700.2	9	10645	16	59510	13
4248.8	25	6791.5	5	10718.16	21	66342	1
4285.3	23	6794.2	5	10921.07	21	69177	4
4294.1	23	6813.6	5	11360	10	80610	19
4310.46	11	6848.4	8	11455	20	96151	17
4314.92	11	6910.8	9	11616.88	21	100780	17
4326.2	23	6955.8	9	12323	7	106790	2
4461.54	11	7054.2	5	12779	21	127250	2
4466.33	11	7078.0	5	12924	16	145330	15
4485.2	25	7102.84	8	13353	20	182360	2
4573.45	11	7255.8	5	13396	20	402820	15
4628.0	23	7307.7	9	16766	14		
5064.2	6	7379.57	22	17245	18		

For this ion, we have selected the data of Nussbaumer and Storey,[1] who calculated M1 and E2 transition probabilities for transitions within the $3d^9$ and $3d^84s$ configurations. These comprise the 17 energetically lowest levels of this spectrum. Nussbaumer and Storey included the $3d^9$, $3d^84s$, and $3d^84d$ configurations as the basis of their calculations, and they derived radial wavefunctions by using adjustable Thomas-Fermi potentials. These authors also applied additional empirical corrections to their coupling coefficients, so that their calculated eigenenergies are in close agreement with observed energy levels. Accuracies for M1 transitions within the same term should be quite good—within 25 percent or better. However, data for other magnetic dipole transitions, as well as for all electric quadrupole transitions, are necessarily much less reliable.

Reference

[1]H. Nussbaumer and P. J. Storey, Astron. Astrophys. **110**, 295 (1982).

Ni II: Forbidden transitions

No.	Transition Array	Multiplet	λ (Å)	E_i (cm^{-1})	E_k (cm^{-1})	g_i	g_k	Type of transition	A_{ki} (s^{-1})	S (at. u.)	Accuracy	Source
1.	$3d^9$–$3d^9$	^2D – ^2D	[66342]	0	1507	6	4	M1	0.0554	2.40	C+	1

Ni II: Forbidden transitions — Continued

No.	Transition Array	Multiplet	λ (Å)	E_i (cm^{-1})	E_k (cm^{-1})	g_i	g_k	Type of transition	A_{ki} (s^{-1})	S (at. u.)	Accu-racy	Source
2.	$3d^8(^3F)4s-$ $3d^8(^3F)4s$	$^4F - ^4F$										
			[106790]	8394	9330	10	8	M1	0.0271	9.79	C+	1
			[127250]	9330	10116	8	6	M1	0.0276	12.7	C+	1
			[182360]	10116	10664	6	4	M1	0.0105	9.44	C+	1
3.		$^4F - ^2F$										
			[19388]	8394	13550	10	8	M1	0.087	0.19	D	1
			[17646]	9330	14996	8	6	M1	0.0028	0.0034	D	1
			[23688]	9330	13550	8	8	M1	0.0055	0.022	D	1
			[20487]	10116	14996	6	6	M1	0.0072	0.014	D	1
			[29106]	10116	13550	6	8	M1	0.014	0.10	D	1
			[23079]	10664	14996	4	6	M1	0.029	0.079	D	1
4.		$^2F - ^2F$										
			[69177]	13550	14996	8	6	M1	0.0471	3.47	C+	1
5.	$3d^8(^3F)4s-$ $3d^8(^3P)4s$	$^4F - ^4P$										
			[6794.2]	8394	23108	10	6	E2	0.0046	0.24	E	1
			[6467.3]	9330	24788	8	4	E2	0.0062	0.17	E	1
			[6791.5]	10116	24836	6	2	E2	0.0049	0.084	E	1
			[7255.8]	9330	23108	8	6	M1	0.16	0.014	E	1
			"	"	"	8	6	E2	0.0010	0.072	E	1
			[6813.6]	10116	24788	6	4	M1	0.092	0.0043	E	1
			"	"	"	6	4	E2	0.0035	0.12	E	1
			[7054.2]	10664	24836	4	2	M1	5.5(−4)[a]	1.4(−5)	E	1
			"	"	"	4	2	E2	0.0060	0.12	E	1
			[7694.6]	10116	23108	6	6	M1	0.014	0.0014	E	1
			"	"	"	6	6	E2	1.6(−4)	0.015	E	1
			[7078.0]	10664	24788	4	4	M1	0.027	0.0014	E	1
			"	"	"	4	4	E2	9.6(−4)	0.041	E	1
			[8033.5]	10664	23108	4	6	M1	0.010	0.0012	E	1
			"	"	"	4	6	E2	1.3(−5)	0.0016	E	1
6.		$^4F - ^2P$ (9F)										
			[5064.2]	9330	29071	8	4	E2	0.0013	0.010	E	1
			[5132.6]	10116	29593	6	2	E2	4.5(−4)	0.0019	E	1
			5274.27	10116	29071	6	4	M1	0.012	2.6(−4)	E	1
			"	"	"	6	4	E2	1.7(−4)	0.0017	E	1
			5281.46	10664	29593	4	2	M1	0.0013	1.4(−5)	E	1
			"	"	"	4	2	E2	5.7(−5)	2.8(−4)	E	1
			5431.39	10664	29071	4	4	M1	0.0019	4.5(−5)	E	1
			"	"	"	4	4	E2	4.3(−5)	4.8(−4)	E	1
7.		$^2F - ^4P$										
			[8896.1]	13550	24788	8	4	E2	4.5(−5)	0.0060	E	1
			[10460]	13550	23108	8	6	M1	0.076	0.019	E	1
			"	"	"	8	6	E2	3.2(−6)	0.0014	E	1
			[10209]	14996	24788	6	4	M1	0.0068	0.0011	E	1
			"	"	"	6	4	E2	1.3(−5)	0.0034	E	1
			[12323]	14996	23108	6	6	M1	0.082	0.034	E	1

Ni II: Forbidden transitions — Continued

No.	Transition Array	Multiplet	λ (Å)	E_i (cm^{-1})	E_k (cm^{-1})	g_i	g_k	Type of transition	A_{ki} (s^{-1})	S (at. u.)	Accuracy	Source
8.		^{2}F – ^{2}P (13F)										
			[6441.3]	13550	29071	8	4	E2	0.031	0.82	E	1
			[6848.4]	14996	29593	6	2	E2	0.029	0.52	E	1
			7102.84	14996	29071	6	4	M1	0.0079	4.2(−4)	E	1
			"	"	"	6	4	E2	0.0026	0.11	E	1
9.	$3d^8(^3$F$)4s-$ $3d^8(^1$D$)4s$	^{4}F – ^{2}D										
			[6007.1]	8394	25036	10	6	E2	0.0074	0.21	E	1
			[6910.8]	9330	23796	8	4	E2	0.0012	0.045	E	1
			[6365.1]	9330	25036	8	6	M1	0.19	0.011	E	1
			"	"	"	8	6	E2	0.0018	0.067	E	1
			[7307.7]	10116	23796	6	4	M1	0.38	0.022	E	1
			"	"	"	6	4	E2	5.8(−4)	0.029	E	1
			[6700.2]	10116	25036	6	6	M1	0.0078	5.2(−4)	E	1
			"	"	"	6	6	E2	3.3(−4)	0.016	E	1
			[7612.7]	10664	23796	4	4	M1	0.18	0.012	E	1
			"	"	"	4	4	E2	1.4(−4)	0.0085	E	1
			[6955.8]	10664	25036	4	6	M1	0.0077	5.8(−4)	E	1
			"	"	"	4	6	E2	3.6(−5)	0.0021	E	1
10.		^{2}F – ^{2}D										
			[9757.4]	13550	23796	8	4	E2	6.0(−5)	0.013	E	1
			[8703.9]	13550	25036	8	6	M1	0.089	0.013	E	1
			"	"	"	8	6	E2	4.3(−5)	0.0077	E	1
			[11360]	14996	23796	6	4	M1	0.057	0.012	E	1
			"	"	"	6	4	E2	2.6(−5)	0.012	E	1
			[9956.6]	14996	25036	6	6	M1	0.062	0.014	E	1
11.	$3d^8(^3$F$)4s-$ $3d^8(^1$G$)4s$	^{4}F – ^{2}G (10F)										
			4147.30	8394	32500	10	10	M1	0.32	0.0085	E	1
			"	"	"	10	10	E2	1.2(−4)	0.0010	E	1
			4310.46	9330	32524	8	8	M1	0.16	0.0038	E	1
			4143.17	8394	32524	10	8	M1	0.0096	2.0(−4)	E	1
			"	"	"	10	8	E2	1.0(−5)	5.8(−5)	E	1
			4314.92	9330	32500	8	10	M1	0.073	0.0022	E	1
			"	"	"	8	10	E2	2.1(−4)	0.0019	E	1
			4461.54	10116	32524	6	8	M1	0.099	0.0026	E	1
			"	"	"	6	8	E2	1.2(−4)	0.0010	E	1
			4466.33	10116	32500	6	10	E2	1.3(−4)	0.0014	E	1
			4573.45	10664	32524	4	8	E2	4.6(−4)	0.0044	E	1
12.		^{2}F – ^{2}G (14F)										
			5711.46	14996	32500	6	10	E2	5.9(−4)	0.021	E	1
			5275.83	13550	32500	8	10	M1	0.082	0.0045	E	1
			"	"	"	8	10	E2	0.0040	0.097	E	1
			5703.64	14996	32524	6	8	M1	0.042	0.0023	E	1
			"	"	"	6	8	E2	0.0020	0.057	E	1
			5269.16	13550	32524	8	8	M1	0.12	0.0052	E	1
			"	"	"	8	8	E2	2.5(−4)	0.0048	E	1

Ni II: Forbidden transitions — Continued

No.	Transition Array	Multiplet	λ (Å)	E_i (cm^{-1})	E_k (cm^{-1})	g_i	g_k	Type of transition	A_{ki} (s^{-1})	S (at. u.)	Accuracy	Source
13.	$3d^8(^3P)4s-$ $3d^8(^3P)4s$	$^4P - {^4P}$										
			[59510]	23108	24788	6	4	M1	0.0192	0.600	C+	1
14.		$^4P - {^2P}$										
			[16766]	23108	29071	6	4	M1	0.080	0.056	D	1
			"	"	"	6	4	E2	2.2(−5)	0.069	D	1
			[20805]	24788	29593	4	2	M1	0.0078	0.0052	D	1
			[23343]	24788	29071	4	4	M1	0.021	0.040	D	1
			[21014]	24836	29593	2	2	M1	0.0061	0.0042	D	1
			[23606]	24836	29071	2	4	M1	0.0014	0.0027	D	1
15.	$3d^8(^3P)4s-$ $3d^8(^1D)4s$	$^4P - {^2D}$										
			[145330]	23108	23796	6	4	M1	0.0051	2.3	E	1
			[51850]	23108	25036	6	6	M1	0.069	2.1	E	1
			[402820]	24788	25036	4	6	M1	1.9(−4)	2.0	E	1
16.	$3d^8(^3P)4s-$ $3d^8(^1G)4s$	$^4P - {^2G}$										
			[10645]	23108	32500	6	10	E2	2.9(−4)	0.24	E	1
			[12924]	24788	32524	4	8	E2	3.7(−5)	0.064	E	1
			[10618]	23108	32524	6	8	M1	8.0(−6)	2.8(−6)	E	1
			"	"	"	6	8	E2	3.8(−5)	0.024	E	1
17.	$3d^8(^1D)4s-$ $3d^8(^3P)4s$	$^2D - {^4P}$										
			[100780]	23796	24788	4	4	M1	0.012	1.8	E	1
			[96151]	23796	24836	4	2	M1	0.0085	0.56	E	1
18.		$^2D - {^2P}$										
			[24779]	25036	29071	6	4	M1	0.0059	0.013	E	1
			"	"	"	6	4	E2	2.1(−6)	0.047	E	1
			[17245]	23796	29593	4	2	M1	0.077	0.029	E	1
			"	"	"	4	2	E2	2.1(−6)	0.0038	E	1
			[18953]	23796	29071	4	4	M1	0.073	0.074	E	1
			"	"	"	4	4	E2	1.5(−5)	0.087	E	1
19.	$3d^8(^1D)4s-$ $3d^8(^1D)4s$	$^2D - {^2D}$										
			[80610]	23796	25036	4	6	M1	0.00163	0.190	C+	1
20.	$3d^8(^1D)4s-$ $3d^8(^1G)4s$	$^2D - {^2G}$										
			[13396]	25036	32500	6	10	E2	9.8(−5)	0.25	E	1
			[11455]	23796	32524	4	8	E2	3.2(−4)	0.30	E	1
			[13353]	25036	32524	6	8	M1	2.3(−6)	1.6(−6)	E	1
			"	"	"	6	8	E2	1.2(−5)	0.024	E	1

Ni II: Forbidden transitions — Continued

No.	Transition Array	Multiplet	λ (Å)	E_i (cm^{-1})	E_k (cm^{-1})	g_i	g_k	Type of transition	A_{ki} (s^{-1})	S (at. u.)	Accuracy	Source
21.	$3d^9$–$3d^8(^3F)4s$	^2D – ^4F (1F)										
			[12779]	1507	9330	4	8	E2	2.1(−5)	0.034	E	1
			10718.16	0	9330	6	8	M1	6.4(−4)	2.3(−4)	E	1
			"	"	"	6	8	E2	7.8(−4)	0.53	E	1
			11616.88	1507	10116	4	6	M1	9.3(−4)	3.2(−4)	E	1
			"	"	"	4	6	E2	2.7(−4)	0.20	E	1
			9885.74	0	10116	6	6	M1	1.2(−4)	2.6(−5)	E	1
			"	"	"	6	6	E2	1.6(−5)	0.0054	E	1
			10921.07	1507	10664	4	4	M1	7.7(−4)	1.5(−4)	E	1
			"	"	"	4	4	E2	6.1(−5)	0.023	E	1
			9377.33	0	10664	6	4	M1	1.3(−5)	1.6(−6)	E	1
			"	"	"	6	4	E2	7.5(−5)	0.013	E	1
22.		^2D – ^2F (2F)										
			8303.23	1507	13550	4	8	E2	0.013	2.4	E	1
			7379.57	0	13550	6	8	M1	3.7(−4)	4.4(−5)	E	1
			"	"	"	6	8	E2	0.23	24	E	1
			7413.33	1507	14996	4	6	M1	3.3(−5)	3.0(−6)	E	1
			"	"	"	4	6	E2	0.18	14	E	1
			6668.16	0	14996	6	6	M1	6.2(−4)	4.1(−5)	E	1
			"	"	"	6	6	E2	0.098	4.6	E	1
23.	$3d^9$–$3d^8(^3P)4s$	^2D – ^4P										
			[4326.2]	0	23108	6	6	M1	0.0025	4.5(−5)	E	1
			"	"	"	6	6	E2	0.35	1.9	E	1
			[4294.1]	1507	24788	4	4	M1	0.0013	1.5(−5)	E	1
			"	"	"	4	4	E2	0.11	0.38	E	1
			[4033.0]	0	24788	6	4	M1	0.0033	3.2(−5)	E	1
			"	"	"	6	4	E2	0.095	0.24	E	1
			[4285.3]	1507	24836	4	2	M1	0.0010	5.8(−6)	E	1
			"	"	"	4	2	E2	0.0033	0.0057	E	1
			[4628.0]	1507	23108	4	6	M1	1.9(−4)	4.2(−6)	E	1
			"	"	"	4	6	E2	0.13	0.99	E	1
			[4025.3]	0	24836	6	2	E2	2.9(−4)	3.6(−4)	E	1
24.		^2D – ^2P (5F)										
			3378.55	0	29593	6	2	E2	4.1	2.1	E	1
			3439.29	0	29071	6	4	M1	5.7(−4)	3.4(−6)	E	1
			"	"	"	6	4	E2	5.5	6.3	E	1
			3559.86	1507	29593	4	2	M1	0.0011	3.7(−6)	E	1
			"	"	"	4	2	E2	4.6	3.1	E	1
			3627.35	1507	29071	4	4	M1	0.0014	9.9(−6)	E	1
			"	"	"	4	4	E2	2.8	4.2	E	1
25.	$3d^9$–$3d^8(^1D)4s$	^2D – ^2D										
			[3993.1]	0	25036	6	6	M1	0.0022	3.1(−5)	E	1
			"	"	"	6	6	E2	0.52	1.9	E	1
			[4485.2]	1507	23796	4	4	M1	8.4(−4)	1.1(−5)	E	1
			"	"	"	4	4	E2	0.30	1.3	E	1
			[4201.2]	0	23796	6	4	M1	0.0015	1.6(−5)	E	1
			"	"	"	6	4	E2	0.66	2.1	E	1
			4248.8]	1507	25036	4	6	M1	0.0015	2.6(−5)	E	1
			"	"	"	4	6	E2	0.17	0.84	E	1

Ni II: Forbidden transitions — Continued

No.	Transition Array	Multiplet	λ (Å)	E_i (cm^{-1})	E_k (cm^{-1})	g_i	g_k	Type of transition	A_{ki} (s^{-1})	S (at. u.)	Accuracy	Source
26.	$3d^9$–$3d^8(^1G)4s$	^2D – ^2G (6F)										
			[3076.1]	0	32500	6	10	E2	4.6	7.5	E	1
			[3223.1]	1507	32524	4	8	E2	3.5	5.8	E	1
			3074.11	0	32524	6	8	E2	0.35	0.46	E	1

ªThe number in parentheses following the tabulated value indicates the power of ten by which this value has to be multiplied.

Ni III

Fe Isoelectronic Sequence

Ground State: $1s^2 2s^2 2p^6 3s^2 3p^6 3d^8\ ^3F_4$

Ionization Energy: 35.19 eV = 283800 cm^{-1}

Allowed transitions

For this spectrum, we have chosen the calculations of Biemont[1] and of Kurucz and Peytremann.[2] Biemont obtained radial wavefunctions by means of the scaled Thomas-Fermi method and calculated individual line strengths in intermediate coupling. Similarly, Kurucz and Peytremann used a semiempirical scaled Thomas-Fermi-Dirac approach with very limited configuration interaction. Generally the agreement between Refs. 1 and 2 was good, particularly for strong lines: 63% of the log gf-values for common lines agreed within ±50 percent. In this compilation, we have included only those lines showing 50 percent or better agreement between Refs. 1 and 2.

As in the case of Fe III, we were able to assess the reliability of Kurucz and Peytremann's (or Biemont's) absolute scale by comparing their theoretical branching ratios to beam-foil lifetime data of Andersen et al.[3] This comparison supports the adopted scale: for the $z\ ^3G_3^\circ$ state (the only level measured by Andersen et al. for Ni III) the inverse sum of the transition probabilities, $(\Sigma_i A_{ki})^{-1}$, taken from Ref. 2 is only 17 percent higher than the corresponding beam-foil lifetime.

References

[1]E. Biemont, J. Quant. Spectrosc. Radiat. Transfer **16**, 137 (1976).

[2]R. L. Kurucz and E. Peytremann, Smithsonian Astrophysical Observatory Special Report 362 (1975).

[3]T. Andersen, P. Petersen, and E. Biemont, J. Quant. Spectrosc. Radiat. Transfer **17**, 389 (1977).

Ni III: Allowed transitions

No.	Transition Array	Multiplet	λ (Å)	E_i (cm^{-1})	E_k (cm^{-1})	g_i	g_k	A_{ki} (10^8 s^{-1})	f_{ik}	S (at. u.)	log gf	Accuracy	Source
1.	$3d^7(^4F)4s$ – $3d^7(^4F)4p$	^5F – $^5F°$											
			1769.64	53704	110213	11	11	6.2	0.29	19	0.50	D	1,2
			1794.90	54658	110371	9	9	2.7	0.13	6.9	0.07	D	1,2
			1791.64	55406	111221	7	7	2.5	0.12	5.0	−0.08	D	1,2
			1786.93	55952	111915	5	5	2.5	0.12	3.5	−0.22	D	1,2
			1782.75	56308	112402	3	3	3.8	0.18	3.2	−0.27	D	1,2
2.		^5F – $^5D°$											
			1724.52	56308	114295	3	1	6.7	0.10	1.7	−0.52	D	1,2
3.		^5F – $^5G°$											
			1692.51	53704	112788	11	13	7.9	0.40	25	0.64	D	1,2
			1709.90	54658	113141	9	11	6.3	0.34	17	0.49	D	1,2
			1719.46	55952	114110	5	7	6.0	0.37	10	0.27	D	1,2
			1722.28	56308	114371	3	5	5.9	0.44	7.5	0.12	D	1,2
			[1666.6]	53704	113705	11	9	0.038	0.0013	0.078	−1.84	D	1,2
4.		^3F – $^3G°$											
			1854.15	61339	115272	9	11	5.4	0.34	19	0.49	D	1,2
			1849.54	62606	116674	7	9	5.3	0.35	15	0.39	D	1,2
5.		^3F – $^3F°$											
			1823.06	61339	116192	9	9	5.6	0.28	15	0.40	D	1,2
			1830.01	62606	117251	7	7	4.6	0.23	9.7	0.21	D	1,2
			1830.08	63472	118115	5	5	5.0	0.25	7.5	0.10	D	1,2
6.		^3F – $^3D°$											
			1741.96	61339	118745	9	7	5.7	0.20	10	0.26	D	1,2
			1752.43	62606	119670	7	5	5.5	0.18	7.3	0.10	D	1,2
			1760.56	63472	120272	5	3	6.5	0.18	5.2	−0.05	D	1,2

Ni III

Forbidden Transitions

List of tabulated lines

Wavelength (Å)	No.	Wavelength (Å)	No.	Wavelength (Å)	No.	Wavelength (Å)	No.
1989.6	5	6000.2	3	7889.9	2	73472	1
2596.6	8	6401.5	3	8499.6	2	109990	1
2787.0	11	6533.8	3	11014	7	316170	9
2811.8	11	6682.2	3	15507	10	395310	9
4326.2	4	6797.1	3	31250	6		
4596.8	4	6946.4	3	33933	6		
4797.3	4	7124.8	2	38012	6		

For this spectrum, we have selected the data of Garstang,[1] who calculated M1 and E2 transition probabilities for lines arising between levels of the $3d^8$ configuration. Garstang's single configuration approximation should be fairly reasonable, since levels of the $3d^8$ configuration are generally well removed from those of the nearest neighboring configuration—$3d^74s$. His calculated energy levels are also in close agreement with observed energy levels. The $3d^8$ ^1S level, however, is close to some levels of the $3d^74s$ configuration. Nevertheless, we feel that configuration interaction should still play a minor role, since there are no nearby $J=0$ levels within the $3d^74s$ configuration. Accuracies for M1 transitions between levels of the same term should be 25 percent or better. However, data for other magnetic dipole transitions, as well as for all E2 transitions, are much less reliable.

Reference

[1]R. H. Garstang, Mon. Not. R. Astron. Soc. **118**, 234 (1958).

Ni III: Forbidden transitions

No.	Transition Array	Multiplet	λ (Å)	E_i (cm^{-1})	E_k (cm^{-1})	g_i	g_k	Type of transition	A_{ki} (s^{-1})	S (at. u.)	Accuracy	Source
1.	$3d^8$–$3d^8$	^3F – ^3F										
			[73472]	0	1361	9	7	M1	0.065	6.7	C+	1
			[109990]	1361	2270	7	5	M1	0.027	6.7	C+	1
2.		^3F – ^1D										
			[7124.8]	0	14032	9	5	E2	0.0045	0.25	E	1
			[7889.9]	1361	14032	7	5	M1	0.48	0.044	E	1
			"	"	"	7	5	E2	5.4(−4)^a	0.049	E	1
			[8499.6]	2270	14032	5	5	M1	0.21	0.024	E	1
			"	"	"	5	5	E2	1.9(−4)	0.025	E	1
3.		^3F – ^3P										
			[6000.2]	0	16662	9	5	E2	0.050	1.2	E	1
			[6401.5]	1361	16978	7	3	E2	0.038	0.73	E	1
			[6682.2]	2270	17231	5	1	E2	0.046	0.36	E	1
			[6533.8]	1361	16662	7	5	M1	0.11	0.0057	E	1
			[6797.1]	2270	16978	5	3	M1	0.0028	9.8(−5)	E	1
			"	"	"	5	3	E2	0.013	0.34	E	1
			[6946.4]	2270	16662	5	5	M1	0.022	0.0014	E	1
			"	"	"	5	5	E2	7.2(−4)	0.035	E	1
4.		^3F – ^1G										
			[4326.2]	0	23109	9	9	M1	0.35	0.0095	E	1
			"	"	"	9	9	E2	2.7(−4)	0.0022	E	1
			[4596.8]	1361	23109	7	9	M1	0.18	0.0058	E	1
			[4797.3]	2270	23109	5	9	E2	3.9(−4)	0.0053	E	1
5.		^3F – ^1S										
			[1989.6]	2270	52532	5	1	E2	0.20	0.0038	E	1n
6.		^1D – ^3P										
			[38012]	14032	16662	5	5	M1	0.098	1.0	E	1
			[33933]	14032	16978	5	3	M1	0.090	0.39	E	1
			"	"	"	5	3	E2	1.8(−6)	0.14	E	1
			[31250]	14032	17231	5	1	E2	2.4(−6)	0.043	E	1
7.		^1D – ^1G										
			[11014]	14032	23109	5	9	E2	6.2(−4)	0.54	E	1

Ni III: Forbidden transitions — Continued

No.	Transition Array	Multiplet	λ (Å)	E_i (cm^{-1})	E_k (cm^{-1})	g_i	g_k	Type of transition	A_{ki} (s^{-1})	S (at. u.)	Accu-racy	Source
8.		$^1D - ^1S$										
			[2596.6]	14032	52532	5	1	E2	10	0.72	E	1n
9.		$^3P - ^3P$										
			[316170]	16662	16978	5	3	M1	5.9(−4)	2.1	C+	1
			[395310]	16978	17231	3	1	M1	8.6(−4)	2.0	C+	1
10.		$^3P - ^1G$										
			[15507]	16662	23109	5	9	E2	2.7(−5)	0.13	E	1
11.		$^3P - ^1S$										
			[2787.0]	16662	52532	5	1	E2	1.9	0.19	E	1n
			[2811.8]	16978	52532	3	1	M1	2.9	0.0024	E	1n

^{a}The number in parentheses following the tabulated value indicates the power of ten by which this value has to be multiplied.

Ni IV

Mn Isoelectronic Sequence

Ground State: $1s^2 2s^2 2p^6 3s^2 3p^6 3d^7 \ ^4F_{9/2}$

Ionization Energy: 54.9 eV = 443000 cm^{-1}

Forbidden Transitions

List of tabulated lines

Wavelength (Å)	No.	Wavelength (Å)	No.	Wavelength (Å)	No.	Wavelength (Å)	No.
1470.8	8	2366.3	7	3987.6	12	5321.6	20
1497.0	8	2390.7	7	3990.3	6	5363.3	3
1511.3	8	2415.0	7	4071.8	29	5517.7	2
1516.4	8	2424.2	7	4084.0	12	5809.1	23
1529.8	8	2444.6	27	4084.6	6	5820.1	2
1531.0	8	2449.3	7	4142.8	29	5905.4	2
1544.7	8	2479.9	24	4160.5	16	5910.0	2
2004.5	13	2482.9	27	4179.0	29	5964.2	26
2014.5	13	2549.5	27	4234.7	16	6117.9	26
2030.2	13	2591.1	27	4253.9	29	6119.3	2
2038.9	13	3612.0	5	4363.5	16	6124.1	2
2040.4	13	3623.7	6	4421.8	4	6178.4	23
2065.4	13	3689.5	6	4445.2	16	6218.7	2
2075.7	17	3739.3	6	4451.3	4	6343.5	23
2125.1	17	3751.4	5	4537.9	4	6349.2	2
2153.9	17	3774.2	5	4627.0	4	6450.9	2
2254.6	21	3822.0	6	4754.3	4	6629.1	26
2279.4	7	3858.9	6	4772.5	3	6819.5	26
2287.0	21	3883.9	12	4946.7	20	9379.1	11
2301.4	7	3899.8	5	5041.6	3	9602.7	11
2306.7	21	3921.7	12	5052.0	20	10181	11
2340.7	21	3926.7	5	5059.9	3	11135	11
2342.9	7	3948.5	12	5288.1	3	11452	11

List of tabulated lines — Continued

Wavelength (Å)	No.	Wavelength (Å)	No.	Wavelength (Å)	No.	Wavelength (Å)	No.
12739	15	18927	10	84032	1	168990	9
14660	15	19492	19	89421	14	172780	1
14855	15	21314	10	97202	22	237410	28
15908	10	24228	19	99723	18	402790	9
17561	10	28998	19	117230	1		
18077	10	59465	25	158740	30		

For this spectrum, we have chosen the work of Hansen et al.,[1] who calculated M1 and E2 transition probabilities for transitions within the $3d^7$ ground configuration. These authors used a single configuration approximation, which should be fairly reliable, since the ground configuration is well separated from other configurations of the same parity. Also, the authors determined eigenvector components by a parametric fitting of theoretical energy expressions to observed energy levels.

Finally, Hartree-Fock calculations were used to determine s_q, the radial electric quadrupole integral, which is needed in the calculation of E2 transition probabilities.

Reference

[1] J. E. Hansen, A. J. J. Raassen, and P. H. M. Uylings, Astrophys. J. **277**, 435 (1984).

Ni IV: Forbidden transitions

No.	Transition Array	Multiplet	λ (Å)	E_i (cm^{-1})	E_k (cm^{-1})	g_i	g_k	Type of transition	A_{ki} (s^{-1})	S (at. u.)	Accuracy	Source
1.	$3d^7$–$3d^7$	^4F – ^4F										
			[84032]	0.0	1189.7	10	8	M1	0.057	10	C+	1
			[117230]	1189.7	2042.5	8	6	M1	0.036	13	C+	1
			[172780]	2042.5	2621.1	6	4	M1	0.013	9.9	C+	1
2.		^4F – ^4P										
			[5517.7]	0.0	18119	10	6	E2	0.068	1.2	E	1
			[5820.1]	1189.7	18367	8	4	E2	0.035	0.56	E	1
			[5910.0]	2042.5	18958	6	2	E2	0.026	0.22	E	1
			[5905.4]	1189.7	18119	8	6	M1	0.0053	2.4(−4)^a	E	1
			"	"	"	8	6	E2	0.015	0.38	E	1
			[6124.1]	2042.5	18367	6	4	M1	0.0018	6.1(−5)	E	1
			"	"	"	6	4	E2	0.020	0.41	E	1
			[6119.3]	2621.1	18958	4	2	E2	0.033	0.34	E	1
			[6218.7]	2042.5	18119	6	6	M1	0.0020	1.1(−4)	E	1
			"	"	"	6	6	E2	0.0027	0.090	E	1
			[6349.2]	2621.1	18367	4	4	M1	3.3(−4)	1.3(−5)	E	1
			"	"	"	4	4	E2	0.0054	0.13	E	1
			[6450.9]	2621.1	18119	4	6	M1	6.9(−4)	4.1(−5)	E	1
			"	"	"	4	6	E2	2.5(−4)	0.010	E	1
3.		^4F – ^2G										
			[5041.6]	0.0	19830	10	10	M1	0.91	0.043	E	1
			"	"	"	10	10	E2	1.3(−4)	0.0025	E	1
			[5059.9]	1189.7	20948	8	8	M1	0.35	0.013	E	1
			[4772.5]	0.0	20948	10	8	M1	0.034	0.0011	E	1
			[5363.3]	1189.7	19830	8	10	M1	0.28	0.016	E	1
			[5288.1]	2042.5	20948	6	8	M1	0.25	0.011	E	1

Ni IV: Forbidden transitions — Continued

No.	Transition Array	Multiplet	λ (Å)	E_i (cm^{-1})	E_k (cm^{-1})	g_i	g_k	Type of transition	A_{ki} (s^{-1})	S (at. u.)	Accuracy	Source
4.		^{4}F – ^{2}P										
			[4451.3]	1189.7	23649	8	4	E2	0.0099	0.041	E	1
			[4421.8]	2042.5	24651	6	2	E2	0.0042	0.0085	E	1
			[4627.0]	2042.5	23649	6	4	M1	0.17	0.0025	E	1
			"	"	"	6	4	E2	0.0043	0.022	E	1
			[4537.9]	2621.1	24651	4	2	M1	0.0039	2.7(−5)	E	1
			"	"	"	4	2	E2	0.0024	0.0055	E	1
			[4754.3]	2621.1	23649	4	4	M1	0.12	0.0019	E	1
			"	"	"	4	4	E2	0.0011	0.0064	E	1
5.		^{4}F – ^{2}H										
			[3926.7]	1189.7	26649	8	12	E2	1.7(−5)	1.1(−4)	E	1
			[3899.8]	2042.5	27678	6	10	E2	1.6(−5)	8.6(−5)	E	1
			[3751.4]	0.0	26649	10	12	M1	0.0016	3.8(−5)	E	1
			"	"	"	10	12	E2	2.5(−4)	0.0013	E	1
			[3774.2]	1189.7	27678	8	10	M1	0.0065	1.3(−4)	E	1
			"	"	"	8	10	E2	6.9(−5)	3.1(−4)	E	1
			[3612.0]	0.0	27678	10	10	M1	0.013	2.3(−4)	E	1
			"	"	"	10	10	E2	3.7(−5)	1.4(−4)	E	1
6.		^{4}F – ^{2}D2										
			[3689.5]	0.0	27097	10	6	E2	0.0011	0.0027	E	1
			[3623.7]	1189.7	28778	8	4	E2	0.0013	0.0019	E	1
			[3858.9]	1189.7	27097	8	6	M1	1.7	0.022	E	1
			"	"	"	8	6	E2	4.0(−4)	0.0012	E	1
			[3739.3]	2042.5	28778	6	4	M1	1.6	0.012	E	1
			"	"	"	6	4	E2	8.4(−4)	0.0015	E	1
			[3990.3]	2042.5	27097	6	6	M1	0.18	0.0025	E	1
			"	"	"	6	6	E2	2.3(−5)	8.3(−5)	E	1
			[3822.0]	2621.1	28778	4	4	M1	0.85	0.0070	E	1
			"	"	"	4	4	E2	4.0(−5)	7.8(−5)	E	1
			[4084.6]	2621.1	27097	4	6	M1	0.081	0.0012	E	1
			"	"	"	4	6	E2	3.1(−6)	1.3(−5)	E	1
7.		^{4}F – ^{2}F										
			[2301.4]	0.0	43438	10	6	E2	0.0010	2.3(−4)	E	1
			[2279.4]	0.0	43859	10	8	M1	0.28	9.8(−4)	D	1
			"	"	"	10	8	E2	0.0066	0.0019	E	1
			[2366.3]	1189.7	43438	8	6	M1	0.072	2.1(−4)	D	1
			"	"	"	8	6	E2	9.9(−4)	2.6(−4)	E	1
			[2342.9]	1189.7	43859	8	8	M1	0.056	2.1(−4)	D	1
			"	"	"	8	8	E2	7.5(−4)	2.5(−4)	E	1
			[2415.0]	2042.5	43438	6	6	M1	0.055	1.7(−4)	D	1
			"	"	"	6	6	E2	0.0013	3.8(−4)	E	1
			[2390.7]	2042.5	43859	6	8	M1	0.12	4.9(−4)	D	1
			"	"	"	6	8	E2	0.0022	8.2(−4)	E	1
			[2449.3]	2621.1	43438	4	6	M1	0.28	9.2(−4)	D	1
			"	"	"	4	6	E2	0.0049	0.0015	E	1
			[2424.2]	2621.1	43859	4	8	E2	4.9(−4)	2.0(−4)	E	1

Ni IV: Forbidden transitions — Continued

No.	Transition Array	Multiplet	λ (Å)	E_i (cm^{-1})	E_k (cm^{-1})	g_i	g_k	Type of transition	A_{ki} (s^{-1})	S (at. u.)	Accuracy	Source
8.		^{4}F – ^{2}D1										
			[1470.8]	0.0	67990	10	6	E2	0.074	0.0018	E	1
			[1511.3]	1189.7	67360	8	4	E2	0.014	2.6(−4)	E	1
			[1497.0]	1189.7	67990	8	6	M1	0.26	1.9(−4)	E	1
			"	"	"	8	6	E2	0.0055	1.5(−4)	E	1
			[1531.0]	2042.5	67360	6	4	M1	0.29	1.5(−4)	E	1
			"	"	"	6	4	E2	3.4(−5)	6.8(−7)	E	1
			[1516.4]	2042.5	67990	6	6	M1	0.040	3.1(−5)	E	1
			"	"	"	6	6	E2	0.0039	1.1(−4)	E	1
			[1544.7]	2621.1	67360	4	4	M1	0.20	1.1(−4)	E	1
			"	"	"	4	4	E2	0.0026	5.4(−5)	E	1
			[1529.8]	2621.1	67990	4	6	M1	0.019	1.5(−5)	E	1
			"	"	"	4	6	E2	0.0016	4.8(−5)	E	1
9.		^{4}P – ^{4}P										
			[402790]	18119	18367	6	4	M1	3.4(−4)	3.3	C+	1
			[168990]	18367	18958	4	2	M1	0.0089	3.2	C+	1
10.		^{4}P – ^{2}P										
			[18077]	18119	23649	6	4	M1	0.33	0.29	D	1
			[15908]	18367	24651	4	2	M1	6.4(−4)	1.9(−4)	D	1
			[18927]	18367	23649	4	4	M1	0.17	0.17	D	1
			[17561]	18958	24651	2	2	M1	0.42	0.17	D	1
			[21314]	18958	23649	2	4	M1	0.064	0.092	D	1
11.		^{4}P – ^{2}D2										
			[11135]	18119	27097	6	6	M1	0.10	0.031	E	1
			[9602.7]	18367	28778	4	4	M1	0.022	0.0029	E	1
			[9379.1]	18119	28778	6	4	M1	0.0090	0.0011	E	1
			[11452]	18367	27097	4	6	M1	0.065	0.022	E	1
			[10181]	18958	28778	2	4	M1	0.021	0.0033	E	1
12.		^{4}P – ^{2}F										
			[3921.7]	18367	43859	4	8	E2	0.0037	0.016	E	1
			[4084.0]	18958	43438	2	6	E2	0.0010	0.0041	E	1
			[3883.9]	18119	43859	6	8	M1	6.0(−4)	1.0(−5)	E	1
			"	"	"	6	8	E2	3.1(−4)	0.0013	E	1
			[3987.6]	18367	43438	4	6	M1	1.3(−4)	1.8(−6)	E	1
			"	"	"	4	6	E2	0.0020	0.0072	E	1
			[3948.5]	18119	43438	6	6	M1	0.0050	6.8(−5)	E	1
			"	"	"	6	6	E2	6.8(−5)	2.3(−4)	E	1
13.		^{4}P – ^{2}D1										
			[2004.5]	18119	67990	6	6	M1	1.9	0.0034	E	1
			"	"	"	6	6	E2	0.0015	1.7(−4)	E	1
			[2040.4]	18367	67360	4	4	M1	0.72	9.1(−4)	E	1
			"	"	"	4	4	E2	0.13	0.011	E	1
			[2030.2]	18119	67360	6	4	M1	0.27	3.4(−4)	E	1
			"	"	"	6	4	E2	2.9(−4)	2.4(−5)	E	1
			[2014.5]	18367	67990	4	6	M1	0.30	5.5(−4)	E	1
			"	"	"	4	6	E2	0.25	0.030	E	1
			[2065.4]	18958	67360	2	4	M1	0.19	2.5(−4)	E	1
			"	"	"	2	4	E2	0.063	0.0056	E	1
			[2038.9]	18958	67990	2	6	E2	0.022	0.0028	E	1

Ni IV: Forbidden transitions — Continued

No.	Transition Array	Multiplet	λ (Å)	E_i (cm^{-1})	E_k (cm^{-1})	g_i	g_k	Type of transition	A_{ki} (s^{-1})	S (at. u.)	Accu-racy	Source
14.		^{2}G – ^{2}G										
			[89421]	19830	20948	10	8	M1	0.020	4.2	C+	1
15.		^{2}G – ^{2}H										
			[14660]	19830	26649	10	12	M1	0.094	0.13	E	1
			[14855]	20948	27678	8	10	M1	0.089	0.11	E	1
			[12739]	19830	27678	10	10	M1	0.31	0.24	E	1
16.		^{2}G – ^{2}F										
			[4234.7]	19830	43438	10	6	E2	0.037	0.18	E	1
			[4160.5]	19830	43859	10	8	M1	0.22	0.0047	E	1
			"	"	"	10	8	E2	0.093	0.55	E	1
			[4445.2]	20948	43438	8	6	M1	0.18	0.0035	E	1
			"	"	"	8	6	E2	0.078	0.48	E	1
			[4363.5]	20948	43859	8	8	M1	0.33	0.0081	E	1
			"	"	"	8	8	E2	0.0075	0.057	E	1
17.		^{2}G – ^{2}D1										
			[2075.7]	19830	67990	10	6	E2	8.5	1.2	E	1
			[2153.9]	20948	67360	8	4	E2	8.5	0.94	E	1
			[2125.1]	20948	67990	8	6	M1	0.0045	9.6(−6)	E	1
			"	"	"	8	6	E2	0.60	0.093	E	1
18.		^{2}P – ^{2}P										
			[99723]	23649	24651	4	2	M1	0.017	1.3	C+	1
19.		^{2}P – ^{2}D2										
			[28998]	23649	27097	4	6	M1	0.037	0.20	E	1
			[24228]	24651	28778	2	4	M1	0.064	0.13	E	1
			[19492]	23649	28778	4	4	M1	0.37	0.41	E	1
20.		^{2}P – ^{2}F										
			[4946.7]	23649	43859	4	8	E2	0.011	0.16	E	1
			[5321.6]	24651	43438	2	6	E2	0.010	0.15	E	1
			[5052.0]	23649	43438	4	6	M1	0.0035	1.0(−4)	E	1
			"	"	"	4	6	E2	0.018	0.21	E	1
21.		^{2}P – ^{2}D1										
			[2306.7]	24651	67990	2	6	E2	0.60	0.14	E	1
			[2254.6]	23649	67990	4	6	M1	0.19	4.8(−4)	E	1
			"	"	"	4	6	E2	2.1	0.44	E	1
			[2340.7]	24651	67360	2	4	M1	0.0055	1.0(−5)	E	1
			"	"	"	2	4	E2	1.2	0.20	E	1
			[2287.0]	23649	67360	4	4	M1	0.052	9.2(−5)	E	1
			"	"	"	4	4	E2	0.66	0.098	E	1
22.		^{2}H – ^{2}H										
			[97202]	26649	27678	12	10	M1	0.016	5.4	C+	1

Ni IV: Forbidden transitions — Continued

No.	Transition Array	Multiplet	λ (Å)	E_i (cm^{-1})	E_k (cm^{-1})	g_i	g_k	Type of transition	A_{ki} (s^{-1})	S (at. u.)	Accuracy	Source
23.		^{2}H – ^{2}F										
			[5809.1]	26649	43859	12	8	E2	0.059	1.9	E	1
			[6343.5]	27678	43438	10	6	E2	0.036	1.3	E	1
			[6178.4]	27678	43859	10	8	M1	0.0016	1.1(−4)	E	1
			"	"	"	10	8	E2	0.0044	0.19	E	1
24.		^{2}H – ^{2}D1										
			[2479.9]	27678	67990	10	6	E2	0.15	0.050	E	1
25.		^{2}D2 – ^{2}D2										
			[59465]	27097	28778	6	4	M1	0.070	2.2	C+	1
26.		^{2}D2 – ^{2}F										
			[6629.1]	28778	43859	4	8	E2	0.0038	0.23	E	1
			[5964.2]	27097	43859	6	8	M1	0.022	0.0014	E	1
			"	"	"	6	8	E2	0.024	0.86	E	1
			[6819.5]	28778	43438	4	6	M1	0.012	8.5(−4)	E	1
			"	"	"	4	6	E2	0.0089	0.47	E	1
			[6117.9]	27097	43438	6	6	M1	0.045	0.0023	E	1
			"	"	"	6	6	E2	0.0047	0.14	E	1
27.		^{2}D2 – ^{2}D1										
			[2444.6]	27097	67990	6	6	M1	0.0099	3.2(−5)	D	1
			"	"	"	6	6	E2	0.54	0.17	E	1
			[2591.1]	28778	67360	4	4	M1	0.0012	3.1(−6)	D	1
			"	"	"	4	4	E2	0.75	0.21	E	1
			[2482.9]	27097	67360	6	4	M1	1.2	0.0027	D	1
			"	"	"	6	4	E2	0.14	0.031	E	1
			[2549.5]	28778	67990	4	6	M1	0.59	0.0022	D	1
			"	"	"	4	6	E2	0.0036	0.0014	E	1
28.		^{2}F – ^{2}F										
			[237410]	43438	43859	6	8	M1	8.6(−4)	3.4	C+	1
29.		^{2}F – ^{2}D1										
			[4253.9]	43859	67360	8	4	E2	0.051	0.17	E	1
			[4142.8]	43859	67990	8	6	M1	0.45	0.0071	E	1
			"	"	"	8	6	E2	0.42	1.8	E	1
			[4179.0]	43438	67360	6	4	M1	0.45	0.0049	E	1
			"	"	"	6	4	E2	0.41	1.2	E	1
			[4071.8]	43438	67990	6	6	M1	0.86	0.013	E	1
			"	"	"	6	6	E2	0.065	0.26	E	1
30.		^{2}D1 – ^{2}D1										
			[158740]	67360	67990	4	6	M1	0.0027	2.4	C+	1

[a]The number in parentheses following the tabulated value indicates the power of ten by which this value has to be multiplied.

Ni v

Cr Isoelectronic Sequence

Ground State: $1s^2 2s^2 2p^6 3s^2 3p^6 3d^6\ ^5D_4$

Ionization Energy: $76.06\ \text{eV} = 613500\ \text{cm}^{-1}$

Forbidden Transitions

List of tabulated lines

Wavelength (Å)	No.	Wavelength (Å)	No.	Wavelength (Å)	No.	Wavelength (Å)	No.
2384.8	5	2904.7	4	3485.6	3	3706.9	2
2436.5	5	2935.0	4	3519.0	3	3726.7	1
2454.0	5	2981.8	4	3540.8	3	3752.7	1
2472.6	5	3006.1	4	3560.1	3	3957.2	1
2486.1	5	3013.7	4	3566.9	3	4053.5	1
2490.7	5	3036.2	4	3588.5	2	4117.2	1
2509.9	5	3380.7	3	3600.2	1		
2514.6	5	3432.6	3	3625.0	2		
2521.7	5	3446.2	3	3674.4	1		

For this spectrum, we have chosen the work of Raassen et al.,[1] who calculated M1 transition probabilities for transitions within the $3d^6$ ground configuration. These authors used a single configuration approximation in intermediate coupling. They determined eigenvector components by a parametric fitting of theoretical energy expressions to observed energy levels.

Reference

[1]A. J. J. Raassen, Th. A. M. van Kleef, and B. C. Metsch, Physica **84C**, 133 (1976).

Ni v: Forbidden transitions

No.	Transition Array	Multiplet	λ (Å)	E_i (cm^{-1})	E_k (cm^{-1})	g_i	g_k	Type of transition	A_{ki} (s^{-1})	S (at. u.)	Accuracy	Source
1.	$3d^6$–$3d^6$	^5D – $^3P2$										
			[3957.2]	889.7	26153	7	5	M1	2.3	0.026	E	1
			[3674.4]	1489.9	28698	5	3	M1	2.9	0.016	E	1
			[3600.2]	1871.5	29640	3	1	M1	3.8	0.0066	E	1
			[4053.5]	1489.9	26153	5	5	M1	7(−4)^a	9(−6)	E	1
			[3726.7]	1871.5	28698	3	3	M1	0.0063	3.6(−5)	E	1
			[4117.2]	1871.5	26153	3	5	M1	0.24	0.0031	E	1
			[3752.7]	2057.6	28698	1	3	M1	0.54	0.0032	E	1
2.		^5D – ^3H										
			[3625.0]	0.0	27578	9	11	M1	1.4(−4)	2.7(−6)	E	1
			[3706.9]	889.7	27859	7	9	M1	0.029	4.9(−4)	E	1
			[3588.5]	0.0	27859	9	9	M1	0.17	0.0026	E	1

Ni v: Forbidden transitions — Continued

No.	Transition Array	Multiplet	λ (Å)	E_i (cm^{-1})	E_k (cm^{-1})	g_i	g_k	Type of transition	A_{ki} (s^{-1})	S (at. u.)	Accuracy	Source
3.		^{5}D – ^{3}F2										
			[3432.6]	0.0	29124	9	9	M1	2.3	0.031	E	1
			[3485.6]	889.7	29571	7	7	M1	1.3	0.014	E	1
			[3519.0]	1489.9	29899	5	5	M1	0.53	0.0043	E	1
			[3380.7]	0.0	29571	9	7	M1	0.23	0.0023	E	1
			[3446.2]	889.7	29899	7	5	M1	0.15	0.0011	E	1
			[3540.8]	889.7	29124	7	9	M1	0.41	0.0061	E	1
			[3560.1]	1489.9	29571	5	7	M1	0.44	0.0052	E	1
			[3566.9]	1871.5	29899	3	5	M1	0.25	0.0021	E	1
4.		^{5}D – ^{3}G										
			[3006.1]	0.0	33257	9	11	M1	0.0030	3.3(−5)	E	1
			[3013.7]	889.7	34062	7	9	M1	0.025	2.3(−4)	E	1
			[3036.2]	1489.9	34416	5	7	M1	0.018	1.3(−4)	E	1
			[2935.0]	0.0	34062	9	9	M1	0.10	8.4(−4)	E	1
			[2981.8]	889.7	34416	7	7	M1	0.046	3.2(−4)	E	1
			[2904.7]	0.0	34416	9	7	M1	0.0070	4.5(−5)	E	1
5.		^{5}D – ^{3}D										
			[2384.8]	0.0	41920	9	7	M1	1.2	0.0042	D	1
			[2454.0]	889.7	41627	7	5	M1	0.12	3.3(−4)	D	1
			[2486.1]	1489.9	41701	5	3	M1	0.018	3.1(−5)	D	1
			[2436.5]	889.7	41920	7	7	M1	0.34	0.0013	D	1
			[2490.7]	1489.9	41627	5	5	M1	0.57	0.0016	D	1
			[2509.9]	1871.5	41701	3	3	M1	0.76	0.0013	D	1
			[2472.6]	1489.9	41920	5	7	M1	0.22	8.6(−4)	D	1
			[2514.6]	1871.5	41627	3	5	M1	0.44	0.0013	D	1
			[2521.7]	2057.6	41701	1	3	M1	0.62	0.0011	D	1

[a]The number in parentheses following the tabulated value indicates the power of ten by which this value has to be multiplied.

Ni VII

Ti Isoelectronic Sequence

Ground State: $1s^2 2s^2 2p^6 3s^2 3p^6 3d^4 \,^5D_0$

Ionization Energy: 133 eV = 1070000 cm^{-1}

Forbidden Transitions

List of tabulated lines

Wavelength (Å)	No.	Wavelength (Å)	No.	Wavelength (Å)	No.	Wavelength (Å)	No.
2676.2	4	2916.6	1	3026.2	1	3168.0	1
2685.5	4	2943.1	3	3048.2	3	3221.6	1
2712.5	4	2962.0	1	3054.7	3	3315.6	2
2728.5	4	2983.0	3	3106.2	3	3344.2	2
2749.9	4	2989.3	3	3131.4	3	3355.0	1
2795.0	4	3024.3	3	3140.2	1	3414.3	2

For this spectrum, we selected the work of Henrichs,[1] who calculated magnetic dipole transition probabilities for 24 lines within the $3d^4$ configuration. The calculations were performed in intermediate coupling, and the eigenvector components were determined by a least-squares fitting of theoretically derived energies to observed energy levels.

Reference

[1] H. F. Henrichs, Astron. Astrophys. **44**, 41 (1975).

Ni VII: Forbidden transitions

No.	Transition Array	Multiplet	λ (Å)	E_i (cm^{-1})	E_k (cm^{-1})	g_i	g_k	Type of transition	A_{ki} (s^{-1})	S (at. u.)	Accuracy	Source
1.	$3d^4$–$3d^4$	^5D – $^3P2$ (1F)										
			[3026.2]	1520	34555	7	5	M1	3.1	0.016	E	1
			[3221.6]	804	31836	5	3	M1	4.7	0.017	E	1
			[3355.0]	279	30077	3	1	M1	6.1	0.0085	E	1
			[2962.0]	804	34555	5	5	M1	0.005	2(−5)[a]	E	1
			[3168.0]	279	31836	3	3	M1	8(−4)	1(−6)	E	1
			[2916.6]	279	34555	3	5	M1	0.11	5.1(−4)	E	1
			[3140.2]	0	31836	1	3	M1	0.51	0.0018	E	1
2.		^5D – ^3H (2F)										
			[3344.2]	2392	32286	9	11	M1	2(−4)	3(−6)	E	1
			[3315.6]	1520	31672	7	9	M1	0.01	1(−4)	E	1
			[3414.3]	2392	31672	9	9	M1	0.05	7(−4)	E	1
3.		^5D – $^3F2$ (3F)										
			[3106.2]	2392	34576	9	9	M1	3.2	0.032	E	1
			[3048.2]	1520	34317	7	7	M1	1.9	0.014	E	1
			[2989.3]	804	34247	5	5	M1	0.87	0.0043	E	1
			[3131.4]	2392	34317	9	7	M1	0.26	0.0021	E	1
			[3054.7]	1520	34247	7	5	M1	0.15	7.9(−4)	E	1
			[3024.3]	1520	34576	7	9	M1	0.70	0.0065	E	1
			[2983.0]	804	34317	5	7	M1	0.75	0.0052	E	1
			[2943.1]	279	34247	3	5	M1	0.48	0.0023	E	1
4.		^5D – ^3G										
			[2712.5]	2392	39247	9	11	M1	0.004	3(−5)	E	1
			[2685.5]	1520	38746	7	9	M1	0.06	4(−4)	E	1
			[2676.2]	804	38160	5	7	M1	0.07	3(−4)	E	1
			[2749.9]	2392	38746	9	9	M1	0.20	0.0014	E	1
			[2728.5]	1520	38160	7	7	M1	0.16	8.4(−4)	E	1
			[2795.0]	2392	38160	9	7	M1	0.02	1(−4)	E	1

[a]The number in parentheses following the tabulated value indicates the power of ten by which the value has to be multiplied.

Ni IX

Ca Isoelectronic Sequence

Ground State: $1s^2 2s^2 2p^6 3s^2 3p^6 3d^2 \; ^3F_2$

Ionization Energy: 193 eV = 1560000 cm^{-1}

Allowed Transitions

For this spectrum, we have chosen the data of Fawcett, Ridgeley, and Ekberg.[1] These authors experimentally observed and classified sixteen Ni IX lines in the $3p^6 3d^2 - 3p^5 3d^3$ transition array. For fifteen of these lines, Fawcett *et al.* calculated oscillator strengths by the Hartree-XR method (self-consistent-field calculations with exchange, configuration interaction, and relativistic effects). We estimate these data to be accurate within fifty percent.

Reference

[1] B. C. Fawcett, A. Ridgeley, and J. O. Ekberg, Phys. Scr. **21**, 155 (1980).

Ni IX: Allowed transitions

No.	Transition Array	Multiplet	λ (Å)	E_i (cm^{-1})	E_k (cm^{-1})	g_i	g_k	A_{ki} (10^8 s^{-1})	f_{ik}	S (at. u.)	log gf	Accuracy	Source
1.	$3p^6 3d^2-$ $3p^5(^2P°)3d^3(^2H)$	$^3F - ^3G°$											
			165.436	4070	608530	9	11	970	0.49	2.4	0.64	D	1
			166.079	1880	604000	7	9	870	0.46	1.8	0.51	D	1
			166.306	0	601300	5	7	430	0.25	0.69	0.10	D	1
2.		$^1G - ^1G°$	147.013			9	9	4100	1.3	5.8	1.08	D	1
3.	$3p^6 3d^2-$ $3p^5(^2P°)3d^3(^2G)$	$^1G - ^1H°$	165.436			9	11	730	0.37	1.8	0.52	D	1
4.	$3p^6 3d^2-$ $3p^5(^2P°)3d^3(^4F)$	$^3F - ^3F°$											
			150.836	4070	667080	9	9	2800	0.95	4.2	0.93	D	1
			151.022	1880	664080	7	7	2200	0.75	2.6	0.72	D	1
			151.281	0	661050	5	5	2300	0.80	2.0	0.60	D	1
			150.32	1880	667080	7	9	180	0.077	0.27	−0.27	D	1
			150.574	0	664080	5	7	180	0.083	0.21	−0.38	D	1
5.		$^3F - ^3D°$											
			141.356	4070	711510	9	7	2600	0.61	2.6	0.74	D	1
			140.917	1880	711520	7	5	2600	0.56	1.8	0.59	D	1
			141.002	0	709210	5	3	1900	0.33	0.77	0.22	D	1
			140.917	1880	711510	7	7	120	0.034	0.11	−0.62	D	1
			140.542	0	711520	5	5	230	0.068	0.16	−0.47	D	1

Ni IX

Forbidden Transitions

For this ion, we selected the work of Warner and Kirkpatrick,[1] who used a single-configuration approximation and calculated radial integrals with scaled Thomas-Fermi wavefunctions. We have tabulated M1 and E2 transition probabilities for 5 lines within the $3d^2$ configuration. Warner and Kirkpatrick also calculated electric quadrupole A-values for transitions within the $3d^2$–$3d4s$ transition array. We have omitted these lines, however, since accurate experimental energy levels within the $3d4s$ configuration were unavailable. For lines within the $3d^2\ ^3F$ term, we have recalculated

Warner and Kirkpatrick's A-values by using observed energy-level data instead of theoretically derived values. Due to the lack of reliable observational material, we did not provide energy level values for the ^1D, ^3P, and ^1G terms.

Reference

[1]B. Warner and R. C. Kirkpatrick, Mon. Not. R. Astron. Soc. **144**, 397 (1969).

Ni IX: Forbidden transitions

No.	Transition Array	Multiplet	λ (Å)	E_i (cm^{-1})	E_k (cm^{-1})	g_i	g_k	Type of transition	A_{ki} (s^{-1})	S (at. u.)	Accuracy	Source
1.	$3d^2$–$3d^2$	^3F – ^3F										
			[45650]	1880	4070	7	9	M1	0.21	6.7	C	1n
			[53180]	0	1880	5	7	M1	0.17	6.6	C	1n
2.		^1D – ^3P										
			[19010]			5	5	M1	0.51	0.65	E	1
3.		^1D – ^1G	[7141.9]			5	9	E2	8.1(−4)[a]	0.081	E	1
4.		^3P – ^1G										
			[11440]			5	9	E2	1.9(−5)	0.020	E	1

[a]The number in parentheses following the tabulated value indicates the power of ten by which this value has to be multiplied.

Ni XI

Ar Isoelectronic Sequence

Ground State: $1s^2 2s^2 2p^6 3s^2 3p^6\ ^1S_0$

Ionization Energy: $321.0\ \text{eV} = 2589000\ \text{cm}^{-1}$

Allowed Transitions

Line strengths for the $3p^6$–$3p^5 3d$ resonance transitions of this argon-like ion were taken from the superposition-of-configurations (SOC) calculations of Weiss,[1] which are expected to be fairly accurate. The remainder of the oscillator strengths were interpolated from the Dirac-Hartree-Fock data of Lin et al.,[2] who included correlation only in the lower state.

References

[1]A. W. Weiss, private communication.
[2]D. L. Lin, W. Fielder, Jr., and L. Armstrong, Jr., Phys. Rev. A **16**, 589 (1977).

Ni XI: Allowed transitions

No.	Transition Array	Multiplet	λ (Å)	E_i (cm^{-1})	E_k (cm^{-1})	g_i	g_k	A_{ki} (10^8 s^{-1})	f_{ik}	S (at. u.)	log gf	Accuracy	Source
1.	$3p^6$–$3p^53d$	^1S – $^3P°$											
			211.428	0	472974	1	3	0.14	2.9(−4)[a]	2.0(−4)	−3.54	E	1
2.		^1S – $^3D°$											
			186.976	0	534828	1	3	4.3	0.0068	0.0042	−2.17	E	1
3.		^1S – $^1P°$	148.374	0	673973	1	3	2340	2.31	1.13	0.364	C+	1
4.	$3p^6$– $3p^5(^2P°_{3/2})4s$	^1S – $(^3/_2,^1/_2)°$											
			78.744	0	1269900	1	3	610	0.17	0.044	−0.77	D	interp.
5.	$3p^6$– $3p^5(^2P°_{1/2})4s$	^1S – $(^1/_2,^1/_2)°$											
			77.393	0	1292100	1	3	850	0.23	0.059	−0.64	D	interp.
6.	$3p^6$– $3p^5(^2P°_{3/2})4d$	^1S – $^2[^3/_2]°$											
			63.641	0	1571300	1	3	2500	0.45	0.094	−0.35	D	interp.
7.	$3p^6$– $3p^5(^2P°_{1/2})4d$	^1S – $^2[^3/_2]°$											
			62.730	0	1594100	1	3	1200	0.22	0.045	−0.66	D	interp.

[a]The number in parentheses following the tabulated value indicates the power of ten by which this value has to be multiplied.

Ni XII

Cl Isoelectronic Sequence

Ground State: $1s^22s^22p^63s^23p^5\ ^2P°_{3/2}$

Ionization Energy: 352 eV = 2840000 cm^{-1}

Allowed Transitions

Line strengths for transitions of the arrays $3s^23p^5$–$3s3p^6$ and $3p^5$–$3p^43d$ are the results of the multiconfiguration Dirac-Fock (MCDF) calculations of Huang et al.[1] These relativistic calculations include a perturbative treatment of the Breit interaction and the Lamb shift. Configuration mixing was limited to some configurations within the $n=3$ complex. Those configurations which were assumed to lie far above $3p^5$ or $3p^43d$ in energy were excluded, as were all configurations outside the complex.

According to the semi-empirical HX (Hartree-Fock with statistical allowance for exchange) calculations of

Bromage et al.[2] for Fe X, some levels of the $3p^43d$ configuration are strongly mixed in the LS basis, and in a few cases the LS designations given in Ref. 2 differed from those of Huang et al. The level designations used in this compilation are in accord with the theoretical results of Refs. 1 and 2 for Fe X. Percentage compositions published by Bromage[3] for the levels of the $3p^43d$ configuration in V VII and Ni XII indicate that the designations for the iron ion are appropriate for the neighboring ions of the chlorine isoelectronic sequence. Transitions involving highly mixed levels have been excluded, as have the very weak transitions.

The calculated wavelengths of Huang *et al*. differ appreciably from the observed ones found in the literature. Thus the available experimentally determined wavelengths were used in making the conversion from line strengths to *f*- and *A*-values. (Otherwise, the calculated wavelengths of Huang *et al*. were used, but they provide only a rough idea of the spectral-line positions.) Bromage *et al*. indicate that it was necessary to scale down some configuration-interaction parameters by a greater amount than usual in order to fit their calculated energy levels for Fe x to the experimental data. This could be an indication that neglecting to take configuration interaction into account on a larger scale yields significant errors in the energy levels and/or *f*-values.

References

[1]K.-N. Huang, Y.-K. Kim, K. T. Cheng, and J. P. Desclaux, At. Data Nucl. Data Tables **28**, 355 (1983).

[2]G. E. Bromage, R. D. Cowan, and B. C. Fawcett, Phys. Scr. **15**, 177 (1977).

[3]G. E. Bromage, Astron. Astrophys., Suppl. Ser. **41**, 79 (1980).

Ni XII: Allowed transitions

No.	Transition Array	Multiplet	λ (Å)	E_i (cm^{-1})	E_k (cm^{-1})	g_i	g_k	A_{ki} (10^8 s^{-1})	f_{ik}	S (at. u.)	log gf	Accuracy	Source
1.	$3s^23p^5$–$3s3p^6$	$^2P°$ – ^2S	*302.35*	*7877*	338615	6	2	73	0.0332	0.198	−0.70	C−	1
			295.321	0	338615	4	2	52	0.0337	0.131	−0.87	C−	1
			317.475	23630	338615	2	2	21	0.032	0.067	−1.19	C−	1
2.	$3p^5$–$3p^4(^3P)3d$	$^2P°$ – ^4F											
			[193]			4	6	0.66	5.5(−4)[a]	0.0014	−2.66	E	1
3.		$^2P°$ – ^4P											
			[189]			4	4	0.49	2.6(−4)	6.5(−4)	−2.98	E	1
			[201]			2	2	1.2	7.2(−4)	9.5(−4)	−2.84	E	1
			[192]			4	2	5.3	0.0015	0.0037	−2.23	E	1
4.		$^2P°$ – ^2F											
			[182]			4	6	0.27	2.0(−4)	4.8(−4)	−3.10	E	1
5.		$^2P°$ – ^2D	*152.20*	*7877*	*664890*	6	10	2070	1.20	3.61	0.86	C−	1
			152.153	0	657233	4	6	2080	1.08	2.17	0.64	C	1
			153.174	23630	676375	2	4	1990	1.40	1.41	0.447	C	1
			147.847	0	676375	4	4	41	0.013	0.026	−1.27	D	1
6.	$3p^5$–$3p^4(^1D)3d$	$^2P°$ – ^2F											
			[172]			4	6	2.6	0.0017	0.0039	−2.16	E	1
7.		$^2P°$ – ^2S	*162.61*	*7877*	622843	6	2	1900	0.25	0.79	0.17	C−	1
			160.554	0	622843	4	2	1400	0.27	0.58	0.04	C−	1
			166.88	23630	622843	2	2	453	0.189	0.208	−0.422	C−	1
8.	$3p^5$–$3p^4(^1S)3d$	$^2P°$ – ^2D											
			[167]			2	4	14	0.012	0.013	−1.63	D	1
			[161]			4	4	5.7	0.0022	0.0047	−2.05	E	1

[a]The number in parentheses following the tabulated value indicates the power of ten by which this value has to be multiplied.

Ni XII

Forbidden Transitions

Line strengths for the magnetic dipole and electric quadrupole contributions to the transition between the two levels of the $3p^5$ configuration are the results of the multiconfiguration Dirac-Fock (MCDF) calculations of Huang *et al.*[1] These relativistic calculations included a perturbative treatment of the Breit interaction and the Lamb shift. Allowance for mixing among odd-parity configurations was limited to the set $3s^23p^5$, $3s\,3p^53d$, $3p^53d^2$, and $3s^23p^33d^2$. The strength of the electric

quadrupole transition as defined in Ref. 1 was multiplied by the factor $^2/_3$ which is needed to bring this value into conformance with the definition of quadrupole strengths used in the NBS tables.

Reference

[1]K.-N. Huang, Y.-K. Kim, K. T. Cheng, and J. P. Desclaux, At. Data Nucl. Data Tables **28**, 355 (1983).

Ni XII: Forbidden transitions

No.	Transition Array	Multiplet	λ (Å)	E_i (cm^{-1})	E_k (cm^{-1})	g_i	g_k	Type of transition	A_{ki} (s^{-1})	S (at. u.)	Accuracy	Source
1.	$3p^5$–$3p^5$	^{2}P° – ^{2}P°										
			4231.4	0	23630	4	2	M1	237	1.33	B	1
			"	"	"	4	2	E2	0.080	0.13	D–	1

Ni XIII

S Isoelectronic Sequence

Ground State: $1s^22s^22p^63s^23p^4\ {}^3P_2$

Ionization Energy: 384 eV = 3100000 cm^{-1}

Allowed Transitions

Oscillator strengths are tabulated for a few transitions of the arrays $3s^23p^4$–$3s\,3p^5$ and $3p^4$–$3p^33d$. These are the results of the Hartree-XR (Hartree-Fock with relativistic effects and statistical allowance for exchange) calculations of Bromage.[1] The percentage compositions are in good agreement with those of Bromage *et al.*[2] for Fe XI. The term designations used here are in accord with the results of these two sources. Transitions involving levels

of low purity in *LS* coupling are omitted, as are very weak transitions.

References

[1]G. E. Bromage, Astron. Astrophys., Suppl. Ser. **41**, 79 (1980).
[2]G. E. Bromage, R. D. Cowan, and B. C. Fawcett, Phys. Scr. **15**, 177 (1977).

Ni XIII: Allowed transitions

No.	Transition Array	Multiplet	λ (Å)	E_i (cm^{-1})	E_k (cm^{-1})	g_i	g_k	A_{ki} (10^8 s^{-1})	f_{ik}	S (at. u.)	log gf	Accuracy	Source
1.	$3s^23p^4$–$3s3p^5$	^{3}P – ^{3}P°											
			302.844	0	330203	5	5	29	0.040	0.20	-0.70	D	1
2.		^{1}D – ^{1}P°	267.468	47033	420910	5	3	100	0.066	0.29	-0.48	D	1
3.	$3p^4$–$3p^3(^2$D°$)3d$	^{3}P – ^{3}P°											
			164.15	0	609200	5	5	1300	0.52	1.4	0.41	D	1
			[165.2]			3	1	1700	0.23	0.38	-0.16	D	1
			[169.59]	19542	609200	3	5	210	0.15	0.25	-0.35	D	1
4.		^{3}P – ^{1}D°											
			[154.69]	19542	666000	3	5	150	0.087	0.13	-0.58	D	1
5.		^{1}D – ^{1}D°	161.56	47033	666000	5	5	1400	0.54	1.4	0.43	C	1
6.		^{1}D – ^{1}F°	157.55	47033	681750	5	7	1920	1.00	2.59	0.70	C	1
7.	$3p^4$–$3p^3(^2$P°$)3d$	^{3}P – ^{3}P°											
			[174.0]			5	5	88	0.040	0.11	-0.70	D	1
8.		^{1}S – ^{1}P°	161.78	97836	715960	1	3	1640	1.93	1.03	0.286	C	1

Ni XIII

Forbidden Transitions

Transition probabilities for magnetic dipole and electric quadrupole lines within the $3p^4$ configuration are the results of the scaled Thomas-Fermi calculations of Mendoza and Zeippen.[1] They included a number of correlation configurations in their basis set and introduced Breit-Pauli relativistic corrections as a perturbation to the nonrelativistic Hamiltonian.

Reference

[1]C. Mendoza and C. J. Zeippen, Mon. Not. R. Astron. Soc. **202**, 981 (1983).

Ni XIII: Forbidden transitions

No.	Transition Array	Multiplet	λ (Å)	E_i (cm^{-1})	E_k (cm^{-1})	g_i	g_k	Type of transition	A_{ki} (s^{-1})	S (at. u.)	Accuracy	Source
1.	$3p^4$–$3p^4$	^3P – ^3P										
			5116.03	0	19542	5	3	M1	157	2.34	C+	1
			"	"	"	5	3	E2	0.019	0.12	D−	1
			[201000]	19542	20040	3	1	M1	0.0063	1.9	D+	1
			[4988.6]	0	20040	5	1	E2	0.036	0.066	E	1
2.		^3P – ^1D										
			2125.50	0	47033	5	5	M1	260	0.46	E	1
			"	"	"	5	5	E2	0.41	0.053	E	1
			[3636.5]	19542	47033	3	5	M1	18	0.16	E	1
			"	"	"	3	5	E2	0.0044	0.0083	E	1
			[3703.6]	20040	47033	1	5	E2	0.0034	0.0071	E	1
3.		^3P – ^1S										
			[1022.1]	0	97836	5	1	E2	3.1	0.0021	E	1
			1277.23	19542	97836	3	1	M1	2500	0.19	E	1
4.		^1D – ^1S	[1968.4]	47033	97836	5	1	E2	12	0.21	D−	1

Ni XIV

P Isoelectronic Sequence

Ground State: $1s^2 2s^2 2p^6 3s^2 3p^3 \, ^4S^\circ_{3/2}$

Ionization Energy: 430 eV = 3470000 cm^{-1}

Allowed Transitions

List of tabulated lines

Wavelength (Å)	No.	Wavelength (Å)	No.	Wavelength (Å)	No.	Wavelength (Å)	No.
157	16	172.16	15	191	14	292.399	3
157.62	13	172.80	15	192	5,12	295.55	3
159	16	177.28	18	196	12	297	1
164.13	10	178	17	198	12	302.264	1
168	18	181	17	213	8	316.113	1
168.12	6	182.14	9	216	8	324	4
169.69	6	185.96	9	225	7	332	4
169.88	15	186.66	9	239	11	336	4
170.50	15	188.69	9	285.88	3	369.58	2
171.37	6	189	5	288.894	3		

Line strengths for transitions of the arrays $3s^23p^3$–$3s3p^4$ and $3p^3$–$3p^23d$ are the results of the multiconfiguration Dirac-Fock (MCDF) calculations of Huang.[1] These relativistic calculations included a perturbative treatment of the Breit interaction and the Lamb shift. Allowance for configuration mixing was limited to configurations within the $n=3$ complex having no more than two electrons in the $3d$ subshell.

Huang published neither an energy-level diagram nor percentage compositions for levels of the $3s^23p^3$, $3s3p^4$, and $3s^23p^23d$ configurations in Ni XIV. We have used the percentages given by Bromage et al.[2] for Fe XII, and by Bromage[3] for V IX and Ni XIV, as a guide to naming the levels; their values resulted from Hartree-Fock calculations with relativistic effects and statistical allowance for exchange (HXR), and incorporated correlation effects due to a few configurations within the $n=3$ complex.

Whenever a term designation of a level in Fe XII, as given in Ref. 1, is different from that indicated in Ref. 2, all transitions involving the corresponding level in Ni XIV are omitted from this compilation.

Transitions involving levels which are indicated to be of low purity in LS coupling are omitted here. Lines which are characterized by very small f-values are assigned lower accuracy ratings; the weakest lines have been excluded.

References

[1]K.-N. Huang, At. Data Nucl. Data Tables **30**, 313 (1984).
[2]G. E. Bromage, R. D. Cowan, and B. C. Fawcett, Mon. Not. R. Astron. Soc. **183**, 19 (1978).
[3]G. E. Bromage, Astron. Astrophys., Suppl. Ser. **41**, 79 (1980).

Ni XIV: Allowed transitions

No.	Transition Array	Multiplet	λ (Å)	E_i (cm^{-1})	E_k (cm^{-1})	g_i	g_k	A_{ki} (10^8 s^{-1})	f_{ik}	S (at. u.)	log gf	Accuracy	Source
1.	$3s^23p^3$–$3s3p^4$	$^4S°$ – ^4P	*308*			4	12	21	0.091	0.37	−0.44	D	1
			316.113	0	316343	4	6	20	0.046	0.19	−0.74	D	1
			302.264	0	330837	4	4	22	0.030	0.12	−0.92	D	1
			[297]			4	2	24	0.016	0.063	−1.19	D	1
2.		$^2D°$ – ^4P											
			[369.58]	45769	316343	4	6	0.34	0.0010	0.0051	−2.38	E	1
3.		$^2D°$ – ^2D	*290.99*	*50449*	*394107*	10	10	40	0.051	0.49	−0.29	D−	1
			292.399	53569	395567	6	6	36	0.047	0.27	−0.55	D	1
			288.894	45769	391917	4	4	46	0.058	0.22	−0.64	D	1
			[295.55]	53569	391917	6	4	0.16	1.4(−4)[a]	8.3(−4)	−3.07	E	1
			[285.88]	45769	395567	4	6	0.11	2.0(−4)	7.7(−4)	−3.09	E	1
4.		$^2P°$ – ^2D	*330*			6	10	6.2	0.017	0.11	−0.99	E	1
			[332]			4	6	7.8	0.019	0.084	−1.11	D	1
			[324]			2	4	3.7	0.012	0.025	−1.63	D	1
			[336]			4	4	0.44	7.5(−4)	0.0033	−2.53	E	1
5.	$3p^3$–$3p^2(^3P)3d$	$^4S°$ – ^4D											
			[189]			4	6	6.0	0.0048	0.012	−1.71	E	1
			[192]			4	4	5.4	0.0030	0.0075	−1.93	E	1
6.		$^4S°$ – ^4P	*170.26*	0	*587340*	4	12	960	1.2	2.8	0.70	D	1
			171.37	0	583530	4	6	940	0.62	1.4	0.39	D	1
			169.69	0	589310	4	4	980	0.43	0.95	0.23	D	1
			168.12	0	594810	4	2	850	0.18	0.40	−0.14	D	1
7.		$^2D°$ – ^4F											
			[225]			4	4	3.2	0.0025	0.0073	−2.01	E	1

Ni XIV: Allowed transitions — Continued

No.	Transition Array	Multiplet	λ (Å)	E_i (cm^{-1})	E_k (cm^{-1})	g_i	g_k	A_{ki} (10^8 s^{-1})	f_{ik}	S (at. u.)	log gf	Accuracy	Source
8.		^{2}D° – ^{4}D											
			[213]			6	6	1.9	0.0013	0.0055	−2.11	E	1
			[216]			6	4	6.5	0.0030	0.013	−1.74	E	1
			[213]			4	2	8.1	0.0027	0.0077	−1.96	E	1
9.		^{2}D° – ^{4}P											
			[188.69]	53569	583530	6	6	19	0.010	0.038	−1.21	E	1
			[186.66]	53569	589310	6	4	9.3	0.0033	0.012	−1.71	E	1
			[182.14]	45769	594810	4	2	150	0.037	0.089	−0.83	E	1
			[185.96]	45769	583530	4	6	10	0.0078	0.019	−1.51	E	1
10.		^{2}D° – ^{2}F											
			164.13	53569	662840	6	8	1200	0.65	2.1	0.59	E	1
11.		^{2}P° – ^{4}D											
			[239]			4	2	4.7	0.0020	0.0064	−2.09	E	1
12.		^{2}P° – ^{4}P											
			[198]			4	4	4.0	0.0023	0.0061	−2.03	E	1
			[192]			2	2	29	0.016	0.020	−1.50	E	1
			[196]			4	2	38	0.011	0.028	−1.36	E	1
13.	$3p^3$–$3p^2(^1$D)$3d$	^{4}S° – ^{2}D											
			[157.62]	0	634430	4	6	2.3	0.0013	0.0027	−2.28	E	1
14.		^{2}D° – ^{2}G											
			[191]			6	8	5.1	0.0037	0.014	−1.65	E	1
15.		^{2}D° – ^{2}D	*171.49*	*50449*	*633570*	10	10	640	0.28	1.6	0.45	D	1
			172.16	53569	634430	6	6	470	0.21	0.71	0.10	D	1
			170.50	45769	632280	4	4	710	0.31	0.69	0.09	D	1
			[172.80]	53569	632280	6	4	140	0.041	0.14	−0.61	D	1
			[169.88]	45769	634430	4	6	17	0.011	0.025	−1.35	D	1
16.		^{2}D° – ^{2}P											
			[159]			6	4	8.4	0.0021	0.0067	−1.89	E	1
			[157]			4	4	22	0.0082	0.017	−1.48	E	1
17.		^{2}P° – ^{2}D	*180*			6	10	83	0.068	0.24	−0.39	E	1
			[181]			4	6	74	0.055	0.13	−0.66	D	1
			[178]			2	4	89	0.084	0.099	−0.77	D	1
			[181]			4	4	7.4	0.0037	0.0087	−1.84	E	1
18.		^{2}P° – ^{2}P											
			177.28			4	4	560	0.27	0.62	0.03	E	1
			[168]			2	4	240	0.20	0.22	−0.40	E	1

[a]The number in parentheses following the tabulated value indicates the power of ten by which this value has to be multiplied.

Ni XIV

Forbidden Transitions

Line strengths for magnetic dipole and electric quadrupole transitions within the $3p^3$ configuration are the results of the multiconfiguration Dirac-Fock (MCDF) calculations of Huang.[1] These relativistic calculations included a perturbative treatment of the Breit interaction and the Lamb shift. Allowance for configuration mixing was limited to configurations within the $n = 3$ complex having no more than two electrons in the $3d$ subshell. Strengths of electric quadrupole transitions as defined in Ref. 1 were multiplied by the factor $2/3$ which is needed to bring these values into conformance with the definition of quadrupole strengths used in the NBS tables. We have excluded from this compilation the electric quadrupole contributions to the $^4S^\circ_{3/2} - {}^2P^\circ_{3/2}$ and $^4S^\circ_{3/2} - {}^2P^\circ_{1/2}$ transitions, since their strengths are very small and thus subject to considerable uncertainty.

Data for these same transitions calculated by Mendoza and Zeippen[2] with the scaled Thomas-Fermi approach with allowance for correlation are generally in very good agreement with the results of Ref. 1. These latter calculations treated relativistic effects by introducing Breit-Pauli corrections as a perturbation to the nonrelativistic Hamiltonian.

References

[1] K.-N. Huang, At. Data Nucl. Data Tables **30**, 313 (1984).
[2] C. Mendoza and C. J. Zeippen, Mon. Not. R. Astron. Soc. **198**, 127 (1982).

Ni XIV: Forbidden transitions

No.	Transition Array	Multiplet	λ (Å)	E_i (cm^{-1})	E_k (cm^{-1})	g_i	g_k	Type of transition	A_{ki} (s^{-1})	S (at. u.)	Accuracy	Source
1.	$3p^3$–$3p^3$	$^4S^\circ - {}^2D^\circ$										
			1866.75	0	53569	4	6	M1	7.6	0.011	D	1
			"	"	"	4	6	E2	0.23	0.019	E	1
			2184.20	0	45769	4	4	M1	160	0.25	C	1
			"	"	"	4	4	E2	0.077	0.0091	E	1
2.		$^4S^\circ - {}^2P^\circ$										
			[1030]			4	4	M1	740	0.12	C	1
			[1180]			4	2	M1	460	0.056	D	1
3.		$^2D^\circ - {}^2D^\circ$										
			[12817]	45769	53569	4	6	M1	4.27	2.00	C+	1
			"	"	"	4	6	E2	5.3(−5)ᵃ	0.066	E	1
4.		$^2D^\circ - {}^2P^\circ$										
			[3170]			6	2	E2	0.34	0.13	D−	1
			[2320]			6	4	M1	210	0.39	C	1
			"			6	4	E2	2.3	0.37	D−	1
			[2540]			4	2	M1	160	0.19	C	1
			"			4	2	E2	1.4	0.17	D−	1
			[1960]			4	4	M1	600	0.67	C	1
			"			4	4	E2	1.3	0.087	E	1
5.		$^2P^\circ - {}^2P^\circ$										
			[8620]			2	4	M1	11.4	1.08	C+	1
			"			2	4	E2	2.6(−4)	0.029	E	1

ᵃThe number in parentheses following the tabulated value indicates the power of ten by which this value has to be multiplied.

Ni xv

Si Isoelectronic Sequence

Ground State: $1s^2 2s^2 2p^6 3s^2 3p^2 \, {}^3P_0$

Ionization Energy: $464 \, \text{eV} = 3740000 \, \text{cm}^{-1}$

Allowed Transitions

List of tabulated lines

Wavelength (Å)	No.	Wavelength (Å)	No.	Wavelength (Å)	No.	Wavelength (Å)	No.
50.249	14	179.28	7	258	2	311.756	1
60.890	15	181	13	268	2	312.03	1
64.635	16	191.45	11	269	2	319.063	1
149	9	206	5	269.05	2	324.35	1
163.64	8	208	5	278	2	324.65	1
173.73	12	212	5	278.386	2	359.78	3
175	6	231	10	298.15	1	367	4

Line strengths for transitions of the arrays $3s^2 3p^2$–$3s \, 3p^3$ and $3p^2$–$3p \, 3d$ are the results of the multiconfiguration Dirac-Fock (MCDF) calculations of Huang.[1] These relativistic calculations included a perturbative treatment of the Breit interaction and the Lamb shift. Allowance for configuration mixing included all configurations within the $n = 3$ complex.

Huang published neither an energy-level diagram nor percentage compositions for levels of the $3s^2 3p^2$, $3s \, 3p^3$, and $3s^2 3p \, 3d$ configurations in Ni xv. We have used the percentages given by Bromage et al.[2] for Fe xiii, and by Bromage[3] for V x and Ni xv, as a guide to naming the levels; their values resulted from Hartree-Fock calculations with relativistic effects and statistical allowance for exchange (HXR), and incorporated correlation effects due to a partial set of configurations within the $n = 3$ complex. Whenever the term designation of a level in Fe xiii, as given in Ref. 1, is different from that indicated in Ref. 2, all transitions involving the corresponding level in Ni xv are omitted from this compilation.

A few f-values for transitions to configurations in which one electron occupies the $n = 4$ shell were interpolated from the results of Kastner et al.[4] for Fe xiii and Zn xvii, which were computed by a multiconfiguration scaled Thomas-Fermi approach.

Transitions involving levels which are indicated to be of low purity in LS coupling are omitted here. Lines which are characterized by very small f-values are assigned lower accuracy ratings; the weakest lines have been excluded.

References

[1] K.-N. Huang, At. Data Nucl. Data Tables **32**, 503 (1985).
[2] G. E. Bromage, R. D. Cowan, and B. C. Fawcett, Mon. Not. R. Astron. Soc. **183**, 19 (1978).
[3] G. E. Bromage, Astron. Astrophys., Suppl. Ser. **41**, 79 (1980).
[4] S. O. Kastner, M. Swartz, A. K. Bhatia, and J. Lapides, J. Opt. Soc. Am. **68**, 1558 (1978).

Ni xv: Allowed transitions

No.	Transition Array	Multiplet	λ (Å)	E_i (cm^{-1})	E_k (cm^{-1})	g_i	g_k	A_{ki} (10^8 s^{-1})	f_{ik}	S (at. u.)	log gf	Accuracy	Source
1.	$3s^2 3p^2$–$3s \, 3p^3$	${}^3P$ – ${}^3D^\circ$	*314.63*	*20181*	*338010*	9	15	18	0.044	0.41	−0.40	E	1
			319.063	27376	340794	5	7	15	0.032	0.17	−0.79	D	1
			311.756	14917	335681	3	5	20	0.049	0.15	−0.84	D	1
			298.15	0	335400	1	3	19	0.074	0.073	−1.13	D	1
			[324.35]	27376	335681	5	5	0.21	3.4(−4)^a	0.0018	−2.77	E	1
			[312.03]	14917	335400	3	3	3.3	0.0049	0.015	−1.84	D−	1
			[324.65]	27376	335400	5	3	0.20	1.9(−4)	0.0010	−3.03	E	1

Ni xv: Allowed transitions — Continued

No.	Transition Array	Multiplet	λ (Å)	E_i (cm^{-1})	E_k (cm^{-1})	g_i	g_k	A_{ki} (10^8 s^{-1})	f_{ik}	S (at. u.)	log gf	Accuracy	Source
2.		^{3}P – ^{3}P°	*272*			9	9	50	0.056	0.45	−0.30	D	1
			278.386	27376	386589	5	5	43	0.050	0.23	−0.60	D	1
			[268]			3	3	28	0.030	0.079	−1.05	D	1
			[278]			5	3	9.4	0.0066	0.030	−1.48	D−	1
			[269]			3	1	53	0.019	0.051	−1.24	C−	1
			[269.05]	14917	386589	3	5	3.7	0.0068	0.018	−1.69	D−	1
			[258]			1	3	16	0.047	0.040	−1.33	D	1
3.		^{1}D – ^{3}D°											
			[359.78]	62849	340794	5	7	1.7	0.0047	0.028	−1.63	E	1
4.		^{1}S – ^{3}P°											
			[367]			1	3	0.41	0.0025	0.0030	−2.61	E	1
5.	$3p^2$–$3p3d$	^{3}P – ^{3}F°											
			[208]			5	7	6.1	0.0055	0.019	−1.56	E	1
			[206]			3	5	2.0	0.0022	0.0044	−2.19	E	1
			[212]			5	5	3.1	0.0021	0.0074	−1.97	E	1
6.		^{3}P – ^{3}P°											
			[175]			3	1	570	0.087	0.15	−0.58	D	1
7.		^{3}P – ^{3}D°											
			179.28	27376	585170	5	7	750	0.51	1.5	0.41	D	1
8.		^{3}P – ^{1}F°											
			[163.64]	27376	638460	5	7	56	0.032	0.085	−0.80	E	1
9.		^{3}P – ^{1}P°											
			[149]			1	3	7.1	0.0071	0.0035	−2.15	E	1
10.		^{1}D – ^{3}F°											
			[231]			5	5	5.9	0.0047	0.018	−1.63	E	1
11.		^{1}D – ^{3}D°											
			[191.45]	62849	585170	5	7	41	0.032	0.10	−0.80	E	1
12.		^{1}D – ^{1}F°	173.73	62849	638460	5	7	760	0.479	1.37	0.379	C	1
13.		^{1}S – ^{1}P°	[181]			1	3	680	1.0	0.60	0.00	D	1
14.	$3p^2$–$3p4d$	^{1}D – ^{1}F°	50.249	62849	2052900	5	7	6800	0.36	0.30	0.26	D	*interp.*
15.	$3p3d$–$3p4f$	^{3}F° – ^{3}G											
			60.890			9	11	1.0(+4)[a]	0.71	1.3	0.81	D	*interp.*
16.		^{1}F° – ^{1}G	64.635	638460	2185600	7	9	9600	0.77	1.1	0.73	E	*interp.*

[a]The number in parentheses following the tabulated value indicates the power of ten by which this value has to be multiplied.

Ni xv

Forbidden Transitions

List of tabulated lines

Wavelength (Å)	No.	Wavelength (Å)	No.	Wavelength (Å)	No.	Wavelength (Å)	No.
335.94	12	720	6	1954	8	2818.2	2
401	10	730	6	1964.3	8	3651.8	1
409.20	11	1030	3	2000	8	6701.7	1
530	9	1100	5	2040	4	8024.1	1
550	9	1190	3	2085.61	2	18500	7
580	9	1200	5	2183.0	8	19550	7
620	9	1900	8	2200	8	360000	7

Line strengths for magnetic dipole and electric quadrupole transitions are the results of the multiconfiguration Dirac-Fock (MCDF) calculations of Huang.[1] These relativistic calculations included a perturbative treatment of the Breit interaction and the Lamb shift. Allowance for configuration interaction encompassed all configurations within the $n=3$ complex. Huang calculated line strengths for transitions within the $3p^2$ configuration, as well as for transitions between pairs of odd-parity levels whose lower level is one of the four lowest-lying odd-parity levels in the $n=3$ complex. Transitions involving odd-parity levels which are indicated by Bromage et al.[2] (for Fe XIII) or Bromage[3] (for V X and Ni XV) to be of low purity in LS coupling in Fe-group species are omitted here, as are lines whose strengths are very small. Strengths of electric quadrupole transitions as reported in Ref. 1 were multiplied by the factor $2/3$ which is needed to bring these values into conformance with the definition of quadrupole strengths used in the NBS tables.

References

[1]K.-N. Huang, At. Data Nucl. Data Tables **32**, 503 (1985).

[2]G. E. Bromage, R. D. Cowan, and B. C. Fawcett, Mon. Not. R. Astron. Soc. **183**, 19 (1978).

[3]G. E. Bromage, Astron. Astrophys., Suppl. Ser. **41**, 79 (1980).

Ni xv: Forbidden transitions

No.	Transition Array	Multiplet	λ (Å)	E_i (cm^{-1})	E_k (cm^{-1})	g_i	g_k	Type of transition	A_{ki} (s^{-1})	S (at. u.)	Accuracy	Source
1.	$3p^2$–$3p^2$	^3P – ^3P										
			8024.1	14917	27376	3	5	M1	22.7	2.17	C+	1
			"	"	"	3	5	E2	9.9(−4)[a]	0.098	E	1
			6701.7	0	14917	1	3	M1	56.5	1.89	C+	1
			[3651.8]	0	27376	1	5	E2	0.030	0.058	E	1
2.		^3P – ^1D										
			[2818.2]	27376	62849	5	5	M1	200	0.85	E	1
			"	"	"	5	5	E2	0.17	0.091	E	1
			2085.61	14917	62849	3	5	M1	200	0.33	E	1
			"	"	"	3	5	E2	0.12	0.014	E	1
3.		^3P – ^1S										
			[1190]			5	1	E2	9.2	0.013	E	1
			[1030]			3	1	M1	2500	0.10	E	1
4.		^1D – ^1S	[2040]			5	1	E2	9.0	0.19	D−	1

Ni xv: Forbidden transitions — Continued

No.	Transition Array	Multiplet	λ (Å)	E_i (cm^{-1})	E_k (cm^{-1})	g_i	g_k	Type of transition	A_{ki} (s^{-1})	S (at. u.)	Accu-racy	Source
5.	$3s\,3p^3$–$3s\,3p^3$	$^5S°$ – $^3D°$										
			[1100]			5	7	M1	6.4	0.0022	E	1
			"	"	"	5	7	E2	1.1	0.0073	E	1
			[1200]			5	5	M1	130	0.043	E	1
			"	"	"	5	5	E2	0.65	0.0048	E	1
			[1200]			5	3	M1	43	0.0083	E	1
			"	"	"	5	3	E2	0.27	0.0012	E	1
6.		$^5S°$ – $^3P°$										
			[720]			5	5	M1	1400	0.097	E	1
			[730]			5	3	M1	830	0.036	E	1
7.		$^3D°$ – $^3D°$										
			[19550]	335681	340794	5	7	M1	2.20	4.27	C−	1
			"	"	"	5	7	E2	3.4(−6)	0.041	E	1
			[360000]	335400	335681	3	5	M1	5.0(−4)	4.3	E	1
			[18500]	335400	340794	3	7	E2	1.3(−6)	0.012	E	1
8.		$^3D°$ – $^3P°$										
			[2200]			7	3	E2	2.2	0.20	D−	1
			[2000]			5	1	E2	6.8	0.13	D−	1
			[2183.0]	340794	386589	7	5	M1	190	0.37	E	1
			"	"	"	7	5	E2	2.3	0.34	D−	1
			[2000]			5	3	E2	0.54	0.031	E	1
			[2000]			3	1	M1	330	0.099	E	1
			[1964.3]	335681	386589	5	5	M1	180	0.25	E	1
			"	"	"	5	5	E2	2.1	0.18	D−	1
			[1900]			3	3	M1	380	0.29	E	1
			"	"	"	3	3	E2	3.6	0.16	D−	1
			[1954]	335400	386589	3	5	M1	51	0.070	E	1
			"	"	"	3	5	E2	0.86	0.073	E	1
9.	$3s\,3p^3$–$3s^2\,3p\,3d$	$^3D°$ – $^3F°$										
			[530]			5	9	E2	23	0.0051	E	1
			[580]			3	7	E2	8.4	0.0023	E	1
			[550]			7	9	M1	1100	0.059	E	1
			[620]			5	5	M1	45	0.0020	E	1
10.		$^3D°$ – $^3P°$										
			[401]			5	1	E2	440	0.0027	E	1
11.		$^3D°$ – $^3D°$										
			[409.20]	340794	585170	7	7	M1	84	0.0015	E	1
12.		$^3D°$ – $^1F°$										
			[335.94]	340794	638460	7	7	M1	130	0.0013	E	1

ᵃThe number in parentheses following the tabulated value indicates the power of ten by which this value has to be multiplied.

Ni XVI

Al Isoelectronic Sequence

Ground State: $1s^2 2s^2 2p^6 3s^2 3p \; ^2P^\circ_{1/2}$

Ionization Energy: $499 \text{ eV} = 4020000 \text{ cm}^{-1}$

Allowed Transitions

List of tabulated lines

Wavelength (Å)	No.	Wavelength (Å)	No.	Wavelength (Å)	No.	Wavelength (Å)	No.
152	25,34	194	22,31,41,42	228	27	266	16
160	37	194.04	20	231	19	288.165	2
161	37	195.27	20	232.475	4	297	11
162	36,37	196	41	233	18,19	299	9,15
163	36	197	29,42	235	18,19	300	8
166	35	198	29,41	236	18	302	32
167	35	199	31	237.875	4	304	8
168	35	200	22	238	6,18	309.179	2
170	24	204	7,40	239.550	3	313.22	2
171	30	206	33	245	6,17	317	15
172	24	209	7	246	10	320	15
175	30	210	33	247	17	328	13
180	39	212	33	249	10,17	335	13
182	39	215	7,21	250	10	338	13
183	22,38	216	28	254	6,16	346	13
185.23	20	217	21,33	255	5	382	1
187	22,23	218	21	256.62	3	385	1
188	23	218.391	4	257	26	407	14
190	29	219	28	259	26	428	12
192	23	221	21	262	26	447	12
193	23,42	223.119	4	263	16		

Line strengths for transitions of the arrays $3s^2 3p$–$3s 3p^2$, $3s 3p^2$–$3p^3$, $3s^2 3d$–$3s 3p 3d$, $3s^2 3p$–$3s^2 3d$, and $3s 3p^2$–$3s 3p 3d$ are the results of the multiconfiguration Dirac-Fock (MCDF) calculations of Huang.[1] These relativistic calculations included a perturbative treatment of the Breit interaction. Allowance for configuration mixing included all configurations within the $n = 3$ complex.

Huang published neither an energy-level diagram nor percentage compositions for levels of the $3s^2 3p$, $3s 3p^2$, $3s^2 3d$, $3p^3$, and $3s 3p 3d$ configurations in Ni XVI. We have used the percentages given by Fawcett[2] as a guide to naming the levels; the latter's values resulted from Hartree-Fock calculations with relativistic effects and statistical allowance for exchange (HXR), and incorporated correlation effects due to all configurations within the $n = 3$ complex.

Transitions involving levels which are indicated to be of low purity in LS coupling are omitted here. Lines which are characterized by very small f-values are assigned lower accuracy ratings; the weakest lines have been excluded. A few wavelengths computed by Huang differ significantly from those which resulted from the fitting and scaling procedure applied by Fawcett[2]; lines for which the wavelengths are in serious disagreement have been omitted.

References

[1] K.-N. Huang, At. Data Nucl. Data Tables **34**, 1 (1986) and private communication.

[2] B. C. Fawcett, At. Data Nucl. Data Tables **28**, 557 (1983).

Ni XVI: Allowed transitions

No.	Transition Array	Multiplet	λ (Å)	E_i (cm^{-1})	E_k (cm^{-1})	g_i	g_k	A_{ki} (10^8 s^{-1})	f_{ik}	S (at. u.)	log gf	Accuracy	Source
1.	$3s^2 3p - 3s 3p^2$	^{2}P° – ^{4}P											
			[385]			4	6	0.57	0.0019	0.0097	−2.12	E	1
			[382]			2	2	0.58	0.0013	0.0032	−2.59	E	1
2.		^{2}P° – ^{2}D	302.10	18507	349528	6	10	26	0.059	0.35	−0.45	E	1
			309.179	27761	351198	4	6	23	0.049	0.20	−0.71	D	1
			288.165	0	347023	2	4	32	0.079	0.15	−0.80	D	1
			[313.22]	27761	347023	4	4	0.43	6.3(−4)[a]	0.0026	−2.60	E	1
3.		^{2}P° – ^{2}S	250.66	18507	417449	6	2	230	0.073	0.36	−0.36	E	1
			[256.62]	27761	417449	4	2	3.8	0.0019	0.0064	−2.12	E	1
			239.550	0	417449	2	2	260	0.22	0.35	−0.35	E	1
4.		^{2}P° – ^{2}P	229.28	18507	454660	6	6	480	0.38	1.7	0.35	E	1
			232.475	27761	457894	4	4	407	0.330	1.01	0.120	C−	1
			223.119	0	448191	2	2	130	0.095	0.14	−0.72	E	1
			237.875	27761	448191	4	2	260	0.11	0.35	−0.35	E	1
			218.391	0	457894	2	4	95	0.136	0.195	−0.57	C−	1
5.	$3s 3p^2 - 3p^3$	^{4}P – ^{2}D°											
			[255]			6	6	1.9	0.0019	0.0094	−1.95	E	1
6.		^{4}P – ^{4}S°	248			12	4	400	0.12	1.2	0.17	D	1
			[254]			6	4	180	0.12	0.59	−0.15	D	1
			[245]			4	4	140	0.13	0.41	−0.29	D	1
			[238]			2	4	75	0.13	0.20	−0.59	D	1
7.		^{4}P – ^{2}P°											
			[215]			6	4	3.0	0.0014	0.0059	−2.08	E	1
			[209]			4	4	8.3	0.0055	0.015	−1.66	E	1
			[204]			2	4	4.2	0.0053	0.0071	−1.98	E	1
8.		^{2}D – ^{2}D°											
			[304]			6	6	40	0.055	0.33	−0.48	E	1
			[300]			4	6	4.5	0.0091	0.036	−1.44	E	1
9.		^{2}D – ^{4}S°											
			[299]			4	4	2.7	0.0036	0.014	−1.85	E	1
10.		^{2}D – ^{2}P°	249			10	6	150	0.083	0.68	−0.08	D	1
			[249]			6	4	120	0.077	0.38	−0.33	D	1
			[250]			4	2	160	0.076	0.25	−0.52	D	1
			[246]			4	4	15	0.014	0.045	−1.26	D	1
11.		^{2}S – ^{2}P°											
			[297]			2	4	25	0.066	0.13	−0.88	E	1
12.		^{2}P – ^{4}S°											
			[447]			4	4	1.2	0.0037	0.022	−1.83	E	1
			[428]			2	4	0.34	0.0019	0.0053	−2.42	E	1

Ni XVI: Allowed transitions — Continued

No.	Transition Array	Multiplet	λ (Å)	E_i (cm^{-1})	E_k (cm^{-1})	g_i	g_k	A_{ki} (10^8 s^{-1})	f_{ik}	S (at. u.)	log gf	Accuracy	Source
13.		^{2}P – ^{2}P°	*337*			6	6	43	0.074	0.49	−0.35	E	1
			[338]			4	4	37	0.063	0.28	−0.60	D	1
			[335]			2	2	46	0.077	0.17	−0.81	E	1
			[346]			4	2	8.3	0.0075	0.034	−1.53	E	1
			[328]			2	4	0.46	0.0015	0.0032	−2.53	E	1
14.	$3s^23d$ – $3s3p(^3$P°$)3d$	^{2}D – ^{4}P°											
			[407]			6	6	0.50	0.0012	0.010	−2.13	E	1
15.		^{2}D – ^{2}F°	*307*			10	14	26	0.050	0.51	−0.30	E	1
			[299]			6	8	27	0.049	0.29	−0.53	E	1
			[317]			4	6	17	0.038	0.16	−0.81	E	1
			[320]			6	6	5.7	0.0087	0.055	−1.28	E	1
16.		^{2}D – ^{2}P°	*262*			10	6	13	0.0082	0.071	−1.08	E	1
			[266]			6	4	7.0	0.0049	0.026	−1.53	E	1
			[254]			4	2	3.3	0.0016	0.0054	−2.19	E	1
			[263]			4	4	11	0.012	0.040	−1.34	E	1
17.	$3s^23d$ – $3s3p(^1$P°$)3d$	^{2}D – ^{2}F°	*247*			10	14	330	0.42	3.4	0.62	E	1
			[249]			6	8	330	0.41	2.0	0.39	E	1
			[245]			4	6	320	0.43	1.4	0.24	E	1
			[247]			6	6	9.2	0.0084	0.041	−1.30	E	1
18.		^{2}D – ^{2}D°	*235*			10	10	250	0.21	1.6	0.32	E	1
			[235]			6	6	250	0.21	0.96	0.09	E	1
			[236]			4	4	120	0.10	0.31	−0.40	E	1
			[238]			6	4	130	0.074	0.35	−0.35	E	1
			[233]			4	6	1.7	0.0020	0.0062	−2.09	E	1
19.		^{2}D – ^{2}P°	*234*			10	6	400	0.19	1.5	0.29	D	1
			[233]			6	4	240	0.13	0.60	−0.11	D	1
			[235]			4	2	380	0.16	0.49	−0.20	D	1
			[231]			4	4	160	0.13	0.39	−0.29	D	1
20.	$3p$–$3d$	^{2}P° – ^{2}D	*191.09*	*18507*	*541820*	6	10	490	0.45	1.7	0.43	D	1
			[194.04]	27761	543120	4	6	460	0.39	0.99	0.19	D	1
			185.23	0	539870	2	4	420	0.43	0.53	−0.06	D	1
			[195.27]	27761	539870	4	4	95	0.054	0.14	−0.66	D	1
21.	$3s3p^2$ – $3s3p(^3$P°$)3d$	^{4}P – ^{4}F°											
			[217]			6	8	5.2	0.0049	0.021	−1.53	E	1
			[215]			4	6	2.8	0.0029	0.0082	−1.94	E	1
			[221]			6	6	1.6	0.0012	0.0052	−2.15	E	1
			[218]			4	4	1.7	0.0012	0.0035	−2.31	E	1

Ni XVI: Allowed transitions — Continued

No.	Transition Array	Multiplet	λ (Å)	E_i (cm^{-1})	E_k (cm^{-1})	g_i	g_k	A_{ki} (10^8 s^{-1})	f_{ik}	S (at. u.)	log gf	Accuracy	Source
22.		^{4}P – ^{4}P°											
			[200]			6	6	32	0.019	0.076	−0.94	E	1
			[183]			2	2	0.31	1.6(−4)	1.9(−4)	−3.50	E	1
			[187]			4	2	330	0.085	0.21	−0.47	D	1
			[194]			4	6	280	0.24	0.61	−0.02	E	1
23.		^{4}P – ^{4}D°											
			[192]			6	8	454	0.335	1.27	0.303	C−	1
			[187]			4	6	120	0.097	0.24	−0.41	D	1
			[192]			6	6	310	0.17	0.64	0.01	D	1
			[188]			2	2	470	0.25	0.31	−0.30	D	1
			[193]			4	2	9.3	0.0026	0.0066	−1.98	E	1
24.		^{4}P – ^{2}F°											
			[170]			6	8	8.2	0.0048	0.016	−1.54	E	1
			[172]			4	6	1.5	0.0010	0.0023	−2.39	E	1
25.		^{4}P – ^{2}P°											
			[152]			2	4	5.6	0.0039	0.0039	−2.11	E	1
26.		^{2}D – ^{4}F°											
			[257]			6	6	1.4	0.0014	0.0072	−2.07	E	1
			[259]			4	4	2.0	0.0021	0.0070	−2.09	E	1
			[262]			6	4	1.6	0.0011	0.0056	−2.19	E	1
27.		^{2}D – ^{4}P°											
			[228]			6	6	14	0.011	0.050	−1.18	E	1
28.		^{2}D – ^{4}D°											
			[219]			6	8	6.8	0.0065	0.028	−1.41	E	1
			[216]			4	6	1.8	0.0019	0.0054	−2.12	E	1
29.		^{2}D – ^{2}F°	193			10	14	190	0.15	0.96	0.18	E	1
			[190]			6	8	200	0.14	0.53	−0.07	E	1
			[197]			4	6	150	0.13	0.33	−0.29	E	1
			[198]			6	6	44	0.026	0.10	−0.81	E	1
30.		^{2}D – ^{2}P°											
			[171]			4	2	2.2	4.9(−4)	0.0011	−2.71	E	1
			[175]			4	4	0.45	2.1(−4)	4.8(−4)	−3.08	E	1
31.		^{2}S – ^{2}P°	197			2	6	370	0.65	0.84	0.11	E	1
			[199]			2	4	490	0.58	0.76	0.06	E	1
			[194]			2	2	110	0.063	0.080	−0.90	E	1
32.		^{2}P – ^{4}P°											
			[302]			4	6	0.51	0.0011	0.0042	−2.37	E	1

Ni XVI: Allowed transitions — Continued

No.	Transition Array	Multiplet	λ (Å)	E_i (cm^{-1})	E_k (cm^{-1})	g_i	g_k	A_{ki} (10^8 s^{-1})	f_{ik}	S (at. u.)	log gf	Accuracy	Source
33.		^{2}P – ^{2}P°	*213*			6	6	210	0.14	0.59	−0.07	E	1
			[217]			4	4	110	0.077	0.22	−0.51	D	1
			[206]			2	2	370	0.24	0.32	−0.33	E	1
			[210]			4	2	49	0.016	0.045	−1.19	D	1
			[212]			2	4	1.1	0.0014	0.0020	−2.54	E	1
34.	$3s\,3p^2-$ $3s\,3p(^1$P°$)3d$	^{4}P – ^{2}F°											
			[152]			6	8	6.6	0.0030	0.0091	−1.74	E	1
35.		^{2}D – ^{2}F°	*167*			10	14	310	0.18	1.0	0.26	E	1
			[168]			6	8	320	0.18	0.59	0.03	E	1
			[166]			4	6	310	0.19	0.42	−0.11	E	1
			[167]			6	6	15	0.0064	0.021	−1.42	E	1
36.		^{2}D – ^{2}D°											
			[162]			4	4	2.1	8.4(−4)	0.0018	−2.47	E	1
			[163]			6	4	2.5	6.5(−4)	0.0021	−2.41	E	1
37.		^{2}D – ^{2}P°	*161*			10	6	6.1	0.0014	0.0075	−1.85	E	1
			[161]			6	4	4.0	0.0010	0.0033	−2.21	E	1
			[162]			4	2	1.2	2.3(−4)	4.9(−4)	−3.04	E	1
			[160]			4	4	4.6	0.0018	0.0037	−2.15	E	1
38.		^{2}S – ^{2}D°											
			[183]			2	4	21	0.021	0.025	−1.38	E	1
39.		^{2}S – ^{2}P°	*181*			2	6	130	0.19	0.23	−0.41	E	1
			[180]			2	4	71	0.069	0.082	−0.86	E	1
			[182]			2	2	250	0.13	0.15	−0.60	E	1
40.		^{2}P – ^{2}F°											
			[204]			4	6	8.0	0.0074	0.020	−1.53	E	1
41.		^{2}P – ^{2}D°	*195*			6	10	660	0.62	2.4	0.57	E	1
			[196]			4	6	670	0.58	1.5	0.37	E	1
			[194]			2	4	550	0.62	0.79	0.09	E	1
			[198]			4	4	52	0.031	0.080	−0.91	E	1
42.		^{2}P – ^{2}P°											
			[194]			4	4	350	0.20	0.51	−0.10	D	1
			[193]			2	2	55	0.031	0.039	−1.21	E	1
			[197]			4	2	120	0.035	0.090	−0.86	C−	1

[a]The number in parentheses following the tabulated value indicates the power of ten by which this value has to be multiplied.

Ni XVI

Forbidden Transitions

Line strengths for magnetic dipole and electric quadrupole transitions within the $3s^2 3p$ $^2P°$ and $3s 3p^2$ ^4P terms are the results of the multiconfiguration Dirac-Fock (MCDF) calculations of Huang.[1] These relativistic calculations included a perturbative treatment of the Breit interaction and the Lamb shift. Allowance for configuration mixing included all configurations within the $n = 3$ complex. Strengths of electric quadrupole transitions as reported in Ref. 1 were multiplied by the factor $^2/_3$ which is needed to bring these values into conformance with the definition of quadrupole strengths used in the NBS tables.

Reference

[1] K.-N. Huang, At. Data Nucl. Data Tables **34**, 1 (1986).

Ni XVI: Forbidden transitions

No.	Transition Array	Multiplet	λ (Å)	E_i (cm^{-1})	E_k (cm^{-1})	g_i	g_k	Type of transition	A_{ki} (s^{-1})	S (at. u.)	Accuracy	Source
1.	$3p-3p$	$^2P° - ^2P°$										
			3601.1	0	27761	2	4	M1	192	1.33	C+	1
			"	"	"	2	4	E2	0.067	0.096	E	1
2.	$3s 3p^2 - 3s 3p^2$	$^4P - ^4P$										
			[7320]			4	6	M1	40.3	3.52	C	1
			"			4	6	E2	0.0019	0.14	D−	1
			[8500]			2	4	M1	35.8	3.26	C	1
			"			2	4	E2	1.1(−4)^a	0.012	E	1
			[3930]			2	6	E2	0.033	0.11	D−	1

^{a}The number in parentheses following the tabulated value indicates the power of ten by which this value has to be mulitplied.

Ni XVII

Mg Isoelectronic Sequence

Ground State: $1s^2 2s^2 2p^6 3s^2\ {}^1S_0$

Ionization Energy: 571.08 eV = 4606000 cm^{-1}

Allowed Transitions

Wavelength (Å)	No.	Wavelength (Å)	No.	Wavelength (Å)	No.	Wavelength (Å)	No.
30.919	16	207	5	269.0	19	294	10
42.855	15	207.50	17	269.39	3	296	10
54.451	26	208.66	17	270.4	19	323	9
55.361	27	209.5	17	272	4	339	9
57.348	28	214	21	277	11	343	9
169	18	215.89	20	278	11	355	9
173	24	216	22,25	279	11	358	9
175	18	217	21	280	10	362	9
197.39	17	226	21	281.50	3	366.7	1
199.87	17	227	23	282	8,10	372.7	6
200.53	17	249.180	2	284	14	412.2	6
204	22	251.97	3	285.59	3	419	7
205	22	263.55	3	289	4	441.3	6
206	21	266.06	3	292	13	465	12

Oscillator strengths for the three transitions $3s^2\ {}^1S_0 - 3snp\ {}^1P_1^\circ$ ($n=3-5$) are the results of the relativistic random phase approximation (RRPA) calculations of Shorer et al.,[1] who allowed for correlation within the context of a frozen core. Oscillator strength data of Fawcett,[2] quoted for most transitions of the arrays $3s\,3p - 3p^2$, $3s\,3d - 3p\,3d$, $3s\,3p - 3s\,3d$, and $3p^2 - 3p\,3d$, were derived by means of Hartree-Fock calculations which included relativistic effects and statistical allowance for exchange (HXR); he incorporated correlation effects due to all configurations in the $n=3$ complex. A-values for all intercombination lines tabulated here were determined by Bhatia and Kastner[3] using a scaled Thomas-Fermi (STF) approach with allowance for configuration mixing; these data are included here, but the A-values were first converted to line strengths and then reconverted to oscillator strengths and transition probabilities which incorporate more accurate wavelength values.

Kastner et al.[4] calculated A-values for a number of lines of the array $3p\,3d - 3p\,4f$ in Fe XV and Zn XIX by application of a multiconfiguration STF approach. These transition probabilities were converted to oscillator strengths, from which f-values for a few transitions of this array in Ni XVII were interpolated.

Transitions involving levels which are indicated in Ref. 2 to be of low purity in LS coupling in Ni XVII, or in Ref. 4 to be of low purity in neighboring Mg-like ions, are omitted here. Lines which are characterized by very small f-values are assigned lower accuracy ratings.

References

[1]P. Shorer, C. D. Lin, and W. R. Johnson, Phys. Rev. A **16**, 1109 (1977).

[2]B. C. Fawcett, At. Data Nucl. Data Tables **28**, 579 (1983).

[3]A. K. Bhatia and S. O. Kastner, J. Quant. Spectrosc. Radiat. Transfer **24**, 53 (1980).

[4]S. O. Kastner, M. Swartz, A. K. Bhatia, and J. Lapides, J. Opt. Soc. Am. **68**, 1558 (1978).

Ni XVII: Allowed transitions

No.	Transition Array	Multiplet	λ (Å)	E_i (cm^{-1})	E_k (cm^{-1})	g_i	g_k	A_{ki} (10^8 s^{-1})	f_{ik}	S (at. u.)	log gf	Accuracy	Source
1.	$3s^2$–$3s3p$	^1S – $^3P°$											
			366.7	0	272700	1	3	0.58	0.0035	0.0042	−2.46	E	3
2.		^1S – $^1P°$	249.180	0	401316	1	3	275	0.767	0.629	−0.115	B	1
3.	$3s3p$–$3p^2$	$^3P°$ – ^3P	*268.2*	*283500*	*656400*	9	9	210	0.23	1.8	0.31	D	2
			266.06	293700	669600	5	5	140	0.15	0.66	−0.12	D	2
			269.39	272700	643900	3	3	58	0.063	0.17	−0.72	C	2
			285.59	293700	643900	5	3	85	0.062	0.29	−0.51	C	2
			281.50	272700	627900	3	1	210	0.083	0.23	−0.60	C	2
			251.97	272700	669600	3	5	50	0.080	0.20	−0.62	D	2
			263.55	264500	643900	1	3	86	0.27	0.23	−0.57	C	2
4.		$^3P°$ – ^1D											
			[289]			5	5	27	0.034	0.16	−0.77	E	3
			[272]			3	5	15	0.028	0.076	−1.07	E	3
5.		$^3P°$ – ^1S											
			[207]			3	1	3.2	6.8(−4)[a]	0.0014	−2.69	E	3
6.		$^1P°$ – ^3P											
			[372.7]	401316	669600	3	5	5.9	0.020	0.075	−1.21	E	3
			[412.2]	401316	643900	3	3	0.14	3.4(−4)	0.0014	−2.99	E	3
			[441.3]	401316	627900	3	1	1.4	0.0014	0.0059	−2.39	E	3
7.		$^1P°$ – ^1D	[419]			3	5	18	0.080	0.33	−0.62	E	2
8.		$^1P°$ – ^1S	[282]			3	1	240	0.097	0.27	−0.54	C	2
9.	$3s3d$–$3p3d$	^3D – $^3F°$	*336*			15	21	64	0.15	2.5	0.35	D	2
			[323]			7	9	75	0.150	1.12	0.021	C	2
			[339]			5	7	50	0.12	0.67	−0.22	C	2
			[355]			3	5	38	0.12	0.42	−0.44	D	2
			[343]			7	7	13	0.023	0.18	−0.79	C	2
			[358]			5	5	10	0.020	0.12	−1.00	D	2
			[362]			7	5	0.10	1.4(−4)	0.0012	−3.01	E	2
10.		^3D – $^3D°$											
			[282]			7	7	84	0.10	0.65	−0.15	C	2
			[294]			3	3	25	0.032	0.093	−1.02	E	2
			[296]			5	3	74	0.058	0.28	−0.54	E	2
			[280]			5	7	26	0.042	0.19	−0.68	C	2
11.		^3D – $^3P°$											
			[279]			5	3	37	0.026	0.12	−0.89	E	2
			[278]			3	1	130	0.050	0.14	−0.82	C	2
			[277]			3	3	96	0.11	0.30	−0.48	E	2
12.		^1D – $^1D°$	[465]			5	5	8.6	0.028	0.21	−0.85	D	2
13.		^1D – $^1F°$	[292]			5	7	220	0.40	1.9	0.30	D	2
14.		^1D – $^1P°$	[284]			5	3	150	0.11	0.51	−0.26	D	2

Ni XVII: Allowed transitions — Continued

No.	Transition Array	Multiplet	λ (Å)	E_i (cm^{-1})	E_k (cm^{-1})	g_i	g_k	A_{ki} (10^8 s^{-1})	f_{ik}	S (at. u.)	log gf	Accuracy	Source
15.	$3s^2$–$3s4p$	^{1}S – ^{1}P°	42.855	0	2333500	1	3	4750	0.392	0.055	−0.407	C	1
16.	$3s^2$–$3s5p$	^{1}S – ^{1}P°	30.919	0	3234300	1	3	2770	0.119	0.0121	−0.92	C	1
17.	$3s3p$–$3s3d$	^{3}P° – ^{3}D	*204.0*	*283500*	*773800*	9	15	263	0.273	1.65	0.390	C	2
			207.50	293700	775600	5	7	250	0.226	0.77	0.053	C	2
			199.87	272700	773000	3	5	210	0.21	0.41	−0.20	C	2
			197.39	264500	771100	1	3	160	0.28	0.18	−0.55	C	2
			208.66	293700	773000	5	5	61	0.040	0.14	−0.70	C	2
			200.53	272700	771100	3	3	120	0.070	0.14	−0.68	C	2
			[209.5]	293700	771100	5	3	6.6	0.0026	0.0090	−1.89	D	2
18.		^{3}P° – ^{1}D											
			[175]	293700	864520	5	5	0.28	1.3(−4)	3.7(−4)	−3.19	E	3
			[169]	272700	864520	3	5	5.1	0.0037	0.0061	−1.96	E	3
19.		^{1}P° – ^{3}D											
			[269.0]	401316	773000	3	5	0.20	3.7(−4)	9.7(−4)	−2.96	E	3
			[270.4]	401316	771100	3	3	0.41	4.5(−4)	0.0012	−2.87	E	3
20.		^{1}P° – ^{1}D	215.89	401316	864520	3	5	480	0.56	1.2	0.23	D	2
21.	$3p^2$–$3p3d$	^{3}P – ^{3}D°											
			[217]			5	7	240	0.24	0.86	0.08	D	2
			[206]			1	3	300	0.57	0.39	−0.24	E	2
			[214]			3	3	48	0.033	0.070	−1.00	E	2
			[226]			5	3	2.6	0.0012	0.0045	−2.22	E	2
22.		^{3}P – ^{3}P°											
			[204]			3	3	180	0.11	0.22	−0.48	E	2
			[216]			5	3	71	0.030	0.11	−0.82	E	2
			[205]			3	1	240	0.050	0.10	−0.82	C	2
23.		^{1}D – ^{1}D°	[227]			5	5	160	0.12	0.45	−0.22	C	2
24.		^{1}D – ^{1}P°	[173]			5	3	4.8	0.0013	0.0037	−2.19	E	2
25.		^{1}S – ^{1}P°	[216]			1	3	270	0.57	0.41	−0.24	C	2
26.	$3p3d$–$3p4f$	^{3}F° – ^{3}G											
			54.451			9	11	1.5(+4)	0.81	1.3	0.86	C	*interp.*
27.		^{3}P° – ^{3}D											
			55.361			1	3	6700	0.93	0.17	−0.03	C	*interp.*
28.		^{1}F° – ^{1}G	57.348			7	9	1.4(+4)	0.90	1.2	0.80	C−	*interp.*

[a]The number in parentheses following the tabulated value indicates the power of ten by which this value has to be multiplied.

Ni XVII

Forbidden Transitions

List of tabulated lines

Wavelength (Å)	No.	Wavelength (Å)	No.	Wavelength (Å)	No.	Wavelength (Å)	No.
115.67	9	861	6	2400	3	6250	3
149.3	10	888	4	3330	5	12000	1
156	11	912	7	3420	1	27700	5
731.0	2	929.4	2	3890	3	38000	8
777.6	2	1160	4	4760	1	53000	8

Transition probabilities for forbidden lines involving pairs of levels belonging to the set of configurations $3s^2$, $3s3p$, $3p^2$, and $3s3d$ were computed by Bhatia and Kastner[1] using a scaled Thomas-Fermi approach with allowance for configuration mixing. These data are quoted here, but we first converted the transition probabilities to line strengths, which we then reconverted to A-values in order to incorporate more accurate wavelength values. The weakest lines were excluded from this compilation.

Reference

[1]A. K. Bhatia and S. O. Kastner, J. Quant. Spectrosc. Radiat. Transfer **24**, 53 (1980).

Ni XVII:　Forbidden transitions

No.	Transition Array	Multiplet	λ (Å)	E_i (cm^{-1})	E_k (cm^{-1})	g_i	g_k	Type of transition	A_{ki} (s^{-1})	S (at. u.)	Accuracy	Source
1.	$3s3p$–$3s3p$	$^3P°$ – $^3P°$										
			[4760]	272700	293700	3	5	M1	125	2.49	C+	1
			[12000]	264500	272700	1	3	M1	10	2.0	D+	1
			[3420]	264500	293700	1	5	E2	0.037	0.051	E	1
2.		$^3P°$ – $^1P°$										
			[929.4]	293700	401316	5	3	M1	210	0.019	E	1
			[777.6]	272700	401316	3	3	M1	210	0.011	E	1
			[731.0]	264500	401316	1	3	M1	350	0.015	E	1
3.	$3p^2$–$3p^2$	^3P – ^3P										
			[3890]	643900	669600	3	5	M1	190	2.1	D	1
			[6250]	627900	643900	1	3	M1	70	1.91	C	1
			[2400]	627900	669600	1	5	E2	0.13	0.031	E	1
4.		^3P – ^1S										
			[1160]			5	1	E2	26	0.032	E	1
			[888]			3	1	M1	2500	0.066	E	1

Ni XVII: Forbidden transitions — Continued

No.	Transition Array	Multiplet	λ (Å)	E_i (cm^{-1})	E_k (cm^{-1})	g_i	g_k	Type of transition	A_{ki} (s^{-1})	S (at. u.)	Accuracy	Source
5.		$^1D - {}^3P$										
			[3330]			5	5	M1	150	1.0	E	1
			[27700]			5	3	M1	0.17	0.41	E	1
6.		$^1D - {}^1S$	[861]			5	1	E2	340	0.097	E	1
7.	$3p^2$–$3s\,3d$	$^1S - {}^1D$	[912]			1	5	E2	37	0.070	E	1
8.	$3s\,3d$–$3s\,3d$	$^3D - {}^3D$										
			[38000]	773000	775600	5	7	M1	0.30	4.3	D+	1
			[53000]	771100	773000	3	5	M1	0.16	4.5	D	1
9.	$3s^2$–$3s\,3d$	$^1S - {}^1D$	[115.67]	0	864520	1	5	E2	2.1(+6)^a	0.13	D−	1
10.	$3s^2$–$3p^2$	$^1S - {}^3P$										
			[149.3]	0	669600	1	5	E2	6.3(+4)	0.014	E	1
11.		$^1S - {}^1D$	[156]			1	5	E2	2.1(+5)	0.059	E	1

^{a}The number in parentheses following the tabulated value indicates the power of ten by which this value has to be multiplied.

Ni XVIII

Na Isoelectronic Sequence

Ground State: $1s^2 2s^2 2p^6 3s\ ^2S_{1/2}$

Ionization Energy: 607.06 eV = 4896200 cm^{-1}

Allowed Transitions

List of tabulated lines

Wavelength (Å)	No.	Wavelength (Å)	No.	Wavelength (Å)	No.	Wavelength (Å)	No.
24.881	17	38.573	20	67.092	53	130.3	51
25.070	17	38.643	20	67.132	53	130.9	43
25.071	17	38.658	20	67.16	53	131.3	43
26.02	15	41.015	2	68.526	37	131.9	43
26.020	4	41.218	2	69.075	37	136.0	70
26.046	4	43.814	9	69.094	37	136.1	70
26.218	15	44.365	9	76.254	46	136.2	70
26.23	15	44.405	9	76.359	46	145.4	74
27.98	27	52.502	41	76.377	46	145.5	74
27.982	13	52.615	19	80.077	45	146	74
28.01	27	52.720	19	80.212	45	187.5	61
28.018	27	52.745	19	80.321	45	189.4	61
28.220	13	52.829	41	81.974	52	189.5	61
28.223	13	52.835	41	82.001	52	211.8	68
29.383	25	57.27	50	82.034	52	212.1	68
29.422	25	57.37	50	94.661	29	212.2	68
29.424	25	57.376	50	95.175	29	220.43	7
29.779	3	57.84	39	99.275	35	233.75	7
29.829	3	58.197	39	100.4	35	236.31	7
31.845	23	58.24	39	100.5	35	239.8	73
31.890	23	59.780	18	102.2	65	240.0	73
31.893	23	59.950	18	102.8	65	244.2	67
32.034	11	60.056	54	110	72	244.7	67
32.340	11	60.064	54	110.5	72	246.5	67
32.350	11	60.089	54	114.46	44	291.983	1
32.493	22	60.212	18	114.6	44	320.56	1
32.533	22	63.512	48	114.74	44	595.2	33
32.543	22	63.589	48	116.0	75	632.5	33
36.990	21	63.597	48	125	63	641.0	33
37.049	21	64.872	30	125.2	63	731.5	28
37.055	21	65.032	30	130.1	51	801.9	28

Strengths of the lines of the 3s–3p and 3p–3d transitions were taken from Edlen's interpolation formulae.[1] These were based on the results of Weiss' Hartree-Fock calculations,[2] in which ratios of relativistic Dirac to nonrelativistic line strengths in hydrogenic ions were applied as scaling factors to the nonrelativistic Hartree-Fock line strengths in the corresponding sodiumlike species. Oscillator strengths for the 4p–4d transitions were derived by Gruzdev and Sherstyuk[3] using the relativistic variant of their effective orbital quantum number method, which utilizes a Coulomb potential in conjunction with a semiempirical orbital quantum number which is determined from experimental energy levels. Strengths of the lines of the 3s–4p and 3p–4d

transitions, as well as f-values of the 3d–4f transitions, were interpolated from the results of the relativistic single-configuration Hartree-Fock calculations of Kim and Cheng[4] for Fe XVI and, depending on the transition, either Kr XXVI or Mo XXXII.

The lifetimes of the $3p_{1/2,3/2}$ and $3d_{5/2}$ levels were measured by Pegg et al.[5] using the beam-foil technique. They used a multiexponential fitting procedure to analyze their results, but did not incorporate a simulation of repopulation from higher levels, so that their method of cascade analysis is not so sophisticated as that reported elsewhere[6,7] for the lifetimes of the corresponding levels in Fe XVI. Nevertheless, the reciprocals of the lifetimes determined by Pegg et al. agree, to within their error estimates, with the theoretical A-values derived from Edlen's interpolation formulae, with our uncertainty estimates of the theoretical results taken into account.

Multiplet f-values calculated by Tull *et al.*[8] using the frozen-core Hartree-Fock approach are quoted for numerous transitions $nl-n'l'$ ($3\leqslant n\leqslant 5$; $4\leqslant n'\leqslant 8$; l, $l'=s,p,d,f$). Data for additional transitions (namely, those for which $n,n' \leqslant 10$, where n,n' are the principal quantum numbers of the lower and upper states, respectively) can be found in Ref. 8. Whenever wavelengths of individual lines within a multiplet either were available directly or could be determined from the energy levels, the multiplet strength was distributed among the lines according to LS-coupling rules, except in the case of the $5f$-$8d$ transition, where the wavelengths for all the lines in the multiplet are identical. The strength of the $3p\ ^2P^\circ$ – $4s\ ^2S$ multiplet was not distributed between the two lines in the multiplet, however, since the calculations of Kim and Cheng indicate that in the corresponding transition in neighboring sodiumlike ions (Fe XVI and

Mo XXXII) the ratio of the two line strengths deviates somewhat from the value that would be obtained in the case of pure LS coupling.

Transitions with small f-values were generally assigned lower accuracy ratings.

References

[1]B. Edlen, Phys. Scr. **17**, 565 (1978).
[2]A. W. Weiss, J. Quant. Spectrosc. Radiat. Transfer **18**, 481 (1977).
[3]P. F. Gruzdev and A. I. Sherstyuk, Opt. Spectrosc. (USSR) **46**, 353 (1979).
[4]Y.-K. Kim and K.-T. Cheng, J. Opt. Soc. Am. **68**, 836 (1978).
[5]D. J. Pegg, P. M. Griffin, B. M. Johnson, K. W. Jones, and T. H. Kruse, Astrophys. J. **224**, 1056 (1978).
[6]J. P. Buchet, M. C. Buchet-Poulizac, A. Denis, J. Desesquelles, and M. Druetta, Phys. Rev. A **22**, 2061 (1980).
[7]E. Träbert, K. W. Jones, B. M. Johnson, D. C. Gregory, and T. H. Kruse, Phys. Lett. A **87**, 336 (1982).
[8]C. E. Tull, R. P. McEachran, and M. Cohen, At. Data **3**, 169 (1971).

Ni XVIII: Allowed transitions

No.	Transition Array	Multiplet	λ (Å)	E_i (cm^{-1})	E_k (cm^{-1})	g_i	g_k	A_{ki} (10^8 s^{-1})	f_{ik}	S (at. u.)	log gf	Accuracy	Source
1.	$3s$-$3p$	^{2}S – ^{2}P°	300.92	0	332310	2	6	90.1	0.367	0.727	−0.134	B	1
			291.983	0	342486	2	4	99.1	0.253	0.487	−0.295	B	1
			320.56	0	311950	2	2	73.8	0.114	0.240	−0.643	B	1
2.	$3s$-$4p$	^{2}S – ^{2}P°	41.078	0	2434400	2	6	3100	0.23	0.063	−0.33	C	interp.
			41.015	0	2438400	2	4	2970	0.150	0.0405	−0.52	C	interp.
			41.218	0	2426400	2	2	3200	0.0814	0.0221	−0.788	C+	interp.
3.	$3s$-$5p$	^{2}S – ^{2}P°	29.796	0	3356200	2	6	1900	0.074	0.015	−0.83	C	8
			29.779	0	3358100	2	4	1900	0.051	0.010	−0.99	C	ls
			29.829	0	3352400	2	2	1900	0.025	0.0050	−1.29	C	ls
4.	$3s$-$6p$	^{2}S – ^{2}P°	26.029	0	3841900	2	6	1100	0.0334	0.0057	−1.175	C	8
			26.020	0	3843200	2	4	1100	0.022	0.0038	−1.35	C	ls
			26.046	0	3839400	2	2	1100	0.011	0.0019	−1.65	C	ls
5.	$3s$-$7p$	^{2}S – ^{2}P°				2	6		0.0183		−1.437	C	8
6.	$3s$-$8p$	^{2}S – ^{2}P°				2	6		0.0113		−1.65	C	8
7.	$3p$-$3d$	^{2}P° – ^{2}D	229.30	332310	768420	6	10	195	0.256	1.16	0.187	B	1
			233.75	342486	770300	4	6	183	0.225	0.694	−0.045	B	1
			220.43	311950	765610	2	4	183	0.266	0.386	−0.274	B	1
			236.31	342486	765610	4	4	29.4	0.0246	0.0766	−1.007	B	1
8.	$3p$-$4s$	^{2}P° – ^{2}S	50.777	332310	2301700	6	2	4600	0.059	0.059	−0.45	C−	8
9.	$3p$-$4d$	^{2}P° – ^{2}D	44.181	332310	2595700	6	10	6800	0.330	0.288	0.297	C	interp.
			44.365	342486	2596500	4	6	6800	0.301	0.176	0.081	C	interp.
			43.814	311950	2594400	2	4	5500	0.32	0.092	−0.20	C	interp.
			44.405	342486	2594400	4	4	1140	0.0337	0.0197	−0.87	C	interp.

Ni XVIII: Allowed transitions — Continued

No.	Transition Array	Multiplet	λ (Å)	E_i (cm^{-1})	E_k (cm^{-1})	g_i	g_k	A_{ki} (10^8 s^{-1})	f_{ik}	S (at. u.)	log gf	Accuracy	Source
10.	3p–5s	^{2}P° – ^{2}S				6	2		0.0119		−1.146	C	8
11.	3p–5d	^{2}P° – ^{2}D	32.238	332310	3434200	6	10	4000	0.104	0.066	−0.205	C	8
			32.340	342486	3434600	4	6	4000	0.094	0.040	−0.43	C	ls
			32.034	311950	3433700	2	4	3400	0.10	0.022	−0.68	C	ls
			[32.350]	342486	3433700	4	4	660	0.010	0.0044	−1.38	D	ls
12.	3p–6s	^{2}P° – ^{2}S				6	2		0.0047		−1.55	D	8
13.	3p–6d	^{2}P° – ^{2}D	28.140	332310	3885900	6	10	2360	0.0466	0.0259	−0.55	C	8
			28.220	342486	3886100	4	6	2330	0.0417	0.0155	−0.78	C	ls
			27.982	311950	3885700	2	4	2000	0.047	0.0086	−1.03	C	ls
			[28.223]	342486	3885700	4	4	380	0.0046	0.0017	−1.74	D	ls
14.	3p–7s	^{2}P° – ^{2}S				6	2		0.0024		−1.84	D	8
15.	3p–7d	^{2}P° – ^{2}D	26.15	332310	4156000	6	10	1490	0.0254	0.0131	−0.82	C	8
			26.218	342486	4156700	4	6	1500	0.023	0.0079	−1.04	C	ls
			26.02	311950	4155000	2	4	1260	0.0255	0.00437	−1.292	C	ls
			[26.23]	342486	4155000	4	4	240	0.0025	8.7(−4)^a	−2.00	D	ls
16.	3p–8s	^{2}P° – ^{2}S				6	2		0.0014		−2.08	D	8
17.	3p–8d	^{2}P° – ^{2}D	25.007	332310	4331200	6	10	1000	0.0156	0.0077	−1.029	C	8
			25.070	342486	4331300	4	6	990	0.014	0.0046	−1.25	C	ls
			24.881	311950	4331100	2	4	860	0.016	0.0026	−1.50	C	ls
			[25.071]	342486	4331100	4	4	160	0.0015	5.1(−4)	−2.21	D	ls
18.	3d–4p	^{2}D – ^{2}P°	60.024	768420	2434400	10	6	1080	0.0349	0.069	−0.457	C−	8
			59.950	770300	2438400	6	4	960	0.035	0.041	−0.68	C−	ls
			60.212	765610	2426400	4	2	1100	0.029	0.023	−0.94	C−	ls
			[59.780]	765610	2438400	4	4	110	0.0058	0.0046	−1.63	D	ls
19.	3d–4f	^{2}D – ^{2}F°	52.679	768420	2666700	10	14	1.60(+4)	0.93	1.62	0.97	C	interp.
			52.720	770300	2667100	6	8	1.6(+4)	0.89	0.93	0.73	C	interp.
			52.615	765610	2666200	4	6	1.5(+4)	0.93	0.64	0.57	C	interp.
			[52.745]	770300	2666200	6	6	1060	0.0444	0.0463	−0.57	C	interp.
20.	3d–5p	^{2}D – ^{2}P°	38.643	768420	3356200	10	6	420	0.0057	0.0073	−1.24	D	8
			[38.643]	770300	3358100	6	4	390	0.0058	0.0044	−1.46	D	ls
			[38.658]	765610	3352400	4	2	420	0.0047	0.0024	−1.72	D	ls
			[38.573]	765610	3358100	4	4	43	9.6(−4)	4.9(−4)	−2.41	E	ls
21.	3d–5f	^{2}D – ^{2}F°	37.026	768420	3469200	10	14	5900	0.170	0.207	0.230	C	8
			37.049	770300	3469400	6	8	5900	0.161	0.118	−0.014	C	ls
			36.990	765610	3469000	4	6	5500	0.17	0.083	−0.17	C	ls
			[37.055]	770300	3469000	6	6	390	0.0081	0.0059	−1.32	D	ls
22.	3d–6p	^{2}D – ^{2}P°	32.536	768420	3841900	10	6	220	0.0021	0.0022	−1.68	D	8
			[32.543]	770300	3843200	6	4	190	0.0020	0.0013	−1.92	D	ls
			[32.533]	765610	3839400	4	2	210	0.0017	7.3(−4)	−2.17	D	ls
			[32.493]	765610	3843200	4	4	22	3.5(−4)	1.5(−4)	−2.85	E	ls

Ni XVIII: Allowed transitions — Continued

No.	Transition Array	Multiplet	λ (Å)	E_i (cm^{-1})	E_k (cm^{-1})	g_i	g_k	A_{ki} (10^8 s^{-1})	f_{ik}	S (at. u.)	log gf	Accuracy	Source
23.	3d–6f	^{2}D – ^{2}F°	31.871	768420	3906000	10	14	3000	0.063	0.066	−0.20	C	8
			31.890	770300	3906100	6	8	3000	0.060	0.038	−0.44	C	ls
			31.845	765610	3905800	4	6	2700	0.062	0.026	−0.61	C	ls
			[31.893]	770300	3905800	6	6	200	0.0030	0.0019	−1.74	D	ls
24.	3d–7p	^{2}D – ^{2}P°				10	6		0.0010		−2.00	D	8
25.	3d–7f	^{2}D – ^{2}F°	29.407	768420	4169000	10	14	1700	0.0308	0.0298	−0.51	C	8
			29.422	770300	4169100	6	8	1690	0.0293	0.0170	−0.76	C	ls
			29.383	765610	4168900	4	6	1580	0.0308	0.0119	−0.91	C	ls
			[29.424]	770300	4168900	6	6	110	0.0015	8.5(−4)	−2.06	D	ls
26.	3d–8p	^{2}D – ^{2}P°				10	6		5.8(−4)		−2.24	E	8
27.	3d–8f	^{2}D – ^{2}F°	28.00	768420	4339000	10	14	1080	0.0177	0.0163	−0.75	C	8
			28.018	770300	4339400	6	8	1100	0.017	0.0093	−1.00	C	ls
			27.98	765610	4340000	4	6	1000	0.018	0.0065	−1.15	C	ls
			[28.01]	770300	4340000	6	6	72	8.5(−4)	4.7(−4)	−2.29	E	ls
28.	4s–4p	^{2}S – ^{2}P°	753.6	2301700	2434400	2	6	19.4	0.496	2.46	−0.003	C	8
			[731.5]	2301700	2438400	2	4	21.2	0.341	1.64	−0.167	C	ls
			[801.9]	2301700	2426400	2	2	16	0.16	0.82	−0.51	C	ls
29.	4s–5p	^{2}S – ^{2}P°	94.832	2301700	3356200	2	6	660	0.265	0.165	−0.276	C	8
			[94.661]	2301700	3358100	2	4	660	0.176	0.110	−0.452	C	ls
			[95.175]	2301700	3352400	2	2	650	0.088	0.055	−0.76	C	ls
30.	4s–6p	^{2}S – ^{2}P°	64.927	2301700	3841900	2	6	430	0.081	0.035	−0.79	C	8
			[64.872]	2301700	3843200	2	4	430	0.054	0.023	−0.97	C	ls
			[65.032]	2301700	3839400	2	2	440	0.028	0.012	−1.25	C	ls
31.	4s–7p	^{2}S – ^{2}P°				2	6		0.0376		−1.124	C	8
32.	4s–8p	^{2}S – ^{2}P°				2	6		0.0210		−1.377	C	8
33.	4p–4d	^{2}P° – ^{2}D	620.0	2434400	2595700	6	10	38	0.37	4.5	0.34	C	3
			[632.5]	2438400	2596500	4	6	37	0.33	2.7	0.12	C	3
			[595.2]	2426400	2594400	2	4	37	0.39	1.5	−0.11	C	3
			[641.0]	2438400	2594400	4	4	5.8	0.036	0.30	−0.84	C	3
34.	4p–5s	^{2}P° – ^{2}S				6	2		0.101		−0.218	C	8
35.	4p–5d	^{2}P° – ^{2}D	100.0	2434400	3434200	6	10	1200	0.301	0.59	0.257	C	8
			[100.4]	2438400	3434600	4	6	1200	0.26	0.35	0.02	C	ls
			[99.275]	2426400	3433700	2	4	1000	0.31	0.20	−0.21	C	ls
			[100.5]	2438400	3433700	4	4	190	0.029	0.039	−0.93	D	ls
36.	4p–6s	^{2}P° – ^{2}S				6	2		0.0203		−0.91	C	8
37.	4p–6d	^{2}P° – ^{2}D	68.894	2434400	3885900	6	10	830	0.098	0.13	−0.23	C	8
			[69.075]	2438400	3886100	4	6	800	0.086	0.078	−0.46	C	ls
			[68.526]	2426400	3885700	2	4	680	0.095	0.043	−0.72	C	ls
			[69.094]	2438400	3885700	4	4	130	0.0096	0.0087	−1.42	D	ls

Ni XVIII: Allowed transitions — Continued

No.	Transition Array	Multiplet	λ (Å)	E_i (cm^{-1})	E_k (cm^{-1})	g_i	g_k	A_{ki} (10^8 s^{-1})	f_{ik}	S (at. u.)	log gf	Accuracy	Source
38.	4p–7s	^{2}P° – ^{2}S				6	2		0.0080		−1.32	D	8
39.	4p–7d	^{2}P° – ^{2}D	58.07	2434400	4156000	6	10	550	0.0460	0.053	−0.56	C	8
			[58.197]	2438400	4156700	4	6	550	0.042	0.032	−0.78	C	ls
			[57.84]	2426400	4155000	2	4	470	0.047	0.018	−1.02	C	ls
			[58.24]	2438400	4155000	4	4	90	0.0046	0.0035	−1.74	D	ls
40.	4p–8s	^{2}P° – ^{2}S				6	2		0.0041		−1.61	D	8
41.	4p–8d	^{2}P° – ^{2}D	52.720	2434400	4331200	6	10	373	0.0259	0.0270	−0.81	C	8
			[52.829]	2438400	4331300	4	6	371	0.0233	0.0162	−1.031	C	ls
			[52.502]	2426400	4331100	2	4	320	0.026	0.0090	−1.28	C	ls
			[52.835]	2438400	4331100	4	4	62	0.0026	0.0018	−1.99	D	ls
42.	4d–4f	^{2}D – ^{2}F°				10	14		0.102		0.009	C	8
43.	4d–5p	^{2}D – ^{2}P°	131.5	2595700	3356200	10	6	510	0.080	0.35	−0.10	C	8
			[131.3]	2596500	3358100	6	4	470	0.081	0.21	−0.31	C	ls
			[131.9]	2594400	3352400	4	2	530	0.069	0.12	−0.56	C	ls
			[130.9]	2594400	3358100	4	4	52	0.013	0.023	−1.27	D	ls
44.	4d–5f	^{2}D – ^{2}F°	114.5	2595700	3469200	10	14	2700	0.74	2.8	0.87	C	8
			114.74	2596500	3469400	6	8	2700	0.71	1.6	0.63	C	ls
			114.46	2594400	3469000	4	6	2500	0.73	1.1	0.47	C	ls
			[114.6]	2596500	3469000	6	6	180	0.035	0.080	−0.67	D	ls
45.	4d–6p	^{2}D – ^{2}P°	80.244	2595700	3841900	10	6	240	0.0139	0.0367	−0.86	C	8
			[80.212]	2596500	3843200	6	4	216	0.0139	0.0220	−1.079	C	ls
			[80.321]	2594400	3839400	4	2	239	0.0115	0.0122	−1.336	C	ls
			[80.077]	2594400	3843200	4	4	24	0.0023	0.0024	−2.04	D	ls
46.	4d–6f	^{2}D – ^{2}F°	76.318	2595700	3906000	10	14	1470	0.180	0.452	0.255	C	8
			[76.359]	2596500	3906100	6	8	1470	0.171	0.258	0.011	C	ls
			[76.254]	2594400	3905800	4	6	1380	0.180	0.181	−0.142	C	ls
			[76.377]	2596500	3905800	6	6	99	0.0086	0.013	−1.29	D	ls
47.	4d–7p	^{2}D – ^{2}P°				10	6		0.0052		−1.28	D	8
48.	4d–7f	^{2}D – ^{2}F°	63.561	2595700	4169000	10	14	870	0.074	0.15	−0.13	C	8
			[63.589]	2596500	4169100	6	8	850	0.068	0.086	−0.39	C	ls
			[63.512]	2594400	4168900	4	6	790	0.072	0.060	−0.54	C	ls
			[63.597]	2596500	4168900	6	6	56	0.0034	0.0043	−1.69	D	ls
49.	4d–8p	^{2}D – ^{2}P°				10	6		0.0026		−1.59	D	8
50.	4d–8f	^{2}D – ^{2}F°	57.37	2595700	4339000	10	14	560	0.0389	0.073	−0.410	C	8
			[57.376]	2596500	4339400	6	8	560	0.037	0.042	−0.65	C	ls
			[57.27]	2594400	4340000	4	6	520	0.038	0.029	−0.81	C	ls
			[57.37]	2596500	4340000	6	6	38	0.0019	0.0021	−1.95	D	ls
51.	4f–5d	^{2}F° – ^{2}D	130.3	2666700	3434200	14	10	100	0.0181	0.109	−0.60	C	8
			[130.3]	2667100	3434600	8	6	95	0.018	0.062	−0.84	C	ls
			[130.3]	2666200	3433700	6	4	100	0.0169	0.0436	−0.99	C	ls
			[130.1]	2666200	3434600	6	6	4.8	0.0012	0.0031	−2.14	D	ls

Ni XVIII: Allowed transitions — Continued

No.	Transition Array	Multiplet	λ (Å)	E_i (cm^{-1})	E_k (cm^{-1})	g_i	g_k	A_{ki} (10^8 s^{-1})	f_{ik}	S (at. u.)	log gf	Accuracy	Source
52.	4f–6d	^{2}F° – ^{2}D	82.021	2666700	3885900	14	10	42	0.0030	0.011	−1.38	D	8
			[82.034]	2667100	3886100	8	6	39	0.0029	0.0063	−1.63	D	ls
			[82.001]	2666200	3885700	6	4	40	0.0027	0.0044	−1.79	D	ls
			[81.974]	2666200	3886100	6	6	1.9	1.9(−4)	3.1(−4)	−2.94	E	ls
53.	4f–7d	^{2}F° – ^{2}D	67.16	2666700	4156000	14	10	23	0.0011	0.0034	−1.81	D	8
			[67.132]	2667100	4156700	8	6	21	0.0011	0.0019	−2.07	D	ls
			[67.16]	2666200	4155000	6	4	23	0.0011	0.0014	−2.20	D	ls
			[67.092]	2666200	4156700	6	6	1.1	7.3(−5)	9.7(−5)	−3.36	E	ls
54.	4f–8d	^{2}F° – ^{2}D	60.078	2666700	4331200	14	10	13	5.1(−4)	0.0014	−2.15	E	8
			[60.089]	2667100	4331300	8	6	12	5.1(−4)	8.0(−4)	−2.39	E	ls
			[60.064]	2666200	4331100	6	4	13	4.7(−4)	5.6(−4)	−2.55	E	ls
			[60.056]	2666200	4331300	6	6	0.62	3.4(−5)	4.0(−5)	−3.69	E	ls
55.	5s–5p	^{2}S – ^{2}P°				2	6		0.64		0.11	C	8
56.	5s–6p	^{2}S – ^{2}P°				2	6		0.287		−0.241	C	8
57.	5s–7p	^{2}S – ^{2}P°				2	6		0.089		−0.75	C	8
58.	5s–8p	^{2}S – ^{2}P°				2	6		0.0415		−1.081	C	8
59.	5p–5d	^{2}P° – ^{2}D				6	10		0.53		0.50	C	8
60.	5p–6s	^{2}P° – ^{2}S				6	2		0.143		−0.067	C	8
61.	5p–6d	^{2}P° – ^{2}D	188.8	3356200	3885900	6	10	321	0.286	1.07	0.235	C	8
			[189.4]	3358100	3886100	4	6	320	0.26	0.64	0.01	C	ls
			[187.5]	3352400	3885700	2	4	274	0.289	0.357	−0.238	C	ls
			[189.5]	3358100	3885700	4	4	53	0.028	0.071	−0.94	D	ls
62.	5p–7s	^{2}P° – ^{2}S				6	2		0.0288		−0.76	C	8
63.	5p–7d	^{2}P° – ^{2}D	125	3356200	4156000	6	10	250	0.096	0.24	−0.24	C	8
			[125.2]	3358100	4156700	4	6	240	0.085	0.14	−0.47	C	ls
			[125]	3352400	4155000	2	4	210	0.097	0.080	−0.71	C	ls
			[125]	3358100	4155000	4	4	41	0.0097	0.016	−1.41	D	ls
64.	5p–8s	^{2}P° – ^{2}S				6	2		0.0114		−1.165	C	8
65.	5p–8d	^{2}P° – ^{2}D	102.6	3356200	4331200	6	10	175	0.0459	0.093	−0.56	C	8
			[102.8]	3358100	4331300	4	6	170	0.041	0.056	−0.78	C	ls
			[102.2]	3352400	4331100	2	4	150	0.046	0.031	−1.04	C	ls
			[102.8]	3358100	4331100	4	4	29	0.0046	0.0062	−1.74	D	ls
66.	5d–5f	^{2}D – ^{2}F°				10	14		0.181		0.258	C	8
67.	5d–6p	^{2}D – ^{2}P°	245.3	3434200	3841900	10	6	238	0.129	1.04	0.111	C	8
			[244.7]	3434600	3843200	6	4	210	0.13	0.62	−0.11	C	ls
			[246.5]	3433700	3839400	4	2	235	0.107	0.347	−0.369	C	ls
			[244.2]	3433700	3843200	4	4	24	0.021	0.069	−1.07	D	ls

Ni XVIII: Allowed transitions — Continued

No.	Transition Array	Multiplet	λ (Å)	E_i (cm^{-1})	E_k (cm^{-1})	g_i	g_k	A_{ki} (10^8 s^{-1})	f_{ik}	S (at. u.)	log gf	Accuracy	Source
68.	5d–6f	^{2}D – ^{2}F°	212.0	3434200	3906000	10	14	700	0.66	4.6	0.82	C	8
			[212.1]	3434600	3906100	6	8	690	0.62	2.6	0.57	C	ls
			[211.8]	3433700	3905800	4	6	640	0.65	1.8	0.41	C	ls
			[212.2]	3434600	3905800	6	6	46	0.031	0.13	−0.73	D	ls
69.	5d–7p	^{2}D – ^{2}P°				10	6		0.0232		−0.63	C	8
70.	5d–7f	^{2}D – ^{2}F°	136.1	3434200	4169000	10	14	458	0.178	0.80	0.250	C	8
			[136.1]	3434600	4169100	6	8	460	0.17	0.46	0.01	C	ls
			[136.0]	3433700	4168900	4	6	430	0.18	0.32	−0.15	C	ls
			[136.2]	3434600	4168900	6	6	31	0.0085	0.023	−1.29	D	ls
71.	5d–8p	^{2}D – ^{2}P°				10	6		0.0087		−1.06	D	8
72.	5d–8f	^{2}D – ^{2}F°	110	3434200	4339000	10	14	310	0.078	0.28	−0.11	C	8
			[110.5]	3434600	4339400	6	8	300	0.073	0.16	−0.36	C	ls
			[110]	3433700	4340000	4	6	280	0.076	0.11	−0.52	C	ls
			[110]	3434600	4340000	6	6	20	0.0037	0.0080	−1.66	D	ls
73.	5f–6d	^{2}F° – ^{2}D	240.0	3469200	3885900	14	10	72	0.0444	0.491	−0.206	C	8
			[240.0]	3469400	3886100	8	6	69	0.0445	0.281	−0.449	C	ls
			[240.0]	3469000	3885700	6	4	72	0.0413	0.196	−0.61	C	ls
			[239.8]	3469000	3886100	6	6	3.4	0.0030	0.014	−1.75	D	ls
74.	5f–7d	^{2}F° – ^{2}D	146	3469200	4156000	14	10	35	0.0079	0.053	−0.96	D	8
			[145.5]	3469400	4156700	8	6	33	0.0078	0.030	−1.20	D	ls
			[146]	3469000	4155000	6	4	34	0.0073	0.021	−1.36	D	ls
			[145.4]	3469000	4156700	6	6	1.6	5.2(−4)	0.0015	−2.50	E	ls
75.	5f–8d	^{2}F° – ^{2}D	[116.0]	3469200	4331200	14	10	20	0.0029	0.016	−1.39	D	8

aThe number in parentheses following the tabulated value indicates the power of ten by which this value has to be multiplied.

Ni XVIII

Forbidden Transitions

List of tabulated lines

Wavelength (Å)	No.	Wavelength (Å)	No.	Wavelength (Å)	No.	Wavelength (Å)	No.
29.115	28	77.567	9	142.8	18	303.5	49
31.981	30	80.671	12	142.9	18	306.8	49
38.513	27	80.710	12	143.0	18	307.0	49
38.545	27	85.027	42	144.5	41	339.2	32
42.477	29	88.269	33	144.7	41	341.6	32
43.018	29	88.339	33	145.7	41	369.0	20
43.035	29	95.914	37	180.7	44	369.5	20
47.028	4	96.993	37	182.5	44	372	20
47.712	4	97.031	37	182.6	44	379.8	22
47.987	4	101	46	200	50	380.1	22
52.604	40	101.9	46	201	50	380.2	22
53.908	35	102	46	201.5	50	380.5	22
53.97	35	107.3	6	203.7	13	417.0	36
54.618	5	108.7	6	206.1	13	437.3	36
54.681	5	109.4	6	207.8	13	439.0	36
54.759	5	111.4	16	221.0	14	568	24
54.822	5	111.5	16	221.2	14	572.7	24
57.389	39	114.9	19	221.5	14	573.4	24
57.780	39	115	19	221.7	14	585	25
63.115	34	119.0	8	224.4	21	586.5	25
63.131	34	119.1	8	224.5	21	587.2	25
64.094	10	119.3	8	224.6	21	857.6	43
65.100	31	119.4	8	224.7	21	898.5	43
65.300	31	122.5	45	228.8	17	901.7	43
67.595	38	123.3	45	228.9	17	1510	48
68.134	38	124.5	11	229.0	17	1590	48
68.148	38	124.6	11	229.1	17	1600	48
70.582	7	124.7	11	230	23	3273	1
71.185	7	129.82	26	230.6	23	8330	3
71.378	7	130.61	26	230.8	23	21300	2
77.417	9	138.3	15	267.2	47		
77.441	9	138.5	15	267.5	47		
77.543	9	139	15	270.0	47		

Electric quadrupole strengths for numerous multiplets in this sodiumlike ion were determined by Tull et al.[1] using the frozen-core Hartree-Fock approach with no allowance for configuration mixing. LS-coupling rules were applied to obtain strengths of lines within multiplets. The strongest lines for which fairly accurate wavelengths could be derived from experimentally determined energy levels are quoted in this compilation.

The strengths given in Ref. 1 for transitions in which both $\Delta n = 0$ and $\Delta l = 0$ (i.e., transitions between the two levels of a given term) are overstated, and had to be reduced as follows:

$$S(np\ ^2P^\circ_{1/2} - np\ ^2P^\circ_{3/2}) = S(\text{Ref. 1}) \times (1/3)$$

$$S(nd\ ^2D_{3/2} - nd\ ^2D_{5/2}) = S(\text{Ref. 1}) \times (3/25)$$

$$S(nf\ ^2F^\circ_{5/2} - nf\ ^2F^\circ_{7/2}) = S(\text{Ref. 1}) \times (3/49).$$

Reference

[1] C. E. Tull, M. Jackson, R. P. McEachran, and M. Cohen, J. Quant. Spectrosc. Radiat. Transfer **12**, 893 (1972).

Ni XVIII: Forbidden transitions

No.	Transition Array	Multiplet	λ (Å)	E_i (cm^{-1})	E_k (cm^{-1})	g_i	g_k	Type of transition	A_{ki} (s^{-1})	S (at. u.)	Accuracy	Source
1.	3p–3p	^{2}P° – ^{2}P°										
			[3273]	311950	342486	2	4	E2	0.11	0.094	D	1
2.	3d–3d	^{2}D – ^{2}D										
			[21300]	7656100	770300	4	6	E2	1.7(−6)^a	0.027	D−	1
3.	4p–4p	^{2}P° – ^{2}P°										
			[8330]	2426400	2438400	2	4	E2	0.013	1.2	D−	1
4.	3p–4p	^{2}P° – ^{2}P°										
			[47.712]	342486	2438400	4	4	E2	6.6(+7)	0.039	D	1,ls
			[47.987]	342486	2426400	4	2	E2	1.3(+8)	0.039	D	1,ls
			[47.028]	311950	2438400	2	4	E2	7.1(+7)	0.039	D	1,ls
5.	3d–4d	^{2}D – ^{2}D										
			[54.759]	770300	2596500	6	6	E2	3.6(+7)	0.063	D	1,ls
			[54.681]	765610	2594400	4	4	E2	3.2(+7)	0.037	D	1,ls
			[54.822]	770300	2594400	6	4	E2	1.4(+7)	0.016	E	1,ls
			[54.618]	765610	2596500	4	6	E2	9.2(+6)	0.016	E	1,ls
6.	4p–5p	^{2}P° – ^{2}P°										
			[108.7]	2438400	3358100	4	4	E2	1.07(+7)	0.387	C	1,ls
			[109.4]	2438400	3352400	4	2	E2	2.07(+7)	0.387	C	1,ls
			[107.3]	2426400	3358100	2	4	E2	1.14(+7)	0.387	C	1,ls
7.	4p–6p	^{2}P° – ^{2}P°										
			[71.185]	2438400	3843200	4	4	E2	6.9(+6)	0.030	D	1,ls
			[71.378]	2438400	3839400	4	2	E2	1.4(+7)	0.030	D	1,ls
			[70.582]	2426400	3843200	2	4	E2	7.2(+6)	0.030	D	1,ls
8.	4d–5d	^{2}D – ^{2}D										
			[119.3]	2596500	3434600	6	6	E2	8.1(+6)	0.70	C	1,ls
			[119.1]	2594400	3433700	4	4	E2	7.1(+6)	0.406	C	1,ls
			[119.4]	2596500	3433700	6	4	E2	3.01(+6)	0.174	C−	1,ls
			[119.0]	2594400	3434600	4	6	E2	2.04(+6)	0.174	C−	1,ls
9.	4d–6d	^{2}D – ^{2}D										
			[77.543]	2596500	3886100	6	6	E2	4.8(+6)	0.048	D	1,ls
			[77.441]	2594400	3885700	4	4	E2	4.2(+6)	0.028	D	1,ls
			[77.567]	2596500	3885700	6	4	E2	1.8(+6)	0.012	E	1,ls
			[77.417]	2594400	3886100	4	6	E2	1.2(+6)	0.012	E	1,ls
10.	4d–7d	^{2}D – ^{2}D										
			[64.094]	2596500	4156700	6	6	E2	3.1(+6)	0.012	D	1,ls
11.	4f–5f	^{2}F° – ^{2}F°										
			[124.6]	2667100	3469400	8	8	E2	4.7(+6)	0.67	C	1,ls
			[124.6]	2666200	3469000	6	6	E2	4.48(+6)	0.481	C	1,ls
			[124.7]	2667100	3469000	8	6	E2	7.4(+5)	0.080	D	1,ls
			[124.5]	2666200	3469400	6	8	E2	5.6(+5)	0.080	D	1,ls

Ni xviii: Forbidden transitions — Continued

No.	Transition Array	Multiplet	λ (Å)	E_i (cm^{-1})	E_k (cm^{-1})	g_i	g_k	Type of transition	A_{ki} (s^{-1})	S (at. u.)	Accuracy	Source
12.	4f–6f	^{2}F° – ^{2}F°										
			[80.710]	2667100	3906100	8	8	E2	2.3(+6)	0.038	D	1,ls
			[80.671]	2666200	3905800	6	6	E2	2.2(+6)	0.027	D	1,ls
13.	5p–6p	^{2}P° – ^{2}P°										
			[206.1]	3358100	3843200	4	4	E2	2.50(+6)	2.21	C	1,ls
			[207.8]	3358100	3839400	4	2	E2	4.79(+6)	2.21	C	1,ls
			[203.7]	3352400	3843200	2	4	E2	2.65(+6)	2.21	C	1,ls
14.	5d–6d	^{2}D – ^{2}D										
			[221.5]	3434600	3886100	6	6	E2	2.13(+6)	4.06	C	1,ls
			[221.2]	3433700	3885700	4	4	E2	1.88(+6)	2.37	C	1,ls
			[221.7]	3434600	3885700	6	4	E2	8.0(+5)	1.02	C–	1,ls
			[221.0]	3433700	3886100	4	6	E2	5.4(+5)	1.02	C–	1,ls
15.	5d–7d	^{2}D – ^{2}D										
			[138.5]	3434600	4156700	6	6	E2	1.46(+6)	0.266	C	1,ls
			[139]	3433700	4155000	4	4	E2	1.25(+6)	0.155	C	1,ls
			[139]	3434600	4155000	6	4	E2	5.3(+5)	0.066	D	1,ls
			[138.3]	3433700	4156700	4	6	E2	3.7(+5)	0.066	D	1,ls
16.	5d–8d	^{2}D – ^{2}D										
			[111.5]	3434600	4331300	6	6	E2	9.9(+5)	0.061	D	1,ls
			[111.4]	3433700	4331100	4	4	E2	8.8(+5)	0.036	D	1,ls
			[111.5]	3434600	4331100	6	4	E2	3.7(+5)	0.015	E	1,ls
			[111.4]	3433700	4331300	4	6	E2	2.4(+5)	0.015	E	1,ls
17.	5f–6f	^{2}F° – ^{2}F°										
			[229.0]	3469400	3906100	8	8	E2	1.63(+6)	4.89	C	1,ls
			[228.9]	3469000	3905800	6	6	E2	1.57(+6)	3.52	C	1,ls
			[229.1]	3469400	3905800	8	6	E2	2.6(+5)	0.59	C–	1,ls
			[228.8]	3469000	3906100	6	8	E2	2.0(+5)	0.59	C–	1,ls
18.	5f–7f	^{2}F° – ^{2}F°										
			[142.9]	3469400	4169100	8	8	E2	1.00(+6)	0.284	C	1,ls
			[142.9]	3469000	4168900	6	6	E2	9.6(+5)	0.205	C	1,ls
			[143.0]	3469400	4168900	8	6	E2	1.6(+5)	0.034	D	1,ls
			[142.8]	3469000	4169100	6	8	E2	1.2(+5)	0.034	D	1,ls
19.	5f–8f	^{2}F° – ^{2}F°										
			[114.9]	3469400	4339400	8	8	E2	6.4(+5)	0.061	D	1,ls
			[115]	3469000	4340000	6	6	E2	6.1(+5)	0.044	D	1,ls
20.	6d–7d	^{2}D – ^{2}D										
			[369.5]	3886100	4156700	6	6	E2	6.8(+5)	16.8	C	1,ls
			[372]	3885700	4155000	4	4	E2	5.8(+5)	9.8	D+	1,ls
			[372]	3886100	4155000	6	4	E2	2.5(+5)	4.2	D	1,ls
			[369.0]	3885700	4156700	4	6	E2	1.72(+5)	4.21	C–	1,ls

Ni xviii: Forbidden transitions — Continued

No.	Transition Array	Multiplet	λ (Å)	E_i (cm^{-1})	E_k (cm^{-1})	g_i	g_k	Type of transition	A_{ki} (s^{-1})	S (at. u.)	Accuracy	Source
21.	6d–8d	^{2}D – ^{2}D										
			[224.6]	3886100	4331300	6	6	E2	5.1(+5)	1.05	C	1,ls
			[224.5]	3885700	4331100	4	4	E2	4.5(+5)	0.61	C	1,ls
			[224.7]	3886100	4331100	6	4	E2	1.93(+5)	0.263	C–	1,ls
			[224.4]	3885700	4331300	4	6	E2	1.29(+5)	0.263	C–	1,ls
22.	6f–7f	^{2}F° – ^{2}F°										
			[380.2]	3906100	4169100	8	8	E2	5.8(+5)	21.9	C	1,ls
			[380.1]	3905800	4168900	6	6	E2	5.6(+5)	15.8	C	1,ls
			[380.5]	3906100	4168900	8	6	E2	9.2(+4)	2.63	C–	1,ls
			[379.8]	3905800	4169100	6	8	E2	7.0(+4)	2.63	C–	1,ls
23.	6f–8f	^{2}F° – ^{2}F°										
			[230.8]	3906100	4339400	8	8	E2	4.04(+5)	1.26	C	1,ls
			[230]	3905800	4340000	6	6	E2	4.0(+5)	0.91	C–	1,ls
			[230]	3906100	4340000	8	6	E2	6.5(+4)	0.15	D+	1,ls
			[230.6]	3905800	4339400	6	8	E2	4.86(+4)	0.151	C–	1,ls
24.	7d–8d	^{2}D – ^{2}D										
			[572.7]	4156700	4331300	6	6	E2	2.5(+5)	56	C	1,ls
			[568]	4155000	4331100	4	4	E2	2.3(+5)	32	D	1,ls
			[573.4]	4156700	4331100	6	4	E2	9.4(+4)	13.9	C–	1,ls
			[568]	4155000	4331300	4	6	E2	6.6(+4)	14	D	1,ls
25.	7f–8f	^{2}F° – ^{2}F°										
			[587.2]	4169100	4339400	8	8	E2	2.3(+5)	75	C	1,ls
			[585]	4168900	4340000	6	6	E2	2.2(+5)	54	D	1,ls
			[585]	4169100	4340000	8	6	E2	3.7(+4)	9.0	D	1,ls
			[586.5]	4168900	4339400	6	8	E2	2.7(+4)	9.0	C–	1,ls
26.	3s–3d	^{2}S – ^{2}D										
			[129.82]	0	770300	2	6	E2	8.4(+5)	0.110	C	1,ls
			[130.61]	0	765610	2	4	E2	8.2(+5)	0.074	D	1,ls
27.	3s–4d	^{2}S – ^{2}D										
			[38.513]	0	2596500	2	6	E2	2.8(+8)	0.086	D	1,ls
			[38.545]	0	2594400	2	4	E2	2.9(+8)	0.058	D	1,ls
28.	3s–5d	^{2}S – ^{2}D										
			[29.115]	0	3434600	2	6	E2	1.5(+8)	0.011	D	1,ls
29.	3p–4f	^{2}P° – ^{2}F°										
			[43.018]	342486	2667100	4	8	E2	4.92(+8)	0.345	C	1,ls
			[42.477]	311950	2666200	2	6	E2	4.07(+8)	0.201	C	1,ls
			[43.035]	342486	2666200	4	6	E2	1.1(+8)	0.057	D	1,ls
30.	3p–5f	^{2}P° – ^{2}F°										
			[31.981]	342486	3469400	4	8	E2	7.5(+7)	0.012	D	1,ls
31.	3d–4s	^{2}D – ^{2}S										
			[65.300]	770300	2301700	6	2	E2	2.7(+7)	0.038	D	1,ls
			[65.100]	765610	2301700	4	2	E2	1.8(+7)	0.025	D	1,ls

Ni XVIII: Forbidden transitions — Continued

No.	Transition Array	Multiplet	λ (Å)	E_i (cm^{-1})	E_k (cm^{-1})	g_i	g_k	Type of transition	A_{ki} (s^{-1})	S (at. u.)	Accuracy	Source
32.	$4s$–$4d$	^{2}S – ^{2}D										
			[339.2]	2301700	2596500	2	6	E2	1.06(+5)	1.70	C	1,ls
			[341.6]	2301700	2594400	2	4	E2	1.02(+5)	1.13	C	1,ls
33.	$4s$–$5d$	^{2}S – ^{2}D										
			[88.269]	2301700	3434600	2	6	E2	3.1(+7)	0.59	C	1,ls
			[88.339]	2301700	3433700	2	4	E2	3.07(+7)	0.393	C	1,ls
34.	$4s$–$6d$	^{2}S – ^{2}D										
			[63.115]	2301700	3886100	2	6	E2	2.1(+7)	0.074	D	1,ls
			[63.131]	2301700	3885700	2	4	E2	2.1(+7)	0.050	D	1,ls
35.	$4s$–$7d$	^{2}S – ^{2}D										
			[53.908]	2301700	4156700	2	6	E2	1.4(+7)	0.022	D	1,ls
			[53.97]	2301700	4155000	2	4	E2	1.4(+7)	0.015	D	1,ls
36.	$4p$–$4f$	^{2}P° – ^{2}F°										
			[437.3]	2438400	2667100	4	8	E2	2.36(+4)	1.80	C	1,ls
			[417.0]	2426400	2666200	2	6	E2	2.33(+4)	1.05	C	1,ls
			[439.0]	2438400	2666200	4	6	E2	5200	0.300	C−	1,ls
37.	$4p$–$5f$	^{2}P° – ^{2}F°										
			[96.993]	2438400	3469400	4	8	E2	5.8(+7)	2.37	C	1,ls
			[95.914]	2426400	3469000	2	6	E2	4.76(+7)	1.38	C	1,ls
			[97.031]	2438400	3469000	4	6	E2	1.28(+7)	0.394	C−	1,ls
38.	$4p$–$6f$	^{2}P° – ^{2}F°										
			[68.134]	2438400	3906100	4	8	E2	2.33(+7)	0.163	C	1,ls
			[67.595]	2426400	3905800	2	6	E2	1.9(+7)	0.095	D	1,ls
			[68.148]	2438400	3905800	4	6	E2	5.1(+6)	0.027	E	1,ls
39.	$4p$–$7f$	^{2}P° – ^{2}F°										
			[57.780]	2438400	4169100	4	8	E2	1.0(+7)	0.032	D	1,ls
			[57.389]	2426400	4168900	2	6	E2	8.5(+6)	0.019	D	1,ls
40.	$4p$–$8f$	^{2}P° – ^{2}F°										
			[52.604]	2438400	4339400	4	8	E2	5.2(+6)	0.010	D	1,ls
41.	$4f$–$5p$	^{2}F° – ^{2}P°										
			[144.7]	2667100	3358100	8	4	E2	1.87(+6)	0.282	C	1,ls
			[145.7]	2666200	3352400	6	2	E2	2.11(+6)	0.165	C	1,ls
			[144.5]	2666200	3358100	6	4	E2	3.1(+5)	0.047	D	1,ls
42.	$4f$–$6p$	^{2}F° – ^{2}P°										
			[85.027]	2667100	3843200	8	4	E2	1.1(+6)	0.012	D	1,ls
43.	$5p$–$5f$	^{2}P° – ^{2}F°										
			[898.5]	3358100	3469400	4	8	E2	6200	17.2	C	1,ls
			[857.6]	3352400	3469000	2	6	E2	6000	10.0	C	1,ls
			[901.7]	3358100	3469000	4	6	E2	1350	2.87	C−	1,ls

Ni XVIII: Forbidden transitions — Continued

No.	Transition Array	Multiplet	λ (Å)	E_i (cm^{-1})	E_k (cm^{-1})	g_i	g_k	Type of transition	A_{ki} (s^{-1})	S (at. u.)	Accuracy	Source
44.	$5p$–$6f$	$^2P°$ – $^2F°$										
			[182.5]	3358100	3906100	4	8	E2	1.06(+7)	10.2	C	1,ls
			[180.7]	3352400	3905800	2	6	E2	8.7(+6)	6.0	C	1,ls
			[182.6]	3358100	3905800	4	6	E2	2.34(+6)	1.70	C−	1,ls
45.	$5p$–$7f$	$^2P°$ – $^2F°$										
			[123.3]	3358100	4169100	4	8	E2	6.3(+6)	0.85	C	1,ls
			[122.5]	3352400	4168900	2	6	E2	5.0(+6)	0.493	C	1,ls
			[123.3]	3358100	4168900	4	6	E2	1.39(+6)	0.141	C−	1,ls
46.	$5p$–$8f$	$^2P°$ – $^2F°$										
			[101.9]	3358100	4339400	4	8	E2	3.61(+6)	0.189	C	1,ls
			[101]	3352400	4340000	2	6	E2	2.93(+6)	0.110	C	1,ls
			[102]	3358100	4340000	4	6	E2	8.1(+5)	0.032	D	1,ls
47.	$5f$–$6p$	$^2F°$ – $^2P°$										
			[267.5]	3469400	3843200	8	4	E2	9.2(+5)	2.99	C	1,ls
			[270.0]	3469000	3839400	6	2	E2	1.02(+6)	1.75	C	1,ls
			[267.2]	3469000	3843200	6	4	E2	1.54(+5)	0.499	C−	1,ls
48.	$6p$–$6f$	$^2P°$ – $^2F°$										
			[1590]	3843200	3906100	4	8	E2	1900	91	C−	1,ls
			[1510]	3839400	3905800	2	6	E2	1900	53	C−	1,ls
			[1600]	3843200	3905800	4	6	E2	400	15	D+	1,ls
49.	$6p$–$7f$	$^2P°$ – $^2F°$										
			[306.8]	3843200	4169100	4	8	E2	2.63(+6)	34.1	C	1,ls
			[303.5]	3839400	4168900	2	6	E2	2.16(+6)	19.9	C	1,ls
			[307.0]	3843200	4168900	4	6	E2	5.9(+5)	5.7	C−	1,ls
50.	$6p$–$8f$	$^2P°$ – $^2F°$										
			[201.5]	3843200	4339400	4	8	E2	1.90(+6)	3.01	C	1,ls
			[200]	3839400	4340000	2	6	E2	1.54(+6)	1.76	C−	1,ls
			[201]	3843200	4340000	4	6	E2	4.3(+5)	0.50	D+	1,ls

[a]The number in parentheses following the tabulated value indicates the power of ten by which this value has to be multiplied.

Ni xix

Ne Isoelectronic Sequence

Ground State: $1s^2 2s^2 2p^6 \, {}^1S_0$

Ionization Energy: 1541 eV = 12430000 cm^{-1}

Allowed Transitions

List of tabulated lines

Wavelength (Å)	No.	Wavelength (Å)	No.	Wavelength (Å)	No.	Wavelength (Å)	No.
9.140	12	13.779	14	45.2	33	227	28
9.153	11	14.043	13	45.3	33	237	28
9.977	22	38.8	27	46.0	39	242	28
10.110	21	38.9	26	46.4	34	256	29
10.157	20	39.1	26	77.32	2	257	30
10.283	19	39.3	26	79.81	1	258	32
10.433	18	40.5	36	80	5	305	24
11.539	10	40.7	36	82	6	348	31
11.599	9	41.132	37	86.36	4	359	23
12.435	17	42.3	38	90	7	736	25
12.656	16	42.6	35	91	7,8		
12.812	15	43.4	40	91.02	3		

A-values for numerous transitions involving an electron jump of the type $2s$–np ($n=2$–4), $2p$–ns, $2p$–nd, $3s$–np, $3p$–nd ($n=3,4$), or $3p$–$4s$ were calculated by Loulergue and Nussbaumer[1] using scaled Thomas-Fermi wavefunctions. The following configurations were included in their basis: $2s^2 2p^6$, $2s^2 2p^5 nl$, and $2s 2p^6 nl$ (for $n=3$: $l=s,p,d$; for $n=4$: $l=s,p,d,f$). Their results are quoted here, but, in cases where better wavelength data were available, their transition probabilities were first converted to line strengths, which were then reconverted to f- and A-values by using the more accurate wavelengths. Data for resonance lines were not modified, as the calculated wavelengths of Ref. 1 for these lines are fairly accurate.

Transition probabilities for a few lines for which Loulergue and Nussbaumer did not report results were taken from the work of Pokleba and Safronova,[2] who used wavefunctions calculated by a charge-expansion perturbation theory approach with allowance for mixing of configurations in which a single $2s$ or $2p$ electron is excited to an $n=3$ orbital but with no inclusion of configurations in which an electron occupies the $n=4$ shell.

Transitions involving levels of the $2p^5 3p$ and $2p^5 3d$

configurations which are indicated by Jupen and Litzen[3] to be of low to moderate purity in LS coupling in Fe XVII are excluded here, as are very weak lines. Transitions involving the corresponding levels in the $2p^5 4l$ configurations are excluded as well, as no percentage composition data were available for these levels. The pattern of levels within the $2s 2p^6 3d$ configuration resulting from the calculations of Loulergue and Nussbaumer is entirely different from that determined by Vainshtein and Safronova,[4] whose energy levels were apparently used by Pokleba and Safronova in their transition probability calculations. We have thus excluded transitions out of these levels from our tabulation.

References

[1]M. Loulergue and H. Nussbaumer, Astron. Astrophys. **45**, 125 (1975).
[2]A. K. Pokleba and U. I. Safronova, Preprint No. 11, Akad. Nauk SSSR, Ot. Ob. Fiz. Astron., Inst. Spektrosk. (Moscow, 1981).
[3]C. Jupen and U. Litzen, Phys. Scr. **30**, 112 (1984).
[4]L. A. Vainshtein and U. I. Safronova, *Spektroskopicheskie Konstanty Atomov*, 5–122 (Ed. V. B. Belyanin, Akad. Nauk SSSR, Ot. Ob. Fiz. Astron., Nauch. Sov. Spektrosk., Moscow, 1977).

Ni XIX: Allowed transitions

No.	Transition Array	Multiplet	λ (Å)	E_i (cm^{-1})	E_k (cm^{-1})	g_i	g_k	A_{ki} (10^8 s^{-1})	f_{ik}	S (at. u.)	log gf	Accuracy	Source
1.	$2s^22p^5(^2P^\circ_{3/2})3s-$ $2s2p^63s$	$(^3/_2,^1/_2)^\circ - {}^3S$											
			[79.81]	7103800	8357000	5	3	1200	0.069	0.090	-0.47	D	1
2.		$(^3/_2,^1/_2)^\circ - {}^1S$											
			77.32	7121000	8415000	3	1	830	0.025	0.019	-1.13	D	1
3.	$2s^22p^5(^2P_{1/2})3s-$ $2s2p^63s$	$(^1/_2,^1/_2)^\circ - {}^3S$											
			91.02	7257400	8357000	3	3	290	0.036	0.032	-0.97	D	1
4.		$(^1/_2,^1/_2)^\circ - {}^1S$											
			86.36	7257400	8415000	3	1	440	0.016	0.014	-1.31	D	1
5.	$2s^22p^53p-$ $2s2p^63p$	$^3S - {}^3P^\circ$											
			[80]			3	3	170	0.016	0.013	-1.31	E	1
			[80]			3	1	670	0.022	0.017	-1.19	D	1
6.		$^3D - {}^3P^\circ$											
			[82]			7	5	960	0.069	0.13	-0.32	D	1
7.	$2s^22p^54p-$ $2s2p^64p$	$^3S - {}^3P^\circ$											
			[90]			3	3	120	0.015	0.013	-1.36	E	1
			[91]			3	1	580	0.024	0.022	-1.14	E	1
8.		$^3D - {}^3P^\circ$											
			[91]			7	5	620	0.055	0.12	-0.41	E	1
9.	$2s^22p^6-2s2p^63p$	$^1S - {}^3P^\circ$											
			11.599	0	8621400	1	3	6300	0.038	0.0015	-1.42	E	1
10.		$^1S - {}^1P^\circ$	11.539	0	8666300	1	3	4.8(+4)^a	0.29	0.011	-0.54	D	1
11.	$2s^22p^6-2s2p^64p$	$^1S - {}^3P^\circ$											
			9.153	0	10930000	1	3	5200	0.020	5.9(-4)	-1.71	E	1
12.		$^1S - {}^1P^\circ$	9.140	0	10940000	1	3	3.1(+4)	0.12	0.0035	-0.93	D	1
13.	$2p^6-$ $2p^5(^2P^\circ_{3/2})3s$	$^1S - (^3/_2,^1/_2)^\circ$											
			14.043	0	7121000	1	3	1.31(+4)	0.116	0.0054	-0.93	C$-$	1
14.	$2p^6-$ $2p^5(^2P^\circ_{1/2})3s$	$^1S - (^1/_2,^1/_2)^\circ$											
			13.779	0	7257400	1	3	1.23(+4)	0.105	0.00476	-0.98	C$-$	1
15.	$2p^6-2p^53d$	$^1S - {}^3P^\circ$											
			12.812	0	7805200	1	3	1100	0.0081	3.4(-4)	-2.09	E	1

Ni XIX: Allowed transitions — Continued

No.	Transition Array	Multiplet	λ (Å)	E_i (cm^{-1})	E_k (cm^{-1})	g_i	g_k	A_{ki} (10^8 s^{-1})	f_{ik}	S (at. u.)	log gf	Accuracy	Source
16.		^{1}S – ^{3}D°											
			12.656	0	7901400	1	3	1.0(+5)	0.72	0.030	−0.14	D	1
17.		^{1}S – ^{1}P°	12.435	0	8041800	1	3	3.66(+5)	2.55	0.104	0.406	C−	1
18.	$2p^6$– $2p^5(^2$P$^°_{3/2})4s$	^{1}S – ($^3/_2$,$^1/_2$)°											
			10.433	0	9585000	1	3	5100	0.025	8.6(−4)	−1.60	D	1
19.	$2p^6$– $2p^5(^2$P$^°_{1/2})4s$	^{1}S – ($^1/_2$,$^1/_2$)°											
			10.283	0	9724800	1	3	4700	0.022	7.6(−4)	−1.65	D	1
20.	$2p^6$–$2p^54d$	^{1}S – ^{3}P°											
			10.157	0	9845400	1	3	700	0.0032	1.1(−4)	−2.49	E	1
21.		^{1}S – ^{3}D°											
			10.110	0	9891200	1	3	9.4(+4)	0.43	0.014	−0.36	D	1
22.		^{1}S – ^{1}P°	9.977	0	10020000	1	3	1.1(+5)	0.49	0.016	−0.31	D	1
23.	$2p^5(^2$P$^°_{3/2})3s$– $2p^53p$	($^3/_2$,$^1/_2$)° – ^{3}S											
			[359]			5	3	44	0.051	0.30	−0.60	D	1
24.		($^3/_2$,$^1/_2$)° – ^{3}D											
			[305]			5	7	80	0.16	0.78	−0.11	D	1
25.	$2p^5(^2$P$^°_{1/2})3s$– $2p^53p$	($^1/_2$,$^1/_2$)° – ^{3}S											
			[736]			1	3	0.060	0.0015	0.0035	−2.84	E	2
26.	$2s2p^63s$– $2s2p^64p$	^{3}S – ^{3}P°	*39.1*			3	9	3800	0.26	0.10	−0.11	D	1
			[39.1]			3	5	3900	0.15	0.058	−0.35	D	1
			[38.9]	8357000	10930000	3	3	3700	0.083	0.032	−0.60	D	1
			[39.3]			3	1	4300	0.033	0.013	−1.00	D	1
27.		^{3}S – ^{1}P°											
			[38.8]	8357000	10940000	3	3	710	0.016	0.0061	−1.32	E	1
28.	$2p^53p$–$2p^53d$	^{3}S – ^{3}P°	*238*			3	9	78	0.20	0.47	−0.22	E	1
			[227]			3	5	59	0.076	0.17	−0.64	E	1
			[237]			3	3	110	0.090	0.21	−0.57	D	1
			[242]			3	1	130	0.038	0.091	−0.94	D	1
29.		^{3}D – ^{3}P°											
			[256]			7	5	5.5	0.0039	0.023	−1.57	E	2

Ni xix: Allowed transitions — Continued

No.	Transition Array	Multiplet	λ (Å)	E_i (cm^{-1})	E_k (cm^{-1})	g_i	g_k	A_{ki} (10^8 s^{-1})	f_{ik}	S (at. u.)	log gf	Accuracy	Source
30.		^{3}D – ^{3}F°											
			[257]			7	9	120	0.16	0.93	0.04	D	1
31.		^{3}P – ^{3}P°											
			[348]			1	3	1.3	0.0071	0.0081	−2.15	E	2
32.		^{3}P – ^{3}D°											
			[258]			1	3	54	0.16	0.14	−0.79	D	2
33.	$2p^53p$– $2p^5(^2P^\circ_{3/2})4s$	^{3}S – $(^3/_2,^1/_2)°$											
			[45.3]			3	5	1000	0.051	0.023	−0.81	D	1
			[45.2]			3	3	45	0.0014	6.2(−4)	−2.38	E	1
34.		^{3}D – $(^3/_2,^1/_2)°$											
			[46.4]			7	5	2600	0.060	0.064	−0.38	D	1
35.	$2p^53p$– $2p^5(^2P^\circ_{1/2})4s$	^{3}S – $(^1/_2,^1/_2)°$											
			[42.6]			3	3	72	0.0020	8.2(−4)	−2.23	E	1
			[42.6]			3	1	93	8.4(−4)	3.5(−4)	−2.60	E	1
36.	$2p^53p$–$2p^54d$	^{3}S – ^{3}P°	*40.6*			3	9	4700	0.35	0.14	0.02	E	1
			[40.5]			3	5	3000	0.12	0.049	−0.43	E	1
			[40.7]			3	3	6400	0.16	0.064	−0.32	D	1
			[40.7]			3	1	8400	0.070	0.028	−0.68	E	1
37.		^{3}D – ^{3}F°											
			41.132			7	9	9400	0.31	0.29	0.33	D	1
38.		^{3}P – ^{3}D°											
			[42.3]			1	3	3800	0.31	0.043	−0.51	D	1
39.		^{1}S – ^{3}D°											
			[46.0]			1	3	1100	0.10	0.016	−0.98	E	1
40.		^{1}S – ^{1}P°	[43.4]			1	3	4000	0.34	0.048	−0.47	D	1

[a]The number in parentheses following the tabulated value indicates the power of ten by which this value has to be multiplied.

Ni XIX

Forbidden Transitions

The A-value for the single magnetic-dipole transition tabulated here is the result of the Hartree-Fock-Relativistic (HFR) calculations of Cowan.[1] The wavelength is the result of these same calculations and may be somewhat uncertain, as the energy of the $J=0$ level has not been determined experimentally. For the magnetic quadrupole resonance transition to the $J=2$ level of the $2p^5 3s$ configuration, we quote the A-value determined by Loulergue and Nussbaumer[2] using scaled Thomas-Fermi wavefunctions with fairly extensive allowance for configuration mixing.

References

[1]R. D. Cowan, Los Alamos Scientific Laboratory Informal Report LA–6679–MS (Jan. 1977).
[2]M. Loulergue and H. Nussbaumer, Astron. Astrophys. **45**, 125 (1975).

Ni XIX: Forbidden transitions

No.	Transition Array	Multiplet	λ (Å)	E_i (cm^{-1})	E_k (cm^{-1})	g_i	g_k	Type of transition	A_{ki} (s^{-1})	S (at. u.)	Accuracy	Source
1.	$2p^5(^2P^\circ_{3/2})3s-$ $2p^5(^2P^\circ_{1/2})3s$	$(^3/_2,^1/_2)^\circ - (^1/_2,^1/_2)^\circ$	[780]			3	1	M1	4.8(+4)^a	0.84	D+	1
2.	$2p^6-$ $2p^5(^2P^\circ_{3/2})3s$	$^1S - (^3/_2,^1/_2)^\circ$	14.077	0	7103800	1	5	M2	4.2(+5)	0.18	D+	2

^{a}The number in parentheses following the tabulated value indicates the power of ten by which this value has to be multiplied.

Ni XX

F Isoelectronic Sequence

Ground State: $1s^2 2s^2 2p^5\ ^2P^\circ_{3/2}$

Ionization Energy: $1648\,\mathrm{eV} = 13290000\,\mathrm{cm}^{-1}$

Allowed Transitions

Oscillator strengths for lines of the multiplet $2s^2 2p^5\ ^2P^\circ - 2s 2p^6\ ^2S$ are the results of the Dirac-Fock calculations of Cheng et al.,[1] which included a perturbative treatment of the Breit interaction and the Lamb shift.

For lines of the arrays $2p^5-2p^4 3s$ and $2p^5-2p^4 3d$, we quote the f-values calculated by Fawcett[2] using Cowan's Hartree-Fock-Relativistic (HFR) method and incorporating scaling of energy parameters on the basis of a least-squares fit to observed energies. Fawcett's calculations included fairly extensive allowance for configuration mixing in both odd- and even-parity states. Transitions involving levels which are indicated by Fawcett to be of low to moderate purity in LS coupling in neighboring fluorinelike ions are excluded from this compilation, as are lines characterized by very small f-values.

The ratio of A-values for the two resonance lines out of the $2s 2p^6\ ^2S_{1/2}$ level as given in Ref. 1 is in reasonably good agreement with the result of Stratton et al.[3] derived from relative-intensity measurements.

References

[1]K. T. Cheng, Y.-K. Kim, and J. P. Desclaux, At. Data Nucl. Data Tables **24**, 111 (1979).
[2]B. C. Fawcett, At. Data Nucl. Data Tables **31**, 495 (1984).
[3]B. C. Stratton, H. W. Moos, S. Suckewer, U. Feldman, J. F. Seely, and A. K. Bhatia, Phys. Rev. A **31**, 2534 (1985).

Ni xx:　Allowed transitions

No.	Transition Array	Multiplet	λ (Å)	E_i (cm^{-1})	E_k (cm^{-1})	g_i	g_k	A_{ki} (10^8 s^{-1})	f_{ik}	S (at. u.)	log gf	Accuracy	Source
1.	$2s^2 2p^5 - 2s\,2p^6$	^{2}P° – ^{2}S	86.66	47990	1202000	6	2	1450	0.0544	0.0932	−0.486	C+	1
			83.17	0	1202000	4	2	1100	0.0571	0.0625	−0.641	C+	1
			94.49	143960	1202000	2	2	368	0.0493	0.0307	−1.006	C+	1
2.	$2p^5 - 2p^4(^3\mathrm{P})3s$	^{2}P° – ^{4}P											
			13.309	0	7513700	4	6	1700	0.0068	0.0012	−1.57	E	2
			13.135	0	7613200	4	2	6000	0.0078	0.0013	−1.51	E	2
3.		^{2}P° – ^{2}P											
			13.282	143960	7673400	2	2	2.0(+4)[a]	0.054	0.0047	−0.97	D	2
			13.032	0	7673400	4	2	1.8(+4)	0.023	0.0039	−1.04	D	2
4.	$2p^5 - 2p^4(^1\mathrm{D})3s$	^{2}P° – ^{2}D	13.003	47990	7738300	6	10	1.8(+4)	0.074	0.019	−0.35	E	2
			12.927	0	7735700	4	6	1.6(+4)	0.060	0.010	−0.62	D	2
			13.161	143960	7742200	2	4	1.9(+4)	0.099	0.0086	−0.70	D	2
			[12.916]	0	7742200	4	4	1000	0.0025	4.3(−4)	−2.00	E	2
5.	$2p^5 - 2p^4(^1\mathrm{S})3s$	^{2}P° – ^{2}S	12.656	47990	7949200	6	2	2.0(+4)	0.016	0.0041	−1.01	E	2
			[12.580]	0	7949200	4	2	2800	0.0033	5.5(−4)	−1.88	E	2
			12.812	143960	7949200	2	2	1.7(+4)	0.042	0.0035	−1.08	D	2
6.	$2p^5 - 2p^4(^1\mathrm{D})3d$	^{2}P° – ^{2}S	11.942	47990	8421800	6	2	2.6(+5)	0.19	0.044	0.05	D	2
			11.874	0	8421800	4	2	2.3(+5)	0.24	0.038	−0.02	D	2
			12.079	143960	8421800	2	2	3.4(+4)	0.074	0.0059	−0.83	D	2
7.	$2p^5 - 2p^4(^1\mathrm{S})3d$	^{2}P° – ^{2}D											
			[11.61]			4	6	9600	0.029	0.0044	−0.94	D	2

[a]The number in parentheses following the value indicates the power of ten by which this value has to be multiplied.

Ni xx

Forbidden Transitions

　Line strengths for the magnetic dipole and electric quadrupole contributions to the transition between the two levels of the $2p^5$ configuration are the results of the Dirac-Fock calculations of Cheng *et al.*[1] These relativistic calculations included a perturbative treatment of the Breit interaction and the Lamb shift. The strength of the electric quadrupole transition as defined in Ref. 1 was multiplied by the factor $^2/_3$ which is needed to bring this value into conformance with the definition of quadrupole strengths used in the NBS tables.

Reference

[1]K. T. Cheng, Y.-K. Kim, and J. P. Desclaux, At. Data Nucl. Data Tables **24**, 111 (1979).

Ni xx:　Forbidden transitions

No.	Transition Array	Multiplet	λ (Å)	E_i (cm^{-1})	E_k (cm^{-1})	g_i	g_k	Type of transition	A_{ki} (s^{-1})	S (at. u.)	Accuracy	Source
1.	$2p^5 - 2p^5$	^{2}P° – ^{2}P°										
			694.64	0	143960	4	2	M1	5.35(+4)[a]	1.33	B	1
			"	"	"	4	2	E2	7.3	0.0014	D	1

[a]The number in parentheses following the tabulated value indicates the power of ten by which this value has to be multiplied.

Ni xxi

O Isoelectronic Sequence

Ground State: $1s^2 2s^2 2p^4\ ^3P_2$

Ionization Energy: $1756\ \mathrm{eV} = 14160000\ \mathrm{cm}^{-1}$

Allowed Transitions

List of tabulated lines

Wavelength (Å)	No.	Wavelength (Å)	No.	Wavelength (Å)	No.	Wavelength (Å)	No.
11.13	25	11.67	18	12.502	14	95.85	1
11.23	26	11.72	18	12.533	13	96.79	1
11.239	21	12.079	15	12.592	10	97.13	6
11.28	24	12.166	15	12.656	10	100.23	1
11.318	19	12.179	12	12.8	9	103.40	8
11.48	20,27	12.208	11	69.62	2	109.29	1
11.517	23	12.209	16	74.431	2	109.44	3
11.539	18	12.245	15	76.45	2	120.33	3
11.54	18	12.345	16	78.28	7	139.05	5
11.596	22	12.370	12	81.69	4		
11.65	17	12.454	10	88.81	1		
11.66	17	12.472	11	93.91	1		

The tabulated oscillator strengths for transitions of the arrays $2s^2 2p^4$–$2s 2p^5$ and $2s 2p^5$–$2p^6$ are the results of the multiconfiguration Dirac-Fock (MCDF) calculations of Cheng et al.[1] These relativistic calculations included a perturbative treatment of the Breit interaction and the Lamb shift. Allowance for configuration mixing was limited to the $n = 2$ complex. The results should be quite accurate, except in the case of weak lines.

Transition probabilities for lines of the $2s^2 2p^4$–$2s 2p^5$ array were calculated by Froese Fischer and Saha[2] using the multiconfiguration Hartree-Fock (MCHF) method with Breit-Pauli corrections. Their basis set included many configurations outside the $n = 2$ complex, but relativistic effects were not treated to the same degree as in Ref. 1. Line strengths derived from these two sources are in reasonably good agreement, particularly for the stronger transitions.

A few experimental data are available for this ion. Stratton et al.[3] measured ratios of transition probabilities for two pairs of transitions, one of these pairs originating from the $2s 2p^5\ ^3P^o_2$ level and the other from the $2s 2p^5\ ^3P^o_1$ level. The former agrees very well with the theoretical

data of Cheng et al.; the latter is nearly a factor of two larger than theory.

For lines of the arrays $2p^4$–$2p^3 3s$ and $2p^4$–$2p^3 3d$, we quote the f-values calculated by Fawcett[4] using Cowan's Hartree-Fock-Relativistic (HFR) method and incorporating scaling of energy parameters on the basis of a least-squares fit to observed energies. Fawcett's calculations included fairly extensive allowance for configuration mixing in both odd- and even-parity states. The weakest lines were not reported, and thus are not tabulated here. Transitions involving levels which are indicated by Fawcett to be of low to moderate purity in LS coupling in neighboring fluorinelike ions are excluded from this compilation.

References

[1]K. T. Cheng, Y.-K. Kim, and J. P. Desclaux, At. Data Nucl. Data Tables **24**, 111 (1979).

[2]C. Froese Fischer and H. P. Saha, J. Phys. B **17**, 943 (1984).

[3]B. C. Stratton, H. W. Moos, S. Suckewer, U. Feldman, J. F. Seely, and A. K. Bhatia, Phys. Rev. A **31**, 2534 (1985).

[4]B. C. Fawcett, At. Data Nucl. Data Tables **34**, 215 (1986).

Ni xxi: Allowed transitions

No.	Transition Array	Multiplet	λ (Å)	E_i (cm^{-1})	E_k (cm^{-1})	g_i	g_k	A_{ki} (10^8 s^{-1})	f_{ik}	S (at. u.)	log gf	Accuracy	Source
1.	$2s^2 2p^4$–$2s 2p^5$	^3P – $^3P°$	*96.671*	*53080*	*1087520*	9	9	640	0.089	0.256	−0.095	C	1
			95.85	0	1043300	5	5	460	0.063	0.099	−0.50	C	1
			100.23	128290	1126000	3	3	143	0.0215	0.0213	−1.190	C	1
			88.81	0	1126000	5	3	419	0.0297	0.0434	−0.83	C	1
			93.91	128290	1193140	3	1	740	0.0325	0.0301	−1.011	C	1
			109.29	128290	1043300	3	5	115	0.0344	0.0371	−0.99	C	1
			96.79	92840	1126000	1	3	190	0.079	0.025	−1.10	C	1
2.		^3P – $^1P°$											
			69.62	0	1436370	5	3	170	0.0076	0.0087	−1.42	E	1
			76.45	128290	1436370	3	3	18	0.0016	0.0012	−2.32	E	1
			[74.431]	92840	1436370	1	3	23	0.0057	0.0014	−2.24	E	1
3.		^1D – $^3P°$											
			120.33	212230	1043300	5	5	32	0.0070	0.014	−1.46	E	1
			[109.44]	212230	1126000	5	3	1.0	1.1(−4)[a]	2.0(−4)	−3.26	E	1
4.		^1D – $^1P°$	81.69	212230	1436370	5	3	1700	0.102	0.137	−0.292	C	1
5.		^1S – $^3P°$											
			[139.05]	406820	1126000	1	3	11	0.0093	0.0043	−2.03	E	1
6.		^1S – $^1P°$	97.13	406820	1436370	1	3	120	0.050	0.016	−1.30	C	1
7.	$2s 2p^5$–$2p^6$	$^3P°$ – ^1S											
			78.28	1126000	2403490	3	1	230	0.0070	0.0054	−1.68	E	1
8.		$^1P°$ – ^1S	103.40	1436370	2403490	3	1	1800	0.098	0.10	−0.53	C	1
9.	$2p^4$ – $2p^3(^4S°)3s$	^3P – $^5S°$											
			[12.8]			5	5	2000	0.0050	0.0011	−1.60	E	4
10.		^3P – $^3S°$	*12.537*	*53080*	*8029700*	9	3	4.8(+4)	0.038	0.014	−0.47	D	4
			[12.454]	0	8029700	5	3	3.3(+4)	0.046	0.0094	−0.64	D	4
			12.656	128290	8029700	3	3	6700	0.016	0.0020	−1.32	D	4
			12.592	92840	8029700	1	3	7400	0.053	0.0022	−1.28	D	4
11.	$2p^4$ – $2p^3(^2D°)3s$	^3P – $^3D°$											
			12.208	0	8191300	5	7	1.41(+4)	0.0442	0.0089	−0.66	C	4
			12.472	128290	8146300	3	3	1.8(+4)	0.042	0.0052	−0.90	D−	4
12.		^3P – $^1D°$											
			[12.179]	0	8210900	5	5	2100	0.0046	9.2(−4)	−1.64	E	4
			12.370	128290	8210900	3	5	3700	0.014	0.0017	−1.38	E	4
13.	$2p^4$ – $2p^3(^2D°)3s$	^1D – $^3D°$											
			[12.533]	212230	8191300	5	7	1800	0.0060	0.0012	−1.52	E	4
14.		^1D – $^1D°$	12.502	212230	8210900	5	5	2.8(+4)	0.066	0.014	−0.48	D−	4

Ni XXI: Allowed transitions — Continued

No.	Transition Array	Multiplet	λ (Å)	E_i (cm^{-1})	E_k (cm^{-1})	g_i	g_k	A_{ki} (10^8 s^{-1})	f_{ik}	S (at. u.)	log gf	Accuracy	Source
15.	$2p^4 - 2p^3(^2$P$)3s$	^{3}P – ^{3}P°											
			12.245	128290	8294900	3	1	1.5(+4)	0.011	0.0013	−1.48	C	4
			12.079	128290	8407100	3	5	9100	0.033	0.0039	−1.00	E	4
			[12.166]	92840	8312600	1	3	1.0(+4)	0.067	0.0027	−1.17	D−	4
16.		^{1}D – ^{3}P°											
			12.209	212230	8407100	5	5	8900	0.020	0.0040	−1.00	E	4
			12.345	212230	8312600	5	3	8000	0.011	0.0022	−1.26	E	4
17.	$2p^4 - 2p^3(^2$P°$)3d$	^{3}P – ^{5}D°											
			[11.66]			5	5	5400	0.011	0.0021	−1.26	E	4
			[11.65]			5	3	6700	0. 0082	0.0016	−1.39	E	4
18.	$2p^4 - 2p^3(^4$S°$)3d$	^{3}P – ^{3}D°											
			11.539	0	8666300	5	7	1.2(+5)	0.33	0.063	0.22	D−	4
			[11.67]			1	3	8.0(+4)	0.49	0.019	−0.31	D	4
			[11.72]			3	3	2.3(+4)	0.048	0.0056	−0.84	D	4
			[11.54]			5	3	4200	0.0050	9.5(−4)	−1.60	D−	4
19.	$2p^4 - 2p^3(^2$D°$)3d$	^{3}P – ^{3}D°											
			11.318	0	8835500	5	7	2.8(+5)	0.76	0.14	0.58	D−	4
20.		^{3}P – ^{3}P°											
			[11.48]			3	1	1.1(+5)	0.075	0.0085	−0.65	D−	4
21.	$2p^4 - 2p^3(^2$D°$)3d$	^{3}P – ^{1}F°											
			11.239	0	8895000	5	7	5.7(+4)	0.15	0.028	−0.12	E	4
22.		^{1}D – ^{3}D°											
			[11.596]	212230	8835500	5	7	6400	0.018	0.0034	−1.05	E	4
23.		^{1}D – ^{1}F°	11.517	212230	8895000	5	7	1.4(+5)	0.39	0.074	0.29	D−	4
24.	$2p^4 - 2p^3(^2$P°$)3d$	^{3}P – ^{3}P°											
			[11.28]			3	1	2.2(+5)	0.14	0.016	−0.38	D−	4
25.	$2p^4 - 2p^3(^2$P°$)3d$	^{3}P – ^{1}P°											
			[11.13]			3	3	1.7(+4)	0.031	0.0034	−1.03	E	4
26.		^{1}D – ^{1}P°	[11.23]			5	3	1.7(+4)	0.019	0.0035	−1.02	D	4
27.		^{1}S – ^{1}P°	[11.48]			1	3	4.0(+5)	2.4	0.091	0.38	D	4

[a]The number in parenthesis following the tabulated value indicates the power of ten by which this value has to be multiplied.

Ni XXI

Forbidden Transitions

Line strengths tabulated for magnetic dipole and electric quadrupole transitions within the $2p^4$ configuration are the results of the multiconfiguration Dirac-Fock (MCDF) calculations of Cheng et al.[1] These relativistic calculations included a perturbative treatment of the Breit interaction and the Lamb shift. Allowance for configuration mixing was limited to the $n=2$ complex. Strengths of electric quadrupole transitions as defined in Ref. 1 were multiplied by the factor $^2/_3$ which is needed to bring these values into conformance with the definition of quadrupole strengths used in the NBS tables.

Transition probabilities for these same lines were calculated by Froese Fischer and Saha[2] using the multiconfiguration Hartree-Fock (MCHF) method with Breit-Pauli corrections. Their basis included many configurations outside the $n=2$ complex, but relativistic effects were not treated to the same degree as in Ref. 1. Line strengths derived from these data are in quite good agreement with the data of Cheng et al. For this ion of the oxygen isoelectronic sequence, correlation effects due to mixing with configurations outside the complex were found by Froese Fischer and Saha to be rather small, as shown by a comparison of the results of their calculations employing an extensive basis to those derived by the same technique but limited to configurations within the $n=2$ complex.

The weakest lines are excluded from this compilation, as their transition probabilities are considered to be very uncertain.

References

[1]K. T. Cheng, Y.-K Kim, and J. P. Desclaux, At. Data Nucl. Data Tables **24**, 111 (1979).

[2]C. Froese Fischer and H. P. Saha, Phys. Rev. A **28**, 3169 (1983).

Ni XXI: Forbidden transitions

No.	Transition Array	Multiplet	λ (Å)	E_i (cm^{-1})	E_k (cm^{-1})	g_i	g_k	Type of transition	A_{ki} (s^{-1})	S (at. u.)	Accuracy	Source
1.	$2p^4$–$2p^4$	^{3}P – ^{3}P										
			779.5	0	128290	5	3	M1	4.14(+4)[a]	2.18	C	1
			"	"	"	5	3	E2	2.5	0.0013	E	1
			2818.2	92840	128290	1	3	M1	560	1.4	D	1
2.		^{3}P – ^{1}D										
			471.15	0	212230	5	5	M1	4.2(+4)	0.82	D	1
			"	"	"	5	5	E2	16	0.0011	E	1
			1191.1	128290	212230	3	5	M1	1000	0.32	D	1
3.		^{3}P – ^{1}S										
			[359.03]	128290	406820	3	1	M1	3.3(+5)	0.57	D	1
4.		^{1}D – ^{1}S	[513.90]	212230	406820	5	1	E2	98	0.0021	E	1

[a]The number in parentheses following the tabulated value indicates the power of ten by which this value has to be multiplied.

Ni XXII

N Isoelectronic Sequence

Ground State:　$1s^2 2s^2 2p^3 \, {}^4S^{\circ}_{3/2}$

Ionization Energy:　$1894 \, \text{eV} = 15280000 \, \text{cm}^{-1}$

Allowed Transitions

List of tabulated lines

Wavelength (Å)	No.	Wavelength (Å)	No.	Wavelength (Å)	No.	Wavelength (Å)	No.
71.48	4	85.02	2	103.31	1	127.30	5
71.54	13	85.86	14	103.43	6	128.86	10
72.52	8	88.00	12	105.88	11	136.30	5
72.837	13	91.20	12	106.04	1	144.80	5
74.37	3	95.604	6	106.16	16	150.26	9
74.49	13	95.95	11	114.45	10	152.89	16
80.16	13	98.16	6	117.91	1	156.56	5
80.55	8	98.58	14	118.21	15	184.19	9
81.04	12	100.12	12	123.38	5	223.22	9
81.794	13	100.38	15	124.31	16		
84.06	8	100.60	6	124.48	10		
84.24	7	101.31	14	126.32	16		

The tabulated oscillator strengths for transitions of the arrays $2s^2 2p^3$–$2s\,2p^4$ and $2s\,2p^4$–$2p^5$ are the results of the multiconfiguration Dirac-Fock (MCDF) calculations of Cheng et al.[1] These relativistic calculations included a perturbative treatment of the Breit interaction and the Lamb shift. The results should be quite accurate, except in the case of weak lines. (A few very weak lines have been omitted from this tabulation.)

Reference

[1]K. T. Cheng, Y.-K. Kim, and J. P. Desclaux, At. Data Nucl. Data Tables **24**, 111 (1979).

Ni XXII: Allowed transitions

No.	Transition Array	Multiplet	λ (Å)	E_i (cm^{-1})	E_k (cm^{-1})	g_i	g_k	A_{ki} (10^8 s^{-1})	f_{ik}	S (at. u.)	log gf	Accuracy	Source
1.	$2s^2 2p^3$–$2s\,2p^4$	$^4S^{\circ}$ – ^4P	*111.15*	0	*899720*	4	12	192	0.107	0.156	−0.370	C	1
			117.91	0	848100	4	6	146	0.0458	0.071	−0.74	C	1
			106.04	0	943040	4	4	236	0.0398	0.056	−0.80	C	1
			103.31	0	967960	4	2	266	0.0213	0.0290	−1.070	C	1
2.		$^4S^{\circ}$ – ^2D											
			85.02	0	1176180	4	4	47	0.0051	0.0057	−1.69	E	1
3.		$^4S^{\circ}$ – ^2S											
			74.37	0	1344630	4	2	39	0.0016	0.0016	−2.19	E	1
4.		$^4S^{\circ}$ – ^2P											
			71.48	0	1398940	4	4	76	0.0058	0.0055	−1.63	E	1

Ni XXII: Allowed transitions — Continued

No.	Transition Array	Multiplet	λ (Å)	E_i (cm^{-1})	E_k (cm^{-1})	g_i	g_k	A_{ki} (10^8 s^{-1})	f_{ik}	S (at. u.)	log gf	Accu-racy	Source
5.		^{2}D° – ^{4}P											
			[156.56]	209380	848100	6	6	4.4	0.0016	0.0049	−2.02	E	1
			[127.30]	157480	943040	4	4	2.9	7.0(−4)^a	0.0012	−2.55	E	1
			[136.30]	209380	943040	6	4	1.1	2.1(−4)	5.7(−4)	−2.90	E	1
			[123.38]	157480	967960	4	2	7.4	8.4(−4)	0.0014	−2.47	E	1
			[144.80]	157480	848100	4	6	13	0.0063	0.012	−1.60	E	1
6.		^{2}D° – ^{2}D	*99.609*	*188620*	*1192550*	10	10	450	0.067	0.22	−0.17	C−	1
			100.60	209380	1203460	6	6	390	0.059	0.12	−0.45	C	1
			98.16	157480	1176180	4	4	520	0.075	0.097	−0.52	C	1
			[103.43]	209380	1176180	6	4	0.48	5.1(−5)	1.0(−4)	−3.51	E	1
			[95.604]	157480	1203460	4	6	1.2	2.4(−4)	3.0(−4)	−3.02	E	1
7.		^{2}D° – ^{2}S											
			84.24	157480	1344630	4	2	560	0.030	0.033	−0.92	E	1
8.		^{2}D° – ^{2}P	*79.607*	*188620*	*1444790*	10	6	1100	0.061	0.16	−0.21	C	1
			84.06	209380	1398940	6	4	1200	0.084	0.14	−0.30	C	1
			72.52	157480	1536480	4	2	284	0.0112	0.0107	−1.349	C	1
			80.55	157480	1398940	4	4	124	0.0121	0.0128	−1.315	C	1
9.		^{2}P° – ^{4}P											
			[223.22]	400120	848100	4	6	0.29	3.3(−4)	9.7(−4)	−2.88	E	1
			[184.19]	400120	943040	4	4	2.8	0.0014	0.0034	−2.25	E	1
			[150.26]	302450	967960	2	2	5.3	0.0018	0.0018	−2.44	E	1
10.		^{2}P° – ^{2}D	*121.21*	*367560*	*1192550*	6	10	57	0.0208	0.0499	−0.90	C−	1
			124.48	400120	1203460	4	6	67	0.0233	0.0382	−1.031	C	1
			114.45	302450	1176180	2	4	30.3	0.0119	0.0090	−1.62	C	1
			[128.86]	400120	1176180	4	4	6.4	0.0016	0.0027	−2.19	D	1
11.		^{2}P° – ^{2}S	*102.35*	*367560*	*1344630*	6	2	390	0.020	0.041	−0.91	C−	1
			105.88	400120	1344630	4	2	14	0.0012	0.0017	−2.32	D	1
			95.95	302450	1344630	2	2	440	0.061	0.039	−0.91	C	1
12.		^{2}P° – ^{2}P	*92.831*	*367560*	*1444790*	6	6	500	0.065	0.119	−0.410	C−	1
			100.12	400120	1398940	4	4	102	0.0153	0.0202	−1.213	C	1
			81.04	302450	1536480	2	2	40	0.0039	0.0021	−2.11	D	1
			88.00	400120	1536480	4	2	1200	0.068	0.079	−0.57	C	1
			91.20	302450	1398940	2	4	119	0.0297	0.0178	−1.226	C	1
13.	$2s\,2p^4$–$2p^5$	^{4}P – ^{2}P°											
			74.49	848100	2190550	6	4	52	0.0029	0.0043	−1.76	E	1
			71.54	943040	2340890	4	2	6.0	2.3(−4)	2.2(−4)	−3.04	E	1
			80.16	943040	2190550	4	4	34	0.0033	0.0035	−1.88	E	1
			[72.837]	967960	2340890	2	2	15	0.0012	5.8(−4)	−2.62	E	1
			[81.794]	967960	2190550	2	4	11	0.0023	0.0012	−2.34	E	1
14.		^{2}D – ^{2}P°	*95.410*	*1192550*	*2240660*	10	6	680	0.056	0.176	−0.252	C	1
			101.31	1203460	2190550	6	4	483	0.0495	0.099	−0.53	C	1
			85.86	1176180	2340890	4	2	490	0.0271	0.0306	−0.96	C	1
			98.58	1176180	2190550	4	4	245	0.0357	0.0463	−0.85	C	1

Ni XXII: Allowed transitions — Continued

No.	Transition Array	Multiplet	λ (Å)	E_i (cm^{-1})	E_k (cm^{-1})	g_i	g_k	A_{ki} (10^8 s^{-1})	f_{ik}	S (at. u.)	log gf	Accuracy	Source
15.		^{2}S – ^{2}P°	*111.60*	1344630	*2240660*	2	6	99	0.056	0.0409	−0.95	C−	1
			118.21	1344630	2190550	2	4	111	0.0463	0.0360	−1.033	C	1
			[100.38]	1344630	2340890	2	2	49	0.0074	0.0049	−1.83	D	1
16.		^{2}P – ^{2}P°	*125.65*	*1444790*	*2240660*	6	6	460	0.11	0.27	−0.19	C	1
			126.32	1398940	2190550	4	4	330	0.080	0.13	−0.49	C	1
			124.31	1536480	2340890	2	2	370	0.085	0.070	−0.77	C	1
			106.16	1398940	2340890	4	2	510	0.0435	0.061	−0.76	C	1
			[152.89]	1536480	2190550	2	4	15.1	0.0106	0.0107	−1.67	C	1

ªThe number in parentheses following the tabulated value indicates the power of ten by which this value has to be multiplied.

Ni XXII

Forbidden Transitions

Line strengths tabulated for magnetic dipole and electric quadrupole transitions within the $2p^3$ configuration are the results of the multiconfiguration Dirac-Fock (MCDF) calculations of Cheng et al.[1] These relativistic calculations included a perturbative treatment of the Breit interaction and the Lamb shift. Allowance for configuration mixing was limited to the $n=2$ complex. Strengths of electric quadrupole transitions as defined in Ref. 1 were multiplied by the factor $^2/_3$ which is needed to bring these values into conformance with the definition of quadrupole strengths used in the NBS tables. The weakest lines are excluded from this compilation, as their strengths are considered to be very uncertain.

A-values for the M1 and E2 components of the single transition within the $2p^5$ configuration were obtained by applying Z-expansion formulas published by Oboladze and Safronova.[2]   Their values for the magnetic dipole contribution to this line are in very good agreement with the results of the scaled Thomas-Fermi calculations of Bhatia et al.[3] and Bhatia[4] for nitrogenlike Ti and Mn, respectively. It is not clear whether Oboladze and Safronova incorporated configuration interaction into their calculations. Thus the A-value for the E2 contribution should be considered rather uncertain.

References

[1]K. T. Cheng, Y.-K. Kim, and J. P. Desclaux, At. Data Nucl. Data Tables **24**, 111 (1979).

[2]N. S. Oboladze and U. I. Safronova, Opt. Spectrosc. (USSR) **48**, 469 (1980).

[3]A. K. Bhatia, U. Feldman, and G. A. Doschek, J. Appl. Phys. **51**, 1464 (1980).

[4]A. K. Bhatia, J. Appl. Phys. **53**, 59 (1982).

Ni XXII: Forbidden transitions

No.	Transition Array	Multiplet	λ (Å)	E_i (cm^{-1})	E_k (cm^{-1})	g_i	g_k	Type of transition	A_{ki} (s^{-1})	S (at. u.)	Accuracy	Source
1.	$2p^3$–$2p^3$	$^4S°$ – $^2D°$										
			477.6	0	209380	4	6	M1	4500	0.11	D	1
			634.8	0	157480	4	4	M1	4.2(+4)^a	1.6	D	1
2.		$^4S°$ – $^2P°$										
			[249.93]	0	400120	4	4	M1	4.8(+4)	0.11	D	1
			[330.63]	0	302450	4	2	M1	7.8(+4)	0.21	D	1
3.		$^2D°$ – $^2D°$										
			[1927]	157480	209380	4	6	M1	1040	1.65	C	1
4.		$^2D°$ – $^2P°$										
			[1074]	209380	302450	6	2	E2	0.88	0.0015	E	1
			[524.27]	209380	400120	6	4	M1	3.0(+4)	0.64	D	1
			"	"	"	6	4	E2	37	0.0035	E	1
			[689.80]	157480	302450	4	2	M1	9000	0.22	D	1
			"	"	"	4	2	E2	8.6	0.0016	E	1
			[412.13]	157480	400120	4	4	M1	1.2(+5)	1.2	D	1
5.		$^2P°$ – $^2P°$										
			[1024]	302450	400120	2	4	M1	5700	0.90	C–	1
6.	$2p^5$–$2p^5$	$^2P°$ – $^2P°$										
			[665.16]	2190550	2340890	4	2	M1	6.0(+4)	1.3	C+	2
			"	"	"	4	2	E2	8.2	0.0013	E	2

^{a}The number in parentheses following the tabulated value indicates the power of ten by which this value has to be multiplied.

Ni XXIII

C Isoelectronic Sequence

Ground State: $1s^2 2s^2 2p^2\ ^3P_0$

Ionization Energy: $2011\ \text{eV} = 16220000\ \text{cm}^{-1}$

Allowed Transitions

List of tabulated lines

Wavelength (Å)	No.	Wavelength (Å)	No.	Wavelength (Å)	No.	Wavelength (Å)	No.
70.752	16	92.32	17	109.06	3	135.42	2
74.07	6	92.75	3	111.23	19	136.47	2
77.027	6	99.812	20	111.78	19	137.55	22
78.21	16	100.42	15	111.86	2	143.89	25
78.751	21	100.50	3	112.55	29	161.15	28
79.99	4	102.08	10	120.51	23	162.30	7
83.707	5	102.50	20	126.54	2	162.74	13
84.721	18	103.07	17	127.14	14	162.85	22
87.50	5	103.23	3	127.21	19	173.86	7
87.66	4	103.67	17	127.46	2	175.59	7
87.77	18	104.70	3	128.22	8	178.51	25
88.11	11	106.02	3	128.30	2	185.33	27
90.49	17	107.00	19	128.87	26	209.98	1
90.96	17	108.03	9	131.60	19	232.37	12
91.094	24	108.27	17	132.69	8	235.61	1
91.83	4	108.59	19	133.54	22	247.04	27

The tabulated oscillator strengths for transitions of the arrays $2s^2 2p^2$–$2s 2p^3$ and $2s 2p^3$–$2p^4$ are the results of the multiconfiguration Dirac-Fock (MCDF) calculations of Cheng et al.[1] These relativistic calculations included a perturbative treatment of the Breit interaction and the Lamb shift. Allowance for configuration mixing was limited to the $n=2$ complex. The results should be quite accurate, except in the case of weak lines. (A few very weak lines have been omitted from this tabulation.)

Transition probabilities for lines of the $2s^2 2p^2$–$2s 2p^3$ array were calculated by Froese Fischer and Saha[2] using the multiconfiguration Hartree-Fock (MCHF) method with Breit-Pauli corrections. Their basis included many configurations outside the $n=2$ complex, but relativistic effects we not treated to the same degree as in Ref. 1.

Line strengths derived from these two sources are in reasonably good agreement, particularly for the stronger transitions.

Stratton et al.[3] measured the ratio of A-values for two lines out of the $2s 2p^3\ ^3S_1^\circ$ level. Their result agrees very well with the theoretical data of Cheng et al.

References

[1]K. T. Cheng, Y.-K. Kim, and J. P. Desclaux, At. Data Nucl. Data Tables **24**, 111 (1979).
[2]C. Froese Fischer and H. P. Saha, Phys. Scr. **32**, 181 (1985).
[3]B. C. Stratton, H. W. Moos, S. Suckewer, U. Feldman, J. F. Seely, and A. K. Bhatia, Phys. Rev. A **31**, 2534 (1985).

Ni XXIII: Allowed transitions

No.	Transition Array	Multiplet	λ (Å)	E_i (cm^{-1})	E_k (cm^{-1})	g_i	g_k	A_{ki} (10^8 s^{-1})	f_{ik}	S (at. u.)	log gf	Accuracy	Source
1.	$2s^2 2p^2$–$2s 2p^3$	^3P – $^5S^\circ$											
			[235.61]	161190	585620	5	5	0.66	5.5(−4)^a	0.0021	−2.56	E	1
			[209.98]	109390	585620	3	5	0.85	9.4(−4)	0.0019	−2.55	E	1

Ni xxiii: Allowed transitions — Continued

No.	Transition Array	Multiplet	λ (Å)	E_i (cm^{-1})	E_k (cm^{-1})	g_i	g_k	A_{ki} (10^8 s^{-1})	f_{ik}	S (at. u.)	log gf	Accuracy	Source
2.		³P – ³D°	*126.32*	*126010*	*917630*	9	15	100	0.040	0.15	−0.44	D	1
			128.30	161190	940610	5	7	74	0.0257	0.054	−0.89	C	1
			126.54	109390	899650	3	5	120	0.0482	0.060	−0.84	C	1
			111.86	0	893970	1	3	170	0.098	0.036	−1.01	C	1
			[135.42]	161190	899650	5	5	1.0	2.8(−4)	6.2(−4)	−2.85	E	1
			127.46	109390	893970	3	3	4.9	0.0012	0.0015	−2.44	D	1
			[136.47]	161190	893970	5	3	2.0	3.3(−4)	7.4(−4)	−2.78	E	1
3.		³P – ³P°	*103.61*	*126010*	*1091190*	9	9	306	0.0492	0.151	−0.354	C−	1
			106.02	161190	1104380	5	5	287	0.0484	0.084	−0.62	C	1
			103.23	109390	1078110	3	3	240	0.0384	0.0391	−0.94	C	1
			109.06	161190	1078110	5	3	29	0.0031	0.0056	−1.81	D	1
			104.70	109390	1064460	3	1	294	0.0161	0.0166	−1.316	C	1
			[100.50]	109390	1104380	3	5	1.1	2.9(−4)	2.9(−4)	−3.06	E	1
			92.75	0	1078110	1	3	45.2	0.0175	0.0053	−1.76	C	1
4.		³P – ³S°	*88.956*	*126010*	*1250160*	9	3	1170	0.0463	0.122	−0.380	C	1
			91.83	161190	1250160	5	3	750	0.057	0.086	−0.55	C	1
			87.66	109390	1250160	3	3	280	0.0322	0.0279	−1.015	C	1
			79.99	0	1250160	1	3	107	0.0308	0.0081	−1.51	C	1
			87.50	161190	1304040	5	5	120	0.014	0.020	−1.15	E	1
5.		³P – ¹D°											
			[83.707]	109390	1304040	3	5	6.9	0.0012	9.9(−4)	−2.44	E	1
6.		³P – ¹P°											
			[77.027]	161190	1459440	5	3	5.1	2.7(−4)	3.4(−4)	−2.87	E	1
			74.07	109390	1459440	3	3	72	0.0059	0.0043	−1.75	E	1
7.		¹D – ³D°											
			[162.30]	324460	940610	5	7	16	0.0086	0.023	−1.37	E	1
			[173.86]	324460	899650	5	5	0.40	1.8(−4)	5.2(−4)	−3.05	E	1
			[175.59]	324460	893970	5	3	2.8	7.8(−4)	0.0023	−2.41	E	1
8.		¹D – ³P°											
			[128.22]	324460	1104380	5	5	1.9	4.7(−4)	9.9(−4)	−2.63	E	1
			[132.69]	324460	1078110	5	3	2.5	3.9(−4)	8.5(−4)	−2.71	E	1
9.		¹D – ³S°	[108.03]	324460	1250160	5	3	7.0	7.3(−4)	0.0013	−2.44	E	1
10.		¹D – ¹D°	102.08	324460	1304040	5	5	530	0.083	0.14	−0.38	C	1
11.		¹D – ¹P°	88.11	324460	1459440	5	3	830	0.058	0.084	−0.54	C	1
12.		¹S – ³D°											
			[232.37]	463620	893970	1	3	0.66	0.0016	0.0012	−2.80	E	1
13.		¹S – ³P°											
			[162.74]	463620	1078110	1	3	2.4	0.0028	0.0015	−2.55	E	1
14.		¹S – ³S°	[127.14]	463620	1250160	1	3	10	0.0075	0.0031	−2.12	E	1
15.		¹S – ¹P°	100.42	463620	1459440	1	3	210	0.096	0.032	−1.02	C	1

Ni XXIII: Allowed transitions — Continued

No.	Transition Array	Multiplet	λ (Å)	E_i (cm^{-1})	E_k (cm^{-1})	g_i	g_k	A_{ki} (10^8 s^{-1})	f_{ik}	S (at. u.)	log gf	Accuracy	Source
16.	$2s\,2p^3$–$2p^4$	^{5}S° – ^{3}P											
			78.21	585620	1864230	5	5	29	0.0027	0.0035	−1.87	E	1
			[70.752]	585620	1999010	5	3	3.8	1.7(−4)	2.0(−4)	−3.07	E	1
17.		^{3}D° – ^{3}P	99.595	917630	1921700	15	9	530	0.0474	0.233	−0.148	C	1
			108.27	940610	1864230	7	5	332	0.0417	0.104	−0.53	C	1
			90.96	899650	1999010	5	3	250	0.0186	0.0278	−1.032	C	1
			92.32	893970	1977150	3	1	439	0.0187	0.0171	−1.251	C	1
			103.67	899650	1864230	5	5	178	0.0286	0.0488	−0.84	C	1
			90.49	893970	1999010	3	3	177	0.0217	0.0194	−1.186	C	1
			103.07	893970	1864230	3	5	60	0.0158	0.0161	−1.324	C	1
18.		^{3}D° – ^{1}D											
			87.77	940610	2079990	7	5	90	0.0074	0.015	−1.29	E	1
			[84.721]	899650	2079990	5	5	9.3	0.0010	0.0014	−2.30	E	1
19.		^{3}P° – ^{3}P	120.41	1091190	1921700	9	9	153	0.0334	0.119	−0.52	C−	1
			131.60	1104380	1864230	5	5	39.7	0.0103	0.0223	−1.288	C	1
			[108.59]	1078110	1999010	3	3	8.5	0.0015	0.0016	−2.35	D	1
			[111.78]	1104380	1999010	5	3	219	0.0246	0.0453	−0.91	C	1
			[111.23]	1078110	1977150	3	1	226	0.0140	0.0154	−1.377	C	1
			127.21	1078110	1864230	3	5	44.0	0.0178	0.0224	−1.272	C	1
			[107.00]	1064460	1999010	1	3	64	0.0329	0.0116	−1.483	C	1
20.		^{3}P° – ^{1}D											
			102.50	1104380	2079990	5	5	57	0.0089	0.015	−1.35	E	1
			[99.812]	1078110	2079990	3	5	23	0.0058	0.0057	−1.76	E	1
21.		^{3}P° – ^{1}S											
			[78.751]	1078110	2347930	3	1	74	0.0023	0.0018	−2.16	E	1
22.		^{3}S° – ^{3}P	148.91	1250160	1921700	3	9	119	0.119	0.175	−0.447	C	1
			[162.85]	1250160	1864230	3	5	72	0.0474	0.076	−0.85	C	1
			133.54	1250160	1999010	3	3	186	0.0497	0.066	−0.83	C	1
			137.55	1250160	1977150	3	1	253	0.0239	0.0325	−1.144	C	1
23.		^{3}S° – ^{1}D	[120.51]	1250160	2079990	3	5	0.39	1.4(−4)	1.7(−4)	−3.38	E	1
24.		^{3}S° – ^{1}S	[91.094]	1250160	2347930	3	1	70	0.0029	0.0026	−2.06	E	1
25.		^{1}D° – ^{3}P											
			[178.51]	1304040	1864230	5	5	9.8	0.0047	0.014	−1.63	E	1
			[143.89]	1304040	1999010	5	3	8.1	0.0015	0.0036	−2.12	E	1
26.		^{1}D° – ^{1}D	128.87	1304040	2079990	5	5	402	0.100	0.212	−0.301	C	1
27.		^{1}P° – ^{3}P											
			[247.04]	1459440	1864230	3	5	1.2	0.0018	0.0044	−2.27	E	1
			[185.33]	1459440	1999010	3	3	11	0.0055	0.010	−1.78	E	1
28.		^{1}P° – ^{1}D	[161.15]	1459440	2079990	3	5	58	0.0378	0.060	−0.95	C	1
29.		^{1}P° – ^{1}S	112.55	1459440	2347930	3	1	1000	0.063	0.070	−0.72	C	1

^{a}The number in parentheses following the tabulated value indicates the power of ten by which this value has to be multiplied.

Ni XXIII

Forbidden Transitions

Line strengths tabulated for magnetic dipole and electric quadrupole transitions within the $2p^2$ configuration are the results of the multiconfiguration Dirac-Fock (MCDF) calculations of Cheng et al.[1] These relativistic calculations included a perturbative treatment of the Breit interaction and the Lamb shift. Allowance for configuration mixing was limited to the $n = 2$ complex. Strengths of electric quadrupole transitions as defined in Ref. 1 were multiplied by the factor $^2/_3$ which is needed to bring these values into conformance with the definition of quadrupole strengths used in the NBS tables. The weakest lines are excluded from this compilation, as their strengths are considered to be very uncertain.

Transition probabilities for these same lines were calculated by Froese Fischer and Saha[2] using the multicon-

figuration Hartree-Fock (MCHF) method with Breit-Pauli corrections. Their basis included many configurations outside the $n = 2$ complex, but relativistic effects were not treated to the same degree as in Ref. 1. Line strengths derived from these data are in good agreement with the data of Cheng et al.

References

[1]K. T. Cheng, Y.-K. Kim, and J. P. Desclaux, At. Data Nucl. Data Tables **24**, 111 (1979).

[2]C. Froese Fischer and H. P. Saha, Phys. Scr. **32**, 181 (1985).

Ni XXIII: Forbidden transitions

No.	Transition Array	Multiplet	λ (Å)	E_i (cm^{-1})	E_k (cm^{-1})	g_i	g_k	Type of transition	A_{ki} (s^{-1})	S (at. u.)	Accuracy	Source
1.	$2p^2$–$2p^2$	^3P – ^3P										
			1915.0	109390	161190	3	5	M1	1320	1.72	C	1
			911.0	0	109390	1	3	M1	2.07(+4)ᵃ	1.74	C	1
2.		^3P – ^1D										
			614.8	161190	324460	5	5	M1	3.7(+4)	1.6	C	1
			"	"	"	5	5	E2	7.3	0.0019	E	1
			465.4	109390	324460	3	5	M1	4.1(+4)	0.77	D	1
3.		^3P – ^1S										
			[282.30]	109390	463620	3	1	M1	3.0(+5)	0.25	D	1
4.		^1D – ^1S	[718.60]	324460	463620	5	1	E2	16	0.0018	E	1

ᵃThe number in parentheses following the tabulated value indicates the power of ten by which this value has to be multiplied.

Ni XXIV

B Isoelectronic Sequence

Ground State: $1s^2 2s^2 2p \ ^2P^\circ_{1/2}$

Ionization Energy: $2131 \ \text{eV} = 17190000 \ \text{cm}^{-1}$

Allowed Transitions

List of tabulated lines

Wavelength (Å)	No.	Wavelength (Å)	No.	Wavelength (Å)	No.	Wavelength (Å)	No.
75.083	7	102.89	6	134.53	16	213.15	11
75.580	7	103.43	5	134.73	9	218.35	1
79.185	7	103.53	4	135.47	9	221.06	15
87.50	3	104.64	3	137.01	13	224.02	1
88.54	4	106.68	10	138.80	2	227.82	12
88.977	6	109.03	13	143.30	9	264.98	1
92.138	6	113.14	5	153.47	16	338.62	14
96.070	6	118.52	2	156.70	13	339.66	1
97.17	10	121.15	13	159.69	12	354.74	11
98.39	6	122.72	5	172.09	8		
101.13	10	126.25	3	184.92	8		
102.11	3	127.78	9	206.88	12		

The tabulated oscillator strengths for transitions of the arrays $2s^2 2p$–$2s 2p^2$ and $2s 2p^2$–$2p^3$ are the results of the multiconfiguration Dirac-Fock (MCDF) calculations of Cheng et al.[1] These relativistic calculations included a perturbative treatment of the Breit interaction and the Lamb shift. The results should be quite accurate, except in the case of weak lines. (A few very weak lines have been omitted from this tabulation.)

According to several sources (see, e.g., introduction to Fe XXII), the lower of the two levels $2s 2p^2 \ ^2P_{1/2}$ and $^2S_{1/2}$ is mostly of ^2P character, having "crossed" the $^2S_{1/2}$ level

at about V XIX or Cr XX. We have thus labeled these two levels accordingly, in contrast to their labeling by Cheng. et al., which is consistent with their ordering at the neutral end of the B sequence.

Reference

[1]K. T. Cheng, Y.-K. Kim, and J. P. Desclaux, At. Data Nucl. Data Tables **24**, 111 (1979).

Ni XXIV: Allowed transitions

No.	Transition Array	Multiplet	λ (Å)	E_i (cm^{-1})	E_k (cm^{-1})	g_i	g_k	A_{ki} (10^8 s^{-1})	f_{ik}	S (at. u.)	log gf	Accuracy	Source
1.	$2s^2 2p$–$2s 2p^2$	$^2P^\circ$ – ^4P											
			[224.02]	163570	609950	4	6	1.4	0.0016	0.0047	−2.19	E	1
			[264.98]	163570	540950	4	4	0.13	1.4(−4)^a	4.9(−4)	−3.25	E	1
			[218.35]	0	457980	2	2	1.7	0.0012	0.0017	−2.62	E	1
			[339.66]	163570	457980	4	2	0.19	1.6(−4)	7.2(−4)	−3.19	E	1
2.		$^2P^\circ$ – ^2D											
			138.80	163570	884030	4	6	72	0.0314	0.057	−0.90	C	1
			118.52	0	843710	2	4	150	0.063	0.049	−0.90	C	1

Ni XXIV: Allowed transitions — Continued

No.	Transition Array	Multiplet	λ (Å)	E_i (cm^{-1})	E_k (cm^{-1})	g_i	g_k	A_{ki} (10^8 s^{-1})	f_{ik}	S (at. u.)	log gf	Accuracy	Source
3.		^{2}P° – ^{2}P	102.94	109050	1080490	6	6	530	0.084	0.17	−0.30	C−	1
			102.11	163570	1142910	4	4	540	0.084	0.11	−0.47	C	1
			104.64	0	955660	2	2	470	0.077	0.053	−0.81	C	1
			[126.25]	163570	955660	4	2	1.8	2.2(−4)	3.7(−4)	−3.06	E	1
			87.50	0	1142910	2	4	67	0.0153	0.0088	−1.51	C	1
4.		^{2}P° – ^{2}S	97.997	109050	1129490	6	2	510	0.0243	0.0471	−0.84	C−	1
			103.53	163570	1129490	4	2	417	0.0335	0.0457	−0.87	C	1
			88.54	0	1129490	2	2	20	0.0024	0.0014	−2.32	D	1
5.	$2s\,2p^2$–$2p^3$	^{4}P – ^{4}S°	115.85	561620	1424810	12	4	500	0.0339	0.155	−0.391	C	1
			122.72	609950	1424810	6	4	217	0.0326	0.079	−0.71	C	1
			113.14	540950	1424810	4	4	165	0.0316	0.0471	−0.90	C	1
			103.43	457980	1424810	2	4	130	0.0418	0.0285	−1.078	C	1
6.		^{4}P – ^{2}D°											
			98.39	609950	1626280	6	6	37	0.0053	0.010	−1.50	E	1
			[96.070]	540950	1581860	4	4	49	0.0068	0.0086	−1.57	E	1
			[102.89]	609950	1581860	6	4	6.6	7.0(−4)	0.0014	−2.38	E	1
			[92.138]	540950	1626280	4	6	0.84	1.6(−4)	1.9(−4)	−3.19	E	1
			[88.977]	457980	1581860	2	4	1.5	3.6(−4)	2.1(−4)	−3.14	E	1
7.		^{4}P – ^{2}P°											
			[79.185]	609950	1872810	6	4	1.8	1.1(−4)	1.7(−4)	−3.18	E	1
			[75.083]	540950	1872810	4	4	4.1	3.5(−4)	3.5(−4)	−2.85	E	1
			[75.580]	457980	1781090	2	2	3.4	2.9(−4)	1.4(−4)	−3.24	E	1
8.		^{2}D – ^{4}S°											
			[184.92]	884030	1424810	6	4	0.38	1.3(−4)	4.7(−4)	−3.11	E	1
			[172.09]	843710	1424810	4	4	3.4	0.0015	0.0034	−2.22	E	1
9.		^{2}D – ^{2}D°	135.02	867900	1608510	10	10	175	0.0477	0.212	−0.322	C	1
			134.73	884030	1626280	6	6	144	0.0393	0.105	−0.63	C	1
			135.47	843710	1581860	4	4	80	0.0221	0.0394	−1.054	C	1
			143.30	884030	1581860	6	4	60	0.0124	0.0351	−1.128	C	1
			127.78	843710	1626280	4	6	52	0.0190	0.0320	−1.119	C	1
10.		^{2}D – ^{2}P°	102.63	867900	1842240	10	6	270	0.026	0.088	−0.58	C−	1
			101.13	884030	1872810	6	4	163	0.0167	0.0334	−1.000	C	1
			106.68	843710	1781090	4	2	367	0.0313	0.0440	−0.90	C	1
			97.17	843710	1872810	4	4	59	0.0084	0.011	−1.47	D	1
11.		^{2}P – ^{4}S°											
			[354.74]	1142910	1424810	4	4	0.42	7.9(−4)	0.0037	−2.50	E	1
			[213.15]	955660	1424810	2	4	3.3	0.0045	0.0063	−2.05	E	1
12.		^{2}P – ^{2}D°	189.39	1080490	1608510	6	10	51	0.045	0.17	−0.56	D	1
			[206.88]	1142910	1626280	4	6	37.4	0.0360	0.098	−0.84	C	1
			159.69	955660	1581860	2	4	89	0.068	0.071	−0.87	C	1
			[227.82]	1142910	1581860	4	4	0.49	3.8(−4)	0.0011	−2.82	E	1

Ni XXIV: Allowed transitions — Continued

No.	Transition Array	Multiplet	λ (Å)	E_i (cm^{-1})	E_k (cm^{-1})	g_i	g_k	A_{ki} (10^8 s^{-1})	f_{ik}	S (at. u.)	log gf	Accuracy	Source
13.		^{2}P – ^{2}P°	131.28	1080490	1842240	6	6	220	0.058	0.15	−0.46	C−	1
			137.01	1142910	1872810	4	4	260	0.073	0.13	−0.53	C	1
			121.15	955660	1781090	2	2	44	0.0097	0.0077	−1.71	D	1
			[156.70]	1142910	1781090	4	2	21	0.0038	0.0078	−1.82	D	1
			109.03	955660	1872810	2	4	36.8	0.0131	0.0094	−1.58	C	1
14.		^{2}S – ^{4}S°	[338.62]	1129490	1424810	2	4	0.18	6.2(−4)	0.0014	−2.91	E	1
15.		^{2}S – ^{2}D°											
			[221.06]	1129490	1581860	2	4	6.3	0.0093	0.014	−1.73	D	1
16.		^{2}S – ^{2}P°	140.30	1129490	1842240	2	6	72	0.064	0.059	−0.89	C	1
			134.53	1129490	1872810	2	4	28.4	0.0154	0.0136	−1.51	C	1
			[153.47]	1129490	1781090	2	2	127	0.0447	0.0452	−1.049	C	1

[a]The number in parentheses following the tabulated value indicates the power of ten by which this value has to be multiplied.

Ni XXIV

Forbidden Transitions

The line strengths tabulated for the single magnetic dipole and single electric quadrupole transition within the $2s^2 2p$ ground state configuration are the results of the multiconfiguration Dirac-Fock (MCDF) calculations of Cheng et al.[1] These relativistic calculations include a perturbative treatment of the Breit interaction and the Lamb shift. Allowance for configuration mixing is limited to the $n = 2$ complex. The strength of the electric quadrupole transition as defined in Ref. 1 was multiplied by the factor $2/3$ in order to bring this value into conformance with the definition of the quadrupole strength used in the NBS tables.

Transition probabilities for the same lines were calculated by Froese Fischer and Saha[2] using the multiconfiguration Hartree-Fock (MCHF) method with Breit-Pauli corrections. Their orbital basis includes many configurations outside the $n = 2$ complex, but relativistic effects were not treated to the same degree as in Ref. 1. The line strengths for both the M1 and E2 transitions, derived from these data by interpolation between appropriately spaced ions of the B sequence, are in very good agreement with the data of Cheng et al.[1]

References

[1]K. T. Cheng, Y.-K. Kim, and J. P. Desclaux, At. Data Nucl. Data Tables **24**, 111 (1979).
[2]C. Froese Fischer and H. P. Saha, Phys. Rev. A **28**, 3169 (1983).

Ni XXIV: Forbidden transitions

No.	Transition Array	Multiplet	λ (Å)	E_i (cm^{-1})	E_k (cm^{-1})	g_i	g_k	Type of transition	A_{ki} (s^{-1})	S (at. u.)	Accuracy	Source
1.	2p–2p	^{2}P° – ^{2}P°										
			609.9	0	163960	2	4	M1	3.95(+4)[a]	1.33	B	1
			"	"	"	2	4	E2	5.1	1.03(−3)	C	1

[a]The number in parentheses following the tabulated value indicates the power of ten by which this value has to be multiplied.

Ni xxv

Be Isoelectronic Sequence

Ground State: $1s^2 2s^2\ {}^1S_0$

Ionization Energy: 2295 eV = 18510000 cm^{-1}

Allowed Transitions

List of tabulated lines

Wavelength (Å)	No.	Wavelength (Å)	No.	Wavelength (Å)	No.	Wavelength (Å)	No.
9.19	11	9.70	13	9.94	26	117.91	2
9.27	10	9.71	29	9.95	19	120.47	4
9.30	11	9.74	24	9.97	25,31	126.73	3
9.31	10	9.75	28	10.02	21	128.85	3
9.32	10	9.76	24,27,29	10.07	21	130.99	7
9.34	9	9.77	12	10.08	21,35	135.95	3
9.39	8	9.78	27	10.09	19,30	151.91	3
9.42	16	9.80	28	10.17	21	158.84	3
9.43	10	9.85	33	10.20	21	165.34	3
9.49	15	9.86	27,34	10.21	22	188.15	6
9.56	14	9.87	27	10.23	21	238.82	1
9.60	24	9.91	19,27	10.32	20	278.01	5
9.63	24	9.92	32	10.46	23	326.58	5
9.64	24	9.93	27	104.07	4	499.50	5

Oscillator strengths for transitions of the arrays $2s^2$–$2s2p$ and $2s2p$–$2p^2$ are taken from the multiconfiguration Dirac-Fock (MCDF) calculations of Cheng et al.[1] These relativistic calculations include the configuration interaction most relevant for the states of these configurations, as well as a perturbative treatment of the Breit interaction and the Lamb shift. The results should be quite accurate, except for the weakest intercombination lines. (The $^3P^\circ_1$ – $^1S_0$ transition of the $2s2p$–$2p^2$ array has been omitted here, since the f-value is considerably smaller than those of the other lines of this array.)

A number of sources of reliable data, from other relativistic calculations, are available for the $2s$–$2p$ transitions. However, with the exception of some of the weaker lines, they all agree well with the results of Cheng et al.[1] The latter are quoted exclusively here since they provide data from a single set of comprehensive calculations, all done at a uniform and reasonably accurate level of approximation, for the valence shell $2s$–$2p$ transitions for all ions of the isoelectronic sequence.

The f-values for the $2s^2$–$2s3p$, $2s2p$–$2p3p$, $2s2p$–$2s3s$, $2p^2$–$2p3s$, $2s2p$–$2s3d$, and $2p^2$–$2p3d$ arrays of transitions are taken from the work of Fawcett,[2] who used Cowan's version of the relativistic Hartree-Fock method with intermediate coupling and configuration interaction. This work provides a comprehensive set of data for the entire isoelectronic sequence, calculated at a uniform level of approximation. Some of these transitions, for some ions of this sequence, have also been calculated by Bhatia et al.[3] using the program SUPERSTRUCTURE, which includes configuration interaction and intermediate coupling. Where they overlap, these two sets of calculations agree to within the uncertainties assigned here. Transitions involving the $J=1$ levels of $2p3p$ ^3S and ^3P have been omitted because of erratic behavior of the f-values along the sequence.

Oscillator strengths for the transition array $2s^2$–$2s4p$ are the results of the relativistic random phase approximation (RRPA) calculations of Lin and Johnson.[4]

A few multiplet f-values for transitions involving the outer electron alone, $2s3s$–$2s3p$ and $2s3p$–$2s3d$, have been interpolated along the isoelectronic sequence and assigned a low accuracy.

References

[1] K. T. Cheng, Y.-K. Kim, and J. P. Desclaux, At. Data Nucl. Data Tables 24, 111 (1979).

[2] B. C. Fawcett, At. Data Nucl. Data Tables 30, 1 (1984); 33, 479 (1985).

[3] A. K. Bhatia, U. Feldman, and J. F. Seely, At. Data Nucl. Data Tables 35, 449 (1986).

[4] C. D. Lin and W. R. Johnson, Phys. Rev. A 15, 1046 (1977).

Ni xxv: Allowed transitions

No.	Transition Array	Multiplet	λ (Å)	E_i (cm^{-1})	E_k (cm^{-1})	g_i	g_k	A_{ki} (10^8 s^{-1})	f_{ik}	S (at. u.)	log gf	Accuracy	Source
1.	$2s^2$–$2s2p$	^{1}S – ^{3}P°											
			[238.82]	0	418720	1	3	0.82	0.0021	0.0017	−2.68	D	1
2.		^{1}S – ^{1}P°	[117.91]	0	848100	1	3	238	0.149	0.0578	−0.827	B	1
3.	$2s2p$–$2p^2$	^{3}P° – ^{3}P	*145.92*	*486870*	*1172200*	9	9	150	0.0479	0.207	−0.366	B	1
			[151.91]	549500	1207800	5	5	76.6	0.0265	0.0663	−0.878	B	1
			[135.95]	418720	1154300	3	3	51.2	0.0142	0.0191	−1.371	B	1
			[165.34]	549500	1154300	5	3	48.8	0.0120	0.0327	−1.222	B	1
			[158.84]	418720	1048300?	3	1	138	0.0174	0.0273	−1.282	B	1
			[126.73]	418720	1207800	3	5	70.0	0.0281	0.0352	−1.074	B	1
			[128.85]	378190	1154300	1	3	83.7	0.0625	0.0265	−1.204	B	1
4.		^{3}P° – ^{1}D											
			[120.47]	549500	1379600	5	5	77	0.0167	0.0331	−1.078	C	1
			[104.07]	418720	1379600	3	5	7.4	0.0020	0.0021	−2.22	D	1
5.		^{1}P° – ^{3}P											
			[278.01]	848100	1207800	3	5	5.7	0.011	0.030	−1.48	D	1
			[326.58]	848100	1154300	3	3	0.13	2.1(−4)^a	6.8(−4)	−3.20	E	1
			[499.50]	848100	1048300?	3	1	0.24	3.0(−4)	0.0015	−3.05	E	1
6.		^{1}P° – ^{1}D	[188.15]	848100	1379600	3	5	59.1	0.0523	0.0972	−0.804	B	1
7.		^{1}P° – ^{1}S	[130.99]	848100	1611500	3	1	399	0.0342	0.0442	−0.989	B	1
8.	$2s^2$–$2s3p$	^{1}S – ^{3}P°											
			[9.39]	0	10650000	1	3	6.6(+4)	0.26	0.0080	−0.59	C−	2
9.		^{1}S – ^{1}P°	[9.34]	0	10707000	1	3	1.1(+5)	0.45	0.014	−0.35	C−	2
10.	$2s2p$–$2p3p$	^{3}P° – ^{3}D											
			[9.31]	549500	11296000	5	7	8.2(+4)	0.15	0.023	−0.12	C−	2
			[9.32]	418720	11153000	3	5	7.8(+4)	0.17	0.016	−0.29	C−	2
			[9.27]	378190	[11162000]	1	3	2.2(+4)	0.084	0.0026	−1.08	D	2
			[9.43]	549500	11153000	5	5	1500	0.0020	3.1(−4)	−2.00	D	2
			[9.31]	418720	[11162000]	3	3	4.1(+4)	0.053	0.0049	−0.80	D	2
11.		^{3}P° – ^{3}P											
			[9.30]	549500	11306000	5	5	6.9(+4)	0.090	0.014	−0.35	C−	2
			[9.30]	418720	[11173000]	3	1	9.3(+4)	0.040	0.0037	−0.92	D	2
			[9.19]	418720	11306000	3	5	1600	0.0033	3.0(−4)	−2.00	D	2
12.		^{1}P° – ^{1}P	[9.77]	848100	[11080000]	3	3	1.7(+4)	0.025	0.0024	−1.12	D	2
13.		^{1}P° – ^{3}D											
			[9.70]	848100	[11162000]	3	3	1.8(+4)	0.025	0.0024	−1.12	D	2
14.		^{1}P° – ^{3}P											
			[9.56]	848100	11306000	3	5	3.4(+4)	0.077	0.0073	−0.64	D	2
			[9.56]	848100	[11313000]	3	3	5.1(+4)	0.070	0.0066	−0.68	C−	2

Ni xxv: Allowed transitions — Continued

No.	Transition Array	Multiplet	λ (Å)	E_i (cm^{-1})	E_k (cm^{-1})	g_i	g_k	A_{ki} (10^8 s^{-1})	f_{ik}	S (at. u.)	log gf	Accuracy	Source
15.		^{1}P° – ^{1}D	[9.49]	848100	[11381000]	3	5	8.9(+4)	0.20	0.019	−0.22	C−	2
16.		^{1}P° – ^{1}S	[9.42]	848100	[11460000]	3	1	9.0(+4)	0.040	0.0037	−0.92	D	2
17.	$2s^2$–$2s\,4p$	^{1}S – ^{3}P°											
						1	3		0.032		−1.49	D	4
18.		^{1}S – ^{1}P°				1	3		0.15		−0.82	D	4
19.	$2s\,2p$–$2s\,3s$	^{3}P° – ^{3}S	10.02	486870	[10465000]	9	3	5.2(+4)	0.026	0.0078	−0.63	D	2
			[10.09]	549500	[10465000]	5	3	2.8(+4)	0.026	0.0043	−0.89	D	2
			[9.95]	418720	[10465000]	3	3	1.8(+4)	0.026	0.0026	−1.11	D	2
			[9.91]	378190	[10465000]	1	3	6100	0.027	8.8(−4)	−1.57	D	2
20.		^{1}P° – ^{1}S	[10.32]	848100	[10536000]	3	1	2.1(+4)	0.011	0.0011	−1.48	D	2
21.	$2p^2$–$2p\,3s$	^{3}P – ^{3}P°	10.11	1172200	[11062000]	9	9	3.5(+4)	0.053	0.016	−0.32	D	2
			[10.08]	1207800	[11130000]	5	5	2.1(+4)	0.032	0.0053	−0.80	D	2
			[10.17]	1154300	[10983000]	3	3	6000	0.0093	9.3(−4)	−1.55	D	2
			[10.23]	1207800	[10983000]	5	3	1.6(+4)	0.015	0.0025	−1.12	D	2
			[10.20]	1154300	[10963000]	3	1	3.1(+4)	0.016	0.0016	−1.32	D	2
			[10.02]	1154300	[11130000]	3	5	1.5(+4)	0.037	0.0037	−0.95	D	2
			[10.07]	1048300?	[10983000]	1	3	1.2(+4)	0.055	0.0018	−1.26	D	2
22.		^{1}D – ^{1}P°	[10.21]	1379600	[11176000]	5	3	2.8(+4)	0.026	0.0044	−0.89	D	2
23.		^{1}S – ^{1}P°	[10.46]	1611500	[11176000]	1	3	1.1(+4)	0.054	0.0019	−1.27	D	2
24.	$2s\,2p$–$2s\,3d$	^{3}P° – ^{3}D	9.69	486870	10805000	9	15	3.09(+5)	0.72	0.208	0.81	C−	2
			[9.74]	549500	10813000	5	7	3.0(+5)	0.60	0.096	0.48	C−	2
			[9.63]	418720	10800000	3	5	2.4(+5)	0.55	0.052	0.22	C−	2
			[9.60]	378190	10794000	1	3	1.8(+5)	0.75	0.024	−0.12	C−	2
			[9.76]	549500	10800000	5	5	7.7(+4)	0.11	0.018	−0.26	C−	2
			[9.64]	418720	10794000	3	3	1.3(+5)	0.18	0.017	−0.27	C−	2
			[9.76]	549500	10794000	5	3	8400	0.0072	0.0012	−1.44	C−	2
25.		^{1}P° – ^{1}D	[9.97]	848100	10880000	3	5	2.5(+5)	0.61	0.060	0.26	C−	2
26.	$2p^2$–$2p\,3d$	^{3}P – ^{3}F°											
			[9.94]	1207800	11271000	5	7	1.29(+5)	0.268	0.0438	0.127	C−	2
27.		^{3}P – ^{3}D°	9.82	1172200	11357000	9	15	2.84(+5)	0.68	0.199	0.79	C−	2
			[9.78]	1207800	11437000	5	7	2.9(+5)	0.59	0.095	0.47	C−	2
			[9.87]	1154300	11283000	3	5	2.03(+5)	0.493	0.0481	0.170	C−	2
			[9.76]	1048300?	11296000	1	3	3.03(+5)	1.30	0.0418	0.114	C−	2
			[9.93]	1207800	11283000	5	5	2.2(+4)	0.032	0.0052	−0.80	D	2
			[9.86]	1154300	11296000	3	3	5.7(+4)	0.083	0.0081	−0.60	C−	2
			[9.91]	1207800	11296000	5	3	3800	0.0034	5.5(−4)	−1.77	D	2
28.		^{3}P – ^{1}D°											
			[9.80]	1207800	11408000	5	5	4.4(+4)	0.064	0.010	−0.49	C−	2
			[9.75]	1154300	11408000	3	5	1.3(+5)	0.30	0.029	−0.05	D	2

Ni xxv: Allowed transitions — Continued

No.	Transition Array	Multiplet	λ (Å)	E_i (cm⁻¹)	E_k (cm⁻¹)	g_i	g_k	A_{ki} (10⁸ s⁻¹)	f_{ik}	S (at. u.)	log gf	Accuracy	Source
29.		³P – ³P°											
			[9.76]	1207800	11456000	5	5	1.3(+5)	0.19	0.031	−0.02	C−	2
			[9.71]	1154300	11456000	3	3	1.8(+5)	0.25	0.024	−0.12	C−	2
			[9.76]	1207800	11456000	5	3	7.5(+4)	0.064	0.010	−0.49	C−	2
			[9.71]	1154300	11456000	3	1	2.3(+5)	0.11	0.011	−0.48	C−	2
			[9.71]	1154300	11456000	3	5	2.7(+4)	0.063	0.0060	−0.72	D	2
30.		¹D – ³F°											
			[10.09]	1379600	[11295000]	5	5	7900	0.012	0.0020	−1.22	D	2
31.		¹D – ¹D°	[9.97]	1379600	11408000	5	5	2.8(+4)	0.042	0.0069	−0.68	C−	2
32.		¹D – ³P°											
			[9.92]	1379600	11456000	5	5	1.3(+5)	0.19	0.031	−0.02	C−	2
33.		¹D – ¹P°	[9.85]	1379600	[11535000]	5	3	1.7(+4)	0.015	0.0024	−1.12	D	2
34.		¹D – ¹F°	[9.86]	1379600	11525000	5	7	4.8(+5)	0.98	0.16	0.69	C−	2
35.		¹S – ¹P°	[10.08]	1611500	[11535000]	1	3	2.80(+5)	1.28	0.0425	0.107	C−	2
36.	2s3s–2s3p	³S – ³P°				3	9		0.11		−0.48	D	interp.
37.		¹S – ¹P°				1	3		0.050		−1.30	E	interp.
38.	2s3p–2s3d	³P° – ³D				9	15		0.025		−0.65	E	interp.
39.		¹P° – ¹D				3	5		0.043		−0.89	E	interp.

ᵃThe number in parentheses following the tabulated value indicates the power of ten by which this value has to be multiplied.

Ni xxv

Forbidden Transitions

Transition probabilities for magnetic dipole and electric quadrupole transitions within the $2s2p$ and $2p^2$ configurations were calculated by Feldman et al.[1] using scaled Thomas-Fermi wavefunctions with allowance for configuration interaction and relativistic effects. We modified their transition probability data by the application of experimental wavelengths, i.e., we first converted their A-values into line strength data utilizing their theoretical transition energies and then reconverted the line strengths into A-values with wavelengths derived from experimental data. This approach should normally yield transition probabilities that are more accurate than those based on theoretically determined wavelengths.

The transition probability for the one electric quadrupole transition listed, which is relatively strong compared to other E2 transitions, has been interpolated from the data of Anderson and Anderson[2] and Glass[3,4] for neighboring ions of the Be sequence. This A-value exhibits a smooth nuclear charge dependence.

References

[1]U. Feldman, G. A. Doschek, Ch.-Ch. Cheng, and A. K. Bhatia, J. Appl. Phys. **51**, 190 (1980).
[2]E. K. Anderson and E. M. Anderson, Opt. Spectrosc. (USSR) **52**, 478 (1982).
[3]R. Glass, Z. Phys. A **320**, 545 (1985).
[4]R. Glass, Astrophys. Space Sci. **92**, 307 (1983).

Ni xxv: Forbidden transitions

No.	Transition Array	Multiplet	λ (Å)	E_i (cm^{-1})	E_k (cm^{-1})	g_i	g_k	Type of transition	A_{ki} (s^{-1})	S (at. u.)	Accuracy	Source
1.	$2s2p$–$2s2p$	$^3P^\circ$ – $^3P^\circ$										
			[764.64]	418720	549500	3	5	M1	2.91(+4)[a]	2.41	C+	1
			[2466.6]	378190	418720	1	3	M1	1160	1.93	C+	1
2.		$^3P^\circ$ – $^1P^\circ$										
			[334.90]	549500	848100	5	3	M1	2.1(+4)	0.087	D	1
			[232.89]	418720	848100	3	3	M1	3.6(+4)	0.050	D	1
			"	"	"	3	3	E2	100	1.2(−4)	D−	interp.
			[212.81]	378190	848100	1	3	M1	6.5(+4)	0.070	D	1
3.	$2p^2$–$2p^2$	^3P – ^3P										
			[1869.2]	1154300	1207800	3	5	M1	1410	1.71	C	1
			[943.40]	1048300	1154300	1	3	M1	1.96(+4)	1.83	C	1
4.		^3P – ^1D										
			[582.07]	1207800	1379600	5	5	M1	4.4(+4)	1.6	C	1
			[443.85]	1154300	1379600	3	5	M1	4.9(+4)	0.79	D+	1
5.		^3P – ^1S										
			[218.72]	1154300	1611500	3	1	M1	4.4(+5)	0.17	D	1

[a]The number in parentheses following the tabulated value indicates the power of ten by which this value has to be multiplied.

Ni xxvi

Li Isoelectronic Sequence

Ground State: $1s^2 2s\ ^2S_{1/2}$

Ionization Energy: 2399.2 eV = 19351000 cm^{-1}

Allowed Transitions

Transition probabilities for the strongest inner-shell transitions to doubly excited $n = 2$ states are taken from the multiconfiguration Dirac-/fock (MCDF) calculations of Hata and Grant.[1] Their results are in good agreement with the Z-expansion perturbation calculations of Vainshtein and Safronova,[2] who included relativistic corrections at the level of the Pauli approximation.

Oscillator strengths for lines of the principal ($2s$–$2p$) resonance multiplet are the results of the MCDF calculations of Cheng *et al.*,[3] which include a perturbative treatment of the Breit interaction and the Lamb shift.

The results of the Hartree-XR (Hartree-Fock with statistical exchange and relativistic effects) calculations

of Fawcett *et al.*[4] are tabulated for the $2p$–$3s$ and $2p$–$3d$ transitions.

The f-value for the $3d$–$4f$ transition was taken from a study of systematic trends along isoelectronic sequences by Smith and Wiese.[5] The tabulated data for the remaining transitions were taken from the theoretical analysis of Martin and Wiese,[6] which was based on a generalized study of systematic trends for several spectral series of the lithium isoelectronic sequence.

Results of the relativistic Hartree-Fock calculations of Kim and Desclaux[7] for several ions of the Li sequence were incorporated into the data of Ref. 6 for the $2s$–$3p$ transitions. For all other transitions for which the results of Ref. 6 are quoted here, no relativistic calculations were available. However, the relativistic calculations of Younger and Weiss[8] for the hydrogen isoelectronic sequence provide a means of assessing the magnitude of relativistic corrections, since the Li sequence is very similar in structure to the H sequence. For those transitions for which relativistic effects were estimated to be significant (specifically, whenever the ratio of the weighted relativistic hydrogenic f-values $g_i f_{ik}$ of any two lines within a multiplet was found to deviate from the corresponding LS-coupling line strength ratio by more than 5% for the appropriate value of the nuclear charge Z), the f-values were excluded from the compilation. A more detailed discussion of this comparison is given in Ref. 6.

Transition probability data are available for numerous transitions involving doubly excited states with the spectator electron occupying the $n = 3$ shell, or higher.[9] These have not been tabluated, however, since they belong to, or are very close to belonging to, the unresolved satellites of the helium-like ion.

References

[1]J. Hata and I. P. Grant, Mon. Not. R. Astron. Soc. **211**, 549 (1984).

[2]L. A. Vainshtein and U. I. Safronova, At. Data Nucl. Data Tables **21**, 49 (1978).

[3]K. T. Cheng, Y.-K. Kim, and J. P. Desclaux, At. Data Nucl. Data Tables **24**, 111 (1979).

[4]B. C. Fawcett, A. Ridgeley, and T. P. Hughes, Mon. Not. R. Astron. Soc. **188**, 365 (1979).

[5]M. W. Smith and W. L. Wiese, Astrophys. J. Suppl. Ser. **23**, No. 196, 103 (1971).

[6]G. A. Martin and W. L. Wiese, J. Phys. Chem. Ref. Data **5**, 537 (1976).

[7]Y.-K. Kim and J. P. Desclaux, Phys. Rev. Lett. **36**, 139 (1976) and private communication.

[8]S. M. Younger and A. W. Weiss, J. Res. Nat. Bur. Stand., Sect. A **79**, 629 (1975).

[9]L. A. Vainshtein and U. I. Safronova, At. Data Nucl. Data Tables **25**, 311 (1980).

Ni XXVI: Allowed transitions

No.	Transition Array	Multiplet	λ (Å)	E_i (cm^{-1})	E_k (cm^{-1})	g_i	g_k	A_{ki} (10^8 s^{-1})	f_{ik}	S (at. u.)	log gf	Accuracy	Source
1.	$1s^2 2s-$ $1s(^2S)2s2p(^3P°)$	$^2S - ^2P°$											
			1.5996	0	62516000	2	2	2.7(+6)^a	0.10	0.0011	−0.68	C	1
2.	$1s^2 2s-$ $1s(^2S)2s2p(^1P°)$	$^2S - ^2P°$											
			1.5935	0	62755000	2	2	4.0(+6)	0.15	0.0016	−0.52	C	1
3.	$1s^2 2p-1s2p^2$	$^2P° - ^2P$											
			1.5973	604520	63211000	4	4	8.1(+6)	0.31	0.0065	0.09	C	1
			1.5982	426990	62997000	2	2	7.3(+6)	0.28	0.0029	−0.25	C	1
			[1.6036]	604520	62997000	4	2	2.1(+6)	0.040	8.5(−4)	−0.79	C	1
4.		$^2P° - ^2D$											
			1.6005	604520	63085000	4	6	2.7(+6)	0.16	0.0033	−0.21	C	1
			1.5977	426990	63017000	2	4	4.4(+6)	0.34	0.0035	−0.17	C	1
5.		$^2P° - ^2S$											
			1.5930	604520	63380000	4	2	3.4(+6)	0.065	0.0014	−0.59	C	1
6.	$2s-2p$	$^2S - ^2P°$	*183.37*	0	*545340*	2	6	42.0	0.0635	0.0767	−0.896	B+	3
			165.42?	0	604520	2	4	57.5	0.0472	0.0514	−1.025	B+	3
			234.20	0	426990	2	2	19.9	0.0164	0.0253	−1.484	B+	3
7.	$2s-3p$	$^2S - ^2P°$	*9.074*	0	*11020000*	2	6	1.01(+5)	0.375	0.0224	−0.125	B+	6
			9.061	0	1104000	2	4	9.99(+4)	0.246	0.0147	−0.308	B+	6
			9.105	0	10980000	2	2	1.04(+5)	0.129	0.00773	−0.588	B+	6
8.	$2s-4p$	$^2S - ^2P°$				2	6		0.101		−0.695	C+	6
9.	$2s-5p$	$^2S - ^2P°$				2	6		0.040		−1.10	C+	6

Ni XXVI: Allowed transitions — Continued

No.	Transition Array	Multiplet	λ (Å)	E_i (cm^{-1})	E_k (cm^{-1})	g_i	g_k	A_{ki} (10^8 s^{-1})	f_{ik}	S (at. u.)	log gf	Accuracy	Source
10.	2s–6p	^{2}S – ^{2}P°				2	6		0.0213		−1.371	C+	6
11.	2s–7p	^{2}S – ^{2}P°				2	6		0.0125		−1.602	C+	6
12.	2p–3s	^{2}P° – ^{2}S	9.676	545340	10880000	6	2	3.8(+4)	0.018	0.0034	−0.97	C	4
			9.732	604520	10880000	4	2	2.5(+4)	0.018	0.0023	−1.14	C	4
			[9.567]	426990	10880000	2	2	1.3(+4)	0.018	0.0011	−1.44	C	4
13.	2p–3d	^{2}P° – ^{2}D	9.483	545340	11090000	6	10	3.02(+5)	0.68	0.127	0.61	C	4
			9.535	604520	11090000	4	6	2.96(+5)	0.605	0.0760	0.384	C+	4
			9.390	426990	11080000	2	4	2.59(+5)	0.685	0.0424	0.137	C+	4
			9.55	604520	11080000	4	4	5.0(+4)	0.068	0.0086	−0.57	C	4
14.	2p–4d	^{2}P° – ^{2}D				6	10		0.12		−0.14	C+	6
15.	2p–5d	^{2}P° – ^{2}D				6	10		0.0450		−0.569	C+	6
16.	2p–6d	^{2}P° – ^{2}D				6	10		0.0220		−0.879	C+	6
17.	2p–7d	^{2}P° – ^{2}D				6	10		0.0125		−1.125	C+	6
18.	3s–4p	^{2}S – ^{2}P°				2	6		0.45		−0.05	C	6
19.	3s–5p	^{2}S – ^{2}P°				2	6		0.108		−0.67	C	6
20.	3s–6p	^{2}S – ^{2}P°				2	6		0.048		−1.02	C	6
21.	3s–7p	^{2}S – ^{2}P°				2	6		0.0250		−1.301	C	6
22.	3p–4d	^{2}P° – ^{2}D				6	10		0.60		0.56	B	6
23.	3p–5d	^{2}P° – ^{2}D				6	10		0.138		−0.082	C+	6
24.	3p–6d	^{2}P° – ^{2}D				6	10		0.0558		−0.475	C+	6
25.	3p–7d	^{2}P° – ^{2}D				6	10		0.0289		−0.761	C+	6
26.	3d–4f	^{2}D – ^{2}F°				10	14		1.00		1.000	B	5
27.	4s–5p	^{2}S – ^{2}P°				2	6		0.483		−0.015	C	6
28.	4s–6p	^{2}S – ^{2}P°				2	6		0.129		−0.59	C	6
29.	4s–7p	^{2}S – ^{2}P°				2	6		0.056		−0.95	C	6
30.	4p–5d	^{2}P° – ^{2}D				6	10		0.586		0.546	C+	6
31.	4p–6d	^{2}P° – ^{2}D				6	10		0.143		−0.067	C+	6
32.	4p–7d	^{2}P° – ^{2}D				6	10		0.0618		−0.431	C+	6

[a]The number in parentheses following the tabulated value indicates the power of ten by which this value has to be multiplied.

Ni xxvi

Forbidden Transitions

The single magnetic dipole transition within the $1s^2 2p$ configuration has the line strength of 1.33 in the absence of relativistic effects in the wavefunctions.[1] It is estimated that these effects are negligible, since comprehensive relativistic calculations by Cheng *et al.*[2] for the analogous transition in the $1s^2 2s^2 2p$ configuration of the boron sequence show that such relativistic corrections are negligible until much more highly charged ions.

The listed transition probability data are also expected to be quite accurate since the energy levels are derived from experimental data.

An electric quadrupole transition at the same wavelength is estimated to be of negligible strength, as calculated by Bhatia[3] for this transition in the case of Mn XXIII. (He obtains a ratio of about 10^{-3} for the ratio of E2 to M1 line strengths).

References

[1]W. L. Wiese, M. W. Smith, and B. M. Miles, "Atomic Transition Probabilities", Vol. II, NSRDS–NBS 22, U.S. Govt. Print. Office, Washington, DC 1969.

[2]K. T. Cheng, Y.–K. Kim, and J. P. Desclaux, At. Data Nucl. Data Tables **24**, 111 (1979).

[3]A. K. Bhatia, private communication (1986).

Ni xxvi: Forbidden transitions

No.	Transition Array	Multiplet	λ (Å)	E_i (cm^{-1})	E_k (cm^{-1})	g_i	g_k	Type of transition	A_{ki} (s^{-1})	S (at. u.)	Accuracy	Source
1.	$2p$–$2p$	$^2P^\circ$ – $^2P^\circ$	[563.38]	[427180]	[604680]	2	4	M1	5.02(+4)[a]	1.33	B	*interp.*

[a]The number in parentheses following the tabulated value indicates the power of ten by which this value has to be multiplied.

Ni XXVII

He Isoelectronic Sequence

Ground State: $1s^2\ {}^1S_0$

Ionization Energy: $10288.8\ \text{eV} = 82984000\ \text{cm}^{-1}$

Allowed Transitions

List of tabulated lines

Wavelength (Å)	No.	Wavelength (Å)	No.	Wavelength (Å)	No.	Wavelength (Å)	No.
1.2534	19	1.543	3,9	6.4225	25	25.500	46
1.2537	18	1.544	9	6.5224	26	25.907	47
1.2824	17	1.546	11	6.5520	33	53.879	57
1.2831	16	1.547	11	6.6779	34	54.177	58
1.3500	15	1.549	5	8.6080	23	55.298	60
1.3516	14	1.550	11	8.7475	24	56.016	61
1.531	4	1.551	7	8.8772	29	168.5	21
1.534	13	1.558	8	9.0740	30	228.0	20
1.537	6,10	1.5883	2	17.084	43	315.5	22
1.538	3	1.5963	1	17.241	44	361.9	20
1.539	9	5.7489	27	17.356	50	388.7	20
1.540	9	5.8352	28	17.590	51		
1.541	12	5.8471	37	24.819	41		
1.542	3,9	5.9524	38	25.036	42		

Oscillator strengths for transitions of the $1s^2$–$1s2p$ array are taken from the results of Drake,[1] who incorporated accurate nonrelativistic matrix elements and Dirac hydrogenic matrix elements into a Z-expansion technique in order to provide f-values which would accurately reflect correlation effects for low-Z ions and relativistic effects for high-Z ions of the helium isoelectronic sequence. The f-values for the $1s^2\ {}^1S$ – $1snp\ {}^3P^\circ$ ($n = 3$–5) transitions were interpolated from results of the relativistic random phase approximation (RRPA) calculations of Johnson and Lin.[2] For other s–p and p–s transitions, we tabulate the published RRPA data of Lin et al.[3,4]

The charge expansion results of Laughlin[5] are given for various p–d and d–p transitions, as well as transitions between $4d$ and $4f$ levels. For those multiplets involving no change in principal quantum number ($3p$–$3d$, $4p$–$4d$, $4d$–$4f$) the f-values should be considered rather uncertain, since they are sensitive to energy differences. Oscillator strengths for the $2p$–$3d$ transitions, and for $1s3p$ ${}^3P^\circ$ – $1s3d$ ${}^3D$, were interpolated from the variational calculations of Weiss.[6] Both of these calculations indicate that, unlike the triplets, the $nd\ {}^1D$ energy levels ($n = 3,4$) lie below the $np\ {}^1P^\circ$ levels, and the $4f\ {}^1F^\circ$ lies below the $4d\ {}^1D$.

Brown and Cortez[7] have provided f-values for numerous d–f and f–d transitions for the isoelectronic sequence by fitting Z-expansion formulas to the results of varia-

tional calculations for the low-Z ions. Their results for transitions between the lower-lying D and F° terms are tabulated here.

Transition probabilities for the stronger transitions involving the doubly excited $n = 2$ states are taken from the comprehensive, charge expansion perturbation theory calculations of Vainshtein and Safronova.[8] Numerous data are also available for transitions involving doubly excited states where the spectator electron has principal quantum number $n = 3$.[9] However, these data are not tabulated here, since most of the transitions are very close to belonging to the unresolved satellites of the H-like ions, if they do not in fact do so.

References

[1]G. W. F. Drake, Phys. Rev. A **19**, 1387 (1979).

[2]W. R. Johnson and C. D. Lin, Phys. Rev. A **14**, 565 (1976).

[3]C. D. Lin, W. R. Johnson, and A. Dalgarno, Astrophys. J. **217**, 1011 (1977).

[4]C. D. Lin, W. R. Johnson, and A. Dalgarno, Phys. Rev. A **15**, 154 (1977).

[5]C. J. Laughlin, J. Phys. B **6**, 1942 (1973).

[6]A. W. Weiss, J. Res. Nat. Bur. Stand., Sect. A **71**, 163 (1967).

[7]R. T. Brown and J.-L. M. Cortez, Astrophys. J. **176**, 267 (1972).

[8]L. A. Vainshtein and U. I. Safronova, At. Data Nucl. Data Tables **21**, 49 (1978).

[9]L. A. Vainshtein and U. I. Safronova, At. Data Nucl. Data Tables **25**, 311 (1980).

Ni XXVII: Allowed transitions

No.	Transition Array	Multiplet	λ (Å)	E_i (cm^{-1})	E_k (cm^{-1})	g_i	g_k	A_{ki} (10^8 s^{-1})	f_{ik}	S (at. u.)	log gf	Accuracy	Source
1.	$1s^2$–$1s2p$	^1S – $^3P°$											
			[1.5963]	0	[62644200]	1	3	7.70(+5)^a	0.0883	4.64(−4)	−1.054	B	1
2.		^1S – $^1P°$	[1.5883]	0	[62961500]	1	3	6.02(+6)	0.683	0.00357	−0.166	B	1
3.	$1s2s$–$2s2p$	^3S – $^3P°$	*1.540*	[62367900]	*[127300000]*	3	9	3.8(+6)	0.41	0.0062	0.09	C	8
			[1.538]	[62367900]	[127380000]	3	5	3.9(+6)	0.23	0.0035	−0.16	C	8
			[1.542]	[62367900]	[127210000]	3	3	3.6(+6)	0.13	0.0020	−0.41	C	8
			[1.543]	[62367900]	[127170000]	3	1	3.8(+6)	0.045	6.9(−4)	−0.87	C	8
4.		^3S – $^1P°$											
			[1.531]	[62367900]	[127690000]	3	3	2.0(+5)	0.0070	1.1(−4)	−1.68	D	8
5.		^1S – $^3P°$											
			[1.549]	[62644500]	[127210000]	1	3	2.0(+5)	0.022	1.1(−4)	−1.67	D	8
6.		^1S – $^1P°$	[1.537]	[62644500]	[127690000]	1	3	3.7(+6)	0.39	0.0020	−0.41	C	8
7.	$1s2p$–$2s^2$	$^3P°$ – ^1S											
			[1.551]	[62644200]	[127130000]	3	1	8.2(+5)	0.0099	1.5(−4)	−1.53	D	8
8.		$^1P°$ – ^1S	[1.558]	[62961500]	[127130000]	3	1	6.5(+5)	0.0079	1.2(−4)	−1.63	D	8
9.	$1s2p$–$2p^2$	$^3P°$ – ^3P	*1.542*	*[62732300]*	*[127600000]*	9	9	6.6(+6)	0.236	0.0108	0.328	C	8
			[1.542]	[62806500]	[127640000]	5	5	3.5(+6)	0.12	0.0032	−0.20	C	8
			[1.540]	[62644200]	[127590000]	3	3	1.7(+6)	0.060	9.2(−4)	−0.74	C	8
			[1.544]	[62806500]	[127590000]	5	3	3.2(+6)	0.069	0.0017	−0.46	C	8
			[1.543]	[62644200]	[127460000]	3	1	6.9(+6)	0.082	0.0013	−0.61	C	8
			[1.539]	[62644200]	[127640000]	3	5	2.6(+6)	0.15	0.0023	−0.34	C	8
			[1.539]	[62625200]	[127590000]	1	3	2.6(+6)	0.28	0.0014	−0.56	C	8
10.		$^3P°$ – ^1D											
			[1.537]	[62806500]	[127860000]	5	5	2.3(+6)	0.081	0.0021	−0.39	C	8
11.		$^1P°$ – ^3P											
			[1.546]	[62961500]	[127640000]	3	5	1.6(+6)	0.096	0.0015	−0.54	C	8
			[1.547]	[62961500]	[127590000]	3	3	2.1(+5)	0.0075	1.2(−4)	−1.65	D	8
			[1.550]	[62961500]	[127460000]	3	1	1.2(+5)	0.0014	2.2(−5)	−2.36	D	8
12.		$^1P°$ – ^1D	[1.541]	[62961500]	[127860000]	3	5	5.5(+6)	0.33	0.0050	−0.01	C	8
13.		$^1P°$ – ^1S	[1.534]	[62961500]	[128140000]	3	1	6.9(+6)	0.081	0.0012	−0.61	C	8
14.	$1s^2$–$1s3p$	^1S – $^3P°$											
			[1.3516]	0	[73985000]	1	3	2.4(+5)	0.020	8.9(−5)	−1.70	E	*interp.*
15.		^1S – $^1P°$	[1.3500]	0	[74076300]	1	3	1.63(+6)	0.134	5.96(−4)	−0.873	B	3
16.	$1s^2$–$1s4p$	^1S – $^3P°$											
			[1.2831]	0	[77938200]	1	3	1.0(+5)	0.0074	3.1(−5)	−2.13	E	*interp.*
17.		^1S – $^1P°$	[1.2824]	0	[77976200]	1	3	6.38(+5)	0.0472	1.99(−4)	−1.326	B	3

Ni XXVII: Allowed transitions — Continued

No.	Transition Array	Multiplet	λ (Å)	E_i (cm^{-1})	E_k (cm^{-1})	g_i	g_k	A_{ki} (10^8 s^{-1})	f_{ik}	S (at. u.)	log gf	Accuracy	Source
18.	$1s^2 - 1s\,5p$	^{1}S $-$ ^{3}P°											
			[1.2537]	0	[79762600]	1	3	5.2(+4)	0.0037	1.5($-$5)	$-$2.43	E	*interp.*
19.		^{1}S $-$ ^{1}P°	[1.2534]	0	[79782000]	1	3	3.35(+5)	0.0237	9.78($-$5)	$-$1.625	B	3
20.	$1s\,2s - 1s\,2p$	^{3}S $-$ ^{3}P°	*274.4*	[62367900]	[*62732300*]	3	9	12.0	0.0406	0.110	$-$0.914	B	4
			[228.0]	[62367900]	[62806500]	3	5	21.7	0.0282	0.0635	$-$1.073	B	4
			[361.9]	[62367900]	[62644200]	3	3	4.82	0.00946	0.0338	$-$1.547	B	4
			[388.7]	[62367900]	[62625200]	3	1	4.35	0.00328	0.0126	$-$2.006	B	4
21.		^{3}S $-$ ^{1}P°											
			[168.5]	[62367900]	[62961500]	3	3	5.95	0.00253	0.00421	$-$2.119	B	4
22.		^{1}S $-$ ^{1}P°	[315.5]	[62644500]	[62961500]	1	3	7.53	0.0337	0.0350	$-$1.472	B	4
23.	$1s\,2s - 1s\,3p$	^{2}S $-$ ^{3}P°											
			[8.6080]	[62367900]	[73985000]	3	3	1.07(+5)	0.119	0.0101	$-$0.447	B	3
24.		^{1}S $-$ ^{1}P°	[8.7475]	[62644500]	[74076300]	1	3	1.03(+5)	0.353	0.0102	$-$0.452	B	3
25.	$1s\,2s - 1s\,4p$	^{2}S $-$ ^{3}P°											
			[6.4225]	[62367900]	[77938200]	3	3	5.2(+4)	0.032	0.0020	$-$1.02	B	3
26.		^{1}S $-$ ^{1}P°	[6.5224]	[62644500]	[77976200]	1	3	4.4(+4)	0.085	0.0018	$-$1.07	B	3
27.	$1s\,2s - 1s\,5p$	^{3}S $-$ ^{3}P°											
			[5.7489]	[62367900]	[79762600]	3	3	2.4(+4)	0.012	6.8($-$4)	$-$1.44	B	3
28.		^{1}S $-$ ^{1}P°	[5.8352]	[62644500]	[79782000]	1	3	2.3(+4)	0.035	6.7($-$4)	$-$1.46	B	3
29.	$1s\,2p - 1s\,3s$	^{3}P° $-$ ^{3}S											
			[8.8772]	[62644200]	[73909000]	3	3	1.1(+4)	0.013	0.0011	$-$1.41	B	3
30.		^{1}P° $-$ ^{1}S	[9.0740]	[62961500]	[73982000]	3	1	3.4(+4)	0.014	0.0013	$-$1.38	B	3
31.	$1s\,2p - 1s\,3d$	^{3}P° $-$ ^{3}D				9	15		0.69		0.79	C+	*interp.*
32.		^{1}P° $-$ ^{1}D				3	5		0.70		0.32	C+	*interp.*
33.	$1s\,2p - 1s\,4s$	^{3}P° $-$ ^{3}S											
			[6.5520]	[62644200]	[77906600]	3	3	4700	0.0030	1.9($-$4)	$-$2.05	C	3
34.		^{1}P° $-$ ^{1}S	[6.6779]	[62961500]	[77936200]	3	1	1.3(+4)	0.0030	2.0($-$4)	$-$2.05	C	3
35.	$1s\,2p - 1s\,4d$	^{3}P° $-$ ^{3}D				9	15		0.12		0.03	C	5
36.		^{1}P° $-$ ^{1}D				3	5		0.12		$-$0.44	C	5
37.	$1s\,2p - 1s\,5s$	^{3}P° $-$ ^{3}S											
			[5.8471]	[62644200]	[79746600]	3	3	2300	0.0012	6.9($-$5)	$-$2.44	C	3
38.		^{1}P° $-$ ^{1}S	[5.9524]	[62961500]	[79761400]	3	1	6800	0.0012	7.1($-$5)	$-$2.44	C	3

Ni XXVII: Allowed transitions — Continued

No.	Transition Array	Multiplet	λ (Å)	E_i (cm^{-1})	E_k (cm^{-1})	g_i	g_k	A_{ki} (10^8 s^{-1})	f_{ik}	S (at. u.)	log gf	Accuracy	Source
39.	$1s3s-1s3p$	^{3}S – ^{3}P°											
						3	3		0.015		−1.35	C	3
40.		^{1}S – ^{1}P°				1	3		0.057		−1.24	C	3
41.	$1s3s-1s4p$	^{3}S – ^{3}P°											
			[24.819]	[73909000]	[77938200]	3	3	1.42(+4)	0.131	0.0321	−0.406	B	3
42.		^{1}S – ^{1}P°	[25.036]	[73982000]	[77976200]	1	3	1.37(+4)	0.387	0.0319	−0.412	B	3
43.	$1s3s-1s5p$	^{3}S – ^{3}P°											
			[17.084]	[73909000]	[79762600]	3	3	7500	0.033	0.0056	−1.00	B	3
44.		^{1}S – ^{1}P°	[17.241]	[73982000]	[79782000]	1	3	7400	0.099	0.0056	−1.00	B	3
45.	$1s3p-1s3d$	^{3}P° – ^{3}D				9	15		0.010		−1.05	D	interp.
46.	$1s3p-1s4s$	^{3}P° – ^{3}S											
			[25.500]	[73985000]	[77906600]	3	3	3200	0.031	0.0078	−1.03	B	3
47.		^{1}P° – ^{1}S	[25.907]	[74076300]	[77936200]	3	1	9800	0.033	0.0084	−1.00	B	3
48.	$1s3p-1s4d$	^{3}P° – ^{3}D				9	15		0.60		0.73	C	5
49.		^{1}P° – ^{1}D				3	5		0.62		0.27	C	5
50.	$1s3p-1s5s$	^{3}P° – ^{3}S											
			[17.356]	[73985000]	[79746600]	3	3	1600	0.0070	0.0012	−1.68	C	3
51.		^{1}P° – ^{1}S	[17.590]	[74076300]	[79761400]	3	1	4700	0.0073	0.0013	−1.66	C	3
52.	$1s3d-1s3p$	^{1}D – ^{1}P°				5	3		0.0019		−2.02	E	5
53.	$1s3d-1s4p$	^{3}D – ^{3}P°				15	9		0.012		−0.74	C	5
54.		^{1}D – ^{1}P°				5	3		0.011		−1.26	C	5
55.	$1s4s-1s4p$	^{3}S – ^{3}P°											
						3	3		0.026		−1.11	E	3
56.		^{1}S – ^{1}P°				1	3		0.062		−1.21	D	3
57.	$1s4s-1s5p$	^{3}S – ^{3}P°											
			[53.879]	[77906600]	[79762600]	3	3	3380	0.147	0.0782	−0.356	B	3
58.		^{1}S – ^{1}P°	[54.177]	[77936200]	[79782000]	1	3	3260	0.431	0.0769	−0.366	B	3
59.	$1s4p-1s4d$	^{3}P° – ^{3}D				9	15		0.018		−0.79	D	5
60.	$1s4p-1s5s$	^{3}P° – ^{3}S											
			[55.298]	[77938200]	[79746600]	3	3	1100	0.051	0.028	−0.82	B	3
61.		^{1}P° – ^{1}S	[56.016]	[77976200]	[79761400]	3	1	3400	0.053	0.029	−0.80	B	3

Ni XXVII: Allowed transitions — Continued

No.	Transition Array	Multiplet	λ (Å)	E_i (cm^{-1})	E_k (cm^{-1})	g_i	g_k	A_{ki} (10^8 s^{-1})	f_{ik}	S (at. u.)	log gf	Accuracy	Source
62.	$1s4d$–$1s4p$	^{1}D – ^{1}P°				5	3		0.0029		−1.84	E	5
63.	$1s4d$–$1s4f$	^{3}D – ^{3}F°				15	21		7.3(−4)		−1.96	E	5
64.	$1s4d$–$1s5f$	^{3}D – ^{3}F°				15	21		0.89		1.13	B	7
65.		^{1}D – ^{1}F°				5	7		0.89		0.65	B	7
66.	$1s4f$–$1s4d$	^{1}F° – ^{1}D				7	5		3.9(−4)		−2.56	E	5
67.	$1s4f$–$1s5d$	^{3}F° – ^{3}D				21	15		0.0089		−0.73	C	7
68.		^{1}F° – ^{1}D				7	5		0.0089		−1.21	C	7
69.	$1s5s$–$1s5p$	^{3}S – ^{3}P°											
						3	3		0.026		−1.11	E	3
70.		^{1}S – ^{1}P°				1	3		0.10		−1.00	E	3

aThe number in parentheses following the tabulated value indicates the power of ten by which this value has to be multiplied.

Ni XXVII

Forbidden Transitions

The results of multi-configuration Dirac-Fock calculations by Hata and Grant[1] have been selected for this tabulation. Their work includes both a very detailed consideration of configuration interaction—with configurational wavefunction sets containing as many as 51 interacting states—as well as a fully relativistic treatment based on the Dirac Hamiltonian. Their calculated wavelengths are in very close agreement with experimental values. For the ions Ti XXI, V XXII and Fe XXV, where accurate experimental lifetime data are available, the agreement between these and the theoretical results of Hata and Grant[1] is excellent, with differences not exceeding a few percent (see the comparison table in the introduction to the forbidden lines of Ti XXI).

Reference

[1] J. Hata and I. P. Grant, Mon. Not. R. Astr. Soc. **211**, 549 (1984).

Ni XXVII: Forbidden transitions

No.	Transition Array	Multiplet	λ (Å)	E_i (cm^{-1})	E_k (cm^{-1})	g_i	g_k	Type of transition	A_{ki} (s^{-1})	S (at. u.)	Accuracy	Source
1.	$1s^2$–$1s2s$	^{1}S – ^{3}S	[1.6036]	0	[62358960]	1	3	M1	4.52(+8)a	2.07(−4)	B	1
2.	$1s^2$–$1s2p$	^{1}S – ^{3}P°										
			[1.5923]	0	[62801270]	1	5	M2	1.22(+10)	0.0942	B	1

aThe number in parentheses following the tabulated value indicates the power of ten by which this value has to be multiplied.

Ni xxviii

H Isoelectronic Sequence

Ground State: $1s\ ^2S_{1/2}$

Ionization Energy: $10775.48\ \text{eV} = 86909400\ \text{cm}^{-1}$

Allowed Transitions

Electric dipole transition probability data for this hydrogen-like ion can be obtained directly, in a non-relativistic approximation, from the data for neutral hydrogen.[1] The oscillator strength is independent of Z along the entire isoelectronic sequence and is therefore identical to the value for the hydrogen atom. Line strengths scale as Z^{-2} and transition probabilities scale as Z^4, i.e.,

$$S_Z = Z^{-2} S_{\text{H}}, \qquad A_Z = Z^4 A_{\text{H}}.$$

For higher nuclear charges in this sequence, relativistic corrections will cause these values to deviate increasingly from the non-relativistic ones. The first effect of relativity will be to alter the transition energies, or wavelengths, from the non-relativistic, even though the line strength itself is still well approximated by the non-relativistic value. In this case, experimental energies should be used in the standard conversion formulas, given in the general introduction to this volume, to calculate the most accurate values of f and A. It should be noted that the relativistic removal of the j-degeneracy introduces dipole transitions which do not occur in the non-relativistic theory, e.g., $2s_{1/2} - 2p_{3/2}$.

For very high Z, it is necessary to use the four-component Dirac spinors rather than two-component Schroedinger functions in theoretical calculations, and this introduces relativistic corrections to the line strengths themselves. Several recent systematic studies of the problem[2,3] indicate that these corrections are not large for stages of ionization in the range 20–30. Corrections for $Z = 30$ are usually no larger than 5–10% and generally substantially less than 5%. If an accuracy greater than this is required, the reader is referred to these papers[2,3] for a more detailed error analysis.

References

[1] W. L. Wiese, M. W. Smith, and B. M. Glennon, Atomic Transition Probabilities – Hydrogen through Neon (A Critical Data Compilation), Vol. I, 157 pp., Nat. Stand. Ref. Data Ser., Nat. Bur. Stand. (U.S.), 4 (May 1966).

[2] S. M. Younger and A. W. Weiss, J. Res. Nat. Bur. Stand., Sect. A **79**, 629 (1975).

[3] S. J. Rose, Rutherford Appleton Laboratory Report RL–82–114 (December 1982).

Journal of Physical and Chemical Reference Data
Cumulative Listing of Reprints and Supplements

Reprints from Volume 1

Reprints from Volume 2

- - - - - - - - - - - - - - - - - -

(Continuation of Cumulative Listing of Reprints)

(Continuation of Cumulative Listing of Reprints)

Reprints from Volume 11

Reprints from Volume 12

Reprints from Volume 13

(Continuation of Cumulative Listing of Reprints)

Reprints from Volume 16

(Continuation of Cumulative Listing of Reprints)

Special Reprints Packages

These special reprints packages offer selected articles in specific subject areas from the JOURNAL OF PHYSICAL AND CHEMICAL REFERENCE DATA, and they are offered at a better rate than when purchased individually. You will have available a complete library of literature for your specific requirements at a fraction of the cost of purchasing back issues of the journal.

Look over the reprints packages available—they are listed by subject area. In the Cumulative Listing of Reprints you will find the titles corresponding to the reprint numbers. You are sure to find building your information bank in this manner to be thorough and economical.

Package C1 (5 Parts) MOLECULAR VIBRATIONAL FREQUENCIES. Consisting of Reprint Nos. 103, 129, 170, 257, NSRD 39.
If purchased individually: $ 33.00
Special package price: **$ 26.00**

Package C2 (22 Parts) ATOMIC ENERGY LEVELS. Consisting of Reprint Nos. 26, 54, 64, 68, 94, 100, 109, 125, 126, 131, 132, 149, 150, 154, 156, 160, 179, 180, 192, 200, 222, 278.
If purchased individually: $121.00
Special package price: **$ 96.00**

Package C3 (6 Parts) ATOMIC SPECTRA. Consisting of Reprint Nos. 33, 56, 77, 78, 110, 132.
If purchased individually: $ 33.00
Special package price: **$ 27.00**

Package C4 (5 Parts) ATOMIC TRANSITION PROBABILITIES. Consisting of Reprint Nos. 20, 63, 82, 118, 182.
If purchased individually: $ 35.00
Special package price: **$ 28.00**

Package C5 (7 Parts) MOLECULAR SPECTRA. Consisting of Reprint Nos. 4, 8, 53, 79, 93, 130, 146.
If purchased individually: $ 51.50
Special package price: **$ 41.00**

Package C6 (9 Parts) THERMODYNAMIC PROPERTIES OF ELECTROLYTE SOLUTIONS. Consisting of Reprint Nos. 15, 95, 111, 151, 152, 174, 184, 185, 186.
If purchased individually: $ 46.00
Special package price: **$ 37.00**

Package C7 (12 Parts) IDEAL GAS THERMODYNAMIC PROPERTIES. Consisting of Reprint Nos. 30, 42, 43, 62, 65, 66, 70, 80, 83, 113, 115, 141.
If purchased individually: $ 38.00
Special package price: **$ 31.00**

Package C8 (7 Parts) RESISTIVITY. Consisting of Reprint Nos. 138, 139, 155, 221, 258, 259, 260.
If purchased individually: $ 47.50
Special package price: **$ 39.00**

Package C9 (7 Parts) MOLTEN SALTS. Consisting of Reprint Nos. 10, 41, 71, 96, 135, 167, 168.
If purchased individually: $ 62.50
Special package price: **$ 44.00**

Package C10 (4 Parts) REFRACTIVE INDEX. Consisting of Reprint Nos. 81, 158, 162, 240.
If purchased individually: $ 32.50
Special package price: **$ 26.00**

Supplements to JPCRD

When the topic demands it, and the quality of the data justifies it, the JOURNAL OF PHYSICAL AND CHEMICAL REFERENCE DATA issues a special Supplement. Each Supplement is a monograph—collected tables of highly significant physical or chemical property data in one complete volume. Listed below are the special Supplements to JPCRD that have been published. Each is a valuable resource for the physical chemist and chemical physicist.

GAS-PHASE ION AND NEUTRAL THERMOCHEMISTRY, by Sharon G. Lias, John E. Bartmess, Joel F. Liebman, John L. Holmes, Rhoda D. Levin, and W. Gary Mallard. (Supplement No. 1 to Volume 17) 1988, 874 pages. Hardcover.
U.S. & Canada: $70.00
Abroad: $84.00

ATOMIC AND IONIC SPECTRUM LINES BELOW 2000 ANGSTROMS: HYDROGEN THROUGH KRYPTON, by Raymond L. Kelly. (Supplement No. 1 to Volume 16) 1987, 1689 pages, 3 volumes. Hardcover.
U.S. & Canada: $75.00
Abroad: $90.00

ATOMIC ENERGY LEVELS OF THE IRON-PERIOD ELEMENTS: POTASSIUM THROUGH NICKEL by J. Sugar and C. Corliss. (Supplement No. 2 to Volume 14) 1985, 664 pages. Hardcover.
U.S. & Canada: $50.00
Abroad: $58.00

JANAF THERMOCHEMICAL TABLES, Third Edition by M. W. Chase, Jr., C. A. Davies, J. R. Downey, Jr., D. J. Frurip, R. A. McDonald, and A. N. Syverud. (Supplement No. 1 to Volume 14) 1985, 1896 pages, 2 volumes. Hardcover.
U.S. & Canada: $130.00
Abroad: $156.00

HEAT CAPACITIES AND ENTROPIES OF ORGANIC COMPOUNDS IN THE CONDENSED PHASE by E.S. Domalski, W.H. Evans, and E.D. Hearing. (Supplement No. 1 to Volume 13) 1984, 288 pages. Hardcover.
U.S. & Canada: $40.00
Abroad: $48.00

THE NBS TABLES OF CHEMICAL THERMODYNAMIC PROPERTIES. SELECTED VALUES FOR INORGANIC AND C_1 AND C_2 ORGANIC SUBSTANCES IN SI UNITS by D.D. Wagman, W.H. Evans, V.B. Parker, R.H. Schumm, I. Halow, S.M. Bailey, K.L. Churney, and R.L. Nuttall. (Supplement No. 2 to Volume 11) 1982, 394 pages. Hardcover.
U.S. & Canada: $40.00
Abroad: $48.00

THERMOPHYSICAL PROPERTIES OF FLUIDS. 1. ARGON, ETHYLENE, PARAHYDROGEN, NITROGEN, NITROGEN TRIFLUORIDE, AND OXYGEN by B.A. Younglove. (Supplement No. 1 to Volume 11) 1982, 368 pages. Hardcover.
U.S. & Canada: $40.00
Abroad: $48.00

EVALUATED KINETIC DATA FOR HIGH TEMPERATURE REACTIONS: VOLUME 4, HOMOGENEOUS GAS PHASE REACTIONS OF HALOGEN- AND CYANIDE-CONTAINING SPECIES by D.L. Baulch, J. Duxbury, S.J. Grant, and D.C. Montague. (Supplement No. 1 to Volume 10) 1981, 721 pages. Hardcover.
U.S. & Canada: $80.00
Abroad: $96.00

THERMAL CONDUCTIVITY OF THE ELEMENTS: A COMPREHENSIVE REVIEW by C.Y. Ho, R.W. Powell, and P.E. Liley. (Supplement No. 1 to Volume 3) 1974, 796 pages.*
U.S. & Canada: $60/$55
Abroad: $72/$66

*Prices are for hardcover/softcover.

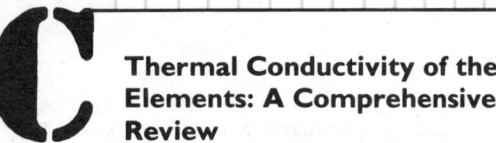